T0140211

Lecture Notes in Computer Science 13623

Founding Editors

Gerhard Goos
Juris Hartmanis

Editorial Board Members

The series Lecture Notes in Computer Science (LNCS), including its subseries Lecture Notes in Artificial Intelligence (LNAI) and Lecture Notes in Bioinformatics (LNBI), has established itself as a medium for the publication of new developments in computer science and information technology research, teaching, and education.

LNCS enjoys close cooperation with the computer science R & D community, the series counts many renowned academics among its volume editors and paper authors, and collaborates with prestigious societies. Its mission is to serve this international community by providing an invaluable service, mainly focused on the publication of conference and workshop proceedings and postproceedings. LNCS commenced publication in 1973.

Mohammad Tanveer · Sonali Agarwal ·
Seiichi Ozawa · Asif Ekbal · Adam Jatowt
Editors

Neural Information Processing

29th International Conference, ICONIP 2022
Virtual Event, November 22–26, 2022
Proceedings, Part I

Springer

Editors
Mohammad Tanveer
Indian Institute of Technology Indore
Indore, India

Seiichi Ozawa
Kobe University
Kobe, Japan

Adam Jatowt
University of Innsbruck
Innsbruck, Austria

Sonali Agarwal ⓘ
Indian Institute of Information Technology -
Allahabad
Prayagraj, India

Asif Ekbal
Indian Institute of Technology Patna
Patna, India

ISSN 0302-9743   ISSN 1611-3349 (electronic)
Lecture Notes in Computer Science
ISBN 978-3-031-30104-9   ISBN 978-3-031-30105-6 (eBook)
https://doi.org/10.1007/978-3-031-30105-6

This Springer imprint is published by the registered company Springer Nature Switzerland AG
The registered company address is: Gewerbestrasse 11, 6330 Cham, Switzerland

Preface

Welcome to the proceedings of the 29th International Conference on Neural Information Processing (ICONIP 2022) of the Asia-Pacific Neural Network Society (APNNS), held virtually from Indore, India, during November 22–26, 2022.

The mission of the Asia-Pacific Neural Network Society is to promote active interactions among researchers, scientists, and industry professionals who are working in neural networks and related fields in the Asia-Pacific region. APNNS has Governing Board Members from 13 countries/regions – Australia, China, Hong Kong, India, Japan, Malaysia, New Zealand, Singapore, South Korea, Qatar, Taiwan, Thailand, and Turkey. The society's flagship annual conference is the International Conference of Neural Information Processing (ICONIP).

The ICONIP conference aims to provide a leading international forum for researchers, scientists, and industry professionals who are working in neuroscience, neural networks, deep learning, and related fields to share their new ideas, progress, and achievements. Due to the current situation regarding the pandemic and international travel, ICONIP 2022, which was planned to be held in New Delhi, India, was organized as a fully virtual conference.

The proceedings of ICONIP 2022 consists of a multi-volume set in LNCS and CCIS, which includes 146 and 213 papers, respectively, selected from 1003 submissions reflecting the increasingly high quality of research in neural networks and related areas. The conference focused on four main areas, i.e., "Theory and Algorithms," "Cognitive Neurosciences," "Human Centered Computing," and "Applications." The conference also had special sessions in 12 niche areas, namely

1. International Workshop on Artificial Intelligence and Cyber Security (AICS)
2. Computationally Intelligent Techniques in Processing and Analysis of Neuronal Information (PANI)
3. Learning with Fewer Labels in Medical Computing (FMC)
4. Computational Intelligence for Biomedical Image Analysis (BIA)
5. Optimized AI Models with Interpretability, Security, and Uncertainty Estimation in Healthcare (OAI)
6. Advances in Deep Learning for Biometrics and Forensics (ADBF)
7. Machine Learning for Decision-Making in Healthcare: Challenges and Opportunities (MDH)
8. Reliable, Robust and Secure Machine Learning Algorithms (RRS)
9. Evolutionary Machine Learning Technologies in Healthcare (EMLH)
10. High Performance Computing Based Scalable Machine Learning Techniques for Big Data and Their Applications (HPCML)
11. Intelligent Transportation Analytics (ITA)
12. Deep Learning and Security Techniques for Secure Video Processing (DLST)

Our great appreciation goes to the Program Committee members and the reviewers who devoted their time and effort to our rigorous peer-review process. Their insightful reviews and timely feedback ensured the high quality of the papers accepted for publication.

The submitted papers in the main conference and special sessions were reviewed following the same process, and we ensured that every paper has at least two high-quality single-blind reviews. The PC Chairs discussed the reviews of every paper very meticulously before making a final decision. Finally, thank you to all the authors of papers, presenters, and participants, which made the conference a grand success. Your support and engagement made it all worthwhile.

December 2022
Mohammad Tanveer
Sonali Agarwal
Seiichi Ozawa
Asif Ekbal
Adam Jatowt

Organization

Program Committee

General Chairs

M. Tanveer   Indian Institute of Technology Indore, India
Sonali Agarwal   IIIT Allahabad, India
Seiichi Ozawa   Kobe University, Japan

Honorary Chairs

Jonathan Chan   King Mongkut's University of Technology Thonburi, Thailand
P. N. Suganthan   Nanyang Technological University, Singapore

Program Chairs

Asif Ekbal   Indian Institute of Technology Patna, India
Adam Jatowt   University of Innsbruck, Austria

Technical Chairs

Shandar Ahmad   JNU, India
Derong Liu   University of Chicago, USA

Special Session Chairs

Kai Qin   Swinburne University of Technology, Australia
Kaizhu Huang   Duke Kunshan University, China
Amit Kumar Singh   NIT Patna, India

Tutorial Chairs

Swagatam Das   ISI Kolkata, India
Partha Pratim Roy   IIT Roorkee, India

Finance Chairs

Shekhar Verma   Indian Institute of Information Technology
Allahabad, India

Hayaru Shouno   University of Electro-Communications, Japan

R. B. Pachori   IIT Indore, India

Publicity Chairs

Jerry Chun-Wei Lin   Western Norway University of Applied Sciences,
Norway

Chandan Gautam   A*STAR, Singapore

Publication Chairs

Deepak Ranjan Nayak   MNIT Jaipur, India

Tripti Goel   NIT Silchar, India

Sponsorship Chairs

Asoke K. Talukder   NIT Surathkal, India

Vrijendra Singh   IIIT Allahabad, India

Website Chairs

M. Arshad   IIT Indore, India

Navjot Singh   IIIT Allahabad, India

Local Arrangement Chairs

Pallavi Somvanshi   JNU, India

Yogendra Meena   University of Delhi, India

M. Javed   IIIT Allahabad, India

Vinay Kumar Gupta   IIT Indore, India

Iqbal Hasan   National Informatics Centre, Ministry of
Electronics and Information Technology, India

Regional Liaison Committee

Sansanee Auephanwiriyakul   Chiang Mai University, Thailand

Nia Kurnianingsih   Politeknik Negeri Semarang, Indonesia

Md Rafiqul Islam	University of Technology Sydney, Australia
Bharat Richhariya	IISc Bangalore, India
Sanjay Kumar Sonbhadra	Shiksha 'O' Anusandhan, India
Mufti Mahmud	Nottingham Trent University, UK
Francesco Piccialli	University of Naples Federico II, Italy

Program Committee

Balamurali A. R.	IITB-Monash Research Academy, India
Ibrahim A. Hameed	Norwegian University of Science and Technology (NTNU), Norway
Fazly Salleh Abas	Multimedia University, Malaysia
Prabath Abeysekara	RMIT University, Australia
Adamu Abubakar Ibrahim	International Islamic University, Malaysia
Muhammad Abulaish	South Asian University, India
Saptakatha Adak	Philips, India
Abhijit Adhikary	King's College, London, UK
Hasin Afzal Ahmed	Gauhati University, India
Rohit Agarwal	UiT The Arctic University of Norway, Norway
A. K. Agarwal	Sharda University, India
Fenty Eka Muzayyana Agustin	UIN Syarif Hidayatullah Jakarta, Indonesia
Gulfam Ahamad	BGSB University, India
Farhad Ahamed	Kent Institute, Australia
Zishan Ahmad	Indian Institute of Technology Patna, India
Mohammad Faizal Ahmad Fauzi	Multimedia University, Malaysia
Mudasir Ahmadganaie	Indian Institute of Technology Indore, India
Hasin Afzal Ahmed	Gauhati University, India
Sangtae Ahn	Kyungpook National University, South Korea
Md. Shad Akhtar	Indraprastha Institute of Information Technology, Delhi, India
Abdulrazak Yahya Saleh Alhababi	University of Malaysia, Sarawak, Malaysia
Ahmed Alharbi	RMIT University, Australia
Irfan Ali	Aligarh Muslim University, India
Ali Anaissi	CSIRO, Australia
Ashish Anand	Indian Institute of Technology, Guwahati, India
C. Anantaram	Indraprastha Institute of Information Technology and Tata Consultancy Services Ltd., India
Nur Afny C. Andryani	Universiti Teknologi Petronas, Malaysia
Marco Anisetti	Università degli Studi di Milano, Italy
Mohd Zeeshan Ansari	Jamia Millia Islamia, India
J. Anuradha	VIT, India
Ramakrishna Appicharla	Indian Institute of Technology Patna, India

He Chen	Hebei University of Technology, China
Hongxu Chen	University of Queensland, Australia
J. Chen	Dalian University of Technology, China
Jianhui Chen	Beijing University of Technology, China
Junxin Chen	Dalian University of Technology, China
Junyi Chen	City University of Hong Kong, China
Junying Chen	South China University of Technology, China
Lisi Chen	Hong Kong Baptist University, China
Mulin Chen	Northwestern Polytechnical University, China
Xiaocong Chen	University of New South Wales, Australia
Xiaofeng Chen	Chongqing Jiaotong University, China
Zhuangbin Chen	The Chinese University of Hong Kong, China
Long Cheng	Institute of Automation, China
Qingrong Cheng	Fudan University, China
Ruting Cheng	George Washington University, USA
Girija Chetty	University of Canberra, Australia
Manoj Chinnakotla	Microsoft R&D Pvt. Ltd., India
Andrew Chiou	CQ University, Australia
Sung-Bae Cho	Yonsei University, South Korea
Kupsze Choi	The Hong Kong Polytechnic University, China
Phatthanaphong Chomphuwiset	Mahasarakham University, Thailand
Fengyu Cong	Dalian University of Technology, China
Jose Alfredo Ferreira Costa	UFRN, Brazil
Ruxandra Liana Costea	Polytechnic University of Bucharest, Romania
Raphaël Couturier	University of Franche-Comte, France
Zhenyu Cui	Peking University, China
Zhihong Cui	Shandong University, China
Juan D. Velasquez	University of Chile, Chile
Rukshima Dabare	Murdoch University, Australia
Cherifi Dalila	University of Boumerdes, Algeria
Minh-Son Dao	National Institute of Information and Communications Technology, Japan
Tedjo Darmanto	STMIK AMIK Bandung, Indonesia
Debasmit Das	IIT Roorkee, India
Dipankar Das	Jadavpur University, India
Niladri Sekhar Dash	Indian Statistical Institute, Kolkata, India
Satya Ranjan Dash	KIIT University, India
Shubhajit Datta	Indian Institute of Technology, Kharagpur, India
Alok Debnath	Trinity College Dublin, Ireland
Amir Dehsarvi	Ludwig Maximilian University of Munich, Germany
Hangyu Deng	Waseda University, Japan

Mingcong Deng	Tokyo University of Agriculture and Technology, Japan
Zhaohong Deng	Jiangnan University, China
V. Susheela Devi	Indian Institute of Science, Bangalore, India
M. M. Dhabu	VNIT Nagpur, India
Dhimas Arief Dharmawan	Universitas Indonesia, Indonesia
Khaldoon Dhou	Texas A&M University Central Texas, USA
Gihan Dias	University of Moratuwa, Sri Lanka
Nat Dilokthanakul	Vidyasirimedhi Institute of Science and Technology, Thailand
Tai Dinh	Kyoto College of Graduate Studies for Informatics, Japan
Gaurav Dixit	Indian Institute of Technology Roorkee, India
Youcef Djenouri	SINTEF Digital, Norway
Hai Dong	RMIT University, Australia
Shichao Dong	Ping An Insurance Group, China
Mohit Dua	NIT Kurukshetra, India
Yijun Duan	Kyoto University, Japan
Shiv Ram Dubey	Indian Institute of Information Technology, Allahabad, India
Piotr Duda	Institute of Computational Intelligence/Czestochowa University of Technology, Poland
Sri Harsha Dumpala	Dalhousie University and Vector Institute, Canada
Hridoy Sankar Dutta	University of Cambridge, UK
Indranil Dutta	Jadavpur University, India
Pratik Dutta	Indian Institute of Technology Patna, India
Rudresh Dwivedi	Netaji Subhas University of Technology, India
Heba El-Fiqi	UNSW Canberra, Australia
Felix Engel	Leibniz Information Centre for Science and Technology (TIB), Germany
Akshay Fajge	Indian Institute of Technology Patna, India
Yuchun Fang	Shanghai University, China
Mohd Fazil	JMI, India
Zhengyang Feng	Shanghai Jiao Tong University, China
Zunlei Feng	Zhejiang University, China
Mauajama Firdaus	University of Alberta, Canada
Devi Fitrianah	Bina Nusantara University, Indonesia
Philippe Fournierviger	Shenzhen University, China
Wai-Keung Fung	Cardiff Metropolitan University, UK
Baban Gain	Indian Institute of Technology, Patna, India
Claudio Gallicchio	University of Pisa, Italy
Yongsheng Gao	Griffith University, Australia

Yunjun Gao	Zhejiang University, China
Vicente García Díaz	University of Oviedo, Spain
Arpit Garg	University of Adelaide, Australia
Chandan Gautam	I2R, A*STAR, Singapore
Yaswanth Gavini	University of Hyderabad, India
Tom Gedeon	Australian National University, Australia
Iuliana Georgescu	University of Bucharest, Romania
Deepanway Ghosal	Indian Institute of Technology Patna, India
Arjun Ghosh	National Institute of Technology Durgapur, India
Sanjukta Ghosh	IIT (BHU) Varanasi, India
Soumitra Ghosh	Indian Institute of Technology Patna, India
Pranav Goel	Bloomberg L.P., India
Tripti Goel	National Institute of Technology Silchar, India
Kah Ong Michael Goh	Multimedia University, Malaysia
Kam Meng Goh	Tunku Abdul Rahman University of Management and Technology, Malaysia
Iqbal Gondal	RMIT University, Australia
Puneet Goyal	Indian Institute of Technology Ropar, India
Vishal Goyal	Punjabi University Patiala, India
Xiaotong Gu	University of Tasmania, Australia
Radha Krishna Guntur	VNRVJIET, India
Li Guo	University of Macau, China
Ping Guo	Beijing Normal University, China
Yu Guo	Xi'an Jiaotong University, China
Akshansh Gupta	CSIR-Central Electronics Engineering Research Institute, India
Deepak Gupta	National Library of Medicine, National Institutes of Health (NIH), USA
Deepak Gupta	NIT Arunachal Pradesh, India
Kamal Gupta	NIT Patna, India
Kapil Gupta	PDPM IIITDM, Jabalpur, India
Komal Gupta	IIT Patna, India
Christophe Guyeux	University of Franche-Comte, France
Katsuyuki Hagiwara	Mie University, Japan
Soyeon Han	University of Sydney, Australia
Palak Handa	IGDTUW, India
Rahmadya Handayanto	Universitas Islam 45 Bekasi, Indonesia
Ahteshamul Haq	Aligarh Muslim University, India
Muhammad Haris	Universitas Nusa Mandiri, Indonesia
Harith Al-Sahaf	Victoria University of Wellington, New Zealand
Md Rakibul Hasan	BRAC University, Bangladesh
Mohammed Hasanuzzaman	ADAPT Centre, Ireland

Takako Hashimoto	Chiba University of Commerce, Japan
Bipan Hazarika	Gauhati University, India
Huiguang He	Institute of Automation, Chinese Academy of Sciences, China
Wei He	University of Science and Technology Beijing, China
Xinwei He	University of Illinois Urbana-Champaign, USA
Enna Hirata	Kobe University, Japan
Akira Hirose	University of Tokyo, Japan
Katsuhiro Honda	Osaka Metropolitan University, Japan
Huy Hongnguyen	National Institute of Informatics, Japan
Wai Lam Hoo	University of Malaya, Malaysia
Shih Hsiung Lee	National Cheng Kung University, Taiwan
Jiankun Hu	UNSW@ADFA, Australia
Yanyan Hu	University of Science and Technology Beijing, China
Chaoran Huang	UNSW Sydney, Australia
He Huang	Soochow University, Taiwan
Ko-Wei Huang	National Kaohsiung University of Science and Technology, Taiwan
Shudong Huang	Sichuan University, China
Chih-Chieh Hung	National Chung Hsing University, Taiwan
Mohamed Ibn Khedher	IRT-SystemX, France
David Iclanzan	Sapientia Hungarian University of Transylvania, Romania
Cosimo Ieracitano	University "Mediterranea" of Reggio Calabria, Italy
Kazushi Ikeda	Nara Institute of Science and Technology, Japan
Hiroaki Inoue	Kobe University, Japan
Teijiro Isokawa	University of Hyogo, Japan
Kokila Jagadeesh	Indian Institute of Information Technology, Allahabad, India
Mukesh Jain	Jawaharlal Nehru University, India
Fuad Jamour	AWS, USA
Mohd. Javed	Indian Institute of Information Technology, Allahabad, India
Balasubramaniam Jayaram	Indian Institute of Technology Hyderabad, India
Jin-Tsong Jeng	National Formosa University, Taiwan
Sungmoon Jeong	Kyungpook National University Hospital, South Korea
Yizhang Jiang	Jiangnan University, China
Ferdinjoe Johnjoseph	Thai-Nichi Institute of Technology, Thailand
Alireza Jolfaei	Federation University, Australia

Ratnesh Joshi	Indian Institute of Technology Patna, India
Roshan Joymartis	Global Academy of Technology, India
Chen Junjie	IMAU, The Netherlands
Ashwini K.	Global Academy of Technology, India
Asoke K. Talukder	National Institute of Technology Karnataka - Surathkal, India
Ashad Kabir	Charles Sturt University, Australia
Narendra Kadoo	CSIR-National Chemical Laboratory, India
Seifedine Kadry	Noroff University College, Norway
M. Shamim Kaiser	Jahangirnagar University, Bangladesh
Ashraf Kamal	ACL Digital, India
Sabyasachi Kamila	Indian Institute of Technology Patna, India
Tomoyuki Kaneko	University of Tokyo, Japan
Rajkumar Kannan	Bishop Heber College, India
Hamid Karimi	Utah State University, USA
Nikola Kasabov	AUT, New Zealand
Dermot Kerr	University of Ulster, UK
Abhishek Kesarwani	NIT Rourkela, India
Shwet Ketu	Shambhunath Institute of Engineering and Technology, India
Asif Khan	Integral University, India
Tariq Khan	UNSW, Australia
Thaweesak Khongtuk	Rajamangala University of Technology Suvarnabhumi (RMUTSB), India
Abbas Khosravi	Deakin University, Australia
Thanh Tung Khuat	University of Technology Sydney, Australia
Junae Kim	DST Group, Australia
Sangwook Kim	Kobe University, Japan
Mutsumi Kimura	Ryukoku University, Japan
Uday Kiran	University of Aizu, Japan
Hisashi Koga	University of Electro-Communications, Japan
Yasuharu Koike	Tokyo Institute of Technology, Japan
Ven Jyn Kok	Universiti Kebangsaan Malaysia, Malaysia
Praveen Kolli	Pinterest Inc, USA
Sunil Kumar Kopparapu	Tata Consultancy Services Ltd., India
Fajri Koto	MBZUAI, UAE
Aneesh Krishna	Curtin University, Australia
Parameswari Krishnamurthy	University of Hyderabad, India
Malhar Kulkarni	IIT Bombay, India
Abhinav Kumar	NIT, Patna, India
Abhishek Kumar	Indian Institute of Technology Patna, India
Amit Kumar	Tarento Technologies Pvt Limited, India

Nagendra Kumar	IIT Indore, India
Pranaw Kumar	Centre for Development of Advanced Computing (CDAC) Mumbai, India
Puneet Kumar	Jawaharlal Nehru University, India
Raja Kumar	Taylor's University, Malaysia
Sachin Kumar	University of Delhi, India
Sandeep Kumar	IIT Patna, India
Sanjaya Kumar Panda	National Institute of Technology, Warangal, India
Chouhan Kumar Rath	National Institute of Technology, Durgapur, India
Sovan Kumar Sahoo	Indian Institute of Technology Patna, India
Anil Kumar Singh	IIT (BHU) Varanasi, India
Vikash Kumar Singh	VIT-AP University, India
Sanjay Kumar Sonbhadra	ITER, SoA, Odisha, India
Gitanjali Kumari	Indian Institute of Technology Patna, India
Rina Kumari	KIIT, India
Amit Kumarsingh	National Institute of Technology Patna, India
Sanjay Kumarsonbhadra	SSITM, India
Vishesh Kumar Tanwar	Missouri University of Science and Technology, USA
Bibekananda Kundu	CDAC Kolkata, India
Yoshimitsu Kuroki	Kurume National College of Technology, Japan
Susumu Kuroyanagi	Nagoya Institute of Technology, Japan
Retno Kusumaningrum	Universitas Diponegoro, Indonesia
Dwina Kuswardani	Institut Teknologi PLN, Indonesia
Stephen Kwok	Murdoch University, Australia
Hamid Laga	Murdoch University, Australia
Edmund Lai	Auckland University of Technology, New Zealand
Weng Kin Lai	Tunku Abdul Rahman University of Management & Technology (TAR UMT), Malaysia
Kittichai Lavangnananda	King Mongkut's University of Technology Thonburi (KMUTT), Thailand
Anwesha Law	Indian Statistical Institute, India
Thao Le	Deakin University, Australia
Xinyi Le	Shanghai Jiao Tong University, China
Dong-Gyu Lee	Kyungpook National University, South Korea
Eui Chul Lee	Sangmyung University, South Korea
Minho Lee	Kyungpook National University, South Korea
Shih Hsiung Lee	National Kaohsiung University of Science and Technology, Taiwan
Gurpreet Lehal	Punjabi University, India
Jiahuan Lei	Meituan-Dianping Group, China

Pui Huang Leong   Tunku Abdul Rahman University of Management
                              and Technology, Malaysia
Chi Sing Leung   City University of Hong Kong, China
Man-Fai Leung   Anglia Ruskin University, UK
Bing-Zhao Li   Beijing Institute of Technology, China
Gang Li   Deakin University, Australia
Jiawei Li   Tsinghua University, China
Mengmeng Li   Zhengzhou University, China
Xiangtao Li   Jilin University, China
Yang Li   East China Normal University, China
Yantao Li   Chongqing University, China
Yaxin Li   Michigan State University, USA
Yiming Li   Tsinghua University, China
Yuankai Li   University of Science and Technology of China,
                              China
Yun Li   Nanjing University of Posts and
                              Telecommunications, China
Zhipeng Li   Tsinghua University, China
Hualou Liang   Drexel University, USA
Xiao Liang   Nankai University, China
Hao Liao   Shenzhen University, China
Alan Wee-Chung Liew   Griffith University, Australia
Chern Hong Lim   Monash University Malaysia, Malaysia
Kok Lim Yau   Universiti Tunku Abdul Rahman (UTAR),
                              Malaysia
Chin-Teng Lin   UTS, Australia
Jerry Chun-Wei Lin   Western Norway University of Applied Sciences,
                              Norway
Jiecong Lin   City University of Hong Kong, China
Dugang Liu   Shenzhen University, China
Feng Liu   Stevens Institute of Technology, USA
Hongtao Liu   Du Xiaoman Financial, China
Ju Liu   Shandong University, China
Linjing Liu   City University of Hong Kong, China
Weifeng Liu   China University of Petroleum (East China),
                              China
Wenqiang Liu   Hong Kong Polytechnic University, China
Xin Liu   National Institute of Advanced Industrial Science
                              and Technology (AIST), Japan
Yang Liu   Harbin Institute of Technology, China
Zhi-Yong Liu   Institute of Automation, Chinese Academy of
                              Sciences, China
Zongying Liu   Dalian Maritime University, China

Jaime Lloret	Universitat Politècnica de València, Spain
Sye Loong Keoh	University of Glasgow, Singapore, Singapore
Hongtao Lu	Shanghai Jiao Tong University, China
Wenlian Lu	Fudan University, China
Xuequan Lu	Deakin University, Australia
Xiao Luo	UCLA, USA
Guozheng Ma	Shenzhen International Graduate School, Tsinghua University, China
Qianli Ma	South China University of Technology, China
Wanli Ma	University of Canberra, Australia
Muhammad Anwar Ma'sum	Universitas Indonesia, Indonesia
Michele Magno	University of Bologna, Italy
Sainik Kumar Mahata	JU, India
Shalni Mahato	Indian Institute of Information Technology (IIIT) Ranchi, India
Adnan Mahmood	Macquarie University, Australia
Mohammed Mahmoud	October University for Modern Sciences & Arts - MSA University, Egypt
Mufti Mahmud	University of Padova, Italy
Krishanu Maity	Indian Institute of Technology Patna, India
Mamta	IIT Patna, India
Aprinaldi Mantau	Kyushu Institute of Technology, Japan
Mohsen Marjani	Taylor's University, Malaysia
Sanparith Marukatat	NECTEC, Thailand
José María Luna	Universidad de Córdoba, Spain
Archana Mathur	Nitte Meenakshi Institute of Technology, India
Patrick McAllister	Ulster University, UK
Piotr Milczarski	Lodz University of Technology, Poland
Kshitij Mishra	IIT Patna, India
Pruthwik Mishra	IIIT-Hyderabad, India
Santosh Mishra	Indian Institute of Technology Patna, India
Sajib Mistry	Curtin University, Australia
Sayantan Mitra	Accenture Labs, India
Vinay Kumar Mittal	Neti International Research Center, India
Daisuke Miyamoto	University of Tokyo, Japan
Kazuteru Miyazaki	National Institution for Academic Degrees and Quality Enhancement of Higher Education, Japan
U. Mmodibbo	Modibbo Adama University Yola, Nigeria
Aditya Mogadala	Saarland University, Germany
Reem Mohamed	Mansoura University, Egypt
Muhammad Syafiq Mohd Pozi	Universiti Utara Malaysia, Malaysia

Anirban Mondal	University of Tokyo, Japan
Anupam Mondal	Jadavpur University, India
Supriyo Mondal	ZBW - Leibniz Information Centre for Economics, Germany
J. Manuel Moreno	Universitat Politècnica de Catalunya, Spain
Francisco J. Moreno-Barea	Universidad de Málaga, Spain
Sakchai Muangsrinoon	Walailak University, Thailand
Siti Anizah Muhamed	Politeknik Sultan Salahuddin Abdul Aziz Shah, Malaysia
Samrat Mukherjee	Indian Institute of Technology, Patna, India
Siddhartha Mukherjee	Samsung R&D Institute India, Bangalore, India
Dharmalingam Muthusamy	Bharathiar University, India
Abhijith Athreya Mysore Gopinath	Pennsylvania State University, USA
Harikrishnan N. B.	BITS Pilani K K Birla Goa Campus, India
Usman Naseem	University of Sydney, Australia
Deepak Nayak	Malaviya National Institute of Technology, Jaipur, India
Hamada Nayel	Benha University, Egypt
Usman Nazir	Lahore University of Management Sciences, Pakistan
Vasudevan Nedumpozhimana	TU Dublin, Ireland
Atul Negi	University of Hyderabad, India
Aneta Neumann	University of Adelaide, Australia
Hea Choon Ngo	Universiti Teknikal Malaysia Melaka, Malaysia
Dang Nguyen	University of Canberra, Australia
Duy Khuong Nguyen	FPT Software Ltd., FPT Group, Vietnam
Hoang D. Nguyen	University College Cork, Ireland
Hong Huy Nguyen	National Institute of Informatics, Japan
Tam Nguyen	Leibniz University Hannover, Germany
Thanh-Son Nguyen	Agency for Science, Technology and Research (A*STAR), Singapore
Vu-Linh Nguyen	Eindhoven University of Technology, Netherlands
Nick Nikzad	Griffith University, Australia
Boda Ning	Swinburne University of Technology, Australia
Haruhiko Nishimura	University of Hyogo, Japan
Kishorjit Nongmeikapam	Indian Institute of Information Technology (IIIT) Manipur, India
Aleksandra Nowak	Jagiellonian University, Poland
Stavros Ntalampiras	University of Milan, Italy
Anupiya Nugaliyadde	Sri Lanka Institute of Information Technology, Sri Lanka

Anto Satriyo Nugroho	Agency for Assessment & Application of Technology, Indonesia
Aparajita Ojha	PDPM IIITDM Jabalpur, India
Akeem Olowolayemo	International Islamic University Malaysia, Malaysia
Toshiaki Omori	Kobe University, Japan
Shih Yin Ooi	Multimedia University, Malaysia
Sidali Ouadfeul	Algerian Petroleum Institute, Algeria
Samir Ouchani	CESI Lineact, France
Srinivas P. Y. K. L.	IIIT Sri City, India
Neelamadhab Padhy	GIET University, India
Worapat Paireekreng	Dhurakij Pundit University, Thailand
Partha Pakray	National Institute of Technology Silchar, India
Santanu Pal	Wipro Limited, India
Bin Pan	Nankai University, China
Rrubaa Panchendrarajan	Sri Lanka Institute of Information Technology, Sri Lanka
Pankaj Pandey	Indian Institute of Technology, Gandhinagar, India
Lie Meng Pang	Southern University of Science and Technology, China
Sweta Panigrahi	National Institute of Technology Warangal, India
T. Pant	IIIT Allahabad, India
Shantipriya Parida	Idiap Research Institute, Switzerland
Hyeyoung Park	Kyungpook National University, South Korea
Md Aslam Parwez	Jamia Millia Islamia, India
Leandro Pasa	Federal University of Technology - Parana (UTFPR), Brazil
Kitsuchart Pasupa	King Mongkut's Institute of Technology Ladkrabang, Thailand
Debanjan Pathak	Kalinga Institute of Industrial Technology (KIIT), India
Vyom Pathak	University of Florida, USA
Sangameshwar Patil	TCS Research, India
Bidyut Kr. Patra	IIT (BHU) Varanasi, India
Dipanjyoti Paul	Indian Institute of Technology Patna, India
Sayanta Paul	Ola, India
Sachin Pawar	Tata Consultancy Services Ltd., India
Pornntiwa Pawara	Mahasarakham University, Thailand
Yong Peng	Hangzhou Dianzi University, China
Yusuf Perwej	Ambalika Institute of Management and Technology (AIMT), India
Olutomilayo Olayemi Petinrin	City University of Hong Kong, China
Arpan Phukan	Indian Institute of Technology Patna, India

Chiara Picardi	University of York, UK
Francesco Piccialli	University of Naples Federico II, Italy
Josephine Plested	University of New South Wales, Australia
Krishna Reddy Polepalli	IIIT Hyderabad, India
Dan Popescu	University Politehnica of Bucharest, Romania
Heru Praptono	Bank Indonesia/UI, Indonesia
Mukesh Prasad	University of Technology Sydney, Australia
Yamuna Prasad	Thompson Rivers University, Canada
Krishna Prasadmiyapuram	IIT Gandhinagar, India
Partha Pratim Sarangi	KIIT Deemed to be University, India
Emanuele Principi	Università Politecnica delle Marche, Italy
Dimeter Prodonov	Imec, Belgium
Ratchakoon Pruengkarn	College of Innovative Technology and Engineering, Dhurakij Pundit University, Thailand
Michal Ptaszynski	Kitami Institute of Technology, Japan
Narinder Singh Punn	Mayo Clinic, Arizona, USA
Abhinanda Ranjit Punnakkal	UiT The Arctic University of Norway, Norway
Zico Pratama Putra	Queen Mary University of London, UK
Zhenyue Qin	Tencent, China
Nawab Muhammad Faseeh Qureshi	SU, South Korea
Md Rafiqul	UTS, Australia
Saifur Rahaman	City University of Hong Kong, China
Shri Rai	Murdoch University, Australia
Vartika Rai	IIIT Hyderabad, India
Kiran Raja	Norwegian University of Science and Technology, Norway
Sutharshan Rajasegarar	Deakin University, Australia
Arief Ramadhan	Bina Nusantara University, Indonesia
Mallipeddi Rammohan	Kyungpook National University, South Korea
Md. Mashud Rana	Commonwealth Scientific and Industrial Research Organisation (CSIRO), Australia
Surangika Ranathunga	University of Moratuwa, Sri Lanka
Soumya Ranjan Mishra	KIIT University, India
Hemant Rathore	Birla Institute of Technology & Science, Pilani, India
Imran Razzak	UNSW, Australia
Yazhou Ren	University of Science and Technology of China, China
Motahar Reza	GITAM University Hyderabad, India
Dwiza Riana	STMIK Nusa Mandiri, Indonesia
Bharat Richhariya	BITS Pilani, India

Pattabhi R. K. Rao	AU-KBC Research Centre, India
Heejun Roh	Korea University, South Korea
Vijay Rowtula	IIIT Hyderabad, India
Aniruddha Roy	IIT Kharagpur, India
Sudipta Roy	Jio Institute, India
Narendra S. Chaudhari	Indian Institute of Technology Indore, India
Fariza Sabrina	Central Queensland University, Australia
Debanjan Sadhya	ABV-IIITM Gwalior, India
Sumit Sah	IIT Dharwad, India
Atanu Saha	Jadavpur University, India
Sajib Saha	Commonwealth Scientific and Industrial Research Organisation, Australia
Snehanshu Saha	BITS Pilani K K Birla Goa Campus, India
Tulika Saha	IIT Patna, India
Navanath Saharia	Indian Institute of Information Technology Manipur, India
Pracheta Sahoo	University of Texas at Dallas, USA
Sovan Kumar Sahoo	Indian Institute of Technology Patna, India
Tanik Saikh	L3S Research Center, Germany
Naveen Saini	Indian Institute of Information Technology Lucknow, India
Fumiaki Saitoh	Chiba Institute of Technology, Japan
Rohit Salgotra	Swansea University, UK
Michel Salomon	Univ. Bourgogne Franche-Comté, France
Yu Sang	Research Institute of Institute of Computing Technology, Exploration and Development, Liaohe Oilfield, PetroChina, China
Suyash Sangwan	Indian Institute of Technology Patna, India
Soubhagya Sankar Barpanda	VIT-AP University, India
Jose A. Santos	Ulster University, UK
Kamal Sarkar	Jadavpur University, India
Sandip Sarkar	Jadavpur University, India
Naoyuki Sato	Future University Hakodate, Japan
Eri Sato-Shimokawara	Tokyo Metropolitan University, Japan
Sunil Saumya	Indian Institute of Information Technology Dharwad, India
Gerald Schaefer	Loughborough University, UK
Rafal Scherer	Czestochowa University of Technology, Poland
Arvind Selwal	Central University of Jammu, India
Noor Akhmad Setiawan	Universitas Gadjah Mada, Indonesia
Mohammad Shahid	Aligarh Muslim University, India
Jie Shao	University of Science and Technology of China, China

Nabin Sharma	University of Technology Sydney, Australia
Raksha Sharma	IIT Bombay, India
Sourabh Sharma	Avantika University, India
Suraj Sharma	International Institute of Information Technology Bhubaneswar, India
Ravi Shekhar	Queen Mary University of London, UK
Michael Sheng	Macquarie University, Australia
Yin Sheng	Huazhong University of Science and Technology, China
Yongpan Sheng	Southwest University, China
Liu Shenglan	Dalian University of Technology, China
Tomohiro Shibata	Kyushu Institute of Technology, Japan
Iksoo Shin	University of Science & Technology, China
Mohd Fairuz Shiratuddin	Murdoch University, Australia
Hayaru Shouno	University of Electro-Communications, Japan
Sanyam Shukla	MANIT, Bhopal, India
Udom Silparcha	KMUTT, Thailand
Apoorva Singh	Indian Institute of Technology Patna, India
Divya Singh	Central University of Bihar, India
Gitanjali Singh	Indian Institute of Technology Patna, India
Gopendra Singh	Indian Institute of Technology Patna, India
K. P. Singh	IIIT Allahabad, India
Navjot Singh	IIIT Allahabad, India
Om Singh	NIT Patna, India
Pardeep Singh	Jawaharlal Nehru University, India
Rajiv Singh	Banasthali Vidyapith, India
Sandhya Singh	Indian Institute of Technology Bombay, India
Smriti Singh	IIT Bombay, India
Narinder Singhpunn	Mayo Clinic, Arizona, USA
Saaveethya Sivakumar	Curtin University, Malaysia
Ferdous Sohel	Murdoch University, Australia
Chattrakul Sombattheera	Mahasarakham University, Thailand
Lei Song	Unitec Institute of Technology, New Zealand
Linqi Song	City University of Hong Kong, China
Yuhua Song	University of Science and Technology Beijing, China
Gautam Srivastava	Brandon University, Canada
Rajeev Srivastava	Banaras Hindu University (IT-BHU), Varanasi, India
Jérémie Sublime	ISEP - Institut Supérieur d'Électronique de Paris, France
P. N. Suganthan	Nanyang Technological University, Singapore

Derwin Suhartono	Bina Nusantara University, Indonesia
Indra Adji Sulistijono	Politeknik Elektronika Negeri Surabaya (PENS), Indonesia
John Sum	National Chung Hsing University, Taiwan
Fuchun Sun	Tsinghua University, China
Ning Sun	Nankai University, China
Anindya Sundar Das	Indian Institute of Technology Patna, India
Bapi Raju Surampudi	International Institute of Information Technology Hyderabad, India
Olarik Surinta	Mahasarakham University, Thailand
Maria Susan Anggreainy	Bina Nusantara University, Indonesia
M. Syafrullah	Universitas Budi Luhur, Indonesia
Murtaza Taj	Lahore University of Management Sciences, Pakistan
Norikazu Takahashi	Okayama University, Japan
Abdelmalik Taleb-Ahmed	Polytechnic University of Hauts-de-France, France
Hakaru Tamukoh	Kyushu Institute of Technology, Japan
Choo Jun Tan	Wawasan Open University, Malaysia
Chuanqi Tan	BIT, China
Shing Chiang Tan	Multimedia University, Malaysia
Xiao Jian Tan	Tunku Abdul Rahman University of Management and Technology (TAR UMT), Malaysia
Xin Tan	East China Normal University, China
Ying Tan	Peking University, China
Gouhei Tanaka	University of Tokyo, Japan
Yang Tang	East China University of Science and Technology, China
Zhiri Tang	City University of Hong Kong, China
Tanveer Tarray	Islamic University of Science and Technology, India
Chee Siong Teh	Universiti Malaysia Sarawak (UNIMAS), Malaysia
Ya-Wen Teng	Academia Sinica, Taiwan
Gaurish Thakkar	University of Zagreb, Croatia
Medari Tham	St. Anthony's College, India
Selvarajah Thuseethan	Sabaragamuwa University of Sri Lanka, Sri Lanka
Shu Tian	University of Science and Technology Beijing, China
Massimo Tistarelli	University of Sassari, Italy
Abhisek Tiwari	IIT Patna, India
Uma Shanker Tiwary	Indian Institute of Information Technology, Allahabad, India

Alex To	University of Sydney, Australia
Stefania Tomasiello	University of Tartu, Estonia
Anh Duong Trinh	Technological University Dublin, Ireland
Enkhtur Tsogbaatar	Mongolian University of Science and Technology, Mongolia
Enmei Tu	Shanghai Jiao Tong University, China
Eiji Uchino	Yamaguchi University, Japan
Prajna Upadhyay	IIT Delhi, India
Sahand Vahidnia	University of New South Wales, Australia
Ashwini Vaidya	IIT Delhi, India
Deeksha Varshney	Indian Institute of Technology, Patna, India
Sowmini Devi Veeramachaneni	Mahindra University, India
Samudra Vijaya	Koneru Lakshmaiah Education Foundation, India
Surbhi Vijh	JSS Academy of Technical Education, Noida, India
Nhi N. Y. Vo	University of Technology Sydney, Australia
Xuan-Son Vu	Umeå University, Sweden
Anil Kumar Vuppala	IIIT Hyderabad, India
Nobuhiko Wagatsuma	Toho University, Japan
Feng Wan	University of Macau, China
Bingshu Wang	Northwestern Polytechnical University Taicang Campus, China
Dianhui Wang	La Trobe University, Australia
Ding Wang	Beijing University of Technology, China
Guanjin Wang	Murdoch University, Australia
Jiasen Wang	City University of Hong Kong, China
Lei Wang	Beihang University, China
Libo Wang	Xiamen University of Technology, China
Meng Wang	Southeast University, China
Qiu-Feng Wang	Xi'an Jiaotong-Liverpool University, China
Sheng Wang	Henan University, China
Weiqun Wang	Institute of Automation, Chinese Academy of Sciences, China
Wentao Wang	Michigan State University, USA
Yongyu Wang	Michigan Technological University, USA
Zhijin Wang	Jimei University, China
Bunthit Watanapa	KMUTT-SIT, Thailand
Yanling Wei	TU Berlin, Germany
Guanghui Wen	RMIT University, Australia
Ari Wibisono	Universitas Indonesia, Indonesia
Adi Wibowo	Diponegoro University, Indonesia
Ka-Chun Wong	City University of Hong Kong, China

Kevin Wong	Murdoch University, Australia
Raymond Wong	Universiti Malaya, Malaysia
Kuntpong Woraratpanya	King Mongkut's Institute of Technology Ladkrabang (KMITL), Thailand
Marcin Woźniak	Silesian University of Technology, Poland
Chengwei Wu	Harbin Institute of Technology, China
Jing Wu	Shanghai Jiao Tong University, China
Weibin Wu	Sun Yat-sen University, China
Hongbing Xia	Beijing Normal University, China
Tao Xiang	Chongqing University, China
Qiang Xiao	Huazhong University of Science and Technology, China
Guandong Xu	University of Technology Sydney, Australia
Qing Xu	Tianjin University, China
Yifan Xu	Huazhong University of Science and Technology, China
Junyu Xuan	University of Technology Sydney, Australia
Hui Xue	Southeast University, China
Saumitra Yadav	IIIT-Hyderabad, India
Shekhar Yadav	Madan Mohan Malaviya University of Technology, India
Sweta Yadav	University of Illinois at Chicago, USA
Tarun Yadav	Defence Research and Development Organisation, India
Shankai Yan	Hainan University, China
Feidiao Yang	Microsoft, China
Gang Yang	Renmin University of China, China
Haiqin Yang	International Digital Economy Academy, China
Jianyi Yang	Shandong University, China
Jinfu Yang	BJUT, China
Minghao Yang	Institute of Automation, Chinese Academy of Sciences, China
Shaofu Yang	Southeast University, China
Wachira Yangyuen	Rajamangala University of Technology Srivijaya, Thailand
Xinye Yi	Guilin University of Electronic Technology, China
Hang Yu	Shanghai University, China
Wen Yu	Cinvestav, Mexico
Wenxin Yu	Southwest University of Science and Technology, China
Zhaoyuan Yu	Nanjing Normal University, China
Ye Yuan	Xi'an Jiaotong University, China
Xiaodong Yue	Shanghai University, China

Aizan Zafar   Indian Institute of Technology Patna, India
Jichuan Zeng   Bytedance, China
Jie Zhang   Newcastle University, UK
Shixiong Zhang   Xidian University, China
Tianlin Zhang   University of Manchester, UK
Mingbo Zhao   Donghua University, China
Shenglin Zhao   Zhejiang University, China
Guoqiang Zhong   Ocean University of China, China
Jinghui Zhong   South China University of Technology, China
Bo Zhou   Southwest University, China
Yucheng Zhou   University of Technology Sydney, Australia
Dengya Zhu   Curtin University, Australia
Xuanying Zhu   ANU, Australia
Hua Zuo   University of Technology Sydney, Australia

Additional Reviewers

Acharya, Rajul   Doborjeh, Maryam
Afrin, Mahbuba   Dong, Zhuben
Alsuhaibani, Abdullah   Dutta, Subhabrata
Amarnath   Dybala, Pawel
Appicharla, Ramakrishna   El Achkar, Charbel
Arora, Ridhi   Feng, Zhengyang
Azar, Joseph   Galkowski, Tomasz
Bai, Weiwei   Garg, Arpit
Bao, Xiwen   Ghobakhlou, Akbar
Barawi, Mohamad Hardyman   Ghosh, Soumitra
Bhat, Mohammad Idrees Bhat   Guo, Hui
Cai, Taotao   Gupta, Ankur
Cao, Feiqi   Gupta, Deepak
Chakraborty, Bodhi   Gupta, Megha
Chang, Yu-Cheng   Han, Yanyang
Chen   Han, Yiyan
Chen, Jianpeng   Hang, Bin
Chen, Yong   Harshit
Chhipa, Priyank   He, Silu
Cho, Joshua   Hua, Ning
Chongyang, Chen   Huang, Meng
Cuenat, Stéphane   Huang, Rongting
Dang, Lili   Huang, Xiuyu
Das Chakladar, Debashis   Hussain, Zawar
Das, Kishalay   Imran, Javed
Dey, Monalisa   Islam, Md Rafiqul

Jain, Samir
Jia, Mei
Jiang, Jincen
Jiang, Xiao
Jiangyu, Wang
Jiaxin, Lou
Jiaxu, Hou
Jinzhou, Bao
Ju, Wei
Kasyap, Harsh
Katai, Zoltan
Keserwani, Prateek
Khan, Asif
Khan, Muhammad Fawad Akbar
Khari, Manju
Kheiri, Kiana
Kirk, Nathan
Kiyani, Arslan
Kolya, Anup Kumar
Krdzavac, Nenad
Kumar, Lov
Kumar, Mukesh
Kumar, Puneet
Kumar, Rahul
Kumar, Sunil
Lan, Meng
Lavangnananda, Kittichai
Li, Qian
Li, Xiaoou
Li, Xin
Li, Xinjia
Liang, Mengnan
Liang, Shuai
Liquan, Li
Liu, Boyang
Liu, Chang
Liu, Feng
Liu, Linjing
Liu, Xinglan
Liu, Xinling
Liu, Zhe
Lotey, Taveena
Ma, Bing
Ma, Zeyu
Madanian, Samaneh

Mahata, Sainik Kumar
Mahmud, Md. Redowan
Man, Jingtao
Meena, Kunj Bihari
Mishra, Pragnyaban
Mistry, Sajib
Modibbo, Umar Muhammad
Na, Na
Nag Choudhury, Somenath
Nampalle, Kishore
Nandi, Palash
Neupane, Dhiraj
Nigam, Nitika
Nigam, Swati
Ning, Jianbo
Oumer, Jehad
Pandey, Abhineet Kumar
Pandey, Sandeep
Paramita, Adi Suryaputra
Paul, Apurba
Petinrin, Olutomilayo Olayemi
Phan Trong, Dat
Pradana, Muhamad Hilmil Muchtar Aditya
Pundhir, Anshul
Rahman, Sheikh Shah Mohammad Motiur
Rai, Sawan
Rajesh, Bulla
Rajput, Amitesh Singh
Rao, Raghunandan K. R.
Rathore, Santosh Singh
Ray, Payel
Roy, Satyaki
Saini, Nikhil
Saki, Mahdi
Salimath, Nagesh
Sang, Haiwei
Shao, Jian
Sharma, Anshul
Sharma, Shivam
Shi, Jichen
Shi, Jun
Shi, Kaize
Shi, Li
Singh, Nagendra Pratap
Singh, Pritpal

Singh, Rituraj
Singh, Shrey
Singh, Tribhuvan
Song, Meilun
Song, Yuhua
Soni, Bharat
Stommel, Martin
Su, Yanchi
Sun, Xiaoxuan
Suryodiningrat, Satrio Pradono
Swarnkar, Mayank
Tammewar, Aniruddha
Tan, Xiaosu
Tanoni, Giulia
Tanwar, Vishesh
Tao, Yuwen
To, Alex
Tran, Khuong
Varshney, Ayush
Vo, Anh-Khoa
Vuppala, Anil
Wang, Hui
Wang, Kai
Wang, Rui
Wang, Xia
Wang, Yansong

Wang, Yuan
Wang, Yunhe
Watanapa, Saowaluk
Wenqian, Fan
Xia, Hongbing
Xie, Weidun
Xiong, Wenxin
Xu, Zhehao
Xu, Zhikun
Yan, Bosheng
Yang, Haoran
Yang, Jie
Yang, Xin
Yansui, Song
Yu, Cunzhe
Yu, Zhuohan
Zandavi, Seid Miad
Zeng, Longbin
Zhang, Jane
Zhang, Ruolan
Zhang, Ziqi
Zhao, Chen
Zhou, Xinxin
Zhou, Zihang
Zhu, Liao
Zhu, Linghui

Contents – Part I

Theory and Algorithms

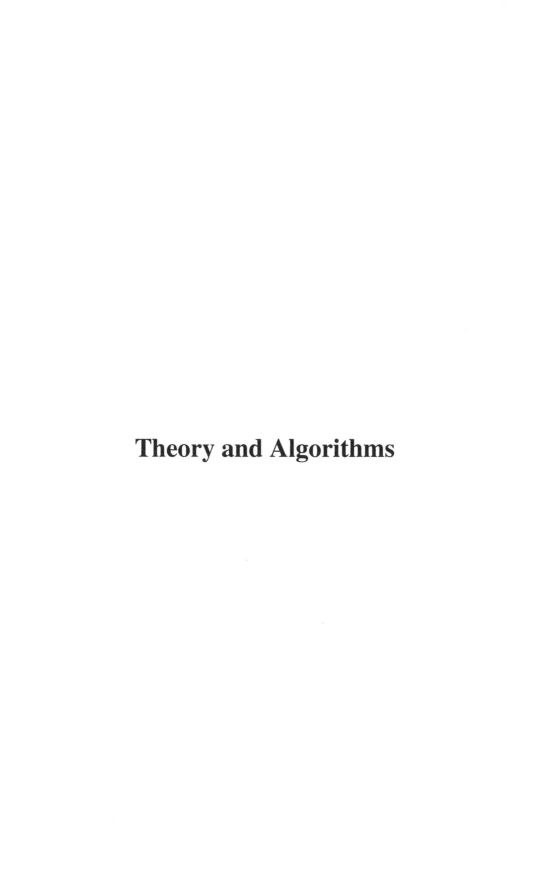

Theory and Algorithms

Solving Partial Differential Equations Using Point-Based Neural Networks

Ning Hua and Wenlian Lu[✉]

School of Mathematical Sciences, Fudan University, Shanghai 200433, China
{nhua19,wenlian}@fudan.edu.cn

Abstract. Recently, solving partial differential equations (PDEs) using neural networks (NNs) has been attracting increasing interests with promising potential to be applied in wide areas. In this paper, we propose a theoretical model that approximates the operator from a parametric function space to a solution function space and prove its universal approximation theorem in the operator space. For practical application, by regarding the domain of a parametric (or solution) function as a discrete point cloud, we propose a novel idea that implements the theoretical model by introducing the point-based NN as the backbone. We show that the present model can approximate the solution operator of static PDEs using the training data generated on unstructured meshes while most existing methods work for the data generated on lattice grid meshes. We conduct experiments to demonstrate the performance of our model on different types of PDEs. The numerical results verify that our model possesses a higher precision and a faster inference speed compared with the existing models for data of unstructured meshes; in addition, our model has competitive performance compared with the existing works dealing with data of lattice grid meshes.

Keywords: Neural network · Partial differential equation · Universal approximation · Point cloud

1 Introduction

Partial differential equations (PDEs) are now employed in various fields. PDEs are used to mathematically model phenomena in physics and chemistry, and can also reveal complex social dynamics. For example, the Navier-Stokes equations describe the motion of viscous fluid substances which are widely used in meteorology and engineering design, among other fields. The Schrödinger equation is the basic equation of quantum mechanics, and it describes the law that governs the state of microscopic particles changing with time. However, it is time-consuming to solve complicated PDEs using conventional numerical methods. With the rapid development of computer simulation and sensor technology, researchers can now easily obtain massive amounts of experimental data. Therefore, data-driven methods are becoming more and more attractive. Among numerous data-driven methods, neural networks (NNs) receive the most attention because of their powerful feature extraction capabilities and efficient computation.

© The Author(s), under exclusive license to Springer Nature Switzerland AG 2023
M. Tanveer et al. (Eds.): ICONIP 2022, LNCS 13623, pp. 3–14, 2023.
https://doi.org/10.1007/978-3-031-30105-6_1

The use of neural networks to solve PDEs can be traced back to the last century. In the 1990s, some researchers tried to represent the solutions to differential equations using NNs [1]. They divided solutions into two components based on the initial/boundary conditions and constructed NNs after eliminating the effects of those conditionalities. In the past few years, NN-based PDE solvers have been introduced to various disciplines. Many approaches embed NNs into classical numerical methods [2–4]. These approaches have been widely studied in computational fluid dynamics (CFD). For example, [5] combines NNs with Reynolds Averaged Navier Stokes (RANS) models by using deep neural networks to learn the Reynolds stress. [6] obtains the closure terms of Large Eddy Simulation (LES) by learning data from the coarse grid quantities. Such approaches usually have comparable performance with conventional methods. There are also many pure neural network approaches. These approaches achieve extremely fast inference speed through end-to-end training without considering too many physical constraints [9–12].

Our study finds that two different perspectives can classify most existing NN-based methods while considering their objects or application scenarios.

Approximating a Single Solution or the Solution Operator. Consider the following PDE:

$$\begin{cases} \mathcal{L}_a(u) &= f, \ x \in \Omega, \ a(\cdot) \in K_1, \ f(\cdot) \in K_2, \\ \mathcal{B}(u) &= g, \ x \in \partial\Omega, \ g(\cdot) \in K_3, \end{cases} \tag{1}$$

where $\mathcal{L}$ is the differential operator and $\mathcal{B}$ is the boundary condition (e.g., for a zero boundary condition, $\mathcal{B}(u) = u \equiv 0$). Given functions $a(\cdot)$, $f(\cdot)$, and $g(\cdot)$, we need to obtain solution $u(\cdot)$. One direct strategy represents the solution, $u(\cdot)$, using an NN, which we call an "approximating a single solution". Based on Monte Carlo sampling and the convenience of automatic differentiation, the NN will approach the true solution during training. PINN treats the residual error between the right- and left-hand sides of (1) as a loss function and impose punishment according to the initial/boundary conditions [7]. In [8], the authors apply an NN in the Galerkin setting and train the network by solving an energy minimization problem. This type of training strategy is usually self-supervised and is more convenient to implement. However, once the initial/boundary conditions or parameters in the equations have changed, the model must be trained again, which is inefficient.

Another type of solution strategy is called the "approximating a solution operator". We assume the solution $u(\cdot) \in H$, where H is a function space. NNs are designed to approximate the solution operator $\mathcal{G} : K_1 \to H$ (more generally, $\mathcal{G} : K_1 \times K_2 \times K_3 \to H$). This task is more difficult and these models usually have lower precision when compared with "approximating a single solution" models. However, once the model has been well-trained, given any $a(\cdot)$ sampled from K_1, it can output the corresponding solution $u(\cdot)$ in less than a second. The neural operator family [9–11] builds a series of powerful and efficient models that learn operators between infinite dimensional spaces. In this paper, we concentrate only on "approximating a solution operator" approaches.

Dependent or Independent on Structured Meshes. Most existing models are suitable only for structured meshes, which means that every training sample is arranged on a Cartesian domain with a lattice grid mesh. Those data can be treated as images. Then, those models use CNNs to learn the image-to-image mapping [11,12]. The convolution operation with small convolution kernels is very fast and can sufficiently utilize the data's spatial structure. However, in many cases, the data is arranged unevenly; this phenomenon occurs with data derived from the finite element method (FEM) on an irregular domain and, data collected from sensors, among other examples. Therefore, these methods have limitations when processing real scenes. To overcome this challenge, [9,10] conceive of the input function as a graph and utilize graph neural networks (GNNs) to extract the features among the nodes and edges.

1.1 Point-Based Neural Networks

In addition to GNNs, another family of networks, point-based NNs, show great potential in scenarios that results in irregular data. Currently, processing 3D irregular data has gradually become a popular topic in deep learning. This type of data has a generic name, point clouds. Typical neural network architectures for image processing cannot adapt well to such data. PointNet [13] is a pioneer in directly processing point clouds. Every point is input into the model independently using weight-sharing multi-layer perceptrons (MLP), and the features are extracted and aggregated using a symmetry function. PointNet preserves the point clouds' spatial characteristics to the greatest extent. Inspired by this work, numerous point-based NNs have been proposed [14–16] and they have demonstrated their capabilities in various fields such as shape classification, object detection and segmentation. One primary difference between point-based NNs and GNNs is that point-based NNs need only the features of points (named nodes in GNNs), while GNNs the need features of both the nodes and edges. In this paper, we aim to solve parametric PDEs using point-based NNs.

1.2 Contributions

We propose a theoretical model that approximates the solution operator of parametric PDEs, and prove a theorem of its universal approximation capability in the nonlinear operator space. Based on the theoretical framework, for practical application, we introduce the point-based NNs as the backbone to approximate the solution operator of parametric PDEs. Unlike most existing models, which apply only to data of a structured mesh, our model works very well with the solution operators using unstructured training data (see Fig. 3(a) and 3(b)). We conduct several experiments to verify that our model has superior performance and a fast inference speed particularly on the data of unstructured meshes.

2   Theoretical Framework

2.1   Descriptions of the Task

In this section, we use the following parametric PDE as an example:

$$\mathcal{L}_a(u) = f, \ x \in \Omega \subset \mathbb{R}^d, \ a(\cdot) \in K. \tag{2}$$

This PDE is combined with a specific boundary condition to ensure the equation has a unique solution and that the space of the solution function $u(\cdot)$ is defined as H. If $f(\cdot)$ is fixed, the solution operator is $\mathcal{G} : K \to H$, $a(\cdot) \mapsto u(\cdot)$. We remark that if $a(\cdot)$ is fixed, then the solution operator is $\mathcal{G}(f) = u$. Additionally, in this paper, we assume that K and H have the same domain Ω.

An NN $\mathcal{F} : \mathbb{R}^{n_1} \to \mathbb{R}^{n_2}$ is a nonlinear function composed of a series of affine transformations and nonlinear scalar activation functions. Given any vector $x \in \mathbb{R}^{n_1}$, $\mathcal{F}(x) = W_L \sigma(\cdots \sigma(W_2 \sigma(W_1 x + b_1) + b_2 \cdots)) + b_L$ is an $L-$layer NN. For any layer $i \in \{1, 2, \cdots, L\}$, $W_i \in \mathbb{R}^{d_{i-1} \times d_i}$ and $b_i \in \mathbb{R}^{d_i}$ ($d_0 = n_1, d_L = n_2$), and σ is a non-polynomial and nonlinear activation function such as the Rectified Linear Unit (ReLU). $W_i, b_i, i = 1, 2, \cdots, L$ are trainable parameters of the NN.

Considering that the dimension of H (or K) is always infinite when any NN has a finite number of output (or input) neurons, we select a point set $\{x_1, x_2, \cdots, x_n\}$ from Ω and use $H_n := \{(u(x_1), u(x_2), \cdots, u(x_n)) : u \in H\}$ to represent H in a finite dimensional space. In Sect. 2.2, we find that when n is large enough, combined with a specific interpolation operator, H_n can approximate H with any given precision. In this way, if $u(\cdot)$ is a scalar function, the original solution operator $\mathcal{G}$ is simplified to $\mathcal{G}_n : K_n \to H_n$, $(a(x_1), \cdots, a(x_n)) \mapsto (u(x_1), \cdots, u(x_n))$, and n neurons output each of those values. For vector functions, the following theory still holds since we only need to multiply the number of output neurons by the corresponding dimension. With some constrains, there exists an NN that can approximate $\mathcal{G}_n$ based on the traditional universal approximation theorems [17,18]. Figure 1 illustrates the pipeline to approximate the solution operator using NNs.

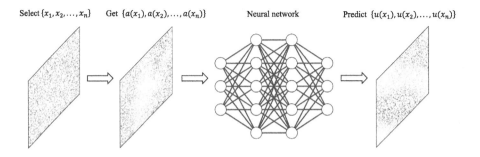

Select $\{x_1, x_2, ..., x_n\}$ Get $\{a(x_1), a(x_2), ..., a(x_n)\}$ Neural network   Predict $\{u(x_1), u(x_2), ..., u(x_n)\}$

Fig. 1. The method overview.

2.2 Discretization Operator and Interpolation Operator

Given a point set $\{x_1, x_2, \cdots, x_n\} \subset \Omega$, we define a mapping

$$\mathcal{D}(a) := (a(x_1), a(x_2), \cdots, a(x_n)), \ \forall a(\cdot) \in K, \tag{3}$$

which is called the discretization operator. As mentioned above, we first use the discretization operator to get the finite-dimensional representation of the input function and then feed it into the NN. The NN predicts the values of the corresponding solution $u(\cdot)$ on $\{x_1, x_2, \cdots, x_n\}$. Then we also need to propose an interpolation operator. Considering that Ω is compact in $\mathbb{R}^d$, for any $\eta > 0$, there is a finite $\eta-$net $\{x_1, x_2, \cdots, x_{n(\eta)}\} \subset \Omega$. For any $a(\cdot) \in K$, we define the interpolation operator as

$$\mathcal{R}_\eta(a(x_1), a(x_2), \cdots, a(x_{n(\eta)})) := \sum_{i=1}^{n(\eta)} a(x_i) T_{\eta,i}(x), \tag{4}$$

$$T_{\eta,i}(x) = \frac{T_{\eta,i}^*(x)}{\sum_{j=1}^{n(\eta)} T_{\eta,j}^*(x)}, T_{\eta,i}^*(x) = \begin{cases} 1 - \frac{\|x - x_i\|_{\mathbb{R}^d}}{\eta} & \text{if } \|x - x_i\|_{\mathbb{R}^d} \le \eta \\ 0 & \text{otherwise.} \end{cases}$$

We follow [19] to define $T_{\eta,i}$ and $T_{\eta,i}^*$. The interpolation operator (4) has three important properties as shown in the following lemma.

Lemma 1. *Suppose Ω is compact in $\mathbb{R}^d$ and K is compact in $C(\Omega)$. For any $\delta > 0$, there exists $\eta > 0$, a finite $\eta-$net $\{x_1, x_2, \cdots, x_{n(\eta)}\} \subset \Omega$, and the corresponding discretization and interpolation operators $\mathcal{D}_\eta$ and $\mathcal{R}_\eta$ such that:*

(1) $\|\mathcal{R}_\eta \circ \mathcal{D}_\eta(a) - a\|_{C(\Omega)} < \delta, \ \forall a(\cdot) \in K$.
(2) $K_\eta := \{\mathcal{R}_\eta \circ \mathcal{D}_\eta(a) : a(\cdot) \in K\}$ is compact in $C(\Omega)$ with $n(\eta)-$dimension.
(3) Given a sequence $\eta_1 > \eta_2 > \cdots > \eta_n \to 0$, $K^* := K \cup (\cup_{i=1}^\infty K_{\eta_i})$ is compact in $C(\Omega)$.

The proof of Lemma 1 can be found in [19]. The first conclusion in Lemma 1 indicates that we can find enough points in Ω and use the discrete function values to reconstruct the original function by the interpolation operator for any given tolerance δ. Additionally, those points are fixed for any $a(\cdot) \in K$ if K is compact. According to the discretization and interpolation operators, we formulate our solution operator approximator as $\mathcal{R} \circ \mathcal{F} \circ \mathcal{D}(a) \approx \mathcal{G}(a)$, where $\mathcal{F}$ is an NN.

2.3 Main Theorem

In this section, we will prove the universal approximation theorem of NNs to learn the solution operator in (2) with the help of the discretization and interpolation operators. Before establishing our theory, we must make some assumptions: **(A1)** The domain Ω is compact in $\mathbb{R}^d$. **(A2)** K is compact in $C(\Omega)$, and $H = C(\Omega)$. **(A3)** The solution operator G is continuous. Based on the above definitions and assumptions, we propose our main theorem as follows:

Theorem 1. *Under assumptions **(A1)-(A3)**, for any $\epsilon > 0$, there exists $n \in \mathbb{N}^+$, $\{x_1, x_2, \cdots, x_n\} \subset \Omega$, the corresponding discretization and interpolation operators $\mathcal{D}$ and $\mathcal{R}$ defined in (3) and (4), and an NN $\mathcal{F} : \mathbb{R}^n \to \mathbb{R}^n$, such that*

$$\|\mathcal{R} \circ \mathcal{F} \circ \mathcal{D}(a) - \mathcal{G}(a)\|_{C(\Omega)} < \epsilon, \ \forall a(\cdot) \in K. \tag{5}$$

Proof. Since $\mathcal{G}$ is continuous and K is compact in $C(\Omega)$, $\mathcal{G}(K) := \{\mathcal{G}(a) : a(\cdot) \in K\}$ is compact in $C(\Omega)$. Then $V := K \cup \mathcal{G}(K)$ is compact in $C(\Omega)$. Now we choose a sequence $\delta_1 > \delta_2 > \cdots > \delta_n \to 0$. According to Lemma 1, there exists a corresponding sequence $\eta_1 > \eta_2 > \cdots > \eta_n \to 0$. For each $\eta_i, i = 1, 2, \cdots$, there is a finite η_i-net $\{x_1^i, x_2^i, \cdots, x_{n(\eta_i)}^i\} \subset \Omega$, and the corresponding discretization and interpolation operators $\mathcal{D}_{\eta_i}$ and $\mathcal{R}_{\eta_i}$ such that

$$\|\mathcal{R}_{\eta_i} \circ \mathcal{D}_{\eta_i}(a) - a\|_{C(\Omega)} < \delta_i, \ \forall a(\cdot) \in V.$$

Recall that we denote $V_{\eta_i} := \{\mathcal{R}_{\eta_i} \circ \mathcal{D}_{\eta_i}(a) : a(\cdot) \in V\}$. Then $V^* := V \cup (\cup_{i=1}^{\infty} V_{\eta_i})$ is compact in $C(\Omega)$ based on Lemma 1. Since K is a closed set, according to a generalization of the Tietze extension theorem [20], we can find an extended $\mathcal{G} : K \to C(\Omega)$ denoted by $\widetilde{\mathcal{G}} : C(\Omega) \to C(\Omega)$ such that $\widetilde{\mathcal{G}}$ is continuous and $\widetilde{\mathcal{G}}(a) = \mathcal{G}(a)$ for all $a(\cdot) \in K$. Considering that V^* is compact and $\widetilde{\mathcal{G}}$ is continuous, there exists $\delta_j, j \in \{1, 2, \cdots\}$, such that

$$\|\widetilde{\mathcal{G}}(a') - \widetilde{\mathcal{G}}(a'')\|_{C(\Omega)} < \frac{\epsilon}{3} \tag{6}$$

whenever $a'(\cdot), a''(\cdot) \in V^*$ and $\|a' - a''\|_{C(\Omega)} < \delta_j$.

Without loss of generality, we assume $\delta_j < \epsilon/3$. We fix j and denote $n = n(\eta_j)$, $\{x_1, x_2, \cdots, x_n\} = \{x_1^j, x_2^j, \cdots, x_{n(\eta_j)}^j\}$, $\mathcal{D} = \mathcal{D}_{\eta_j}$, and $\mathcal{R} = \mathcal{R}_{\eta_j}$. Then

$$\|\mathcal{R} \circ \mathcal{D}(a) - a\|_{C(\Omega)} < \delta_j < \frac{\epsilon}{3}, \ \forall a(\cdot) \in V. \tag{7}$$

Now let us consider the operator $\mathcal{N} := \mathcal{D} \circ \widetilde{\mathcal{G}} \circ \mathcal{R} : \mathbb{R}^n \supset \mathcal{D}(K) \to \mathbb{R}^n$. Since $\mathcal{D}, \mathcal{R}$ are Lipschitz continuous with Lipschitz constants $\text{Lip}(\mathcal{D}), \text{Lip}(\mathcal{R}) \leq 1$ by the property of unity partition and $\widetilde{\mathcal{G}}$ is continuous, $\mathcal{N}$ is continuous. Additionally, the domain $\mathcal{D}(K)$ is compact. Then based on the traditional universal approximation theorems [17,18], there exists an NN $\mathcal{F} : \mathbb{R}^n \to \mathbb{R}^n$ such that

$$\|\mathcal{F}(a(x_1), \cdots, a(x_n)) - \mathcal{N}(a(x_1), \cdots, a(x_n))\|_{l^{\infty}} < \frac{\epsilon}{3}, \tag{8}$$

for any $a(\cdot) \in K$.

We decompose the approximation error into three parts. According to the triangle inequality and the Lipschitz continuity, we have

$$\|\mathcal{R} \circ \mathcal{F} \circ \mathcal{D}(a) - \mathcal{G}(a)\|_{C(\Omega)}$$
$$\leq \|\mathcal{R} \circ \mathcal{F} \circ \mathcal{D}(a) - \mathcal{R} \circ \mathcal{D} \circ \widetilde{\mathcal{G}} \circ \mathcal{R} \circ \mathcal{D}(a)\|_{C(\Omega)} +$$
$$\|\mathcal{R} \circ \mathcal{D} \circ \widetilde{\mathcal{G}} \circ \mathcal{R} \circ \mathcal{D}(a) - \mathcal{R} \circ \mathcal{D} \circ \mathcal{G}(a)\|_{C(\Omega)} + \|\mathcal{R} \circ \mathcal{D} \circ \mathcal{G}(a) - \mathcal{G}(a)\|_{C(\Omega)} \tag{9}$$
$$\leq \|\mathcal{F} \circ \mathcal{D}(a) - \mathcal{D} \circ \widetilde{\mathcal{G}} \circ \mathcal{R} \circ \mathcal{D}(a)\|_{l^{\infty}} +$$
$$\|\widetilde{\mathcal{G}} \circ \mathcal{R} \circ \mathcal{D}(a) - \mathcal{G}(a)\|_{C(\Omega)} + \|\mathcal{R} \circ \mathcal{D} \circ \mathcal{G}(a) - \mathcal{G}(a)\|_{C(\Omega)}$$
$$:= T_1 + T_2 + T_3.$$

Equations (8), (6), and (7) indicate that $T_1 < \epsilon/3$, $T_2 < \epsilon/3$, and $T_3 < \epsilon/3$ respectively. Therefore, $\|\mathcal{R} \circ \mathcal{F} \circ \mathcal{D}(a) - \mathcal{G}(a)\|_{C(\Omega)} < \epsilon$ for any $a(\cdot) \in K$.

3 Framework of the Practical Model

In practical situations, we only access to data pairs:

$$\{(a_k(x_1), \cdots, a_k(x_n), u_k(x_1), \cdots, u_k(x_n))\}_{k=1}^{N},$$

where N is the dataset size, $a_k(\cdot)$ is sampled from K and $u_k(\cdot)$ is the corresponding solution. Such discretized data exactly match our theory. If the points $\{x_1, \cdots, x_n\}$ are arranged on a Cartesian domain with a lattice grid mesh, based on this inductive bias, the CNN is a good choice to learn the mapping between two discretized domains. This, however, is not always the case. When PDE solutions are generated through FEM on an irregular domain or collected from sensors, the points $\{x_1, \cdots, x_n\}$ are always arranged on the unstructured meshes. CNNs are not suitable for these data and fully-connected networks cannot fit the data well when the number of points n is large. If we treat $\{x_1, \cdots, x_n\}$ as a point cloud with additional features $\{a(x_1), \cdots, a(x_n)\}$, a point-based deep learning framework shows great potential. Point-based NNs have been widely used in computer vision and computer graphics for two and three dimensional tasks. In this paper, we build our framework of approximating solution operators based on the point-based NN. There are various point-based NNs proposed in recent years. We choose PointConv [16] as our model since its convolution operation is similar to the traditional CNN and it demonstrates outstanding performance in the point cloud segmentation task (a type of point-to-point prediction task that is similar to ours). We will briefly introduce PointConv in the next paragraph. One may refer to [16] for more details.

PointConv includes a novel convolution for data on the unstructured mesh. Convolution is defined as a Monte Carlo integral. The convolution kernels are represented by MLPs and reweighted by a kernelized density estimation. The architecture of the PointConv is similar to U-Net [21] that has been widely used in image segmentation tasks. PointConv consists of convolution layers and deconvolution layers. The convolutional part and the deconvolutional part are symmetrical and share latent features using skip connections. Figure 2 briefly demonstrates the model structure. Suppose $a(\cdot)$ is a scalar function on the domain $\Omega \subset \mathbb{R}^2$. Then $\{(x_j, a(x_j))\}_{j=1,2,\cdots,n}$ are fed into the input layer. Thus the symbol n in Fig. 2 represents the number of points and $k_0 = 3$. The first three layers are convolution layers. The number four to six layers are deconvolution layers. The last two layers are point-wise fully connected layers. Except for the last layer, we impose the ReLU activation function on each layer. Some other hyperparameters will be discussed in the next section.

4 Experiments

In this section, we test our models on two different PDEs: the 2D Poisson's equation and the Darcy Flow equation. We use numerical methods to sample $a(\cdot)$

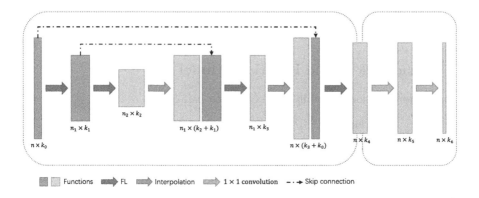

Fig. 2. The structure of the PointConv.

($f(\cdot)$ in the 2D Poisson's equation) and get the corresponding solution. Unless otherwise stated, the size of the training set, validation set, and testing set are 1000, 100, and 1000 respectively. We choose the PointConv with four convolution (deconvolution) layers. The kernel size of the convolution (deconvolution) layers in the PointConv is 32 (16). During training, all data are first normalized to $[0, 1]$. We choose the L2 relative error as the loss function and minimize it using Adam. We train the model with 500 epochs with an initial learning rate of 0.001 which decays by half every 200 epochs. Recall that $\mathcal{G}_n$ is the discretized version of $\mathcal{G}$. In practice, to train and compare our models, we do not interpolate the discretized predicted values using $\mathcal{R}$. Thus we actually use a discretized version of the loss function

$$l(\mathcal{G}_n, \mathcal{F}) := \sum_{k=1}^{N} \frac{\sqrt{\sum_{j=1}^{n}(\widehat{u}_k(x_j) - u_k(x_j))^2}}{\sqrt{\sum_{j=1}^{n} u_k^2(x_j)}}. \tag{10}$$

4.1   The 2D Poisson's Equation

Equation (11) is Poisson's equation with the Dirichlet boundary condition. $f(\cdot)$ is a source function, and we predict the corresponding solution $u(\cdot)$ by approximating $\mathcal{G}_n$ with our point-based model.

$$\begin{cases} \Delta u(x) &= f(x), \ x \in \Omega, f(\cdot) \in K, \\ u(x) &= 0, \ x \in \partial\Omega. \end{cases} \tag{11}$$

In this experiment, we consider two datasets of different types of meshes.

Case One. We set $\Omega = [-1, 1] \times [-1, 1]$ and K refers to a Gaussian random field (GRF) with length scale $= 3$. We divide the area Ω evenly into an 80×80 grid and generate the dataset using the finite difference method (FDM). Then we unevenly sample 3200 points from the 80×80 mesh. The points in the center are densely sampled while the surrounding points are sparsely sampled. Therefore, the input is $\{(x_j, f(x_j))\}_{j=1,2,\cdots,3200}$ and the output is the predicted $\{\widehat{u}(x_j)\}_{j=1,2,\cdots,3200}$. Finally, we choose the model with the lowest validation loss during 500 epochs and test it using the testing set. We compare our model with two graph-based models, the graph neural operator (GNO) [9] and the multipole graph neural operator (MGNO) [10], which are also designed for unstructured meshes. Table 1 shows that our model has the lowest testing loss while the other two models fail to approximate the solution operator using only 1000 training samples. Inference time per sample is another important indicator, especially for unstructured-mesh-based models, since they require more intensive computation. We record the inference times per sample for these three models on a single Titan Xp GPU. According to Table 1, our model is significantly faster than GNO and MGNO. We also select a testing result and illustrate it in Fig. 3(a). The result reveals that our point-based model can approximate the solution operator in high precision using data on the unstructured mesh.

Case 2. We also consider the Poisson's equation on the irregular domain. As Fig. 3(b) shows, the domain Ω is a rectangle with three holes of different shapes. $f(\cdot)$ is sampled from the following space

$$f(x) = \sum_{i=0}^{4} \sum_{j=0}^{4} \beta_{ij} \sin([i\pi, j\pi] \cdot x^T),$$

where β_{ij} is sampled from the standard normal distribution. We generate the dataset using the FEM and the number of points $n = 1780$. We select a testing result and illustrate them in Fig. 3(b). The numerical results in Table 1 further verify that our point-based model can efficiently and effectively approximate the solution operator on the unstructured mesh.

Table 1. Testing loss and inference time on the unstructured meshes

Model	Testing loss (Case One/Two)	Inference time (s)
GNO [9]	0.2796/0.0997	0.1922/0.1498
MGNO [10]	0.1976/0.0657	0.4081/0.3008
Ours	**0.0397/0.0435**	**0.0622/0.0509**

4.2 The 2D Darcy Flow Equation

The Darcy Flow equation is applied in numerous fields such as hydromechanics and the mechanics of materials. Equation (12) describes this equation with the Dirichlet boundary condition. $f(\cdot)$ is the source function and $a(\cdot)$ is the diffusion coefficient. We fix $f(\cdot)$ and utilize our model to approximate the solution operator from $a(\cdot)$ to $u(\cdot)$. [9–12] have tested their models on this equation, so it is wise to treat the 2D Darcy Flow equation as a benchmark to compare our model with other NN-based PDE solvers.

$$\begin{cases} -\nabla \cdot (a(x)\nabla u(x)) & = f(x), \ x \in \Omega, a \in K, \\ u(x) & = 0, \ x \in \partial\Omega. \end{cases} \quad (12)$$

Instructed by [11], we set $\Omega = (0,1) \times (0,1)$ and keep $f(x) \equiv 1$. We generate $a(x)$ based on the distribution $a \sim \mu$ where $\mu = \psi_\# \mathcal{N}\left(0, (-\Delta + 9I)^{-2}\right)$ with a Neumann boundary condition on the operator $-\Delta + 9I$. The mapping $\psi_\#$

Table 2. Relative L2 error on the testing set (2D Darcy Flow Equation)

	Model	Testing loss
Structured-mesh-based	NN	0.1716
	FCN	0.0253
	PCANN [12]	0.0299
	FNO [11]	0.0108
Graph-based	GNO [9]	0.0346
	MGNO [10]	0.0416
Point-based	Ours	**0.0088**

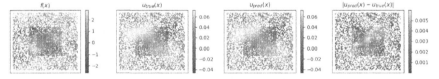

(a) Poisson's Equation on the unstructured mesh (Case 1).

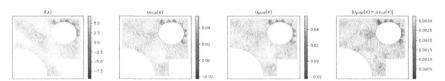

(b) Poisson's Equation on the unstructured mesh (Case 2).

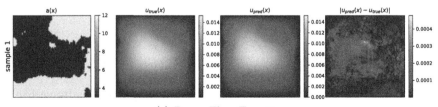

(c) Darcy Flow Equation.

Fig. 3. Visualization of results sampled from the testing set.

converts the positive values to 12 and the negative values to 3. The equation is solved by the FDM on a 421×421 grid and downsampled to 85×85.

The performance comparison is shown in Table 2. Our point-based model has the best performance when compared to the others. We note that the FNO in Table 2 uses ReLU activation function. However, if one replaces the ReLU activation by GeLU [22], the testing loss is reduced from 0.0108 to 0.0082, which is slightly better than our model. It is worth noting that our model is significantly better than the graph-based models that can also handle unstructured meshes. A visualization of the result can be found in Fig. 3(c).

5 Conclusions

In this paper, we propose a theoretical framework to approximate the solution operator of parametric PDEs using NNs. Combining the traditional universal approximation theorems with the specific discretization and interpolation operators, we prove the universal approximation capability in the operator space of NNs. To implement the framework in practice, we introduce the point-based NN, PointConv, as the backbone. We demonstrate that our point-based model has a good approximation to the real solution operator after training with data of structured or unstructured meshes. Our model has significantly higher precision and faster inference speed compared with the graph-based methods which can also handle data of unstructured meshes. Besides, our model also has competitive performance compared with the methods including the structured-mesh-based methods. We are convinced that while we primarily tested our model on synthetic data sets, point-based models are capable of handling real-world data collected from experiments and sensors. Our code is available at https://github.com/hanCi422/point-based-model-for-PDEs.

Acknowledgments. This work is jointly supported by the National Key R&D Program of China (No. 2018AAA0100303), the Shanghai Municipal Science and Technology Major Project (No.2018SHZDZX01) and the ZHANGJIANG LAB, the National Natural Science Foundation of China under Grant 62072111.

References

1. Lagaris, I.E., Likas, A., Fotiadis, D.I.: Artificial neural networks for solving ordinary and partial differential equations. IEEE Trans. Neural Networks **9**(5), 987–1000 (1998)
2. He, J., Xu, J.: Mgnet: a unified framework of multigrid and convolutional neural network. Sci. China Math. **62**(7), 1331–1354 (2019)
3. Kochkov, D., Smith, J.A., Alieva, A., Wang, Q., Brenner, M.P., Hoyer, S.: Machine learning-accelerated computational fluid dynamics. Proc. Natl. Acad. Sci. **118**(21), e2101784118 (2021)
4. Ramuhalli, P., Udpa, L., Udpa, S.S.: Finite-element neural networks for solving differential equations. IEEE Trans. Neural Networks **16**(6), 1381–1392 (2005)

5. Ling, J., Kurzawski, A., Templeton, J.: Reynolds averaged turbulence modelling using deep neural networks with embedded invariance. J. Fluid Mech. **807**, 155–166 (2016)
6. Beck, A., Flad, D., Munz, C.D.: Deep neural networks for data-driven les closure models. J. Comput. Phys. **398**, 108910 (2019)
7. Raissi, M., Perdikaris, P., Karniadakis, G.E.: Physics-informed neural networks: a deep learning framework for solving forward and inverse problems involving nonlinear partial differential equations. J. Comput. Phys. **378**, 686–707 (2019)
8. Weinan, E., Yu, B.: The deep Ritz method: a deep learning-based numerical algorithm for solving variational problems. Commun. Math. Stat. **6**(1), 1–12 (2018)
9. Anandkumar, A., et al.: Neural operator: graph kernel network for partial differential equations. In: ICLR 2020 Workshop on Integration of Deep Neural Models and Differential Equations (2020)
10. Li, Z., Kovachki, N., Azizzadenesheli, K., Liu, B., Stuart, A., Bhattacharya, K., Anandkumar, A.: Multipole graph neural operator for parametric partial differential equations. In: Advances in Neural Information Processing Systems, vol. 33 (2020)
11. Li, Z., et al.: Fourier neural operator for parametric partial differential equations. In: International Conference on Learning Representations (2020)
12. Bhattacharya, K., Hosseini, B., Kovachki, N.B., Stuart, A.M.: Model reduction and neural networks for parametric PDEs. The SMAI J. Comput. Math. **7**, 121–157 (2021)
13. Qi, C.R., Su, H., Mo, K., Guibas, L.J.: Pointnet: Deep learning on point sets for 3d classification and segmentation. In: Proceedings of the IEEE Conference on Computer Vision and Pattern Recognition, pp. 652–660 (2017)
14. Qi, C.R., Yi, L., Su, H., Guibas, L.J.: Pointnet++: deep hierarchical feature learning on point sets in a metric space. In: Advances in Neural Information Processing Systems, vol. 30 (2017)
15. Thomas, H., Qi, C.R., Deschaud, J.E., Marcotegui, B., Goulette, F., Guibas, L.J.: Kpconv: flexible and deformable convolution for point clouds. In: Proceedings of the IEEE/CVF International Conference on Computer Vision, pp. 6411–6420 (2019)
16. Wu, W., Qi, Z., Fuxin, L.: Pointconv: deep convolutional networks on 3D point clouds. In: Proceedings of the IEEE/CVF Conference on Computer Vision and Pattern Recognition, pp. 9621–9630 (2019)
17. Hornik, K., Stinchcombe, M., White, H.: Multilayer feedforward networks are universal approximators. Neural Netw. **2**(5), 359–366 (1989)
18. Leshno, M., Lin, V.Y., Pinkus, A., Schocken, S.: Multilayer feedforward networks with a nonpolynomial activation function can approximate any function. Neural Netw. **6**(6), 861–867 (1993)
19. Chen, T., Chen, H.: Universal approximation to nonlinear operators by neural networks with arbitrary activation functions and its application to dynamical systems. IEEE Trans. Neural Networks **6**(4), 911–917 (1995)
20. Dugundji, J.: An extension of Tietze's theorem. Pac. J. Math. **1**(3), 353–367 (1951)
21. Ronneberger, O., Fischer, P., Brox, T.: U-Net: convolutional networks for biomedical image segmentation. In: Navab, N., Hornegger, J., Wells, W.M., Frangi, A.F. (eds.) MICCAI 2015. LNCS, vol. 9351, pp. 234–241. Springer, Cham (2015). https://doi.org/10.1007/978-3-319-24574-4_28
22. Hendrycks, D., Gimpel, K.: Gaussian error linear units (gelus). arXiv preprint arXiv:1606.08415 (2016)

Patch Mix Augmentation with Dual Encoders for Meta-Learning

Hong Yu and Fanzhang Li[✉]

School of Computer Science and Technolgy, Soochow University, Suzhou, China
`20205227028@stu.suda.edu.cn`, `lfzh@suda.edu.cn`

Abstract. Meta-learning aims to learn models that can make quick adaptations to new tasks. However, due to the lack of data, the further improvement of meta-learning can be severely constrained. Since, data augmentation has been a commonly used method to help models reach state-of-art performance in various image classification tasks. It is wise to use data augmentation methods in meta-learning. Different strategies for applying data augmentation to meta-learning have emerged. One common combination of data augmentation and meta-learning is performing different transformations on images. Other methods use generative models, such as GAN, VAE, or AE, to generate samples and expand the data set. In this paper, we proposed a novel data augmentation method aiming to enlarge the number of samples in the support sets. Our approach uses wavelet transform, a widely used method in signal analysis and processing and style mix from AdaIn. Furthermore, we use both ResNet and ViT as our feature encoder. Combining with the idea of contrastive learning, we train our ViT in an unsupervised way. Experimental results show that we achieve a decent performance improvement.

Keywords: Data Augmentation · Meta-Learning · InfoNCE · Wavelet Transform · AdaIN

1 Introduction

Meta-learning [1,2] or learning to learn, has become an area of great interest. Although traditional deep learning has promoted huge progress in computer vision [3–5] using large-scale datasets like ImageNet [6]. However, in the real world, it can be difficult to collect a sufficient amount of data in certain areas. Meta-learning aims to solve this problem. Through training a robust meta-learner on base classes, meta-learning methods aim to make fast adaptation to novel classes with learning from few labeled samples. Meta-learning leverages the experience from prior learning to help reach its goal. In meta-learning, data are grouped into different N-way, K-shot tasks and meta learner uses different tasks $\{\mathcal{T}_1, \mathcal{T}_2, \mathcal{T}_3, \cdots\}$ to train and test. Each task samples from N classes and then divides these samples into support sets and query sets. The support sets of task $\mathcal{T}_i$ can be defined as $S_{\mathcal{T}_i} = \{(X_i, Y_i)\}^{N \times K}$ and query sets as $Q_{\mathcal{T}_i} = \{(X_i, Y_i)\}^{N \times L}$,

M. Tanveer et al. (Eds.): ICONIP 2022, LNCS 13623, pp. 15–26, 2023.
https://doi.org/10.1007/978-3-031-30105-6_2

where L stands for the size of each query set. Meta-learner learns a classifier from the support sets and computes the classification loss on the query sets. Tasks used during training and testing are sampled from completely different sets of categories. In addition, we use $S_{\mathcal{C}_i}$ and $Q_{\mathcal{C}_i}$ to denote the support set and query set of a certain class.

Similar to conventional image classification, we apply data augmentation strategies as a part of the meta-learning pipeline to help meta-learner improve its performance. In meta-learning, commonly used data augmentation methods are different image transformations [7] like random rotation, horizontal flip, color jitter, and random crop, which help improve the performance and robustness of the model. More sophisticated methods like CutMix [8], Mixup [9], and Manifold mixup [10]. Although these methods are proposed within the traditional image classification, they are still able to improve the performance of meta-learning models. Others, use GAN like MetaGAN [11] or VAE like META-GMVAE [12] as a combination to help improve their learning performance.

In this paper, we proposed a novel data augmentation method for meta-learning which focuses on adding the number of samples in the support sets. We use the wavelet transform combined with AdaIn [13] style mix as the augmentation. Unlike Wave-SAN [14], we only apply our augmentation to the patches from different positions of the image instead of the whole picture and the feature maps. To make better use of these mixed patches, we use Vision Transformer(ViT) [15] to encoder these augmented images. To train our ViT module, we borrow the idea from contrastive learning and use InfoNCE [16,17] as the loss to update the ViT. Apart from ViT, we use ResNet to encode the original images and query images. The feature vectors generated by ViT are taken to form new support sets together with those from ResNet.

- We use the wavelet transform combined with AdaIn style mix to augment patches from randomly chosen positions of the image.
- We introduce Vision Transformer as an additional encoder to make better use of those mixed patches. The feature vectors output from ViT are directly used as support samples together with those from ResNet. In this way, we expand the available samples for each support set.
- We use the idea of contrastive learning to train our ViT module in an unsupervised approach by using InfoNCE loss.

2   Related Work

2.1   Meta Learning

Meta-learning intends to train the meta-learner, a model that can adapt to new classes quickly. To achieve this goal, in meta-learning, datasets are organized into many N-way, K-shot tasks. N-way means we sample from N classes and K-shot means from each class we sample K examples to form its support set, the remaining samples are used as query set. Thus, we can simulate fast adaptation to the new environment after seeing limited samples in the real world. The classes used during training and testing are separated. Meta-learning methods can be roughly grouped into three major categories: optimization-based [18, 19], model-based [20,21], and metric-based [22,23]. Prototypical networks [24] is the representative work of metric-based meta-learning methods. In Prototypical networks, the mean of each support set is computed and taken as the prototype of its class. The goal of prototypical networks is to find the best prototypes, which help cluster the samples from their respective class and separate from others of other classes.

2.2 Data Augmentation

Data augmentation methods have been widely used in image classification. A commonly used data augmentation strategy is performing different transformations on images before feeding them to the feature extractor. Frequently used transformations are random rotation, horizontal flip, random crop, color jitter, etc. More complex strategies like CutMix [8], cut and pastes patches among training samples. Mixup [9] mixes two images from different classes and Manifold mixup [10], on the other hand, does this process in the feature space. META-MAXUP [25] incorporates the idea of adversarial learning, for each task, a method is selected that maximizes the classification loss from a set of data augmentation methods, and then this augmentation method will be applied to the task's support sets or query sets. Generative models such as GAN, VAE, and AE, can also be used to perform data augmentation. Delta-encoder [26] modifies the Auto Encoder module to learn the deformation between images from the same class to generate new samples for unseen classes. Example generation methods are especially suitable for few-shot learning. The key point is to choose good side information such as the semantic information from the labels.

3 Method

In meta-learning, we divide all classes into C_{base} and C_{novel}. To test whether our model can adapt to the new environment quickly, we specify that $C_{base} \cap C_{novel} = \emptyset$. The data used for training D_{train} and testing D_{test} are sampled separately: $D_{train} \subset D_{C_{base}}, D_{test} \subset D_{C_{novel}}$. Our goal is to expand the amount of available image samples in the support sets. Inspired by Wave-SAN [14], we proposed a novel method using wavelet transform and style mix to do patch-level data augmentation. We also apply ViT as the secondary encoder to make better use of the patch-level information. We use prototypical networks as the few-shot classifier in our model and baseline model. The whole process is shown in Fig. 1

3.1 Patch Mix Module

Images selected from the support sets will be fed into the patch mix module for data augmentation. The workflow of the patch mix module is illustrated in Fig. 2. The workflow can be divided into three main steps: segmentation and selection, augmentation, and reassembling. The first and last steps are relatively

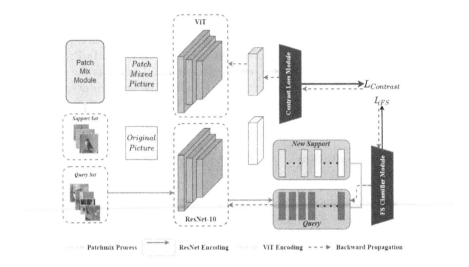

Fig. 1. The full procedure of our Patch Mix Augmentation method. We applied two encoders: ResNet-10 and ViT. ViT is responsible for encoding those augmented images. ResNet encodes those original images and query images. Feature vectors encoded by ViT will be added back to their corresponding support set to form a new support set.

simple. For segmentation and selection, we split the two images into patches of the equal size. Then we randomly select patches in different positions. Two images share this position selection. As for reassembling, we just need to paste these mixed patches into their corresponding positions on the original image. The augmentation step is most important and it consists of two key components.

Wavelet Transform. Wavelet Transform has two directions: wavelet Transform (DWT) and inverse wavelet transform (IDWT). DWT decomposes the given patch P into a low-frequency component P_{low} and three high-frequency components P_{high}. The low-frequency component retains the style of the image. And the high-frequency components contain the textures and edge contours of objects in the image. Three high-frequency components correspond to three different directions: horizontal, vertical, and diagonal. Both P_{low} and P_{high} are is $1/2$ the size of P. To avoid inconsistency in size after transformation, we set the size of A to an even number.

Correspondingly, IDWT takes low-frequency component P_{low} and high-frequency components and reconstructs the original image. DWT and IDWT are each other's inverse transforms and neither DWT nor IDWT transform will lead to information loss.

Using the features of DWT, we can handle P_{low} and P_{high} separately with different strategies. Given the processed P_{low} and P_{high}, IDWT make us enable to reconstruct a new image.

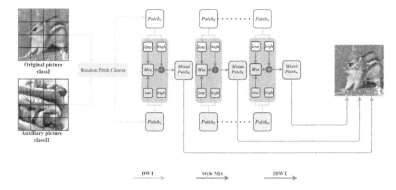

Fig. 2. The workflow of the Patch Mix module. The patches chosen from the original picture and auxiliary picture are from the same position. Each patch is first transformed using DWT to get the high and low-frequency parts. And then we mix their low-frequency parts. Finally inverse wavelet transform (IDWT) is performed.

Style Mix. Our style mix method is directly from AdaIn [13]. Given two low-frequency component of image from different classes P_{low}^{auxl} and P_{low}^{orig}. Superscript *auxl* and *orig* correspond to the auxiliary picture and the original picture in Fig. 2. We take P_{low}^{auxl} as the content image and P_{low}^{orig} as the style image ($P_{low}^{auxl}, P_{low}^{orig} \in \mathcal{R}^{C \times H \times W}$). Since low-frequency components are taken as the input, we actually mix the style of two patch images and generate P_{low}^{mix}:

$$P_{low}^{mix} = \mathbf{AdaIN}(P_{low}^{auxl}, P_{low}^{orig}) = \sigma(P_{low}^{orig}) \frac{\mu(P_{low}^{auxl})}{\sigma(P_{low}^{auxl})} + \mu(P_{low}^{orig}). \qquad (1)$$

The $\mu(\cdot)$ and $\sigma(\cdot)$ stand for the calculation of the mean and variance of input image. They are defined as follows:

$$\mu_c(x) = \frac{1}{HW} \sum_{h=1}^{H} \sum_{w=1}^{W} x_{c,h,w} \qquad (2)$$

$$\sigma_c(x) = \sqrt{\frac{1}{HW} \sum_{h=1}^{H} \sum_{w=1}^{W} (x_{c,h,w} - \mu_c(x))^2 + \epsilon}, \qquad (3)$$

where ϵ is a very small value to prevent the denominator in (1) from becoming 0. After style mix, P_{low}^{mix} together with P_{high}^{orig} will be taken as the input of IDWT to reconstruct a mixed patch.

3.2 Contrastive Training

In our method, we introduce ViT as our second feature encoder. The traditional method of training ViT needs a large number of labeled data, which is not available in meta-learning. Thus we borrow the idea from contrastive learning to train our ViT without labeled data and let our ViT learns from comparison.

We use $f_{ResNet}(\cdot)$ and $f_{ViT}(\cdot)$ to stand for the feature encoding of ResNet and ViT. We note the original image as x and its augmented one as $\tilde{x}$. Correspondingly, z and $\tilde{z}$ represent their feature vectors encoded by ResNet and ViT:

$$z = f_{ResNet}(x), \tilde{z} = f_{ViT}(\tilde{x}). \tag{4}$$

Given a $N-way, K-shot$ task $\mathcal{T}$, we treat sample pairs consisting of augmented images and their origin ones $\{\tilde{x}_i^c, x_i^c\}, c \in \{1, 2, \cdots N\}$ as positive pairs. Pairs with samples from different classes $\{\tilde{x}_i^{c\prime}, x_j^{c\prime}\}, c, c\prime \in \{1, 2, \cdots N\}$, where $c \neq c\prime$, are viewed as negative pairs. We denote positive and negative pairs in the feature space as $\{z^+\}$ and $\{z^-\}$. Note that $\{z^+\} = \{f_{ViT}(\tilde{x}_i^c), f_{ResNet}(x_i^c)\}$ and $\{z^-\} = \{f_{ViT}(\tilde{x}_i^{c\prime}), f_{ResNet}(x_j^{c\prime})\}$. We use InfoNCE [17] to train ViT:

$$\mathcal{L}_v = -log \frac{exp(d(\{z^+\}))}{exp(d(\{z^+\})) + \sum_{n=1}^{(N-1) \times K} exp(d(\{z^-\}_n))}, \tag{5}$$

where $d(\cdot)$ measures the cosine distance of two samples in the sample pair. Each augmented sample has one positive pair and $(N-1) \times K$ negative pairs.

4   Experiments

4.1   DataSet

mini-ImageNet. Mini-Imagenet [22] samples images from ILSVRC-2012, including 100 classes with 60000 images and each class contains 600 samples. The whole dataset is divided into training sets with 64 classes, validating sets with 16 classes, and testing sets with 20 classes. All images are resized into 224×224 in our experiments.

CUB-200-2011. CUB dataset [27] is offen used for evaluating cross-domain few-shot classification performance. It consists of 11788 images, belonging to 200 different classes. We followed [28] and split these classes into 100 for training, 50 for validation, and 50 for testing.

Cars. Introduced by [29] is also used for our cross-domain evaluation. The dataset is mainly used for fine-grained classification tasks, including 16185 images of 196 types of cars. The dataset is divided into two parts, 8144 images for training and 8041 for testing.

4.2   Implementation Details

Encoder Pre-Traing. We choose ResNet-10 as our primary encoder and ViT as the secondary encoder. The same as [28] we pre-trained the ResNet-10 network for 400 epochs and each batch is of size 16 on mini-ImageNet. By minimizing the standard cross-entropy classification loss. We removed the final layer used for classification after the pre-training is completed. Our baseline model is pre-trained in the same way.

Optimizing and Testing. We use the Adam as the optimizer for both encoders with the learning rate $\alpha = 0.001$. We trained both our model and the baseline for a maximum of 400 epochs on mini-ImageNet. It is worth noting that the ViT is not pre-trained on any dataset. The query set size is set as 16. During the few-shot classification test, we test our model with 2000 novel tasks on mini-ImageNet to see its performance. Differently cross-domain few-shot classification (CD-FS) aims to test our model on unseen domains. Therefore we test its performance on two unseen datasets: CUB and Cars. The baseline model is tested in the same way.

Patch Mix Augmentation. We use three hyperparameters to control our patch mix augmentation. *naug* is the number of fake samples to be generated for each support set. *patchsize* controls the size of each patch. And *mixpercent* decides how many patches in an image will be selected for augmentation. In all experiments except for 4.4, we set *patchsize* and *mixpercent* to 16 and 0.3 in both 1-shot and 5-shot. While *naug* is set to 5 for 1-shot and 10 for 5-shot.

Table 1. Few-shot Classification Accuracies (%) on mini-ImageNet

Model	Backbone	mini-ImageNet	
		1-shot	5-shot
Optimization-based			
MAML [18]	ConvNet-4	48.70 ± 1.75	63.15 ± 0.91
LEO [19]	WRN-28-10	61.76 ± 0.08	77.59 ± 0.12
MTL [30]	ResNet-12	61.20 ± 1.80	75.50 ± 0.80
Model-based			
Meta-learner LSTM [20]	ConvNet-4	43.44 ± 0.77	60.60 ± 0.71
Meta-SGD [21]	ConvNet-4	54.24 ± 0.03	70.86 ± 0.04
SNAIL [31]	ResNet-12	55.71 ± 0.99	68.88 ± 0.92
Metrics-based			
Matching network [22]	ConvNet-4	43.56 ± 0.84	55.31 ± 0.73
Relation network [23]	ConvNet-4	50.44 ± 0.82	65.32 ± 0.70
MetaOptNet [32]	ResNet-12	62.64 ± 0.61	78.63 ± 0.46
Meta-Baseline [33]	ResNet-12	63.17 ± 0.23	$\mathbf{79.26 \pm 0.17}$
Baseline (Metrics-based)	ResNet-10	58.53 ± 0.46	78.70 ± 0.32
Ours (Metrics-based)	ResNet-10	$\mathbf{63.45 \pm 0.45}$	79.15 ± 0.33

4.3 Result

Few-shot Classification. As introduced in 2.1, we classified existing methods into three major categories and we compare our work with these mainstream meta-learning methods. It's worth noting that both our model and baseline use prototypical networks as the few-shot classifier. The result in Table 1 shows that The performance of our model surpasses other models in 1-shot classification. When compared with our baseline, we can see significant performance improvement, nearly 5%. However, only about 0.45% accuracy improvement is gained in 5-shot classification. Even so, our model's performance can still match or exceed other models in the 5-shot classification.

Cross-domain Few-shot Classification. Although it is a challenging task, we still get decent performance boosts in cross-domain few-shot classification. As can be seen in Table 2, we reach the state-of-the-art performance and gain huge improvement compared to existing state-of-the-art models. To be clear, these state-of-the-art models use ResNet-10 as the backbone network. When compared to our baseline we can still have significant performance improvement, except in the 5-shot classification on CUB. Compared with our baseline, our model's classification accuracy increased by nearly 5% in 1-shot on Cars and 2–3% in other scenarios. The result strongly proves that our augmentation method does strengthen the ability to generalize.

Table 2. Cross-domain Few-shot Classification Accuracies (%) on CUB and CARS

Model	CUB		Cars	
	1-shot	5-shot	1-shot	5-shot
RelationNet+FT [34]	43.33 ± 0.40	59.77 ± 0.40	30.45 ± 0.30	40.18 ± 0.40
RelationNet+LRP [35]	41.57 ± 0.40	57.70 ± 0.40	30.48 ± 0.30	41.21 ± 0.40
RelationNet+ATA [36]	43.02 ± 0.40	59.36 ± 0.40	31.79 ± 0.30	42.95 ± 0.40
GNN+FT [34]	45.50 ± 0.50	64.97 ± 0.50	32.25 ± 0.40	46.19 ± 0.40
GNN+LRP [35]	43.89 ± 0.50	62.86 ± 0.50	31.46 ± 0.40	46.07 ± 0.40
GNN+ATA [36]	45.00 ± 0.50	66.22 ± 0.50	33.61 ± 0.40	49.14 ± 0.40
Baseline	46.78 ± 0.42	**67.86 ± 0.40**	31.46 ± 0.32	49.18 ± 0.40
Ours	**49.12 ± 0.42**	66.79 ± 0.38	**36.26 ± 0.38**	**51.20 ± 0.40**

4.4   Study of Path Mix Augmentation.

In order to analyze how each hyperparameter of patch mix augmentation will affect the performance, we carry out three experiments in 1-shot image classification on mini-ImageNet. Mix percent, the number of fake samples, and patch size are initialized to 0.3, 5, and 16 respectively. Each experiment changes only the hyperparameter it is studying while keeping the other two fixed. The results are shown in Fig. 3

The figure on the left in Fig. 3 shows the performance going up and then down and reaches a peak at 0.3. While the mix percent is small, ViT can learn useful information from the comparison of the original image with the augmented one. However, when the mix percent grows too big, too much information has been changed in the original image. This makes it difficult for ViT to learn from the comparison of the original image with the augmented one. Ultimately leads to a decrease in the usefulness of fake samples.

The middle figure in Fig. 3 shows an overall trend of increasing performance with more fake samples are generated. Under a reasonable hyperparameter setting, it is reasonable that the more fake samples the better the performance. But considering the complexity of the calculation, it should not be set to a value that is too large.

The right figure in Fig. 3 shows a clear downward trend in performance as the patch size becomes bigger. Just as mix percent, increasing the size of each patch will significantly expand the augmented area on the image. And leads to performance degradation.

Considering the above experiments and the computational complexity, our configuration in 4.2 is reasonable.

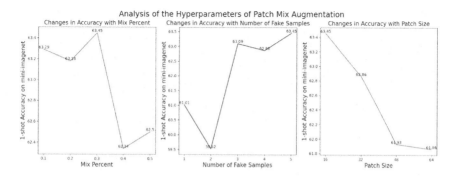

Fig. 3. The analysis of the hyperparameters of the Patch Mix module. We study the effect of Mix Percent, Number of Fake Samples and, Patch Size on the performance of our model. We test and evaluate different settings with 1-shot accuracy on mini-ImageNet

4.5 Ablation Study

For a more in-depth study of wavelet transform and style mix, we conduct the ablation experiments. The result is shown in Table 3. By comparing (I) and (V), we can conclude that performing our augmentation to the patches is better than just pasting directly. For the wavelet transform, comparing (I) and (II) can prove its effectiveness. As for the style mix, the problem is a little more complicated. While (V) only shows a slight performance increase compared to (II). By comparing (II) and (III) yet prove the effectiveness of style mix. We believe that both wavelet transform and style mix are capable of improving performance. But when the two come together, the combination is not very good and wavelet transform contributes more to performance improvement.

Table 3. Ablation Experiments on mini-ImageNet ('✓' With; '-' Without)

	Our Model($naug = 5, patchsize = 16, mixpercent = 0.3$)		Accuracy (%)
	Wavelet Transform	Style Mix	1-shot
(I)	-	-	62.39 ± 0.44
(II)	✓	-	63.37 ± 0.43
(III)	-	✓	62.70 ± 0.43
(V)	✓	✓	63.45 ± 0.45

5   Conclusion

In this paper, we proposed a novel data augmentation method with dual encoders. We augment images at patch-level using wavelet transform and style mix. To make better use of patch-level information, we introduce ViT as the second encoder in addition to ResNet. By using InfoNCE, ViT is trained in an unsupervised way. Multiple experiments have been carried out. The results can strongly prove that our method can help improve the generalization capability.

Acknowledgment. This work is supported by the National Key R&D Program of China (2018YFA0701700; 2018YFA0701701), and the National Natural Science Foundation of China under Grant No. 61672364, No. 62176172 and No. 61902269.

References

1. Vilalta, R., Drissi, Y.: A perspective view and survey of meta-learning. Artif. Intell. Rev. **18**(2), 77–95 (2002)
2. Vanschoren, J.: Meta-learning: A survey. arXiv preprint arXiv:1810.03548 (2018)
3. Krizhevsky, A., Sutskever, I., Hinton, G.E.: Imagenet classification with deep convolutional neural networks. Adv. Neural. Inf. Process. Syst. **25**, 1097–1105 (2012)
4. He, K., Zhang, X., Ren, S., Sun, J.: Deep residual learning for image recognition. In: Proceedings of the IEEE Conference on Computer Vision and Pattern Recognition, pp. 770–778 (2016)
5. Szegedy, C., et al.: Going deeper with convolutions. In: Proceedings of the IEEE Conference on Computer Vision and Pattern Recognition, pp. 1–9 (2015)
6. Deng, J., Dong, W., Socher, R., Li, L.J., Li, K., Fei-Fei, L.: Imagenet: a large-scale hierarchical image database. In: 2009 IEEE Conference on Computer Vision and Pattern Recognition, pp. 248–255. IEEE (2009)
7. Ni, R., Goldblum, M., Sharaf, A., Kong, K., Goldstein, T.: Data augmentation for meta-learning. In: International Conference on Machine Learning, pp. 8152–8161. PMLR (2021)
8. Yun, S., Han, D., Oh, S.J., Chun, S., Choe, J., Yoo, Y.: Cutmix: regularization strategy to train strong classifiers with localizable features. In: Proceedings of the IEEE/CVF International Conference on Computer Vision, pp. 6023–6032 (2019)

9. Zhang, H., Cissé, M., Dauphin, Y.N., Lopez-Paz, D.: mixup: beyond empirical risk minimization. In: 6th International Conference on Learning Representations, ICLR 2018, Vancouver, BC, Canada, April 30 - May 3, 2018, Conference Track Proceedings. OpenReview.net (2018)

10. Verma, V., et al.: Manifold mixup: Better representations by interpolating hidden states. In: Chaudhuri, K., Salakhutdinov, R. (eds.) Proceedings of the 36th International Conference on Machine Learning, ICML 2019. Proceedings of Machine Learning Research, vol. 97, pp. 6438–6447. PMLR (2019)

11. Zhang, R., Che, T., Ghahramani, Z., Bengio, Y., Song, Y.: Metagan: an adversarial approach to few-shot learning. In: NeurIPS, vol. 2, p. 8 (2018)

12. Lee, D.B., Min, D., Lee, S., Hwang, S.J.: Meta-GMVAE: mixture of gaussian VAE for unsupervised meta-learning. In: International Conference on Learning Representations (2020)

13. Huang, X., Belongie, S.J.: Arbitrary style transfer in real-time with adaptive instance normalization. In: 5th International Conference on Learning Representations, ICLR 2017, Toulon, France, April 24–26, 2017, Workshop Track Proceedings (2017)

14. Fu, Y., Xie, Y., Fu, Y., Chen, J., Jiang, Y.G.: Wave-san: Wavelet based style augmentation network for cross-domain few-shot learning. arXiv preprint arXiv:2203.07656 (2022)

15. Dosovitskiy, A., et al.: An image is worth 16x16 words: transformers for image recognition at scale. In: 9th International Conference on Learning Representations, ICLR 2021 (2021)

16. van den Oord, A., Li, Y., Vinyals, O.: Representation learning with contrastive predictive coding. CoRR abs/1807.03748 (2018)

17. He, K., Fan, H., Wu, Y., Xie, S., Girshick, R.B.: Momentum contrast for unsupervised visual representation learning. In: 2020 IEEE/CVF Conference on Computer Vision and Pattern Recognition, CVPR 2020, Seattle, WA, USA, June 13–19, 2020, pp. 9726–9735 (2020)

18. Finn, C., Abbeel, P., Levine, S.: Model-agnostic meta-learning for fast adaptation of deep networks. In: International Conference on Machine Learning, pp. 1126–1135. PMLR (2017)

19. Rusu, A.A., et al.: Meta-learning with latent embedding optimization. arXiv preprint arXiv:1807.05960 (2018)

20. Ravi, S., Larochelle, H.: Optimization as a model for few-shot learning (2016)

21. Li, Z., Zhou, F., Chen, F., Li, H.: Meta-SGD: learning to learn quickly for few-shot learning. arXiv preprint arXiv:1707.09835 (2017)

22. Vinyals, O., Blundell, C., Lillicrap, T., Wierstra, D., et al.: Matching networks for one shot learning. Adv. Neural. Inf. Process. Syst. **29**, 3630–3638 (2016)

23. Sung, F., Yang, Y., Zhang, L., Xiang, T., Torr, P.H., Hospedales, T.M.: Learning to compare: Relation network for few-shot learning. In: Proceedings of the IEEE Conference on Computer Vision and Pattern Recognition, pp. 1199–1208 (2018)

24. Snell, J., Swersky, K., Zemel, R.: Prototypical networks for few-shot learning. In: Guyon, I., et al. (eds.) Advances in Neural Information Processing Systems. vol. 30. Curran Associates, Inc. (2017)

25. Ni, R., Goldblum, M., Sharaf, A., Kong, K., Goldstein, T.: Data augmentation for meta-learning. CoRR abs/2010.07092 (2020)

26. Schwartz, E., et al.: Delta-encoder: an effective sample synthesis method for few-shot object recognition. In: Advances in Neural Information Processing Systems, vol. 31 (2018)

27. Wah, C., Branson, S., Welinder, P., Perona, P., Belongie, S.: The caltech-ucsd birds-200-2011 dataset. Technical report CNS-TR-2011-001, California Institute of Technology (2011)
28. Wang, H., Deng, Z.H.: Cross-domain few-shot classification via adversarial task augmentation. arXiv preprint arXiv:2104.14385 (2021)
29. Krause, J., Stark, M., Deng, J., Fei-Fei, L.: 3D object representations for fine-grained categorization. In: 4th International IEEE Workshop on 3D Representation and Recognition (3dRR-13). Sydney, Australia (2013)
30. Sun, Q., Liu, Y., Chua, T.S., Schiele, B.: Meta-transfer learning for few-shot learning. In: Proceedings of the IEEE/CVF Conference on Computer Vision and Pattern Recognition, pp. 403–412 (2019)
31. Mishra, N., Rohaninejad, M., Chen, X., Abbeel, P.: A simple neural attentive meta-learner. arXiv preprint arXiv:1707.03141 (2017)
32. Lee, K., Maji, S., Ravichandran, A., Soatto, S.: Meta-learning with differentiable convex optimization. In: Proceedings of the IEEE/CVF Conference on Computer Vision and Pattern Recognition, pp. 10657–10665 (2019)
33. Chen, Y., Wang, X., Liu, Z., Xu, H., Darrell, T.: A new meta-baseline for few-shot learning. arXiv preprint arXiv:2003.04390 (2020)
34. Tseng, H.Y., Lee, H.Y., Huang, J.B., Yang, M.H.: Cross-domain few-shot classification via learned feature-wise transformation. In: International Conference on Learning Representations (2020)
35. Sun, J., Lapuschkin, S., Samek, W., Zhao, Y., Cheung, N.M., Binder, A.: Explanation-guided training for cross-domain few-shot classification. In: 2020 25th International Conference on Pattern Recognition (ICPR), pp. 7609–7616 (2021)
36. Wang, H., Deng, Z.H.: Cross-domain few-shot classification via adversarial task augmentation. In: Zhou, Z.H. (ed.) Proceedings of the Thirtieth International Joint Conference on Artificial Intelligence, IJCAI-21, pp. 1075–1081, August 2021. main Track

Tacit Commitments Emergence in Multi-agent Reinforcement Learning

Boyin Liu[1,2], Zhiqiang Pu[1,2(✉)], Junlong Gao[3], Jianqiang Yi[1,2], and Zhenyu Guo[3]

[1] School of Artificial Intelligence, University of Chinese Academy of Sciences, Beijing 100049, China
{liuboyin2019,zhiqiang.pu,jianqiang.yi}@ia.ac.cn
[2] Institute of Automation, Chinese Academy of Sciences, Beijing 100190, China
[3] Alibaba Group, Hangzhou, China
{JunlongGao,ZhenyuGuo}@alibaba-inc.com

Abstract. Tacit commitments have been widely seen as a crucial underpinning for real-world cooperation. Similarly, it could also be a key to multi-agent cooperation. This paper proposes a novel tacit commitment emergence multi-agent reinforcement learning (MARL) framework (TCEM). In MARL, we define commitment as the unique state that the agent will exhibit through its action. TCEM first equips each agent with a commitment inference module (CIM) to infer its neighbor's commitments. Then, TCEM proposes that commitments influence intrinsic motivation (CIR) to encourage agents to have casual influence on others' actions. Finally, commitment acceptance intrinsic (CAI) motivation is constructed to guide the agent in behaving considering neighbors' commitments. CIR and CAI calculate intrinsic reward using counterfactual reasoning deriving from causal inference. Empirical results show that our method can effectively improve learning performance and deliver better cooperation among agents, which helps our method show superior performance on the Google Research Football benchmark.

Keywords: Casual Inference · Counterfactual Reasoning · Intrinsic Reward · Multi-agent Systems · Reinforcement Learning

1 Introduction

In recent years, cooperative multi-agent reinforcement learning (MARL) has achieved meaningful progress, and many deep approaches have been proposed [3,8,12,15]. However, learning complicated and effective coordination policies among agents is still a challenge in the MARL field.

The human ability to make and stick to commitments is crucial to human social cooperation [1,9]. Tacit commitment induces effective cooperation behavior, which improves the productivity and efficiency of human society [10]. Analogically, the emergence of tacit commitment should also be essential for multi-agent cooperation. Tacit commitments mean that the transmission of intention

M. Tanveer et al. (Eds.): ICONIP 2022, LNCS 13623, pp. 27–36, 2023.
https://doi.org/10.1007/978-3-031-30105-6_3

among agents is implicit without explicit communication. For the commitments receiving agent, it is necessary to understand other agents' commitments through historical observation. Then, its behaviors must be influenced by others' commitments and also make corresponding commitments.

In this paper, we propose a novel tacit commitment emergence MARL framework (TCEM) to develop effective cooperation among agents. To equip agents with the ability to know others' commitments, TCEM first learns a commitment inference module (CIM) to predict neighbors' commitments. Humans usually infer others' commitments according to their historical behavior and current observation. CIM is trained using neighbors' past behavior data and outputs their commitments according to current observation. To further encourage tacit commitment emergence, we propose two terms of intrinsic motivation, i.e., commitments influence intrinsic motivation (CIR) and commitments acceptance intrinsic motivation (CAI). CIR gives an agent an additional reward for having a casual commitment influence towards others. With this intrinsic reward, we hope the agent makes meaningful commitments. CAI encourages the agent to be influenced by its inferring neighbors' commitments. CIR and CAI form a closed circle, one for making commitments and another for accepting commitments.

We benchmark our approach on Google Research Football (GRF) benchmark. The superior performance of our approach on challenging benchmarking tasks shows that our approach achieves significantly higher coordination capacity than baselines while using tacit commitment as a catalyst for more robust talent policies.

2 Related Work

A fully cooperative multi-agent task can be formulated as a Decentralized Partially Observable Markov Decision Process (Dec-POMDP) [11]. There are n agents in the environment, where each agent i receives a local observation o_i^t and then executes its action a_i^t and gets a shared reward r^t. All agents aim to maximize their expected return $E[\sum_{t=0}^{T} \gamma^t r^t]$, where γ denotes the discount factor and T the time horizon. Many methods have been proposed for Dec-POMDP, most of which follow the paradigm of centralized training and decentralized execution (CTDE). One of the promising ways to implement the CTDE framework is value function factorization [12–14,19]. QMIX [12] proposes monotonicity for factorization structures. QPLEX [15] uses a duplex dueling network architecture for factorization. Except for value decomposition methods, policy gradient method [3,4,7,17,18] is also popular in MARL field. COMA [3] proposes a counterfactual baseline for multi-agent credit assignment.

To develop MARL methods qualified for complex tasks, many concepts in human collaboration are introduced into MARL, such as role [16], diversity [2,8], individuality [5], etc. ROMA [16] constructs a stochastic role embedding space by introducing two novel regularizers and conditioning individual policies on roles. MAVEN [8] learns a diverse ensemble of monotonic approximations with the help of a latent space to explore. In this paper, we introduce tacit commitments into MARL to help agents construct more effective cooperation.

3   Method

In this section, we propose a novel tacit commitments emergence MARL framework (TCEM) that guides agents to learn high-level cooperation. TCEM adopts the CTDE paradigm. As shown in Fig. 1, to enable tacit commitments emergence among agents, TCEM firstly equips each agent with a commitments inference module (CIM). CIM allows agents to speculate on other agents' commitments based on the observed historical trajectory. Then, the inferred commitments combined with the agent's hidden states are fed into a multi-layer perception network to output the Q function of the agent. More importantly, TCEM proposes two terms of intrinsic reward, one for encouraging the agent to exert its commitment to others and another for guiding the agent to learn to be influenced by others' commitments.

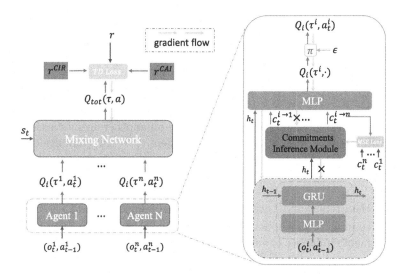

Fig. 1. Schematics of our approach. $c_t^{i \to n}$ denotes the inferred n's commitments by i. c_t^n represents n's real commitments, observed by other agents after several steps. Additionally, the TD Loss gradient does not influence the parameters in CIM, which updates its parameters isolately.

3.1   Commitments Inference Module

In high-level human cooperation, such as professional football, people usually make decisions based on their teammates' unspoken commitments. Superior coordination is shown when the decision-maker infers the teammates accurately. To learn tacit commitments among agents, we first need to define the agent's commitments.

Definition 1. *The **controllable states** of an agent is the states that can be directly and completely achieved by agent itself through corresponding actions.*

Remark 1. For example, in football game, the controllable states refer to the states such as position and direction of the agent itself but not the positions of the ball, neighbors or opponents.

Definition 2. *The **commitments** c_i of the agent i is its expected performing controllable states at next step.*

To construct tacit commitments among agents, each agent must learn to infer the possible commitments of its neighbors. As agents constantly learn and improve their policies, they must also constantly learn to infer their neighbors' commitments. Therefore, commitments need to be learned by neural networks. We adopt a MLP to construct the CIM, with extracted hidden states as input and inferred neighbors' commitments as output. CIM can be updated when the agent sees the real controllable states of its neighbors. It is noted that the gradients of TD Loss do not influence the parameters of CIM.

3.2 Commitments Influence Intrinsic Motivation

Based on CIM, we then propose commitments influence intrinsic reward (CIR) to motivate the agent having a casual influence on another agent's actions. Suppose there are two agents i and j, agent i infers $c_t^{i \to j}$ the commitments c_t^j of agent j using CIM. The prediction of $c_t^{i \to j}$ is based on the history observation of agent j. Thus, the agent i computes the probability of its next action as $p(a_t^i | o_t^i, c_t^{i \to j}, c_t^{i \to \sim(i,j)})$ where $\sim(i,j)$ denotes agents $1, 2, ..., n$ except agents i and j. We then replace $c_t^{i \to j}$ with counterfactual commitments $\tilde{c}_t^{i \to j}$, and calculating the counterfactual policy distribution $p(a_t^i | o_t^i, \tilde{c}_t^{i \to j}, c_t^{i \to \sim(i,j)})$. The agent j's commitments influence towards i is then computed by asking a question: How would agent i's policy distribution change if i had only inferred j's commitments different? If agent i's policy change as the inferred agent j's commitments change, we can say that agent j has commitments influence towards agent i.

The marginal distribution of $p(a_t^i | o_t^i, \tilde{c}_t^{i \to j}, c_t^{i \to \sim(i,j)})$ can not be directly computed. We achieve it by sampling several counterfactual commitments of agent j:

$$p\left(a_t^i \mid c_t^{i \to \sim(i,j)}, o_t^i\right) = \sum_{\tilde{c}_t^{i \to j}} p\left(a_t^i \mid \tilde{c}_t^{i \to j}, c_t^{i \to \sim(i,j)}, o_t^i\right) p\left(\tilde{c}_t^{i \to j} \mid c_t^{i \to \sim(i,j)}, o_t^i\right). \quad (1)$$

Thus, the causal influence reward c_{CIR} for agent j is defined as:

$$c_{CIR}^j = \sum_{i=0, i \neq j}^{N} \left[D_{KL} \left[p\left(a_t^i \mid c_t^{i \to j}, c_t^{i \to \sim(i,j)}, o_t^i\right) \| p\left(a_t^i \mid c_t^{i \to \sim(i,j)}, o_t^i\right) \right] \right]. \quad (2)$$

3.3   Commitments Acceptance Intrinsic Motivation

CIR encourages agents to be influencers. To show tacit commitments emergence among agents, the agent that is influenced by inferred other agents' commitments, could also be rewarded. Therefore, we propose commitments acceptance intrinsic motivation (CAI) to give an agent an additional reward for being influenced by inferred other agents' commitments. For agent j, it calculates CAI by asking a retrospective question: How would agent j's policy distribution change if i had inferred neighbors' commitments different?

Thus, the causal influence reward c_{CAI} for agent j is defined as:

$$c^j_{CAI} = \left[D_{KL} \left[p \left(a^j_t \mid c^{j \to \sim j}_t, o^j_t \right) \| p \left(a^j_t \mid o^j_t \right) \right] \right]. \tag{3}$$

3.4   Overall Learning Objective

In this paper, the proposed learning framework TCEM adopts QMIX [12] style mixing network. With CIR and CAI, the final intrinsic reward function is calculated as:

$$r_I = \beta_1 r_{CIR} + \beta_2 r_{CAI}, \tag{4}$$

where β_1 and β_2 are weight coefficient of CIR and CAI respectively.

We add r_I to extrinsic rewards r and use the following TD loss:

$$\mathcal{L}_{TD}(\theta) = \left[r + \beta r_I + \gamma \max_{a'} Q_{tot} \left(s', a'; \theta^- \right) - Q_{tot}(s, a; \theta) \right]^2, \tag{5}$$

where θ is the parameters of the whole framework, θ^- is periodically fixed parameters copied from θ for a stable update, and β is the weight of intrinsic rewards.

4   Experiments Setup

To clearly interpret the mechanism and show the effectiveness of TCEM, we evaluate our method in the scenarios in Google Research Football (GRF) [6]. In this section, we describe the environment and experimental setup.

4.1   Environments

Football is a game that needs high-level cooperation among agents, and players essentially require tacit commitment to score goals. As shown in Fig. 2, we chose three challenging football tasks i.e., *academy 3vs1 with keeper*, *academy 3vs3*, and *academy counterattack hard* to test the learning performance of TCEM and other baselines. In these tasks, agents need to choose an action from 19 actions at each step, including move, pass, shot, etc. In our experiments, we control left-side players (in yellow) except the goalkeeper. The right-side players are rule-based bots controlled by the game engine. All the agents must coordinate well, and

academy 3vs1 with keeper   academy 3vs3   academy counterattack hard

Fig. 2. Initial snapshot of three GRF tasks.

then it is possible to overcome the opponent and score a goal. Except for the goal reward of +80, a reward +5 is given for the ball being first controlled by agents.

To speed up training and reduce useless exploration, the episodes are terminated either when some events happen (including score, ball possession loss, and game stops) or the time steps exceeding 400).

4.2 Baseline and Ablation Methods Setup

In this section, we compare our methods with QMIX [12], MAVEN [8] and COMA [3]. In addition, we carry out the following ablation studies: (1) QMIX-CIM. QMIX with CIM module to infer other agents' commitments. (2)QMIX-CIR (QMIX-CIM-CIR). Based on QMIX-CIM, we add commitments influence intrinsic motivation to encourage agents to behave with influence. (3) QMIX-CAI (QMIX-CIM-CAI). Based on QMIX-CIM, we add commitments acceptance and intrinsic motivation to encourage agents to be influenced by others' commitments.

The content of controllable states are essential to effectively tacit commitment emergence. In GRF task, the agent's controllable state is its position and direction. For agent i, the agent j's commitments for him is its relative position and direction after m steps towards current i and ball's position and direction. Football is a game about the ball. Therefore, we also add the ball's information into the commitments design. Agent j's commitments about the ball are its expected position or direction after m steps relative to the current ball, so it is still controllable for agent j. We choose $m = 2$ in GRF tasks.

For all experiments, the optimization is conducted using RMSprop with a learning rate of 5×10^4, α of 0.99, and with no momentum or weight decay. For exploration, we use ϵ-greedy, with ϵ annealed linearly from 1.0 to 0.05 over 500K time steps and kept constant for the rest of the training for both TCME and all the baselines and ablations. We introduce three important hyperparameters: β, β_1, β_2. The $(\beta, \beta_1, \beta_2)$ of our methods is shown in Table 1. The intrinsic motivation stops after 600 million steps of training. In addition, to reduce randomness, we show the average and variance of the performance for our method, baselines, and ablations tested with three random seeds.

Table 1. Parameters setting of our methods for all tasks.

Methods	Parameters Setting $(\beta, \beta_1, \beta_2)$
TCEM	$(0.08, 1, 2)$
QMIX-CIM	$(0, 0, 0)$
QMIX-CIR	$(0.08, 1, 0)$
QMIX-CAI	$(0.08, 0, 2)$

5   Experiments Results

5.1   Comparison Results of Baseline Methods

As illustrated by Fig. 3, TCEM outperforms all other methods for each task with acceptable variance across random seeds. With tacit commitments introduced, TCEM evidently learns better strategy and forms effective coordination faster. The baseline QMIX can achieve satisfactory performance compared with MAVEN and COMA. However, at *academy 3vs1 with keeper* and *academy 3vs3* tasks, QMIX converges to the strategy with a lower wining rate compared with TCEM. In addition, TCEM always rises faster than QMIX without falling into the local optimum. At *academy 3vs1 with keeper* task, although TCEM's learning curve rises later than QMIX, it rises faster and achieves better performance finally. Among these baselines, MAVEN and COMA fail to show meaningful strategy learning.

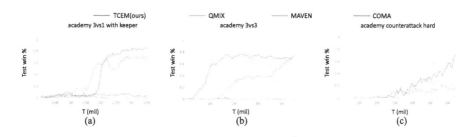

Fig. 3. Comparison of our method against baseline algorithms.

5.2   Comparison Results of Ablation Methods

In this section, we conduct ablation studies, comparing with the ablations explained in Sect. 4.2 at three GRF tasks. As shown in Fig. 4, TCEM offers the best performance among ablation methods. The ablation of each part of our method will induce an evident decrease in learning performance.

The superiority of TCEM against QMIX-CIM highlights the contribution of CAI and CIR. By comparing TCEM with QMIX-CAI, we can conclude that the CIM effectively improves learning performance both in speed and quality.

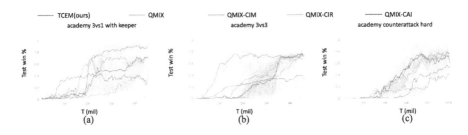

Fig. 4. Comparison of our method against ablation methods.

In *academy 3vs1 with keeper*, there is only two right team players to defend MARL agents. It is noted that QMIX also shows superior learning performance in this easy task, even better than QMIX-CIM, QMIX-CIR, and QMIX-CAI. However, when it comes to *academy 3vs3*, a more challenging task, QMIX-CIR and QMIX-CAI both outperform QMIX. Finally, in *academy counterattack hard*, the most challenging task, even QMIX-CIM, shows better performance than QMIX. These results correspond to the intuition that tacit commitments have broader potential in a complex task. CIM infers commitments and extracts information from observations. The inferred commitments do not provide additional information for agents. Therefore, we observe that QMIX-CIM fails to show better performance than QMIX. Based on QMIX-CIM, adding CIR or CAI will bring noticeable performance promotion for complex tasks. This is because CAI and CIR will guide the agents to notice inferred commitments and form better cooperation. The comparison of the performance of TCEM with QMIX-CIR and QMIX-CAI proves that the commitments influence and acceptance intrinsic motivation loop can effectively speed up training and increase stability.

The performance gap between TCME and ablations is more evident on harder tasks. This observation supports the previous discussion—tacit commitment is more likely to improve labor efficiency in complex tasks.

5.3   Policy Visualization

We further visualize the final learning policy by TCEM at *academy counterattack hard* task, which shows evident tacit commitment emergence between agents.

As shown in Fig. 5, we control the yellow team players. At the start, shown in Fig. 5(a), the yellow player 1 dribbles the ball and runs down to draw the attention of blue player 2. It is noted that the movement of blue player 2 makes space for yellow player 2, who also runs below to bypass the defender. At the same time, yellow player 3 also runs to the backcourt to distract defenders. As shown in Fig. 5(b), from $t = 11$ to $t = 30$, yellow player 1 spends 19 steps dribbling the ball and distracting blue players 1 and 2 with yellow player 3's support. With the help of teammates, yellow player 2 successfully runs to the backfield. At $t = 30$, yellow player 1 suddenly kicks the ball to the penalty arc. The ignored yellow player 2 then runs fast to the penalty arc and successfully receives the ball. Finally, yellow player 2 dribbles the ball to the penalty area and overcomes the goalkeeper's defense. These mutual commitments among yellow players have led to efficient cooperation and reflect the tacit commitments learned by TCEM.

(a) $t = 11$ (b) $t = 30$

Fig. 5. Visualization of learning policies by TCEM at *academy counterattack hard* task, which achieve complex cooperation with impressive off-the-ball moving strategies.

Throughout the yellow team players' behaviors, there are tacit commitments among agents. First, at the dribbling ball phase, both yellow team players 1 and 3 move to distract two blue players, making commitments that they will draw the blue players' attention. The yellow team player 2 receives the commitments of their teammates and also makes a commitment that he will directly run to the backfield. When it comes to the pass, it is noted that yellow player 1 does not pass the ball straight to yellow player 2, but kicks the ball to the penalty arc. Left team player 2 has made a commitment. Yellow player 1 kicks the ball to receive the commitment from yellow player 2 that he will run to the penalty arc and receive the ball.

6   Conclusion

Many cooperation concepts in human society have shown meaningful potential to improve performance in cooperative MARL. In this paper, we propose TCEM to introduce tacit commitments into MARL to promote cooperation performance among agents. Experimental results demonstrate that our method accelerates the learning and improves the learning performance, which performs best in all GRF tasks. When compared to the baseline methods, the results confirm that TCEM is significantly superior to QMIX, COMA, and MAVEN. Ablation results suggest that each part in TECM makes a meaningful contribution towards ultimate superior performance. Finally, the critical snapshot analysis confirms that TCEM indeed learns tacit commitments among agents.

Acknowledgments. This work was supported by the National Key Research and Development Program of China under Grant 2020AAA0103404, the National Natural Science Foundation of China under Grant 62073323 and Alibaba Group through Alibaba Innovative Research (AIR) Program.

References

1. Agranov, M., Potamites, E., Schotter, A., Tergiman, C.: Beliefs and endogenous cognitive levels: an experimental study. Games Econom. Behav. **75**(2), 449–463 (2012)
2. Chenghao, L., Wang, T., Wu, C., Zhao, Q., Yang, J., Zhang, C.: Celebrating diversity in shared multi-agent reinforcement learning. In: Advances in Neural Information Processing Systems 34 (2021)
3. Foerster, J., Farquhar, G., Afouras, T., Nardelli, N., Whiteson, S.: Counterfactual multi-agent policy gradients. In: Proceedings of the AAAI Conference on Artificial Intelligence, vol. 32 (2018)
4. Iqbal, S., Sha, F.: Actor-attention-critic for multi-agent reinforcement learning. In: International Conference on Machine Learning, pp. 2961–2970. PMLR (2019)
5. Jiang, J., Lu, Z.: The emergence of individuality. In: International Conference on Machine Learning, pp. 4992–5001. PMLR (2021)
6. Kurach, K., et al.: Google research football: a novel reinforcement learning environment. arXiv preprint arXiv:1907.11180 (2019)
7. Lowe, R., Wu, Y.I., Tamar, A., Harb, J., Pieter Abbeel, O., Mordatch, I.: Multi-agent actor-critic for mixed cooperative-competitive environments. In: Advances in Neural Information Processing Systems 30 (2017)
8. Mahajan, A., Rashid, T., Samvelyan, M., Whiteson, S.: Maven: Multi-agent variational exploration. In: Advances in Neural Information Processing Systems 32 (2019)
9. Melkonyan, T., Zeitoun, H., Chater, N.: The cognitive foundations of tacit commitments. Available at SSRN 3168669 (2018)
10. Nguyen, N.L.: Tacit knowledge sharing within project teams: an application of social commitments theory. VINE Journal of Information and Knowledge Management Systems (2021)
11. Oliehoek, F.A., Amato, C.: A concise introduction to decentralized POMDPs. Springer (2016). https://doi.org/10.1007/978-3-319-28929-8
12. Rashid, T., Samvelyan, M., Schroeder, C., Farquhar, G., Foerster, J., Whiteson, S.: QMIX: monotonic value function factorisation for deep multi-agent reinforcement learning. In: International Conference on Machine Learning, pp. 4295–4304. PMLR (2018)
13. Son, K., Kim, D., Kang, W.J., Hostallero, D.E., Yi, Y.: QTRAN: learning to factorize with transformation for cooperative multi-agent reinforcement learning. In: International Conference on Machine Learning, pp. 5887–5896. PMLR (2019)
14. Sunehag, P., et al.: Value-decomposition networks for cooperative multi-agent learning. arXiv preprint arXiv:1706.05296 (2017)
15. Wang, J., Ren, Z., Liu, T., Yu, Y., Zhang, C.: QPLEX: duplex dueling multi-agent Q-learning. arXiv preprint arXiv:2008.01062 (2020)
16. Wang, T., Dong, H., Lesser, V., Zhang, C.: Roma: Multi-agent reinforcement learning with emergent roles. arXiv preprint arXiv:2003.08039 (2020)
17. Wang, Y., Han, B., Wang, T., Dong, H., Zhang, C.: DOP: Off-policy multi-agent decomposed policy gradients. In: International Conference on Learning Representations (2020)
18. Wen, Y., Yang, Y., Luo, R., Wang, J., Pan, W.: Probabilistic recursive reasoning for multi-agent reinforcement learning. arXiv preprint arXiv:1901.09207 (2019)
19. Yang, Y., et al.: Qatten: a general framework for cooperative multiagent reinforcement learning. arXiv preprint arXiv:2002.03939 (2020)

Saccade Direction Information Channel

Qiaohong Hao[1], Mateu Sbert[2(✉)] ⓘ, Miquel Feixas[2] ⓘ, Yi Zhang[1],
Marius Vila[2] ⓘ, and Jiawan Zhang[1(✉)] ⓘ

[1] College of Intelligence and Computing, Tianjin University, Tianjin 300350, China
jwzhang@tju.edu.cn
[2] Institute of Informatics and Applications, University of Girona,
Girona 17003, Spain
mateu.sbert@udg.edu

Abstract. Eye tracking has become an increasingly important technology in many fields of research, such as marketing, human computer interaction, psychology, and also in human cognition. Understanding the human eye movements, while viewing specific scenarios, can be of great support for improving visual stimuli. However, the challenging problem with this kind of spatio-temporal data is to find quantitative links between eye movements and human cognition. This paper introduces the information channel based on saccade direction. The gaze transition between different saccade directions is modeled as a discrete information channel, which we call *saccade direction information channel*. The channel is applied to an eye-tracking dataset on Van Gogh's paintings observation. In our results, horizontal saccades are more frequent than vertical saccades, and the information conveyed in horizontal/vertical displacements, measured as the mutual information of the channel, is higher than in diagonal displacements. By comparing the results to our previous spatial gaze channel between Areas of Interest (AOIs) we constate that the spatial channel discriminates better between the observed images, while the direction channel discriminates better between observers.

Keywords: Gaze information channel · Eye tracking · Saccade Direction · Markov chain · Entropy

1   Introduction

Eyes can reflect the human thoughts and reveal the way in which observers view the scene. Eye tracking is a widely used method of recording eye positions and movements of human for further explanation and application. Eye tracking hardware sample gaze locations at $50 - 2000$Hz. Samples are then reduced to fixations, periods of visual attention at particular positions. The eye movements between fixations are generally referred to as *saccades*, and a scanpath is a

Supported by the National Natural Science Foundation of China under grant No. 61702359, and by Grant PID2019-106426RB-C31 funded by MCIN/AEI/10.13039/501100011033.

repetitive sequence of fixations and saccades [1,2]. In recent years, with the increased portability and accessibility of eye tracker, a great deal of interest has been generated in using eye tracking data, which has become an increasingly important technology in many disciplines like medicine [3], visual search and attention analysis [4], driving [5], media and marketing [6], human computer interaction [7], psychology [8], and human cognition [9,10]. Research leveraging eye tracking as an instructional tool, however, is still in its infancy. Quantitative comparison of eye movement metrics is badly needed [11]. Two representative eye movement metrics, scanpaths [12] and heatmaps [13], have been developed. Recently, deep learning techniques have been introduced to predict position and duration of visual scanpaths [14,15], inferring human age [16], and for generating new Emojis for scanpath classification [17]. Ponsoda et al. [18] introduced the study of transitions between saccade directions, and Tatler and Vincent [19] and Smith and Henderson [20] considered the correlation between successive saccade directions when free viewing images of natural scenes. They found an overall bias to make saccades either in the same direction as the previous saccade (saccade momentum), or 180°C reversals [19].

In this paper, we model eye tracking fixation sequences between saccade directions as a discrete information channel, which we call *saccade direction information channel* (in brief saccade direction channel). To illustrate the usefulness of this channel, we apply it to formerly conducted eye tracking studies focusing on Van Gogh's paintings observation [21,22]. The main contributions of this work can be summarized as follows:

- Following our previous work [21–23] where we defined the gaze information channel based on transitions between AOIs, we introduce here a novel information channel based on saccade direction.
- As in the gaze information channel, in the saccade direction channel clustering works well to study group or cluster properties, for instance grouping for a single observer all her/his observed images results, or grouping for a single image all its observers results.
- Differently to the gaze information channel, in the saccade direction channel the mutual information is much smaller than the entropy of the channel, i.e., randomness is much bigger than determinism.
- Horizontal saccades tend to be more frequent than vertical saccades. This was already observed in [24] for photographs and in [25] for fractal images, but it is different to the result obtained in [25] for natural scene images.
- The information transferred, measured as mutual information, is in general higher in horizontal/vertical displacements than in diagonal ones, except in the case where the visual content is structured along a diagonal.
- The information channel measures based on saccade direction discriminate better between observers than between images, differently than for the gaze channel, which distinguishes better between images. Thus, the saccade direction channel might be useful to help customizing the way the information is presented according to the observer particular behavior pattern.

The rest of the paper is organized as follows. In Sect. 2 we present previous work on eye tracking data analysis based on transition matrices, in Sect. 3 we model the saccade direction sequences as an information channel, in Sect. 4 we discuss our results, and conclusions and future work are presented in Sect. 5.

2   Background: Modeling Saccades as a Random Walk

Since the pioneering work by Ellis and Stark [26] several researchers have modeled the eye movement data as a Markov chain transition matrix. The key for this modeling is how to construct a reasonable and appropriate transition matrix.

Ellis and Stark [26], to study dynamic displays of air traffic, divided cockpit display traffic information (CDTI) into eight AOIs. They introduced first-order fixation transition matrices firstly, and converted them to conditional probability matrices. And then, they computed the conditional entropy by employing the conditional probabilities or transition matrices. Ellis and Stark did not consider the self-transitions between AOIs, although their work provided a measure of the statistical dependency of fixations represented by the transition matrix.

Ponsoda et al. [18] conducted an eye tracking study during visual search, and introduced probability vectors and transition matrices by classifying the saccade directions. Unlike Ellis and Stark's work [26], Ponsoda et al.'s transition matrices were built based on saccade directions rather than between the AOIs. Interestingly, although their matrices are compared with a statistical method, the sequence of saccade directions was not modeled as a Markov chain. The Markov model, comparatively, is used and emphasized in this article.

Besag and Mondal [27] verified the feasibility of modeling gaze transition as a first-order Markov process. Subsequently, Krejtz et al. [28–30] asserted stationary entropy H_s and transition entropy H_t to measure the complexity of the Markov process through modeling gaze transitions between AOIs as a Markov chain. In this paper, the gaze transition modeling is based on saccade directions rather than AOIs.

A recent survey that reviews the use of the entropies H_s and H_t of a Markov chain between AOIs can be found in [31].

Ma et al. [21] and Hong et al. [22,23] built the gaze information channel between AOIs, that extends the Markov chain model of visual paths, by interpreting the transition matrix as the conditional matrix of a communication or information channel [32,33].

3   Saccade Direction Information Channel

3.1   Saccade Direction

Different from our previous work where the gaze information channel was based on AOIs [21–23], here we model the eye tracking gaze transition between the saccade directions S, discretized to 8 in this paper, i.e. $S = \{1, 2, \ldots, 8\}$. Next, we describe how we build the saccade direction transitions.

As in Ponsoda et al. [18], we introduce saccade direction, where a saccade can be described as a movement between two fixations. The angle between them is used to obtain the direction. We consider eight directions: East (E), SouthEast (SE), South (S), SouthWest (SW), West (W), NorthWest (NW), North (N) and NorthEast (NE), see Fig. 1(b). In Fig. 1(a) (note that the fixation duration time was not considered) we show an example of a saccade direction sequence.

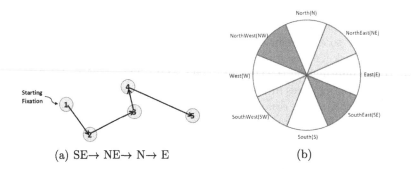

(a) SE→ NE→ N→ E   (b)

Fig. 1. (a) an example of a saccade with 5 fixations and saccade direction sequence SE→ NE→ N→ E, (b) the 8 saccade directions considered in this paper.

Then, we build the transition matrix, which has 64 elements, corresponding to the $8 \times 8 = 64$ different consecutive pairs of saccade direction ($E, E; E, SE; E, S;$ ...; $E, NE;$ ...; NE, NE). In general, N^2 pairs for N saccade directions.

Next, we go further than Ponsoda et al.'s approach by extending the Markov chain model for gaze transition between saccade directions to an information channel $X \rightarrow Y$, where X and Y are discrete random variables with alphabet $S = \{E, SE, S, SW, W, NW, N, NE\}$. The basic elements of the *saccade direction information channel* are as follows:

- The conditional probabilities $p_{ij} = p(j|i)$ represent the probability of transition from i saccade direction to j saccade direction. To obtain the matrix elements p_{ij}, we take the number of transitions from i source or incoming saccade direction, to destination or outgoing saccade direction j, and then we normalize relative to each source saccade direction (i.e., per row), $p_{ij} = \frac{n_{ij}}{\sum_{j=1}^{s} n_{ij}}$, $i, j \in S$. Conditional probabilities hold that $\sum_{j=1}^{s} p_{ij} = 1, \forall i \in S$.
- The marginal probabilities of input X and output Y, $p(X)$ and $p(Y)$, are both given by the equilibrium probability $\pi = (\pi_1, \pi_2, \ldots, \pi_s)$, representing the proportion of times a saccade has arrived at a given direction from any direction. The equilibrium distributions can be obtained by normalizing the totals per row (or per column) by the total number of transitions.

3.2 Measures in the Saccade Direction Information Channel

In this Section, we define Shannon's information measures [33] for the saccade direction information channel. The entropy of the input (and also output) random variables, with probability distribution equal to the equilibrium distribution, is

$$H_s = H(X) = H(Y) = -\sum_{i=1}^{s} \pi_i \log \pi_i \qquad (1)$$

As the equilibrium distribution describes the proportion of times a saccade has arrived at a given direction from any direction (or alternatively, the proportion of times the saccade has exited from a given direction to any direction), H_s measures the average uncertainty of gaze direction between the eight saccade directions.

The conditional entropy of ith row, $H(Y|i)$, is defined as

$$H(Y|i) = -\sum_{j=1}^{s} p_{ij} \log p_{ij} \qquad (2)$$

It depicts the uncertainty that the next saccade would be in the j_{th} saccade direction if it had arrived at the i_{th} saccade direction.

The conditional entropy H_t of the information channel is the weighted average of row entropies,

$$H_t = H(Y|X) = \sum_{i=1}^{s} \pi_i H(Y|i) = -\sum_{i=1}^{s} \pi_i \sum_{j=1}^{s} p_{ij} \log p_{ij} \qquad (3)$$

which indicates the average uncertainty of gaze transition between two saccade directions, or average uncertainty about the destination saccade direction when the source saccade direction is known.

The joint entropy $H(X, Y)$ of the information channel is the entropy of the joint distribution of X and Y, $\{\pi_i p_{ij}\}, 1 \le i, j \le s$,

$$H(X,Y) = H(X) + H(Y|X) = H_s + H_t = \sum_{i=1}^{s} \sum_{j=1}^{s} \pi_i p_{ij} \log (\pi_i p_{ij}) \qquad (4)$$

and measures the total uncertainty of the information channel.

The mutual information $I(X; Y)$ is given by the Kullback-Leibler distance [33] between the actual joint distribution of X and Y, $\{\pi_i p_{ij}\}$ and their joint distribution in case they were independent, $\{\pi_i \pi_j\}$,

$$I(X;Y) = H(X) + H(Y) - H(Y|X)$$
$$= \sum_{i=1}^{s} \sum_{j=1}^{s} \pi_i p_{ij} \log \frac{\pi_i p_{ij}}{\pi_i \pi_j} = \sum_{i=1}^{s} \sum_{j=1}^{s} \pi_i p_{ij} \log \frac{p_{ij}}{\pi_j} \qquad (5)$$

and indicates the total correlation, or information shared, between the saccade directions. Observe that if X and Y are independent, this is, if output saccade direction would be independent of input saccade direction, then for all i, j, $\pi_i p_{ij} = \pi_i \pi_j$ and the mutual information would be 0.

Figure 2 shows in a Venn diagram the relationship between the channel measures.

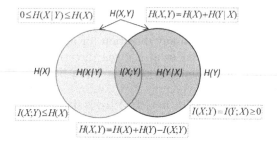

Fig. 2. Venn diagram showing the relationship of information channel measures.

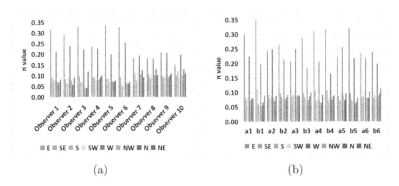

Fig. 3. (a)&(b) show the equilibrium distribution π of channels grouped by observers and paintings in Van Gogh dataset respectively.

4 Experimental Results: The Van Gogh Paintings Dataset

The Van Gogh dataset is an eye tracking dataset from our previous work [22]. It contains 12 paintings of Vincent Van Gogh in digital format observed by 10 observers, where observers randomly viewed each painting for 45 s in free viewing mode. The paintings are divided in two groups (a and b), as shown in Fig. 4.

Both groups include 6 representative paintings of each period of Van Gogh's painting trajectory (periods numbered from 1 to 6) [34,35].

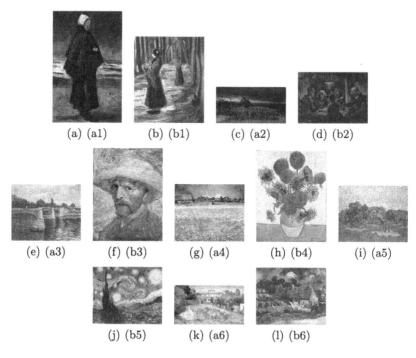

(a) (a1)   (b) (b1)   (c) (a2)   (d) (b2)

(e) (a3)   (f) (b3)   (g) (a4)   (h) (b4)   (i) (a5)

(j) (b5)   (k) (a6)   (l) (b6)

Fig. 4. Representative paintings of each Van Gogh's period (chronologically ordered from period 1 to 6. Credits for images except a3&b2: The Vincent Van Gogh Gallery of David Brooks, http://www.Vggallery.com, Copyright 1996–2022 David Brooks. Credits for a3&b2: Van Gogh Museum, Amsterdam (Vincent van Gogh Foundation).

Transition Probability Matrices. In Table 1 we show the unnormalized transition matrices values clustered by observer. The (normalized) transition matrices or transition probabilities matrices (not shown) are obtained by normalizing each row by its total. We see that the majority of the transitions from all directions go to E direction. We also observe the saccadic momentum [20] specially for E direction, and even for some observers there are more $E \rightarrow E$ transitions than $E \rightarrow W$ ones. A similar behaviour is observed when clustering values of all observers for each painting Table 2. The preferred directions for all paintings are E and W, and the other directions are more or less evenly distributed. Horizontal saccades are much more frequent than vertical ones.

In Figs. 3 a&b we plot the equilibrium distributions for observers and paintings, respectively, which represent the relative frequency of arriving from (or exiting to) a given direction, and obtained by normalizing the totals in Tables 1 and 2. We see that in general the preferred directions are E and W, even for paintings with aspect ratio less than 1.

Table 1. Unnormalized transition matrices and equilibrium distribution of paintings grouped by observers, with equilibrium distribution highest values highlighted.

Observer 1

Transition matrix	E	SE	S	SW	W	NW	N	NE	Total
E	220	43	24	38	173	33	9	29	569
SE	52	15	4	18	38	15	10	16	168
S	30	10	18	14	12	11	24	14	133
SW	41	16	12	14	18	7	19	21	148
W	127	42	32	21	70	21	33	34	380
NW	35	17	18	11	14	13	6	11	125
N	26	12	10	15	18	11	18	13	123
NE	35	13	15	16	40	13	8	7	147
Total	566	168	133	147	383	124	127	145	1793
Equilibrium distribution	0.317	0.094	0.074	0.083	0.212	0.070	0.069	0.082	1.000

Observer 2

Transition matrix	E	SE	S	SW	W	NW	N	NE	Total
E	133	24	10	35	176	36	8	14	436
SE	27	5	7	14	36	16	10	12	127
S	21	11	8	8	12	8	14	16	98
SW	38	7	7	19	22	12	16	21	142
W	136	36	34	11	49	19	27	44	356
NW	36	16	13	13	14	10	4	9	115
N	17	14	8	13	13	7	4	9	85
NE	24	15	12	29	34	9	4	12	139
Total	432	128	99	142	356	117	87	137	1498
Equilibrium distribution	0.291	0.085	0.065	0.095	0.238	0.077	0.057	0.093	1.000

Observer 3

Transition matrix	E	SE	S	SW	W	NW	N	NE	Total
E	301	98	59	46	146	16	19	96	781
SE	88	37	7	14	32	3	8	43	232
S	43	11	13	28	20	8	8	32	163
SW	38	4	5	22	60	25	10	14	178
W	184	21	31	43	170	22	13	48	532
NW	25	16	6	0	10	18	6	16	97
N	12	10	13	15	29	4	15	2	100
NE	91	36	29	10	61	1	21	33	282
Total	782	233	163	178	528	97	100	284	2365
Equilibrium distribution	0.330	0.098	0.069	0.075	0.225	0.041	0.042	0.119	1.000

Observer 4

Transition matrix	E	SE	S	SW	W	NW	N	NE	Total
E	74	20	20	18	93	20	17	24	286
SE	28	14	13	4	28	11	4	10	112
S	20	8	19	9	15	12	20	10	113
SW	17	6	7	13	17	7	15	13	95
W	81	32	16	13	63	17	22	29	273
NW	21	11	11	11	15	6	9	9	93
N	19	9	14	10	22	12	15	11	112
NE	22	12	14	17	22	10	11	11	119
Total	282	112	114	95	275	98	113	119	1208
Equilibrium distribution	0.237	0.093	0.094	0.079	0.226	0.081	0.093	0.099	1.000

Observer 5

Transition matrix	E	SE	S	SW	W	NW	N	NE	Total
E	484	96	21	68	321	49	17	45	1101
SE	111	20	10	16	66	36	9	22	290
S	65	27	35	5	27	44	38	43	284
SW	40	23	41	26	17	18	30	36	231
W	216	41	70	28	131	46	63	59	654
NW	69	22	35	26	23	22	28	24	249
N	52	25	38	30	21	13	38	13	230
NE	57	36	36	33	45	23	10	21	261
Total	1094	290	286	232	651	251	233	263	3300
Equilibrium distribution	0.334	0.088	0.086	0.070	0.198	0.075	0.070	0.079	1.000

Observer 6

Transition matrix	E	SE	S	SW	W	NW	N	NE	Total
E	123	37	20	23	156	20	15	25	419
SE	42	10	1	14	29	9	9	6	120
S	18	6	9	4	6	6	5	10	64
SW	34	8	3	17	19	5	4	7	97
W	119	37	13	21	69	24	20	26	329
NW	31	8	8	8	9	9	9	6	88
N	26	4	5	4	19	5	12	6	81
NE	24	10	5	7	19	10	7	8	90
Total	417	120	64	98	329	86	82	92	1288
Equilibrium distribution	0.325	0.093	0.050	0.075	0.255	0.068	0.063	0.070	1.000

Observer 7

Transition matrix	E	SE	S	SW	W	NW	N	NE	Total
E	52	33	23	31	50	23	23	21	256
SE	30	10	14	14	33	20	16	18	155
S	22	10	10	10	23	17	13	5	110
SW	29	13	12	16	26	17	25	20	158
W	46	36	14	24	68	22	44	21	275
NW	26	19	11	20	30	15	15	12	148
N	28	16	17	23	21	19	32	22	178
NE	19	16	10	21	27	16	10	10	129
Total	252	153	111	159	278	149	178	129	1409
Equilibrium distribution	0.182	0.110	0.078	0.112	0.195	0.105	0.126	0.092	1.000

Observer 8

Transition matrix	E	SE	S	SW	W	NW	N	NE	Total
E	38	25	23	17	31	21	25	22	202
SE	21	12	11	10	19	14	16	14	117
S	17	9	11	13	17	8	14	8	97
SW	22	8	7	10	20	16	22	14	119
W	39	24	11	23	40	20	22	18	197
NW	22	13	7	13	20	14	10	15	114
N	24	8	12	19	28	14	27	12	144
NE	16	15	13	14	23	10	10	11	112
Total	199	114	95	119	198	117	146	114	1102
Equilibrium distribution	0.183	0.106	0.088	0.108	0.179	0.103	0.131	0.102	1.000

Observer 9

Transition matrix	E	SE	S	SW	W	NW	N	NE	Total
E	52	22	22	24	67	16	13	17	233
SE	19	11	18	8	25	10	7	11	109
S	13	16	10	11	15	11	13	13	102
SW	19	6	6	14	20	12	14	16	107
W	62	20	18	15	48	16	25	28	232
NW	25	8	8	8	19	8	8	11	105
N	21	9	10	8	21	12	16	14	111
NE	19	18	11	19	19	9	15	14	124
Total	230	110	103	107	234	94	111	124	1123
Equilibrium distribution	0.208	0.097	0.091	0.095	0.207	0.094	0.099	0.110	1.000

Observer 10

Transition matrix	E	SE	S	SW	W	NW	N	NE	Total
E	40	18	18	13	31	15	18	22	175
SE	21	11	10	14	19	14	14	14	117
S	16	11	25	12	24	10	22	19	139
SW	12	12	10	15	22	10	15	10	106
W	27	20	19	21	72	20	28	21	228
NW	12	15	16	11	26	14	8	12	114
N	21	19	14	11	20	17	26	21	149
NE	23	13	24	10	14	18	19	7	128
Total	172	119	136	107	228	114	150	130	1156
Equilibrium distribution	0.151	0.101	0.120	0.092	0.197	0.099	0.129	0.111	1.000

Averaging Results Versus Clustering for All Observers and Paintings.

Figures 5(a,c) give the stacked H_t and H_s values, and Figs. 5(b,d) the stacked H_t and $I(X;Y)$ values grouped by observer, Figs. 5(a,b), and grouped by painting, Figs. 5(c,d). These figures confirm the validity of the clustering to analyze collective behaviour, as for every observer and painting, the values of joint entropy $H(X,Y)$ (red dots) are approximately equal to the sum of H_s and H_t, and that the value of H_s (blue dots) is approximately equal to the sum of H_t and $I(X;Y)$.

We observe that the values grouped by painting, in Figs. 5(c,d), present less variation than the values grouped by observer, in Figs. 5(a,b).

The equilibrium entropy H_s means how balanced the saccade directions have been visited. From Fig. 5(a), observers 3 and 6 have the lowest H_s values, while 7, 8, 9, 10 have the largest H_s values, telling us that observers 3, 6 were the ones to spend more time focused on certain saccade directions, while observers 7, 8, 9, 10 distributed more evenly their time.

Table 2. Unnormalized transition matrices and equilibrium distribution of observers grouped by painting, with equilibrium distribution highest values highlighted.

Painting name: **a1**

Transition matrix	E	SE	S	SW	W	NW	N	NE	Total
E	107	19	11	15	89	8	7	14	270
SE	22	6	4	5	14	11	11	4	77
S	12	6	8	9	6	9	10	6	66
SW	14	4	6	11	8	7	14	12	76
W	70	19	11	11	86	9	12	15	203
NW	18	11	8	5	5	8	2	7	64
N	12	4	8	5	10	8	13	12	72
NE	11	8	9	16	14	4	5	6	73
Total	266	77	65	77	202	64	74	76	901
Equilibrium distribution	**0.300**	0.085	0.073	0.084	0.225	0.071	0.080	0.081	1.000

Painting name: **b1**

Transition matrix	E	SE	S	SW	W	NW	N	NE	Total
E	125	31	8	16	77	7	5	22	291
SE	24	10	7	4	14	10	8	12	89
S	7	5	6	2	12	5	8	4	49
SW	13	4	2	3	6	2	3	10	43
W	77	18	6	5	19	6	12	15	158
NW	13	5	5	2	6	6	2	5	44
N	9	5	9	5	6	4	11	1	52
NE	19	11	7	5	18	4	4	4	72
Total	287	89	50	42	160	44	53	73	798
Equilibrium distribution	**0.365**	0.112	0.061	0.054	0.198	0.055	0.065	0.090	1.000

Painting name: **a2**

Transition matrix	E	SE	S	SW	W	NW	N	NE	Total
E	32	14	6	16	70	14	6	11	169
SE	9	9	4	6	18	11	1	7	65
S	13	5	9	2	5	10	6	5	55
SW	14	4	4	9	8	6	9	9	63
W	62	15	12	7	38	13	11	15	173
NW	16	10	6	6	10	3	3	3	57
N	11	4	5	4	11	1	11	3	50
NE	10	5	9	13	19	2	3	7	62
Total	167	66	55	63	173	60	50	60	694
Equilibrium distribution	0.244	0.094	0.079	0.091	**0.249**	0.082	0.072	0.089	1.000

Painting name: **b2**

Transition matrix	E	SE	S	SW	W	NW	N	NE	Total
E	191	46	27	28	151	29	21	36	529
SE	52	23	7	16	50	16	12	21	197
S	35	11	24	13	20	12	26	27	168
SW	30	16	20	20	23	13	18	22	162
W	101	45	28	35	103	34	40	38	424
NW	46	16	21	12	19	16	13	12	155
N	33	15	17	18	28	18	24	15	168
NE	35	24	24	20	32	17	16	15	183
Total	523	196	168	162	426	155	170	186	1986
Equilibrium distribution	**0.266**	0.099	0.085	0.082	0.214	0.078	0.085	0.092	1.000

Painting name: **a3**

Transition matrix	E	SE	S	SW	W	NW	N	NE	Total
E	47	23	23	32	109	27	11	18	287
SE	21	13	11	13	31	12	7	10	118
S	24	9	17	14	17	13	13	13	131
SW	33	12	7	10	23	13	14	20	132
W	82	25	28	26	95	25	36	29	346
NW	32	18	15	13	12	10	9	12	121
N	26	12	16	9	23	10	22	10	128
NE	21	8	13	17	36	8	9	12	124
Total	286	117	130	134	346	121	129	124	1387
Equilibrium distribution	0.207	0.085	0.094	0.095	**0.250**	0.087	0.092	0.089	1.000

Painting name: **b3**

Transition matrix	E	SE	S	SW	W	NW	N	NE	Total
E	167	50	20	40	117	19	12	26	451
SE	49	9	9	9	36	13	8	20	153
S	26	17	8	10	10	16	24	12	123
SW	16	10	18	11	13	11	24	12	121
W	93	28	28	12	53	18	27	28	287
NW	29	8	14	11	15	14	10	17	118
N	33	8	13	18	20	12	19	15	138
NE	38	23	20	12	23	15	12	14	156
Total	451	153	130	121	287	118	138	156	1554
Equilibrium distribution	**0.290**	0.098	0.086	0.077	0.185	0.076	0.088	0.100	1.000

Painting name: **a4**

Transition matrix	E	SE	S	SW	W	NW	N	NE	Total
E	161	48	22	27	78	18	9	36	399
SE	46	14	8	11	25	15	8	10	137
S	19	13	8	13	13	4	7	14	91
SW	24	6	7	9	24	15	8	11	104
W	79	24	13	18	70	14	23	23	264
NW	20	11	11	5	19	10	5	6	87
N	18	9	10	7	11	10	12	6	83
NE	30	11	12	14	25	6	11	11	120
Total	397	136	91	104	265	92	83	117	1285
Equilibrium distribution	**0.311**	0.107	0.071	0.081	0.205	0.068	0.065	0.093	1.000

Painting name: **b4**

Transition matrix	E	SE	S	SW	W	NW	N	NE	Total
E	212	64	26	27	119	23	28	38	537
SE	74	15	14	15	27	13	6	21	185
S	30	10	10	14	17	13	15	14	123
SW	26	18	12	18	15	14	20	18	141
W	90	24	23	18	45	24	33	24	281
NW	29	12	8	17	18	15	18	12	129
N	26	16	14	17	25	12	21	18	149
NE	46	27	19	15	16	18	9	9	154
Total	533	186	126	141	282	127	150	154	1699
Equilibrium distribution	**0.316**	0.109	0.072	0.083	0.165	0.076	0.088	0.091	1.000

Painting name: **a5**

Transition matrix	E	SE	S	SW	W	NW	N	NE	Total
E	78	24	20	22	94	30	17	27	310
SE	29	16	8	11	29	8	6	14	121
S	15	11	14	13	17	10	8	11	99
SW	29	10	8	16	32	12	10	16	133
W	101	26	25	30	98	21	22	31	354
NW	20	13	7	12	25	20	10	18	125
N	13	12	6	20	18	6	19	8	102
NE	24	8	11	9	42	15	10	15	134
Total	307	120	99	133	355	122	102	140	1378
Equilibrium distribution	0.225	0.088	0.072	0.097	**0.257**	0.091	0.074	0.097	1.000

Painting name: **b5**

Transition matrix	E	SE	S	SW	W	NW	N	NE	Total
E	195	49	22	26	134	23	16	34	499
SE	54	10	9	11	30	12	10	16	152
S	37	9	9	6	19	13	14	13	120
SW	31	2	9	15	23	7	10	13	110
W	108	38	30	18	66	18	30	19	337
NW	22	12	15	10	17	6	9	6	97
N	22	11	13	15	21	5	10	8	105
NE	32	23	14	11	27	12	7	3	129
Total	499	154	121	110	337	96	106	106	1549
Equilibrium distribution	**0.322**	0.098	0.077	0.071	0.218	0.063	0.068	0.063	1.000

Painting name: **a6**

Transition matrix	E	SE	S	SW	W	NW	N	NE	Total
E	110	32	24	31	117	26	18	27	385
SE	38	15	12	16	19	8	15	18	141
S	30	12	23	15	19	9	17	16	141
SW	29	10	9	30	41	19	21	13	172
W	94	23	32	34	90	25	18	40	356
NW	27	19	14	14	14	23	9	15	135
N	19	15	13	14	19	16	21	15	132
NE	37	15	15	18	34	11	15	22	167
Total	384	141	142	172	353	137	134	166	1629
Equilibrium distribution	**0.236**	0.087	0.087	0.106	0.219	0.083	0.081	0.103	1.000

Painting name: **b6**

Transition matrix	E	SE	S	SW	W	NW	N	NE	Total
E	94	19	31	33	80	25	14	34	331
SE	21	5	2	9	32	15	11	17	112
S	17	11	22	3	19	16	18	14	123
SW	28	7	8	13	25	10	17	18	127
W	82	24	22	8	47	20	33	37	273
NW	30	10	9	14	20	11	13	14	121
N	24	15	17	18	18	12	20	12	136
NE	30	21	16	26	24	12	14	14	157
Total	326	112	127	125	274	121	138	159	1382
Equilibrium distribution	**0.240**	0.081	0.092	0.090	0.198	0.088	0.098	0.114	1.000

Comparison of Saccade Direction Information Channel over Directions with Gaze Information Channel over AOIs.

In Fig. 6 we compare, for direction channel and gaze channel, the normalized $I(X;Y)$ by H_s of all tested Van Gogh's paintings by participant Fig. 6(a), and per tested painting from all participants Fig. 6(b). We observe that the mutual information of direction channel is much smaller than for the gaze channel, both by observer and by painting. From Fig. 6(b) we see that the gaze channel discriminates more the participant than the direction channel does, as the values of normalized $I(X;Y)$ for direction channel are very similar. On the other hand we can see that, although the absolute variations in Fig. 6(a) for the direction channel are smaller than for the gaze channel, the relative variations are higher, and thus the gaze channel can be useful to discriminate by observer . Finally, let us note from Fig. 6(b) that

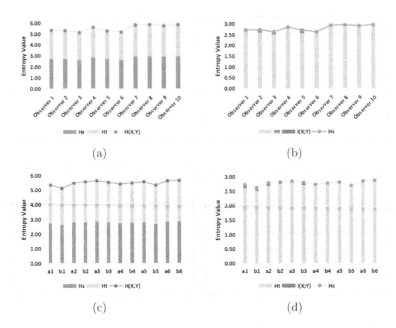

Fig. 5. (a) the stacked H_t, H_s, (b) the stacked H_t, $I(X;Y)$, for all paintings grouped by observer, (c) the stacked H_t, H_s, and (d) the stacked H_t, $I(X;Y)$, for all observers grouped by painting. Observe that the stacked values, that are obtained by clustering the individual transition matrices, almost coincide with the line graph values, obtained from averaging the measure values from the individual channels.

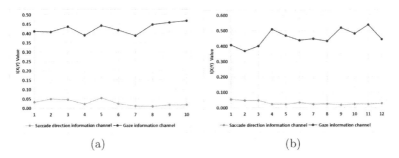

Fig. 6. Normalized $I(X;Y)$ by H_s for saccade direction information channel (in blue) compared with gaze information channel [22] (in orange) of paintings grouped by observer (a), and observers grouped by painting (b). Observe that for saccade direction information channel the relative variation of values is higher for observers (a) than for paintings (b), while in gaze information channel it is the contrary. (Color figure online)

the saccade direction channel values for a1, b1 and a2 are higher than for the rest of paintings. The paintings a1, b1, a2 are simpler than the other paintings, they correspond to the earlier periods in Van Gogh's life, where the composition

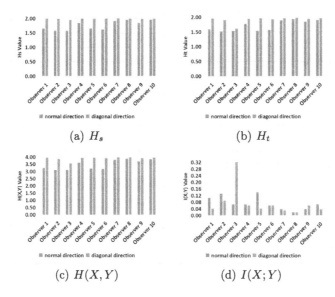

Fig. 7. Measures for normal & diagonal directions for all information channels clustered by paintings. Observe that the values of $I(X;Y)$ (d) are equal or higher for normal directions (in blue) than for diagonal directions (orange), except for observer 3. (Color figure online)

and palette were simpler, and not much visual exploration in different directions would be needed to understand the painting, leading to less randomness and higher mutual information. At the same time, the more complex nature of the other paintings allows for the discovery of spatial correlations between the AOIs that make the mutual information in the spatial gaze channel to be higher than for the more simpler paintings a1, b1 and a2.

Results from 4 Saccade Directions. We extracted two groups: *normal direction* and *diagonal direction* in Van Gogh dataset, respectively. We display the values H_s, H_t, $H(X,Y)$ and $I(X;Y)$ in Figs. 7 and 8 from the perspective of observers and paintings, respectively.

We can see that the normal direction group has the lower $H_s, H_t, H(X,Y)$ (see blue bars from Figs. 7 and 8(a)–(c)), and in general higher mutual information $I(X;Y)$ (see blue bars from Figs. 7 and 8(d)) whereas diagonal direction group has the opposite pattern with higher $H_s, H_t, H(X,Y)$ and lower mutual information $I(X;Y)$. We observe some exceptions to mutual information being higher in normal directions. From the perspective of observers, observer 3 extracts more information from diagonal saccades than from normal ones. From the perspective of paintings, for painting b5 the diagonal mutual information is much higher. By observing the painting we see that it is mostly organized about the diagonal, from the biggest star on the upper right corner to the trunk on the lower left side, from the village in the lower right corner to the two shining

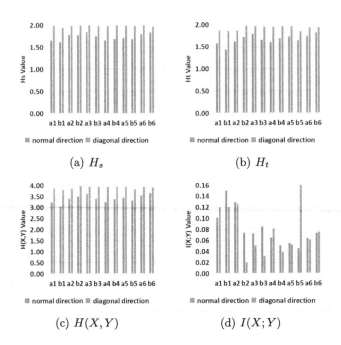

Fig. 8. Measures for normal & diagonal directions for all information channels clustered by paintings. Observe that the values of $I(X;Y)$ (d) are similar or higher for normal directions (in blue) than for diagonal directions (orange), except for painting b5. (Color figure online)

stars on the upper left corner, thus this structure forces the observation to run rather in diagonal.

5   Conclusion and Future Work

To find quantitative links between eye movements and human cognition, this paper models saccade direction transitions as a discrete information channel, which we call *saccade direction information channel*. We then examine the usefulness of the proposed approach with eye tracking data on a set of Van Gogh's paintings. Our experiment confirm the previous finding that human horizontal saccades always tend to be more frequent than vertical saccades. Also, mutual information is explored to measure the relative importance of displacement in diagonal or in horizontal/vertical, with the result that the information transfer for horizontal/vertical directions is higher than for diagonal ones. The saccade information channel seems mainly to discriminate observer behaviour, rather than the image observed, in opposite behaviour to our previous spatial gaze information channel. The better discrimination by observer might lead in the future to customize in an optimized way the presentation of information to the specific observer. While preliminary results with a different dataset [23] confirm our findings, more participants in the eye tracking experiments need to be considered in the future. We also plan to integrate the fixation time into the channel.

References

1. Noton, D., Stark, L.: Scanpaths in eye movements during pattern perception. Science **171**(3968), 308–311 (1971)
2. Liman, T.G., Zangemeister, W.H.: Scanpath eye movements during visual mental imagery in a simulated hemianopia paradigm. J. Eye Mov. Res. **5**(1), 308–311 (2012). https://doi.org/10.16910/jemr.5.1.2
3. Lévêque, L., Bosmans, H., Cockmartin, L., Liu, H.: State of the art: eye-tracking studies in medical imaging. IEEE Access **6**, 37023–37034 (2018)
4. Huckauf, A., Urbina, M.H.: Object selection in gaze controlled systems: what you don't look at is what you get. ACM Trans. Appl. Percept. **8**(2), 1–14 (2011)
5. Silva, Y., Simoes, W., Naves, E., Filho, T., Lucena, V.: Teleoperation training environment for new users of electric powered wheelchairs based on multiple driving methods. IEEE Access **6**, 55099–55111 (2018)
6. Breeden, K., Hanrahan, P.: Gaze data for the analysis of attention in feature films. ACM Trans. Appl. Percept. **14**(4), 23 (2017)
7. Rienks, R., Poppe, R., Heylen, D.: Differences in head orientation behavior for speakers and listeners: an experiment in a virtual environment. ACM Trans. Appl. Percept. **7**(1), 2 (2016)
8. van Wermeskerken, M., van Gog, T.: Seeing the instructor's face and gaze in demonstration video examples affects attention allocation but not learning. Comput. Educ. **113**, 98–107 (2017)
9. Was, C., Sansosti, F., Morris, B.: Eye-Tracking Technology Applications in Educational Research. IGI Global (2016). https://doi.org/10.4018/978-1-5225-1005-5
10. Ellis, E.M., Borovsky, A., Elman, J.L., Evans, J.L.: Novel word learning: an eye-tracking study. are 18-month-old late talkers really different from their typical peers? IEEE Trans. Affect. Comput. **58**, 143–157 (2015)
11. Duchowski, A., Driver, J., Jolaoso, S., Tan, W., Ramey, B., Robbins, A.: Scanpath comparison revisited. In: Proceedings of the 2010 Symposium on Eye-Tracking Research and Applications, pp. 219–226 (2010)
12. de Bruin, J.A., Malan, K.M., Eloff, J.H.P.: Saccade deviation indicators for automated eye tracking analysis. In: Proceedings of the 2013 Conference on Eye Tracking South, pp. 47–54 (2013)
13. Duchowski, A.T., Price, M.M., Meyer, M., Orero, P.: Aggregate gaze visualization with real-time heatmaps. In: Proceedings of the Symposium on Eye Tracking Research and Applications, pp. 13–20, June 2012
14. Sun, W., Chen, Z., Wu, F.: Visual scanpath prediction using IOR-ROI recurrent mixture density network. IEEE Trans. Pattern Anal. Mach. Intell. **43**(6), 2101–2118 (2021)
15. Bao, W., Chen, Z.: Human scanpath prediction based on deep convolutional saccadic model. Neurocomputing **404**(3), 154–164 (2020)
16. Zhang, A.T., Le Meur, B.O.: How old do you look? Inferring your age from your gaze. In: International Conference on Image Processing. Athènes, Greece, October 2018

17. Fuhl, W., et al.: Encodji: encoding gaze data into emoji space for an amusing scan-path classification approach;). In: Proceedings of the 11th ACM Symposium on Eye Tracking Research and Applications. ETRA 2019. Association for Computing Machinery, New York (2019)

18. Ponsoda, V., Scott, D., Findlay, J.M.: A probability vector and transition matrix analysis of eye movements during visual search. Acta Physiol. (Oxf) **88**(2), 167–185 (1995)

19. Tatler, B.W., Vincent, B.T.: Systematic tendencies in scene viewing. J. Eye Mov. Res. **2**(2), 1–18 (2008)

20. Smith, T.J., Henderson, J.M.: Facilitation of return during scene viewing. In: Visual Cognition, vol. 17, pp. 1083–1108 (2009)

21. Ma, L., Sbert, M., Feixas, M.: Gaze information channel. In: Proceedings of Pacific Rim Conference on Multimedia, pp. 575–585 (2018)

22. Hao, Q., Ma, L., Sbert, M., Feixas, M., Zhang, J.: Gaze information channel in van Gogh's paintings. Entropy **22**(5), 540 (2020)

23. Hao, Q., Sbert, M., Ma, L.: Gaze information channel in cognitive comprehension of poster reading. Entropy **21**(5), 444 (2019)

24. Le Meur, O., Liu, Z.: Saccadic model of eye movements for free-viewing condition. Vision. Res. **116**, 152–164 (2015)

25. Foulsham, T., Kingstone, A.: Asymmetries in the direction of saccades during perception of scenes and fractals: effects of image type and image features. Vision. Res. **50**(8), 779–795 (2010)

26. Ellis, S.R., Stark, L.: Statistical dependency in visual scanning. Hum. Factors **28**(4), 421–438 (1986)

27. Besag, J., Mondal, D.: Exact goodness-of-fit tests for Markov chains. Biometrics **69**(2), 488–496 (2013)

28. Krejtz, K., Szmidt, T., Duchowski, A.T., Krejtz, I.: Entropy-based statistical analysis of eye movement transitions. In: Proceedings of the Symposium on Eye Tracking Research and Applications, pp. 159–166 (2014)

29. Krejtz, K., et al.: Gaze transition entropy. ACM Trans. Appl. Percept. **13**(1), 4 (2015)

30. Raptis, G.E., Fidas, C.A., Avouris, N.M.: On implicit elicitation of cognitive strategies using gaze transition entropies in pattern recognition tasks. In: Proceedings of the 2017 CHI Conference Extended Abstracts on Human Factors in Computing Systems, pp. 1993–2000 (2017)

31. Shiferaw, B., Downey, L., Crewther, D.: A review of gaze entropy as a measure of visual scanning efficiency. Neurosci. Biobehav. Rev. **96**, 353–366 (2019)

32. Shannon, C.E.: A mathematical theory of communication. Bell Syst. Tech. J. **27**(3), 379–423 (1948)

33. Cover, T.M., Thomas, J.A.: Elements of Information Theory. Wiley Series in Telecommunications. Wiley, Hoboken (1991)

34. Rigau, J., Feixas, M., Sbert, M.: Informational dialogue with van Gogh's paintings. In: Computational Aesthetics in Graphics, Visualization, and Imaging. The Eurographics Association (2008)

35. Rigau, J., Feixas, M., Sbert, M., Wallraven, C.: Toward Auvers period: Evolution of van Gogh's style. In: Computational Aesthetics, pp. 99–106 (2010)

Shared-Attribute Multi-Graph Clustering with Global Self-Attention

Jianpeng Chen[1], Zhimeng Yang[1], Jingyu Pu[1], Yazhou Ren[1(✉)], Xiaorong Pu[1], Li Gao[2,3], and Lifang He[4]

[1] School of Computer Science and Engineering, University of Electronic Science and Technology of China, Chengdu, China
{yazhou.ren,puxiaor}@uestc.edu.cn
[2] Chengdu Third People's Hospital, Chengdu, China
[3] Affiliated Hospital of Southwest Jiaotong University, Chengdu, China
[4] Department of Computer Science and Engineering, Lehigh University, Bethlehem, PA, USA
lih319@lehigh.edu

Abstract. Recently, multi-view attributed graph clustering has attracted lots of attention with the explosion of graph-structured data. Existing methods are primarily designed for the form in which every graph has its attributes. We argue that a more natural form of multi-view attributed graph data contains shared node attributes and multiple graphs, which we called "multi-graph". When simply applying existing methods to multi-graph clustering, the information of shared attributes is not well exploited to eliminate the large variances among different graphs. Therefore, we propose a Shared-Attribute Multi-Graph Clustering with global self-attention (SAMGC) method for multi-graph clustering. The main ideas of SAMGC are: 1) Global self-attention is proposed to construct the supplementary graph from shared attributes for each graph. 2) Layer attention is proposed to meet the requirements for different layers in different graphs. 3) A novel self-supervised weighting strategy is proposed to de-emphasize unimportant graphs. Our experiments on four benchmark datasets show the superiority of SAMGC over 14 SOTA methods. The source code is available at https://github.com/cjpcool/SAMGC.

Keywords: Multi-graph learning · multi-view clustering · self-attention

1 Introduction

With the increased size and applications of graph-structured data, graph clustering methods [1,21] have emerged from unsupervised data analysis to mine the topological information of the graph. Generally graphs can be established with different relations among the nodes. For example, the graphs of films can be established with their leading roles, themes, or film distribution corporations. To utilize the topological information and nodes information of multiple graphs,

© The Author(s), under exclusive license to Springer Nature Switzerland AG 2023
M. Tanveer et al. (Eds.): ICONIP 2022, LNCS 13623, pp. 51–63, 2023.
https://doi.org/10.1007/978-3-031-30105-6_5

several multi-view graph clustering methods are proposed [13, 19, 20, 33]. These methods particularly focus on the data in which every graph has its attributes. However, in the real world, a more natural form of multi-view attributed graph data is shared node attributes with multiple graphs, which we called "multi-graph". For example, in academic network, each paper has an attribute of its content but the graphs can be constructed from citation or co-author, and in social network, everyone has his characteristics while there are followers graph and visitors graph. When simply applying existing methods to multi-graph clustering (*e.g.*, copy the shared attributes for every graph), the information of shared attributes is not well exploited to eliminate the variances among different graphs.

In general, there are three challenges for multi-graph clustering. 1) Different graphs may have different edges. For instance, the graph constructed by co-subject contains 2,210,761 edges, but the graph constructed by co-paper only contains 29,281 edges in the ACM dataset. This means that the same node may play different roles and contributes differently in different graphs. So some important parameters such as aggregation order should be individually different for each graph. 2) In a specific graph, some edges are noisy or some important edges may be lost, these redundant or missing relations between nodes should be considered with caution. 3) Multi-graph data can provide complementary information, however, it is difficult to capture both common and distinct latent components of variation across different graphs. To cope with these questions, [7] proposed an auto-encoder based model to learn the common information of all graphs. However, it uses only one graph in its encoding term, thus, some individual information has been lost. [13] filtered out noises by introducing a graph filter. However, in essence, the graph filter is to smooth the node representation, which can not filter out noises. [17] proposed reconstructing a consensus graph by contrastive learning, to some extent, this method could ignore the noises and avoid the large variance between different graphs. However, the clustering results are largely dependent on the quality of the consensus graph, and some noisy graphs may dominant the contrastive learning process.

In this paper, we propose a Shared-Attribute Multi-Graph Clustering with global self-attention (SAMGC) method which conducts the shared and complementary information to resolve the three issues. In SAMGC, we introduce the layer attention and global self-attention mechanisms to solve the questions (1) and (2). The aggregation orders of different graphs are trained with the layer attention mechanism to consider different weights from neighbor nodes. On the other hand, according to the shared features, the global self-attention mechanism can generate a supplementary graph for every view, which can enhance the weights of important relations and meanwhile eliminate noisy relations. Except for this, the global-self attention mechanism can also offer help to question (3). Finally, we apply a collaborative training algorithm for every view, which helps to learn both common and individual information from different graphs.

Our main contributions can be summarized as follows:

– To deal with the identified three challenges in the multi-graph clustering task, we propose a novel multi-graph clustering method called SAMGC.

- A novel global self-attention is proposed for multi-graph clustering, which can effectively mitigate the influence of noisy relations while complementing the variances among different graphs. Moreover, layer attention is introduced to satisfy different graphs' requirements of different aggregation orders.
- A collaborative training method with self-supervised weighting strategy is proposed to re-weight each view iteratively and learn the common and individual information of all graphs.

2   Related Works

Multi-view clustering methods have attracted much attention with the boosting of the application scenarios [4, 19, 28–31, 34]. Multi-view clustering strategy can naturally extend to graph-structured data which can make use of both nodes information and correlation information. For example, [33] used a disagreement cost function for regularizing graphs from different views. [25] learned a unified graph matrix from all views and used the unified graph matrix for clustering. [9] performed graph fusion and spectral clustering simultaneously. So far, most multi-view graph clustering methods aim at computing a new graph matrix that unified all views' information. There are also some other works employing the strategy of traditional clustering methods such as self-supervised learning and contrastive learning. In particular, [8,16] explored weights for each view. [17] used contrastive learning strategy and it performs well in the reported results.

Graph Neural Networks (GNNs) can efficiently integrate the information of nodes with topological structure [26,32]. Different single-graph clustering strategies have been proposed based on GNNs. For example, [3] proposed min-cut pooling layer as the regularization which can be jointly optimized with a task-specific loss. [22] proposed an unsupervised pooling method DMoN which allows optimization of cluster assignments in an end-to-end way. To the best of our knowledge, none of existing works have focused on multi-graph clustering.

3   Methodology

In Sect. 3.1, we introduce the preliminaries. In Sect. 3.2, we propose the shared-attribute multi-graph clustering with global self-attention (SAMGC). In Sect. 3.3, we present the collaborative optimizing mechanism of SAMGC. The inference process is shown in Sect. 3.4.

3.1   Preliminaries

Graph Neural Networks. Let $\mathcal{G} = (V, E)$ be a connected undirected graph, where V and E denote the sets of nodes and edges. $\mathbf{X} \in \mathbb{R}^{N \times D}$ denotes the attributed features of the graph $\mathcal{G}$, where N and D represent the number of nodes and the dimension of features respectively. Using $\mathbf{A} = \{a_{ij} | a_{ij} \in \{0, 1\}\}$ to denote the corresponding adjacency matrix (without self-loops), the normalized matrix is defined as $\widetilde{\mathbf{A}} = \mathbf{D}^{-1/2}(\mathbf{A} + \mathbf{I}_n)\mathbf{D}^{-1/2}$, where $\mathbf{I}_n$ is the identify matrix and $\mathbf{D}$ is the degree matrix, which is a diagonal matrix with $\mathbf{D}_{ii} = \sum_j a_{ij} + 1$. Referring to previous works [11,36], the general graph convolutional layer is:

$$\mathbf{E}^{(l+1)} = \sigma(\widetilde{\mathbf{A}}\mathbf{E}^{(l)}\mathbf{W}^{(l)}), \tag{1}$$

where σ represents a non-linear activation function, $\{\mathbf{E}^{(l)}|\mathbf{E}^{(0)} = \mathbf{X}\}$ denote the learned embeddings of lth layer, and $\mathbf{W}^{(l)}$ are learnable parameters in l^{th} layer. The differences among different GNNs are mainly dependent on the different definitions of $\widetilde{\mathbf{A}}$ and $\mathbf{W}$.

Auto-Encoder. Given an input data $\mathbf{X} = \{\mathbf{x}_1, \mathbf{x}_2, \ldots, \mathbf{x}_N\} \in \mathbb{R}^{N \times D}$, the encoder $\mathbf{Z} = f(\mathbf{X}; \theta)$ is to map $\mathbf{X}$ to latent features $\mathbf{Z} = \{\mathbf{z}_1, \mathbf{z}_2, \ldots, \mathbf{z}_N\} \in \mathbb{R}^{N \times d}$, the decoder $\hat{\mathbf{X}} = g(\mathbf{Z}; \psi)$ is to reconstruct features $\mathbf{X}$ from $\mathbf{Z}$. Where θ and ψ are the trainable parameters of $f(\cdot)$ and $g(\cdot)$.

Shared-Attribute Multi-Graph Clustering. Define a multi-view graph as $\mathcal{G} = \{\mathcal{G}^1, \ldots, \mathcal{G}^V\}$ with the shared features $\mathbf{X}$, shared-attribute multi-graph clustering is to cluster the N samples by utilizing multiple graphs $\{\mathcal{G}\}_{v=1}^V$ with their shared attributes $\mathbf{X}$.

3.2 Shared-Attribute Multi-Graph Clustering with Global Self-Attention

In this section, we first present our method for feature learning on every single graph. Then, we propose global self-attention and layer attention mechanisms for multi-graph fusion with shared attributes. The overall architecture of SAMGC is shown in Fig. 1.

Feature Learning on Single Graph. The process of feature learning on a single graph is consists of three steps: The first step is "graph encoding", the second step is "latent space mapping", and the third step is "decoding", as shown in the first and third lines in Fig. 1.

Primal Graph Encoding. Graph encoding aims to extract graph structure information and aggregate it to each node. We propose a simple GNN which removes non-linear transformation function because the representation space (of $\mathbf{E}^v$) needs not to be transformed after graph encoding. Thus, Eq. (1) can be simplified as:

$$\mathbf{E}^{(l+1),v} = \widetilde{\mathbf{A}}^v \mathbf{E}^{(l),v}, \tag{2}$$

where $\mathbf{E}^{(l),v}$ is the learned node embeddings of lth layer in v^{th} view. Referring to [12], we compute the mean value of all layers to avoid over-smoothing:

$$\mathbf{E}^v = [\frac{1}{M} \sum_{m=0}^{M-1} (\widetilde{\mathbf{A}}^v)^m]\mathbf{X}, \tag{3}$$

where M represents the aggregation orders and equals to the number of graph convolution layers.

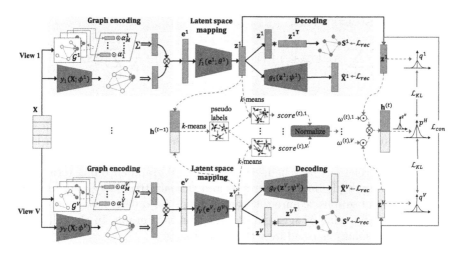

Fig. 1. The overall architecture of SAMGC. $\otimes$ and $\odot$ denote concatenation and dot production operation respectively. The blue dotted line and red line represent the collaborative training process and losses' backpropagation process respectively. For the vth view, the vth global self-attention module y_v encodes shared features to catch its hidden topology structure, meanwhile, the vth graph is encoded by a GNN with layer attention. In latent space mapping term, the concatenated node embeddings $\mathbf{E}^v$ are mapped to latent space $\mathbf{Z}^v$ which will be used for collaborative training. In the decoding term, the latent features $\mathbf{Z}^v$ are decoded to reconstruct view specific adjacency matrix of $\mathcal{G}^v$ and shared features $\mathbf{X}$. Finally, the latent features $\mathbf{Z}^v$ are re-weighted in the supervision of pseudo-labels. At the end of the training process, the final representation $\mathbf{H}$ can be fed into k-means to obtain the clustering results. (Color figure online)

Latent Space Mapping. The goal is to project the embeddings of all graphs to latent space with non-linear transformation. Formally, we use $f_v(\cdot)$ to represent the non-linear transformation of $\mathcal{G}^v$, the process can be formulated as:

$$\mathbf{Z}^v = f_v(\mathbf{E}^v; \theta^v), \tag{4}$$

where θ^v is the trainable parameters of f_v, and $\mathbf{Z}^v = \{\mathbf{z}_1^v, \mathbf{z}_2^v, ..., \mathbf{z}_N^v\} \in \mathbb{R}^{N \times d}$ represents the latent features of the vth view.

Decoding. In the decoding term, $\mathcal{G}^v$ and $\mathbf{X}$ are reconstructed from $\mathbf{Z}^v$. Based on the assumption that the more similar two nodes are, the more probably there exists an edge between them, we introduce a similarity matrix $\mathbf{S}^v = \mathbf{Z}^v(\mathbf{Z}^v)^{\mathsf{T}}$, where $\mathbf{S}^v = \{s_{ij}^v | s_{ij}^v \in [0, 1]\}$, to calculate the probability of whether there is an edge between two nodes. Additionally, to reconstruct the shared features $\mathbf{X}$, we use a non-linear transformation decoder $g_v(\cdot)$ to decode $\mathbf{Z}^v$, *i.e.*, $\hat{\mathbf{X}}^v = g_v(\mathbf{Z}^v; \psi^v)$ where ψ^v are trainable parameters, $\hat{\mathbf{X}}^v$ is reconstructed features from the vth view, and g_v has an approximate symmetrical structure of f_v. Finally, we obtain

v^{th} view's reconstruction loss:

$$\mathcal{L}_{rec}^v = \frac{1}{N^2} \sum_{i=0}^{N-1} \sum_{j=0}^{N-1} [a_{ij}^v \log(s_{ij}^v) + (1 - a_{ij}^v) \log(1 - s_{ij}^v)] + \frac{1}{N} \sum_{i=0}^{N-1} \|\mathbf{x}_i - \hat{\mathbf{x}}_i^v\|_2^2. \quad (5)$$

Multi-Graph Fusion. Till now we have introduced how we learn the latent representations for every single graph. To fuse the information of multi-graph with shared attributes, there are two issues to be addressed: 1) There may exist large variances between different graphs. Some important relations may be ignored and some noisy relations may be used in the primal graph encoding term. How to make use of multiple imperfect graphs and shared attributes to address this issue. 2) Different graphs require different aggregation orders, how to automatically compute the value of aggregation order for each graph. In the following, we propose our solutions to address these two issues.

Global Self-Attention. To address the first issue, we propose global self-attention, which constructs a weight matrix for each view from shared attributes $\mathbf{X}$ directly. Each entry in this weight matrix represents the weight between two nodes. Meanwhile, the backpropagation of graph reconstruction loss (Eq. (5)) and collaborative training loss (which will be introduced in Sect. 3.3) can promote the global self-attention module to learn the individual topological information of the specific graph and supplementary topological information from other graphs. As thus, in each view, the important common topological information could be emphasized and some lost topological information could be supplemented. To better understand this, let's take an example.

Example 1. Considering there is an important relationship between two nodes but a specific graph has lost this edge. As long as the information of this relationship is included in the shared features or other graphs, the global self-attention has a chance to learn this lost information by constructing a supplementary graph from the shared features directly or by collaborative learning other graphs.

To catch the hidden global topology structure of $\mathbf{X}$, we propose our simple yet effective self-attention mechanism [23]. First, using $\sigma(\cdot)$ to denote activation function (ELU [6] in this paper), we generate queries and keys by nonlinearly projecting $\mathbf{X}$, *i.e.*, queries $\mathbf{Q} = \sigma(\mathbf{X}\mathbf{\Phi}_1)$, and keys $\mathbf{K} = \sigma(\mathbf{X}\mathbf{\Phi}_2)$, where $\mathbf{\Phi}_1, \mathbf{\Phi}_2$ are trainable parameters. Then, we construct the supplementary graph via $Scale(\mathbf{Q}\mathbf{K}^\mathsf{T})$, where $Scale(\cdot)$ presses the attention coefficient to [-1,1], *e.g.*, using $\mathbf{v}$ to denote a vector, $Scale(\mathbf{v}) = \mathbf{v}/\|\mathbf{v}\|_2$. Notably, the negative attention coefficient is kept to eliminate the noisy edges in given graph.

Finally, the simplified graph convolution is conducted in the supplementary graph to generate final embeddings. Using $y_v(\cdot)$ to represent the function of global self-attention in the vth view, the whole process is formulated as follow:

$$y_v(\mathbf{X}; \phi^v) = Scale(\sigma(\mathbf{X}\mathbf{\Phi}_1^v)\sigma(\mathbf{X}\mathbf{\Phi}_2^v)^\mathsf{T})\mathbf{X}, \quad (6)$$

where ϕ^v is trainable parameters, *i.e.*, the union of $\mathbf{\Phi}_1^v$ and $\mathbf{\Phi}_2^v$.

Layer Attention. To solve the second issue, layer attention is proposed to self-adaptively learn the proper orders for different graphs. Layer attention re-weights every layer of GNN, so the aggregation orders of different graphs do not need to be designed elaborately. We can improve Eq. (3) with the layer attention:

$$\mathbf{E}^v = (\frac{1}{M} \sum_{m=0}^{M-1} \alpha_m^v (\widetilde{\mathbf{A}}^v)^m) \mathbf{X}, \tag{7}$$

where α_m^v is the trainable parameter representing the attention coefficient for the mth order in the vth view's graph.

Eventually, by combining global self-attention (Eq. (6)) and layer attention (Eq. (7)), we can obtain the final graph embeddings $\mathbf{E}^v$:

$$\mathbf{E}^v = \gamma y_v(\mathbf{X}; \phi^v) \otimes [(\frac{1}{M} \sum_{m=0}^{M-1} \alpha_m^v (\widetilde{\mathbf{A}}^v)^m) \mathbf{X}], \tag{8}$$

where $\otimes$ denotes concatenation operation, γ is a hyper-parameter to control the trade-off of global self-attention, ϕ^v and α^v are trainable parameters.

After that, the encoded embeddings $\mathbf{E}^v = \{\mathbf{e}_1^v, \mathbf{e}_2^v, ..., \mathbf{e}_N^v\} \in \mathbb{R}^{N \times (2D)}$ will be mapped to latent features $\mathbf{Z}^v = \{\mathbf{z}_1^v, \mathbf{z}_2^v, ..., \mathbf{z}_N^v\} \in \mathbb{R}^{N \times d}$ via Eq. (4).

3.3 Collaborative Training and Objective Function

In this section, we propose a novel collaborative training method with self-supervised weighting strategy for multi-graph clustering. The overall final objective function consists of reconstruction loss Eq. (5), consistency loss Eq. (10) and clustering loss Eq. (14):

$$\mathcal{L} = \mathcal{L}_{rec} + \lambda_1 \mathcal{L}_{con} + \lambda_2 \mathcal{L}_{KL}, \tag{9}$$

where $\mathcal{L}_{rec} = \sum_{v=1}^V \mathcal{L}_{rec}^v$, λ_1 and λ_2 are the trade-off parameters.

Consistency Loss. Consistency loss is introduced to inject common information to each specific view in feature level. First, we average all views' latent space features, *i.e.*, $\overline{\mathbf{Z}} = \frac{1}{V} \sum_{v=1}^V \mathbf{Z}^v$. Then, we sharpen its distributions with a temperature T [2] which is fixed to 0.5 in SAMGC, *i.e.*, $\overline{z}_{ij}' = \overline{z}_{ij}^{\frac{1}{T}} / \sum_{j=0}^{d-1} \overline{z}_{ij}^{\frac{1}{T}}$, where d is the number of dimensions of each latent feature, and $\overline{z}_{ij}'$ is the sharpened average feature of ith node in j^{th} dimension. Thus, the common information is emphasized. Finally, the consistency loss can be writen as :

$$\mathcal{L}_{con} = \sum_{v=1}^V \left\| \mathbf{Z}^v - \overline{\mathbf{Z}}' \right\|_2^2. \tag{10}$$

Clustering Loss. To present the clustering loss, we first introduce KL divergence with t-distribution for clustering. After that, we improve this and propose our collaborative training with a self-supervised graph re-weighting method.

KL Divergence for Clustering. By following previous works [27,29], the clustering loss is computed by KL divergence:

$$KL(\mathbf{P}^v \| \mathbf{Q}^v) = \sum_{i=1}^{N} \sum_{j=1}^{K} p_{ij}^v \log \frac{p_{ij}^v}{q_{ij}^v}, \tag{11}$$

where K denotes the number of clusters, and $\mathbf{Q}^v = \{q_{ij}^v\}$ calculated by Student's t-distribution [14] is defined as a soft assignment of ith node of the vth view in j^{th} cluster, and $\mathbf{P}^v = \{p_{ij}^v\}$ derived from the soft assignment q_{ij}^v is target distribution. Furthermore, $\mathbf{P}^v$ can be viewed as a sharpened distribution of $\mathbf{Q}^v$, so it is used to guide $\mathbf{Q}^v$ by KL divergence in Eq. (11). Next, we improve this and propose our collaborative training with self-supervised re-weighting method.

Collaborative Training. In SAMGC, an aggregation function $\mathcal{H}(\cdot)$ is conducted to aggregate the information of all graphs:

$$\mathbf{H}^{(t)} = \mathcal{H}(\omega^{(t),1}\mathbf{Z}^1, ..., \omega^{(t),V}\mathbf{Z}^V), \tag{12}$$

where $\mathcal{H}(\cdot)$ can be concatenating operation. $\omega^{(t),v}$ denotes the weight of the vth graph computed from t^{th} epoch, and $\mathbf{H}^{(t)} = \{\mathbf{h}_1^{(t)}, \mathbf{h}_2^{(t)}, ..., \mathbf{h}_N^{(t)}\} \in \mathbb{R}^{N \times (V \cdot d)}$ denotes N output representations from t^{th} epoch. Thus, finding a way to define the weight ω^v of each graph is imperative.

First, fixing other parameters, we obtain $\mathbf{H}^{(t-1)}$ using the weight $\omega^{(t-1),v}$ computed from previous epoch, *i.e.*, $\mathbf{H}^{(t-1)} = \mathcal{H}(\omega^{(t-1),1}\mathbf{Z}^1, ..., \omega^{(t-1),V}\mathbf{Z}^V)$, where all weights are initialized to 1, *i.e.*, $\omega^{(0),v} = 1$. Then, k-means [15] is implemented on $\mathbf{H}^{(t-1)}$ to generate pseudo labels. Meanwhile, all latent features $\{\mathbf{Z}^v\}_{v=1}^V$ are clustered by k-means. After that, we use the pseudo labels from $\mathbf{H}^{(t-1)}$ as ground-truth, and the predictions from $\mathbf{Z}^v$ to compute the prediction scores, such as normalized mutual information (NMI), of each view $score^{(t),v}$. Finally, we normalize $\{score^{(t),1}, ..., score^{(t),V}\}$ with a soft parameter $\rho \in [0,1]$:

$$w^{(t),v} = \left(\frac{score^{(t),v}}{\max(score^{(t),1}, ..., score^{(t),V})}\right)^\rho. \tag{13}$$

Therefore, we can generate a better common representation $\mathbf{H}^{(t)}$ via Eq. (12), and $\{\omega^v\}_{v=1}^V$ are optimized in a EM-like method.

After obtaining $\mathbf{H}$, we can compute its corresponding soft assignment distribution $\mathbf{Q}^H$ and target distribution $\mathbf{P}^H$. As thus, every view can be optimized by the common representation $\mathbf{H}$:

$$\mathcal{L}_{KL} = \sum_{v=1}^{V} KL(\mathbf{P}^H \| \mathbf{Q}^v) + KL(\mathbf{P}^H \| \mathbf{Q}^H). \tag{14}$$

The whole procedure of the collaborative training term is shown as the middle blue dotted line and red line in Fig. 1.

Table 1. The overall clustering results on both multi-graph and single-graph datasets. **Bold** denotes the best results, and "-" means the results not reported in their papers.

Method	Cora				Citeseer			
	NMI%	ARI%	ACC%	F1%	NMI%	ARI%	ACC%	F1%
VGAE [10] (2016)	40.8	34.7	59.2	45.6	16.3	10.1	39.2	27.8
ARVGA [18] (2018)	42.6	32.9	58.1	56.0	32.3	32.2	59.8	57.0
AGC [35] (2019)	53.7	44.8	68.9	65.6	41.4	42.0	67.2	62.7
MAGCN [5] (2020)	55.3	47.6	71.0	-	41.8	40.3	69.8	-
DNENC [24] (2022)	52.8	49.6	70.4	68.2	42.6	44.9	69.2	63.9
SAMGC	**58.2**	**51.1**	**73.5**	**72.7**	**45.1**	**46.0**	**69.9**	**65.3**
Method	ACM				DBLP			
GAE [10] (2016)	49.1	54.4	82.2	82.3	69.3	74.1	88.6	87.4
SwMC [16] (2017)	8.4	4.0	41.6	47.1	37.6	38.0	65.4	56.0
O2MAC [7] (2020)	69.2	73.9	90.4	90.5	72.9	77.8	90.7	90.1
MvAGC [13] (2021)	67.4	72.1	89.8	89.9	77.2	82.8	92.8	92.3
MCGC [17] (2021)	71.3	76.3	91.5	91.6	**83.0**	77.5	93.0	92.5
SAMGC	**76.8**	**82.3**	**93.8**	**93.8**	78.2	**84.0**	**93.3**	**92.8**

3.4 Inference

In this section, we introduce the inference phase of SAMGC. Inputting shared features and multi-graphs, SAMGC first generates graph embeddings $\mathbf{E}^v$ via Eq. (8) for each graph in graph encoding term; Then, it maps the embeddings to latent space by latent space mapping via Eq. (4); Finally, SAMGC uses the optimal group of $\{\omega^v\}_{v=1}^V$ to re-weight the latent features $\mathbf{Z}^v$ and aggregates these latent features by the aggregation function $\mathcal{H}(\cdot)$. The final representation $\mathbf{H}$ is fed into k-means to obtain the clustering results.

4 Experiments

4.1 Experimental Settings

Datasets. To evaluate SAMGC, we select four benchmark datasets for experiments, *i.e.*., ACM, DBLP, Cora and Citeseer. ACM and DBLP [7] consist of multiple graphs and shared attributes. Specifically, ACM is a paper network dataset, which contains two graphs, *i.e.*, co-paper relationships and co-subject relationships. The paper features are the elements of bag-of-words represented of keywords. DBLP is an author network, which contains three graphs constructed from co-author, co-conference, and co-term. The author features are the representation of authors' keywords. Cora and Citeseer both consist of a single graph with auxiliary features.

Baselines. For ACM and DBLP datasets, five methods are chosen for comparison. GAE [10] is a classical single-view clustering method. SwMC [16] is a multi-view graph embedding method. O2MAC [7] is a SOTA GNN based deep multi-view graph clustering method. MvAGC [13] and MCGC [17] are two

Table 2. Ablation studies. "-" denotes it does not contain this graph.

Component & Input	ACM				DBLP			
	NMI%	ARI%	ACC%	F1%	NMI%	ARI%	ACC%	F1%
w/o global self-att	68.5	72.3	89.6	89.6	70.5	74.2	88.4	86.5
w/o layer att	69.2	74.2	90.5	90.4	76.0	81.3	92.2	91.7
w/o $\mathcal{L}_{con}$	69.1	73.6	90.2	90.1	76.0	81.8	92.3	91.7
w/o $\mathcal{L}_{KL}$	74.1	77.6	91.2	92.0	76.1	81.7	92.3	91.7
$\mathbf{X}$	47.5	43.8	63.1	52.3	73.7	79.5	91.4	90.6
$\mathbf{X}\&\mathcal{G}^1$	66.9	73.1	90.2	90.1	18.4	14.1	49.1	46.5
$\mathbf{X}\&\mathcal{G}^2$	51.9	46.5	63.7	55.4	75.5	81.2	92.2	91.6
$\mathbf{X}\&\mathcal{G}^3$	-	-	-	-	16.1	14.7	43.1	41.2
SAMGC	**76.8**	**82.3**	**93.8**	**93.8**	**78.2**	**84.0**	**93.3**	**92.8**

SOTA graph-filter based multi-view graph clustering methods. For Cora and Citeseer datasets, because they are single-view graph data, our method simply copies their original graph as the second graph to make the collaborative training strategy work. We choose five GNN based single-view graph clustering methods, including VGAE [10], ARVGA [18], AGC [35], MAGCN [5] and DNENC [24]. Specifically, MAGCN is designed for a single graph with multiple features. To be fair, we report the results of MAGCN inputting a graph with single features.

Metrics. We adopt four widely used metrics: Normalized Mutual Information (NMI), Adjusted Rand Index (ARI), Accuracy (ACC), and F1-Score (F1). To have a fair comparison, all the results are drawn from the best in literature.

4.2   Comparison with State-of-the-Arts

Table 1 shows the comparison results of SAMGC and other SOTAs on four datasets. Based on the results, we have the following observations. SAMGC outperforms other SOTAs on all four datasets. Specifically, for ACM and DBLP, SAMGC achieves the ACC improvements of 2.3% and 0.3% compared with MCGC, which demonstrate that SAMGC can better mine the multi-graph information than other multi-view graph clustering methods. For Cora and Citeseer, SAMGC improves the NMI by 5.4% and 2.5% compared with the second-best SOTA respectively. As we can see, from these results on single-graph datasets, SAMGC achieves large improvements by simply copying the original graph as a second view, which shows the powerful generalization ability of SAMGC.

4.3   Ablation Study

Effect of Each Component. We remove each proposed component from SAMGC to demonstrate their effect on the clustering quality. As shown in the upper part of Table 2, all components are effective. Specifically, the global self-attention module (Eq. (6)) contributes most and the KL divergency loss (Eq. (14)) contributes least. This might because that the global self-attention module has already complemented the variances among graphs to some extents.

Results on Each Single Graph. As shown in last four rows of Table 2, we run SAMGC on each single graph. These results demonstrate that SAMGC could well integrate multiple graphs with shared attributes. For example, on ACM dataset, although the ACC of $\mathbf{X}\&\mathcal{G}^2$ only improves 0.6% compared to that of $\mathbf{X}$, the ACC of origional SAMGC can still improve about 3.6%, which demonstrates SAMGC's impressive multi-graph fusion ability.

4.4 Parameter Sensitivity Analysis

We investigate two important parameters in the proposed method, *i.e.*, the weight soft coefficient ρ which controls the variance of $\{\omega^v\}_{v=1}^V$ and the parameter γ in Eq. (8) which represents the importance of global self-attention module in graph encoding term and depends on the importance of the hidden topology from shared attributes. We fix the parameters of λ_1 and λ_2 to 10 and 1 respectively for all experiments according to their numerical scales.

Figure 2 shows the sensitivity of the two parameters on the ACM dataset. Overall, we can observe that our model is not very sensitive to both ρ and γ on ACM. In addition, in Fig. 2, with the increasing of γ, the NMI would increase, but be more sensitive to ρ, this shows global self-attention module is more sensitive to ρ. Moreover, we can see the results are stable and good when $0.2 \leq \rho \leq 0.6$, $0.6 \leq \gamma \leq 0.8$.

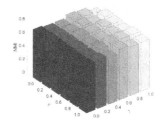

Fig. 2. NMI on ACM.

5 Conclusion

In this paper, we concentrate on a more natural but neglected multi-view clustering task, *i.e.*, multi-graph clustering. Due to inadequate consideration, most existing works can not well adapt to this task. Therefore, we propose Shared-Attribute Multi-Graph Clustering with global self-attention (SAMGC) for the task. In SAMGC, global self-attention and layer attention are proposed to address the potential three issues of multi-graph clustering. Besides, the traditional clustering loss is improved by the collaborative training method with a self-supervised weighting strategy to better integrate all graphs. The experiments on both multi and single-graph datasets show the effectiveness of SAMGC. Finally, we expect our work could inspire more researches about multi-view clustering.

Acknowledgements. This work was supported in part by Sichuan Science and Technology Program (Nos. 2021YFS0172, 2022YFS0047, and 2022YFS0055), Medico-Engineering Cooperation Funds from University of Electronic Science and Technology of China (No. ZYGX2021YGLH022), Guangzhou Science and Technology Program (No. 202002030266), Opening Funds from Radiation Oncology Key Laboratory of Sichuan Province (No. 2021ROKF02), and Major Science and Technology Application Demonstration Project of Chengdu Science and Technology Bureau (No. 2019-YF09-00086-SN).

References

1. Agrawal, S., Patel, A.: Sag cluster: an unsupervised graph clustering based on collaborative similarity for community detection in complex networks. Phys. A **563**, 125459 (2021)
2. Berthelot, D., Carlini, N., Goodfellow, I., Papernot, N., Oliver, A., Raffel, C.: Mixmatch: a holistic approach to semi-supervised learning. In: NeurIPS, pp. 5050–5060 (2019)
3. Bianchi, F.M., Grattarola, D., Alippi, C.: Spectral clustering with graph neural networks for graph pooling. In: ICML, pp. 874–883 (2020)
4. Chen, K., Pu, X., Ren, Y., Qiu, H., Li, H., Sun, J.: Low-dose CT image blind denoising with graph convolutional networks. In: ICONIP, pp. 423–435 (2020)
5. Cheng, J., Wang, Q., Tao, Z., Gao, Q.: Multi-view attribute graph convolution networks for clustering. In: IJCAI, pp. 1–7 (2020)
6. Clevert, D., Unterthiner, T., Hochreiter, S.: Fast and accurate deep network learning by exponential linear units (elus). In: ICLR (2016)
7. Fan, S., Wang, X., Shi, C., Lu, E., Lin, K., Wang, B.: One2multi graph autoencoder for multi-view graph clustering. In: WWW. pp. 3070–3076 (2020)
8. Huang, S., Kang, Z., Tsang, I.W., Xu, Z.: Auto-weighted multi-view clustering via kernelized graph learning. PR 88, 174–184 (2019)
9. Kang, Z., et al.: Multi-graph fusion for multi-view spectral clustering. KBS **189**, 105102 (2020)
10. Kipf, T.N., Welling, M.: Variational graph auto-encoders. In: NeurIPS Workshop on Bayesian Deep Learning (2016)
11. Kipf, T.N., Welling, M.: Semi-supervised classification with graph convolutional networks. In: ICLR (2017)
12. Klicpera, J., Weißenberger, S., Günnemann, S.: Diffusion improves graph learning. In: NeurIPS, pp. 13354–13366 (2019)
13. Lin, Z., Kang, Z.: Graph filter-based multi-view attributed graph clustering. In: IJCAI, pp. 19–26 (2021)
14. Maaten, L.V.D., Hinton, G.: Visualizing data using t-sne. JMLR **9**(2605), 2579–2605 (2008)
15. Macqueen, J.: Some methods for classification and analysis of multivariate observations. In: 5-th Berkeley Symposium on Mathematical Statistics and Probability, pp. 281–297 (1967)
16. Nie, F., Li, J., Li, X.: Self-weighted multiview clustering with multiple graphs. In: IJCAI, pp. 2564–2570 (2017)
17. Pan, E., Kang, Z.: Multi-view contrastive graph clustering. In: NeurIPS, pp. 2148–2159 (2021)
18. Pan, S., Hu, R., Long, G., Jiang, J., Yao, L., Zhang, C.: Adversarially regularized graph autoencoder for graph embedding. In: IJCAI, pp. 2609–2615 (2018)
19. Peng, X., Huang, Z., Lv, J., Zhu, H., Zhou, J.T.: Comic: Multi-view clustering without parameter selection. In: ICML, pp. 5092–5101 (2019)
20. Ren, Y., Yan, X., Hu, Z., Xu, Z.: Self-paced multi-task multi-view capped-norm clustering. In: ICONIP, pp. 205–217 (2018)
21. Schaeffer, S.E.: Graph clustering. Comput. Sci. Rev. **1**(1), 27–64 (2007)
22. Tsitsulin, A., Palowitch, J., Perozzi, B., Müller, E.: Graph clustering with graph neural networks. arXiv preprint arXiv:2006.16904 (2020)
23. Vaswani, A., et al.: Attention is all you need. In: NeurIPS, pp. 6000–6010 (2017)

24. Wang, C., Pan, S., Yu, C.P., Hu, R., Long, G., Zhang, C.: Deep neighbor-aware embedding for node clustering in attributed graphs. PR 122, 108230 (2022)
25. Wang, H., Yang, Y., Liu, B., Fujita, H.: A study of graph-based system for multi-view clustering. KBS **163**, 1009–1019 (2019)
26. Wu, D., Luo, X., Guo, X., Chen, C., Deng, M., Ma, J.: Concordant contrastive learning for semi-supervised node classification on graph. In: ICONIP, pp. 584–595 (2021)
27. Xie, J., Girshick, R., Farhadi, A.: Unsupervised deep embedding for clustering analysis. In: ICML, pp. 478–487 (2016)
28. Xu, J., et al.: Deep incomplete multi-view clustering via mining cluster complementarity. In: AAAI (2022)
29. Xu, J., Ren, Y., Li, G., Pan, L., Zhu, C., Xu, Z.: Deep embedded multi-view clustering with collaborative training. Inf. Sci. **573**, 279–290 (2021)
30. Xu, J., et al.: Self-supervised discriminative feature learning for deep multi-view clustering. In: TKDE, pp. 1–12 (2022)
31. Xu, J., Tang, H., Ren, Y., Peng, L., Zhu, X., He, L.: Multi-level feature learning for contrastive multi-view clustering. In: CVPR, pp. 16051–16060, June 2022
32. Xu, K., Hu, W., Leskovec, J., Jegelka, S.: How powerful are graph neural networks? In: ICLR (2019)
33. Zhan, K., Nie, F., Wang, J., Yang, Y.: Multiview consensus graph clustering. TIP **28**(3), 1261–1270 (2018)
34. Zhang, C., Cui, Y., Han, Z., Zhou, J.T., Fu, H., Hu, Q.: Deep partial multi-view learning. TPAMI **44**(5), 2402–2415 (2022)
35. Zhang, X., Liu, H., Li, Q., Wu, X.M.: Attributed graph clustering via adaptive graph convolution. In: IJCAI, pp. 4327–4333, July 2019
36. Zhu, H., Koniusz, P.: Simple spectral graph convolution. In: ICLR (2021)

Mutual Diverse-Label Adversarial Training

Mingzhe Li, Sizhe Chen, and Xiaolin Huang[✉]

Department of Automation, Shanghai Jiao Tong University, Shanghai, China
{lmz1313159,sizhe.chen,xiaolinhuang}@sjtu.edu.cn

Abstract. Adversarial training is validated to be the most effective method to defend against adversarial attacks. In adversarial training, stronger capacity networks can achieve higher robustness. Mutual learning is plugged into adversarial training to increase robustness by improving model capacity. Specifically, two deep neural networks (DNNs) are trained together with two adversarial examples. Each DNN's prediction not only fits the right label but also aligns with the other DNN's prediction. To take full advantage of mutual learning, each DNN needs to learn more extra information about different incorrect class from the other. To achieve it, we propose diverse-label attack to help with training. Concretely, we generate two adversarial examples for the two DNNs making DNNs predict not only incorrectly but also differently. Combining the above two stages, we propose a novel adversarial training method called *mutual diverse-label adversarial training* (MDLAT). Experiments on CIFAR-10 and CIFAR-100 indicate that our method is effective in improving model robustness under different settings, and our method achieves state-of-the-art (SOTA) robustness under ℓ_∞ attack.

Keywords: adversarial training · mutual learning · label diversity

1 Introduction

Recently, deep neural networks (DNNs) have achieved great success in many fields, including computer vision [12], natural language processing [21], etc. However, DNNs are found vulnerable to small perturbations in the real world [19]. Attackers inject carefully designed imperceptible perturbations into images, generating the so-called adversarial examples that can mislead a well-trained DNN to incorrect predictions.

Under such threat, lots of methods are proposed to defend against adversarial examples [18,25,26]. Among them, the most effective way is adversarial training [16], which solves a min-max problem and trains DNNs by the adversarial examples crafted on-the-fly. In adversarial training, model robustness is positively correlated with model capacity [6]. Intuitively, a larger model has more capacity but slower inference speed in applications. A way to improve model capacity without increasing model complexity is mutual learning [30]. A natural intuition is to combine mutual learning with adversarial training. However, this

© The Author(s), under exclusive license to Springer Nature Switzerland AG 2023
M. Tanveer et al. (Eds.): ICONIP 2022, LNCS 13623, pp. 64–75, 2023.
https://doi.org/10.1007/978-3-031-30105-6_6

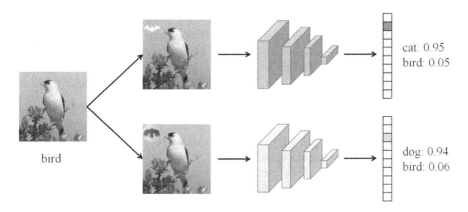

Fig. 1. Diverse-label attack for MDLAT. The original example is in class "bird". The adversarial example for network f_1 is attacked to class "cat", and the adversarial example for network f_2 is attacked to class "dog".

combination is not easy, e.g., Liu et al. [14] do not exploit the extra adversarial information well. Hence, we combine mutual learning with adversarial training in a novel way in order to improve robustness. Specifically, during the training stage, we train two DNNs simultaneously with adversarial examples by not only minimizing the supervised learning loss function but also making two DNNs predict similarly. Mutual adversarial training can learn a better representation by transferring adversarial knowledge of the two DNNs [7].

The key to improving model capacity in mutual learning is providing each DNN with extra information from the others. In order to let mutual learning work better in adversarial training, on the one hand, we expect each DNN to learn more knowledge that it does not have from the other DNN, so we generate samples to encourage the two DNNs to respond differently. On the other hand, when facing unseen attacks, it is difficult to know which class attackers may attack to. So the adversarial knowledge among the right label and multiple incorrect labels is needed. However, most methods based on adversarial training use Project Gradient Descent (PGD) attack [23,24], which attacks a sample to one wrong class. They do not pay enough attention to other incorrect classes. Therefore, we propose diverse-label attack to help transfer adversarial knowledge. In the attack stage, we generate two adversarial examples, which leads two DNNs to predict not only incorrectly but also differently for two DNNs. In other words, the two adversarial examples are attacked to two different classes and attached to more wrong labels. Figure 1 shows our attack strategy.

To sum up, during the attack stage, we apply diverse-label attack to generate two adversarial examples. During the training stage, DNNs are mutually trained on such examples. We call our method mutual diverse-label adversarial training (MDLAT). The models trained by MDLAT become more robust against unseen attacks. Based on standard adversarial training [16] with both ℓ_∞ bound and ℓ_2 bound, we evaluate our method on CIFAR-10/100 [11] using ResNet18 [9]. Our method improves robust accuracy under PGD-20 attack and AutoAttack [5] and achieves state-of-the-art (SOTA) performance. For example, on CIFAR-100 dataset, our method improves robust accuracy by about 4% under $\ell_\infty = 8/255$ bound and by about 6% under $\ell_2 = 128/255$ bound, respectively. We also plug our methods into MART [23] to achieve high PGD-20 attack robust accuracy. Our method improves robust accuracy under PGD-20 attack by about 0.5% on CIFAR-10 and CIFAR-100.

Overall, our contributions can be summarized in three folds:

- We combine mutual learning with adversarial training in a novel way to improve model representation capacity.
- We put forward diverse-label attack for mutual adversarial training, which generates two adversarial examples to different wrong labels. Each DNN obtains extra information about another wrong class from the other DNN.
- We perform extensive experiments on CIFAR-10/100 with different baselines under different settings to show robustness improvement and achieve SOTA robustness under ℓ_∞ attack.

2   Related Work

2.1   Adversarial Attack

As DNNs are vulnerable to malicious perturbation, many studies focus on adversarial attack methods, e.g. black-box attack [15], white-box attack [3,8,13,16,17]. White-box attack, where attackers know all the parameters of the threat model, is an essential tool to measure model robustness. The methods can be divided into gradient-based and optimization-based attacks. Based on gradient, Goodfellow et al. [8] propose a classic attack method named fast gradient sign method (FGSM), which uses the gradient of input data for gradient ascent. One-step Target Class Method (OTCM) [13] is a target attack version of FGSM, which minimizes the cross entropy loss with the target label. Madry et al. [16] utilize an iterative version of FGSM and add random initialization to it, which is called PGD. Deepfool [17] is a boundary based attack method solving the problem of parameter selection in FGSM. Based on optimization, C&W attack [3] is put forward to attack the samples to the highest confidence incorrect class.

However, Gradient obfuscation [2], which means attackers cannot get the right gradient, is a serious problem. Some models can only defend against a few types of attacks and give a false sense of security. It is necessary to ensemble the strongest attacks to measure model robustness. AutoAttack [5] is an ensemble of APGD, APGD-T, FAB [4] and square attack [1]. Besides, it is a good benchmark for white-box robustness and is not subject to gradient obfuscation.

2.2   Adversarial Training

Adversarial training [16] is a useful way to improve model robustness without suffering from gradient obfuscation. Adversarial training can be regarded as a special data augmentation method that feeds adversarial examples using PGD attack to DNNs. The inner maximum problem and the outer minimum problem are equally important to enhance performance [22].

Lots of works are presented to improve robustness based on adversarial training. TRADES [27] focuses on the trade-off between nature error and robust error by regularizing the output of clean samples and their corresponding adversarial samples using Kullback-Leibler (KL) divergence. Misclassification aware adversarial training (MART) [23] concentrates on misclassified examples by using KL-divergence to regularize the predictions of misclassified clean samples and their corresponding adversarial samples. Ensemble adversarial training [20] is proposed to feed adversarial examples generated from both the training DNN and a well-trained DNN to enhance the robustness of transfer attacks. Adversarial weight perturbation (AWP) [24] is declared to perturb weights to get a relatively smooth loss landscape resulting in reducing the robust generalization gap. Friendly adversarial training (FAT) [28] finds adversarial examples closest to the decision boundary to improve clean accuracy. Geometry-aware instance-reweighted adversarial training (GAIRAT) [29] reweights adversarial samples according to the number of attack steps to focus on samples near the decision boundary.

2.3   Mutual Learning

Knowledge distillation [10] is a well-known way to transfer knowledge. An online knowledge transfer method named deep mutual learning [30] is presented, which trains two DNNs in parallel and makes them learn from each other. Mutual learning is a useful way to improve model capacity. Due to the next most likely class varying, each DNN can get extra information from the other and learn the other DNN's representation.

Though Liu et al. [14] present mutual adversarial training (MAT), it leads each DNN's output of adversarial examples to align with the other DNN's output of natural examples. MAT does not guide the two DNNs to learn extra information of different labels, which doesn't take full advantage of mutual learning. Therefore, we combine mutual learning with adversarial training in a new way.

3   Method

3.1   Mutual Adversarial Training

Standard adversarial training solves a min-max problem. In the attack stage, the network generates adversarial examples by maximizing cross entropy loss iteratively. In the training stage, the network updates trainable weights by minimizing cross entropy loss. The following equations show the procedure:

$$\arg\min_{\theta} \mathcal{L}_{\mathrm{CE}}(f(x', \theta), y), \tag{1}$$

$$x' = \arg\max_{\|x'-x\|_p \leq \epsilon} \mathcal{L}_{\mathrm{CE}}(f(x', \theta), y), \tag{2}$$

where $\mathcal{L}_{CE}$ is cross entropy loss, f is a DNN model, ϵ is the upper bound for perturbations, and x' is called an adversarial example.

We propose mutual adversarial training to improve capacity in adversarial training. In our framework, two DNNs are trained simultaneously with adversarial examples. We lead the two DNNs' outputs not only to fit the right labels of adversarial examples but also to align with each other's predictions. It can be regarded as the following three indicator functions:

$$(a)\mathbb{1}(f_1(\theta_1, x') = y), (b)\mathbb{1}(f_2(\theta_2, x'') = y), (c)\mathbb{1}(f_1(\theta_1, x') = f_2(\theta_2, x'')), \quad (3)$$

where x' and x'' are adversarial examples for f_1 and f_2, y is label. However, the indicator function is a hard decision and can not be directly optimized. Surrogate losses are needed. The surrogate loss function for the first two indicator functions is cross entropy loss. KL-divergence is used to measure the distance between two predictions' distributions. The surrogate loss function for the third indicator is KL-divergence to measure different outputs. Since the two networks are expected to learn together and train symmetrically but KL-divergence is asymmetric, we use a symmetric version of KL-divergence:

$$\mathcal{L}_{KL} = KL(f_1(\theta_1, x'), f_2(\theta_2, x'')) + KL(f_2(\theta_2, x''), f_1(\theta_1, x')), \quad (4)$$

where $f_i(\theta_i, :)$ stands for the output predictions of DNN, KL stands for KL-divergence.

To sum up, the training stage is to minimize the following loss function:

$$\mathcal{L}_{MDALT} = \mathcal{L}_{CE}(f_1(\theta_1, x'), y) + \mathcal{L}_{CE}(f_2(\theta_2, x''), y) + \lambda\mathcal{L}_{KL}(\theta_1, \theta_2, x', x''), \quad (5)$$

with it, we obtain two more powerful DNNs. the whole training framework is presented in Fig. 2.

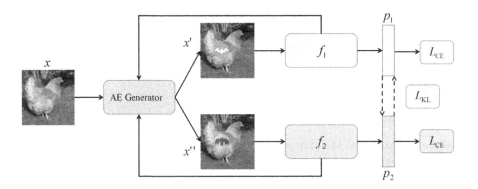

Fig. 2. The framework of mutual adversarial training. Two DNNs' f_1 and f_2 are adversarially trained together with a supervised loss, and a KL loss to learn from each other. AE generator uses diverse-label attack.

3.2   Diverse-Label Attack

In the mutual adversarial training framework, we wish the two DNNs to respond differently to get more extra adversarial information. We also encourage the two DNNs' predictions to different classes to get adversarial knowledge about more labels. Models trained with diverse-label attack can defend against various types of attacks. The two goals are not in conflict with each other. Since normal PGD attack is an untargeted attack and does not guarantee to achieve these, a regular term is needed to guide it. Specially, we encourage the two DNNs to predict not only incorrectly but also away from each other during the generation of adversarial examples. The indicator function for the regular term is $\mathbb{1}(f_1(\theta_1, x') \neq f_2(\theta_2, x''))$. In the same way, we use symmetric KL-divergence as the surrogate loss function. The attack strategy is shown in the following:

$$x' = \underset{\|x'-x\|_p \leq \epsilon}{\arg\max} \left(\mathcal{L}_{\mathrm{CE}}(f(\theta_1, x'), y) + \lambda \mathcal{L}_{\mathrm{KL}}(\theta_1, \theta_2, x', x'') \right), \tag{6}$$

$$x'' = \underset{\|x''-x\|_p \leq \epsilon}{\arg\max} \left(\mathcal{L}_{\mathrm{CE}}(f(\theta_2, x''), y) + \lambda \mathcal{L}_{\mathrm{KL}}(\theta_1, \theta_2, x', x'') \right), \tag{7}$$

We combine diverse-label attack with mutual adversarial training, more details are listed in Algorithm 1. Our attack stage and training stage are both symmetrical, so the capacity of the two networks is the same. Our algorithm only increases the cost in the training phase. During the testing phase, each of the two DNNs could be used to make predictions.

Algorithm 1. Mutual Diverse-label Adversarial Training (MDLAT)

Input: Training dataset S, two DNN classifiers $f_1(\theta_1, :)$ and $f_2(\theta_2, :)$ with initial parameters θ_{01}, θ_{02}, maximum perturbation ϵ, number of training epochs N, batch size m, number of batches M, number of adversarial attack steps K, adversarial attack step size α, and learning rate η.
Output: Robust classifiers f_1 and f_2
Initialization: $\theta_1 \leftarrow \theta_{01}$, $\theta_2 \leftarrow \theta_{02}$
1: **for** $t = 1, 2, ..., N$ **do**
2:    **for** $i = 1, 2, ..., M$ **do**
3:       Read a minibatch $\{(x_i, y_i)\}_{i=1}^m$ from S
4:       $x_i' \leftarrow x_i + \delta_1$, $\delta_1 \sim Uniform(-\epsilon, \epsilon)$
5:       $x_i'' \leftarrow x_i + \delta_2$, $\delta_2 \sim Uniform(-\epsilon, \epsilon)$
6:       **for** $k = 1, 2, ..., K$ **do**
7:          $x_i' \leftarrow \prod_{x_i' \in \mathbb{B}_\epsilon(x)}(\alpha \mathrm{sign}(x_i' + \nabla_{x_i'} \mathcal{L}_{\mathrm{MDALT}}(\theta_1, \theta_2, x_i', x_i'', y_i)))$
8:          $x_i'' \leftarrow \prod_{x_i'' \in \mathbb{B}_\epsilon(x)}(\alpha \mathrm{sign}(x_i'' + \nabla_{x_i''} \mathcal{L}_{\mathrm{MDALT}}(\theta_1, \theta_2, x_i', x_i'', y_i)))$
9:       **end for**
10:      $\theta_1 \leftarrow \theta_1 - \eta \frac{1}{m}(\nabla_{\theta_1} \mathcal{L}_{\mathrm{MDALT}}(\theta_1, \theta_2, x_i', x_i'', y_i))$
11:      $\theta_2 \leftarrow \theta_2 - \eta \frac{1}{m}(\nabla_{\theta_2} \mathcal{L}_{\mathrm{MDALT}}(\theta_1, \theta_2, x_i', x_i'', y_i))$
12:   **end for**
13: **end for**

3.3   Combination with MART

Our approach provides a new method to improve adversarial training. It is different from the current methods based on adversarial training and doesn't conflict with them. MART [23] manipulates misclassified examples with a regularized adversarial risk. It achieves high robustness on PGD-20 attack but doesn't improve much on AutoAttack. In order to improve PGD-20 attack robustness and maintain AutoAttack robustness, we combine our method MDLAT with MART.

Specifically, we train two DNNs mutually, during the attack stage, Diverse-label attack is used for two DNNs, and during the training stage, we use the MART loss to replace cross entropy loss in mutual adversarial training. The loss functions are listed below:

$$\mathcal{L}_{\mathrm{MART}} = \mathcal{L}_{\mathrm{BCE}}(f(\theta, x'), y) + \eta \mathrm{KL}(f(\theta, x), f(\theta, x'))(1 - f_y(\theta, x)), \qquad (8)$$

$$\mathcal{L}_{\mathrm{MART\_MDLAT}} = \mathcal{L}_{\mathrm{MART}}(\theta_1, x') + \mathcal{L}_{\mathrm{MART}}(\theta_2, x'') + \lambda \mathcal{L}_{\mathrm{KL}}(\theta_1, \theta_2, x', x''), \quad (9)$$

where $\mathcal{L}_{\mathrm{BCE}}$ is a combination of cross entropy loss and margin loss between the right class and the highest confidence wrong class.

4   Experiments

We evaluate the model robustness on CIFAR-10/100 under different settings. The code is released at https://github.com/mingzheli814/MDLAT.

4.1   Experimental Setup

For training, we use ResNet18 for all experiments on CIFAR-10/100. We select two types of adversarial perturbations: ℓ_∞ norm attack and ℓ_2 norm attack. Experiments are performed with NVIDIA GeForce RTX 2080Ti GPUs.

– For ℓ_∞ adversarial training, we generate adversarial examples with radius $\epsilon = 8/255$, step size $\alpha = 2/255$ and step numbers 10, we train DNNs for 100 epochs with an initial learning rate of 0.05 and decaying by ten at 75 epoch and 90 epoch. In our method, we use $\lambda = 1$ on CIFAR-10 and $\lambda = 1.5$ on CIFAR-100, respectively.
– For ℓ_2 adversarial training, we generate adversarial examples with radius $\epsilon = 128/255$, step size $\alpha = 15/255$ and step numbers 10, we train DNNs for 120 epochs with an initial learning rate of 0.01 and decaying by ten at 75 epoch, 90 epoch, 100 epoch. In our method, we use $\lambda = 1$ on CIFAR-10/100.

We use SGD to train DNNs with 0.9 momentum, and add random cropping and random horizontalflip as data augmentation methods. We use two kinds of criteria to evaluate robustness: 20-steps PGD attack accuracy and AutoAttack accuracy.

We also plug our method into ℓ_∞ adversarial training method MART [23]. We set the learning rate as 0.01. DNNs are trained for 120 epochs and the learning rate is divided by 10 at epoch 75, 90, and 100. The hyper-parameter η is set to 5.0. For our method, we set our hyper-parameter $\lambda = 3$ on CIFAR-10 and $\lambda = 2$ on CIFAR-100, respectively.

4.2   Comparison with Standard Adversarial Training

We utilize our method in both ℓ_∞ adversarial training and ℓ_2 adversarial training to defend against adversarial attacks with different norm bounds. Figure 3 shows the run-time test PGD-20 attack robust accuracy of different methods. Our models achieve high robustness after the learning rate decays and do not suffer from overfitting. And all numerical results are shown in Table 1.

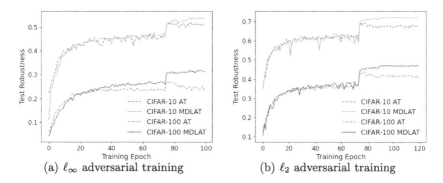

(a) ℓ_∞ adversarial training   (b) ℓ_2 adversarial training

Fig. 3. Robust accuracy under PGD-20 attack with ℓ_∞ norm and ℓ_2 norm on CIFAR-10/100 using standard adversarial training and MDLAT during the training process.

Table 1. Classification accuracy (%) of ℓ_∞ and ℓ_2 adversarial training using ResNet18 on CIFAR-10/100. The maximum perturbation is $\varepsilon = 8/255$ for ℓ_∞ attack and $\varepsilon = 128/255$ for ℓ_2 attack.

Dataset	Norm	Method	Clean	PGD-20	AutoAttack
CIFAR-10	ℓ_∞	AT	**83.36**	52.14	48.28
		MDLAT	82.23	**54.54**	**50.61**
	ℓ_2	AT	**89.92**	67.86	65.65
		MDLAT	89.21	**71.97**	**70.13**
CIFAR-100	ℓ_∞	AT	56.05	27.08	23.49
		MDLAT	**56.28**	**31.84**	**26.73**
	ℓ_2	AT	**66.74**	40.43	36.57
		MDLAT	65.84	**46.86**	**42.99**

Our model improves robustness ranging from 2% to 6%. It can be found that our method gains more robustness on CIFAR-100 dataset. Due to CIFAR-100 having more classes and being easier to attack to different incorrect classes, the two DNNs can learn adversarial knowledge of various incorrect labels well.

Figure 4 shows three training loss curves during training. The two cross entropy losses may be different at the beginning due to different initialization. As the training goes on, they get closer and go down together, which means they learn robust features together. Instead of dropping down, the KL loss function fluctuates around a certain value, which means the models have been passing on knowledge. This indicates that we succeed in achieving our purpose: two DNNs predict differently and learn extra information during the whole training process.

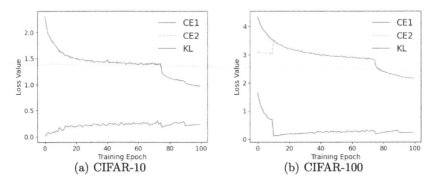

Fig. 4. Loss curves during the training procedure, CE1 and CE2 stand for cross entropy loss for f_1 and f_2, and KL stands for symmetric KL-divergence.

4.3 Comparison with SOTA Baselines

We also plug our method into MART [23], which is called MDLAT-MART. Our methods can defend against attacks with bigger bounds. Figure 5 shows the results.

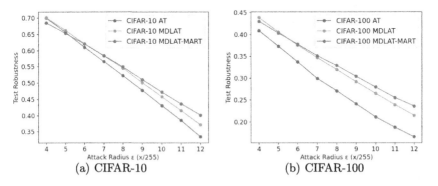

Fig. 5. Test robust accuracy under PGD-20 attack concerning different perturbation radius.

We evaluate our methods MDLAT and MDLAT-MART with several baselines under ℓ_∞ attack: AT [16], TRADES [27], MART [23], FAT [28], GAIRAT [29] and

Table 2. Classification accuracy (%) of different methods on ℓ_∞ adversarial training using ResNet18 on CIFAR-10/100. The maximum perturbation is $\varepsilon = 8/255$.

Dataset	Method	Clean	PGD-20	AutoAttack
CIFAR-10	AT	83.36	52.14	48.28
	TRADES	81.87	53.35	48.34
	MART	80.87	56.00	48.21
	FAT	**87.61**	45.95	43.05
	GAIRAT	81.85	54.62	39.57
	MAT	81.18	55.69	50.10
	MDLAT	82.23	54.54	**50.61**
	MDLAT-MART	80.18	**56.39**	49.59
CIFAR-100	AT	56.05	27.08	23.49
	TRADES	55.08	28.30	23.77
	MART	55.06	32.86	26.05
	FAT	**63.68**	21.00	19.14
	GAIRAT	55.71	24.02	19.50
	MAT	55.58	31.19	26.49
	MDLAT	56.28	31.84	26.73
	MDLAT-MART	53.74	**33.43**	**27.05**

MAT [14]. Our methods achieve the highest robustness under PGD-20 attack and AutoAttack on CIFAR-100/100. All results are shown in Table 2.

Compared with the highest baseline, on CIFAR-10, our methods improve by about 0.4% robust accuracy under PGD-20 attack and by about 0.5% under AutoAttack, respectively. On CIFAR-100, our methods improve by about 0.6% robust accuracy under PGD-20 attack and AutoAttack.

4.4 Ablation Study

Our method combines diverse-label attack with mutual learning using two DNNs. We explore the importance of the following three elements: using two DNNs, applying mutual learning, and applying diverse-label attack. So we do three experiments. First, we use one DNN to generate diverse-label adversarial examples and train them mutually. Second, two DNNs are trained just with mutual learning. Third, two DNNs are trained just with diverse-label attack. The results are shown in Tabel 3.

Table 3. Classification accuracy (%) of ℓ_∞ adversarial training on CIFAR-10. DL is diverse-label attack. ML is mutual learning. 1NN stands for just using one DNN.

Method	Clean	PGD-20
AT	83.36	52.14
MDLAT w/o DL	82.76	53.68
MDLAT w/o ML	**84.49**	52.53
MDLAT w/ 1NN	83.03	53.29
MDLAT	82.23	**54.54**

It can see that the gain of robustness comes from three aspects. The number of DNNs and mutual learning method are equally important. Diverse-label attack is an essential way to help mutual adversarial training.

5   Conclusion

We propose a novel adversarial training system called MDLAT. Two DNNs are trained together. During the attack stage, we generate an adversarial example for each DNN and lead the predictions of them away from each other to attach to more incorrect labels. During the training stage, these samples are trained mutually to improve model capacity. A model which can better defend against invisible attacks is obtained. We conduct experiments on CIFAR-10 and CIFAR-100 under different baselines and different settings.

References

1. Andriushchenko, M., Croce, F., Flammarion, N., Hein, M.: Square attack: a query-efficient black-box adversarial attack via random search. In: Vedaldi, A., Bischof, H., Brox, T., Frahm, J.-M. (eds.) ECCV 2020. LNCS, vol. 12368, pp. 484–501. Springer, Cham (2020). https://doi.org/10.1007/978-3-030-58592-1_29
2. Athalye, A., Carlini, N., Wagner, D.: Obfuscated gradients give a false sense of security: circumventing defenses to adversarial examples. In: International Conference on Machine Learning, pp. 274–283. PMLR (2018)
3. Carlini, N., Wagner, D.: Towards evaluating the robustness of neural networks. In: IEEE Symposium on Security and Privacy, pp. 39–57. IEEE (2017)
4. Croce, F., Hein, M.: Minimally distorted adversarial examples with a fast adaptive boundary attack. In: International Conference on Machine Learning, pp. 2196–2205. PMLR (2020)
5. Croce, F., Hein, M.: Reliable evaluation of adversarial robustness with an ensemble of diverse parameter-free attacks. In: International Conference on Machine Learning, pp. 2206–2216. PMLR (2020)
6. Gao, R., Cai, T., Li, H., Hsieh, C.J., Wang, L., Lee, J.D.: Convergence of adversarial training in overparametrized neural networks. In: Advances in Neural Information Processing Systems, vol. 32 (2019)
7. Goldblum, M., Fowl, L., Feizi, S., Goldstein, T.: Adversarially robust distillation. In: Association for the Advancement of Artificial Intelligence, pp. 3996–4003 (2020)
8. Goodfellow, I.J., Shlens, J., Szegedy, C.: Explaining and harnessing adversarial examples. In: International Conference on Learning Representations (2015)
9. He, K., Zhang, X., Ren, S., Sun, J.: Deep residual learning for image recognition. In: IEEE Conference on Computer Vision and Pattern Recognition, pp. 770–778 (2016)
10. Hinton, G., Vinyals, O., Dean, J.: Distilling the knowledge in a neural network. Comput. Sci. **14**(7), 38–39 (2015)
11. Krizhevsky, A., Hinton, G., et al.: Learning multiple layers of features from tiny images (2009)
12. Krizhevsky, A., Sutskever, I., Hinton, G.E.: Imagenet classification with deep convolutional neural networks. In: Advances in Neural Information Processing Systems, vol. 25 (2012)

13. Kurakin, A., Goodfellow, I., Bengio, S.: Adversarial machine learning at scale. arXiv preprint arXiv:1611.01236 (2016)
14. Liu, J., Lau, C.P., Souri, H., Feizi, S., Chellappa, R.: Mutual adversarial training: Learning together is better than going alone. arXiv preprint arXiv:2112.05005 (2021)
15. Liu, Y., Chen, X., Liu, C., Song, D.: Delving into transferable adversarial examples and black-box attacks. In: International Conference on Learning Representations (2017)
16. Madry, A., Makelov, A., Schmidt, L., Tsipras, D., Vladu, A.: Towards deep learning models resistant to adversarial attacks. In: International Conference on Learning Representations (2018)
17. Moosavi-Dezfooli, S.M., Fawzi, A., Frossard, P.: Deepfool: a simple and accurate method to fool deep neural networks. In: IEEE Conference on Computer Vision and Pattern Recognition, pp. 2574–2582 (2016)
18. Papernot, N., McDaniel, P., Wu, X., Jha, S., Swami, A.: Distillation as a defense to adversarial perturbations against deep neural networks. In: IEEE Symposium on Security and Privacy, pp. 582–597. IEEE (2016)
19. Szegedy, C., et al.: Intriguing properties of neural networks. In: International Conference on Learning Representations (2014)
20. Tramèr, F., Kurakin, A., Papernot, N., Goodfellow, I., Boneh, D., McDaniel, P.: Ensemble adversarial training: attacks and defenses. In: International Conference on Learning Representations (2018)
21. Vaswani, A., et al.: Attention is all you need. In: Advances in Neural Information Processing Systems, vol. 30 (2017)
22. Wang, Y., Ma, X., Bailey, J., Yi, J., Zhou, B., Gu, Q.: On the convergence and robustness of adversarial training. In: International Conference on Machine Learning, pp. 6586–6595. PMLR (2019)
23. Wang, Y., Zou, D., Yi, J., Bailey, J., Ma, X., Gu, Q.: Improving adversarial robustness requires revisiting misclassified examples. In: International Conference on Learning Representations (2019)
24. Wu, D., Xia, S.T., Wang, Y.: Adversarial weight perturbation helps robust generalization. Advances in Neural Information Processing Systems, vol. 33 (2020)
25. Xie, C., Wang, J., Zhang, Z., Ren, Z., Yuille, A.: Mitigating adversarial effects through randomization. In: International Conference on Learning Representations (2018)
26. Xu, W., Evans, D., Qi, Y.: Feature squeezing: Detecting adversarial examples in deep neural networks. In: Network and Distributed System Security Symposium (2017)
27. Zhang, H., Yu, Y., Jiao, J., Xing, E., El Ghaoui, L., Jordan, M.: Theoretically principled trade-off between robustness and accuracy. In: International Conference on Machine Learning, pp. 7472–7482. PMLR (2019)
28. Zhang, J., et al.: Attacks which do not kill training make adversarial learning stronger. In: International Conference on Machine Learning, pp. 11278–11287. PMLR (2020)
29. Zhang, J., Zhu, J., Niu, G., Han, B., Sugiyama, M., Kankanhalli, M.S.: Geometry-aware instance-reweighted adversarial training. In: International Conference on Learning Representation (2021)
30. Zhang, Y., Xiang, T., Hospedales, T.M., Lu, H.: Deep mutual learning. In: IEEE Conference on Computer Vision and Pattern Recognition, pp. 4320–4328 (2018)

Multi-Agent Hyper-Attention Policy Optimization

Bin Zhang, Zhiwei Xu, Yiqun Chen, Dapeng Li, Yunpeng Bai, Guoliang Fan,
and Lijuan Li[✉]

Institute of Automation, Chinese Academy of Sciences, School of Artificial
Intelligence, University of Chinese Academy of Sciences, Beijing, China
{zhangbin2020,xuzhiwei2019,lidapeng2020,baiyunpeng2020,
guoliang.fan,lijuan.li}@ia.ac.cn

Abstract. Policy-based methods like MAPPO have exhibited amazing results in diverse test scenarios in multi-agent reinforcement learning. Nevertheless, current actor-critic algorithms do not fully leverage the benefits of the centralized training with decentralized execution paradigm and do not effectively use global information to train the centralized critic, as seen by IPPO's superior performance in some scenarios compared to MAPPO. To address this problem, we propose a game abstraction technique based on a state-conditioned hyper-attention network. It can help agents integrate important data and refine complex game interactions to achieve efficient policy optimization. In addition, to improve the stability of the trust-region methods, we introduce a point probability distance penalty in addition to the clipping operation in PPO. Experimental results demonstrate the advantages of our method in various cooperative environments.

Keywords: Multi-Agent Reinforcement Learning · Game
Abstraction · Trust Region

1 Introduction

Many real-world challenges, such as autonomous intelligent transportation systems [13], network packet routing [24], and computer games [26], need collective intelligence, in which agents in a multi-agent system (MAS) collaborate to achieve a shared objective. And it has piqued academics' interest in multi-agent reinforcement learning (MARL). In contrast to the single-agent deep reinforcement learning (DRL), which focuses on strategy learning in a static environment, MARL requires agents to simultaneously grasp the dynamics of the environment and other agents in order to build efficient cooperative policies. As a result, non-stationarity and scalability issues arise, posing additional challenges for MARL.

This Project was Supported by National Defence Foundation Reinforcement Fund.
B. Zhang, Z. Xu, Y. Chen and L. Li—These Authors Contributed Equally.

Instead of just training all agents as a group or training a reinforcement learning model for each agent independently, the centralized training with decentralized execution (CTDE) paradigm [12] tackles both of the above challenges. To fulfill this paradigm, value-based approaches such as VDN [22], QMIX [14], and QPLEX [27], require constructing a value function decomposition framework. Furthermore, the actor-critic methods are well-suited to the CTDE paradigm. We just need to learn a critic with a global perspective to guide the actor during training so that agents can act based only on their local observations during the test phase. These approaches can cope with multi-agent competitive or mixed tasks, stochastic policies, and applications in non-monotonic environments that value decomposition methods cannot manage due to their unique structure. However, given their performance in the StarCraft II benchmark, researchers believe that value-based methods outperform policy-based methods, and that off-policy approaches have higher sample efficiency than on-policy approaches in multi-agent applications, despite the fact that these algorithms have achieved revolutionary results in single-agent reinforcement learning.

The performance disparity may be traced in that policy-based approaches make insufficient use of the centralized training program in the CTDE paradigm. For example, MADDPG [12] simply concatenates local observations of all agents as input to the critic. To this end, many algorithms adopt game abstraction mechanism, such as MAAC's use of soft attention [9], and MAPPO's use of feature pruning [31]. However, the processing of these algorithms can be characterized as the integration of all agents' local observations, without considering the global state perspective. Moreover, the success of MAPPO in the StarCraft II challenge is partly due to the inclusion of well-designed prior knowledge of certain situations. But this operation may still result in information redundancy in some situations, making the algorithm's performance even worse than IPPO that relies entirely on agents' local observation.

In this paper, we take full advantage of CTDE and propose a novel game abstraction method called the state-conditioned hyper-attention network, which can assist agents in extracting local characteristics from other agents and making full use of global information to build the value function, thereby improving the efficiency and accuracy of value function estimation. In addition, we propose adding a point probability distance penalty term [1] to the trust-region methods to increase their robustness in more complex multi-agent environments. Finally we provide the complete Multi-Agent Hyper-Attention Policy Optimization (MAHAPO) algorithm. Experiments in the StarCraft II micromanagement tasks show that our approach has obvious advantages over the fine-tuning QMIX [7] and the feature-pruned MAPPO, especially in some Super Hard scenarios.

2 Background

2.1 Dec-POMDP

We consider the decentralized partially observable Markov decision process (Dec-POMDP) framework. A special type of the partially-observable stochastic games

(POSG) under the branch of Markov games. In Dec-POMDP, each agent has its own local observation and shares the same global rewards. It can be defined by a tuple $\mathcal{G} = <\mathcal{N}, \mathcal{S}, \Omega, \mathcal{A}, \mathcal{P}, \mathcal{O}, \nabla, \gamma>$, with $\mathcal{N} = \{1, ..., n\}$, $\mathcal{S}$, Ω and $\mathcal{A} = \prod_{i=1}^{n} \mathcal{A}^i$ corresponding to the sets of agents, global states, local observations, and joint actions. At each time step, agent $i \in \mathcal{N}$ chooses an action $a_i \in \mathcal{A}^i$ according to its local observation $o_i \in \Omega$ (based on the observation function $\mathcal{O} : \mathcal{S} \times \mathcal{A} \to \Omega$) in the global state $s \in \mathcal{S}$. Then the environment gives the global reward $r : \mathcal{S} \times \mathcal{A} \to \mathbb{R}$ according to the joint action $\mathbf{a} = [a_i]_{i=1}^n$, and transfers to the next state s' based on the state transition probability $\mathcal{P} : \mathcal{S} \times \mathcal{A} \times \mathcal{S} \to [0, 1]$. Agent i maintains a local action-observation history $\tau_i \in T \equiv (\Omega \times \mathcal{A})$, and its policy can be expressed as $\pi_i(a_i|\tau_i) : T \times \mathcal{A} \to [0, 1]$. Similar to single agent reinforcement learning, our goal is to maximize the expectation of cumulative discount rewards $\mathcal{J}(\pi) \triangleq \mathbb{E}_{\mathbf{a} \sim \pi, s \sim P} \left[\sum_{t=0}^{\infty} \gamma^t r_t \right]$. The value function and the action-state value function are expressed as: $V_\pi(s) \triangleq \mathbb{E}_{\mathbf{a} \sim \pi, s \sim P} \left[\sum_{t=0}^{\infty} \gamma^t r_t \mid s_0 = s \right]$ and $Q_\pi(s, \mathbf{a}) \triangleq \mathbb{E}_{\mathbf{a} \sim \pi, s \sim P} \left[\sum_{t=0}^{\infty} \gamma^t r_t \mid s_0 = s, \mathbf{a}_0 = \mathbf{a} \right]$, and the advantage function can be written as: $A_\pi(s, \mathbf{a}) \triangleq Q_\pi(s, \mathbf{a}) - V_\pi(s)$.

2.2   Centralized Training with Decentralized Execution

Centralized training with decentralized execution (CTDE) paradigm is widely used in multi-agent reinforcement learning tasks. Because there is usually no limitation on information communication throughout the training process, all agents can utilize the global information. Hence the centralized training is employed. Moreover, each agent can determine the next action based only on the local observation during the actual execution process because of the partially-observable setting. As a result, the approach of decentralized execution is used. Value function decomposition is commonly used in value-based multi-agent reinforcement learning methods. Furthermore, policy-based methods are naturally suitable for the CTDE paradigm. During the training stage, a centralized critic is trained to guide the update of each agent's policy module. During the execution stage, the actor receives local observations and outputs cooperative behaviors consistent with collective intelligence.

2.3   Proximal Policy Optimization

Proximal Policy Optimization (PPO) [19] is a simplified variant of the Trust Region Policy Optimization (TRPO) [17]. TRPO is a policy-based technique that employs KL divergence to restrict the update step in the trust region during the policy update process. It aims to maximize the payoff function by finding a policy that is close to the present policy throughout each iteration, which can be described as the following form:

$$\text{maximize } \mathbb{E}_{s, a \sim \pi_{\theta_{old}}} \left[\frac{\pi_\theta(a|s)}{\pi_{\theta_{old}}(a|s)} A_{\theta_{old}}(s, a) \right]$$
$$\text{subject to } \mathbb{E}_s \left[D_{KL} \left(\pi_{\theta_{old}} (\cdot \mid s) \| \pi_\theta(\cdot \mid s) \right) \right] \le \delta,$$

Table 1. Gaming abstraction mechanism in different algorithms

Algorithm	Base Algorithm	Game Abstraction
MADDPG	DDPG	oncatenation of all agents' observations and actions
MAAC	SAC	Soft attention
GA-AC	SAC	Two-stage attention
MFAC	A2C	Information of neighborhoods
MAPPO	PPO	Expert agent-specific global state
HGAC	SAC	Hypergraph convolution
MAHAPO	PPO	**State-conditioned hyper-attention network**

where θ and θ_{old} are the policy's parameters prior to and after the update respectively, $D_{KL}(\cdot\|\cdot)$ represents the KL divergence.

In order to minimize computational complexity, PPO replaces the trust region restriction with a more succinct optimization target, that is, clipping the ratio of the old and new policies $r(\theta) = \frac{\pi_\theta(a|s)}{\pi_{\theta_{old}}(a|s)}$ to be no more than $1 + \epsilon$ and no less than $1 - \epsilon$. So the objective function can be written as:

$$\mathcal{L}^{\text{CLIP}}(\theta) = \mathbb{E}_{s,a\sim\pi_{\theta_{old}}} \left[\min\left(r(\theta)A(s,a), \text{clip}\left(r(\theta), 1 - \epsilon, 1 + \epsilon\right) A(s,a)\right)\right].$$

2.4 Game Abstraction

Game abstraction is known as the process of refining complicated game interactions in multi-agent systems into simpler models, enhancing algorithm's performance and reducing algorithm's complexity. Many algorithms, either expressly or indirectly, employ the game abstraction mechanism. For the actor-critic methods, the completely decentralized IPPO [29] just employs the agents' original local observations, and the earlier MADDPG [12] merely concatenates each agent's local observations and actions as the input to the centralized critic in the Multi-Agent Particle Environment (MPE). Subsequently, the performance of the algorithms has been significantly enhanced because of the application of the game abstraction mechanism. As illustrated in Table 1, MAAC employs a soft attention mechanism to selectively focus on information from other agents, and G2ANet employs a two-stage attention mechanism to further optimize it. MAPPO uses a well-designed feature pruning method, and HGAC [32] utilizes a hypergraph neural network [4] to enhance cooperation. To handle large-scale dynamic games, MFAC [30] employs the mean-field approach, in which each agent gets information exclusively from agents in its near neighborhood.

3 Multi-Agent Hyper-Attention Policy Optimization

In this section, we first present a new game abstraction mechanism based on the state-conditioned hyper-attention neural network, then introduce a computationally friendly point probability distance penalty term to trust-region methods

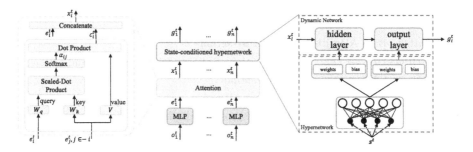

Fig. 1. The overall architecture of State-Conditioned Hyper-Attention Neural Network. *Left*: use the attention mechanism to generate information encoding x_i^t that incorporates all the agents' local observations o_i^t. *Right*: use the state-conditioned hypernetwork to process the global state s^t and agents's encoding x_i^t to generate the final embedding g_i^t.

to improve their robustness in dealing with multi-agent reinforcement learning problems. Finally, we propose the Multi-Agent Hyper-Attention Policy Optimization (MAHAPO) algorithm based on the two methods.

3.1 State-Conditioned Hyper-Attention Neural Network

Our state-conditioned hyper-attention game abstraction technique highlights the use of the attention mechanism [25] and state-conditioned hypernetwork [6] to achieve information aggregation of local perspective and information processing of global perspective, as well as improving the accuracy and efficiency of value function estimation. Its overall architecture is shown in Fig. 1.

At each time step t, the local observation o_i^t received by agent i is firstly encoded by a one-layer multilayer perceptron (MLP) as $e_i^t = MLP(o_i^t)$ and then the weight of agent j's contribution to agent i is obtained through the attention mechanism: $\alpha_j^t \propto exp(e_j^{t^T} W_k^T W_q e_i^t)$, where W_k and W_q are learnable parameters encoding the two vectors as Key and Query, respectively. Finally, the embedding of agent i can be expressed as the concatenation of e_i and other agents' contribution (a weighted sum of each agent's encoding):

$$x_i^t = Concatenate(e_i^t, \sum_{j \neq i} \alpha_j^t e_j^t).$$

After that, in order to further refine the game interaction, rather than mapping x_i^t directly to the value function V_i^t of agent i through a neural network, we learn a dynamic hypernetwork, which is a deep neural network model that generates the weights of the primary (dynamic) network. In single-agent reinforcement learning, hypernetworks have been used to enable the agent to acquire the capacity of continuous learning in model-based RL [8] and reduce the training variance of meta-policies in meta-RL [16]. In the field of MARL, hypernetworks are used for fitting joint action-state value functions Q_{tot} in value function decomposition methods.

In this paper, we use a state-conditioned hypernetwork H_{ϕ_h} with parameters ϕ_h as the weight generator for the dynamic network $f_{\phi_{tar}}$. The global state s^t is mapped to the weights of the dynamic network $f_{\phi_{tar}} = f_{H_{\phi_h}(s^t)}$ via H_{ϕ_h}, so the output of the dynamic network is determined by its input x_i^t and the global state s^t. Both networks are trained simultaneously, and the weights of the dynamic network will change dynamically according to the global state s^t during the test phase. Finally, as shown in Fig. 1, the output of our state-conditioned hyper-attention network can be expressed as: $g_i^t = f_{\phi_{tar}}(x_i^t)$.

3.2   Point Probability Distance Penalty

Trust-region methods have shown their powerful effects on multi-agent tasks, and the success of MAPPO and HATRPO [10] has shown their comparable sample efficiency and experimental performance to off-policy methods. Even the entirely decentralized IPPO benefits from constraint processing for policy update, leading to better results than Independent Q-Learning [23]. However, the debate over the PPO algorithm's effect is still continuing. The effect of the PPO, according to the study of Engstrom et al. [3], is more derived from code-optimization techniques such as value function clipping, reward scaling and Adam learning rate annealing. In contrast, the policy ratio clipping does not have the expected effect. TR-PPO-RB [28] also demonstrates that PPO could not rigorously perform the necessary trust region limitation. Consequently, PPO still risks performance instability, which will be more severe in more complicated multi-agent environments. It might be one of the reasons why HAPPO [10] is worse than HATRPO.

To this end, we suggest introducing a new penalty term of point probability Euclidean distance constraint to the optimization problem as proposed in [1]. At time t, for example, the point probability distance penalty for agent i's old and new policies at the action a_m^t is given as: $D_{pp}\left(\pi_{\theta_{old}}^i\left(\cdot \mid o^t\right) \| \pi_\theta^i\left(\cdot \mid o^t\right)\right) = \left(\pi_{\theta_{old}}^i\left(o_i^t, a_m^t\right) - \pi_\theta^i\left(o_i^t, a_m^t\right)\right)^2$, which can be proved to be the symmetric lower bound of the total variation divergence square of the old and new policies $D_{TV}^2 = \left(\frac{1}{2}\sum_k \left|\pi_{\theta_{old}}^i\left(o_i^t, a_k^t\right) - \pi_\theta^i\left(o_i^t, a_k^t\right)\right|\right)^2$ for improving the algorithm's robustness and performance without increasing computational complexity.

$$D_{TV}^2(\pi_{\theta_{old}}^i(\cdot|o^t)\|\pi_\theta^i(\cdot|o^t)) = \left(\frac{1}{2}\sum_k \left|\pi_{\theta_{old}}^i(o_i^t, a_k^t) - \pi_\theta^i(o_i^t, a_k^t)\right|\right)^2$$

$$= \left(\frac{1}{2}\sum_{k \neq m} \left|\pi_{\theta_{old}}^i(o_i^t, a_k^t) - \pi_\theta^i(o_i^t, a_k^t)\right| + \frac{1}{2}\left|\pi_{\theta_{old}}^i(o_i^t, a_m^t) - \pi_\theta^i(o_i^t, a_m^t)\right|\right)^2$$

$$\geq \left(\frac{1}{2}\left|\sum_{k \neq m} (\pi_{\theta_{old}}^i(o_i^t, a_k^t) - \pi_\theta^i(o_i^t, a_k^t))\right| + \frac{1}{2}(\pi_{\theta_{old}}^i(o_i^t, a_m^t) - \pi_\theta^i(o_i^t, a_m^t))\right)^2$$

$$= \left(\frac{1}{2}|1 - \pi_{\theta_{old}}^i(o_i^t, a_m^t) - (1 - \pi_\theta^i(o_i^t, a_m^t))| + \frac{1}{2}(\pi_{\theta_{old}}^i(o_i^t, a_m^t) - \pi_\theta^i(o_i^t, a_m^t))\right)^2$$

$$= D_{pp}(\pi_{\theta_{old}}^i(\cdot|o^t)\|\pi_\theta^i(\cdot|o^t)).$$

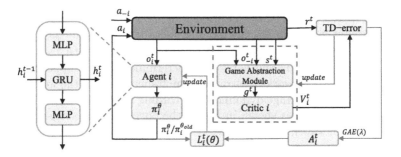

Fig. 2. The architecture of Multi-Agent Hyper-Attention Policy Optimization. The environment receives all agents' actions $< a_i, a_{-i} >$ and provides current observations o_i^t, states s^t, and reward r^t. Agent i maps o_i^t to an action at the next moment. Game Abstraction Module and the Critic i gets all of the data and generates the current time value function V_i^t. The red line depicts the process of gathering data and calculating the loss function to update the network during training.

We apply the point probability distance penalty to MARL, and it only works on the current action for each agent at each update, as opposed to the KL divergence penalty, which affects all action probabilities. This is in line with intuition, and all that is required is for the old and new policy networks to have the same high probability of taking the proper actions.

3.3   Multi-Agent Hyper-Attention Policy Optimization

Combined with the above two improvements, we present Multi-Agent Hyper-Attention Policy Optimization algorithm (MAHAPO), an actor-critic architecture method with completely end-to-end training.

As shown in the Fig. 2, MAHAPO trains two independent networks. The actor network, or the policy network, is represented as π_θ, receiving the agent's local observations, outputting the categorical distribution of each action (discrete action space) or the mean and standard deviation of the multivariate Gaussian distribution (continuous action space), and sampling actions from the distribution. The critic network, or the value network, only works in the training phase, mapping local observations and global state information to the value function $V_\phi(s)$ of the current state. If all agents are homogenous, they can share the same actor and critic network or train their own set of networks. The critic network is updated by minimizing $\mathcal{L}(\phi)$:

$$\mathcal{L}(\phi) = \sum_{i=1}^{n} \max \left[(V_\phi(s_i) - R)^2, (\text{clip}(V_\phi(s_i), V_{\phi_{old}}(s_i) - \varepsilon, V_{\phi_{old}}(s_i) + \varepsilon) - R)^2 \right],$$

where R is the cumulative return, n is the number of agents, and ε is the clipping ratio of the value function. The actor network can be updated by maximizing the target function $\mathcal{L}(\theta)$:

$$\mathcal{L}(\theta) = \mathbb{E}_{s,a} \left[\sum_{i=1}^{n} \min \left(\frac{\pi_{\theta_{otd}}^{i}(a_i \mid o_i)}{\pi_{\theta}^{i}(a_i \mid o_i)} A_i, \text{clip} \left(\frac{\pi_{\theta_{old}}^{i}(a_i \mid o_i)}{\pi_{\theta}^{i}(a_i \mid o_i)}, 1 - \epsilon, 1 + \epsilon \right) A_i \right) \right.$$
$$\left. - \sigma \sum_{i=1}^{n} D_{pp} \left(\pi_{\theta_{old}}^{i}(a_i \mid o_i) \| \pi_{\theta}^{i}(a_i \mid o_i) \right) + \eta \sum_{i=1}^{n} S \left(\pi_{\theta}(o_i) \right) \right],$$

where A_i represents the advantage function calculated by Generalized Advantage Estimate (GAE) [18], $S(\cdot)$ is the Shannon entropy of policy used to encourage exploration, and ϵ, σ, η are the coefficient hyperparameters of ratio clipping, distance penalty and the policy entropy, respectively.

4   Experiments

In this section, we evaluate MAHAPO and other MARL baselines in the Star-Craft Multi-Agent Challenge (SMAC) [15], which includes a variety of test scenarios and stringent control requirements. Furthermore, we also design ablation experiments to demonstrate the effectiveness of state-conditioned hyper-attention network and point probability distance penalty.

4.1   Settings

We choose SMAC as the test platform. It is one of the most prominent MARL benchmarks and the best choice for measuring the performance of algorithms. StarCraft II micromanagement tasks fall into three categories: Easy, Hard, and Super Hard scenarios, in which each agent takes action to complete a cooperative task based on its own local observations. To demonstrate the effects of our approach, we focus on evaluating all Super Hard scenarios as well as some representative Easy and Hard scenarios. We compare MAHAPO with the current state-of-the-art algorithms, including fine-tuned QMIX and feature-pruned MAPPO, as well as IPPO, VDAC [21], and COMA [5] based on the actor-critic architecture and QTRAN [20] based on the value decomposition framework. Following the evaluation metric proposed in MAPPO, we compute and record the win rates over 32 test games after each training iteration (10,000 parallel time steps). Finally, we show the results of five random seeds for each algorithm on each task.

4.2   MAHAPO Performance

The median win rates of all the algorithms evaluated in 9 situations are shown in Fig. 3. It is not hard to see that MAHAPO performs excellently and significantly outperforms the other algorithms, particularly in super hard scenarios. This shows MAHAPO's ability to fully exploit global information in very complicated circumstances during training. Even in the scenarios where IPPO outperforms MAPPO, such as *3s5z_vs_3s6z* and *corridor*, MAHAPO still achieves optimal results. It demonstrates that the usage of carefully constructed agent-specific

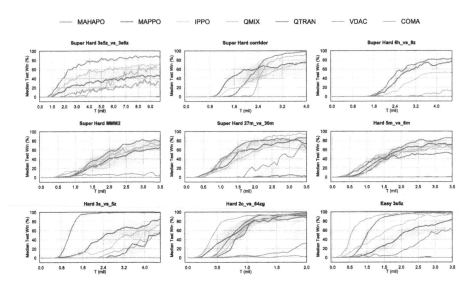

Fig. 3. Learning curves on 9 cooperative scenarios in SMAC. The solid line denotes the median win rates and the shadow part represents 25–75% percentiles.

characteristics in MAPPO cannot grasp critical information in some conditions, and cannot accomplish the desired results in the scenarios that necessitate close attention to individual data. The game abstraction method in MAHAPO, on the other hand, successfully overcomes this difficulty, and the self-adaptive extraction of game relations can assist agents in learning strategies efficiently. Moreover, MAHAPO still exhibits strong performance and similar sample efficiency compared to the value decomposition methods. Though it converges slightly slower than the fine-tuned QMIX in some simple scenarios like *3s5z*, there are clear advantages in more complex scenarios such as *3s5z_vs_3s6z* and *6h_vs_8z*. Finally, as a trust-region method, MAHAPO also outperforms MAPPO in terms of performance stability in most scenarios.

4.3   Ablation Studies

We conduct two ablation experiments to test the effect of the state-conditioned hyper-attention network and the point probability distance penalty on the algorithm. Firstly, we disassemble the hyper-attention network into two variants. One of them solely employs the attention method, and agents' embedded information comprises their own local observations as well as the extraction of other agents' local observations via the attention network. The other variant just considers the state-conditioned hypernetwork. It takes the agent's local observation and maps it to the value function through the dynamic network, while the hypernetwork is used to map the global information to the dynamic network's weight. We express these two variants as MAHAPO-att and MAHAPO-hyper, respectively.

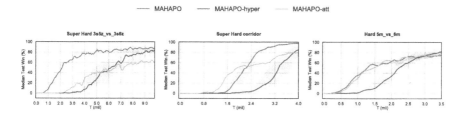

Fig. 4. Learning curves with MAHAPO, MAHAPO-att and MAHAPO-hyper on three scenarios in SMAC.

After that, we look at the impact of the penalty term on the trust region restriction by evaluating the effects of different penalty coefficients on the algorithm's performance.

As shown in the Fig. 4, algorithms that lack part of the structure lose some input information correspondingly so that their effects are naturally inferior to the complete MAHAPO. In addition, because MAHAPO-hyper employs the global state, it has more information and more intricate game interactions to deal with, sacrificing the convergence speed of the algorithm without losing much of the final performance. In contrast, MAHAPO-att converges relatively quickly but has a worse effect since it does not realize information processing under the global perspective, which means that it simplifies the game interaction while reducing accuracy. In short, the hyper-attention network not only considers all available data, but also significantly simplifies the game relationship, therefore optimal results are obtained.

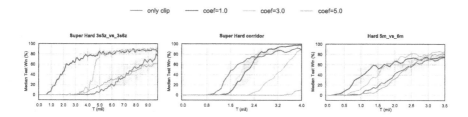

Fig. 5. Performance curves with MAHAPO for different Punishment Coefficient σ, 0 (clip), 1.0, 3.0, 5.0, in SMAC.

Figure 5 depicts the performance curves for various penalty term coefficients σ, where *clip* denotes the absence of a penalty term. On the whole, it can be observed that the algorithm utilizing the penalty term outperforms the simple ratio clipping approach (PPO). At the same time, as the penalty coefficient increases, the algorithm's convergence rate slows down, which can be inferred to be the effect of the trust region constraint. Furthermore, Fig. 5 shows that although setting σ to 3 converges slower, it yields similar or better results than setting it to 1 in all scenarios. As a result, we recommend that in a complicated circumstance, the smaller penalty coefficient can be chosen first, and then the penalty coefficient should be increased properly in the acceptable range of convergence rate.

5 Conclusion

We present a new game abstraction method for speeding training that integrates local observation and global information to generate self-adaptive centralized interactive information, called state-conditioned hyper-attention network. It is worth emphasizing that it is independent of the fundamental RL algorithm and has strong portability. Furthermore, we also introduce a point probability distance penalty term to increase the stability of the trust-region MARL. Based on the two improvements, we present the Multi-Agent Hyper-Attention Policy Optimization method. The experimental results in the StarCraft challenge demonstrate the MAHAPO's strong competitiveness against other baselines in a variety of scenarios. The structure ablation study and the penalty term coefficient analysis both demonstrate the efficiency of our approach.

References

1. Chu, X.: Policy optimization with penalized point probability distance: an alternative to proximal policy optimization. arXiv preprint arXiv:1807.00442 (2018)
2. Defferrard, M., Bresson, X., Vandergheynst, P.: Convolutional neural networks on graphs with fast localized spectral filtering. In: NIPS (2016)
3. Engstrom, L., Ilyas, A., Santurkar, S., Tsipras, D., Janoos, F., Rudolph, L., Madry, A.: Implementation matters in deep policy gradients: A case study on ppo and trpo. arXiv preprint arXiv:2005.12729 (2020)
4. Feng, Y., You, H., Zhang, Z., Ji, R., Gao, Y.: Hypergraph neural networks. In: Proceedings of the AAAI Conference on Artificial Intelligence (2019)
5. Foerster, J., Farquhar, G., Afouras, T., et al.: Counterfactual multi-agent policy gradients. In: Proceedings of the AAAI Conference on Artificial Intelligence (2018)
6. Ha, D., Dai, A., Le, Q.V.: Hypernetworks. arXiv preprint arXiv:1609.09106 (2016)
7. Hu, J., Jiang, S., Harding, S.A., Wu, H., Liao, S.: Rethinking the implementation tricks and monotonicity constraint in cooperative multi-agent reinforcement learning. arXiv preprint arXiv:2102.03479 (2021)
8. Huang, Y., Xie, K., Bharadhwaj, H., Shkurti, F.: Continual model-based reinforcement learning with hypernetworks. In: 2021 IEEE International Conference on Robotics and Automation (ICRA), pp. 799–805. IEEE (2021)
9. Iqbal, S., Sha, F.: Actor-attention-critic for multi-agent reinforcement learning. In: International Conference on Machine Learning, pp. 2961–2970. PMLR (2019)
10. Kuba, J.G., Chen, R., Wen, M., Wen, Y., et al.: Trust region policy optimisation in multi-agent reinforcement learning. arXiv preprint arXiv:2109.11251 (2021)
11. Liu, Y., Wang, W., Hu, Y., Hao, J., Chen, X., Gao, Y.: Multi-agent game abstraction via graph attention neural network. In: Proceedings of the AAAI Conference on Artificial Intelligence (2020)

12. Lowe, R., Wu, Y.I., Tamar, A., Harb, J., Pieter Abbeel, O., Mordatch, I.: Multi-agent actor-critic for mixed cooperative-competitive environments. In: Advances in Neural Information Processing Systems (2017)
13. Peng, Z., Li, Q., Hui, K.M., Liu, C., Zhou, B.: Learning to simulate self-driven particles system with coordinated policy optimization. In: Advances in Neural Information Processing Systems, vol. 34, pp. 10784–10797 (2021)
14. Rashid, T., Samvelyan, M., Witt, C.S., Farquhar, G., Foerster, J.N., Whiteson, S.: Qmix: monotonic value function factorisation for deep multi-agent reinforcement learning. arXiv:abs/1803.11485 (2018)
15. Samvelyan, M., et al.: The starcraft multi-agent challenge. arXiv:abs/1902.04043 (2019)
16. Sarafian, E., Keynan, S., Kraus, S.: Recomposing the reinforcement learning building blocks with hypernetworks. In: International Conference on Machine Learning, pp. 9301–9312. PMLR (2021)
17. Schulman, J., Levine, S., Abbeel, P., et al.: Trust region policy optimization. In: International Conference on Machine Learning, pp. 1889–1897. PMLR (2015)
18. Schulman, J., Moritz, P., Levine, S., et al.: High-dimensional continuous control using generalized advantage estimation. arXiv preprint arXiv:1506.02438 (2015)
19. Schulman, J., Wolski, F., Dhariwal, P., Radford, A., Klimov, O.: Proximal policy optimization algorithms. arXiv preprint arXiv:1707.06347 (2017)
20. Son, K., Kim, D., et al.: Qtran: learning to factorize with transformation for cooperative multi-agent reinforcement learning. arXiv:abs/1905.05408 (2019)
21. Su, J., Adams, S., Beling, P.A.: Value-decomposition multi-agent actor-critics. In: Proceedings of the AAAI Conference on Artificial Intelligence (2021)
22. Sunehag, P., et al.: Value-decomposition networks for cooperative multi-agent learning. arXiv:abs/1706.05296 (2018)
23. Tampuu, A., Matiisen, T., Kodelja, D., Kuzovkin, I., et al.: Multiagent cooperation and competition with deep reinforcement learning. PLoS ONE (2017)
24. Tao, N., Baxter, J., Weaver, L.: A multi-agent, policy-gradient approach to network routing. In: Proceedings of the 18th International Conference on Machine Learning. Citeseer (2001)
25. Vaswani, A., Shazeer, N., Parmar, N., Uszkoreit, J., Jones, L., et al.: Attention is all you need. In: Advances in Neural Information Processing Systems (2017)
26. Vinyals, O., Babuschkin, I., Czarnecki, W.M., et al.: Grandmaster level in starcraft II using multi-agent reinforcement learning. Nature 575(7782), 350–354 (2019)
27. Wang, J., Ren, Z., Liu, T., et al.: QPLEX: duplex dueling multi-agent q-learning. In: International Conference on Learning Representations, ICLR (2021)
28. Wang, Y., He, H., Tan, X.: Truly proximal policy optimization. In: Uncertainty in Artificial Intelligence, pp. 113–122. PMLR (2020)
29. de Witt, C.S., Gupta, T., Makoviichuk, D., Makoviychuk, V., Torr, P.H., Sun, M., Whiteson, S.: Is independent learning all you need in the starcraft multi-agent challenge? arXiv preprint arXiv:2011.09533 (2020)
30. Yang, Y., Luo, R., Li, M., et al.: Mean field multi-agent reinforcement learning. In: International Conference on Machine Learning. PMLR (2018)
31. Yu, C., Velu, A., Vinitsky, E., Wang, Y., et al.: The surprising effectiveness of PPO in cooperative, multi-agent games. arXiv preprint arXiv:2103.01955 (2021)
32. Zhang, B., Bai, Y., Xu, Z., Li, D., Fan, G.: Efficient cooperation strategy generation in multi-agent video games via hypergraph neural network. arXiv preprint arXiv:2203.03265 (2022)

Filter Pruning via Similarity Clustering for Deep Convolutional Neural Networks

Kang Song[1], Wangshu Yao[1,2(✉)], and Xuan Zhu[1]

[1] School of Computer Science and Technology, Soochow University, Suzhou, China
{20205227062,20205227093}@stu.suda.edu.cn, wshyao@suda.edu.cn
[2] Collaborative Innovation Center of Novel Software Technology and Industrialization, Nanjing, Jiangsu, China

Abstract. Network pruning is a technique to obtain a smaller lightweight model by removing the redundant structure from pre-trained models. However, existing methods are mainly based on the importance of filters in the whole network. Unlike previous methods, in this paper, we propose a filter pruning strategy, called Filter Pruning via Similarity Clustering(FPSC). FPSC uses the Euclidean distance between filters to measure their similarity, and then selects the filter with the smaller sum of k-nearest neighbor distances among the similar filters for removal. We consider that the selected filter is more likely to be replaced by neighbor filters. FPSC is applied to a variety of different networks, and compared with the existing filter pruning approaches. The experimental results show that FPSC has better pruning performance. On CIFAR-10, it is worth noting that FPSC reduces more than 70% FLOPs and parameters on GoogLeNet, and the accuracy is even 0.09% higher than the baseline model. Moreover, on ImageNet, FPSC reduces more than 43.1% FLOPs and 42.2% parameters, the accuracy only dropped 0.66% on ResNet-50.

Keywords: Filter pruning · Clustering · Convolutional neural network

1 Introduction

CNN have been great success in many fields in the past decade, such as image classification [1] and object detection [2]. However, the improvement of network performance is accompanied by the number of parameters increasing exponentially. Nowadays, models have millions of parameters, resulting in expensive computational costs. It is difficult for the models to be deployed on resource-constrained devices. Therefore, model compression and acceleration has quickly become a hot research topic. In recent years, researchers have proposed various model compression and acceleration methods, such as knowledge distillation [3], network quantization [4,5] and network pruning [5,6].

According to the pruning granularity, network pruning can be divided into unstructured pruning [5–7] and structured pruning [8–10]. The typical of unstructured pruning is weight pruning, which reduces network parameters by

© The Author(s), under exclusive license to Springer Nature Switzerland AG 2023
M. Tanveer et al. (Eds.): ICONIP 2022, LNCS 13623, pp. 88–99, 2023.
https://doi.org/10.1007/978-3-031-30105-6_8

removing unimportant weights in the filters. Unstructured pruning must use specialized software or hardware to achieve model acceleration, while structured pruning does not have this problem. Therefore, structured pruning has received more attention in recent years. Structured pruning mainly refers to filter pruning.

The key of filter pruning is how to select the filter to be removed. As shown in Fig. 1(a), previous works utilized "smaller-norm-less-important" criterion to prune filters. In general, filters with small norm are removed according to threshold and then fine-tuning is used to recover the network performance after pruning. He et al. [10] proposed that this criterion need to satisfy that the variance of the filter norm is large enough and the minimum norm is small enough. However, the actual network model does not completely meet this condition. Therefore, they proposed filter pruning based on the geometric median. Shao et al. [11] used Cosine distance between filters to cluster the filters of the same layer, and evaluated the importance of filters with eigenvalues of feature maps. Then, the most important filters in the same cluster were kept. Inspired by the above papers, we propose a novel filter pruning strategy, called Filter Pruning via Similarity Clustering(FPSC). As shown in Fig. 1(b), the removed filter is replaceable, not the unimportant.

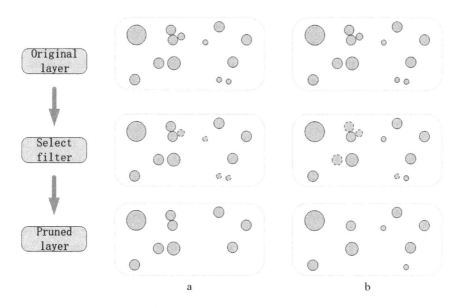

Fig. 1. The geometric distribution of filters in the same layer. The larger the circle is, the larger the filter norm is. (a) Filter pruning based on small norm. Removal of filters with small norm. (b) FPSC. Removal of filters that are close to each other.

The main contributions of this paper are as follows:

(1) The Euclidean distance between filters is proposed to represent the similarity between filters in the same layer. The smaller the distance is, the more similar the filters are.

(2) A method of selecting pruning filter based on clustering centrality is proposed. For similar filter pairs in the same layer, the sum of the Euclidean distances between the k-nearest neighbor filters other than each other is calculated, and then the filter with the smaller sum of distance is pruned. We think that filters with higher cluster centrality are more suitable for co-representation with their neighbor filters.

(3) Extensive comparative experiments were conducted on two image classification benchmarks (CIFAR10, ImageNet), and the results show that FPSC can select more appropriate filters to be removed.

2   Related Work

Network pruning achieves a more simplified model by removing redundant structures in the original network. Before 2017, researchers focused on weight pruning to remove the redundant parameters of the model, resulting in unstructured models that could not be directly accelerated using efficient BLAS libraries. In recent years, structured pruning with filters as pruning units has received much attention. Filter pruning methods can be classified into importance-based pruning, similarity-based pruning and other pruning.

Importance-Based Pruning. Li et al. [8] calculated the ℓ_1-norm of filter and then removed the filter with smaller ℓ_1-norms. Refs. [9,12] used the ℓ_2-norm of filter to represent the importance of the filters and invoked the masking technique to zero the filters with smaller ℓ_2-norm at the end of a training epoch. These zeroed filters could be recovered in subsequent training. Hu et al. [13] used APoZ (Average Percentage of Zeros) to denote the percentage of zero activation values and removed the filters with large APoZ. Lin et al. [14] used the rank of the output feature maps to indicate the importance of the corresponding filter and then removed the filters that produced low rank feature maps.

Similarity-Based Pruning. He et al. [10] proposed a filter pruning method based on the geometric median. The filter closest to the geometric median would be removed. He considered that the pruning filter can be replaced by other filters. Refs. [11] used a clustering algorithm to classify all filters in the same layer, and then one filter with the largest eigenvalue of the corresponding feature maps is reserved in each category. The number of clusters is determined by the pruning rate. Luo et al. [15] considered that pruning filters do not depend on the current layer, but the next layer. If an approximation of the original output can be derived using part of the filters in the same layer, the other filters can be removed.

Other Pruning. Xie et al. [16] proposed compensation-aware pruning to remove redundant channels by minimizing information loss. Refs. [17] used random pruning and found similar results to other pruning methods with no significant performance degradation. Meng et al. [18] proposed stripe pruning, which prunes the weights at the same position in all channels of a filter.

Discussion. The feature extraction of the input image can be accomplished by similarity filters, which is the basic idea of similarity filter pruning. At present, researchers have proposed various similarity filter pruning methods [10,11]. Inspired by [10,11], we propose Filter Pruning via Similarity Clustering (FPSC). Compared with [10], FPSC does not require the complicated computational process to calculate the geometric median and not need to calculate the sum of distances when selecting filters to delete. This avoids the problem that when only two filters are similar, the sum of distances between filters is too large and the filter pruning fails. In [11], if there are too few categories, two filters with large differences will be clustered into one category, which leads to inappropriate pruning of filters. FPSC selects the filter pair with the minimum distance and prunes the filter with the smaller sum of k-nearest neighbor distances, which effectively avoids the above problems.

3   The Proposed Method

3.1   Preliminaries

In this section, we formally introduce symbols and notations. Suppose a CNN model M has L convolution layers. C_i represents the i-th convolution layer. N_i and N_{i+1} represent the number of input channels and output channels of the i-th convolution layer, respectively, where $1 \leq i \leq L$. $F_{i,j}$ represents the j-th filter at the i-th layer, whose weight can be expressed as $\{W_{i,j} \in \mathbb{R}^{N_i \times K \times K}, 1 \leq j \leq N_{i+1}\}$, K is the size of the convolution kernel in the filter at the i-th layer. The weights in the i-th convolution layer can be expressed as $\{W_{C_i} \in \mathbb{R}^{N_{i+1} \times N_i \times K \times K}\}$. The global pruning rate is expressed by α.

In this paper, the cluster centrality of filter is represented by the sum of the distances of the k-neighbor filters of the filter. The smaller the sum of the distances, the higher the cluster centrality.

3.2   Filter Pruning via Similarity Clustering

Different from the existing filter pruning methods, in this paper, we focus on using the similarity filters to implement the feature extraction function of the pruned filters on the input image. Inspired by [10,11], we propose to use the Euclidean distance between filters to present the similarity between filters. A smaller distance indicates that two filters are more similar, and then one of the filters is removed. Once the most similar filter pair is found, a filter in the similar filter pair, which has smaller sum of distance from k-nearest neighbor filters, is removed. We think that the filter with the smaller sum of distance from k-nearest neighbor filters is more suitable to be replaced by the nearest neighbor filter than the other filter.

3.3   Filter Similarity Calculation

In the same layer, each filter has the same size. In this paper, inspired by [11], these filters are converted into a matrix to calculate the distance. The individual convolution kernel of the filters in the i-th layer can be expressed as:

$$w = \begin{pmatrix} w_{1,1} & \cdots & w_{1,K} \\ \vdots & \ddots & \vdots \\ w_{K,1} & \cdots & w_{K,K} \end{pmatrix} \tag{1}$$

Then, after conversion into a 1-D vector, the n-th kernel in the j-th filter of the i-th layer can be expressed as:

$$w_{i,j}^n = \left(w_{1,1}^n, \cdots w_{1,K}^n, \cdots w_{K,1}^n, \cdots w_{K,K}^n \right) \tag{2}$$

After the weights $W_{i,j}$ of the filter $F_{i,j}$ are converted from 3-D tensor to 1-D tensor, it is denoted by $\overline{W}_{i,j}$ and can be expressed as:

$$\overline{W}_{i,j} = \left(w_{i,j}^1, w_{i,j}^2, \cdots w_{i,j}^{N_i} \right) \tag{3}$$

The weights of all filters in the i-th convolution layer can be expressed as:

$$\overline{W}_i = \begin{pmatrix} w_{i,1}^1 & \cdots & w_{i,1}^{N_i} \\ \vdots & \ddots & \vdots \\ w_{i,N+1}^1 & \cdots & w_{i,N+1}^{N_i} \end{pmatrix} \tag{4}$$

Each row in the matrix indicates all weights in a filter that converted to 1-D vector, and the length is $N_i \times K \times K$. The number of rows in the matrix indicates the number of filters in i-th layer, and the number of rows is N_{i+1}.

The distance matrix between different filters in i-th layer is defined as:

$$D_i = \overline{W}_i \times \overline{W}_i^T \tag{5}$$

where $\times$ denotes the multiplication of two matrices. Distance mainly includes Euclidean distance and cosine distance. After experimental verification, the Euclidean distance gets slightly better results. Therefore, FPSC uses Euclidean distance. The smaller the distance between the filters, the more similar they are.

3.4   Pruning Filter Selection Strategy

After obtaining the distance between all filters in the same layer, we select one filter from the minimum distance filter pair to prune. Then, we select the filter pair with the minimum distance among the remaining filters and select one filter from them to prune. Iterate like this until the number of pruning filters reaches the pruning rate. In the minimum distance filter pair, which filter do we choose to prune? In this paper, Filter Pruning via Similarity Clustering(FPSC) is proposed.

Suppose filters A and B are minimum distance filter pair. First, the sum of the distances of the k-nearest neighbor filters to A and B are calculated, respectively.

The sum of distances denotes as DisSumA and DisSumB. Then, according to the value of DisSumA and DisSumB, the filter with smaller sum of distance will be removed. We think that the k-nearest neighbor filters can replace the removed filter. In addition, the number of filters in different convolution layers is different, and each layer will take different k, the k can be set as αN_{i+1}. The pruning filter can be expressed as:

$$F_{i,*} = min\left(DisSum\left(A\right), \left(DisSum\left(B\right)\right)\right) \tag{6}$$

$DisSum\left(X\right)$ can be expressed as:

$$DisSum\left(X\right) = \sum_{t=0}^{k} min\left(D_{i,X}, t\right) \tag{7}$$

where, $min\left(D_{i,X}, t\right)$ denotes the t-th smallest Euclidean distance between the X-th filter and other filters in the i-th convolution layer.

It is worth noting that when some of the smaller distance filters form a ring, it is important to ensure that at least one filter is retained so that this kind of features can be extracted.

The pruning algorithm of Filter Pruning via Similarity Clustering (FPSC) is shown in Algorithm 1.

Algorithm 1: Algorithm Description of FPSC

Input   : The parameters of pre-trained model $W = \{W_1, W_2, \cdots W_L\}$
         pruning rate α
Output: Select pruned filters subset P

1 *Initialize $P \leftarrow \emptyset$;*
2 **for** $i \leftarrow 1$ **to** L **do**
3 $\quad$ *Compute distance D_i by using Eq.(5);*
4 $\quad$ **for** $p \leftarrow 1$ **to** αN_{i+1} **do**
5 $\quad\quad$ *Find most small distance filters F1 and F2;*
6 $\quad\quad$ *Calculate the distance sum of the k-neighbor filter of F1 and F2;*
7 $\quad\quad$ *Select a filter F to be pruned that satisfy Eq.(6);*
8 $\quad\quad$ *Let $P \leftarrow P \cup F$;*
9 $\quad$ **end**
10 **end**
11 *Obtain the compact model parameters W^* from W;*
12 *Fine-tuning;*

---

4   Experiments

This experiment is based on PyTorch and verifies the effectiveness of FPSC on two different datasets: CIFAR-10 and ImageNet. The CIFAR-10 contains 60,000 images with 32×32 pixels in 10 classes, including 50,000 training images and

10,000 test images. ImageNet dataset is a large-scale dataset containing 1.28 million training images and 50k validation images of 1,000 classes. All experiments in this paper are performed on a Tesla V100-SXM2-32GB.

4.1   Experimental Settings

Pruning Setting. On the CIFAR-10, the batch size is set to 128, the weight decay is set to 5e−3, and the momentum is 0.9. The model is trained by SGD for 200 epochs, the initial learning rate is 0.01, and the learning rate is divided by 5 at 60, 120, and 160 epochs, respectively. On ImageNet dataset, the batch size is set to 256, the weight decay is set to 1e−4, and the momentum is 0.9. The model is trained by SGD for 100 epochs, the initial learning rate is 0.1, and the learning rate is divided by 10 at the 30, 60, and 90 epochs.

Evaluation Protocols. We adopt the widely-used criteria, parameters and FLOPs, to evaluate the pruned model. In addition, we provide Top-1 accuracy of pruned models and the pruning rate (denoted as PR) on CIFAR-10. Top-1 and Top-5 accuracy of pruned models on ImageNet are provided. The FPSC method is compared with other pruning methods, including but not limited to PFEC [8], SFP [9], FPGM [10], HRank [14] and Hinge [19]. The experimental results of the comparison methods are from the original references.

4.2   Results and Analysis

Results on CIFAR-10. For the CIFAR-10 dataset, its validity is verified on several mainstream models, including VGGNet, GoogLeNet, DenseNet-40 and ResNet-20/56. VGGNet only uses one fully connection layer. The output of GoogLeNet has been changed to accommodate the CIFAR-10. The growth rate of DenseNet-40 is set to 12. Table 1 shows the comparison results between FPSC and other network pruning methods on these models.

VGGNet. Two different rates of pruning is performed on FPSC, which are 42% and 52%, respectively. When α is set to 0.42, our method reduces up to 65.7% FLOPs and 66.3% parameters while the accuracy reaches 93.71%. FPSC is leading among many methods. Compared with other pruning methods, FPSC achieves the best accuracy regardless of whether the pruning rate is 0.42 or 0.52. Although FPSC removes few parameters than some methods, it has fewer FLOPs, faster inference time and better effect on model acceleration.

GoogLeNet. FPSC achieves superior performance on GoogLeNet. Instead of decreasing the accuracy of the pruned model when the pruning rate is set to 34%, the accuracy is 0.2% higher than baseline model. FPSC still achieves 0.09% higher accuracy than the baseline model when pruning 70.1% of FLOPs. The performance of FPSC is obviously superior to other methods. It has been proved that FPSC is more suitable for models with inception blocks like GoogLenet.

DenseNet-40. When the pruning rate is set to 34%, FPSC achieves better results in terms of accuracy and FLOPs. When the pruning rate is set to 42%, the

Table 1. Pruning results on CIFAR-10

Model	Method	Top-1 %	FLOPs(PR)	Parameters(PR)
VGGNet	PFEC [8]	93.4	206M(34.2%)	5.4M(64.0%)
	Hinge [19]	93.59	191.16M(39.1%)	2.99M(80.1%)
	CHSE [11]	93.08	122.44M(61.0%)	3.11M(79.2%)
	HRank [14]	92.34	108.61M(65.3%)	**2.64M(82.1%)**
	Ours(FPSC-42%)	**93.71**	**107.53M(65.7%)**	4.97M(66.3%)
	CHSE [11]	92	97.18M(69.0%)	2.64M(82.1%)
	ABCPruner [20]	93.08	82.81M(73.7%)	**1.67M(88.7%)**
	HRank [14]	91.23	73.70M(76.5%)	1.78M(92.0%)
	Ours(FPSC-52%)	93.36	**73.41M(76.6%)**	3.40M(76.9%)
GoogLenet	CHSE [11]	94.51	715.87M(53.0%)	2.92M(52.5%)
	HRank [14]	94.53	690M(54.9%)	2.74M(55.4%)
	Ours(FPSC-34%)	**95.25**	**679.26M(55.6%)**	**2.73M(55.8%)**
	FSketch [21]	94.88	590M(61.1%)	2.61M(57.6%)
	CHSE [11]	94.06	450.06M(70.4%)	1.86M(69.8%)
	HRank [14]	94.07	**450M(70.4%)**	1.86M(69.8%)
	Ours(FPSC-46%)	**95.14**	456.62M(70.1%)	**1.83M(70.3%)**
DenseNet-40	CHSE [11]	93.81	137.19M(51.4%)	0.55M(47.0%)
	Ours(FPSC-34%)	**93.95**	**128.47M(55.3%)**	**0.47M(55.4%)**
	HRank [14]	93.68	110.15M(61.0%)	0.48M(53.8%)
	FPGM [10]	93.48	99.13M(65.5%)	0.37M(65.5%)
	Ours(FPSC-42%)	93.68	**99.13M(65.5%)**	**0.37M(65.5%)**
ResNet-20	FPGM [10]	91.09	24.3M(42.2%)	-
	SFP [9]	90.83	24.3M(42.2%)	-
	Ours(FPSC-25%)	**91.72**	**23.34M(43.4%)**	**0.15M(43.6%)**
	DSA [24]	**91.38**	20.49M(50.3%)	-
	FPGM [10]	90.44	18.7M(54.0%)	-
	Ours(FPSC-35%)	90.97	**18.59M(54.9%)**	**0.12M(56.5%)**
ResNet-56	DSA [24]	93.08	88.72M(29.3%)	-
	HRank [14]	93.52	88.72M(29.3%)	0.71M(16.8%)
	CHSE [11]	93.13	80.30M(36.1%)	0.55M(35.3%)
	FSketch [21]	93.19	73.36M(41.5%)	0.50M(41.2%)
	Ours(FPSC-25%)	**93.52**	**71.64M(43.5%)**	**0.48M(43.9%)**
	CHSE [11]	93.07	62.72M(50.0%)	0.49M(42.4%)
	HRank [14]	**93.17**	62.72M(50.0%)	0.49M(42.4%)
	DSA [24]	92.91	60.63M(52.2%)	-
	FPGM [10]	92.93	59.4M(52.6%)	-
	Ours(FPSC-33%)	93.14	**59.4M(52.6%)**	**0.39M(54.2%)**

accuracy is the same as HRank, but FPSC removes more FLOPs (65.5% vs. 61.0%). This result indicates that when pruning rate increases, the accuracy of FPSC decreases less than HRank. Therefore, when the pruning rate continues to increase, FPSC can achieve better results than HRank.

ResNet-20/56. On ResNet-20, FPSC outperforms the other pruning methods when the pruning rate is set to 25% and 33%. When more FLOPs are removed, FPSC is still better than FPGM (90.97% vs. 90.44%). Two different pruning ratios are set on ResNet-56. When 25% of the filter is removed, the FLOPs and parameters are reduced by 43.5% and 43.9%, while the accuracy is only 0.09% lower than the baseline model. FPSC achieves the same accuracy as HRank, but removes much more FLOPs (43.5% vs. 29.3%). When the pruning rate is between 25% and 33%, only HRank achieves comparable performance (93.17% vs. 93.14%). All other methods are less accuracy than FPSC.

Results on ImageNet. Two residual networks with different depths, ResNet-18 and ResNet-50, are experimentally compared on Imagenet. ResNet-18 and ResNet-50 use basicblock and bottleneck structures respectively. The experimental results are shown in Table 2.

Table 2. Pruning results on ImageNet

Model	Method	Top-1 %	Top-5 %	FLOPs(PR)	Parameters(PR)
ResNet-18	MIL [23]	66.33	86.94	1189.5M(34.6%)	-
	SFP [9]	67.25	87.76	1058.7M(41.8%)	-
	FPGM [10]	**67.78**	**88.01**	1058.7M(41.8%)	-
	ASFP [12]	67.41	87.89	1058.7M(41.8%)	-
	Ours(FPSC-26%)	67.43	87.99	**1031.26M(43.3%)**	**6.52M(44.2%)**
ResNet-50	C-SGD [25]	75.27	92.46	2600.53M(36.8%)	-
	DSA [24]	75.1	92.45	2466.91M(40%)	-
	ASFP [12]	74.88	92.39	2392.9M(41.8%)	-
	FPGM [10]	**75.50**	**92.63**	2376.43M(42.2%)	-
	Ours(FPSC-25%)	75.36	92.49	**2339.43M(43.1%)**	**14.77M(42.2%)**
	CHSE [11]	72.25	-	1985.23M(51.5%)	18.43M(27.8%)
	FPGM [20]	74.13	91.94	1911.85M(53.5%)	-
	ABCPruner [20]	73.86	91.69	1890.6M(54.3%)	**11.75M(54.0%)**
	Ours(FPSC-33%)	**74.14**	**92.01**	**1886.46M(54.4%)**	11.99M(53.1%)
	HRank [14]	71.98	**91.01**	1550M(62.1%)	13.77M(46.1%)
	Comp [16]	71.13	90.96	1538.53M(62.6%)	-
	CHSE [11]	71.02	-	1425.34M(65.1%)	13.08M(48.7%)
	Ours(FPSC-42%)	**72.69**	90.96	**1425.34M(65.1%)**	**9.13M(64.3%)**

ResNet-18. When the pruning rate is 26%, FPSC achieves superior accuracy of Top-1 and Top-5. It is slightly lower to FPGM in accuracy (67.43% vs. 67.78%), but it removes more FLOPs(43.8% vs. 41.8%). Besides, FPSC is superior to other methods in all aspects.

ResNet-50. Three different pruning ratios of 0.25, 0.33 and 0.42 are set on ResNet-50, and the results can be viewed in Table 2. With the increase of pruning ratio, the error increases faster. When the pruning ratio is set to 25%, FPSC surpasses most methods and is slightly lower than FPGM (75.36% vs. 75.50%). When the pruning rate is set to 33%, FPSC outperforms FPGM in all aspects. In addition, FPSC achieves more obvious results than other methods in terms of Top-1 accuracy, but not in Top-5 accuracy. When the pruning rate is set to 42%, it also shows this feature. We analyze the whole experimental process of FPSC, and found that the accuracy of Top-5 has been increasing slowly. It indicates that FPSC can achieve better accuracy if the number of training epoch is increased.

4.3   Ablation Study

Euclidean or Cosine. Cosine distance is used instead of Euclidean distance to calculate the distance between filters. ResNet-56 and VGGNet are taken as the baseline model, and the pruning rate is set at 25% and 42% respectively. The two different distance are compared, and the results are shown in Fig. 2(a). As can be seen from the figure, the results of Euclidean distance are better than Cosine distance in both ResNet-56 and VGGNet.

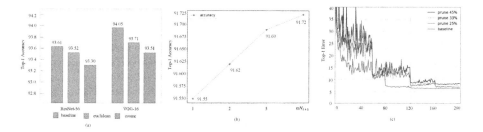

Fig. 2. (a) Experimental comparison of Euclidean distance and Cosine distance. (b) Pruning results of different k on ResNet-20. (c) Error at different pruning ratios on ResNet-56.

Filter Selection Criteria. In the similar filter pairs, pruning filters are selected by the clustering centrality, and the effects of different k on pruning performance are verified by experiments. The experiments are conducted on ResNet-20, and the range of k is $(1, 2, 5, \alpha N_{i+1})$. The results are shown in Fig. 2(b). It can be seen that the best results are obtained when k is automatically taken according to the number of filters in each layer and the pruning rate.

Varying Pruned FLOPs. To comprehensively understand FPSC, we test the error rate of ResNet-56 with different pruning rates, as shown in Fig. 2(c). When the pruning rate is 25%, the error only increases 0.09% compared with the baseline model. When the pruning ratio is 33% and 45%, the error increased by 0.39% and 1.44%, respectively. Obviously, the accuracy decreases slowly when about 50% of FLOPs are removed, and decreases sharply when 68% is reached. This phenomenon suggests that each network have pruning threshold, and when threshold is exceeded, the accuracy rate decreases rapidly.

5   Conclusion and Future Work

In order to speed up the inference of convolutional neural networks, we propose Filter Pruning via Similarity Clustering(FPSC). Unlike the previous norm-based filter pruning, FPSC considers the relationship between all filters in the same layer. The Euclidean distance between filters is used to measure their similarity and the pruning filter is selected based on the cluster centrality. FPSC achieves the state-of-the-art performance on GoogLeNet. However, FPSC still has some shortcomings. A global uniform pruning ratio is difficult to maximize the network pruning. It is worth investigating that different convolutional layers should set different pruning ratios to guide the network pruning so as to better improve the network compression rate.

Acknowledgments. This work was partially supported by Collaborative Innovation Center of Novel Software Technology and Industrialization, and Project Funded by the Priority Academic Program Development of Jiangsu Higher Education Institutions.

References

1. Durand, T., Mehrasa, N., Mori, G.: Learning a deep convnet for multi-label classification with partial labels. In: Proceedings of the IEEE/CVF Conference on Computer Vision and Pattern Recognition, pp. 647–657 (2019)
2. Zhu, C., Chen, F., Ahmed, U., Shen, Z., Savvides, M.: Semantic relation reasoning for shot-stable few-shot object detection. In: Proceedings of the IEEE/CVF Conference on Computer Vision and Pattern Recognition, pp. 8782–8791 (2021)
3. Hinton, G., Vinyals, O., Dean, J., et al.: Distilling the knowledge in a neural network. arXiv preprint arXiv:1503.02531, **2**(7) (2015)
4. Zhou, S., Wu, Y., Ni, Z., Zhou, X., Wen, H., Y. Zou: Dorefa-net: training low bitwidth convolutional neural networks with low bitwidth gradients. arXiv preprint arXiv:1606.06160 (2016)
5. Han, S., Mao, H., Dally, W.J.: Deep compression: compressing deep neural networks with pruning, trained quantization and huffman coding. arXiv preprint arXiv:1510.00149 (2015)
6. Han, S., Pool, J., Tran, J., Dally, W.: Learning both weights and connections for efficient neural network. In: Advances in Neural Information Processing Systems, vol. 28 (2015)
7. Ye, S., et al.: Progressive weight pruning of deep neural networks using ADMM. arXiv preprint arXiv:1810.07378 (2018)
8. Li, H., Kadav, A., Durdanovic, I., Samet, H., Graf, H.P.: Pruning filters for efficient convnets. arXiv preprint arXiv:1608.08710 (2016)
9. He, Y., Kang, G., Dong, X., Fu, Y., Yang, Y.: Soft filter pruning for accelerating deep convolutional neural networks. arXiv preprint arXiv:1808.06866 (2018)

10. He, Y., Liu, P., Wang, Z., Hu, Z., Yang, Y.: Filter pruning via geometric median for deep convolutional neural networks acceleration. In: Proceedings of the IEEE/CVF Conference on Computer Vision and Pattern Recognition, pp. 4340–4349 (2019)

11. Shao, M., Dai, J., Wang, R., Kuang, J., Zuo, W.: Cshe: network pruning by using cluster similarity and matrix eigenvalues. Int. J. Mach. Learn. Cybern. **13**(2), 371–382 (2022)

12. He, Y., Dong, X., Kang, G., Fu, Y., Yan, C., Yang, Y.: Asymptotic soft filter pruning for deep convolutional neural networks. IEEE Trans. Cybern. **50**(8), 3594–3604 (2019)

13. Hu, H., Peng, R., Tai, Y.-W., Tang, C.-K.: Network trimming: a data-driven neuron pruning approach towards efficient deep architectures. arXiv preprint arXiv:1607.03250 (2016)

14. Lin, M., et al.: Hrank: filter pruning using high-rank feature map. In: Proceedings of the IEEE/CVF Conference on Computer Vision and Pattern Recognition, pp. 1529–1538 (2020)

15. J.-H. Luo, J. Wu, and W. Lin, "Thinet: A filter level pruning method for deep neural network compression," in Proceedings of the IEEE international conference on computer vision, 2017, pp. 5058–5066

16. Xie, Z., Fu, Y., Tian, S., Zhou, J., Chen, D.: Pruning with compensation: efficient channel pruning for deep convolutional neural networks. arXiv preprint arXiv:2108.13728 (2021)

17. Liu, Z., Sun, M., Zhou, T., Huang, G., Darrell, T.: Rethinking the value of network pruning. arXiv preprint arXiv:1810.05270 (2018)

18. Meng, F., et al.: Pruning filter in filter. In: Advances in Neural Information Processing Systems, vol. 33, pp. 17 629–17 640 (2020)

19. Li, Y., Gu, S., Mayer, C., Gool, L.V., Timofte, R.: Group sparsity: the hinge between filter pruning and decomposition for network compression. In: Proceedings of the IEEE/CVF Conference on Computer Vision and Pattern Recognition, pp. 8018–8027 (2020)

20. Lin, M., Ji, R., Zhang, Y., Zhang, B., Wu, Y., Tian, Y.: Channel pruning via automatic structure search. arXiv preprint arXiv:2001.08565 (2020)

21. Lin, M., et al.: Filter sketch for network pruning. IEEE Trans. Neural Networks Learn. Syst. (2021)

22. Li, Y., Gu, S., Zhang, K., Van Gool, L., Timofte, R.: DHP: differentiable meta pruning via HyperNetworks. In: Vedaldi, A., Bischof, H., Brox, T., Frahm, J.-M. (eds.) ECCV 2020. LNCS, vol. 12353, pp. 608–624. Springer, Cham (2020). https://doi.org/10.1007/978-3-030-58598-3_36

23. Dong, X., Huang, J., Yang, Y., Yan, S.: More is less: a more complicated network with less inference complexity. In: Proceedings of the IEEE Conference on Computer Vision and Pattern Recognition, pp. 5840–5848 (2017)

24. Ning, X., Zhao, T., Li, W., Lei, P., Wang, Yu., Yang, H.: DSA: more efficient budgeted pruning via differentiable sparsity allocation. In: Vedaldi, A., Bischof, H., Brox, T., Frahm, J.-M. (eds.) ECCV 2020. LNCS, vol. 12348, pp. 592–607. Springer, Cham (2020). https://doi.org/10.1007/978-3-030-58580-8_35

25. Ding, X., Ding, G., Guo, Y., Han, J.: Centripetal SGD for pruning very deep convolutional networks with complicated structure. In: Proceedings of the IEEE/CVF Conference on Computer Vision and Pattern Recognition, pp. 4943–4953 (2019)

FPD: Feature Pyramid Knowledge Distillation

Qi Wang[1], Lu Liu[1], Wenxin Yu[1(✉)], Zhiqiang Zhang[2], Yuxin Liu[1], Shiyu Cheng[1], Xuewen Zhang[1], and Jun Gong[3]

[1] Southwest University of Science and Technology, Mianyang, Sichuan, China
yuwenxin@swust.edu.cn
[2] Hosei Univeristy, Tokyo, Japan
[3] Southwest Automation Research Institute, Chengdu, China

Abstract. Knowledge distillation is a commonly used method for model compression, aims to compress a powerful yet cumbersome model into a lightweight model without much sacrifice of performance, giving the accuracy of a lightweight model close to that of the cumbersome model. Commonly, the efficient but bulky model is called the teacher model and the lightweight model is called the student model. For this purpose, various approaches have been proposed over the past few years. Some classical distillation methods are mainly based on distilling deep features from the intermediate layer or the logits layer, and some methods combine knowledge distillation with contrastive learning. However, classical distillation methods have a significant gap in feature representation between teacher and student, and contrastive learning distillation methods also need massive diversified data for training. For above these issues, our study aims to narrow the gap in feature representation between teacher and student and obtain more feature representation from images in limited datasets to achieve better performance. In addition, the superiority of our method is all validated on a generalized dataset (CIFAR-100) and a small-scale dataset (CIFAR-10). On CIFAR-100, we achieve 19.21%, 20.01% of top-1 error with Resnet50 and Resnet18, respectively. Especially, Resnet50 and Resnet18 as student model achieves better performance than the pre-trained Resnet152 and Resnet34 teacher model. On CIFAR-10, we perform 4.22% of top-1 error with Resnet-18. Whether on CIFAR-10 or CIFAR-100, we all achieve better performance, and even the student model performs better than the teacher.

Keywords: Knowledge distillation · Feature pyramid network · Feature pyramid distillation

1 Introduction

Knowledge distillation is a commonly method for compressing models. In 2015, Geoffrey Hinton et al. added "Temperature (T)" in softmax as the cross-entropy loss function. The initial distillation methods are based on the probability of training samples in the logits layer, such as KD [5]. And then, some classical

M. Tanveer et al. (Eds.): ICONIP 2022, LNCS 13623, pp. 100–111, 2023.
https://doi.org/10.1007/978-3-031-30105-6_9

features representation distillation methods such as Finets [10] added "hints layer", AT [17] used the attention map of feature maps to transfer knowledge, FSP [16] defined the process of image processing between teacher model and student model as an additional matrix, VID [1] maximized the mutual information between teacher model and student model, and other methods such as AB [4], SP [13]. These methods mainly study the features representation among intermediate layer between the teacher model and student model to improve accuracy of the student model and make the student model as similar as possible to the teacher model. Later, to further improve the accuracy of the student model, some methods combine knowledge distillation with contrastive learning, such as CRD [12]. However, classical distillation methods have a significant gap in feature representation and model ability between teacher model and student model, and contrastive learning distillation method also need massive diversified data for training. The three knowledge distillation methods described above are shown in Fig. 1.

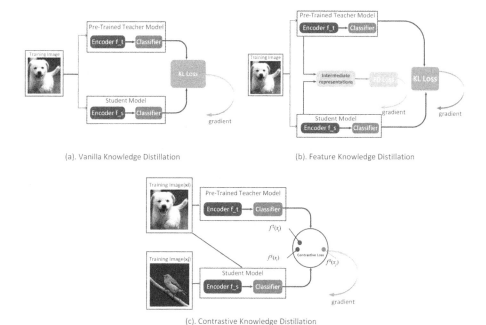

Fig. 1. Three kinds of knowledge distillation methods. (a) Vanilla KD calculates the gradient using final class predictions by pre-trained teacher and student, such as KD [5]. (b) Feature knowledge distillation gathers more gradient information from the intermediate layers through various knowledge representations such as Finets [10], AT [17], FSP [16]. (c) Contrastive knowledge distillation methods such as CRD [12].

Among the previously proposed knowledge distillation methods, the teacher's feature maps differ significantly from the student's due to differences of the net architecture. Using the feature maps of the teacher model directly to learn will lead the student model to learn insufficiently. In order to solve these existing issues, we propose feature pyramid knowledge distillation (named FPD) to fuse the teacher's and student's feature maps at each stage, which can narrow the gap in feature maps between teacher model and student model during training to help the student model get more knowledge from the teacher model to get better performance. As we expected, our propose method is superior to these methods mentioned above, since, our method can be approximated as combining the logits distillation with the feature representation distillation rather than simply adding another loss function. Although the feature pyramid knowledge distillation reduces the gap of feature representation between teachers and students, but the error will be increased by feature pyramid when the teacher's prediction is wrong, so we use guided knowledge distillation [19] to avoid these errors. As expected, our experiments show that our method outperforms even the use of contrastive distillation methods on small-scale dataset (CIFAR-10), where the accuracy of using contrastive distillation methods significantly decreases in limited sample data. From experimental studies, it is shown that our method is more suitable in some special domains, such as medical and military, where massive and diverse samples are unavailable easily.

Overall, our contributions are summarized as follows:

- We provide an novel view to study feature representation distillation by feature pyramid to help student model have a better performance.
- We reveal limitations of the contrastive distillation methods in some special domain cause by the insufficient of datasets.
- We propose an efficient feature representation distillation method (named FPD) to overcome these issues and limitations. Our method achieves better performance and empirically demonstrates its superiority on CIFAR-100 and CIFAR-10.

2   Related Work

2.1   Knowledge Distillation

KD [5] transferred the teacher's output distribution as a kind of knowledge to the students, prompting student model to approach the teacher's output distribution, it used soft target to learn more about the teacher's predictions rather than learn about ground truth directly. FitNet [10] added some additional monitoring signals in the middle of the network to help the output of the intermediate student layer to be as close to the teacher as possible. AT [17] used the attention map of the teacher model and the student model to distillation. FSP [16] defined a solution process matrix for feature mapping between intermediate layers, representing how the teacher processes information between the two layers and allowing students to learn how the teacher processes information.

2.2   Feature Pyramid

Targets of different sizes all undergo the same downsampling ratio. They will have a sizeable semantic gap, and the most common performance is the relatively low accuracy of small target detection. The feature pyramid has different resolutions at different scales, and targets of different sizes can have suitable feature representations at the corresponding scales. The model's performance can be improved by fusing multi-scale information to predict targets of different sizes at different scales. ASPP [2] proposes to build by multiple branches with different dilation rates of dilated convolution, and after multiple branches concate and then 1×1 convolution, FPN [8] proposes to make the shallow feature map with better semantic information by top-down stacking.

2.3   Attention Mechanisms

The study of attention mechanisms originates from human observation habits and is widely used in natural language processing and computer vision tasks. In SENet [6], global average pool and fully connected is used to obtain weights on feature map channels to obtain channel attention. Attention is paid to channels and spatial locations in the feature map considering global average pool and maximum global pool information in CBAM [14]. The detailed structure of the core module of SENet [6] (denote as **SEBlock**) is shown in Fig. 2.

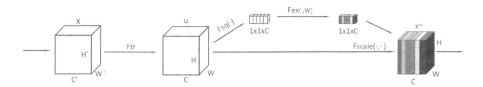

Fig. 2. A core Squeeze-and-Excitation block in SENet [6].

3   Method

In this section, we will introduce the theory behind feature pyramid distillation (named **FPD**), then explain why FPD is performed, and why we use guided knowledge distillation [19], and finally introduce the design of our loss function.

3.1   Feature Pyramid Knowledge Distillation

The FPN [8] consists of two parts: The first part is a bottom-up process, and the second part is a top-down and lateral connection fusion process. The bottom-up process is ordinary convolutional process. The top-down process scales up the small feature map of the top layer to the same size as the feature map of the previous stage by upsampling. This method takes advantage of the more robust semantic features of the top layer and the detailed information of the bottom layer. The lateral connection fusion process takes the features of the previous layer that have been upsampled to the exact resolution as the current layer and fuses them by summation. The detailed architecture of FPN [8] is shown in Fig. 3.

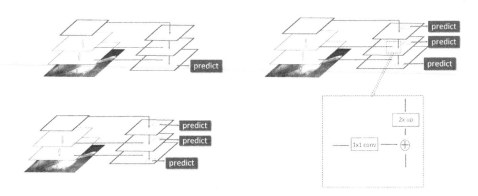

Fig. 3. Left-Top: a top-down architecture with skip connections, where predictions are made on the first level. Left-Bottom: our model that has a similar structure but leverages it as a feature pyramid, with predictions made independently at all levels. Right: A building block illustrating the lateral connection and the top-down pathway, merged by addition.

Our method FPD is derived from the FPN [8]. We believe that feature maps carried by the top layer can be approximated to the logits layer. Meanwhile, the model's bottom and intermediate feature maps carry more details of the original images, such as texture information. Therefore, we think of using the feature pyramid method, which will allow students to learn semantic and texture information at the same time during the training process, this way will narrow the gap of feature maps between teacher and student. In our paper, we take ResNet [3] as the base model, and select the feature map which is derived by residual block of the $conv^1$, $conv^2$, ..., $conv^i$ layers as the distillation feature representation of our FPD, using $\{f_1, f_2, ..., f_i\}$ denotes correspondingly. We put these feature maps into the feature pyramid block to fuse the feature information of each residual block, and feature maps after fusion are represented by $\{p_1, p_2, ..., p_i\}$. And then, we calculate correspondingly the loss value of each pair of feature maps after the feature pyramid between teacher and student. The detailed architecture of our FPD is shown in Fig. 4.

We use the mean squared error (called **MSE**) as the loss function to calculate the gap of feature maps between the teacher and student for each pair after the feature pyramid, divide the obtained loss values for each pair by the sum of the loss values for all pair, then we use the softmax function to calculate the **Weight**

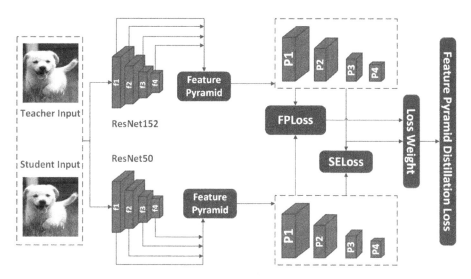

Fig. 4. The structure of our FPD. (1) FPLoss: Mean squared error loss function to calculate the loss value of each pair of feature maps after the feature pyramid. (2) SELoss: feature pyramid loss value of each pair feature maps after SEBlock (the structure of SEBlock is shown in Fig. 2).

for the corresponding pair, and multiply the loss value by the corresponding **Weight** to obtain the total loss, called **FPLoss**.

$$Weight(t,s) = SoftMax(\frac{MSE(p_t^i, p_s^i)}{\sum_{i=1}^{N} MSE(p_t^i, p_s^i)}) = [w_1 ..., w_N] \qquad (1)$$

$$FPLoss(t,s) = \sum_{i=1}^{N} Weight[i] * MSE(p_t^i, p_s^i) \qquad (2)$$

The **t**, **s** in all equation denotes feature maps from teacher model and student model respectively. The p_t^i, p_s^i denotes i-th feature map of all feature maps after feature pyramid block from teacher and student respectively. The **Weight[i]** denotes the i-th weight of loss each pair of the p_t^i, p_s^i calculate by softmax. To gain more knowledge from the teacher model, we put feature maps for each pair after the feature pyramid into **SEBlock** (shown in Fig. 2) to get feature maps channel-wise attention, use $p_{se_t}^i$, $p_{se_s}^i$ to denote respectively, and use the same method as **FPLoss** to calculate the **AT_Weight**, and calculate the gap of channel-wise attention between the teacher and student, called **SELoss**.

$$SE(t,s) = SEBlock(p_t^i, p_s^i) = (p_{se_t}^i, p_{se_s}^i) \qquad (3)$$

$$SELoss(t,s) = \sum_{i=1}^{N} MSE(p_{se_t}^i, p_{se_s}^i) * AT\_Weight[i] \qquad (4)$$

Finally to calculate the weight of **FPLoss** and **SELoss** use above-mentioned same method. The **FP_weight, SE_weight** denotes the weight of FPLoss and SELoss respectively. We call the final Loss as **FPDLoss**, the equation is as follows:

$$FPDLoss(t, s) = FPLoss * FP\_weight + SELoss * SE\_weight \qquad (5)$$

3.2 Guided Knowledge Distillation

During the experiment, we observe that the feature maps after the feature pyramid help students get more information from the teacher during the training process. Nevertheless, we observe the feature pyramid method will extend the gap of the feature map when the teacher's prediction is wrong. Because the prediction of the teacher's is not always correct, if the prediction of the teacher is wrong, the feature maps of the teacher at the same time are not correct, and error feature maps will lead the student to learn error information from the teacher and lead students to learn insufficiently. For this issue, we try to use guided knowledge distillation [19] to rectify the error feature maps from the teacher to help student perform better. The equation of guided knowledge distillation is defined as follows:

$$GKD(t, s) = \frac{\sum_{i=1}^{N} I(p_t^i, y_i) * KL(p_t^i, p_s^i)}{\sum_{i=1}^{N} I(p_t^i, y_i)} \qquad (6)$$

The p_t^i, p_s^i denotes final probability distribution of teacher and student respectively, and y_i denotes true label .The I is an indicator function and $I(p_t^i, y_i)$ is **1** when the output of the teacher equals true label, or else $I(p_t^i, y_i)$ is **0**. The n is the batch size and the $KL(p_t^i, p_s^i)$ is the mean of Kullback-Leibler divergence (KL) between p_t^i and p_s^i.

3.3 Loss Function

We follow the common approach to designing our final loss function, using our **FPDLoss** as the primary loss function, the guided knowledge distillation loss function as the primary auxiliary to rectify the error outputs of teachers, the cross-entropy loss function as a secondary auxiliary. Then, three kinds of loss function is multiplied by the corresponding weight factors, and add them to obtain the final loss function. Our final loss function is as following equation:

$$Loss(t, s) = \alpha * CE(s, y) + \beta * GKD(t, s) + \gamma * FPD(t, s) \qquad (7)$$

In above equation, the t, s denote the final probability distribution of teacher and student, and the y denotes the true label in $CE(s, y), GKD(t, s)$, but the t, s denote feature maps after feature pyramid block in $FPD(t, s)$. Besides, the α, β, γ weight value is set based on the ratio of teacher model's error rate to the correct rate.

4 Experiments

We perform experiments on CIFAR-100 [7] and CIFAR-10 [7] datasets and compare with other networks. We use the pre-trained resnet32 × 4 [15], vgg13 [11], wrn-40-2 [18], resnet152 [3], resnet34 [3] as our teacher models, the pre-trained resnet32 × 4, vgg13, wrn-40-2 teacher model are publicly available from CRD [12] and resnet152, resnet34 teacher model is trained by ourselves.

4.1 Datasets and Baselines

We validated our distillation method on CIFAR-100 [7] and CIFAR-10 [7]. Besides the vanilla KD [5], various approaches are reproduced for comparison, including FitNet [10], AT [17], VID [1], PKT [9], SP [13], CRD [12]. To ensure the fairness of the experiments, we use the same data sampling and data augmentation methods as CRD [12], and such as learning rate, epoch and other hyperparameters, are also the same as CRD [12].

4.2 Implement Details

For all experiments on CIFAR-100 and CIFAR-10 datasets, since the image size is 32 × 32, in order to make the model learn better feature maps on low-resolution images, we use a random crop with padding=4, random horizontal flip and the standard data augmentation and normalize all images by channel means and standard deviations. We still use the SGD optimizer; the weight decay is set to 5e–4, and momentum is set to 0.9. The initial learning rate is set to 0.05, and the decay rate for the learning rate is 0.1, and when epoch = 150, 180, 210, the learning rate decays. The entire model is trained with 240 epochs. All initial settings above are the same with CRD [12]. In addition, for ourselves design, the weight of $CE(s,y)$, $\alpha = 1$, the weight of $GKD(t,s)$, $\beta = 5$, the weight of $FPD(t,s)$, $\gamma = 20$. In order to accommodate small networks with few layers, we use the feature maps of the first four stages of the model to compute uniformly. For the feature maps of the first four stages passing through the feature pyramid, we first use the convolution operation with padding =1, kernel_size = 3 to shift the number of feature map channels to 32, 64, 128, 256. If the input is smaller than the set number of channels, and if it is larger than the set number of channels, we use the original number of channels and uniformly expand the channels to 256 in the upsampling process of the feature pyramid. According to the previous upsampling experience, we use the "bilinear" interpolation method in the upsampling process.

4.3 Comparison of Test Accuracy

From the Table 1, we can see that our method consistently outperforms all other knowledge distillation methods on CIFAR-100 and the improvements are pretty significant, and the student model perform better than the teacher such as "ResNet50 & ResNet18". Especially, the performance of ResNet18 [3] even better than ResNet152 [3]. It is shown that our method can help the student to learn more knowledge from the teacher, and can rectify wrong information, making the student more robust. In Table 2, we can see that performance of the contrastive distillation is poor in small-scale datasets. However, our method outperforms in small-scale datasets.

Table 1. Results on the **CIFAR-100** validation. All results are the average over 3 trials.

Distillation	Teacher	wrn-40-2	resnet32 × 4	vgg13	resnet152	resnet34
	Acc.	75.61	79.42	74.64	79.90	79.12
Mechanism	Student	wrn-16-2	resnet8 × 4	vgg8	resnet50	resnet18
	Acc.	73.26	72.50	70.36	79.26	78.36
Logits	KD [5]	74.92	73.33	72.98	80.32	79.53
Intermediate Layer	FitNet [10]	73.58	73.50	71.02	79.85	78.22
	AT [17]	74.08	73.44	71.43	80.46	78.69
	VID [1]	74.11	73.09	71.23	79.12	79.09
	PKT [9]	74.54	73.64	72.88	80.56	79.81
	SP [13]	73.83	72.94	72.68	80.42	79.90
Contrastive Learning	CRD [12]	**75.48**	**75.51**	**73.94**	**80.62**	**79.72**
Ours	FPD	**75.67**	**75.71**	**74.23**	**80.79** (↑)	**79.99** (↑)

Table 2. Results on the **CIFAR-10** validation. All results are the average over 3 trials.

Distillation	Teacher	wrn-40-2	resnet32 × 4	vgg13	resnet34
	Acc.	94.73	95.52	94.05	95.47
Mechanism	Student	wrn-16-2	resnet8 × 4	vgg8	resnet18
	Acc.	93.83	92.61	91.95	95.13
Logits	KD [5]	94.66	93.80	92.95	95.36
Intermediate Layer	VID [1]	94.02	93.10	92.49	95.26
Contrastive Learning	CRD [12]	**88.41** (↓)	**88.89** (↓)	**84.75** (↓)	**89.56** (↓)
Ours	FPD	**94.70**	**94.16**	**93.13**	**95.78** (↑)

4.4   Ablation Study

In this section, we use two ablation experiments to verify that student performance will be better during knowledge distillation by using our methods. Meanwhile, verifying the feature map of different levels fuse with feature pyramids will help the student to learn more knowledge from the teacher during training. The feature pyramid magnifies the error when prediction of the teacher is wrong and can affect students performance. Using our methods can narrow the gap in feature maps between teachers and students, reinforce the correct information learned by students, and help students rectify the wrong information learned from the teacher. When we only using FPD, the performance of all student network perform better. It shows that our proposed FPD makes the student learn more knowledge from the teacher. Next, we add guided knowledge distillation as an auxiliary loss function for our FPD, helping students learn more positive knowledge and improve performance. Results of ablation study see in Table 3.

Table 3. Results of ablation study on the **CIFAR-100** validation. All results are the average over 3 trials.

Distillation	Teacher	wrn-40-2	resnet32 × 4	vgg13	resnet34
	Acc.	75.61	79.42	74.64	79.12
Mechanism	Student	wrn-16-2	resnet8 × 4	vgg8	resnet18
	Acc.	73.26	72.50	70.36	78.36
Logits	KD [5]	74.92	73.33	72.98	79.53
Intermediate Layer	VID [1]	74.11	73.09	71.23	79.09
	SP [13]	73.83	72.94	72.68	79.90
Contrastive Learning	CRD [12]	**75.48**	**75.51**	**73.94**	**79.72**
	FPD	74.62	73.94	72.65	78.95
Ours	FPD+GKD	**75.76**	**75.71**	**74.23**	**79.99** (↑)

5   Conclusion

We propose feature pyramid knowledge distillation (named FPD): a novel knowledge distillation method to narrow the gap of feature maps between teachers and students to help get more information. And, the effectiveness and superiority of our method was demonstrated through experiments. Moreover, by using our FPD method allows students to learn both semantic information and texture details during training, which seems to serve as a novel method to learn both the teacher's predicted probability distribution and the teacher's intermediate feature map representations for students during training.

Acknowledgements. This research is supported by the Sichuan Science and Technology Program(No.2020YFS0307), Mianyang Science and Technology Program(2020YFZJ016), Sichuan Provincial M. C. Integration Office Program, and IEDA laboratory of SWUST.

References

1. Ahn, S., Hu, S.X., Damianou, A., Lawrence, N.D., Dai, Z.: Variational information distillation for knowledge transfer. In: Proceedings of the IEEE/CVF Conference on Computer Vision and Pattern Recognition, pp. 9163–9171 (2019)
2. Chen, L.C., Papandreou, G., Kokkinos, I., Murphy, K., Yuille, A.L.: DeepLab: semantic image segmentation with deep convolutional nets, atrous convolution, and fully connected CRFs. IEEE Trans. Pattern Anal. Mach. Intell. **40**(4), 834–848 (2017)
3. He, K., Zhang, X., Ren, S., Sun, J.: Deep residual learning for image recognition. In: Proceedings of the IEEE Conference on Computer Vision and Pattern Recognition, pp. 770–778 (2016)
4. Heo, B., Lee, M., Yun, S., Choi, J.Y.: Knowledge transfer via distillation of activation boundaries formed by hidden neurons. In: Proceedings of the AAAI Conference on Artificial Intelligence, vol. 33, pp. 3779–3787 (2019)
5. Hinton, G., Vinyals, O., Dean, J., et al.: Distilling the knowledge in a neural network. arXiv preprint arXiv:1503.02531 (2015)
6. Hu, J., Shen, L., Sun, G.: Squeeze-and-excitation networks. In: Proceedings of the IEEE Conference on Computer Vision and Pattern Recognition, pp. 7132–7141 (2018)
7. Krizhevsky, A., Hinton, G., et al.: Learning multiple layers of features from tiny images (2009)
8. Lin, T.Y., Dollár, P., Girshick, R., He, K., Hariharan, B., Belongie, S.: Feature pyramid networks for object detection. In: Proceedings of the IEEE Conference on Computer Vision and Pattern Recognition, pp. 2117–2125 (2017)
9. Passalis, N., Tefas, A.: Probabilistic knowledge transfer for deep representation learning. CoRR, abs/1803.10837 **1**(2), 5 (2018)
10. Romero, A., Ballas, N., Kahou, S.E., Chassang, A., Gatta, C., Bengio, Y.: Fitnets: Hints for thin deep nets. arXiv preprint arXiv:1412.6550 (2014)
11. Simonyan, K., Zisserman, A.: Very deep convolutional networks for large-scale image recognition. arXiv preprint arXiv:1409.1556 (2014)
12. Tian, Y., Krishnan, D., Isola, P.: Contrastive representation distillation. arXiv preprint arXiv:1910.10699 (2019)
13. Tung, F., Mori, G.: Similarity-preserving knowledge distillation. In: Proceedings of the IEEE/CVF International Conference on Computer Vision, pp. 1365–1374 (2019)
14. Woo, S., Park, J., Lee, J.-Y., Kweon, I.S.: CBAM: convolutional block attention module. In: Ferrari, V., Hebert, M., Sminchisescu, C., Weiss, Y. (eds.) ECCV 2018. LNCS, vol. 11211, pp. 3–19. Springer, Cham (2018). https://doi.org/10.1007/978-3-030-01234-2_1
15. Xie, S., Girshick, R., Dollár, P., Tu, Z., He, K.: Aggregated residual transformations for deep neural networks. In: Proceedings of the IEEE Conference on Computer Vision and Pattern Recognition, pp. 1492–1500 (2017)

16. Yim, J., Joo, D., Bae, J., Kim, J.: A gift from knowledge distillation: fast optimization, network minimization and transfer learning. In: Proceedings of the IEEE
Conference on Computer Vision and Pattern Recognition, pp. 4133–4141 (2017)
17. Zagoruyko, S., Komodakis, N.: Paying more attention to attention: Improving the
performance of convolutional neural networks via attention transfer. arXiv preprint
arXiv:1612.03928 (2016)
18. Zagoruyko, S., Komodakis, N.: Wide residual networks. arXiv preprint
arXiv:1605.07146 (2016)
19. Zhou, Z., Zhuge, C., Guan, X., Liu, W.: Channel distillation: channel-wise attention
for knowledge distillation. arXiv preprint arXiv:2006.01683 (2020)

An Effective Ensemble Model Related to Incremental Learning in Neural Machine Translation

Pumeng Shi[✉]

The University of British Columbia, Vancouver, BC V6T1Z4, Canada
zodiac12@students.cs.ubc.ca

Abstract. In recent years, machine translation has made great progress with the rapid development of deep learning. However, there still exists a problem of catastrophic forgetting in the field of neural machine translation, namely, a decrease in overall performance will happen when training with new data added incrementally. Many methods related to incremental learning have been proposed to solve this problem in the tasks of computer vision, but few for machine translation. In this paper, firstly, several prevailing methods relevant to incremental learning are applied into the task of machine translation, then we proposed an ensemble model to deal with the problem of catastrophic forgetting, at last, some important and authoritative metrics are used to evaluate the model performances in our experiments. The results can prove that the incremental learning is also effective in the task of neural machine translation, and the ensemble model we put forward is also capable of improving the model performance to some extent.

Keywords: ensemble model · incremental learning · neural machine translation

1 Introduction

In recent years, with the rapid development of artificial intelligence and computational systems, deep learning models have achieved great performances in multiple tasks, such as computer vision, recommendation system and natural language processing. However, this kind of models based on deep neural networks have to face the problem of catastrophic forgetting or catastrophic interference [1], which means that the performance of a model with previously learned knowledge on a task or domain tends to decrease significantly as the new tasks or domains are added. In terms of biological systems, human beings can store learned knowledge effectively in diverse continuous learning tasks, and use previous knowledge to better understand new information. But it is difficult for existing deep learning models to learn the information from a new task without dramatically forgetting previously acquired knowledge at the same time. Therefore, the phenomenon of catastrophic forgetting is regarded as an inevitable feature of deep neural networks to some extent.

M. Tanveer et al. (Eds.): ICONIP 2022, LNCS 13623, pp. 112–127, 2023.
https://doi.org/10.1007/978-3-031-30105-6_10

In real-world application, it is extremely complicated to retrain the model from scratch by using the data over different tasks, although training deep learning models with the whole data is capable of achieving better performances. In most instances, data is usually obtained in the form of continuous streams, and sometimes a few data may be only available temporarily according to privacy problems or data loss [2]. Besides, a large amount of data poses a severe challenge to the storage systems, and it also takes more time to carry out the training process for the model. Given that the deep learning models are prone to forget previously learned information significantly when learning a new task, it has been a research hotspot and emphasis to make sure the model can accumulate knowledge continually from different tasks over time without losing the original information obviously.

Incremental learning, which is also referred to as lifelong learning [3], continual learning [4] or sequential learning [1], is a learning paradigm that makes the model to continually learn over time from dynamic data distributions of multiple tasks, while alleviating the phenomenon of catastrophic forgetting. In order to deal with the issue of catastrophic forgetting, the model needs to have a strong ability to continually acquire new knowledge and refine the learned multiple knowledge, in addition, it also tends to be vital for the model to prevent existing important knowledge from being forgotten, which is known as the stability-plasticity dilemma [5].

Nowadays, more and more researchers have paid attention to the incremental learning and its practical applications. The majority of existing works focus on the tasks related to computer vision. Progressive Neural Network [6] is a typical model-growth based method that retains a part of pretrained models throughout training, and uses lateral connections between all existing models to leverage prior knowledge of previous features while acquiring new information from a new task. Similarly, PackNet [7] was put forward to identify redundant parameters in the model and utilize them to train the model on the new tasks due to a specific pruning technique. Roy et al. [8] proposed a novel model named Tree-CNN, which is mainly consisted of multiple nodes connected in a tree-like manner, and each node is a hierarchical deep convolutional neural network. The Tree-CNN model can simplify the training process with the new tasks and maintains competitive accuracy in real applications. In addition, rehearsal methods are also the essential approaches in solving the problem of catastrophic forgetting. Rebuffi et al. [9] put forward an influential model called iCaRL on incremental learning, which aims to select and store a subset of representative samples per class, and only these training data for a small amount of classes are used at the same time, while new classes will be added progressively. Besides, as proposed in GEM [10], the model update of a new task will not interfere with the learned knowledge of previous tasks, which can be achieved through projecting the estimated gradient direction in a feasible area outlined by the gradients of previous tasks through a first order Taylor series approximation.

However, there are relatively few research works on incremental learning in the field of natural language processing. Recently, several approaches have been proposed or applied to mitigate the issue of catastrophic forgetting for multiple NLP tasks. Wang et al. [11] introduced an embedding alignment technique to implement continual learning for the task of relation extraction, which tends to align the sentence embeddings using anchoring to avoid the problem of embedding vector space distortion when the model is trained on a new task. Chen et al. [12] proved that the incremental learning methods are capable of improving traditional sentiment analysis models by adaptively adding extra information of new domains. In addition, Monaikul et al. [13] applied incremental learning into the application of named entity recognition, here a novel knowledge distillation framework was presented to preserve the required knowledge previously learned by the model, and the refined model learned new entity types effectively by utilizing new training materials at the same time.

Machine translation is regarded as an indispensable and significant task in the field of natural language processing, which has been widely applied into a large number of scenarios. In terms of neural machine translation, some researchers have tried to explore relevant incremental learning methods and made some desirable progress. Thompson et al. [14] used a special kind of elastic weight consolidation method to carry out the continued training in neural machine translation, and proved that the model can translate domain-specific sentences without forgetting how to translate the sentences in the previous domains. Khayrallah et al. [15] argued that some fine-tuning based approaches are easily inclined to cause the phenomenon of catastrophic forgetting, and proposed a regularized training objective to increase the model ability to learn the out-of-domain knowledge. Recently, Cao et al. [16] introduced a new incremental learning framework for NMT models, which is mainly consisted of a dynamic knowledge distillation-based module and a bias-correction module. The former is used to alleviate the catastrophic forgetting while the latter tends to deal with the problem of biased weights.

The works shown above can mitigate the issue of catastrophic forgetting to some extent in the field of computer vision and nature language processing, but there exist several apparent differences between these two kinds of tasks. When it comes to neural machine translation, it is difficult to apply the rehearsal methods into real-world scenarios due to its limited scalability. The storage and computation cost of source data for each task tends to become intractable with an ever increasing amount of tasks, although the samples are selected partially from all the tasks. In addition, regularization-based methods are also beneficial to implement the incremental learning in neural machine translation, however, it needs to take more time and space to compute the regularization when a great deal of parameters are involved in the model. It is worth noting that the majority of research works pay attention to the single model with the continual learning methods, few researchers conduct in-depth study on the ensemble model. Many experiments in other research areas have proved that ensemble models usually achieve better performances than a single model. In terms of metrics in machine translation, the BLEU score has been the widely accepted metric to evaluate the performances of NMT models, but sometimes other authoritative metrics are needed to make the experimental results more convincing.

Here, the main contributions of this paper are as follows:

- Two kinds of typical regularization-based methods, namely, EWC [17] and R-EWC [18], are applied into the task of neural machine translation to prove the effectiveness and suitability of incremental learning approaches, both in the in-domain incremental training and multi-domain incremental learning.
- An effective ensemble model related to incremental learning is proposed in this paper, which is able to take advantage of the learned knowledge in the previous tasks, and also can obtain the added information in the new tasks or domains from the model according to the specific circumstances.
- In order to make the experimental results become more persuasive, the BLEU score and the TER score are utilized here to evaluate the model performances.

2 Related Work

2.1 Neural Machine Translation

As an important field of natural language processing, machine translation aims to translate a text from one natural language to another by a computing device, which is committed to providing high quality translation between different pairs of languages. With regard to the methodology, machine translation mainly falls into three fundamental categories, namely, rule-based machine translation, statistical machine translation (SMT) and neural machine translation (NMT). In recent years, due to the rapid progress of deep neural networks, neural machine translation gradually plays a vital role in the field of modern machine translation.

The traditional NMT model depends on a typical sequence-to-sequence structure, which is mainly composed of two essential components, namely, an encoder network maps the source sentence into a real-valued vector, and a decode network produces the translation result according to this vector. However, the emergence of attention mechanism has brought the development and application of machine translation into a new era, laying a solid foundation for its future research and progress. Bahdanau et al. [19] firstly introduced the attention mechanism into neural machine translation, which is also regarded as soft attention. The model based on LSTM module with attention mechanism [20] has been widely used in many applications until the appearance of Transformer architecture. Vaswani et al. [21] proposed a novel and influential model structure named Transformer, which is built only on self-attention mechanism without any convolutional or recurrent module. Compared with traditional model structures, Transformer shows stronger parallelization and representation ability to capture the potential features in the sentences, and achieved superb performances on machine translation.

Therefore, in this paper, the model in the experiments is based on the Transformer architecture proposed in [21].

2.2   Incremental Learning Methods

The majority of existing methods related to incremental learning can be divided into three categories depending on how the specific information of a task is stored and utilized throughout the continual learning process:

- Replay-based methods
- Regularization-based methods
- Parameter isolation-based methods

Replay-based methods tend to make use of the data samples in previous tasks while learning a new task to mitigate the phenomenon of catastrophic forgetting, which are mainly consisted of rehearsal methods and pseudo-rehearsal methods. Rehearsal methods usually retain some valuable training data from prior tasks, while these examples are generated by pseudo-rehearsal models instead of selecting from source data in pseudo-rehearsal methods. Given that the added training data from new tasks are prone to increase the storage consumption and time cost, replay-based methods are rarely applied into the tasks of natural language processing.

Regularization-based methods aim to alleviate the issue of catastrophic forgetting by adding an extra regularization term in the loss function or taking advantage of knowledge distillation-based approaches, which can preserve the important learned information from original tasks to some extent. Compared with other methods, regularization-based methods are more simple and more feasible in a variety of applications. Here two kinds of typical regularization-based approaches are utilized in this paper.

Parameter isolation-based methods refer to those techniques used to modify model architectures or adjust model parameters in the process of dealing with a new task. Sometimes the model size can be expanded in learning new knowledge, even making a complete copy of the model in some extreme cases. Therefore, the parameter isolation-based methods are usually suitable for the scenarios that model size tends to be small and the number of new tasks is limited.

3   Model

Here the EWC method and R-EWC method are introduced in detail respectively, the Transformer model applied in the experiments is also presented, including scaled dot-product attention and multi-head attention mechanism. In addition, the ensemble model related to incremental learning in neural machine translation is elaborated in this section.

3.1   EWC

EWC (Elastic Weight Consolidation) [17] is a vital regularization-based method in incremental learning, which tends to consolidate previous knowledge by constraining significant parameters to stay close to the prior values. This approach introduces an extra quadratic penalty item into the loss function to slow down the learning on task-related weights represented for old tasks.

Due to the Bayes theory, given some data $\mathcal{D}$, the conditional probability $p(\theta|\mathcal{D})$ can be obtained based on the prior probability of the parameters $p(\theta)$ and the probability of these data $p(\mathcal{D}|\theta)$:

$$logp(\theta|\mathcal{D}) = logp(\mathcal{D}|\theta) + logp(\theta) - logp(\mathcal{D}) \tag{1}$$

Assuming a scenario with two independent tasks $A(\mathcal{D}_A)$ and $B(\mathcal{D}_B)$, then the log value of the posterior probability of the parameters given the entire dataset can be computed as follows:

$$logp(\theta|\mathcal{D}) = logp(\mathcal{D}_B|\theta) + logp(\theta|\mathcal{D}_A) - logp(\mathcal{D}_B) \tag{2}$$

where $logp(\mathcal{D}_B|\theta)$ represents the loss function for task B, and $p(\theta|\mathcal{D}_A)$ is the posterior probability for task A, which contains all the essential information about task A. Due to the fact that the true posterior probability is intractable, here EWC method approximates it as a Gaussian distribution with mean given by the parameters θ_A^* and a diagonal precision matrix based on the diagonal of the Fisher information matrix F. The modified loss function is:

$$\mathcal{L}(\theta) = \mathcal{L}_B(\theta) + \sum_i \frac{\lambda}{2} F_i \left(\theta_i - \theta_{A,i}^*\right)^2 \tag{3}$$

where $\mathcal{L}_B(\theta)$ stands for the loss for task B, and λ represents the importance between the previous task and a new one, i denotes each parameter in the model.

3.2   R-EWC

R-EWC [18], which is short for Rotated Elastic Weight Consolidation, is an elegant method in solving the problem of catastrophic forgetting. In essence, the R-EWC approach is an effective improvement of the typical EWC method. EWC tends to use the Fisher Information Matrix (FIM) to identify directions in feature space related to the important learned knowledge, and assumes that the Fisher Information Matrix can be always diagonal. However, If the FIM is not diagonal, EWC may fail to achieve the desirable performances in some specific applications.

Here, R-EWC aims to find a reparameterization of the parameter space θ and receive an appropriate FIM in this space. An indirect rotation technique is also proposed to obtain the rotated weight matrix in the model.

Assuming a scenario with a linear model $y = Wx$, input $x \in R^{d_1}$, output $y \in R^{d_2}$, and the weight matrix $W \in R^{d_2 \times d_1}$. Here the parameter space λ equals to W, the FIM in this simple linear case can be calculated as follows:

$$F_W = \mathbb{E}_{\substack{x \sim \pi \\ y \sim p}} \left[\left(\frac{\partial L}{\partial y}\frac{\partial y}{\partial W}\right) \left(\frac{\partial L}{\partial y}\frac{\partial y}{\partial W}\right)^T \right] \tag{4}$$

$$= \mathbb{E}_{p \sim \pi} \left[\left(\frac{\partial L}{\partial y} \right) x x^T \left(\frac{\partial L}{\partial y} \right)^T \right] \tag{5}$$

In general, $\partial L / \partial y$ and x are regarded as unrelated random variables, therefore the equation (5) can be factorized as follows:

$$F_W = \mathbb{E}_{\substack{x \sim \pi \\ y \sim p}} \left[\left(\frac{\partial L}{\partial y} \right) \left(\frac{\partial L}{\partial y} \right)^T \right] \mathbb{E}_{x \sim \pi} \left[x x^T \right] \tag{6}$$

which means that the FIM is decided by two basic factors, namely, the input x and the backpropagated gradient at the output $\partial L / \partial y$. Due to the SVD decompositions, the rotation matrices U_1 and U_2 are shown as below:

$$\mathbb{E}_{x \sim \pi} \left[x x^T \right] = U_1 S_1 V_1^T \tag{7}$$

$$\mathbb{E}_{\substack{x \sim \pi \\ y \sim p}} \left[\left(\frac{\partial L}{\partial y} \right) \left(\frac{\partial L}{\partial y} \right)^T \right] = U_2 S_2 V_2^T \tag{8}$$

The new rotated weight matrix W^* is computed then:

$$W^* = U_2^T W U_1^T \tag{9}$$

which will be utilized in the process of approximating the FIM in the model.

The R-EWC approach is capable of improving the model performances on incremental learning by refining the Fisher Information Matrix, which has been a common regularization-based method applied in a variety of scenarios.

3.3 Multi-head Attention

The Transformer architecture has been the mainstream model in modern neural machine translation, which uses stacked self-attention and fully connected layers for both encoders and decoders. The encoder module is consisted of 6 identical layers, and each layer owns two sub-layers. The first sub-layer is the multi-head self-attention mechanism, and the second one is a common, position-wise fully connected feed-forward network. In terms of the decoder module, it is also composed of 6 same layers, in addition to the multi-head self-attention mechanism sub-layer and the simple fully connected feed-forward network, the decoder has its unique third sub-layer, which is called as masked multi-head self-attention mechanism [21].

Here the self-attention is calculated by a scaled dot-product method, the input consists of three essential parts, namely, queries, keys and values, and the dot products of the query with all keys are calculated, then the softmax function is applied to get the weights on the values:

$$Attention\,(Q, K, V) = softmax\left(\frac{QK^T}{\sqrt{d}}\right)V \qquad (10)$$

where Q, K and V stand for the query, key, and value vectors respectively, K^T is the transpose of matrix K, and d is a scaling factor.

In order to make the model to jointly attend to all kinds of information from different representation subspaces at different positions, the multi-head attention mechanism is also introduced, which can be described as follows:

$$MultiHead\,(Q, K, V) = Concat\,(head_1, \ldots, head_h)\,W^O \qquad (11)$$

$$head_i = Attention\left(QW_i^Q, KW_i^K, VW_i^V\right) \qquad (12)$$

where h denotes the number of heads, W^O, W_i^Q, W_i^K and W_i^V are parameter matrices, and $head_i$ represents different self-attention spaces. The potential information involved in all these heads will be concatenated at last.

3.4   Ensemble Model

Ensemble learning has been extensively proved effective in a great deal of applications. In essence, ensemble learning methods tend to reduce the variance and bias of prediction results by integrating the decision results of multiple learners [22].

Given that a single Transformer model is sensitive to the fluctuation of hyperparameters, such as the number of layers, the dimension of word embeddings and hidden states [23], the translation result generated by a single model may not be the appropriate answer. In addition, when it refers to the incremental learning, the regularization-based methods can be used to mitigate the phenomenon of catastrophic forgetting, however, the model still drops part of the previous information when learning the new knowledge.

Therefore, the ensemble learning technique is applied into the task of neural machine translation here to achieve a better model performance than a single model.

The ensemble model proposed here combines the original model and the incremental learning model, which can be obtained through three phases:

1. **Model parameters averaging**: In the process of model training, the model parameters usually are saved at regular intervals. In order to get a more robust model, A new parameter matrix is generated by averaging the values in the corresponding position in each candidate matrix, which can produce an optimized original model trained on the previous task.
2. **Incremental model generation**: Based on the optimized original model produced in the first phase, the regularization-based methods are utilized to train a model with the added data from a new task or a specific domain.

3. **Prediction results fusion**: Due to the fact that the original model on the previous task has the same neural network structure as the incremental learning model, the final translation result can be depended on both these two models. With regard to the prediction results based on the ensembled models, the common applied methods usually include arithmetic average, geometric average, weighted average and voting. In the process of model inference here, the probability distribution of the word which will be produced is computed by averaging the results of source model and incremental learning model.

The procedures of ensemble model is shown in Fig. 1 as below.

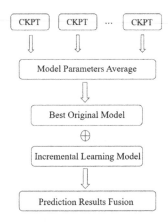

Fig. 1. The general procedure of ensemble model

In general, the ensemble model can take advantage of the merits of the original model and the model on incremental learning, which is more likely to achieve better translation result in real-world applications.

4   Experiments

In this section, more details about the experiments are introduced here, including the experimental setting, data preparation, implementation methods and the evaluation metrics in neural machine translation.

4.1   Experimental Setting

In terms of training scenarios in neural machine translation, in-domain training and multi-domain training are the two most representative tasks, which will be carried out in our experiments to explore the results of incremental learning.

1. **In-domain incremental training:** In-domain training means that the training data comes from the same source or a specialized domain, when it is involved in the field of incremental learning, here the entire training data is split into several subsets, some of them are selected as the validation and test sets while the rest are used to train the model. The in-domain incremental learning tends to be simpler compared with other scenarios.
2. **Multi-domain incremental training:** With regard to the multi-domain training, the training dataset comes not only from one domain, but also from a variety of other fields. The model is first trained on a general-domain corpus, then the incremental data from other domains will be added to fine-tune this model. Correspondingly, the validation and test sets are consisted of the data samples in the previous task, as well as the data from a new domain.

4.2   Data Preparation

In this paper, all the experiments are based on the task of German-English translation. In order to facilitate the parameter adjustment and problems solving in the experiments, and make the experimental results comparable, some widely used dataset are chosen here to carry out the model training and testing.

Here the WMT 2015 de-en training data [24] is selected as the in-domain incremental learning data, which is consisted of 3.9M parallel sentences. The data of OPUS multi-domain dataset [25] released by Aharoni and Goldberg is used to implement the multi-domain incremental training. The part of OPUS multi-domain dataset is involved in a variety of categories, such as Koran, medical and law. Three important categories, namely, medical, IT and law, are selected in the task of multi-domain incremental learning, and 3000 data samples are chosen randomly as the validation and test dataset while the rest is utilized for model training.

4.3   Implementation Methods

1. *Fine-tuning:* The neural machine translation model is fine-tuned directly based on the incremental data of specific tasks.
2. *Combined:* The entire datasets, including the previous data and the new data, are simply employed to train the model. The result of this combined approach tends to be regarded as the upper bound of the tasks relevant to incremental learning.
3. *EWC:* The model is trained with an optimized loss function, which is a weighted result of the standard NLL loss and an extra EWC term.
4. *R-EWC:* The model is trained based on the R-EWC approach, the improved parameters in the model can be helpful in storing the existing knowledge.
5. *Ensemble Model:* The ensemble model, which aims to integrate the important previous knowledge with the refined new information, is applied to generate translation results. In detail, the desirable model trained on the previous task is obtained by ensembling several different saved models, and the incremental

learning model, which is trained on specific tasks, will be added to produce the appropriate translation results in the process of prediction results fusion.

4.4 Evaluation Metrics

Recently, many metrics have been proposed to evaluate the model performances in neural machine translation, and each one owns its advantages and drawbacks. In this paper, two kinds of widely-used evaluation metrics, including BLEU score [26] and TER score [28], are applied into our experiments.

- **BLEU** (Bilingual Evaluation Understudy) BLEU was first put forward by IBM research team in 2002, which was utilized to implement the automatic evaluation for machine translation. The BLEU metric ranges from 0 to 1, and the higher score usually represents the better translation performance of the model. The BLEU score can be computed as follow:

$$BP = 1 \qquad if \ c > r \tag{13}$$

$$BP = e^{(1-r)/c} \qquad if \ c \le r \tag{14}$$

$$BLEU = BP \bullet exp\left(\sum_{n=1}^{N} w_n log p_n\right) \tag{15}$$

where BP is the penalty item, p_n denotes the geometric average of the modified n-gram precisions, and w_n is the positive weights. In addition, c represents the length of the candidate translation sentence, which is usually generated by the model, and r stands for the length of effective reference corpus.
- **TER** (Translation Edit Rate) As a novel metric in the field of machine translation, TER was introduced by Snover et al. in 2006. The main purpose of TER is to measure the number of editing that a human would have to modify the model output to make sure it exactly matches a reference translation. The TER score is calculated as below:

$$TER = \frac{\# \ of \ edits}{average \ \# \ of \ reference \ words} \tag{16}$$

Briefly, The TER is defined as the minimum amount of edits to the closest reference, normalized by the average length of the references.

5   Results and Analysis

Our experiments are carried out on a Nvidia 2080Ti GPU with 4 CPU. The model structure employed here is the vanilla Transformer, which has been widely applied in a great deal of tasks related to NLP. In addition, the experiments are all based on the OpenNMT [28], which is an open source ecosystem for neural machine translation and neural sequence learning.

The procedure of data preprocessing is implemented first to improve the dataset quality, including removing useless symbols and error codes, length ratio filtering for the bilingual corpus and punctuation conversion.

Then the subword-nmt is used here to carry out the word tokenization for the datasets, which is based on the BPE algorithm. The vocabulary size of BPE word tokenization is 32K.

In terms of model training, the main parameters of these Transformer model in the experiments are shown in the Table 1.

Table 1. Model Parameters

Model Parameters	Value
Heads	8
Encoders/Decoders	6
word vector size	768
Initial learning rate	768
Dropout	0.3
Label smoothing	0.1
Batch size	2048
Batch type	tokens
Optimizer	Adam
Early stopping	10

Here the early stopping method is used to help to find out the best model in the training process and avoid the overfitting at the same time. This best model and its two adjacent saved models will be selected to implement the model parameters averaging in the phase of model ensemble. Then the incremental learning model can be obtained by the similar training process according to the special task. At last, the translation results will be produced based on these two models in the process of prediction results fusion.

The BLEU scores of different models under in-domain incremental training setting on the WMT 15 dataset are shown in Table 2.

Table 2. BLEU Scores on In-Domain Incremental Learning

WMT 15	60%	80%	100%
Combined	23.39	24.83	26.52
Fine-tuning	21.55	22.76	24.31
EWC	21.93	23.21	24.63
R-EWC	22.05	24.33	24.70
Ensemble Model	**22.09**	**24.42**	**24.78**

According to Table 2, it is obvious that the combined method has the highest BLEU score regardless of the amount of data, and there exists an apparent decrease in BLEU score in the fine-tuning approach. In addition, the EWC and R-EWC methods tend to improve the model performances to some extent. It is worth noting that the ensemble model proposed in this work owns a higher BLEU score than other techniques related to incremental learning, proving that the ensemble model is more preferable in solving the problem of catastrophic forgetting.

When it comes to multi-domain incremental learning, the BLEU scores of different models under multi-domain incremental training setting with different dataset is shown in Table 3 as below.

Table 3. BLEU Scores on Multi-Domain Incremental Learning

Methods	+Medical	+IT	+Law
Fine-tuning	52.86	42.10	56.82
EWC	51.92	41.56	55.79
R-EWC	51.94	41.49	55.87
Ensemble Model	52.38	**42.17**	55.90

We can find that the results in Table 3 is slightly different from the ones in in-domain incremental learning. Compared with the fine-tuning method, here EWC and R-EWC methods tend to be more difficult to retain the previous knowledge in the model. However, the ensemble model can outperform these two approaches in facing with the task of multi-domain incremental learning.

In addition, the model performances are also relevant with the specific domains or areas. Given that the corpus of medical and law fields has specialized sentences and words, the model trained from these data are prone to have a better translation performance.

In order to make the experiment results more convincing and persuasive, TER metric is also applied here. The TER scores of in-domain incremental learning and multi-domain incremental learning are shown in Table 4 and Table 5 respectively.

Table 4. TER Scores on In-Domain Incremental Learning

WMT 15	60%	80%	100%
Combined	59.11	56.02	51.75
Fine-tuning	63.17	60.24	56.88
EWC	61.49	58.78	55.23
R-EWC	61.25	56.91	54.62
Ensemble Model	**61.10**	**56.10**	**54.58**

Table 5. TER Scores on Multi-Domain Incremental Learning

Methods	+Medical	+IT	+Law
Fine-tuning	28.77	35.38	24.97
EWC	30.35	37.11	25.89
R-EWC	30.41	37.56	25.57
Ensemble Model	29.51	**35.26**	25.30

As shown in Table 4, similarly, the combined approach has the best results than other applied techniques. Compared with fine-tuning, EWC and R-EWC, the ensemble model also achieves the most desirable performance in in-domain incremental learning.

In terms of multi-domain incremental learning, according to the TER scores, the conclusion can be the same as in Table 3. The approaches applied in the in-domain incremental learning usually become more effective than the ones involved in the multi-domain fields.

6 Conclusion

In this paper, the EWC and R-EWC are applied into the task of neural machine translation to prove the effectiveness and suitability of incremental learning approaches, both in the in-domain incremental training and multi-domain incremental learning. Then a typical ensemble model related to incremental learning is proposed here, which aims to take advantage of the learned knowledge in the previous tasks, and also can obtain the added information in the new tasks. In addition, the BLEU score and TER score are introduced to evaluate the model performances. The experiment results can prove that the incremental learning can achieve successes in the task of neural machine translation, and the ensemble model tends to have a desirable performance in incremental learning.

References

1. McCloskey, M., Cohen, N.J.: Catastrophic interference in connectionist networks: the sequential learning problem. In: Psychology of Learning and Motivation, vol. 24, pp. 109–165. Elsevier (1989)
2. Daems, J., Macken, L.: Interactive adaptive SMT versus interactive adaptive NMT: a user experience evaluation. Mach. Transl. 1–18 (2019)
3. Chen, Z., Liu, B.: Lifelong machine learning. Synth. Lect. Artif. Intell. Mach. Lear. **12**(3), 1207 (2018)
4. Mark, B.: Ring. Continual learning in reinforcement environments. In: GMD-Bericht (1994)
5. Grossberg, S.: Studies of Mind and Brain: Neural Principles of Learning, Perception, Development, Cognition, and Motor Control. Boston Studies in the Philosophy of Science, vol. 70. Reidel, Dordrecht (1982)

6. Rusu, A.A., et al.: Progressive neural networks. CoRR (2016)
7. Mallya, A., Lazebnik, S.: Packnet: adding multiple tasks to a single network by iterative pruning. In: 2018 IEEE Conference on Computer Vision and Pattern Recognition, CVPR 2018, Salt Lake City, UT, USA, 18–22 June 2018, pp. 7765–7773 (2018)
8. Roy, D., Panda, P., Roy, K.: Tree-CNN: a hierarchical deep convolutional neural network for incremental learning. Neural Netw. **121**, 148–160 (2020)
9. Rebuffi, A., Kolesnikov, A., Sperl, G., Lampert, C.H.: ICARL: incremental classifier and representation learning. In: CVPR, pp. 2001–2010 (2017)
10. Chang, M., Gupta, A., Levine, S., Griffiths, T.L.: Automatically composing representation transformations as a means for generalization. In: ICML workshop Neural Abstract Machines and Program Induction vol. 2 (2018)
11. Wang, H., Xiong, W., Yu, M., Guo, X., Chang, S., Wang, W.Y.: Sentence embedding alignment for lifelong relation extraction. In: Proceedings of the 2019 Conference of the North American Chapter of the Association for Computational Linguistics: Human Language Technologies, Volume 1 (Long and Short Papers), pp. 796–806, Minneapolis, Minnesota. Association for Computational Linguistics (2019b)
12. Chen, Z., Liu, B.: Lifelong machine learning. Synth. Lect. Artif. Intell. Mach. Learn. **12**(3), 1–207 (2018)
13. Monaikul, N., Castellucci, G., Filice, S.: Continual learning for named entity recognition. In: Proceedings of the AAAI Conference on Artificial Intelligence, vol. 35, no. 15, pp. 13570–13577 (2021)
14. Thompson, B., Gwinnup, J., Khayrallah, H., Duh, K., Koehn, P.: Overcoming catastrophic forgetting during domain adaptation of neural machine translation. In: Proceedings of the 2019 Conference of the North American Chapter of the Association for Computational Linguistics: Human Language Technologies, NAACL-HLT 2019, Minneapolis, MN, USA, 2–7 June 2019, Volume 1 (Long and Short Papers), pp. 2062–2068. Association for Computational Linguistics (2019)
15. Khayrallah, H., Thompson, B., Duh, K., Koehn, P.: Regularized training objective for continued training for domain adaptation in neural machine translation. In: Proceedings of the 2nd Workshop on Neural Machine Translation and Generation, ACL, Melbourne, Australia, 20 July 2018, pp. 36–44. Association for Computational Linguistics (2018)
16. Cao, Y., Wei, H.R., Chen, B., et al.: Continual learning for neural machine translation. In: Proceedings of the 2021 Conference of the North American Chapter of the Association for Computational Linguistics: Human Language Technologies (2021)
17. Kirkpatrick, J., Pascanu, R., Rabinowitz, N., et al.: Overcoming catastrophic forgetting in neural networks. Proc. Natl. Acad. Sci. U S A **114**(13), 3521–3526 (2016)
18. Liu, X., Masana, M., Herranz, L., et al.: Rotate your networks: better weight consolidation and less catastrophic forgetting. IEEE (2018)
19. Bahdanau, D., Cho, K., Bengio, Y.: Neural machine translation by jointly learning to align and translate. Comput. Sci. (2014)
20. Luong, T., Pham, H., Manning, C.D.: Effective approaches to attention-based neural machine translation. In: Proceedings of the 2015 Conference on Empirical Methods in Natural Language Processing, EMNLP 2015, Lisbon, Portugal, 17–21 September 2015, pp. 1412–1421. The Association for Computational Linguistics (2015)
21. Vaswani, A., et al.: Attention is all you need. In: Proceedings of the 31st Conference on Neural Information Processing Systems (NIPS 2017), 4–9 December 2017, Long Beach, CA, USA (2017)

22. Tan, L., Li, L., Han, Y., et al.: An empirical study on ensemble learning of multi-modal machine translation. In: IEEE Sixth International Conference on Multimedia Big Data. IEEE (2020)
23. Tefánik, M., Novotn, V., Sojka, P.: Regressive ensemble for machine translation quality evaluation (2021)
24. Bojar, O., et al.: In: Proceedings of the Tenth Workshop on Statistical Machine Translation, pp. 1–46. Association for Computational Linguistics, Lisbon (2015)
25. Aharoni, R., Goldberg, Y.: Unsupervised domain clusters in pretrained language models. In: Proceedings of the 58th Annual Meeting of the Association for Computational Linguistics, ACL, pp. 7747–7763. Association for Computational Linguistics (2020)
26. Papineni, K., Roukos, S., Ward, T., Zhu, J.: Bleu: a method for automatic evaluation of machine translation. In: Proceedings of the 40th Annual Meeting on Association for Computational Linguistics, pp. 311–318. Association for Computational Linguistics (2002)
27. Snover, M., Dorr, B., Schwartz, R., Micciulla, L., Makhoul, J.: A study of translation edit rate with targeted human annotation. In: Proceedings of Association for Machine Translation in the Americas, pp. 223–231 (2006)
28. Harvard NLP group and SYSTRAN. The OpenNMT ecosystem (2016). https://opennmt.net/

Local-Global Semantic Fusion Single-shot Classification Method

Jianwei Cai, Kun Fang, Weihao Yu, Jie Yang$^{(\boxtimes)}$, and Yu Qiao$^{(\boxtimes)}$

Institute of Image Processing and Pattern Recognition, Shanghai Jiao Tong University, Shanghai, China
{caijianwei,fanghenshao,yuweihao,jieyang,qiaoyu}@sjtu.edu.cn

Abstract. In few-shot learning tasks, a series of semantic-based methods have shown excellent performance due to the modality fusion of both visual and semantic modalities. However, in single-shot learning tasks, the fused visual modality fails to comprehensively capture the class information since only one image is available. To address this issue, we propose a semantic-based single-shot method which considers from both local and global perspectives. Specifically, we fully exploit local visual features to replace the traditional image-level features in the modality fusion in those semantic-based methods. Moreover, a global classification loss is introduced to enlarge the encoding space for accurate and distinguishable local embeddings. Through a series of experiments, we show that by exploiting local features from a global classification perspective, our model boosts the performance of semantic-based approaches by a large margin on two different data sets and global classification loss is effective on both metrics.

Keywords: Single-shot learning · Modality fusion · Local feature · Global classification

1 Introduction

Deep neural networks (DNNs) have achieved tremendous success in many computer vision tasks [7] due to the availability to sufficient labeled high-quality data for training, e.g., the famous ImageNet data set [21]. However, in many real-world applications, data acquisition and annotation might be impractical or expensive for fully-supervised training, e.g., cold-start recommendation [27] and rare disease diagnosis [34]. As a consequence, Few-shot Learning (FSL) has recently drawn growing interests [13] to improve the performance of models trained with very little data.

Different from conventional machine learning, FSL aims to learn a classifier from a set of base classes with abundant labeled samples, then adapts to a set of novel classes with little labeled data [30]. Most FSL approaches [23,24,28] train a few-shot model by constructing massive few-shot tasks from the base classes

M. Tanveer et al. (Eds.): ICONIP 2022, LNCS 13623, pp. 128–139, 2023.
https://doi.org/10.1007/978-3-031-30105-6_11

sets (for training) and then perform evaluation on the set of novel classes. These few-shot tasks are usually formed as N-way K-shot in which each task consists of N classes with K labeled samples per class (the support set) and Q unlabeled query samples per class (the query set). The goal is to classify these $N \times Q$ query samples based on the $N \times K$ support samples. The core principle is to employ the same N-way K-shot tasks for training and test. In this paper, we focus on the single-shot image classification problem, i.e., $K = 1$.

Recently, lots of researchers [2,3,26] realized that it is necessary and effective to pre-train model on abundant labeled base class data. It was found [26] that training models on the whole base classes provides good embeddings for classification in novel classes since more class information is involved in the training. Aside from those methods, another type of semantic-based approach [31] in FSL introduces semantic modality to alleviate data scarcity and obtain better class prototypes for classification. However, in the single-shot setting, a single labeled image can not accurately represent its class due to its unique feature or lack of class commonality so that detailed visual information is needed.

To address these issues, we propose to exploit local features from a global classification perspective to further boost the semantic-based classification methods in single-shot settings. Specifically, we propose to fully utilize the local feature maps, i.e., visual modality, to replace the traditional image-level features in the modality fusion in those semantic-based methods. The fused semantic prototypes generated from the proposed feature maps are much more beneficial in single-shot tasks since the detailed information in the visual modality is fully excavated. Aside from the local visual modality, we further take a global perspective, which is inspired by the pre-training methods. To be specific, we propose a global classification loss based on an additional linear classifier to categorize local prototypes over the entire base classes. Such a global classification loss enlarges the encoding space so as to achieve accurate and distinguishable embeddings, which is also beneficial in few-shot tasks. In this way, by considering from both the local visual modality and the global classification loss, more informative prototypes of images are learned, which is of great significance in single-shot tasks and further boosts those semantic-based methods. Extensive empirical results show the superiority of our methods compared with semantic-based methods Adaptive Modality Mixture Mechanism (AM3, [31]) and Task-Relevant Additive Margin Loss (TRAML, [12]) in single-shot tasks by a 5% improvement on two few-shot data sets, miniImageNet [28] and tieredImageNet [20].

In summary, our contributions are summarized as follows:

- We propose to focus on local visual modality to learn more informative image prototypes in semantic-based methods on single-shot classification tasks.
- A global classification loss is introduced to enlarge the encoding space to learn accurate and distinguishable embeddings.
- Extensive empirical results illustrate the effectiveness of the proposed method in single-shot learning tasks in different data sets over other state-of-the-art methods.

2   Related Work

2.1   Few-shot Learning

Most recent approaches for few-shot image classification are based on meta-learning framework [23] or episodic-training mechanism [24,28]. The various meta-learning methods for few-shot learning can be roughly grouped into five categories. **Optimization-based methods** [5,18] aim at training models that can fast adapt to new tasks with only a few fine-tuning updates. **Memory-based methods** [15,23] use external memory bank or LSTM [8] to store meta knowledge from base classes to learn novel concepts. **Metric-based methods** are based on the idea that we can represent each class by the mean of its examples in a representation space learned by a neural network. Following this core idea, metric-based approaches usually learn a good embedding space through an episodic training procedure where a new sample from the novel classes can be easily classified by computing the distance to given labeled samples. However, when data from visual modality is limited, semantic features from text can be a powerful source of information in the context of few-shot image classification. **Semantic-based methods** like AM3 [31] combines information from two modalities, visual and semantic, for few-shot image classification to get better results. With the study of graph networks, **graph-based methods** [33] introduce graph neural networks to few-shot classification and learn how to construct a good graph structure for label propagation.

Recently, following the standard transfer learning procedure of network pre-training and fine-tuning, [2] trains a feature extractor and classifier with base class data in the pre-training stage, in fine-tuning stage, the parameters of the feature extractor are fixed while a new classifier is trained with the given labeled examples in novel classes. [4] proposed a transductive fine-tuning baseline for few-shot classification and they used information from the test data. [26] finds that training classification tasks on the whole base classes provides good embedding that can be easily classified by nearest neighbor classifier or logistic regression in novel classes. [3] combines pre-training and meta-learning: in the classification training stage, they train a classification model on all base classes and remove its last FC layer to get the encoder; then in meta-learning stage, it further optimizes a cross entropy loss with scaling cosine similarity distance of average class feature in support set and query feature.

2.2   Zero-shot Learning

Zero-shot learning (ZSL) is closely related to FSL, which does not have access to any visual information when learning new concepts. The key idea is to align the two modalities, semantic and visual. The semantic spaces in ZSL are typically word-vector based [6], text description-based [19] and attribute-based [29]. For example, [19] proposed a model that train end-to-end to align with fine-grained and category-specific content of images by training neural language models without pre-training and only consuming words and characters. In this article, we use a word embedding model to extract semantic information from each class name.

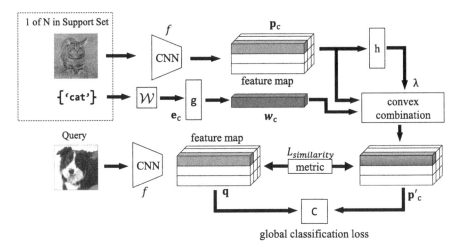

Fig. 1. Overview of Model. f: local features extractor, $\mathcal{W}$: pre-trained word-embedding model, g: semantic transformation network, h: adaptive fusion network, C: global classification FC layer. When evaluating, C is invalid.

3   Method

3.1   Problem Definition

Few-shot learning models are trained on a labeled dataset of base classes C_{base} with a large number of images and tested on a class-disjoint dataset of novel classes C_{novel}. For N-way K-shot episodes, we are given a small support set ($\mathcal{S}_e = \{(s_i, y_i)\}_{i=1}^{N \times K}$) of $N \times K$ labeled images and a query set ($\mathcal{Q}_e = \{(q_j, y_j)\}_{j=1}^{N \times Q}$) of $N \times Q$ unlabeled images for test. N indicates the number of class in $\mathcal{S}_e$ and $\mathcal{Q}_e$, K denotes the number of images of each class in $\mathcal{S}_e$, and Q denotes the number of images of each class in $\mathcal{Q}_e$. Episodic-training mechanism [24,28] constructs massive few-shot tasks from the base classes to simulate evaluation scheme on the novel classes. Focusing on single-shot tasks, i.e., $K = 1$, the support set can be simplified as $\mathcal{S}_e = \{(s_i, y_i)\}_{i=1}^{N}$.

3.2   Semantic Local Prototype

In single-shot image classification tasks, a single image can not accurately represent the characteristics of its class due to its unique feature or lack of class commonality, while high-shot tasks can use mean feature vectors to obtain accurate class representation. Semantic-based methods introduce semantic modality to alleviate data scarcity while ignoring local details from visual modality. In order to better utilize the visual details, we remove the last average pooling layer of the convolutional neural network to obtain a feature map with a large number of local features.

As shown in Fig. 1, we use a convolutional neural network (CNN) [11] $f : \mathbb{R}^{n_v} \to \mathbb{R}^{k \times n_p}$, parameterized by θ_f, to extract k n_p-dimensional local features from one image. For each class c in the support set, we use $\mathbf{p}_c$ to denote one of local features in the feature map. Following [31], we use a word-embedding model $\mathcal{W}$ (pre-trained on unsupervised large text corpora) to obtain label embeddings $\mathbf{e}_c$ of class c. To keep the semantic feature and visual feature dimensions consistent, we use a semantic transformation network to convert the label embedding $\mathbf{e}_c$ to $\mathbf{w}_c$. This transformation $g : \mathbb{R}^{n_w} \to \mathbb{R}^{n_p}$, parameterized by θ_g, is important to guarantee that both modalities lie on the space $\mathbb{R}^{n_p}$ of the same dimension and can be combined. Then we can get semantic local prototype $\mathbf{p}'_c$ as a convex combination of the transformed semantic feature $\mathbf{w}_c$ and the local visual feature $\mathbf{p}_c$:

$$\mathbf{p}'_c = \lambda \cdot \mathbf{w}_c + (1 - \lambda) \cdot \mathbf{p}_c , \qquad (1)$$

where λ is the adaptive fusion coefficient conditioned on the local visual feature and $\mathbf{w}_c = g(\mathbf{e}_c)$. The coefficient λ is calculated as follows:

$$\lambda = \frac{1}{1 + \exp\left(-h\left(\mathbf{p}_c\right)\right)} , \qquad (2)$$

where h is the adaptive fusion network, with parameters θ_h, outputs one-dimensional coefficient. Different from [31], local features are the input of the network, in order to ensure different fusion coefficients for each local feature.

For every episode e, we can get k semantic local prototypes from each class c in the support set and k local prototypes for one query image in the query set. The episodic training for single-shot classification is achieved by minimizing the negative log likelihood loss of the prediction on examples in the query set, given the support set. The metric-based episodic classification loss, also known as similarity loss, can be computed as follows:

$$\mathcal{L}_{similarity}(\theta) = \underset{(\mathcal{S}_e, \mathcal{Q}_e)}{\mathbb{E}} - \sum_{j=1}^{N \times Q} \log p_\theta\left(y_j \mid q_j, \mathcal{S}_e\right), \qquad (3)$$

where $(q_j, y_j) \in \mathcal{Q}_e$ and $\mathcal{S}_e$ are, respectively, query and support set at the same episode e and θ are model parameters. $p_\theta\left(y_j \mid q_j, \mathcal{S}_e\right)$ is the conditional probability of predicting q_j as true class y_j.

We use mean negative Euclidean distances d between each pair of semantic local prototype $\mathbf{p}'_c$ and local prototype $\mathbf{q}$ as logits for softmax [1] function to produce conditional probabilities:

$$p_\theta\left(y = c \mid q, S_e, \theta, \mathcal{W}\right) = \frac{\exp\left(\frac{1}{n^2} \sum_i^n \sum_j^n -d\left(\mathbf{q}_i, \mathbf{p}'_{cj}\right)\right)}{\sum_k^K \exp\left(\frac{1}{n^2} \sum_i^n \sum_j^n -d\left(\mathbf{q}_i, \mathbf{p}'_{kj}\right)\right)}, \qquad (4)$$

where q is one query image, n is the number of local features, $\mathcal{W}$ is the pre-trained word-embedding model.

We also follow [3] to use a scaling cosine similarity as distance metric to compute logits:

$$p_\theta \left(y = c \mid q, S_e, \theta, \mathcal{W}\right) = \frac{\exp\left(\sigma \cdot \frac{1}{n^2} \sum_i^n \sum_j^n \left\langle \mathbf{q}_i, \mathbf{p}'_{cj}\right\rangle\right)}{\sum_k^K \exp\left(\sigma \cdot \frac{1}{n^2} \sum_i^n \sum_j^n \left\langle \mathbf{q}_i, \mathbf{p}'_{kj}\right\rangle\right)}, \quad (5)$$

where σ is a learnable scalar factor, which adjusts original value range of $[-1, 1]$ to be more appropriate for logits computing, $\langle \cdot, \cdot \rangle$ is the cosine similarity of two vectors. Note that θ includes $\theta_f, \theta_h, \theta_g, \sigma$.

3.3   Global Classification Loss

Many recent works [26] find that training classification task on the whole base classes provides good embedding that can be easily classified in novel classes. To introduce the whole base classes embeddings to episodic training mechanism, we propose a global classification loss based on an extra FC layer ($C : \mathbb{R}^{n_p} \rightarrow \mathbb{R}^{N_{base}}$ in Fig. 1, N_{base} is the number of base classes) to obtain global class logits. Different from original image classification, we classify semantic local prototypes from support set and local prototypes from query set into the whole base classes for better embeddings and fusion effect. The global classification loss for support set and query set is computed as negative log likelihood loss based on the whole base classes:

$$\mathcal{L}_{support\_cls}(\theta) = -\mathbb{E} \log p_\theta \left(y \mid s\right),$$
$$\mathcal{L}_{query\_cls}(\theta) = -\mathbb{E} \log p_\theta \left(y \mid q\right), \quad (6)$$

where (s, y) and (q, y) are respectively image-label pair in support set and query set, $\theta = \{\theta_f, \theta_h, \theta_g, \theta_C\}$ are model parameters. Note that label y is global index in the whole base classes. The possibility can be calculated by mean logits through C:

$$p_\theta \left(y \mid s\right) = softmax \left(\frac{1}{n} \sum_{i=1}^n C\left(\mathbf{p}'_i\right)\right),$$
$$p_\theta \left(y \mid q\right) = softmax \left(\frac{1}{n} \sum_{i=1}^n C\left(\mathbf{q}_i\right)\right). \quad (7)$$

Finally, the total loss can be defined as:

$$\mathcal{L}\left(\theta\right) = \mathcal{L}_{similarity} + \mathcal{L}_{support\_cls} + \mathcal{L}_{query\_cls}, \quad (8)$$

which is minimized during episodic training.

In short, we fuse semantic information and local features in a distinguishable encoding space and use global classification loss further strengthen the classification ability of these semantic local prototypes. Besides, the entire training process is done in episodic training mechanism using only the base class data, which makes parameter tuning simple and avoids fine-tuning on novel class.

4   Experiments

4.1   Datasets

We conduct single-shot classification experiments on two commonly used few-shot data sets: *miniImageNet* dataset [28] and *tieredImageNet* dataset [20]. They are both a subset of *ImageNet* ILSVRC-2012 [21]. The *miniImageNet* consists of 100 classes and each class includes 600 images of size 84×84. For fair comparison with other methods, we follow [18] to split the dataset into 64 classes for meta-training, 16 classes for meta-validation and 20 classes for meta-testing, respectively. The *tieredImageNet* includes 608 classes from 34 high-level categories and each class contains more than 1000 images of size 84×84. According to [20], we split the dataset into 20,6,8 high-level categories, resulting in 351,97,160 classes for training, validation, test respectively. This construction method ensures that the base classes and novel classes are totally class-disjoint and makes the problem more realistic and challenging.

4.2   Implementation Details

Following previous works [3,9,26,31,32,35], we use a ResNet12 [7] as our visual feature extractor f. Different from methods that extract one prototype from one image, we remove the last average pooling layer of CNN and get many local prototypes instead. Following [31], we use GloVe [17] to extract the word embeddings from the class labels. The embedding transformation g and the adaptive fusion network h are both two-layer MLP with a 300-dimensional hidden layer, both of them contain ReLU non-linearity and dropout (dropout coefficient is 0.7 on miniImageNet and 0.9 on tieredImageNet). Pytorch is used to implement all our experiments on two NVIDIA 3090 GPUs. The model is trained with SGD with a momentum of 0.9 and a weight decay of $5e^{-4}$. The learning rate is initialized to 0.1 and decays by a factor of 0.1 at some epochs. During training, we adopt color jittering, random crop, random horizontal flip and normalization. Training Details are listed in Table 1.

Table 1. Training Details

Dateset	Metric	Batch size	Q	Total epochs	Learning rate decay epoch
miniImageNet	Euclidean distance	12	5	10000	6000
miniImageNet	Cosine Similarity	12	6	5000	3000
tieredImageNet	Euclidean distance	22	8	10000	6000
tieredImageNet	Cosine Similarity	16	6	10000	6000

When evaluating, we randomly construct 5000 5-way 1-shot tasks and each task contains 20 unlabeled images to be classified. Then we calculate the average accuracies with 95% confidence intervals for fair comparison.

4.3   Discussion of Results

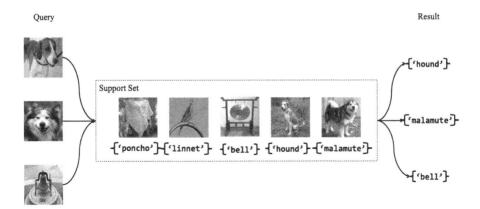

Fig. 2. Example of Classification in 5-way single-shot task.

Figure 2 illustrates an example of query image classification with our approach. There are five labeled images with class name in the support set. Three input query images are correctly classified by measuring mean distance between local prototypes of query images and semantic local prototypes from support set images. The input query images of hound and malamute are close in semantic space. Therefore semantic-based methods may not discriminate their correct class prototypes. In contrast our method fully exploits detailed information from visual modality and hence can accurately classify these two semantically similar categories. Moreover, the query bell image is quite different from the bell image in support set. But our method still distinguishes the category precisely due to good embeddings brought by global classification loss.

Table 2 presents classification accuracies of 5-way 1-shot tasks on miniImageNet and tieredImageNet. We compare our methods with existing inductive few-shot classification approaches, which are roughly divided into four types, i.e., metric-based, semantic-based, optimization-based and pre-training based methods. It can be found that our approach outperforms the state-of-the-art approaches, by around 2% on miniImageNet and 1% on tieredImageNet. The results show that Euclidean distance metric is better than cosine similarity by 2.7% on miniImageNet and 2.1% on tieredImageNet. Compared with metric-based methods, our method makes full use of semantic features to obtain better prototypes, which are vital in metric learning. The results show our method is effective with an improvement of 5.7% on miniImageNet and 1.5% on tieredImageNet. Compared with semantic-based methods, especially AM3-ProtoNets++ baseline, our method achieves better embeddings with global classification loss and pays more attention to details through local prototypes, creating a 5% boost on both data sets. As for pre-training-based methods, our method uses global classification loss to replace the pre-training stage similarly in episodic training scheme. Our method exceeds RFS-Simple by a huge margin of 10% on miniImageNet due to semantic modality. The result shows that only training a standard classification model on base classes for feature extraction is insufficient for few-shot tasks. We need to construct some few-shot tasks to simulate evaluating environment for better performance.

Table 2. Comparison to prior work on miniImageNet and tieredImageNet. Average few-shot classification accuracies (%) with 95% confidence intervals on mini-ImageNet and tieredImageNet test splits.

Method	Type	Backbone	miniImageNet 5way-1shot	tieredImageNet 5way-1shot
Matching Networks [28]	Metric	ConvNet	43.56±0.84	-
Prototypical Networks [24]	Metric	ConvNet	49.42±0.78	53.31±0.89
Relation Network [25]	Metric	ConvNet	51.38±0.82	54.48±0.93
DN4 [14]	Metric	ConvNet	51.24±0.74	-
TADAM [16]	Metric	ResNet12	58.50±0.30	-
CAN [9]	Metric	ResNet12	63.85±0.48	69.89±0.51
ConstellationNet [32]	Metric	ResNet12	64.89±0.23	-
AM3-ProtoNets++ [31]	Semantic	ResNet12	65.21±0.30	67.23±0.34
AM3-TADAM [31]	Semantic	ResNet12	65.30±0.49	69.08±0.47
TRAML-ProtoNets++ [12]	Semantic	ResNet12	60.31±0.48	-
TRAML-AM3-ProtoNets++ [12]	Semantic	ResNet12	67.10±0.52	-
MAML [5]	Optimization	ConvNet	48.70±0.84	51.67±1.81
CAML [10]	Optimization	ResNet12	59.23±0.99	-
LEO [22]	Optimization	WRN-28	61.76±0.08	66.33±0.05
Baseline++ [2]	Pre-training	ResNet18	51.87±0.77	-
MetaBaseline [3]	Pre-training	ResNet12	63.17±0.23	68.62±0.27
RFS-Simple [26]	Pre-training	ResNet12	62.02±0.63	69.74±0.72
Ours-Euclidean distance	Semantic	ResNet12	**68.58±0.34**	**70.93±0.39**
Ours-Cosine Similarity	Semantic	ResNet12	66.75±0.34	69.46±0.38

4.4   Ablation Study

We conduct an ablation study on miniImageNet and tieredImageNet to compare the impact of local feature and global classification loss on the model under two metrics, i.e., Euclidean distance and cosine similarity. Specifically, (i) we remove all components, i.e., classifying each sample by the semantic fusion prototype; (ii) we remove the last average pooling layer of feature extractor to obtain local prototypes for adaptive semantic fusion and classify each sample by computing metric between every pair of semantic local prototypes and local prototypes; (iii) we add global classification loss to (i); (iv) full model. The results are shown in Table 3.

When using Euclidean distance as metric function, local feature and global classification both can improve the performance on both data sets. Especially on tieredImageNet, local feature and global classification exceed the former by 7.8% and 10.3% respectively, proving their effectiveness. While using cosine similarity as metric function, local feature and global classification loss can improve the performance on miniImageNet. However, on tieredImageNet, using local feature is invalid and reduces classification accuracy (a tiny boost on miniImageNet). We consider that cosine similarity is not suitable for local features.

Table 3. Ablation study on miniImageNet and tieredImageNet. Average few-shot classification accuracies (%) with 95% confidence intervals on miniImageNet and tieredImageNet test splits. Notation : No Local feature means keeping last average pooling layer of ResNet12, Eu.d means Euclidean distance while Cos means cosine similarity for metric function.

Local feature	Global classification loss	Metric		miniImageNet	tieredImageNet
		Eu.d	Cos	5way-1shot	5way-1shot
✗	✗	✔		65.85±0.35	61.41±0.40
✔	✗	✔		66.68±0.36	66.18±0.39
✗	✔	✔		68.28±0.34	67.75±0.39
✔	✔	✔		**68.58±0.34**	**70.93±0.39**
✗	✗		✔	64.54±0.36	65.81±0.40
✔	✗		✔	64.74±0.36	63.94±0.39
✗	✔		✔	65.03±0.34	**70.18±0.38**
✔	✔		✔	**66.75±0.34**	69.46±0.38

5   Conclusion and Discussion

In this article, we propose a semantic-based single-shot method which adaptively and effectively makes full use of semantic information and visual local features. The proposed method boosts the performance of semantic-based and metric-based approaches by a large margin on different data sets. Moreover, by introducing global classification loss, our method learns better embeddings and outperforms state-of-the-art approaches on single-shot classification. We also conduct ablation experiments to show that local feature and global classification loss can reasonably and effectively improve metric-based few-shot classification methods. Our code will be available online soon.

Acknowledgement. This research is partly supported by Ministry of Science and Technology, China (No. 2019YFB1311503) and Committee of Science and Technology, Shanghai, China (No.19510711200).

References

1. Bridle, J.S.: Probabilistic interpretation of feedforward classification network outputs, with relationships to statistical pattern recognition. In: Soulié, F.F., Hérault, J. (eds.) Neurocomputing, pp. 227–236. Springer (1990). https://doi.org/10.1007/978-3-642-76153-9_28
2. Chen, W.Y., Liu, Y.C., Kira, Z., Wang, Y.C.F., Huang, J.B.: A closer look at few-shot classification. In: ICLR (2019)
3. Chen, Y., Liu, Z., Xu, H., Darrell, T., Wang, X.: Meta-baseline: exploring simple meta-learning for few-shot learning. In: Proceedings of the IEEE/CVF International Conference on Computer Vision, pp. 9062–9071 (2021)
4. Dhillon, G.S., Chaudhari, P., Ravichandran, A., Soatto, S.: A baseline for few-shot image classification. ICLR (2020)
5. Finn, C., Abbeel, P., Levine, S.: Model-agnostic meta-learning for fast adaptation of deep networks. In: ICML (2017)
6. Frome, A., Corrado, G.S., Shlens, J., Bengio, S., Dean, J., Ranzato, M., Mikolov, T.: Devise: A deep visual-semantic embedding model. Advances in neural information processing systems 26 (2013)
7. He, K., Zhang, X., Ren, S., Sun, J.: Deep residual learning for image recognition. In: Proceedings of the IEEE Conference on Computer Vision and Pattern Recognition, pp. 770–778 (2016)
8. Hochreiter, S., Schmidhuber, J.: Long short-term memory. Neural Comput. **9**(8), 1735–1780 (1997)
9. Hou, R., Chang, H., Ma, B., Shan, S., Chen, X.: Cross attention network for few-shot classification. In: Advances in Neural Information Processing Systems, vol. 32 (2019)
10. Jiang, X., Havaei, M., Varno, F., Chartrand, G., Chapados, N., Matwin, S.: Learning to learn with conditional class dependencies. In: International Conference on Learning Representations (2018)
11. LeCun, Y., Bottou, L., Bengio, Y., Haffner, P.: Gradient-based learning applied to document recognition. Proc. IEEE **86**(11), 2278–2324 (1998)
12. Li, A., Huang, W., Lan, X., Feng, J., Li, Z., Wang, L.: Boosting few-shot learning with adaptive margin loss. In: Proceedings of the IEEE/CVF Conference on Computer Vision and Pattern Recognition, pp. 12576–12584 (2020)
13. Li, W., et al.: Libfewshot: a comprehensive library for few-shot learning. arXiv preprint arXiv:2109.04898 (2021)
14. Li, W., Wang, L., Xu, J., Huo, J., Gao, Y., Luo, J.: Revisiting local descriptor based image-to-class measure for few-shot learning. In: Proceedings of the IEEE/CVF Conference on Computer Vision and Pattern Recognition, pp. 7260–7268 (2019)
15. Liu, X., et al.: Learn from concepts: towards the purified memory for few-shot learning. In: Zhou, Z.H. (ed.) Proceedings of the Thirtieth International Joint Conference on Artificial Intelligence, IJCAI-21, pp. 888–894. International Joint Conferences on Artificial Intelligence Organization, August 2021. https://doi.org/10.24963/ijcai.2021/123, main Track
16. Oreshkin, B., Rodríguez López, P., Lacoste, A.: Tadam: task dependent adaptive metric for improved few-shot learning. In: Advances in Neural Information Processing Systems, vol. 31 (2018)
17. Pennington, J., Socher, R., Manning, C.D.: Glove: Global vectors for word representation. In: Proceedings of the 2014 Conference on Empirical Methods in Natural Language Processing (EMNLP), pp. 1532–1543 (2014)

18. Ravi, S., Larochelle, H.: Optimization as a model for few-shot learning. In: ICLR (2017)
19. Reed, S., Akata, Z., Lee, H., Schiele, B.: Learning deep representations of fine-grained visual descriptions. In: Proceedings of the IEEE Conference on Computer Vision and Pattern Recognition, pp. 49–58 (2016)
20. Ren, M., et al.: Meta-learning for semi-supervised few-shot classification. In: ICLR (2018)
21. Russakovsky, O., et al.: ImageNet large scale visual recognition challenge. Int. J. Comput. Vis. **115**(3), 211–252 (2015). https://doi.org/10.1007/s11263-015-0816-y
22. Rusu, A.A., et al.: Meta-learning with latent embedding optimization. In: International Conference on Learning Representations (2018)
23. Santoro, A., Bartunov, S., Botvinick, M., Wierstra, D., Lillicrap, T.: Meta-learning with memory-augmented neural networks. In: International Conference on Machine Learning, pp. 1842–1850. PMLR (2016)
24. Snell, J., Swersky, K., Zemel, R.: Prototypical networks for few-shot learning. In: Advances in Neural Information Processing Systems, vol. 30 (2017)
25. Sung, F., Yang, Y., Zhang, L., Xiang, T., Torr, P.H., Hospedales, T.M.: Learning to compare: relation network for few-shot learning. In: Proceedings of the IEEE Conference on Computer Vision and Pattern Recognition, pp. 1199–1208 (2018)
26. Tian, Y., Wang, Y., Krishnan, D., Tenenbaum, J.B., Isola, P.: Rethinking few-shot image classification: a good embedding is all you need? In: Vedaldi, A., Bischof, H., Brox, T., Frahm, J.-M. (eds.) ECCV 2020. LNCS, vol. 12359, pp. 266–282. Springer, Cham (2020). https://doi.org/10.1007/978-3-030-58568-6_16
27. Vartak, M., Thiagarajan, A., Miranda, C., Bratman, J., Larochelle, H.: A meta-learning perspective on cold-start recommendations for items. In: Advances in Neural Information Processing Systems, vol. 30 (2017)
28. Vinyals, O., Blundell, C., Lillicrap, T., Wierstra, D., et al.: Matching networks for one shot learning. In: Advances in Neural Information Processing Systems, vol. 29 (2016)
29. Wan, Z., et al.: Transductive zero-shot learning with visual structure constraint. In: Advances in Neural Information Processing Systems, vol. 32 (2019)
30. Wang, Y., Yao, Q., Kwok, J.T., Ni, L.M.: Generalizing from a few examples: a survey on few-shot learning. ACM Comput. Surv. (CSUR) **53**(3), 1–34 (2020)
31. Xing, C., Rostamzadeh, N., Oreshkin, B., O Pinheiro, P.O.: Adaptive cross-modal few-shot learning. In: Advances in Neural Information Processing Systems, vol. 32 (2019)
32. Xu, W., Wang, H., Tu, Z., et al.: Attentional constellation nets for few-shot learning. In: International Conference on Learning Representations (2021)
33. Yang, L., Li, L., Zhang, Z., Zhou, X., Zhou, E., Liu, Y.: DPGN: distribution propagation graph network for few-shot learning. In: Proceedings of the IEEE/CVF Conference on Computer Vision and Pattern Recognition, pp. 13390–13399 (2020)
34. Yoo, T.K., Choi, J.Y., Kim, H.K.: Feasibility study to improve deep learning in oct diagnosis of rare retinal diseases with few-shot classification. Med. Biol. Eng. Comput. **59**(2), 401–415 (2021)
35. Zhang, B., Li, X., Ye, Y., Huang, Z., Zhang, L.: Prototype completion with primitive knowledge for few-shot learning. In: Proceedings of the IEEE/CVF Conference on Computer Vision and Pattern Recognition, pp. 3754–3762 (2021)

Self-Reinforcing Feedback Domain Adaptation Channel

Yan Jia, Xiang Zhang$^{(\boxtimes)}$, Long Lan, and Zhigang Luo

College of Computer Science and Technology, National University of Defense
Technology, Changsha, China
{jia.yan20,zhangxiang08,long.lan,zgluo}@nudt.edu.cn

Abstract. Unsupervised domain adaptation methods utilize feature representations of instances in the source and target domains to eliminate domain shifts. It is worth noting that the instance features are closely related to the entire distribution of the domain, and the current information after adaptation in the execution of the domain adaptation task is closely related to the original features. Common methods are based on only one of these pieces of information and do not make sufficient use of them. We develop the Self-Reinforcing Feedback Domain Adaptation Channel (SRFC). Pioneeringly, on the feature representation of the network, SRFC fuses global and instance information simultaneously, and utilizes the past history and current information in domain adaptation, so that the information can be effectively enhanced to better complete the domain adaptation. Through the designed self-reinforcing feedback mechanism, SRFC skillfully integrates multi-level information in a robust way in the process of domain adaptation, and actively enhances the availability and comprehensive value of features in domain adaptation with manageable continuous feedback. Experiments on benchmark datasets verify the advantages of SRFC fusion information for instance feature enhancement and domain adaptation. The modular Self-Reinforcing Feedback Domain Adaptation Channel has scalability and R&D potential, and we hope that it can be extended to more domain adaptation networks using enhanced instance representations to better accomplish different tasks.

Keywords: Unsupervised domain adaptation · Feedback mechanism · Attention mechanism

1 Introduction

When the trained model is directly transferred, the inherent generalization problem can lead to drastic performance degradation. This generalization loss comes from the domain shift present in the entire dataset [4,23]. Domain adaptation is the task of learning a discriminative model in the presence of a domain offset between such source and target datasets. Unsupervised domain adaptation

---

This work was funded by Haihe Laboratory in Tianjin, Grants No. 22HHXCJC00007.

M. Tanveer et al. (Eds.): ICONIP 2022, LNCS 13623, pp. 140–152, 2023.
https://doi.org/10.1007/978-3-031-30105-6_12

methods [9,15,16] attempt to transfer knowledge from label-rich source domains to unlabeled target domains.

Many existing domain adaptation methods [13,20] align the features of the source distribution with the target distribution. Adversarial learning methods have shown impressive performance in reducing the discrepancy between source and target domains, such as DANN [7] and ADDA [23]. They utilize adversarial techniques to align the source domain and target distributions. Due to the success of contrastive methods [2,3,10] in self-representation learning, some recent works in [12,22] have turned to instance-based contrastive learning to reduce the differences between different domains. CLDA [21] applies contrastive learning for domain adaptation, class contrastive learning for reducing inter-domain gaps. In addition to these types of methods, we note some other methods based on attention mechanisms. In TADA [25] a self-attention mechanism is implemented on source domain samples and target domain samples, respectively, to enhance their own features. ABMSD [28] focuses on Multi-Source Domain Adaptation and proposes an attention mechanism to reweight the importance of source domains.

Instance features are closely related to the entire distribution of the domain. However, in adversarial domain adaptation and contrastive domain adaptation, most of these difference-based or adversary-based methods are at the instance level, and they use instance representations to directly align the inter-domain distribution. For the whole network, they often perform tasks based on only one kind of information, such as using the feature information of the instance, while ignoring other equally valuable information such as the global features of the domain. With attention-based methods, they only focus on the enhancement of their own features, ignoring the synthesis of the global information of the entire domain. At the same time, we also noticed that under the feedback mechanism, the historical information of the system is closely related to the current information, and the deviation of the information can assist the system and promote it to optimize towards the goal. For domain adaptation, there is no method that directly utilizes feedback information bias to enhance domain adaptation ability in feature representation. In the application of instance features, previous domain adaptation methods lacked the attention and synthesis of domain global information and historical information during task execution.

Based on the above analysis, we have developed the Self-Reinforcing Feedback Domain Adaptation Channel (SRFC), which has scalability and R&D potential. Pioneeringly, on the feature representation of the network, SRFC fuses global and instance information simultaneously, and utilizes the past history and current information in domain adaptation, so that the information can highlight inter-domain differences to better complete the domain adaptation task. Specifically, SRFC consists of a self-reinforcing feedback mechanism and domain adaptation constraints. The self-reinforcing feedback mechanism is formed through the tight association of the four components: the embedder, the attention block, the adaptor, and the feedback. The attention block enhances the preliminarily integrated comprehensive features with a self-attention mechanism. The feedback will use the deviation between the output after the initial adaptation of the upper layer and the original feature. It continuously feeds back the original feature through the proportional-integral-derivative PID function for feedback enhancement. Under the control of feedback, the original features move through the entire superstructure towards the desired optimal output. It skillfully fuses multi-level information in a robust manner to actively enhance the availability and comprehensive value of features in domain adaptation with manageable continuous feedback. For domain adaptation constraints, SRFC will use the contrastive learning paradigm to directly use the enhanced instance features. The conditional discriminator in SRFC uses the original features and the enhanced classifier prediction to indirectly use the enhanced features

SRFC performs well in the domain adaptation task under the combination of self-reinforcing feedback mechanism and domain adaptation constraints. The modular design of SRFC gives it the potential for extensibility and in-depth research and development. We hope that SRFC can be extended to more domain adaptation networks, using enhanced instance representations to facilitate networks to perform their different tasks better.

The main contributions of this paper are as follows: 1) The self-reinforcing feedback mechanism combines the multi-level information skillfully in a robust way. It actively enhances feature availability and comprehensive value with manageable continuous feedback. 2) We developed a modular SRFC. Pioneeringly, the feature representation in the domain adaptation network combines both global and instance information, and can utilize the past history and current information to better complete the domain adaptation task. 3) Experiments on benchmark datasets verify the advantages of SRFC fusion information for instance feature enhancement and domain adaptation.

2 Related Works

2.1 Unsupervised Domain Adaptation

Many existing unsupervised domain adaptation methods [13,20] align the features of the source distribution with the target distribution. Adversarial learning methods [7,15,23] have shown impressive performance in reducing the discrepancy between source and target domains. CDAN [15] proposes conditional adversarial domain adaptation to better align the distributions of different domains by utilizing adversarial techniques. Due to the success of contrastive methods Contrastive methods [2,3,10] succeeded in self-representation learning, [12] turned to instance-based contrastive learning to reduce differences between domains. CLDA [21] applies contrastive learning to domain adaptation and class-contrastive learning to reduce inter-domain gaps, CDA [22] was proposed to extend contrastive learning to domain adaptation problems without access to labels.

2.2   Attention Mechanism

In recent years, self-attention mechanisms have made remarkable progress in a wide range of tasks such as vision and text. The attention mechanism was first proposed by [24]. It takes natural language processing (NLP) research to a new level. ViT [5] applies Transformer to the image classification model. The model is efficient and scalable. When there is enough data for pre-training, the effect is remarkable, breaking through the limitations of Transformer. DANet [6] proposes a dual attention network to adaptively integrate local features and global dependencies. The self-attention mechanism is good at capturing the internal correlation of features, and at the same time, it can obtain larger receptive field and contextual information by capturing the global information. In the domain adaptation problem, we hope that instance features are enhanced by incorporating global information with the help of a self-attention mechanism.

2.3   PID Controller

No matter in industry or computer field [18], the feedback mechanism uses the error to update the system state appropriately, and promotes the system to develop to a better state. What information to use and in what way are the two key points to be considered in the feedback mechanism. The PID controller utilizes the current, past and future information of the prediction error to control the feedback system [1]. Mathematically, this process takes the form:

$$u(t) = K_p e(t) + K_i \int_0^t e(t)dt + K_d \frac{d}{dt}e(t) \tag{1}$$

According to the formula, the PID algorithm first calculates the error $e(t)$ between the system output and the expected value; then implements three error terms of proportional, integral and derivative to update the system state, so that the system can obtain the optimal performance index. The coefficients K_p, K_i and K_d determine the contribution of current, past and future errors to the current correction. The PID algorithm considers sufficiently comprehensive information flow with the properties of simplicity, robustness and ease of use [29]. The error of the feature before and after the domain adaptation is performed has a high similarity with the error utilized in the PID algorithm. This just happens to be exploited in our goal of augmenting existing features with historical information. As we will see later in this article, when we apply it to the SRFC, a properly designed error will allow the correction $u(t)$ of the system to be used to enhance the original instance features, allowing the entire channel to better complete the domain adaptation tasks.

3   Methodology

The goal of unsupervised domain adaptation is to use labeled source domain samples $S = \{(x_1^s, y_1^s), \cdots, (x_N^s, y_N^s)\}$ and unlabeled target domain samples

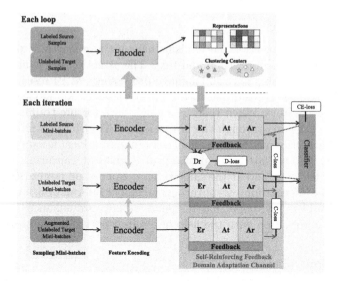

Fig. 1. Outline of our SRFC-Network Framework. In SRFC, enhancements are performed on each link, but they share the same set of components, that is, each enhanced link shares parameters and optimizes at the same time in SRFC. (Er: embedder, At: attention block, Ar: adaptor, Dr: domain discriminator.)

$T = \{x_1^t, \cdots, x_{N'}^t\}$, to reduce the distribution shift between domains. The source domain samples are sampled from $P_s(X, Y)$, and the target domain samples are sampled from $P_t(X, Y)$. $y^s \in \{0, 1, \cdots, M-1\}$ denotes the source data labels of M classes. The target data labels $y^t \in \{0, 1, \cdots, M-1\}$ are unknown. But we consider the case of having the same classes as the source domain. Domain adaptation improves the generalization performance of the model when applied to the target domain.

3.1 Model Overview

We show the entire structural framework of the SRFC network in Fig. 1. The model can be divided into an upper structure that executes once per loop and a lower structure that iterates continuously in the loop. The model is mainly composed of encoder $f(\cdot)$, SRFC and classifier $c(\cdot)$. Parameters are shared between each encoder $f(\cdot)$ while optimizing. In SRFC, it is specifically composed of five components: an embedder Er, an attention block At, an adapter Ar, a feedback device Feedback, and a domain discriminator $d(\cdot)$. The first four are closely combined to build self-reinforcing feedback mechanism in Fig. 2. The domain discriminator $d(\cdot)$ is used for domain adaptation constraints. In SRFC, enhancements are performed on each link, but they share the same set of components, that is, each enhanced link shares parameters and optimizes at the same time. We will cover SRFC in detail in 3.2. In each iteration, the input data are minibatches of data sampled with the same labels from source domain data and target domain data with pseudo-labels.

3.2   Self-Reinforcing Feedback Domain Adaptation Channel

Self-Reinforcing Feedback Mechanism. SRFC consists of the self-reinforcing feedback mechanism and domain adaptation constraints. The whole process of the self-reinforcing feedback mechanism is clearly shown in Fig. 2. The embedder will format the global feature information and instance feature information to input the attention block. The attention block enhances the preliminarily integrated comprehensive features with a self-attention mechanism. The attention block then passes the information to the adaptor for better feature alignment. The feedback will use the deviation between the output after the initial adaptation of the upper layer and the original feature. It continuously feeds back the original feature through the proportional-integral-derivative PID function for feedback enhancement.

The input of SRFC is instance feature representation and global feature information extracted by spherical K-means clustering. The cluster centers O^s of the source domain are used to initialize the clustering O^t of the target domain: $O^t \leftarrow O^s$, we use cosine Dissimilarity $\text{dist}(\boldsymbol{a}, \boldsymbol{b}) = \frac{1}{2}\left(1 - \frac{\langle \boldsymbol{a}, \boldsymbol{b} \rangle}{\|\boldsymbol{a}\|\|\boldsymbol{b}\|}\right)$, $\hat{y}_i^t = \text{argmin}_c \text{dist}(f(x_i^t), O^t)$ and $O^t = \sum_{i=1}^{N_t} \mathbf{1}_{\hat{y}_i^t = c} \frac{f(x_i^t)}{\|f(x_i^t)\|}$ in the feature space to generate pseudo-labels for the target domain $\hat{T} = \{(x_1^t, y_1^t), \cdots, (x_{N'}^t, y_{N'}^t)\}$.

We use the cluster centers (O^s, O^t) to reflect the global information, and generate the global information once before the start of each loop through the upper structure. In subsequent iterations, the features of minibatch sample sampled by the same class, $f(x_{m(i,c)}^s), f(x_{m(j,c)}^t)$ are fed together with (O^s, O^t) to SRFC.

First, the data list $[O^s, f(x_{m(i,c)}^s)], [O^t, f(x_{m(j,c)}^t)]$ input into the embedder together. The global features are transformed into the (Batchsize,1,768) matrix and concatenated with the instance feature matrix transformed into (Batchsize,128,768). Put the matrix corresponding to the global feature at the head of the instance feature. After concatenation, the new matrix is the initial fusion of information obtained under the embedder. The detailed operations and dimensional changes are detailed in Fig. 2.

Next, in the attention block, it will be combined with classtoken and position embedding, and passed to the subsequent self-attention layer to complete self-attention enhancement. The output augmented feature matrix of (Batchsize, 130, 768) dimension is an instance representation that is more responsive to domain features after integrating global information and instance information. As shown in Fig. 2, the output matrix will be divided into a three-layer structure. Only the first and third layer results will be passed into the adapter.

From the discussion in Sect. 2, we know that the difference in features before and after domain adaptation is performed has a high similarity to the error exploited in the PID algorithm. At the same time, we cannot limit the scalability of the channel for the feedback mechanism, which requires the feedback

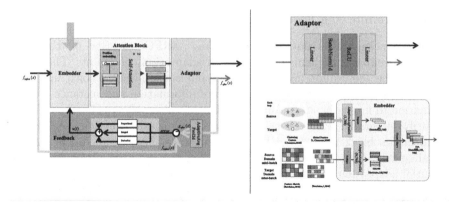

Fig. 2. Illustration of Self-Reinforcing Feedback Mechanism. The SRFC is mainly constructed from a self-attention contrastive adaptor and a domain conditional discriminator.

mechanism to exist in a modular form. The PID feedback algorithm implementation in Eq. (1) starts with a system error. Therefore, as shown in Fig. 2, we first upsample the input so that the domain-adapted information is the same as the original feature dimension, and then construct an error $e(t)$ between $f_{after}(x)$ and $f_{before}(x)$. The feature difference before and after task execution can be described as $f_{before}(x) = g_{after}(x) + shift$, and the relationship between the original feature and the perfect feature can be described as $f_{before}(x) = \mathbf{g_{ideal}}(x) + \mathbf{shift}$. In an ideal state, each item should be fixed. Therefore, under the constraints of the mathematical relationship, the deviation between the current state and the ideal state $Error = \mathbf{g_{ideal}}(x) - g_{after}(x)$ is equivalent to describing the difference between $f_{before}(x)$ and $g_{after}(x)$. Therefore, we can directly use $f_{before}(x) - g_{after}(x) = Error + \mathbf{shift} = error$ to calculate the increment in Eq. (1), and continuously feedback $u(t)$ to promote the channel to continuously approach the ideal state. This achieves the goal of incremental feedback reinforcement through a proportional-integral-derivative algorithm using historical and current information. At the same time, the ideal state can only be approached infinitely but is difficult to achieve. The original control constraints make the length of the feedback mechanism controllable, avoiding model overfitting and negative optimization. The experimental results demonstrate the rationality of our conjecture.

The information used for feedback always comes from two epochs output before and after domain adaptation. This makes the upper and lower structures closely integrated to jointly realize the self-reinforcing feedback mechanism, which can integrate multi-level information and achieve effective feature enhancement.

Domain Adaptation Constraints. Domain adaptation constraints will apply the augmented instance features in both direct and indirect ways. The adaptor outputs the first and third layer results after attention block $f'_{after}(x), f_{after}(x)$. $f'_{after}(x)$ is passed to the classifier to get the prediction. $f_{after}(x)$ is used for computing the contrastive loss.

$$L_C = \sum_{m \in I} - \log \frac{S(f_a(x^s_{m(i,c)}), f_a(x^t_{rifm(j,c)}))}{S(f_a(x^s_{m(i,c)}), f_a(x^t_{rifm(j,c)})) + \sum_{\substack{n \neq m \in \mathcal{I} \\ q \in s,t}} S(f_a(x^s_{m(i,c)}), f_a(x^q_{n(j,c')}))}$$

(2)

$$S(\mathbf{u}, \mathbf{v}) = sim(\mathbf{u}, \mathbf{v}) = \exp\left(\frac{1}{\tau} \frac{(\mathbf{u} \cdot \mathbf{v})}{\|\mathbf{u}\| \|\mathbf{v}\|}\right)$$

(3)

Contrastive learning in SRFC will directly utilize augmented features for inter-domain as well as intra-domain distribution alignment. The data index of the mini-batches is m, n, and the path index of the sample is i, j.

The domain-conditional discriminator $d(\cdot)$ exploits the augmented information in an indirect way by exploiting the prediction of augmented features, and directly eliminates distributional differences in the original features. It enables the encoder to obtain higher quality domain global features and stabilize the whole channel performance.

$$\mathcal{L}_D = -\frac{1}{n_s} \sum_{i=1}^{n_s} \log\left[D\left(f_i^s, y_{ci}^s\right)\right] - \frac{1}{n_t} \sum_{j=1}^{n_t} \log\left[1 - D\left(f_j^t, y_{cj}^t\right)\right]$$

(4)

Domain discriminator has a similar principle as in [15], but here the labels are obtained from reinforcement features.

Our SRFC has attractive features. First, SRFC combines the self-reinforcing feedback mechanism and domain adaptation constraints, which enables the features in the domain adaptation network to combine global and instance information at the same time, and can use the past history and current information in the domain adaptation to strengthen itself. The SRFC network utilizes effectively enhanced information to better accomplish the task of domain adaptation. Second, modular SRFC has more possibilities in the future, and can be extended to more domain adaptation methods, using feature enhancement to better accomplish different tasks. And its own components can also conduct more in-depth research, and integrate more levels of information more efficiently. Finally, based on the characteristics of the feedback algorithm, SRFC can complete the enhancement of features in the process of domain adaptation in a controllable form.

3.3   Objective Function

Outside the SRFC, the classifier is optimized by computing the standard cross-entropy loss $\mathcal{L}_{CE}$.

$$\mathcal{L}_{CE} = -\sum_{m=1}^{n} (y_m^{\text{s}}) \log(f(x_{\text{c}m}^{\text{s}})) \tag{5}$$

Therefore, the overall training objective uses a combination of three losses, which can be formulated as follows:

$$\mathcal{L} = \mathcal{L}_{\text{CE}} + \alpha * \mathcal{L}_{\text{C}} - \beta * \mathcal{L}_{\text{D}} \tag{6}$$

4   Experiments

4.1   Datasets and Setups

We validate our method on two public benchmarks Office-31 and Office-Home. They are general datasets for domain adaptation tasks. The Office-31 dataset contains three distinct domains and consists of 4110 images belonging to 31 categories. Office-Home contains 15,588 images in 65 categories across 4 domains. Where similar to the Amazon category in Office-31, Real World's images were collected using conventional cameras. This is a challenging dataset with large domain shift between domains, so that it is hard to perform domain adaptation well on it.

For the encoder of SRFC-net, we choose to use ResNet-50 [11] pretrained on ImageNet. It should be noted here that the attention block in SRFC does not need to load any pre-training parameters like ViT [5], but is optimized synchronously with the network after random initialization. The adaptor is a standard MLP structure. We train the network using mini-batch stochastic gradient descent (SGD) with momentum 0.9. We follow the same learning rate schedule described in [14,17]. The temperature τ in Eq. (3) is set to 0.5.

4.2   Performance Analysis

We test the effectiveness of our method on two datasets, Office-31 and Office-Home. The performance of SRFC-net is shown in Table 1 and Table 2. Some typical performances of domain adaptation methods are listed in the table, and we compare the performance with them. On the six tasks on office31, compared with DANN, SRFC improves by 9.2% on A→W tasks and 13.4% on A→D tasks, which is a significant improvement. Compared with CDAN, it improves by 7.3% on the W→A task and 4.9% on the D→A task. Table 2 shows the performance of SAC on 12 tasks on the officehome dataset. The domain adaptation task on this dataset is more challenging. We can see that most of the tasks have been improved compared with CDAN, especially the Ar→Cl task has improved by 5.5%, the Pr→Ar task has improved by 4.9%, and the Cl→Ar task has improved by 5.6%. Table 3 shows the accuracy (%) decline without self-reinforcing feedback mechanism about some key tasks on both datasets.

Table 1. Accuracy (%) on Office-31 dataset (based on ResNet50)

Method	A→W	A→D	W→ A	W→D	D→A	D→W	Ave
ResNet50 [11]	68.4	68.9	60.7	99.3	62.5	96.7	76.1
DAN [14]	80.5	78.6	62.8	99.6	63.6	62.8	80.4
DANN [8]	82.0	79.7	67.4	99.1	68.2	96.9	82.2
JAN [17]	85.4	84.7	70.0	99.8	68.6	97.4	84.3
GSM [27]	85.9	84.1	75.5	97.2	73.6	97.1	85.3
CDAN [15]	94.1	92.9	69.3	100	71.0	98.6	87.7
SymNets [26]	90.8	93.9	72.5	100	74.6	98.8	88.4
SRFC-net(ours)	91.2	93.1	**76.6**	99.8	**75.9**	98.3	**89.1**

Table 2. Accuracy (%) on Office-Home dataset (based on ResNet50)

Method	A→C	A→P	A→R	C→A	C→P	C→R	P→A	P→C	P→R	R→A	R→C	R→P	Ave
ResNet50	34.9	50.0	58.0	37.4	41.9	46.2	38.5	31.2	60.4	53.9	41.2	59.9	46.1
DAN	43.6	57.0	67.9	45.8	56.5	60.4	44.0	43.6	67.7	63.1	51.5	74.3	56.3
DANN	45.6	59.3	70.1	47.0	58.5	60.9	46.1	43.7	68.5	63.2	51.8	76.8	57.6
JAN	45.9	61.2	68.9	50.4	59.7	61.0	45.8	43.4	70.3	63.9	52.4	76.8	58.3
CDAN	50.7	70.6	76.0	57.6	70.0	70.0	57.4	50.9	77.3	70.9	56.7	81.6	65.8
GSM	49.4	75.5	80.2	62.9	70.6	70.3	65.6	50.0	80.8	72.4	50.4	81.6	67.5
ours	**56.2**	71.1	77.3	**67.9**	**73.4**	**75.8**	**67.8**	**53.9**	80.5	71.3	**60.1**	**82.5**	**69.8**

Table 3. Accuracy (%) decline without self-reinforcing feedback mechanism

Office-31	A→W	W→ A	A→D	D→A
w/o.SRF	89.9	76.6	90.5	75.3
decline	1.3↓	–	2.6↓	0.6↓
Office-Home	A→P	C→ P	P→A	R→P
w/o.SRF	69.6	72.5	67.3	81.5
decline	1.5↓	0.9↓	0.6↓	1.0↓

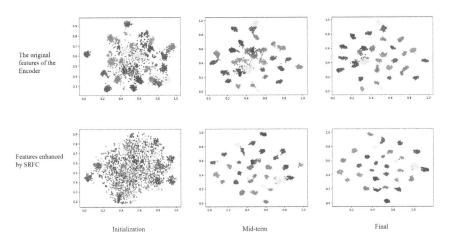

Fig. 3. Visualization of the SRFC on W→A with t-SNE.

To better reflect the performance of SRFC in the network, we perform t-SNE visualizations of the original and augmented expressions in model optimization. We choose three special stage visualizations to show the details. Figure 3 shows t-SNE visualizations for the initial, mid and final stages. The upper layer is the raw feature visualization of the encoder output. The lower layer is SRFC-enhanced feature visualization. Comparing the visualizations of different periods, we can find that in the initial stage of initialization, SRFC is randomly initialized and does not play a role. With the continuous optimization of the model, SRFC can enhance the original feature information and project it into a more compact space. The edges between each category become clearer, especially the blurred centers of some original feature information can be well separated. At the end of model optimization, the original features can already achieve a similar effect as the enhanced features. The obvious difference in t-SNE visualization of upper and lower layer features also shows that the features through SRFC are better enhanced to reflect the differences in domains with higher quality information, which is more conducive to eliminating domain shifts. The self-reinforcing feedback mechanism in the SRFC works well. In SRFC, the self-reinforcing feedback mechanism and the domain adaptation paradigm are closely integrated, complement each other and achieve each other. Only under the constraints of the domain adaptation paradigm, SRFC can continue to advance towards excellence. The self-reinforcing feedback mechanism is what makes SRFC go further.

Fig. 4. Feature Attention Maps of SRFC

To further observe how the model works, we visualize the feature attention map obtained after going through SRFC. The attention map in Fig. 4 is used to reflect the attention of the entire network to each region of the data image. The hotter the area, the higher the attention of the reaction network. This figure shows part of a clear typical model attention map. We can see that the enhanced feature attention is at the core key parts.

5   Conclusion

We developed SRFC, a modular, scalable and potentially domain-adaptive component. SRFC combines the self-reinforcing feedback mechanism with domain adaptation constraints, which enables features in domain adaptation networks to incorporate both global and instance information, and can utilize past history and current information in domain adaptation to augment itself. The SRFC network effectively utilizes the enhanced information to better accomplish domain adaptation. The core self-reinforcing feedback mechanism actively improves the usability and comprehensiveness of features through manageable continuous feedback, which can be further generalized. Modular SRFC has more possibilities in the future and can be extended to more domain adaptation methods, using feature enhancement to better accomplish different tasks. And its own components can also conduct more in-depth research and integrate more layers of information more efficiently.

References

1. Ang, K.H., Chong, G., Li, Y.: PID control system analysis, design, and technology. IEEE Trans. Control Syst. Technol. **13**(4), 559–576 (2005)
2. Bachman, P., Hjelm, R.D., Buchwalter, W.: Learning representations by maximizing mutual information across views. In: Advances in Neural Information Processing Systems, vol. 32 (2019)
3. Chen, T., Kornblith, S., Norouzi, M., Hinton, G.: A simple framework for contrastive learning of visual representations. In: International Conference on Machine Learning, pp. 1597–1607. PMLR (2020)
4. Donahue, J., et al.: Decaf: a deep convolutional activation feature for generic visual recognition. In: International Conference on Machine Learning, pp. 647–655. PMLR (2014)
5. Dosovitskiy, A., et al.: An image is worth 16x16 words: transformers for image recognition at scale. arXiv preprint arXiv:2010.11929 (2020)
6. Fu, J., et al.: Dual attention network for scene segmentation. In: Proceedings of the IEEE/CVF Conference on Computer Vision and Pattern Recognition, pp. 3146–3154 (2019)
7. Ganin, Y., Ustinova, E., Ajakan, H., Germain, P., Larochelle, H., Laviolette, F., Marchand, M., Lempitsky, V.: Domain-adversarial training of neural networks. The journal of machine learning research **17**(1), 2096–2030 (2016)
8. Ghifary, M., Kleijn, W.B., Zhang, M.: Domain adaptive neural networks for object recognition. In: Pham, D.-N., Park, S.-B. (eds.) PRICAI 2014. LNCS (LNAI), vol. 8862, pp. 898–904. Springer, Cham (2014). https://doi.org/10.1007/978-3-319-13560-1_76
9. Gopalan, R., Li, R., Chellappa, R.: Domain adaptation for object recognition: an unsupervised approach. In: 2011 International Conference on Computer Vision, pp. 999–1006. IEEE (2011)
10. He, K., Fan, H., Wu, Y., Xie, S., Girshick, R.: Momentum contrast for unsupervised visual representation learning. In: Proceedings of the IEEE/CVF Conference on Computer Vision and Pattern Recognition, pp. 9729–9738 (2020)

11. He, K., Zhang, X., Ren, S., Sun, J.: Deep residual learning for image recognition. In: Proceedings of the IEEE Conference on Computer Vision and Pattern Recognition, pp. 770–778 (2016)
12. Kim, D., Saito, K., Oh, T.H., Plummer, B.A., Sclaroff, S., Saenko, K.: Cross-domain self-supervised learning for domain adaptation with few source labels. arXiv preprint arXiv:2003.08264 (2020)
13. Li, S., et al.: Domain conditioned adaptation network. In: Proceedings of the AAAI Conference on Artificial Intelligence, vol. 34, pp. 11386–11393 (2020)
14. Long, M., Cao, Y., Wang, J., Jordan, M.: Learning transferable features with deep adaptation networks. In: International Conference on Machine Learning, pp. 97–105. PMLR (2015)
15. Long, M., Cao, Z., Wang, J., Jordan, M.I.: Conditional adversarial domain adaptation. In: Advances in Neural Information Processing Systems, vol. 31 (2018)
16. Long, M., Zhu, H., Wang, J., Jordan, M.I.: Unsupervised domain adaptation with residual transfer networks. In: Advances in Neural Information Processing Systems, 29 (2016)
17. Long, M., Zhu, H., Wang, J., Jordan, M.I.: Deep transfer learning with joint adaptation networks. In: International Conference on Machine Learning, pp. 2208–2217. PMLR (2017)
18. Medsker, L.R., Jain, L.: Recurrent neural networks. Des. Appl. **5**, 64–67 (2001)
19. Minorsky, N.: Directional stability of automatically steered bodies. J. Am. Soc. Naval Eng. **34**(2), 280–309 (1922)
20. Pei, Z., Cao, Z., Long, M., Wang, J.: Multi-adversarial domain adaptation. In: Thirty-Second AAAI Conference on Artificial Intelligence (2018)
21. Singh, A.: CLDA: contrastive learning for semi-supervised domain adaptation. In: Advances in Neural Information Processing Systems, vol. 34 (2021)
22. Thota, M., Leontidis, G.: Contrastive domain adaptation. In: Proceedings of the IEEE/CVF Conference on Computer Vision and Pattern Recognition, pp. 2209–2218 (2021)
23. Tzeng, E., Hoffman, J., Saenko, K., Darrell, T.: Adversarial discriminative domain adaptation. In: Proceedings of the IEEE Conference on Computer Vision and Pattern Recognition, pp. 7167–7176 (2017)
24. Vaswani, A., et al.: Attention is all you need. In: Advances in Neural Information Processing Systems, vol. 30 (2017)
25. Wang, X., Li, L., Ye, W., Long, M., Wang, J.: Transferable attention for domain adaptation. In: Proceedings of the AAAI Conference on Artificial Intelligence, vol. 33, pp. 5345–5352 (2019)
26. Zhang, Y., Tang, H., Jia, K., Tan, M.: Domain-symmetric networks for adversarial domain adaptation. In: Proceedings of the IEEE/CVF Conference on Computer Vision and Pattern Recognition, pp. 5031–5040 (2019)
27. Zhang, Y., Xie, S., Davison, B.D.: Transductive learning via improved geodesic sampling. In: BMVC, p. 122 (2019)
28. Zuo, Y., Yao, H., Xu, C.: Attention-based multi-source domain adaptation. IEEE Trans. Image Process. **30**, 3793–3803 (2021)

General Algorithm for Learning from Grouped Uncoupled Data and Pairwise Comparison Data

Masahiro Kohjima[✉], Yuta Nambu, Yuki Kurauchi, and Ryuji Yamamoto

NTT Human Informatics Laboratories, NTT Corporation, Yokosuka, Japan
{masahiro.kohjima.ev,yuta.nambu.fs,yuuki.kurauchi.mv,
ryuji.yamamoto.sv}@hco.ntt.co.jp

Abstract. Uncoupled regression is the problem of learning a regression model from uncoupled data that consists of a set of input values (unlabeled data) and a set of output values where the correspondence between the input and output is unknown. A recent study showed that a method using both uncoupled data and pairwise comparison data can learn the optimal model under some assumptions. However, this method cannot use grouped information and may require the implementation (almost) from scratch for some models. In this study, we extend the above existing method for handling group information and derive a general algorithm that can learn a model using the standard regression method by approximating the loss function. The effectiveness of the proposed method is confirmed by synthetic and benchmark data experiments.

Keywords: Uncoupled data · Regression · Bregman divergence

1 Introduction

Due to limits of data collection, e.g., privacy-aware or non-centralized data collection, the data to be analyzed are often represented by *uncoupled data* (UD) consisting of (a) a set of input values (unlabeled data) and (b) a set of output values where the correspondence between the input and output is unknown [1–3]. Since we do not know the true output value for each input, uncoupled data can be collected even if outputs represent sensitive information such as age and annual income while protecting anonymity. It is clearly impossible to use standard regression methods, which assume standard *coupled data* (Fig. 1a), to learn a model using uncoupled data.

Uncoupled regression (UR) is the problem of estimating a model from uncoupled data [1–4], see Fig. 1b. Among the various UR studies, we follow the approach of Xu et al. [4] which also uses *pairwise comparison data* (PCD). PCD include ranking information indicating which input value yields the larger output value for given two input values. This data can be collected by asking users "who is older than you?", "who has a higher income than you?" etc. Even if the question is about sensitive matters such as age and income, we can collect the

© The Author(s), under exclusive license to Springer Nature Switzerland AG 2023
M. Tanveer et al. (Eds.): ICONIP 2022, LNCS 13623, pp. 153–164, 2023.
https://doi.org/10.1007/978-3-031-30105-6_13

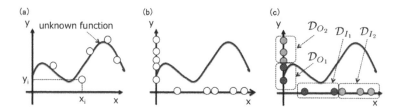

Fig. 1. Comparison of (a) (standard) coupled data, (b) uncoupled data used in [4] and (c) grouped uncoupled data used in our study. The number of data in $\mathcal{D}_{I_k}$ and $\mathcal{D}_{O_k}$ can vary.

information while protecting privacy since the user does not disclose the actual value itself. By the use of uncoupled data and pairwise comparison data, the method proposed in [4] can learn an optimal model (under some assumptions).

However, techniques to process pairwise comparison data for UR are not mature so the existing method [4] still has two limitations: (a) inability to handle *grouped uncoupled data* (GUD) and (b) necessity of implementing algorithm from scratch (for some representative models).

Limitation-(a): in practical data collection scenarios such as survey data collection, data are often collected multiple times from different questionnaire respondents. Therefore, even if the correspondence between the input and output is obscured to protect privacy, we can know which data collection event the data was gathered in. So the data take the form of grouped uncoupled data where the group to which the input and output belong are known (See Fig. 1c). We expect that the use of group information improves the performance.

Limitation-(b): the existing method [4] requires the use of a loss function designed for UR; the loss is computed using both uncoupled data and pairwise comparison data. Implementing this algorithm may be not difficult for some models such as linear-in-parameter model and deep neural network (DNN) model [5] (since the optimization module using automatic differentiation is provided in the ML toolkit such as PyTorch[1]), but it is a burden for other representative models such as Gaussian process (GP) [6] and a variant of decision tree (such as random forest [7] and gradient boosting decision tree [8]) since the learning of these models adopts many additional techniques including hyperparameter optimization, computation cost reduction and training speed acceleration [6,9,10]. The modification of the loss function may impact these techniques and require the implementation (almost) from scratch, so it is desirable to avoid such labor.

In this study, we tackle grouped uncoupled regression (GUR), the problem of learning regression models from grouped uncoupled data and pairwise comparison data; we propose two algorithms; 1st algorithm (GUR-1) is a natural extension of the existing method [4], which is a special case of our proposal, for handling grouped coupled data. 2nd algorithm (GUR-2) allows GUR to be addressed by applying (any) standard regression method to *quasi coupled*

---

[1] https://pytorch.org/.

Table 1. Bregman divergence with various convex functions

	domain	$\varphi(x)$	$\psi(x) = \nabla\varphi(x)$	$d_\varphi(y, x)$
Squared Error	$\mathbb{R}$	$x^2/2$	x	$(y-x)^2/2$
Generalized KL-divergence	$\mathbb{R}_+$	$x\log(x)$	$\log(x) + 1$	$y\log\frac{y}{x} - y + x$
Itakura-Saito distance	$\mathbb{R}_{++}$	$-\log(x)$	$-1/x$	$\frac{y}{x} - \log\frac{y}{x} - 1$

data made from grouped uncoupled data and paired comparison data. GUR-2 is derived by approximating the loss function used in GUR-1. GUR-1 and GUR-2 do not suffer from limitation (a) and GUR-2 overcomes limitation (b). We conduct experiments on both synthetic and benchmark datasets that confirm the effectiveness of the proposals.

The contributions of this paper are summarized below:

- We tackle GUR, the problem of learning regression models from grouped uncoupled data and pairwise comparison data.
- We develop two algorithms for GUR (GUR-1 and GUR-2) that can handle grouped uncoupled data. GUR-2 solves GUR by the (standard) regression method using quasi coupled data.
- We confirm the effectiveness of the proposed algorithms by numerical experiments on synthetic data and benchmark data.

2   Related Work

Uncoupled regression (UR) [1,4], which is also called "regression without correspondence" [2], "regression with shuffled labels" [3,11,12], and "unlabeled sensing" [13], is the problem of estimating a model from uncoupled data. There are two main classes of UR methods. (i) The approach that explicitly estimates the correspondence between input and output [2,11–13] and (ii) the approach that does not estimate the correspondence [1,3,4]. Although the approach (i) is intuitive, the cost of estimating the correspondence is very high since the maximum likelihood estimation of correspondence is NP-hard in general [11]. Thus, we adopt the approach (ii). Among the methods using approach (ii) [1,3,4], the methods [1,3] can be applied to models with some specific structure; the model needs to be monotone in [1] or linear in [3]. So we consider extending the method [4] that can be applied without assuming model structure.

3   Preliminary

3.1   Regression with Bregman Divergence

Let $\mathcal{X} \subset \mathbb{R}^d$ and $\mathcal{Y} \subset \mathbb{R}$ be a d-dimensional input space and output space, respectively. In the (standard) regression problem, training or empirical sample pairs of input and output variables, $\{(x_i \in \mathcal{X}, y_i \in \mathcal{Y})\}_{i=1}^n$, which are independently and identically taken from some probability distribution, are given. n is the number of samples. Regression model $h : \mathcal{X} \to \mathcal{Y}$ in hypothesis space $\mathcal{H}$ is then estimated by empirical risk minimization; when the loss function is the Bregman divergence (BD) [14] d_φ, the empirical risk is given by

$$R_{emp}(h) = \sum_{i=1}^{n} d_\varphi(y_i, h(x_i)) = \sum_{i=1}^{n} [\varphi(y_i) - \varphi(h(x_i)) - \{y_i - h(x_i)\} \psi(h(x_i))], \quad (1)$$

where φ is a convex function and ψ is its 1st derivative $\psi = \nabla\varphi$. It is well known that BD can generate various loss functions by varying φ [14,15]. When $\varphi(x) = x^2/2$, for example, BD corresponds to the squared error, see Table 1. In next section, we use BD to develop the general algorithms that can handle various divergences.

4 Proposed Method

4.1 Formulation of Grouped Uncoupled Regression

The grouped uncoupled regression (GUR) is the problem of estimating a model $h \in \mathcal{H}$ without requiring coupled data. Instead, problem is tackled by three types of data, $\mathcal{D}_I, \mathcal{D}_O$ and $\mathcal{D}_C$, made from grouped uncoupled data and pairwise comparison data. Below, we describe the details of data.

Grouped Uncoupled Data (GUD): GUD consists of Input Value Data (IVD) $\mathcal{D}_I$ and Output Value Data (OVD) $\mathcal{D}_O$. IVD is a set of input values for each group and is defined as $\mathcal{D}_I = \{\mathcal{D}_{I_k}\}_{k=1}^{n_K} = \{\{x_{ki}\}_{i=1}^{n_{I_k}}\}_{k=1}^{n_K}$ where n_K is the number of groups and n_{I_k} is the number of input values in the k-th group. Similarly, OVD is a set of output values for each group and is defined as $\mathcal{D}_O = \{\mathcal{D}_{O_k}\}_{k=1}^{n_K} = \{\{y_{kj}\}_{j=1}^{n_{O_k}}\}_{k=1}^{n_K}$ where n_{O_k} is the number of output values in k-th group. Note that, in general, $n_{I_k} \neq n_{O_k}$ since e.g., some invalid answers like "N/A" or "400 years old" may be weeded from OVD. We also denote the total number of IVD and that of OVD as $n_I = \sum_{k=1}^{n_K} n_{I_k}$, $n_O = \sum_{k=1}^{n_K} n_{O_k}$, respectively.

Pairwise Comparison Data (PCD): PCD is defined by $\mathcal{D}_C = \{\mathcal{D}_{C_k}\}_{k=1}^{n_K} = \{\{(x_{km}^+, x_{km}^-)\}_{m=1}^{n_{C_k}}\}_{k=1}^{n_K}$ where n_{C_k} is the number of PCD in the k-th group. (x_{km}^+, x_{km}^-) indicates that the output value is, given input x_{km}^+, larger than the output given x_{km}^-. We denote the total number of PCD as $n_C = \sum_{k=1}^{n_K} n_{C_k}$.

4.2 Loss Function

To derive the loss functions for GUR, we first define random variables and probability distributions used in this study. Let $\mathcal{K} = \{1, \cdots, n_K\}$ be a set of indexes of the group. We use the symbols K, X and Y as random variables on $\mathcal{K}$, $\mathcal{X}$ and $\mathcal{Y}$ following some joint distribution $P_{K,X,Y}$. We denote the marginal distribution on $\mathcal{K}$ (marginalized over X, Y) as P_K. Its probability mass function (PMF) is given by the symbol p_K. We also denote the conditional distribution and its probability density function (PDF) given $K = k$ as $P_{X,Y|k}$ and $f_{X,Y|k}$, respectively. Its marginal distribution on $\mathcal{X}$ ($\mathcal{Y}$) is denoted as $P_{X|k}$ ($P_{Y|k}$).

Using the above symbols, we define the loss function for GUR as the following expected BD:

$$\mathcal{R}(h) = \mathbb{E}_{K,X,Y}[d_\varphi(Y, h(X))] = \mathbb{E}_K[\mathcal{R}_k(h)] = \sum_{k=1}^{n_K} p_K(k)\mathcal{R}_k(h), \tag{2}$$

where $\mathbb{E}_{K,X,Y}$ and $\mathbb{E}_K$ is the expectation over $P_{I,J,X}$ and that over P_K, respectively. $\mathcal{R}_k$ is defined as follows.

$$\mathcal{R}_k(h) = \mathfrak{C}_k - \mathbb{E}_{X|k}[\varphi(h(X)) - h(X)\psi(h(X))] - \mathbb{E}_{X,Y|k}[Y\psi(h(X))], \tag{3}$$

where $\mathfrak{C}_k$ is a constant term and $\mathbb{E}_{X|k}, \mathbb{E}_{X,Y|k}$ is the expectation over $P_{X|k}$, $P_{X,Y|k}$, respectively.

The difficulty of evaluating $\mathcal{R}_k$ comes from the final term in Eq. (3) which involves the expectation over $P_{X,Y|k}$, $\mathbb{E}_{X,Y|k}[Y\psi(h(X))]$. This term cannot be evaluated even by sample averaging since input X and output Y are not observed simultaneously in (grouped) uncoupled data.

4.3   Approximation by Pairwise Comparison Data

To (approximately) evaluate the above problematic term, we introduce a pair of random variables on $\mathcal{X}$, (X^+, X^-); it indicates that the output value is, given X^+, larger than the output given X^-. The formal definition is given by

$$X^+ = \begin{cases} X & (Y \geq Y') \\ X' & (Y < Y') \end{cases}, \quad X^- = \begin{cases} X' & (Y \geq Y') \\ X & (Y < Y') \end{cases}, \tag{4}$$

where (X, Y) and (X', Y') are two independent random variables following $P_{X,Y|k}$. We denote the conditional distribution and its PDF of X^+ (X^-) given $K = k$ as $P_{X^+|k}$ ($P_{X^-|k}$) and $f_{X^+|k}$ ($f_{X^-|k}$), respectively. The expectation over $P_{X^+|k}$ ($P_{X^-|k}$) is denoted as $\mathbb{E}_{X^+|k}$ ($\mathbb{E}_{X^-|k}$). Later, we will use the fact that sample (x^+_{km}, x^-_{km}) in PCD is regarded as a realization of (X^+, X^-) given $K = k$.

The use of random variable (X^+, X^-) yields the following equation connecting terms using expectation over $P_{X^+|k}$, $P_{X^-|k}$ and $P_{X,Y|k}$:

Lemma 1. *Let $f_{Y|k}$ and $F_{Y|k}$ be PDF and cumulative density function (CDF) of probability distribution $P_{Y|k}$, respectively. Then,*

$$\mathbb{E}_{X^+|k}[\psi(h(X^+))] = 2\mathbb{E}_{X,Y|k}[F_{Y|k}(Y)\psi(h(X))], \tag{5}$$

$$\mathbb{E}_{X^-|k}[\psi(h(X^-))] = 2\mathbb{E}_{X,Y|k}[\{1 - F_{Y|k}(Y)\}\psi(h(X))]. \tag{6}$$

Proof. From the definition of X^+, $f_{X^+|k}$ is written as

$$f_{X^+|k}(x) = \frac{1}{Z} \int \int \int f_{X,Y|k}(x, y) f_{X,Y|k}(x', y') \mathbb{I}(y > y') dy dy' dx'$$

$$= \frac{1}{Z} \int\int f_{X,Y|k}(x, y) \left[\int f_{Y|k}(y') \mathbb{I}(y > y') dy' \right] dy = \frac{1}{Z} \int f_{X,Y|k}(x, y) F_{Y|k}(y) dy,$$

where Z is a normalizing factor and $Z = 1/2$ is obtained by integration by parts. Therefore, we get $\mathbb{E}_{X^+|k}[\psi(h(X^+))] = \int f_{X^+|k}(x)\psi(h(x))dx = \int 2\{\int f_{X,Y|k}(x,y) \, F_{Y|k}(y)dy\}\psi(h(x))dx = 2\mathbb{E}_{X,Y|k}[F_{Y|k}(Y)\psi(h(X))]$. Equation (6) is obtained in an analogous manner. $\square$

The term $\mathbb{E}_{X,Y|k}[F_{Y|k}(Y)\psi(h(X))]$ in Lemma 1 is different from the problematic term $\mathbb{E}_{X,Y|k}[Y\psi(h(X))]$ in Eq. (3). However, it can be used as its approximation allowing us to approximate the loss function $\mathcal{R}$ (Eq. (2)) as follows:

Theorem 1. *Suppose that there exists a constant M such that $\psi(x) < M$ for all $x \in \mathcal{X}$. The loss $\mathcal{R}(h)$ is approximated by $\tilde{\mathcal{R}}(h;\theta) = \mathbb{E}_K[\tilde{\mathcal{R}}_k(h;\theta_k)]$ and its approximation error $\mathbb{E}_K[|\mathcal{R}_k(h) - \tilde{\mathcal{R}}_k(h;\theta_k)|]$ is bounded by $\mathbb{E}_K[\mathrm{Err}_k(\alpha_k,\beta_k)]$ where*

$$\tilde{\mathcal{R}}_k(h;\theta_k=\{\alpha_k,\beta_k,\lambda_k\}) = \mathfrak{C}_k - \mathbb{E}_{X|k}[\varphi(h(X)) - (h(X) - \lambda_k)\psi(h(X))]$$
$$- \left(\alpha_k - \frac{\lambda_k}{2}\right)\mathbb{E}_{X^+|k}[\psi(h(X^+))] - \left(\beta_k - \frac{\lambda_k}{2}\right)\mathbb{E}_{X^-|k}[\psi(h(X^-))].$$
$$\mathrm{Err}_k(\alpha_k,\beta_k) = \int f_{Y|k}(y)\big|g_k(y;\alpha_k,\beta_k)\big|dy, \tag{7}$$
$$g_k(y;\alpha_k,\beta_k) = y - 2\alpha_k F_{Y|k}(y) - 2\beta_k\{1 - F_{Y|k}(y)\}.$$

Proof. Taking the sum of the equations in Lemma 1, we get $\mathbb{E}_{X,Y|k}[\psi(h(X)] = \frac{1}{2}\mathbb{E}_{X^+|k}[\psi(h(X^+))] + \frac{1}{2}\mathbb{E}_{X^-|k}[\psi(h(X^-))]$. Then, the value of $\tilde{\mathcal{R}}_k$ does not depend on λ_k, so we focus on $\tilde{\mathcal{R}}_k(h;\theta' = \{\alpha_k,\beta_k,0\})$. The bound is obtained as follows:

$$|\mathcal{R}_k(h) - \tilde{\mathcal{R}}_k(h;\theta')|$$
$$= \left|\mathbb{E}_{X,Y|k}[Y\psi(h(X))] - \alpha_k\mathbb{E}_{X^+|k}[\psi(h(X^+))] - \beta_k\mathbb{E}_{X^-|k}[\psi(h(X^-))]\right|$$
$$= \left|\iint f_{X,Y|k}(x,y)\psi(h(x))g_k(y;\alpha_k,\beta_k)dxdy\right|$$
$$\leq \iint f_{X,Y|k}(x,y)\big|\psi(h(x))\big|\big|g_k(y;\alpha_k,\beta_k)\big|dxdy \leq M\mathrm{Err}_k(\alpha_k,\beta_k). \square$$

The approximation bound can be exact when $F_{Y|k}$ is a uniform distribution, i.e., $F_{Y|k}(y) = (y - a_k)/(b_k - a_k)$ for all $y \in [a_k, b_k]$, since $g_k(y; b_k/2, a_k/2) = 0$. For general non-uniform distributions we can optimize α_k, β_k by minimizing the bound (explained in Sect. 5).

By removing constant terms and substituting the expectation over K, X, X^+, X^- by sample averaging from the approximated loss $\tilde{\mathcal{R}}$, we obtain the following objective function $\mathcal{L}$ that can be computed using GUD and PCD:

$$\mathcal{L}(h) = -\frac{1}{n_I}\sum_{x_{km}\in\mathcal{D}_I}\left\{\varphi(h(x_{km})) - (h(x_{km}) - \lambda_k)\psi(h(x_{km}))\right\} \tag{8}$$
$$- \frac{1}{n_C}\sum_{(x_{km}^+,x_{km}^-)\in\mathcal{D}_C}\left\{\left(\alpha_k - \frac{\lambda_k}{2}\right)\psi(h(x_{km}^+)) + \left(\beta_k - \frac{\lambda_k}{2}\right)\psi(h(x_{km}^-))\right\}.$$

This objective function reduces to the one used for UR in [4] when the number of groups, n_K equals 1. Thus $\mathcal{L}$ can be seen as a natural extension for handling grouped uncoupled data. Note that λ_k can take an arbitrary value (see the proof of Theorem 1). In experiment, we set $\lambda_k = (\alpha_k + \beta_k)/2$, similar to [4]. In Sect. 5, we construct Algorithm 1 based on this objective.

4.4 Re-approximation by Nearest Neighbor in Unlabeled Data

In this subsection, we consider converting the objective $\mathcal{L}$ (Eq. (8)) for deriving Algorithm 2. For the approximation, we make the following assumption.

Assumption 1 (Lipschitz continuity on ψ and h). *There exists a constant A and B such that $|\psi(x) - \psi(x')| \leq A|x - x'|$ and $|h(x) - h(x')| \leq B|x - x'|$ for all $x, x' \in \mathcal{X}$.*

This assumption holds in many divergence and models. From assumption 1, we obtain $|\psi(h(x)) - \psi(h(x'))| \leq AB|x - x'|$. So if sample point $x_{k\ell}^+$(or $x_{k\ell}^-$) $\in \mathcal{D}_{C_k}$ (PCD) and its (nearest) point $x' \in \mathcal{D}_{I_k}$ (IVD) are sufficiently close, $\psi(h(x_{k\ell}^+))$ can be approximated by $\psi(h(x'))$. We denote the nearest point $x' \in \mathcal{D}_{I_k}$ from point x as $\text{Nearest}_{D_{I_k}}(x)$, and use the symbol δ_+ and δ_- to represent the maximum the distance between a sample in GUD and its nearest sample in PCD in the following proof. The formal definitions are given as follows:

$$\delta_+ = \max_{k \in \mathcal{K}} \max_{x_{k\ell}^+ \in \mathcal{D}_{C_k}} \min_{x_{km} \in \mathcal{D}_{I_k}} |x_{k\ell}^+ - x_{km}|, \quad \delta_- = \max_{k \in \mathcal{K}} \max_{x_{k\ell}^+ \in \mathcal{D}_{C_k}} \min_{x_{km} \in \mathcal{D}_{I_k}} |x_{k\ell}^- - x_{km}|.$$

Theorem 2. *Under the assumption 1, the objective function $\mathcal{L}(\theta)$ is approximated by $\hat{\mathcal{L}}$ and its approximation error $|\mathcal{L}(h) - \hat{\mathcal{L}}(h)|$ is bounded by $\mathcal{E}$ where*

$$\hat{\mathcal{L}}(h) = -\frac{1}{n_I} \sum_{x_{km} \in \mathcal{D}_I} \left[\varphi(h(x_{km})) - \{h(x_{km}) - \tilde{y}_{km}\} \psi(h(x_{km})) \right], \quad (9)$$

$$\tilde{y}_{km} = \lambda_k + n_I \gamma_{km}^+ (\alpha_k - \lambda_k/2) + n_I \gamma_{km}^- (\beta_k - \lambda_k/2), \quad (10)$$

$\gamma_{km}^+ = \frac{1}{n_C} \sum_{(x_{k\ell}^+, x_{k\ell}^-) \in \mathcal{D}_C} \mathbb{I}(x_{km} = \text{Nearest}_{D_{I_k}}(x_{k\ell}^+))$, $\gamma_{km}^- = \frac{1}{n_C} \sum_{(x_{k\ell}^+, x_{k\ell}^-) \in \mathcal{D}_C}$ $\mathbb{I}(x_{km} = \text{Nearest}_{D_{I_k}}(x_{k\ell}^-))$, *and* $\mathcal{E} = AB \sum_{k=1}^{n_K} \frac{n_{C_k}}{n_C} \{|\alpha_k - \lambda_k/2|\delta^+ + |\beta_k - \lambda_k/2|\delta^-\}$.

Proof.

$$|\mathcal{L}(h) - \hat{\mathcal{L}}(h)|$$

$$= \frac{1}{n_C} \left| \sum_{(x_{km}^+, x_{km}^-) \in \mathcal{D}_C} \left[(\alpha_k - \frac{\lambda_k}{2}) \psi(h(x_{km}^+)) + (\beta_k - \frac{\lambda_k}{2}) \psi(h(x_{km}^-)) \right. \right.$$

$$\left. \left. - (\alpha_k - \frac{\lambda_k}{2}) \psi(h(\text{Nearest}_{D_{I_k}}(x_{km}^+))) - (\beta_k - \frac{\lambda_k}{2}) \psi(h(\text{Nearest}_{D_{I_k}}(x_{km}^-))) \right] \right|$$

$$\leq \frac{1}{n_C} \sum_{(x_{km}^+, x_{km}^-) \in \mathcal{D}_C} \left\{ \left| \alpha_k - \frac{\lambda_k}{2} \right| \left| \psi(h(x_{km}^+)) - \psi(h(\text{Nearest}_{D_{I_k}}(x_{km}^+))) \right| \right.$$

$$\left. + \left| \beta_k - \frac{\lambda_k}{2} \right| \left| \psi(h(x_{km}^-)) - \psi(h(\text{Nearest}_{D_{I_k}}(x_{km}^-))) \right| \right\}$$

$$\leq \frac{1}{n_C} \sum_{(x_{km}^+, x_{km}^-) \in \mathcal{D}_C} \left\{ \left| \alpha_k - \frac{\lambda_k}{2} \right| AB\delta^+ + \left| \beta_k - \frac{\lambda_k}{2} \right| AB\delta^- \right\} = \mathcal{E} \qquad \square$$

The value $\tilde{y}_{km}$ can be viewed as a quasi-output given input $x_{km} \in \mathcal{D}_I$. This is because, ignoring constant terms, the approximated objective $\hat{\mathcal{L}}(h)$ is equivalent to the empirical loss used for standard coupled regression (cf. Eq. (1)) given *quasi coupled data* (QCD) $\mathcal{D}_Q = \{\{x_{km}, \tilde{y}_{km}\}_{m=1}^{n_{I_k}}\}_{k=1}^{n_K}$, i.e.,

$$\hat{\mathcal{L}}(h) = \frac{1}{n_I} \sum_{(x_{km}, \tilde{y}_{km}) \in \mathcal{D}_Q} d_\varphi(\tilde{y}_{km}, h(x_{km})) + Const. \qquad (11)$$

Thus we can use (any) coupled regression method in minimizing this loss function $\hat{\mathcal{L}}$. In Sect. 5, we construct Algorithm 2 based on this viewpoint.

Algorithm 1. GUR-1

Input: $\mathcal{D}_I, \mathcal{D}_O, \mathcal{D}_C$ **Output:** $\hat{h}$
1: Estimate $(\hat{\alpha}_k, \hat{\beta}_k) \leftarrow \arg\min_{\alpha_k, \beta_k} \widehat{Err}_k$.

2: Estimate model $\hat{h} \leftarrow \arg\min_{h} \mathcal{L}$.

Algorithm 2. GUR-2

Input: $\mathcal{D}_I, \mathcal{D}_O, \mathcal{D}_C$ **Output:** $\hat{h}$
1: Estimate $(\hat{\alpha}_k, \hat{\beta}_k) \leftarrow \arg\min_{\alpha_k, \beta_k} \widehat{Err}_k$.
2: Compute $\tilde{y}_{km}$ for each $x_{km} \in \mathcal{D}_I$.
3: $\hat{h} \leftarrow$ FitRegressionModel($\mathcal{D}_Q$).

---

5   Algorithm

In this section, we construct the algorithms for GUR. The algorithms are derived using the loss function $\mathcal{L}$ (Eq. (8)) or $\hat{\mathcal{L}}$ (Eq. (9)), and the upper bound of its approximation error Err_k (Eq. (7)) shown in previous section. The algorithms consist of two steps: (i) the estimation of α_k, β_k by minimizing the upper bound, and (ii) the estimation of model h by solving the optimization problem $\min_h \mathcal{L}$ or $\min_h \hat{\mathcal{L}}$. Pseudo codes of the algorithms are shown in Alg. 1 and Alg. 2. We first state step (ii) of the two algorithms. Step (i), which is common to the two algorithms, is detailed at the end of this section.

5.1   Algorithm Using Eq. (8) as the Objective (GUR-1)

Here we explain the algorithm using $\mathcal{L}$ (Eq. (8)) as the the objective function, which we call GUR-1. When using models such as the linear-in-parameter model and deep neural network, objective $\mathcal{L}$ can be minimized in the following manner.

Linear Model: A linear-in parameter model is expressed by $h(x) = w^T \phi(x)$ where w is a model parameter and ϕ is a feature vector. When $\varphi(x) = x^2/2$ (i.e., BD is squared error), the loss function $\mathcal{L}$ is expanded to

$$\mathcal{L}(w) = \frac{1}{2} w^T \left[\frac{1}{n_I} \sum_{x_{km} \in \mathcal{D}_I} \phi(x_{km}) \phi^T(x_{km}) \right] w - w^T \left[\frac{1}{n_I} \sum_{x_{km} \in \mathcal{D}_I} \lambda_k \phi(x_{km}) \right.$$
$$\left. + \frac{1}{n_C} \sum_{(x_{km}^+, x_{km}^-) \in \mathcal{D}_C} \left\{ (\alpha_k - \lambda_k/2) \phi(x_{km}^+) + (\beta_k - \lambda_k/2) \phi(x_{km}^-) \right\} \right].$$

This loss function is quadratic w.r.t parameter, and can be solved analytically.

DNN Model: By using the representative framework for DNN such as PyTorch, we can minimize the loss function by the optimization module. Although we can derive the algorithm for the other model, it will require a model-dependent derivation and so is omitted.

5.2   Algorithm Using Eq. (9) as Loss Function (GUR-2)

Next, we explain the algorithm using $\hat{\mathcal{L}}$ (Eq. (9)) as the objective function, which we call GUR-2. As shown in Eq. (11), the loss function of GUR-2 can be seen as that of standard regression (if the loss function belongs to BD). Thus the learning of the model (minimization of the objective function) can be done by using (any) algorithm designed for the regression problem; we may use scikit-learn[2] for training the linear-in-parameter model and PyTorch for training DNN. Moreover, we can also use other representative supervised regression methods including Gaussian process (GP) [6] and decision tree variants (such as random forest [7] and gradient boosting decision tree (GBDT) [8]). Although these methods adopt additional techniques for e.g., hyperparameter optimization, computation cost reduction and training speed acceleration [6,9,10], no modifications are needed.

GUR-1 and GUR-2 have different strong points. GUR-1 does not rely on Assumption 1 and can be applied easily for the linear-model and deep neural network model. However, it may require some implementation for some models such as GP [6] and decision tree variants [7,8]. In contrast, although GUR-2 relies on Assumption 1, arbitrary models can be trained using algorithms designed for (standard) regression. Note that arbitrary regularization terms (such as L_2-norm and many others) can be added to the objective in both algorithms.

5.3   Hyperparameter Estimation

This subsection explains hyperparameter estimation, which is a common step in both GUR-1 and GUR-2. The hyperparameter $\{\alpha_k, \beta_k\}$ is estimated by minimizing the upper bound of the approximation. The upper bound Err_k can be evaluated by sample approximation using OVD $\mathcal{D}_O$ as follows:

$$\widehat{\mathrm{Err}}_k(\alpha_k, \beta_k) = \sum\nolimits_{y_{km} \in \mathcal{D}_{O_k}} \left| y_{km} - 2\alpha_k \hat{F}_{Y|k}(y_{km}) - 2\beta_k \{1 - \hat{F}_{Y|k}(y_{km})\} \right|, \quad (12)$$

where $\hat{F}_{Y|k}$ is the empirical approximation of $F_{Y|k}$, i.e., $\hat{F}_{Y|k}(y) = \frac{1}{n_{O_k}} \sum_{y_{km} \in \mathcal{D}_{O_k}} \mathbb{I}(y_{km} \leq y)$. Thus, we can estimate the hyperparameter by using arbitrary numerical optimization techniques to minimize $\widehat{\mathrm{Err}}_k$.

Remark: The above estimation algorithm shows that the availability of OVD $\mathcal{D}_Y$ is not essential. This is because, if knowledge of distribution $\{f_{Y|k}\}_{k=1}^{n_K}$ is available, we can estimate the hyperparameters by directly minimizing Err_k.

---

[2] https://scikit-learn.org/.

6 Validation Experiments

We conducted experiments on synthetic and benchmark datasets to confirm that the proposed two algorithms (GUR-1 and GUR-2) can well handle GUD.

6.1 Setting

For the synthetic data (Synth), we set the number of groups $n_K = 2$ and generated the parameters of linear-in-parameter model w from a uniform distribution on $[-1, 1]$. We first prepared coupled data as follows: input sample x is generated following a mixture of 1D-gaussian distribution (GMM) with mixing ratio $\pi = (1/2, 1/2)$, mean $\mu = (-2, 2)$, variance $\sigma^2 = 1.0$; output y is computed as $y = xw + \epsilon$ where ϵ is white noise that follows zero-mean Gaussian distribution with variance 0.1^2. If sample x generated from the k-th component, we treat x and output y as belonging to the k-th group, i.e., $x \in \mathcal{D}_{I_k}$ and $y \in \mathcal{D}_{O_k}$. We generated both training data and test data, and obtained the grouped uncoupled data by shuffling the index values of the outputs in training data. For PCD, we randomly selected a group k following mixing ratio π and generated pairs of variables (x^+, x^-) following Eq. (4) where X, X' follow the k-th component distribution of GMM.

For the benchmark data, we used three representative data for regression problem provided in the UCI machine learning repository[3], Boston housing data (Boston), concrete compressive strength data (Concrete) and auto-MPG data (MPG). Boston records 506 samples of house price with 13 features including average number of rooms per dwelling. Concrete records 1030 samples of the compressive strength of concrete with 8 features including amount of cement, water and so on. MPG records 398 samples of the city-cycle fuel consumption in miles per gallon of automobiles with 7 features including weight and horsepower. We excluded samples with missing value and converted the categorical feature in MPG to a one-hot vector. We divided the sample in each data into two groups depending on whether the average number of rooms (Boston)/amount of cement (Concrete)/weight of automobiles (MPG) is larger than the average or not. We prepared five data sets by randomly dividing the data and using 80% for training and 20% for the test. We obtained the grouped uncoupled data by shuffling the index of the outputs in training data. PCD is generated by randomly drawing two samples in each group.

We use test mean squared error (test MSE) as the performance metric. Test MSE is defined as $\frac{1}{|\mathcal{D}_{\text{test}}|} \sum_{(x_m, y_m) \in \mathcal{D}_{\text{test}}} \{y_m - h(x_m)\}^2$, where $\mathcal{D}_{\text{test}}$ is the test (coupled) data and $|\cdot|$ indicates the number of elements in the set. We used GUR-1 and GUR-2 algorithms with linear-in parameter model and $\varphi(x) = x^2/2$ (i.e., squared error). For the benchmark data, we also used GUR-2 algorithm with Gaussian process model using linear (dot-product) kernel (GUR-2(GP)). Their performances are compared with that of an existing method [4] (UR) that doesn't (cannot) use group information and standard (linear-)regression (SR)

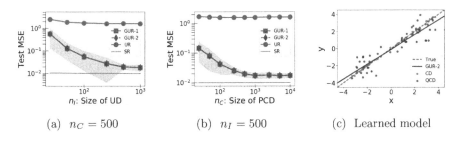

(a) $n_C = 500$ (b) $n_I = 500$ (c) Learned model

Fig. 2. MSE performance on synthetic datasets varying the number of (a) UD (b) PCD. Average and standard deviation are shown (log-scale). Lower values are better. (c) shows the true and estimated regression line by GUR-2, and coupled data (CD) used for SR and QCD used for GUR-2 when $n_I = 50, n_C = 100$.

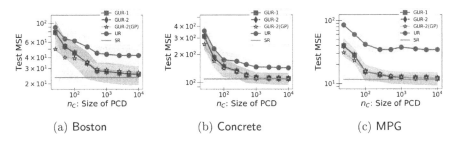

(a) Boston   (b) Concrete   (c) MPG

Fig. 3. MSE performance on three benchmark datasets. Average and standard deviation are shown (log-scale). SR is not a method for (G)UR since it uses coupled data. Lower values are better.

using coupled data. SR do not use PCD. Uncoupled data without group information (used in UR) are made by removing group information from GUD.

6.2   Results

Figure 2 shows the results of the synthetic data experiments. It is confirmed that both GUR-1 and GUR-2 outperform UR and approaches the performance of SR as the number of UD or PCD increases. This implies that our proposed algorithms for GUR well handle the grouped uncoupled data. The performance of GUR-1 and GUR-2 are almost same, and so the negative impact of Assumption 1 on performance was not observed in this experiment. From Fig. 2c, we can also see that quasi coupled data well capture the structure of the true regression line. These results validates the effectiveness of GUR-2. Figure 3 shows the results of the benchmark data experiments. These results also show that our two algorithms outperform UR and approaches SR. This supports the effectiveness of our algorithms. Moreover, we can confirm that the performance of GUR-2 (GP) is better than that of GUR-2 when the number of PCD is small. This imply the use of other models such as GP may improve the performance.

7   Conclusion

In this paper, we proposed the two algorithms for GUR. The algorithms are constructed by deriving the loss functions that can be evaluated using grouped uncoupled data and pairwise comparison data. The effectiveness of the proposal with the linear-in-parameter model and Gaussian process was confirmed by experiments on both synthetic and benchmark data sets. Future work in this research includes examining GUR performance with other representative models such as deep neural network [5] and a variant of decision tree [7,8].

References

1. Carpentier, A., Schlüter, T.: Learning relationships between data obtained independently. In: Artificial Intelligence and Statistics, pp. 658–666 (2016)
2. Hsu, D., Shi, K., Sun, X.: Linear regression without correspondence. In: Proceedings of the 31st International Conference on Neural Information Processing Systems, pp. 1530–1539 (2017)
3. Abid, A., Poon, A., Zou, J.: Linear regression with shuffled labels. arXiv preprint arXiv:1705.01342 (2017)
4. Xu, L., Niu, G., Honda, J., Sugiyama, M.: Uncoupled regression from pairwise comparison data. In: Advances in Neural Information Processing Systems, pp. 3992–4002 (2019)
5. Goodfellow, I., Bengio, Y., Courville, A.: Deep learning. MIT Press Cambridge (2016)
6. Williams, C.K., Rasmussen, C.E.: Gaussian processes for machine learning, vol. 2. MIT press Cambridge, MA (2006)
7. Breiman, L.: Random forests. Mach. Learn. **45**(1), 5–32 (2001)
8. Friedman, J.H.: Greedy function approximation: a gradient boosting machine. Ann. Statist. **29**(5), 1189–1232 (2001)
9. Titsias, M.: Variational learning of inducing variables in sparse Gaussian processes. In: Artificial intelligence and statistics, pp. 567–574. PMLR (2009)
10. Ke, G., et al.: LightGBM: a highly efficient gradient boosting decision tree. Adv. Neural. Inf. Process. Syst. **30**, 3146–3154 (2017)
11. Pananjady, A., Wainwright, M.J., Courtade, T.A.: Linear regression with shuffled data: statistical and computational limits of permutation recovery. IEEE Trans. Inf. Theory **64**(5), 3286–3300 (2017)
12. Slawski, M., Rahmani, M., Li, P.: A sparse representation-based approach to linear regression with partially shuffled labels. In: Proceedings of The 35th Uncertainty in Artificial Intelligence Conference, pp. 38–48 (2020)
13. Unnikrishnan, J., Haghighatshoar, S., Vetterli, M.: Unlabeled sensing with random linear measurements. IEEE Trans. Inf. Theory **64**(5), 3237–3253 (2018)
14. Bregman, L.M.: The relaxation method of finding the common point of convex sets and its application to the solution of problems in convex programming. USSR Comput. Math. Math. Phys. **7**(3), 200–217 (1967)
15. Banerjee, A., Merugu, S., Dhillon, I.S., Ghosh, J., Lafferty, J.: Clustering with Bregman divergences. J. Mach. Learn. Res. **6**(10), 1705–1749 (2005)

Additional Learning for Joint Probability Distribution Matching in BiGAN

Jiazhou Zheng$^{(\boxtimes)}$, Hiroaki Aizawa, and Takio Kurita

Hiroshima University, Higashihiroshima, Japan
`zhengjiazhou36@gmail.com`

Abstract. Bidirectional Generative Adversarial Networks (BiGANs) is a generative model with an invertible mapping between latent and image space. The mapping allows us to encode real images into latent representations and reconstruct input images. However, from preliminary experiments, we found that the joint probability distributions learned by the generator and the encoder are inconsistent, leading to poor-quality mapping. Therefore, to solve this issue, we propose an architecture-agnostic additional learning method to make the two joint probability distributions closer. In the experiments, we evaluated the reconstruction quality on synthetic and natural image datasets and found that our additional learning improves the invertible mapping of BiGAN.

Keywords: Bidirectional Generative Adversarial Networks · Joint Probability Distribution · Additional Learning

1 Introduction

Generative Adversarial Networks (GANs) [1] are capable of learning the latent distribution of arbitrarily complex data but do not have the ability to inverse mapping. Bidirectional Generative Adversarial Networks (BiGANs) [2,3] have been proposed a new unsupervised feature learning framework by adding an encoder E to the standard GAN framework to enable the mapping of data $\mathbf{x}$ to latent representations $\mathbf{z}$.

The goal of BiGAN is to make the encoder learn to invert the generator by training the joint probability distribution $Q(\mathbf{x}, E(\mathbf{x}))$ of the encoder E to match the joint probability distribution $P(G(\mathbf{z}), \mathbf{z})$ of the generator G. However, it turns out that the encoder of pre-trained BiGAN does not invert the generator very well.

In order to improve the reconstruction capabilities of BiGAN, previous works have been proposed by adding various losses to the original BiGAN framework. However, as the overall training of BiGAN follows an adversarial game, any change in the generator or encoder has harmful effects on the training phase and the architecture design.

In this work, to improve the reconstruction performance of the BiGAN encoder without changing the representational power of the generator, we propose an architecture-agnostic method to make the two joint probability distributions

M. Tanveer et al. (Eds.): ICONIP 2022, LNCS 13623, pp. 165–176, 2023.
https://doi.org/10.1007/978-3-031-30105-6_14

within BiGAN more closer, i.e. additional learning. We use the pre-trained BiGAN (generator and encoder) for additional learning, during which the parameters of the generator are fixed and only the encoder is additionally trained so that the encoder can better match the generator. In the experiments, our results demonstrate that our additional learning can improve the similarity of the joint probability distributions and improve the inverse mapping capability of the encoder. Moreover, our method is equally applicable to various BiGAN-related frameworks.

2   Related Work

2.1   Generative Adversarial Networks

The Generative Adversarial Networks (GANs) framework consists of two interacting neural networks, the generator G and the discriminator D, which are trained to learn the latent distribution of the data by iteratively performing adversarial game. The generator generates fake data $G(\mathbf{z})$ from the latent variable $\mathbf{z}$, while the discriminator determines the truthfulness of the input data($G(\mathbf{z})$ or $\mathbf{x}$). GAN's objective function $V_{GAN}(D, G)$ is as follows:

$$\min_{G} \max_{D} V_{GAN}(D, G) = \mathbb{E}_{\mathbf{x} \sim p(\mathbf{x})}[\log D(\mathbf{x})] + \mathbb{E}_{\mathbf{z} \sim p(\mathbf{z})}[1 - \log D(G(\mathbf{z}))], \quad (1)$$

where $D(\mathbf{x}) \in [0, 1]$ indicates truthfulness of the input. When the discriminator reaches an optimal solution, it will not be able to distinguish between real and fake samples, i.e. $D(\mathbf{x}) = 0.5$.

2.2   Wasserstein GAN

The training of GAN is unstable and difficult to achieve Nash equilibrium, and there are problems such as the loss not reflecting the good or bad training and the lack of diversity of the generated samples. Wasserstein GAN (WGAN) [4] introduces the Wasserstein distance to measure the distance between real data and fake data distribution, thus solving the problems of vanishing gradients and lack of generation diversity that have plagued GAN for a long time. Its objective function $V_{WGAN}(D, G)$ is as follows:

$$\min_{G} \max_{D} V_{WGAN}(D, G) = \mathbb{E}_{\mathbf{x} \sim p(\mathbf{x})}[D(\mathbf{x})] - \mathbb{E}_{\mathbf{z} \sim p(\mathbf{z})}[D(G(\mathbf{z}))], \quad (2)$$

where $D(\mathbf{x}) \in \mathbb{R}$. In WGAN, the function of the discriminator D is changed from judging to scoring the truthfulness of the input.

To better satisfy the requirement of Lipschitz continuity in WGAN, WGAN-GP [5] uses a gradient penalty instead of weight clipping. At this point, the new objective is as follows:

$$L_{WGAN-GP} = L_{WGAN} + \lambda \mathbb{E}_{\hat{\mathbf{x}} \sim p(\hat{\mathbf{x}})}[(\| \bigtriangledown_{\hat{x}} D(\hat{\mathbf{x}})\|_2 - 1)^2]. \quad (3)$$

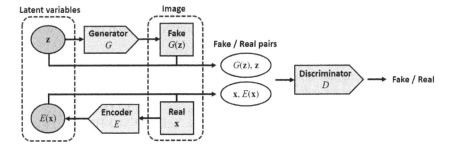

Fig. 1. Network structure of BiGAN

2.3 Bidirectional Generative Adversarial Networks

GAN allows us to map the latent variable to the generated data but cannot project the data back into the latent space. Bidirectional Generative Adversarial Networks (BiGANs) [2] or Adversarially Learned Inference model (ALIGAN) [3] addresses its lack of inverse mapping capability by adding an encoder to the original GAN framework. The discriminator will identify both the data and its corresponding latent component, i.e. $(\mathbf{x}, E(\mathbf{x}))$ and $(G(\mathbf{z}), \mathbf{z})$. Throughout the process, the generator G and the encoder E are trained simultaneously with the discriminator D for the adversarial game, as shown in Fig. 1. The objective function $V_{BiGAN}(D, E, G)$ is as follows:

$$\min_{G,E} \max_{D} V_{BiGAN}(D, E, G) = \mathbb{E}_{\mathbf{x} \sim p(\mathbf{x})}[\log D(\mathbf{x}, E(\mathbf{x}))] + \mathbb{E}_{\mathbf{z} \sim p(\mathbf{z})}[1 - \log D(G(\mathbf{z}), \mathbf{z})].$$

$$(4)$$

Improvement Techniques for BiGAN's Mapping. Although BiGAN gives GAN the ability to learn representations, BiGAN does not have good reconstruction capabilities and often reconstructs incorrect samples.

To improve the reconstruction capabilities of BiGAN, Rubenstein et al. [6] attempted to add various losses to the training of BiGAN that could be used to train the encoder to invert the generator. They found that adding additional loss to the training did improve the reconstruction ability of BiGAN, but it took a lot of time to select the right loss and tune its parameter for different datasets in order to get the best results.

Pablo Sánchez-Martín et al. [7] adopted the proposal of Multi-Discriminator Generative Adversarial Networks (MDGAN) [8] to add a reconstruction loss to the original BiGAN to drive better reconstruction (although they still refer to it by MDGAN, it is used to differ from the original MDGAN framework, and here we refer to the current model as MD-BiGAN). Their updated objective function is:

$$L_{MD-BiGAN} = L_{BiGAN} + \lambda_1 \mathbb{E}_{\mathbf{x} \sim p(\mathbf{x})}[(G(E(\mathbf{x})), \mathbf{x})], \qquad (5)$$

where λ_1 is the additional hyperparameter that controls the quality of the reconstruction. This term measures the reconstruction error based on some distance metric $d(\cdot, \cdot)$, typically minimum squared error.

Moreover, they found that although MD-BiGAN can provide a realistic reconstruction of the original image $\mathbf{x} \approx G(E(\mathbf{x}))$, $E(\mathbf{x})$ may not be typical of $P(\mathbf{z})$, where $P(\mathbf{z})$ is often regarded as a Gaussian distribution (or other simple distribution). Therefore, they added a regularization of $E(\mathbf{x})$ to the objective function of MD-BiGAN, called P-MDGAN (here, we call it P-MD-BiGAN), with an objective function:

$$L_{P-MD-BiGAN} = L_{MD-BiGAN} + \lambda_2 \mathbb{E}_{\mathbf{x} \sim p(\mathbf{x})}[(\|E(\mathbf{x})\|_2 - \sqrt{dim(\mathbf{z})})^2], \quad (6)$$

where $\| \cdot \|_2$ is the L_2 norm, $dim(\mathbf{z})$ is the latent dimension, and λ_2 is an automatically-tuned hyperparameter during training that ensures the distribution of the encoded latent vectors matches the prior distribution.

Consistency BiGAN (CBiGAN) [9] added cyclic consistency regularization to the generator and the encoder. The objective function is:

$$L_{CBiGAN} = (1 - \alpha)L_{BiGAN} + \alpha(\|\mathbf{x} - G(E(\mathbf{x}))\|_1 + \|\mathbf{z} - E(G(\mathbf{z}))\|_1), \quad (7)$$

where $\| \cdot \|_1$ is the L_1 norm and α controls the weight of each contribution.

3 Proposed Method

3.1 Visualization of BiGAN's Latent and Image Space

First, as a preliminary experiment to reveal underlying issues in BiGAN, we visualized the latent space and the reconstructed images obtained from pretrained BiGAN's generator and encoder which has been trained in advance on MNIST for 200 epochs. The visualizations are shown in Fig. 2.

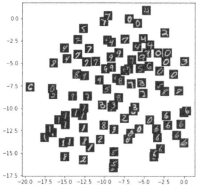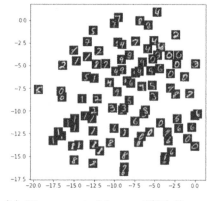

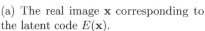

(a) The real image $\mathbf{x}$ corresponding to the latent code $E(\mathbf{x})$.

(b) The generated image $G(E(\mathbf{x}))$ corresponding to the latent code $E(\mathbf{x})$.

Fig. 2. Visualization of the latent space for image embedding.

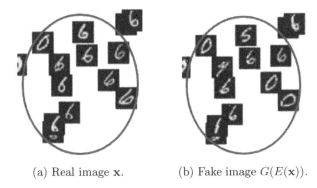

(a) Real image **x**.   (b) Fake image $G(E(\mathbf{x}))$.

Fig. 3. Visualization of inconsistent reconstructions.

As it can be observed in Fig. 2, we can see that although both the generator and the encoder learn the latent distribution of the MNIST data during training, the probability distribution $P(G(\mathbf{z}), \mathbf{z})$ of the generator G is not the same as the probability distribution $Q(\mathbf{x}, E(\mathbf{x}))$ of the encoder E. For the same locations in the latent space, the real image is inconsistent with the corresponding reconstructed image. Although most real images have the same labels as the corresponding reconstructed images, there are still differences in shape and detail. In addition, some images were incorrectly reconstructed. However, Fig. 3 shows that a tendency for semantic similarity can be observed in the distribution of the reconstructed images, despite the errors in the reconstruction.

The training goal of BiGAN is to enable the encoder E to be the inverse of the generator G, i.e., such that $P(G(\mathbf{z}), \mathbf{z}) = Q(\mathbf{x}, E(\mathbf{x}))$, but this is only theoretical and needs to be realized in the limit of the optimal discriminator.

In practice, although the semantics of the images are consistent in the two distributions $P(G(\mathbf{z}), \mathbf{z})$ and $Q(\mathbf{x}, E(\mathbf{x}))$ learned by BiGAN, the reconstructed images are not consistent with the original ones. In other words, the distribution between them does not match. Our insight is that it is an essential issue for learning better BiGAN to alleviate the misalignment of the distribution between the original and reconstructed images. This will be achieved by making the reconstructed image semantically and pixel-wise close to the original image.

3.2 Additional Learning

To improve the reconstruction capabilities of BiGAN, many existing improvements have been made by adding various consistency regularisation functions to the original BiGAN framework to constrain the encoder and the generator, thus enabling the encoder to 'communicate' directly with the generator and bring the probability distributions of the two into the same.

However, since the overall training of BiGAN still follows the adversarial game of the original GAN, any change in the generator or encoder has harmful effects on the training phase and the architecture design. Generator's loss will

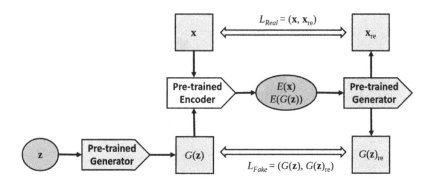

Fig. 4. Overview of our additional learning

change the representational power of the generator. Even if the loss is only for the encoder, it will likewise have some effect on the training of the generator as the training of the generator is directly influenced by the encoder.

According to the comprehensive experimental results of Rubenstein et al. [6], there are no optimal hyperparameters and architecture design for the generative capability, which depends on the capacity of the model and the objective function. Moreover, the training behaviour is sensitive to hyperparameters. Therefore, it requires laborious tuning on each dataset, which is computationally expensive.

Hence, in order not to impact the generator's representational power, our idea is to perform additional learning on the pre-trained encoder without changing the model.

From our preliminary visualizations, since BiGAN fails to reconstruct real images, we train the encoder using real images by the following loss function:

$$L_{Real}(\mathbf{x}, \mathbf{x}_{re}) = \frac{1}{N} \sum_{i=1}^{N} (\mathbf{x} - \mathbf{x}_{re})^2, \tag{8}$$

where N is the mini-batch size and $\mathbf{x}_{re} = G(E(\mathbf{x}))$ is the reconstructed image of $\mathbf{x}$. We use $L_{Real}(\mathbf{x}, \mathbf{x}_{re})$ to improve the match of the true data distribution in BiGAN, i.e. so that $P(\mathbf{x}_{re}, E(\mathbf{x})) = Q(\mathbf{x}, E(\mathbf{x}))$.

However, the latent space of the GAN is richer in images than the limited dataset. Alec Radford et al. [10] demonstrates that the GAN does not simply memorize the training samples. The rich semantic variation in the latent space allows the GAN to generate images beyond the original dataset. Therefore, using real images alone would result in the probability distributions of encoder and generation remaining at local optima.

To emphasize the association between encoder and generator, we also train the encoder by the following loss function:

$$L_{Fake}(G(\mathbf{z}), G(\mathbf{z})_{re}) = \frac{1}{N} \sum_{i=1}^{N} (G(\mathbf{z}) - G(\mathbf{z})_{re})^2, \tag{9}$$

where $G(\mathbf{z})_{re}$ is the reconstructed image of $G(\mathbf{z})$. We use $L_{Fake}(G(\mathbf{z}), G(\mathbf{z})_{re})$ to improve the matching of the fake data distribution in BiGAN, i.e. so that $P(G(\mathbf{z})_{re}, E(G(\mathbf{z}))) = Q(G(\mathbf{z}), E(G(\mathbf{z})))$.

Hence, the complete loss function L_{AL} is

$$L_{AL} = L_{Real}(\mathbf{x}, \mathbf{x}_{re}) + L_{Fake}(G(\mathbf{z}), G(\mathbf{z})_{re}). \tag{10}$$

As shown in Fig. 4, we perform additional learning on the encoder by using a pre-trained generator. In this case, the pre-trained generator has two functions: to acquire random fake images and to reconstruct the real and fake images that have been encoded by the encoder.

4 Evaluation

4.1 Evaluation Protocol

To validate the effectiveness of the proposed additional learning, we evaluated BiGAN, and its follow-up works with and without the additional learning scheme on both synthetic and natural image benchmarks. For the model architecture, we selected BiGAN [2], MD-BiGAN [8], P-MD-BiGAN [8], and CBiGAN [9] as baseline. All models were optimized with Adam [11] and stopped after 200 epochs. The dataset we used and the details are shown in Table 1. The architectures of Encoder, Generator, and Discriminator are based on the convolutional structure. To make the training more stable, we applied a gradient penalty of WGAN-GP to the training scheme. We set the other model-dependent hyperparameters as follows. For MD-BiGAN, $\lambda_1 = 9, 8, 7$ (FashionMNIST, CIFAR10, CelebaA). For P-MD-BiGAN, $\lambda_2 = 9, 7, 7$ (FashionMNIST, CIFAR10, CelebaA). For CBiGAN, $\alpha = 0.00001$.

Table 1. The Datasets used in our experiments.

Dataset	Image size	Mini-batch size	Pre-processing
FashionMNIST	$28 \times 28 \times 1$	64	None
CIFAR10	$32 \times 32 \times 3$	64	None
CelebaA	$64 \times 64 \times 3$	256	Area downsampled

In the additional learning step, we used the generator and the encoder of each pre-trained model, all parameters of the generator were fixed during training and only the encoder was subjected to additional learning with the 20 epochs.

To quantitatively validate the role of the additional learning, we used PSNR (Peak Signal-to-Noise Ratio), SSIM (Structural Similarity), and LPIPS (Learned Perceptual Image Patch Similarity) [12] as evaluation metrics. For PSNR and SSIM, the higher the score the better, while for LPIPS, the lower the score the better. These metrics allows us to measure the reconstruction quality based on similarity between the two images. We calculated these scores from the real image $P_{\mathbf{x}}$ and the corresponding reconstructed image $P_{\mathbf{x}_{re}}$, and on the randomly generated image $P_{G(\mathbf{z})}$ and the corresponding reconstructed image $P_{G(\mathbf{z})_{re}}$.

4.2   The Representational Power of BiGAN and Its Variants

We investigated the representational power of BiGAN and its variants of the pre-trained generator before performing additional learning. To measure the performance of their generators, We used the FID (Frechet Inception Distance score) [13] to calculate the distance between feature vectors of real images $\mathbf{x}$ and generated images $G(\mathbf{z})$.

As shown in Table 2, although the models with the best FID scores varied across the different datasets, by combining the evaluations of each dataset, the original BiGAN model achieved better FID scores compared with BiGAN variants. It can be seen that each BiGAN variant incorporates a different consistency loss during training to obtain good reconstruction capability, but at the cost of instability in the generator's representational power.

Table 2. FID score results for BiGAN and its variants. Lower FID is better.

Model	Dataset		
	FashionMNIST	CIFAR10	CelebaA
BiGAN	10.05	**57.70**	98.46
MD-BiGAN	18.85	74.47	108.34
P-MD-BiGAN	16.89	67.45	**91.55**
CBiGAN	**9.78**	66.35	101.94

4.3   Reconstruction Results on Synthetic Images

As shown in Table 3, after additional learning, BiGAN's ability to reconstruct real images and random fake images was substantially improved, with both PSNR scores increasing by several times and average SSIM scores increasing by 55%.

Although the BiGAN variants already showed significant improvements in reconstruction capability over the original BiGAN, the additional learning still allowed them to improve further in terms of reconstruction capability. The PSNR scores of all variants improved, while the average SSIM scores also improved by about 3.8% to 12.6%.

Table 3. Reconstruction results. We evaluated convolutional BiGAN and its variants on FashionMNIST. The "real" indicates the similarity between $(P_{\mathbf{x}}, P_{\mathbf{x}_{re}})$ and the "fake" indicates the similarity between $(P_{G(\mathbf{z})}, P_{G(\mathbf{z})_{re}})$.

Model	w/ AL	PSNR		SSIM	
		real	fake	real	fake
BiGAN		3.0417	3.0349	0.1747	0.1833
	✓	**13.2874**	**13.7659**	**0.7133**	**0.7509**
MD-BiGAN		15.5700	17.3853	0.7894	0.8496
	✓	**16.7484**	**19.6225**	**0.8250**	**0.8943**
P-MD-BiGAN		15.4344	16.9487	0.7899	0.8477
	✓	**16.8011**	**19.3884**	**0.8248**	**0.8894**
CBiGAN		10.9508	11.7047	0.6313	0.6808
	✓	**14.4497**	**15.8151**	**0.7509**	**0.8124**

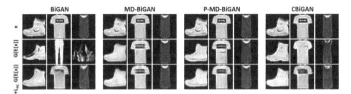

(a) The real image $\mathbf{x}$ and the reconstructed image $\mathbf{x}_{re}$.

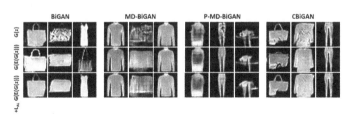

(b) The fake image $G(\mathbf{z})$ and the reconstructed fake image $G(\mathbf{z})_{re}$.

Fig. 5. Qualitative results on FashionMNIST. The first row is the real or fake image, the second row is the original reconstructed image and the third row is the reconstructed image after additional learning.

As shown in Fig. 5, after additional learning, the reconstruction errors in the pre-trained BiGAN also ceased to appear, and the tendency to reconstruct semantically close images was resolved. Moreover, for each pre-trained BiGAN variant model, the additional learning improved the detailed similarity of the corresponding reconstructed images.

Table 4. Reconstruction results. We evaluated convolutional BiGAN and its variants on CIFAR10.

Model	w/ AL	PSNR		SSIM		LPIPS	
		real	fake	real	fake	real	fake
BiGAN		4.6591	4.9101	0.0893	0.1035	0.1766	0.1763
	✓	**13.3993**	**15.2819**	**0.5526**	**0.7070**	**0.1033**	**0.0735**
MD-BiGAN		14.6478	17.5625	0.6513	0.8579	**0.0832**	0.0295
	✓	**15.5717**	**20.8723**	**0.6834**	**0.9162**	0.0932	**0.0210**
P-MD-BiGAN		14.1335	16.7885	0.6184	0.8234	**0.0734**	0.0360
	✓	**15.3736**	**20.4430**	**0.6708**	**0.9038**	0.0794	**0.0219**
CBiGAN		13.5123	16.7193	0.5752	0.8023	0.0863	0.0407
	✓	**14.7991**	**19.9099**	**0.6379**	**0.8871**	**0.0826**	**0.0239**

Table 5. Reconstruction results. We evaluated convolutional BiGAN and its variant on CelebA.

Model	w/ AL	PSNR		SSIM		LPIPS	
		real	fake	real	fake	real	fake
BiGAN		0.8557	0.7118	0.1233	0.1210	0.3402	0.3446
	✓	**12.6331**	**16.3348**	**0.5383**	**0.8261**	**0.1821**	**0.0618**
MD-BiGAN		13.1888	14.1777	0.5585	0.6587	0.1828	0.1710
	✓	**15.2102**	**21.7175**	**0.6449**	**0.9421**	**0.1461**	**0.0262**
P-MD-BiGAN		12.9391	14.9890	0.5487	0.7440	0.1670	0.1007
	✓	**15.1351**	**22.5889**	**0.6368**	**0.9409**	**0.1520**	**0.0255**
CBiGAN		2.3185	1.9134	0.1217	0.1119	0.2916	0.3060
	✓	**12.9370**	**16.0808**	**0.5416**	**0.8199**	**0.1905**	**0.0686**

4.4   Reconstruction Results on Natural Images

As shown in Tables 4 and 5, for both CIFAR10 and CelebaA, for BiGAN, the PNSR scores improved significantly after additional learning, with average SSIM scores improving by 53% and 56%, and LPIPS also decreasing by an average of 8.9% and 22%. For the BiGAN variants, PSNR scores improved for all variants, with average SSIM scores improving by 4.5% to 7.4% and 14.3% to 56.4% respectively, and LPIPS also decreasing respectively (except for LPIPS scores for MD-BiGAN and P-MD-BiGAN about $(P_\mathbf{x}, P_{\mathbf{x}_{re}})$ in CIFAR10).

The generated images of MD-BiGAN and P-MD-BiGAN in Fig. 6 indicates that although the LPIPS scores of these two variants concerning $(P_\mathbf{x}, P_{\mathbf{x}_{re}})$ have increased by about 1%, the detailed similarity of their reconstructed images has been improved.

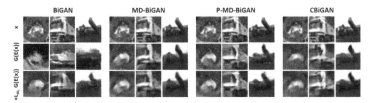

(a) The real image **x** and the reconstructed image $\mathbf{x}_{re}$.

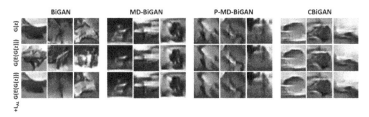

(b) The fake image $G(\mathbf{z})$ and the reconstructed fake image $G(\mathbf{z})_{re}$.

Fig. 6. Qualitative results on CIFAR10. The first row is the real or fake image, the second row is the original reconstructed image and the third row is the reconstructed image after additional learning.

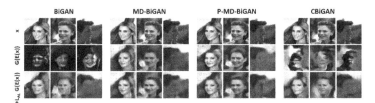

(a) The real image **x** and the reconstructed image $\mathbf{x}_{re}$.

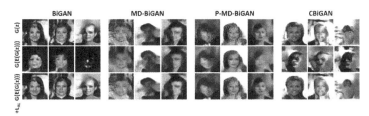

(b) The fake image $G(\mathbf{z})$ and the reconstructed fake image $G(\mathbf{z})_{re}$.

Fig. 7. Qualitative results on CelebaA. The first row is the real or fake image, the second row is the original reconstructed image and the third row is the reconstructed image after additional learning.

5   Conclusion

We proposed additional learning to improve the similarity of the generator and encoder probability distributions in the BiGAN model without affecting the capability of the generator in BiGAN. Our experimental results showed that our proposed method effectively improves the reconstruction capability of BiGAN and its variants for both real and fake images, and is independent of the structure design of BiGAN.

However, our proposed method is an additional process during training and thus requires additional computational resources and time. Moreover, for more complex data, we must consider the difficulty of matching their probability distributions. As part of future work, we will continue to work on how to further deal with these issues of inconsistent probability distributions in BiGAN.

References

1. Goodfellow, I., et al.: Generative adversarial nets. In: Advances in Neural Information Processing Systems 27 (2014)
2. Donahue, J., Krähenbühl, P., Darrell, T.: Adversarial feature learning. In: ICLR (2017)
3. Dumoulin, V., et al.: Adversarially learned inference. In: ICLR, Alex Lamb (2017)
4. Arjovsky, M., Chintala, S., Bottou, L.: Wasserstein generative adversarial networks. In: International Conference on Machine Learning, pp. 214–223. PMLR (2017)
5. Gulrajani, I., Ahmed, F., Arjovsky, M., Dumoulin, V., Courville, A.C.: Improved training of Wasserstein GANs. In: Advances in Neural Information Processing Systems 30 (2017)
6. Rubenstein, P.K., Li, Y., Roblek, D.: An empirical study of generative models with encoders. arXiv preprint arXiv:1812.07909 (2018)
7. Sánchez-Martín, P., Olmos, P.M., Perez-Cruz, F.: Improved BiGAN training with marginal likelihood equalization. arXiv preprint arXiv:1911.01425 (2019)
8. Intrator, Y., Katz, G., Shabtai, A.: MDGAN: boosting anomaly detection using multi-discriminator generative adversarial networks. arXiv preprint arXiv:1810.05221 (2018)
9. Carrara, F., Amato, G., Brombin, L., Falchi, F., Gennaro, C.: Combining GANs and autoencoders for efficient anomaly detection. In: 2020 25th International Conference on Pattern Recognition (ICPR), pp. 3939–3946. IEEE (2021)
10. Radford, A., Metz, L., Chintala, S.: Unsupervised representation learning with deep convolutional generative adversarial networks. arXiv preprint arXiv:1511.06434 (2015)
11. Kingma, D.P., Ba, J.: Adam: a method for stochastic optimization. In: ICLR (2015)
12. Zhang, R., Isola, P., Efros, A.A., Shechtman, E., Wang, O.: The unreasonable effectiveness of deep features as a perceptual metric. In: Proceedings of the IEEE Conference on Computer Vision and Pattern Recognition, pp. 586–595 (2018)
13. Heusel, M., Ramsauer, H., Unterthiner, T., Nessler, B., Hochreiter, S.: GANs trained by a two time-scale update rule converge to a local Nash equilibrium. In: Advances in Neural Information Processing Systems 30 (2017)

Multi-view Self-attention for Regression Domain Adaptation with Feature Selection

Mehdi Hennequin[1,2(✉)] (iD), Khalid Benabdeslem[2], Haytham Elghazel[2], Thomas Ranvier[2] (iD), and Eric Michoux[1]

[1] Galilé Group, 28 Bd de la République, 71100 Chalon-sur-Saône, France
m.hennequin@fondation.galile.fr
[2] Université Lyon 1, LIRIS, UMR CNRS 5205, 69622 Lyon, France
{khalid.benabdeslem,haytham.elghazel,thomas.ranvier}@univ-lyon1.fr

Abstract. In this paper, we address the problem of unsupervised domain adaptation in a regression setting considering that source data have different representations (multiple views). In this work, we investigate an original method which takes advantage of different representations using a discrepancy distance while using attention-based neural networks mechanism to estimate feature importance in domain adaptation. For this purpose, we will begin by introducing a novel formulation of the optimization objective. Then, we will develop an adversarial network domain adaptation algorithm adjusting weights given to each feature, ensuring that those related to the target receive higher weights. Finally, we will evaluate our method on public dataset and compare it to other domain adaptation baselines to demonstrate the improvement for regression tasks.

Keywords: Domain Adaptation · Feature Selection · Multi-view · Regression

1 Introduction

In most predictive maintenance problems, data are collected from various production lines, assembly lines, or are captured by different devices. Those industrial processes there define several domains each with different distribution. In this context, an algorithm trained for predictive maintenance for a specific domain (referred as source domain) cannot be correctly generalized to another domain (referred as target domain). Therefore, it is common practice to retrain the predictive maintenance models. However, this retraining leads to delayed forecast actions until enough data are available for accurate prediction. To address this issue, predictive models, trained with a specific domain, have to adapt to data with different distributions and limited or non-existing fault information. In machine learning, this situation is often referred to as *domain adaptation* or *covariate shift* [28]. In general, domain adaptation methods attempt to solve the learning problem when the main learning task is the same but the domains have different feature spaces or different marginal conditional probabilities [22,25,35].

M. Tanveer et al. (Eds.): ICONIP 2022, LNCS 13623, pp. 177–188, 2023.
https://doi.org/10.1007/978-3-031-30105-6_15

On the other hand, data can be represented by several independent sets of features. For instance, in the example of the aforementioned predictive maintenance, data are collected from diverse sensors and exhibit heterogeneous properties. Thus, data from different sensors can be naturally partitioned into independent groups [37]. Each group is referred to as a particular view. Multi-view learning [15,37] aims to improve predictors by taking advantage of the redundancy and consistency between these multiple views.

In the domain adaptation context, views are generally concatenated into one single view to adapt to the learning task. However, this concatenation might cause negative transfer [39], (*i.e.* introduce source domain data/knowledge undesirably) because each view has a specific statistical property. This will result in a decreased learning performance in the target domain. Furthermore, the risk of negative transfer might also come from one or several features being prejudicial to adaptation [35]. It is particularly true with adversarial methods trying to match source and target domains. Therefore, to avoid negative transfer we want to find a way to give high weights to features most related to the target domain. We can find little research on multi-view domain adaptation [36,38] where considerable attention has been given on the classification problem, while regression task and selection features remains largely under-studied.

In this paper, we propose a novel approach for multi-view domain adaptation using self-attention for regression tasks. This work makes two main contributions: first, we propose to extend the measure between distributions Source-guided Discrepancy [13] to multi-views learning concept, and we also adapt this measure to Adversarial method. The second main contribution is the introduction of self-attention to select important features to avoid negative transfer. We conduct experiments on real-world datasets and improve on state-of-the-art results for multi-view adversarial domain adaptation for regression.

2   Learning Scenario

This section introduces the definitions and concepts needed for the following sections. For the most part, we follow the definitions and notations of Cortes and Mohri [5]. Let $\mathcal{X} \in \mathbb{R}^p$ and $\mathcal{Y} \in \mathbb{R}$, denote respectively input and output spaces. We define a domain as a pair formed by a distribution over $\mathcal{X}$ and a target labeling function mapping from $\mathcal{X}$ to $\mathcal{Y}$. Throughout the paper, (Q, f_Q) denotes the source domain and (P, f_P) the target domain with Q the source and P the target distribution over $\mathcal{X}$ and with $f_Q, f_P : \mathcal{X} \to \mathcal{Y}$ the source and target labeling functions, respectively. In the scenario of *multi-view domain adaptation* the learning algorithm receives a labeled sample $\mathcal{S}$ of m points from the source domain, and the data instances can be represented in M different views. More formally, for $v \in \{1, ..., M\}$, $\mathcal{S}_v = \{(x_1^{(v)}, y_1^{(v)}), ..., (x_m^{(v)}, y_m^{(v)})\} \in (\mathcal{X} \times \mathcal{Y})^m$ where $\mathcal{S}_{x_v} = \{x_1^{(v)}, ..., x_m^{(v)}\}$ is supposed to be drawn i.i.d. according to distribution Q and $y_i = f_Q(x_i)$ for all $i \in [1, m]$. In the same way, we define unlabeled samples from the target domain, $\mathcal{T} = \{x_1', ..., x_n'\} \in \mathcal{X}^n$ where $\mathcal{T}_x = \{x_1', ..., x_n'\}$ is assumed drawn iid according to P and $y_i = f_P(x_i')$ for all $i \in [1, n]$. We denote by $\hat{Q}$ and

$\hat{P}$ the empirical distributions of the respective samples $\mathcal{S}_x$ and $\mathcal{T}_x$. We consider in the following that the covariate shift assumption holds, i.e. $f = f_Q = f_P$.

We also consider a loss function $L : \mathcal{Y} \times \mathcal{Y} \rightarrow \mathbb{R}_+$ jointly convex in its two arguments. The L_p loss functions commonly used in regression and defined by $L_p(y, y') = |y - y'|$ for $p > 1$ are special instances of this definition. We define a hypothesis class H of hypotheses $h : X \rightarrow Y$ .For any two functions $h, h' : \mathcal{X} \rightarrow \mathcal{Y}$ and any distribution D over $\mathcal{X}$, we denote by $L_D(h, h')$, the expected loss of $h(x)$ and $h'(x) : L_D(h, h') = \mathbb{E}_{x \sim D}[L(h(x), h'(x))]$.

Objectif. The goal of Domain Adaptation is to minimize the target risk $\mathcal{L}_P(h, f_p) = \mathbb{E}_{x \sim P}[L(h(x), f_P(x))]$. In unsupervised domain adaptation, no label is available in the target task and we cannot directly estimate f_P. Consequently we want to leverage the information about the labels in the source domains f_Q to adapt to the target domain.

3   Adversarial Algorithm for Multi-view Domain Adaptation

3.1   Source-guided Discrepancy (S-disc)

The Source-guided Discrepancy (S-disc) introduced in [13], is defined as the maximal difference between source and target risk over a set of hypotheses. We recall below its definition.

Definition 1. *(Source-guided discrepancy). Let $\mathcal{H}$ be a hypothesis class and $h, h^* \in \mathcal{H}$. S-disc between two distributions Q and P is defined as:*

$$\varsigma_{\mathcal{H}}^l(P, Q) = \max_{h \in \mathcal{H}} |\mathcal{L}_P(h, h_S^*) - \mathcal{L}_Q(h, h_S^*)|. \tag{1}$$

where $h_Q^* = \arg\min_{h \in \mathcal{H}} \mathcal{L}_Q(h, f_Q)$ in the source domain. Here, note that the risk minimizer h_Q^* is not necessarily equal to the labeling function f_Q as we consider a restricted hypothesis class. S-disc offers several advantages compared to existing discrepancy measures [13], nevertheless in the context of multi-view learning S-disc is not adapted. Consider M hypotheses, $h_v^* \in \mathcal{H}_v$, for $v \in \{1...M\}$ (for reasons of simplification we abbreviate $h_1 \in \mathcal{H}_1, ..., h_M \in \mathcal{H}_M$ with $h_v \in \mathcal{H}_v$), such as $h_v^* = \arg\min_{h_v \in \mathcal{H}_v} \mathcal{L}_{Q_v}(h_v, f_{Q_v})$, with $\bigcup_{v \in M}(Q_v, f_{Q_v}) \subseteq (Q, f_Q)$, where the pair (Q_v, f_{Q_v}) is the vth subset of (Q, f_Q) (we note that $f_{Q_v} = f_Q$). In our context a view is considered as a subset of source domain. In this case, it is difficult to choose the appropriate predictor h_v^* to measure the difference between the two domains with S-disc. To overcome this problem we define a novel discrepancy measure Multi-Views-guided Discrepancy (MV-Disc):

Definition 2. *(Multi-Views-guided Discrepancy) For any $h_v^* \in \mathcal{H}_v$:*

$$MV\text{-}Disc(P, Q) = \max_{h \in \mathcal{H}} |\frac{1}{M} \sum_{v=1}^{M} \mathcal{L}_P(h, h_v^*) - \frac{1}{M} \sum_{v=1}^{M} \mathcal{L}_{Q_v}(h, h_v^*)| \tag{2}$$

3.2 Propositional Self-attention Feature Importance

In this section we explore the use of attention-based neural networks mechanism for estimating feature importance in domain adaptation. This section is inspired from the seminal works on the attention mechanism [3,29,34]. We took inspiration from [29] regarding feature importance, with a different implementation of the attention mechanism, we defined it as follows:

$$\Omega(X) = \frac{1}{k} \bigoplus_k \left[\text{softmax}\left(f(q^k(W^k X + b^k))\right) \right] \tag{3}$$

Input vectors $X \in \mathcal{X}$ are first used as input to a softmax-activated layer containing the number of neurons equal to the number of features p, where the softmax function applied to the j_i-th element of a weight vector v is defined as:

$$\text{softmax}(v_{j_i}) = \frac{\exp(v_{j_i})}{\sum_{j=1}^{p} \exp(v_j)} \tag{4}$$

where $v \in \mathbb{R}^p$. Note that k represents the number of attention heads distinct matrices representing relations between the input features. The $\otimes$ sign corresponds to the Hadamard product, the $\oplus$ refers to the Hadamard summation across individual heads and f corresponds to the activation function. For the activation function f we use a $tanh$ as proposed in [1]. We extend the idea of integrating a weight vector q following the attention layer as proposed in [24]. The proposed architecture maintains a bijection between the set of features p and the set of weights in W, thereby the weights in the head can be understood as relations between features [29]. Ω is a mapping from the feature space to the space of non-negative real values, i.e. $\Omega : p \to \mathbb{R}_0^+$, to obtain the pondered features we multiply the output of Ω by the input space X, and we define $\Omega_R = \Omega(X) \otimes X$.

3.3 Propositional Algorithm

A Min-Max Problem. Since the Multi-Views-guided Discrepancy is defined as a maximum on a functional space, we propose to use adversarial training to align domains. We introduce a feature extractor called generator $G : \mathcal{X} \to \mathcal{Z}$, typically a neural network parametrized by ϕ. The generator aims to produce a latent space $\mathcal{Z}$ where domains are not distinguished by any predictor $h \in \mathcal{H}_{\mathcal{Z}}$, such as $\mathcal{H}_{\mathcal{Z}} : \mathcal{Z} \to \mathcal{Y}$. Using the proposed attention mechanism Ω_R and the definition of MV-Disc, we formulate the following objective function for our Adversarial Multi-Views Self Attention-guided Discrepancy (AMVSAD). For any $h_v^* \in \mathcal{H}_v$:

$$\min_{\Omega_{R_v} \in \mathcal{H}_v, g_\phi \in \mathcal{H}, \Omega_{R_T} \in \mathcal{H}} \max_{h \in \mathcal{H}} |\frac{1}{M} \sum_{v=1}^{M} \mathcal{L}_P(h \circ G_\phi \circ \Omega_{R_T}, h_v^* \circ G_\phi \circ \Omega_{R_T})$$

$$- \sum_{v=1}^{M} \beta_v \mathcal{L}_{Q_v}(h \circ G_\phi \circ \Omega_{R_v}, h_v^* \circ G_\phi \circ \Omega_{R_v})| + \lambda_1 \sigma(h) + \lambda_2 ||\beta||_2 \tag{5}$$

where σ is the spectral norm, λ_1 and λ_2 are hyperparameters. We add spectral normalization [20] to control the discriminator and avoid instability during the training. We also propose to attribute weights to each view [16,26], β ensuring that the most related views to the target receive higher weights. Ω_{R_v} is the vth attention mechanism for the vth view and Ω_{R_T} is the target attention mechanism. For any given $h \in \mathcal{H}, h_v^* \in \mathcal{H}_v$ the discrepancy term constrains all three representations ϕ_θ, weights β and Ω_R to align domains.

While computing the true solution of this min-max problem is still impossible in practice, we derive an alternate optimization algorithm. Similarly to most other adversarial methods, we sequentially optimize differents parameters of our networks according to different objectives. At a given iteration, losses are minimized/maximized sequentially (the general structure of our algorithm is available at the following github repository[1]):

Step 1. First, we train the predictors and generator on labeled source data with the different views. Our aim is for the vth predictor to predict correctly the vth view to obtain h_v^*. For $v \in \{1, ..., M\}$:

$$\min_{h_v \in \mathcal{H}_v, \Omega_{R_v} \in \mathcal{H}_v, G_\phi \in \mathcal{H}} \mathcal{L}_{Q_v}(h_v \circ G_\phi \circ \Omega_{R_v}, f_{Q_v}). \tag{6}$$

Step 2. Thus, we update the predictor h as a discriminator to increase the MV-Disc loss for a fixed generator:

$$\max_{h \in \mathcal{H}} |\frac{1}{M} \sum_{v=1}^{M} \mathcal{L}_P(h \circ G_\phi \circ \Omega_{R_T}, h_v^* \circ G_\phi \circ \Omega_{R_T})$$
$$- \sum_{v=1}^{M} \beta_v \mathcal{L}_{Q_v}(h \circ G_\phi \circ \Omega_{R_v}, h_v^* \circ G_\phi \circ \Omega_{R_v})| + \lambda_1 \sigma(h) \tag{7}$$

Step 3. We train the generator, the attention mechanism, and β to minimize the MV-Disc loss for fixed predictor h:

$$\min_{\Omega_{R_v} \in \mathcal{H}_v, \Omega_{R_T} \in \mathcal{H}, g_\phi \in \mathcal{H}} |\frac{1}{M} \sum_{v=1}^{M} \mathcal{L}_P(h \circ G_\phi \circ \Omega_{R_T}, h_v^* \circ G_\phi \circ \Omega_{R_T})$$
$$- \sum_{v=1}^{M} \beta_v \mathcal{L}_{Q_v}(h \circ G_\phi \circ \Omega_{R_v}, h_v^* \circ G_\phi \circ \Omega_{R_v})| + \lambda_2 ||\beta||_2 \tag{8}$$

---

[1] https://github.com/HennequinMehdi/Adversarial-Multi-View-Attention-guided-Discrepancy.git.

4   Related Work

Adversarial Domain Adaptation. Adversarial techniques, for Domain Adaptation was introduced in [8]. Based on the $\mathcal{H}\Delta\mathcal{H}$-divergence [2], authors found a new representation of the input features where source and target instances cannot be distinguished by any discriminative hypothesis. [27,32,40], follow a similar idea. The above mentioned papers give considerable attention to classification setting, while regression task and selection features remains largely under-studied. Nonetheless, the authors in [19,26] propose methods tailored for regression task. Compared to above mentioned methods we add an attention mechanism to assist the generator to find a subspace shared by domains selecting the most relevant features.

Discrepancy Minimization. The present work is in line with discrepancy minimization methods, which were first introduced in [17], and further developed in [4,5,13,21,26,40]. More specifically, our algorithm aims at minimizing the empirical S-disc introduced in [13]. Discrepancy is the key measure of the difference between two distributions in the context of domain adaptation and has several advantages over other common divergence measures such as the l_1 distance. Besides, several generalizations bound for adaptation in terms of discrepancy were proposed [4–6,21]. In comparison to others methods, we introduce the concept of subset in the source domain, in this way we can use different views to compare two distribution.

Feature Selection Domain Adaptation. Classical feature selection methods [10] are not designed for domain adaptation. For instance, in [14], the authors searched a latent subspace and deploys $l_{2,1}$-norm to select common features shared by the domains. Another example of feature selection methods in domain adaptation are [33] and [9]. The contribution of the former paper consists in the use of parametric maximum mean discrepancy distance in order to find a weight matrix that allows to identify invariant and shifting features in the original space. The method described in the latter paper proposes a similar idea using optimal transport to find a shared feature representation. The biggest advantage using our method over the above mentioned ones, is the search of domains shared features during the training of the regression task (Table 1).

5   Experiments

In this section, we evaluate our AMVSAD method. It should be noted that unsupervised Domain Adaptation with multi-view for regression is hard to evaluate as we have no real public database that corresponds entirely to the problem we described in the introduction. Consequently, we build scenarios and, for each one, we will describe the protocol. We report the results of the AMVSAD algorithm compared to other domain adaptation methods. The experiments are conducted

Table 1. Superconductivity experiments MSE

Expe.	l →ml	l →mh	l →h	ml →l	ml →mh	ml→ h	mh →l
WANN	0.0844	0.0469	0.0343	0.0404	0.0544	**0.0276**	0.0391
KLIEP	0.0619	0.0418	0.0446	0.0377	0.0400	0.0372	0.0268
KMM	0.0667	0.0694	0.0273	0.0513	0.0428	0.0282	0.0342
DANN	0.0885	0.0501	0.0333	**0.0335**	0.1134	0.3368	0.0578
ADDA	0.0450	0.0448	0.1155	0.0340	**0.0310**	0.1626	0.0478
DeepCORAL	0.0672	0.0431	0.0493	0.0502	0.0553	0.0324	0.0538
MDD	0.0691	0.0450	0.0446	0.0395	0.0483	0.0325	0.0343
TrAdaBoostR2	0.0627	0.0499	0.0417	0.0480	0.0538	0.0284	0.0410
AHD-MSDA	0.0801	0.0324	**0.0264**	0.0679	0.0559	0.0592	**0.0259**
AMVSAD	**0.0281**	**0.0252**	0.0275	0.0780	0.0570	0.0772	0.0496

Expe.	mh →ml	mh →h	h →l	h →ml	h →mh	Avg.
WANN	0.0630	0.0661	0.0300	0.0712	0.0395	0.0497
KLIEP	0.0685	0.0337	0.0273	0.0656	0.0429	0.0440
KMM	0.0587	0.0955	0.0350	0.0680	0.0410	0.0515
DANN	0.1052	0.0262	0.0498	0.1235	0.0472	0.0888
ADDA	0.0815	**0.0256**	**0.0264**	0.1877	0.0322	0,0695
DeepCORAL	0.0769	0.0642	0.0586	0.0694	0.0507	0.0559
MDD	0.0667	0.0499	0.0477	0.0762	0.0578	0.0510
TrAdaBoostR2	0.0654	0.0744	0.0427	0.0664	0.0466	0.0517
AHD-MSDA	0.0514	0.0386	0.0292	0.0662	0.0325	0.0471
AMVSAD	**0.0204**	0.0528	0.0368	**0.0206**	**0.0299**	**0.0419**

on public dataset. The following competitors are selected to compare the performance of our algorithm:

- Weighting Adversarial Neural Network (WANN) [19] is a semi-supervised domain adaptation method based on the empirical $\mathcal{Y}$-discrepancy [21]. It is used for regression tasks.
- Discriminative Adversarial Neural Network (DANN) [8] is an unsupervised domain adaptation method. It is used here for regression tasks by considering the mean squared error as task loss instead of the binary cross-entropy proposed in the original algorithm.
- Adversarial Discriminative Domain Adaptation (ADDA) [32] performs a DANN algorithm in two-stage: it first learns a source encoder and a task hypothesis using labeled data and then learns the target encoder with adversarial training.
- Deep Correlation Alignment (DeepCORAL) [31] is an unsupervised domain adaptation method that aligns the second-order statistics of the source and target distributions with a linear transformation.
- Margin Disparity Discrepancy (MDD) [41] is an unsupervised domain adaptation, it learns a new feature representation by minimizing the disparity discrepancy.

- TrAdaBoostR2 [23] is a semi-supervised domain adaptation method for regression tasks. The method is based on a reverse-boosting principle where the weight of source instances poorly predicted are decreased at each boosting iteration.
- Kullback-Leibler Importance Estimation Procedure (KLIEP) [30] is a sample bias correction method minimizing the KL-divergence between a reweighted source and target distributions.
- Kernel Mean Matching (KMM) [12] reweights source instances in order to minimize the MMD between domains.
- Adversarial Hypothesis-Discrepancy Multi-Source Domain Adaptation (AHD-MSDA) [26] is a multi-source unsupervised domain adaptation.

We propose here to demonstrate the efficiency of AMVSAD on the UCI dataset superconductivity [7,11]. The goal is to predict the critical temperature of superconductors. This is a common regression problem in the industry, as industrialists are particularly interested in modeling the relationship between a material and its properties. The dataset contains two views: the first view contains 81 features extracted from 21263 superconductors, while the second view contains the chemical formula broken up for all the 21263 superconductors, whose format is binary. We divide this dataset into separate domains as per the setup of [23]. We select an input feature with a moderate correlation factor with the output (0.3). We then sort the set according to this feature and split it into four parts: low (l), midle-low (ml), midle-high (mh), high (h). Each part defines a domain with around 5000 instances. We conduct an experiment for each pair of domains which leads to 12 experiments. For each pair of domains we also randomly select different features from the two views. Therefore, the source domain and the target domain do not have the same features. 10 target labeled instances are used in the training except for our method, AHD-MSDA, which benefit multi-view/multi-source learning method. The other target data are used to compute the results. For the multi-source methods such as AHD-MSDA, we consider a view to be a source, while for the other baseline methods that do not consider multi-source learning, we merge the views. We reported the results in tables, We also report the average MSE over the 12 experiments. For all base line methods implementation except AHD-MSDA, the python library ADAPT is used [41]. The optimization parameters used in the presented experiments for baseline methods are $lr = 0.01/0.001$, and the loss function is the mean squared error (MSE). The base hypothesis used to learn the task is a neural network with two hidden fully-connected layers of 100 neurons each, ReLU activation functions, weights clipping $C = 1$ and Adam optimizer; 250/350 epochs with a batch size of 128 are performed. Cross-validation is also applied to select best parameters and best scores for each baseline method. Our method and AHD-MSDA have been implemented using the Pytorch library [18], and the network architecture in table. For more detail, the codes and experiments are available at the following github repository[2] (Table 2).

---

[2] https://github.com/HennequinMehdi/Adversarial-Multi-View-Attention-guided-Discrepancy.git.

Table 2. Architectures AMVSAD

Generator	Discriminator
Dense(size(features),50, LeakyReLU)	Spectral_Norm(Dense(25, 9, ReLU))
Dropout(0.1)	Spectral_Norm(Dense(9, 1, ReLU))
Dense(50, 25, Tanh)	$\lambda_1 = 0.001$ for Spectral Norm
Predictors	**Optimization parameters**
Dense(25, 9, ReLU)	Adam lr = 0.001, epochs = 100, $\lambda_2 = 0.01$
Dense(9, 1, ReLU)	**Attention Mechanism Head**:
	Dense(size(features), size(features))
	Attention Mechanism q:
	Dense(size(features), size(features))
	Num Head and q = 2

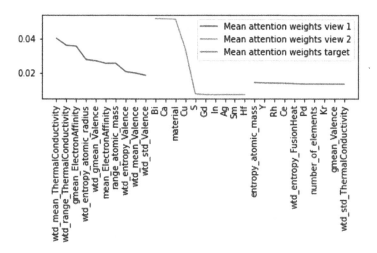

Fig. 1. Features importance of experiments $l \rightarrow ml$

Discussion. Overall, we find that our method performs better than state-of-the-art methods in the target domain. However, for a pair of domains our method performs as well or less than some methods. The reason for this is that some methods leverage information from a few target labeled instances during training, thus penalizing our performance. Nevertheless, the advantage of our method is that we can access the attention level associated with each feature averaged over predictions (see Fig. 1). Since we compute feature importance shared by domains, we can visualize the feature's ranking that contributes to adaptation.

6   Conclusion

In this work, we proposed an adversarial domain adaptation algorithm based on a new discrepancy, MV-Disc, tailored for multi-view regression. We demonstrated the efficiency of our method in real dataset especially with feature importance. For our future work, we aim to extend our MV-disc to classification problems. In the future we hope to access to more real database in our problematic to perform more exhaustive experiments. We also intend to investigate the self-supervised learning and active learning settings, to try labeling target data with a high degree of confidence.

References

1. Bahdanau, D., Cho, K., Bengio, Y.: Neural machine translation by jointly learning to align and translate. CoRR abs/1409.0473 (2015)
2. Ben-David, S., Blitzer, J., Crammer, K., Pereira, F.: Analysis of representations for domain adaptation. In: Schölkopf, B., Platt, J., Hoffman, T. (eds.) Advances in Neural Information Processing Systems, vol. 19. MIT Press, Cambridge (2006)
3. Chorowski, J.K., Bahdanau, D., Serdyuk, D., Cho, K., Bengio, Y.: Attention-based models for speech recognition. In: Cortes, C., Lawrence, N., Lee, D., Sugiyama, M., Garnett, R. (eds.) Advances in Neural Information Processing Systems, vol. 28. Curran Associates, Inc., Red Hook (2015)
4. Cortes, C., Mohri, M.: Domain adaptation in regression. In: Kivinen, J., Szepesvári, C., Ukkonen, E., Zeugmann, T. (eds.) ALT 2011. LNCS (LNAI), vol. 6925, pp. 308–323. Springer, Heidelberg (2011). https://doi.org/10.1007/978-3-642-24412-4_25
5. Cortes, C., Mohri, M.: Domain adaptation and sample bias correction theory and algorithm for regression. Theor. Comput. Sci. **519** (2014)
6. Cortes, C., Mohri, M., Medina, A.M.: Adaptation based on generalized discrepancy. J. Mach. Learn. Res. **20**(1), 1–30 (2019)
7. Dua, D., Graff, C.: UCI machine learning repository (2017). http://archive.ics.uci.edu/ml
8. Ganin, Y., et al.: Domain-adversarial training of neural networks. J. Mach. Learn. Res. (2016)
9. Gautheron, L., Redko, I., Lartizien, C.: Feature selection for unsupervised domain adaptation using optimal transport. CoRR abs/1806.10861 (2018)
10. Guyon, I., Elisseeff, A.: An introduction to variable and feature selection. J. Mach. Learn. Res. **3**, 1157–1182 (2003)
11. Hamidieh, K.: A data-driven statistical model for predicting the critical temperature of a superconductor. Comput. Mater. Sci. **154**, 346–354 (2018)
12. Huang, J., Gretton, A., Borgwardt, K., Schölkopf, B., Smola, A.: Correcting sample selection bias by unlabeled data. In: Schölkopf, B., Platt, J., Hoffman, T. (eds.) Advances in Neural Information Processing Systems, vol. 19. MIT Press, Cambridge (2007)
13. Kuroki, S., Charoenphakdee, N., Bao, H., Honda, J., Sato, I., Sugiyama, M.: Unsupervised domain adaptation based on source-guided discrepancy (2018)
14. Li, J., Zhao, J., Lu, K.: Joint feature selection and structure preservation for domain adaptation. In: Proceedings of the Twenty-Fifth International Joint Conference on Artificial Intelligence. IJCAI 2016, pp. 1697–1703. AAAI Press (2016)

15. Li, Y., Yang, M., Zhang, Z.: Multi-view representation learning: a survey from shallow methods to deep methods. CoRR abs/1610.01206 (2016)
16. Mansour, Y., Mohri, M., Rostamizadeh, A.: Domain adaptation with multiple sources. In: Advances in Neural Information Processing Systems 21, Proceedings of the Twenty-Second Annual Conference on Neural Information Processing Systems, Vancouver, British Columbia, Canada, 8–11 December 2008, pp. 1041–1048. Curran Associates, Inc. (2008)
17. Mansour, Y., Mohri, M., Rostamizadeh, A.: Domain adaptation: learning bounds and algorithms. In: COLT 2009 - The 22nd Conference on Learning Theory, Montreal, Quebec, Canada, 18–21 June 2009 (2009)
18. de Mathelin, A., Deheeger, F., Richard, G., Mougeot, M., Vayatis, N.: ADAPT: awesome domain adaptation python toolbox. CoRR abs/2107.03049 (2021)
19. de Mathelin, A., Richard, G., Mougeot, M., Vayatis, N.: Adversarial weighting for domain adaptation in regression. CoRR abs/2006.08251 (2020)
20. Miyato, T., Kataoka, T., Koyama, M., Yoshida, Y.: Spectral normalization for generative adversarial networks. CoRR abs/1802.05957 (2018)
21. Mohri, M., Muñoz Medina, A.: New analysis and algorithm for learning with drifting distributions. In: Bshouty, N.H., Stoltz, G., Vayatis, N., Zeugmann, T. (eds.) ALT 2012. LNCS (LNAI), vol. 7568, pp. 124–138. Springer, Heidelberg (2012). https://doi.org/10.1007/978-3-642-34106-9_13
22. Pan, S.J., Yang, Q.: A survey on transfer learning. IEEE Trans. Knowl. Data Eng. 22(10), 1345–1359 (2010)
23. Pardoe, D., Stone, P.: Boosting for regression transfer. In: Proceedings of the 27th International Conference on International Conference on Machine Learning. ICML 2010, pp. 863–870. Omnipress, Madison (2010)
24. Ranvier, T., Benabdeslem, K., Bourhis, K., Canitia, B.: Deep multi-view learning for tire recommendation. In: 2021 International Joint Conference on Neural Networks (IJCNN), pp. 1–8 (2021)
25. Redko, I., Morvant, E., Habrard, A., Sebban, M., Bennani, Y.: A survey on domain adaptation theory. CoRR abs/2004.11829 (2020)
26. Richard, G., Mathelin, A., Hébrail, G., Mougeot, M., Vayatis, N.: Unsupervised multi-source domain adaptation for regression. In: Hutter, F., Kersting, K., Lijffijt, J., Valera, I. (eds.) ECML PKDD 2020. LNCS (LNAI), vol. 12457, pp. 395–411. Springer, Cham (2021). https://doi.org/10.1007/978-3-030-67658-2_23
27. Saito, K., Watanabe, K., Ushiku, Y., Harada, T.: Maximum classifier discrepancy for unsupervised domain adaptation. In: 2018 IEEE Conference on Computer Vision and Pattern Recognition, CVPR 2018, Salt Lake City, UT, USA, 18–22 June 2018. pp. 3723–3732. IEEE Computer Society (2018)
28. Shimodaira, H.: Improving predictive inference under covariate shift by weighting the log-likelihood function. J. Stat. Plann. Infer. 90(2), 227–244 (2000)
29. Skrlj, B., Dzeroski, S., Lavrac, N., Petkovic, M.: Feature importance estimation with self-attention networks. CoRR abs/2002.04464 (2020)
30. Sugiyama, M., Nakajima, S., Kashima, H., Buenau, P., Kawanabe, M.: Direct importance estimation with model selection and its application to covariate shift adaptation. In: Platt, J., Koller, D., Singer, Y., Roweis, S. (eds.) Advances in Neural Information Processing Systems, vol. 20. Curran Associates, Inc., Red Hook (2008)
31. Sun, B., Saenko, K.: Deep CORAL: correlation alignment for deep domain adaptation. In: Hua, G., Jégou, H. (eds.) ECCV 2016. LNCS, vol. 9915, pp. 443–450. Springer, Cham (2016). https://doi.org/10.1007/978-3-319-49409-8_35
32. Tzeng, E., Hoffman, J., Saenko, K., Darrell, T.: Adversarial discriminative domain adaptation. CoRR abs/1702.05464 (2017)

33. Uguroglu, S., Carbonell, J.: Feature selection for transfer learning. In: Gunopulos, D., Hofmann, T., Malerba, D., Vazirgiannis, M. (eds.) ECML PKDD 2011. LNCS (LNAI), vol. 6913, pp. 430–442. Springer, Heidelberg (2011). https://doi.org/10.1007/978-3-642-23808-6_28

34. Vaswani, A., et al.: Attention is all you need. In: Guyon, I., et al. (eds.) Advances in Neural Information Processing Systems, vol. 30. Curran Associates, Inc., Red Hook (2017)

35. Wilson, G., Cook, D.J.: A survey of unsupervised deep domain adaptation. ACM Trans. Intell. Syst. Technol. **11**(5) (2020)

36. Xia, Y., et al.: Uncertainty-aware multi-view co-training for semi-supervised medical image segmentation and domain adaptation. Med. Image Anal. **65**, 101766 (2020)

37. Xu, C., Tao, D., Xu, C.: A survey on multi-view learning. CoRR abs/1304.5634 (2013)

38. Yang, P., Gao, W., Tan, Q., Wong, K.F.: Information-theoretic multi-view domain adaptation. In: Proceedings of the 50th Annual Meeting of the Association for Computational Linguistics (Volume 2: Short Papers), pp. 270–274. Association for Computational Linguistics, Jeju Island, July 2012

39. Zhang, W., Deng, L., Wu, D.: Overcoming negative transfer: a survey. CoRR abs/2009.00909 (2020)

40. Zhang, Y., Liu, T., Long, M., Jordan, M.: Bridging theory and algorithm for domain adaptation. In: Chaudhuri, K., Salakhutdinov, R. (eds.) Proceedings of the 36th International Conference on Machine Learning. Proceedings of Machine Learning Research, vol. 97, pp. 7404–7413. PMLR, 09–15 June 2019

41. Zhang, Y., Liu, T., Long, M., Jordan, M.: Bridging theory and algorithm for domain adaptation. CoRR **abs/1904.05801** (2019). http://arxiv.org/abs/1904.05801

EigenGRF: Layer-Wise Eigen-Learning for Controllable Generative Radiance Fields

Zhiyuan Yang[✉][iD] and Qingfu Zhang[iD]

Department of Computer Science, City University of Hong Kong,
Kowloon Tong, Hong Kong
zhiyuan.yang@my.cityu.edu.hk

Abstract. Neural Radiance Fields (NeRF) learn a model for the high-quality 3D-view reconstruction of a single object. Category-specific representation makes it possible to generalize to the reconstruction and even generation of multiple objects. Existing efforts mainly focus on the reconstruction performance including speed and quality. The steerability of generation processes has not been well studied while semantic attributes still exist in 3D neural representations. Inspired by interpreting underlying factors of GANs, this paper proposes a novel method named Eigen-GRF to disentangle the latent semantic subspace in an unsupervised manner. By learning a set of eigenbasis, we can readily control the process and the result of object synthesis accordingly. Concretely, our method brings a mapping network to NeRF by conditioning on a FiLM-SIREN layer. Then we use a component analysis method for discovering steerable latent subspaces. Our experiments reveal that the proposed method is powerful for the 3D-aware generation with steerability by both synthetic and real-world datasets.

Keywords: Neural Radiance Fields · Generative Model · Disentanglement Learning · 3D-Aware Generation

1 Introduction

A fundamental task of computer vision and computer graphics is to learn how to represent the 3D shape and appearance of a scene given only 2D observations. Recent neural radiance fields (NeRF) [22] have reached state-of-the-art performance by mapping a spatial coordinate and a camera view to the volume density and the light field using a neural network. This efficient and powerful method with many other variants has made a figure in a number of applications, such as novel view synthesis [2,27,35], and scene relighting [24,34] (Fig. 1).

Vanilla NeRF can only render novel views of a single object or a scene. Following works [3,30] condition NeRF-like network on the latent code to form category-specific implicit representations, which learns the shape and appearance of multiple objects of the same class from images leveraging a GAN-based [9] structure. Utilizing generative models to map latent codes into the radiance

© The Author(s), under exclusive license to Springer Nature Switzerland AG 2023
M. Tanveer et al. (Eds.): ICONIP 2022, LNCS 13623, pp. 189–199, 2023.
https://doi.org/10.1007/978-3-031-30105-6_16

Fig. 1. Selected examples synthesized by EigenGRF with CelebA [20]. The first row is a random sample generated by EigenGRF under five camera viewpoints (angle $= -50$, -25, 0, 25, 50) of one person in a 3D representation. We show the attribute change by traversing the sampled latent point along two specific bases of the learning subspace in the last two rows.

fields, these methods typically concentrate on the generation capability and the 3D consistency rendering. Other efforts [15,19] also try to disentangle the variation of object shapes and appearances as different latent codes, respectively. By optimizing the optimal codes in the latent space, unseen objects can be inferred from a few or even single images.

Previous efforts often care about the result of reconstruction. While controlling and editing an implicit continuous volumetric representation is challenging but valuable. For 3D editing, prior works primarily focus on editing an explicit representation such as mesh and voxel, and they do not apply to implicit neural representations.

2D generative models [9,14,29,41] have shined for many years and they have achieved photorealistic image synthesis with high fidelity. In this domain, there have been many approaches for generative adversarial networks [9] to dig and try to understand the black-box synthesis process. The generative models have a common structure that utilizes a generator to map the low-dimensional latent space into the high-dimensional data space. We could define a prior distribution for the latent space, while this prior may not match the faithful and agnostic data manifold. It is an obstacle yielding less accurate generation. Recent works [6, 10,12,13,18,42] have uncovered a lot of approaches to manipulate the latent codes and bring the steering of some attributes of the generation results. This steerability should freeze the other parts or attributes and only influence our care parts.

Inspired by 2D disentanglement and controllability, we hope to enable 3D shape controlling and editing with NeRF representation in an unsupervised manner. Our method can be classified as a category-specific 3D semantic editing method. The recent work [16] also pays attention to the editing task. They first disentangle the latent shape and color codes and then interactively update the network and latent code parameters to fit the desired object. Their main contribution is that individual images can be manually edited and transferred to the entire 3D model. However, we find an unsupervised way to exploit the potentialities of the NeRF-like model and provide a more principled way to control the generation.

In this paper, we first design a generative radiance field for 3D-aware synthesis. Then we mine interpretable and controllable possibilities on 3D structures. Specifically, we condition the NeRF model on an embedding latent vector comprised of a set of linear subspaces with orthogonal bases. These principal variations learn to capture the semantic attributes. The condition is operated by a FILM-SIREN [26,33] layer. We hope this will inspire the community for further research and pave the way for the 3D-aware generation with full steerability.

2 Related Work

We first review the controllability in 2D generative models and then transfer this mechanism into the 3D domain. Utilizing neural representation and neural rendering, we concentrate on the branch of category-specific synthesis.

2.1 Controllability in 2D Generative Models

Conventional generative models [9] excel at generating realistic random samples with statistics resembling the training set. However, controllable and interactive matters rather than random. GANs do not provide an inherent way of comprehending or controlling the underlying generative factors. But some research shows that a well-trained GAN is able to encode different semantics inside the latent space. Therefore, a fundamental problem of generative models is to gain explicit control of the synthesis process or results. The goal is to generate or modify images satisfying users' specific requirements. The requirement should be semantical meaningful interpretable and easy to distinguish without affecting other attributes. The manipulation could be single or multi attributes of interest.

There are two paradigms in the domain of 2D controllable generative models, one is the conditional GANs [4,23] which uses a variable connecting to the main networks and the other is to find a meaningful direction to manipulate the generation latent manifold. The latter mechanism could also help interpret the black box of generation models [12,32] by disentangling the latent space and discovering meaningful image manipulation directions.

GLO [1] pioneered optimizing the latent code and a decoder simultaneously. Many non-disentangled GAN-based methods [31,32,43] supervised or unsupervised discover semantically meaningful directions in the latent space of existing state-of-the-art models (e.g., StyleGAN [17,18]) by dissecting the layer weights and feature outputs. LowRankGAN [42] and EigenGAN [13] both introduce several subspaces that enable more precise control of GAN generation. EigenGAN [13] especially inspects the layers of a generator and learns explicit dimensions for semantic attribute control simultaneously. Editing 3D scenes have received much attention in the computer graphics community. This paper combines existing 2D knowledge with 3D reconstruction knowledge organically.

2.2   3D Shape Representation

Recently, a new prominent direction for 3D shape and appearance representation has emerged. Since photorealistic synthesis is the destination of researchers in computer vision. In the field of 3D representation, the goal is to infer the scene structure and view-dependent appearance given a set of 2D images. It's more difficult than 2D synthesis because the real 3D world contains a lot of complex elements (lights, reflections, depth, etc.).

Implicit Neural Representations [5,11,21,25] have exploded in recent years by learning a continuous function (neural network) to represent 3D geometry objects or scenes. Compared with the conventional approaches based on voxel grids [39,40] or meshes [28,38] which discretize space and are restricted in topology, implicit representations provide a compact and continuous mechanism to capture impressive levels of geometric details. The advantage of continuity also makes it possible to be trained with only 2D images using differentiable rendering. Neural rendering projects a 3D neural representation into multiple 2D images which could backpropagate the reconstruction error to optimize. DeepSDF [25] proposes to learn an implicit function where the output of the network represents the signed distance of the point to its nearest surface. NeRF [22] approximates a continuous 5D scene representation with multilayer perceptron (MLP) networks.

2.3   3D-Aware Image Generation

Given unstructured 2D image collections, 3D-aware image generation aims to learn a generative model that can explicitly control the camera viewpoint of the synthesis content. 2D datasets lack knowledge of 3D structure, we usually learn a 3D prior from a collection within a category. 3D-aware synthesis is highly needed because our real-world objects are three-dimensional and 3D information also brings richer perception. Methods to solve can be categorized into two types. The first line of work [36,37] have been tried to disentangle pose and identity with an encoder-decoder structure, but these 2D generations still struggle to synthesize novel view images with identity consistent at high quality. Another line [3,30] generates a 3D shape model and outputs view-consistent images simultaneously. Directly learning a 3D model with 3D shape representation and synthesizing images under a physical-based rendering process (volumetric rendering) achieves more strict 3D consistency. These methods optimize the whole model using an adversarial loss from the discriminators. To reduce the expensive computational cost of volumetric representation learning, [7] learns a generative radiance field on 2D manifolds, which efficiently achieves more realistic image generation with finer details.

3   Methods

In this section, we first give a statement of such a 3D-aware synthesis task in Sect. 3.1. Then we introduce the central part of EigenGRF in Sect. 3.2, a

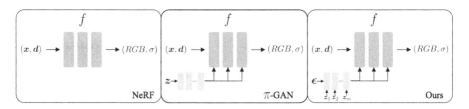

Fig. 2. Comparison of vanilla NeRF [22], π-GAN [3] and our method. (1) Vanilla NeRF use a sequence MLPs f for mapping 3-dimension sparse location x and camera parameters d into 3-channel color RGB and density σ. Then a full image or raster is reconstructed by this implicit radiance field with volume rendering. (2) π-GAN produces a neural radiance field conditioned on a latent vector z for generalization in category-level synthesis. (3) Instead of directly conditioning on a single latent vector, our method embeds a series of subspaces with orthogonal bases into each mapping network. Multiple different latent vectors $z_{1,...,m}$ control the corresponding interpretable attribute variations.

neural radiance field conditioned on a modified FiLM-ed SIREN. The training process and an additional discriminator are also included. Finally, we describe a new layer-wise disentanglement architecture with orthogonal subspace design in Sect. 3.4.

3.1 Problem Statement

Given a set of unstructured 2D photographs within a category (e.g., CelebA [20] and CARLA [20]), high-quality 3D-aware image synthesis aims to learn a generative model that can explicitly control the camera view of the generated content. Besides, a meaningful semantic control of such generated contents is our ending point. To achieve the goal, we build upon the recent neural radiance fields(NeRF) representation. We utilize the implicit neural representation as an intermediate shape to achieve more strict 3D identity consistency. While the NeRF can only render novel views of a particular scene, we extend the NeRF representation to multiple objects with latent vectors. Then we disentangle the attributes of subspaces to enable more precise control of 3D-aware generation.

Let $x = (x, y, z) \in \mathbb{R}^3$ be a queried 3D spatial location in a cubic space, $d = (\theta, \phi) \in \mathbb{S}^2$ be a camera direction. The NeRF model build a mapping $(c, \sigma) = \mathcal{F}(x, d)$ and returns a radiance $c = (r, g, b)$ and a scale density σ. The network $\mathcal{F}$ is parametrized as a multi-layer perceptron (MLP) such that the density output σ is independent of the viewing direction, while the radiance c depends on both position and camera direction. In practice, a subset of points is sampled and the rendered color of the ray is approximated by using the quadrature rule as follows:

$$\hat{C}(\mathbf{r}) = \sum_{i=1}^{N} T_i \left(1 - \exp\left(-\sigma_i \delta_i\right)\right) c_i, \qquad (1)$$

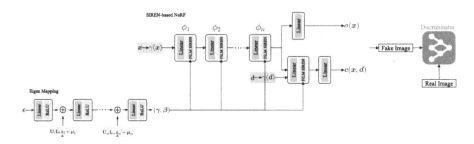

Fig. 3. The framework of the proposed EigenGRF. (1) The Eigen Mapping part is a sequence of fully connection networks embedding of m linear subspaces with orthonormal bases $U_i = [u_{i1}, u_{i2}, \cdots u_{iq},]$, and each basis vector u_{ij} is an "eigen-dimension" capturing meaningful semantic variation. The SIREN-based NeRF part forms a mapping from sparse location x and camera viewpoint d into density σ and color c. Each layer of the NeRF model condition the SIREN on multiple sub-latent vectors z through FiLM conditioning. Leveraging volume rendering, we integrate the fake image and then train the whole parameters in an adversarial manner.

where $\mathbf{r}$ denotes a ray, N is the number of samples, and δ_i is the distance between the ith point and its adjacent sample. Note that T_i indicates the accumulated transmittance along the ray until the ith point, which is given by

$$T_i = \exp(-\sum_{j=1}^{i-1} \sigma_j \delta_j).$$ (2)

This process can be seen as each point in the space emitting radiance in each direction. As the whole process is fully differentiable, the neural networks can be optimized with the following objective:

$$\mathcal{L} = \frac{1}{|\mathcal{R}|} \sum_{r \in \mathcal{R}} \|C(r) - \hat{C}(r)\|_2^2,$$ (3)

where $\mathcal{R}$ denotes a set of sampled rays.

3.2   Generative Radiance Fields

The vanilla NeRF can only represent a single scene or object, we first generalize it into representing multiple objects within one model. Building a new mapping $(c, \sigma) = \mathcal{F}(x, d, z)$, generative neural radiance fields methods [3,30] use a latent code z to embed a 3D representation like GANs. As shown in Fig. 2 (1–2), we leverage an extra mapping neural network to condition the NeRF model on a latent code through FiLM [26] which will be introduced in the next section. Following [22], we also use positional encoding $\gamma(\cdot)$ for both scene coordinates x and camera viewpoint to achieve high-quality generation.

$$\gamma(x) = (\sin(2^0 \pi x), \cos(2^0 \pi x), \sin(2^1 \pi x), \cos(2^1 \pi x), \cdots).$$ (4)

To transform the original reconstruction process into a generation process and improve the synthesis quality, we utilize an adversarial framework to train this generative model. The discriminator D is implemented as a convolutional neural network to distinguish the real image I and the generated image I'.

The whole training process of the generative radiance fields is shown in Fig. 3. Without the ground truth of the camera viewpoint, we randomly sample it from a Gaussian distribution. Obtaining the color and volume density of all sampled rays, we can render the full image I' from a given camera viewpoint. The arbitrary camera viewpoints will produce 3D-aware images with view-consistent. We use the non-saturating GAN loss with R1 regularization:

$$\mathcal{L} = \mathbb{E}[f(D(I'))] + \mathbb{E}[f(-D(I)) + \lambda\|\nabla D(I)\|^2],$$
$$\text{where} f(x) = -\log(1 + \exp(-x)). \tag{5}$$

3.3 Modification of FiLM-SIREN Layer

Following [3], to account for different objects, we condition the emitted radiance using Feature-wise Linear Modulation (FiLM) [26] of SIREN [33] layer, which is termed the FiLM-SIREN layer:

$$\phi(\mathbf{h}_i) = \sin(\boldsymbol{\gamma}_i \times \mathbf{h}_i + \boldsymbol{\beta}_i), \tag{6}$$

where $\mathbf{h}_i$ denotes the layer's input, $\times$ denotes the element-wise multiplication of two equally-sized vectors, $\boldsymbol{\gamma}_i$ denote the frequencies and $\boldsymbol{\beta}_i$ denotes the phase shifts, both of which are conditioned on the latent $\boldsymbol{z}$ via a mapping network, similar with StyleGAN [17]. This mapping network is more expressive than concatenation-based conditioning.

In practice, we find that directly multiplying frequencies vectors $\boldsymbol{\gamma}$ with the layer input $\mathbf{h}_i$ would lead to gradient vanishing, thus we pass the frequencies $\boldsymbol{\gamma}$ through element-wise exponential function $f(x) = e^x$ before the multiplication with the layer input:

$$\phi(\mathbf{h}_i) = \sin(e^{\gamma_i} \times \mathbf{h}_i + \boldsymbol{\beta}_i), \tag{7}$$

which has achieved significant training speed and convergence boost.

3.4 Learning Layer-Wise Subspace

To endow our method with the ability to learn disentanglement factors, we adapt layer-wise latent variables. Our target is to learn the generated image $I' = G(\boldsymbol{z}_1, \boldsymbol{z}_2, \cdots, \boldsymbol{z}_m)$ dominated by multiple orthonormal sub-latent vectors rather than $I' = G(\boldsymbol{z})$. As shown in Fig. 3, in the ith layer of mapping network, we embed a linear subspace model $S_i = (\mathbf{U}_i, \mathbf{L}_i, \boldsymbol{\mu}_i)$, where $\mathbf{U} = [\mathbf{u}_1, \cdots, \mathbf{u}_q]$ is the orthonomal basis of the subspace, $\mathbf{L} = \text{diag}(l_1, \cdots, l_q)$ is a diagonal matrix, and $\boldsymbol{\mu}$ is the origin. Then we can sample a point from the subspace:

$$\psi_i = \mathbf{U}_i \mathbf{L}_i \boldsymbol{z}_i + \boldsymbol{\mu}_i. \tag{8}$$

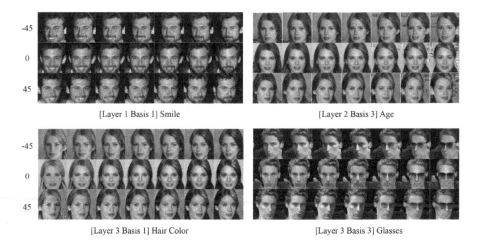

[Layer 1 Basis 1] Smile   [Layer 2 Basis 3] Age

[Layer 3 Basis 1] Hair Color   [Layer 3 Basis 3] Glasses

Fig. 4. Discovered semantic attributes at different layers and different bases for CelebA dataset. The column shows different camera viewpoints (angle $= -45$, 0, 45) of one person in a 3D representation. Each row denotes the changing of semantic attributes along a basic direction.

The sampled point ψ_i will be directly added to the mapping network. We also inject a random noise $\epsilon \sim \mathcal{N}(\mathbf{0}, \mathbf{I})$ to capture the rest variations missed by the subspaces. We constrain the orthogonality of $\mathbf{U}_i$ by a regularization of $\|\mathbf{U}_i^T\mathbf{U}_i - \mathbf{I}\|_F^2$. Our goal is to learn a series of orthonormal vectors to form a latent basis for determining different semantic attributes. After training this model, we can directly traverse such a subspace to control the 3D shape of a specific object.

4 Experiment

We evaluate our proposed method using a moving average of parameters on CeleA [20] and CARAR [8] dataset. CelebA contains more than 200,000 high-resolution celebrity face images. CARLA contains more than 10,000 car images rendered by the Carla driving simulator. Our generative radiance field is parameterized as an MLP with eight modified FiLM-SIREN layers. All these layers are conditioned on the same output of the mapping network. The mapping network consists of nine linear subspaces and six eigenbases in each subspace.

4.1 Discovered Semantic Attributes

As shown in Fig. 4, our model has learned multiple semantic attributes in different subspaces achieving the steerability of 3D generation processes. These disentangled subspaces help us make semantic image editing with 3D consistency. The changes in the concepts of humans are consistent with all views while we only select three angles. Different basis of different layers determines an interpretable direction for semantic editing. For example, L1 B1 learns "Smile" given

Fig. 5. Selected examples synthesized by EigenGRF with CARLA. We achieve the controllability of the car shape and the type.

expression to the mouth. L2 B3 learns "Age" from young to old. L3 B1 learns to change "Hair Color" from brown to black. L3 B3 learns to add sunglasses. Moving along an eigenbasis, the synthesized images continuously change by a specific meaning without interfering with other attributes. In Fig. 5, we also conduct our experiment on a dataset of cars. Appearance colors, shape types, and some minor attributes (e.g., the ceiling size) are captured by our model.

5   Conclusion

We introduce an approach for learning generative radiance fields along with interpretable attribute disentanglement. Furthermore, we have shown how to perform intuitive editing using the orthonormal bases of the subspace. We endow 3D-aware generation with better steerability and give a more strict view-inconsistency to interpretable GAN controls. The limitation of our approach is that we cannot foresee the extent of change. Trying to identify the specific layer and dimension which control semantic attributes is overhead. Despite these limitations, our approach opens up new windows for more enriched neural radiance fields (NeRF).

References

1. Bojanowski, P., Joulin, A., Lopez-Paz, D., Szlam, A.: Optimizing the latent space of generative networks. In: Proceedings of ICML (2017)
2. Cai, S., Obukhov, A., Dai, D., Van Gool, L.: Pix2Nerf: unsupervised conditional pi-GAN for single image to neural radiance fields translation. In: Proceedings of CVPR (2022)
3. Chan, E.R., Monteiro, M., Kellnhofer, P., Wu, J., Wetzstein, G.: pi-GAN: periodic implicit generative adversarial networks for 3D-aware image synthesis. In: Proceedings of CVPR (2021)
4. Chen, S.A., Li, C.L., Lin, H.T.: A unified view of CGANs with and without classifiers. In: Proceedings of NeurIPS (2021)
5. Chen, Z., Zhang, H.: Learning implicit fields for generative shape modeling. In: Proceedings of CVPR (2019)
6. Chong, M.J., Lee, H.Y., Forsyth, D.: Stylegan of all trades: image manipulation with only pretrained styleGAN. arXiv preprint arXiv:2111.01619 (2021)
7. Deng, Y., Yang, J., Xiang, J., Tong, X.: Gram: generative radiance manifolds for 3D-aware image generation. In: Proceedings of CVPR (2022)

8. Dosovitskiy, A., Ros, G., Codevilla, F., Lopez, A., Koltun, V.: CARLA: an open urban driving simulator. In: Proceedings of CORL (2017)

9. Goodfellow, I.J., et al.: Generative adversarial nets. In: Proceedings of NeurIPS (2014)

10. Grigoryev, T., Voynov, A., Babenko, A.: When, why, and which pretrained GANs are useful? In: Proceedings of ICLR (2022)

11. Gropp, A., Yariv, L., Haim, N., Atzmon, M., Lipman, Y.: Implicit geometric regularization for learning shapes. In: Proceedings of ICML (2020)

12. Härkönen, E., Hertzmann, A., Lehtinen, J., Paris, S.: GANspace: discovering interpretable GAN controls. In: Proceedings of NeurIPS (2021)

13. He, Z., Kan, M., Shan, S.: EigenGAN: layer-wise eigen-learning for GANs. In: Proceedings of ICCV (2021)

14. Hinton, G.E., Salakhutdinov, R.R.: Reducing the dimensionality of data with neural networks. Science (2006)

15. Jang, W., Agapito, L.: CodeNerf: disentangled neural radiance fields for object categories. In: Proceedings of ICCV (2021)

16. Kania, K., Yi, K.M., Kowalski, M., Trzciński, T., Tagliasacchi, A.: CoNerf: controllable neural radiance fields. In: Proceedings of ICCV (2021)

17. Karras, T., Laine, S., Aila, T.: A style-based generator architecture for generative adversarial networks. In: Proceedings of CVPR (2019)

18. Karras, T., Laine, S., Aittala, M., Hellsten, J., Lehtinen, J., Aila, T.: Analyzing and improving the image quality of styleGAN. In: Proceedings of CVPR (2020)

19. Liu, S., Zhang, X., Zhang, Z., Zhang, R., Zhu, J.Y., Russell, B.: Editing conditional radiance fields. In: Proceedings of ICCV (2021)

20. Liu, Z., Luo, P., Wang, X., Tang, X.: Deep learning face attributes in the wild. In: Proceedings of ICCV (2015)

21. Mescheder, L., Oechsle, M., Niemeyer, M., Nowozin, S., Geiger, A.: Occupancy networks: learning 3D reconstruction in function space. In: Proceedings of CVPR (2019)

22. Mildenhall, B., Srinivasan, P.P., Tancik, M., Barron, J.T., Ramamoorthi, R., Ng, R.: NeRF: representing scenes as neural radiance fields for view synthesis. In: Vedaldi, A., Bischof, H., Brox, T., Frahm, J.-M. (eds.) ECCV 2020. LNCS, vol. 12346, pp. 405–421. Springer, Cham (2020). https://doi.org/10.1007/978-3-030-58452-8_24

23. Mirza, M., Osindero, S.: Conditional generative adversarial nets. arXiv preprint arXiv:1411.1784 (2014)

24. Oechsle, M., Niemeyer, M., Reiser, C., Mescheder, L., Strauss, T., Geiger, A.: Learning implicit surface light fields. In: Proceedings of 3DV (2020)

25. Park, J.J., Florence, P., Straub, J., Newcombe, R., Lovegrove, S.: DeepSDF: learning continuous signed distance functions for shape representation. In: Proceedings of CVPR (2019)

26. Perez, E., Strub, F., De Vries, H., Dumoulin, V., Courville, A.: Film: Visual reasoning with a general conditioning layer. In: Proceedings of AAAI (2018)

27. Ramirez, P.Z., Tonioni, A., Tombari, F.: Unsupervised novel view synthesis from a single image. arXiv preprint arXiv:2102.03285 (2021)

28. Ranjan, A., Bolkart, T., Sanyal, S., Black, M.J.: Generating 3D faces using convolutional mesh autoencoders. In: Ferrari, V., Hebert, M., Sminchisescu, C., Weiss, Y. (eds.) ECCV 2018. LNCS, vol. 11207, pp. 725–741. Springer, Cham (2018). https://doi.org/10.1007/978-3-030-01219-9_43

29. Rezende, D.J., Mohamed, S.: Variational inference with normalizing flows (2016)

30. Schwarz, K., Liao, Y., Niemeyer, M., Geiger, A.: Graf: generative radiance fields for 3d-aware image synthesis. In: Proceedings of NeurIPS (2020)
31. Shen, Y., Gu, J., Tang, X., Zhou, B.: Interpreting the latent space of GANs for semantic face editing. In: Proceedings of CVPR (2020)
32. Shen, Y., Zhou, B.: Closed-form factorization of latent semantics in GANs. In: Proceedings of CVPR (2021)
33. Sitzmann, V., Martel, J.N., Bergman, A.W., Lindell, D.B., Wetzstein, G.: Implicit neural representations with periodic activation functions. In: Proceedings of NeurIPS (2020)
34. Srinivasan, P.P., Deng, B., Zhang, X., Tancik, M., Mildenhall, B., Barron, J.T.: Nerv: neural reflectance and visibility fields for relighting and view synthesis. In: Proceedings of CVPR (2021)
35. Sun, S.-H., Huh, M., Liao, Y.-H., Zhang, N., Lim, J.J.: Multi-view to novel view: synthesizing novel views with self-learned confidence. In: Ferrari, V., Hebert, M., Sminchisescu, C., Weiss, Y. (eds.) ECCV 2018. LNCS, vol. 11207, pp. 162–178. Springer, Cham (2018). https://doi.org/10.1007/978-3-030-01219-9_10
36. Tian, Y., Peng, X., Zhao, L., Zhang, S., Metaxas, D.N.: CR-GAN: learning complete representations for multi-view generation. In: Proceedings of IJCAI (2018)
37. Tran, L., Yin, X., Liu, X.: Disentangled representation learning GAN for pose-invariant face recognition. In: Proceedings of CVPR (2017)
38. Wang, N., Zhang, Y., Li, Z., Fu, Y., Liu, W., Jiang, Y.-G.: Pixel2Mesh: generating 3D mesh models from single RGB images. In: Ferrari, V., Hebert, M., Sminchisescu, C., Weiss, Y. (eds.) ECCV 2018. LNCS, vol. 11215, pp. 55–71. Springer, Cham (2018). https://doi.org/10.1007/978-3-030-01252-6_4
39. Wu, J., Zhang, C., Zhang, X., Zhang, Z., Freeman, W.T., Tenenbaum, J.B.: Learning shape priors for single-view 3D completion and reconstruction. In: Ferrari, V., Hebert, M., Sminchisescu, C., Weiss, Y. (eds.) ECCV 2018. LNCS, vol. 11215, pp. 673–691. Springer, Cham (2018). https://doi.org/10.1007/978-3-030-01252-6_40
40. Wu, Z., et al.: 3D shapenets: a deep representation for volumetric shapes. In: Proceedings of CVPR (2015)
41. Zhao, J.J., Mathieu, M., LeCun, Y.: Energy-based generative adversarial network. In: Proceedings of ICLR (2017)
42. Zhu, J., et al.: Low-rank subspaces in GANs. In: Proceedings of NeurIPS (2021)
43. Zhuang, P., Koyejo, O., Schwing, A.G.: Enjoy your editing: controllable GANs for image editing via latent space navigation. In: Proceedings of ICLR (2021)

Partial Label Learning with Gradually Induced Error-Correction Output Codes

Yu-Xuan Shi, Deng-Bao Wang, and Min-Ling Zhang[✉]

School of Computer Science and Engineering, Southeast University,
Nanjing 210096, China
zhangml@seu.edu.cn

Abstract. Partial label learning (PLL) is a specific weakly supervised learning problem, where each training example is associated with a set of candidate labels while only one of them is the ground truth. Recently, a disambiguation-free partial label learning method based on error-correcting output codes has been proposed, and achieves outstanding performance among existing partial label learning methods. Despite its good performance, it cannot deal with high ambiguity scenario and large candidate label size. To tackle this issue, we propose a new partial label learning method called PL-GECOC that gradually induces error-correction output codes during iterative model training. Experiments show that PL-GECOC outperforms most of the existing methods, especially in high ambiguity and large candidate label size scenarios.

Keywords: partial label learning · weakly supervised learning · error-correcting output codes

1 Introduction

Classical supervised learning requires large-scale accurately labeled training data, which is expensive to collect, thus it creates a critical bottleneck in many tasks. In contrast, weakly supervised learning methods can induce predictive models from less expensive training data. Partial label learning (PLL) is an emerging weakly-supervised learning framework for learning from ambiguous supervision, where each training example is associated with a set of candidate labels while only one of them is the ground truth. For example, in online image annotation systems, users may annotate an image differently, which associates an image with different painting styles, yet only one of them is correct.

Formally speaking, let $\mathcal{X} = \mathbb{R}^d$ represent the d-dimensional instance space and $\mathcal{Y} = \{y_1, y_2, \ldots, y_q\}$ represent the label space with q class labels. Let $\mathcal{D} = \{(\boldsymbol{x}_i, S_i) \,|\, 1 \leq i \leq m\}$ be the training set which contains m PL training examples, where each instance $\boldsymbol{x}_i \in \mathcal{X}$ is a d-dimensional vector $[x_{i1}, x_{i2}, \ldots, x_{id}]^\top$. $S_i \subset \mathcal{Y}$ is the set of candidate labels associated with x_i, in which the ground truth label y_i is concealed. The final goal of PLL is to train a multi-class classifier $f : \mathcal{X} \mapsto \mathcal{Y}$ from $\mathcal{D}$ that correctly predict the ground truth label of unknown instance.

M. Tanveer et al. (Eds.): ICONIP 2022, LNCS 13623, pp. 200–211, 2023.
https://doi.org/10.1007/978-3-031-30105-6_17

Existing PLL methods can be roughly divided into two types: disambiguation-based and disambiguation-free methods. The former requires identification of the ground truth labels among candidate label sets, which it is quite complicated and difficult to accurately identify them. The latter, represented by PL-ECOC which is recently proposed by Zhang et al. [1], is efficient while it can achieve outstanding performance. The key of PL-ECOC lies in adapting Error-Correcting Output Coding (ECOC) matrix to convert the multi-classification problem into multiple binary problems. Then a series of binary classifier could be induced from all the generated binary datasets. After that, a decoding phase is conducted to convert the outputs of binary classifiers into the original multi-class label space. Adapting ECOC in PLL successfully avoids the influence of false positive labels, but a significant number of instances are discarded after the encoding phase. Any partially labeled instance will be regarded as a positive or negative training example only if its candidate label set entirely falls into the coding dichotomy. Therefore, PL-ECOC shows instability in the absence of sufficient instances.

In this paper, we propose a new PLL method called *Partial Label learning with Gradually induced Error-Correction Output Codes* PL-GECOC. Based on the observation that existing ECOC framework for partial label learning fails to deal with large candidate label size, we propose to use the soft error-correction output coding strategy to transform the original partial label learning problem into a series of regression problems. As this procedure would induce noises into the generated targets, we further introduce the gradually induced coding strategy that iteratively trains the regressors and then uses their own outputs to reduce the generated target noise. Our experiments on eight real world datasets and five artificial datasets show that PL-GECOC outperforms most of the existing methods.

The rest of this paper is organized as follows. Section 2 briefly reviews the related work. Section 3 discusses the technical details of the proposed PL-GECOC method. Section 4 reports comparative experimental results, and Sect. 5 concludes this paper.

2   Related Work

In fully supervised learning, each instance is associated with a label to identify its class. In semi-supervised learning, all labels of label set are likely to become the ground truth label of each instance. In contrast, partial label learning problem induces the candidate label set in which the ground truth label of each training example is concealed. The challenge of partial label learning problems lies in that the ground truth label of the training examples is not directly accessible by the training model. To solve partial label learning problem, two types of methods are proposed, namely disambiguation-based and disambiguation-free partial label learning.

Traditional PLL methods focused on label disambiguation, i.e. to distinguish the ground truth label from many confusing labels in candidate label set. This strategy induces two basic methods: averaging-based disambiguation and identification-based disambiguation. The averaging-based disambiguation assigns same weight to each label [3,4,6]. The average output values on candidate labels are used to calculate the positive loss, while the average output values on the non-candidate labels are used to calculate the negative loss. The final loss function is defined based on a specific convex loss function as the mixture of both positive and negative loss [8]. However, treating the set of candidate labels equally lacks consideration of the differences between ground truth label and false positive labels, which motivates the proposal of the identification-based disambiguation. In identification-based disambiguation methods, the ground truth label is considered as a latent variable to be optimized during the model iteration [2,5], in which the model outputs can be used as the disambiguated label set to guide the model training in the next iteration. The process of label disambiguation is usually complex and even intractable when the label size is large.

Disambiguation-free methods was proposed by Zhang et al. [1] firstly. By adopting Error-Correcting Output Code framework, the label set is divided into positive and negative dichotomies, and the original partial label learning problem would be transformed into a series of binary problems. Any partially labeled instance will be regarded as a positive or negative training example only if its candidate label set entirely falls into the dichotomy. PL-ECOC outperforms numbers of PLL methods due to the error correction mechanism of ECOC. We will further discuss the details of PL-ECOC in Sect. 3.

3 The PL-GECOC Approach

3.1 Preliminary

ECOC is a widely studied decomposition solution for multi-class classification problems [9,10], and can be naturally used to deal with partial label learning problems. Generally, PL-ECOC consists of three phases: encoding, binary classifier induction, and a decoding.

In encoding phase, a coding matrix $M \in \{0,1\}^{q \times L}$ is generated to transform the partial label learning into a number of binary classification tasks. As is shown in Fig. 1(a), each row of the coding matrix $M(i,:)$ represents a code word of a certain class, while each column of the matrix $M(:,j)$ divides the label space into two parts. Let $\mathcal{Y}_j^+$ and $\mathcal{Y}_j^-$ denote positive and negative part of $M(:,j)$:

$$\mathcal{Y}_j^+ = \{y_j | M(i,j) = +1, \ 1 \le i \le q, \ 1 \le j \le L\}$$

$$\mathcal{Y}_j^- = \mathcal{Y} \setminus \mathcal{Y}_j^+, \ 1 \le j \le L$$

To avoid bias, we require that the quantities of $|\mathcal{Y}_j^+|$ and $|\mathcal{Y}_j^-|$ remain same (or try to keep it same when q is odd), i.e. $|\mathcal{Y}_j^+| = |\mathcal{Y}_j^-|$. With this coding matrix, the original partial label dataset could be transformed into a series of binary datasets. Figure 1 (b) and (c) show an intuitive example of constructing ECOC matrix and datasets regeneration.

$y_1 \to$	1	0	1	1	1	0
$y_2 \to$	0	1	0	0	1	1
$y_3 \to$	1	1	0	1	1	0
$y_4 \to$	1	0	0	0	0	1
$y_5 \to$	0	0	1	0	0	1
$y_6 \to$	0	1	1	1	0	0

$$\downarrow \ \downarrow \ \downarrow \ \downarrow \ \downarrow \ \downarrow$$
$$h_1 \ h_2 \ h_3 \ h_4 \ h_5 \ h_6$$
(a)

$$\mathcal{D} = \{(x_i, S_i) | 1 \leq i \leq 5\}$$
$$S_1 = \{y_1, y_2\}$$
$$S_2 = \{y_2, y_5\}$$
$$S_3 = \{y_1, y_3, y_6\}$$
$$S_4 = \{y_4\}$$
$$S_5 = \{y_5, y_6\}$$
(b)

	Positive Set	Negative Set
h_1	$\{x_4\}$	$\{x_2, x_5\}$
h_2	ϕ	$\{x_4\}$
h_3	$\{x_4, x_5\}$	$\{x_2\}$
h_4	$\{x_3\}$	$\{x_2, x_4\}$
h_5	$\{x_1\}$	$\{x_4, x_5\}$
h_6	$\{x_2, x_4\}$	$\{x_3\}$

(c)

Fig. 1. An illustrative example. (a) Demonstration of coding matrix. (b) An example of partial label dataset. (c) The transformed binary datasets, where the original dataset $\mathcal{D}$ is transformed into 6 binary datasets.

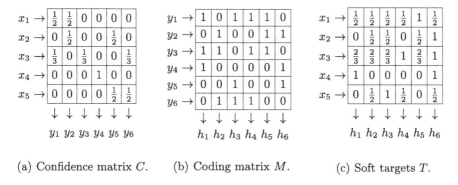

(a) Confidence matrix C. (b) Coding matrix M. (c) Soft targets T.

Fig. 2. Example of coding matrix and soft targets.

For each binary classification problem, a classifier is built by treating training examples whose candidate label set completely falls into $\mathcal{Y}_j^+$ as the positive and those training examples whose candidate label set completely falls into $\mathcal{Y}_j^-$ as the negative. Given a transformed dataset, one can use a binary model like logistic regression model to fit it. One drawback of the encoding process in PL-ECOC is that a number of instances fall in neither the positive part nor negative part, and hence a lot of instances would be discarded during training phase.

In inference phase, a L-bit signed output vector from L binary classifiers $\boldsymbol{h}(x) = [\text{sign}(h_1(x)), \text{sign}(h_2(x)), \ldots, \text{sign}(h_L(x))]$ is generated, where $\text{sign}(z)$ returns $+1$ if $z > 0$ and 0 otherwise. The predicted class labels is determined by choosing the most closest codeword between all class label and the L-bit signed output:

$$y^* = \arg\min_{y_j (1 \leq j \leq q)} \text{dist}(\boldsymbol{h}(x), M(j,:)),$$

where the distance function $\text{dist}(\cdot)$ could be chosen from Hamming distance [9], Euclidean distance [13], loss-based distance [14], etc.

3.2   Soft Error-Correction Output Codes

As we discussed in above, PL-ECOC fails to deal with large candidate label size. As shown in Fig. 1(c), classifier h_2 receives zero positive training example, while other classifiers also suffer from low utilization rate of training examples. To tackle this issue, we introduce soft error-correction output coding strategy. Firstly, we normalize the partial labels in S_i to a confidence vector c_i. For example, the samples in Fig. 1(b) could be normalized as shown in Fig. 2(a), where each row could be produced as follows:

$$c_{ij} = \begin{cases} 1/|S_i|, & \text{if } y_j \in S_i \\ 0, & \text{otherwise} \end{cases} \tag{1}$$

Then, we determine the soft target of each instance by calculating the proportion of label confidence that lies in the positive part of each codeword. This can be implemented by the matrix multiplication, i.e. the soft encoded target matrix $T = C \cdot M$, where C denotes the label confidence matrix. Instead of a series of binary classification datasets, we now have constructed a set of regression datasets $\mathcal{B}_l, 1 \leq l \leq L$ for all the codeword of the encoding matrix:

$$\mathcal{B}_l = \{(x_i, t_{il}) | 1 \leq i \leq m\} \tag{2}$$

Then we use dataset B_l to learn a regressor h_l instead of a binary classifier that are learned in existing PL-ECOC framework. As we discussed, in Fig. 2(c) h_2 would fail to be learned since there is no positive example for training but only a negative example x_4. With our soft coding strategy, h_2 could be learned with all examples with the target values $\{\frac{1}{2}, \frac{1}{2}, \frac{2}{3}, 0, \frac{1}{2}\}$.

3.3   Gradually Induced Encoding

One drawback of the above soft ECOC strategy is that many target noises may be generated and hence harms the performance of the learned model. To handle this issue, we further propose a gradually induced encoding strategy that iteratively trains the regressors for all codewords and then use the model outputs to refine the quality of soft ECOC. Inspired by the self-training strategy, we use the decoded outputs of L regressors as the soft pseudo labels after each iteration, and then use the same encoding procedure in Subsect. 3.2 to generate the soft ECOC for further training. For a training instance x_i, an L-bit real value output codeword is generated by the L classifiers. Then we obtain the normalized confidence vector $f(x_i)$ for each instance x_i with a softmax process:

$$f_j(x_i) = \frac{e^{d_{ij}/T}}{\sum_{k=1}^q e^{d_{ik}/T}} \tag{3}$$

where $d_{ij} = \text{dist}(h(x_i), M(j,:))$ denotes the distance between output $h(x_i)$ and the codeword of label y_j. Then we use the similar form in Eq.(1) to generate the label confidence:

$$f(x_i) = \begin{cases} \frac{f_j(x_i)}{\sum_{y_j \in S_i} f_j(x_i)}, & \text{if } y_j \in S_i \\ 0, & \text{otherwise} \end{cases} \tag{4}$$

---

input : $\mathcal{D}$: partial label training set
$\{(x_i, S_i) \,|\, 1 \leq i \leq m\} \,(x_i \in \mathcal{X}, S_i \subseteq \mathcal{Y}, \mathcal{X} = \mathbb{R}^d, \mathcal{Y} = \{y_1, y_2, \ldots, y_q\})$
x^*: unseen instance
L: code word length
Iter: Max epoch
$\mathfrak{L}$: regression learner
output: y^*:predicted class label for instance x^*

1 **Procedure:**
2 Generate coding matrix M;
3 Construct initial confidence level matrix C according to Eq.(1);
4 Compute soft target matrix T;
5 **for** $i \leftarrow 1$ **to** *Iter* **do**
6   Regenerate coding matrix M;
7   Feed B_l to learner according to Eq.(2), $h_l \leftarrow \mathfrak{L}(B_l)$;
8   Use Eq.(3) to predict all training instances;
9   Update soft targets and confidence levels;
10 **end**
11 Obtain distance matrix according to Eq.(5);
12 Return predicted class label according to Eq.(6);

Algorithm 1: The pseudo-code of PL-GECOC

Notice that T in Eq.(3) denotes the temperature used to adjust the smoothness of label confidence. We repeat the above procedure multiple times and update the encoded soft targets in each iteration. By iteration, classifiers will gradually refine the quality of generated codes while reduce the label noise.

3.4 Decoding Based on Cross-Entropy Loss

Given a testing instance x^*, firstly, we obtain its L-bit output $h(x^*) = [\mathtt{sign}(h_1(x^*)), \mathtt{sign}(h_2(x^*)), \ldots, \mathtt{sign}(h_L(x^*))]$. Then, we need to determine the predicted label based on $h(x^*)$. Different from previous PL-ECOC that uses hamming distance, we adopt cross-entropy loss, i.e.,

$$L_{\mathtt{ce}}(p, q) = -\sum_{i=1}^{k} (p_i \cdot \ln q_i) \tag{5}$$

to determine the prediction, where p and q are two k-dimensional probabilities. Accordingly, we choose the label whose corresponding codeword is the closest to $h(x^*)$ as the final output label:

$$y^* = \arg\min_{y_i (1 \leq i \leq q)} L(M(i, :), H(x^*)) \tag{6}$$

4 Experiments

In this section, we present two series of comparative experiments on both controlled UCI datasets and real-world partial label datasets. We firstly introduce

Table 1. Characteristics of real-world datasets.

dataset	#examples	#features	#classes	avg.#CLs	Task Domain
Lost	1122	108	16	2.23	*automatic face naming* [15]
MSRCv2	1758	48	23	3.16	*object classification* [17]
Mirflickr	2780	1536	14	2.76	*web image classification* [20]
Bird Song	4998	38	13	2.18	*bird song classification* [18]
Malagasy	5303	384	44	8.35	*POS tagging* citech17refspsmalagasy

the experimental setup and then report the detailed experimental results with statistical performance comparisons. Finally, we conduct ablation study to validate the contribution of soft ECOC and gradually induced coding strategy.

4.1 Experimental Setup

Following the controlling protocol widely used in partial label learning studies, a partially labeled data can be generated from a multi-class UCI dataset by controlling three basic parameters p, r and ϵ, where p denotes the proportion of examples which are corrupted to be partially labeled(i.e. $|S_i| > 1$), r controls the number of false positive labels in candidate label set of each partially labeled example(i.e. $|S_i| = r + 1$) and ϵ represents the co-occurring probability between coupling candidate labels and the ground truth label.

In addition, several real-world partial label datasets have been collected from several domains including *Lost* [15], *MSRCv2* [17] from object classification, *BirdSong* [18] from bird song classification, *Malagasy* [19] from POS tagging and *Mirflickr* [20] from web image classification. For the automatic face naming task, faces cropped from images are considered to be instances while the associated captions or subtitles are considered as corresponding candidate labels. For the object classification task, image segmentations are regarded as instances while objects appearing within the origin image are regarded as the candidate labels. For bird song classification task, singing syllables of birds are considered as instances while bird species jointly singing during a 10-seconds period are considered as candidate labels. For POS tagging task, a given word with contextual information can be considered as an instance and all possible POS tags are regarded as the candidate labels. Characteristics of five real-world datasets are also presented in Table 1. Avg.#CLs denotes the average number of candidate labels for each real-world partial label dataset.

Five well-established partial label learning methods are employed for comparative studies, including:

- PL-KNN: A k-nearest neighbor approach to partial label learning which conducts averaging-based disambiguation;
- IPAL: An instance-based approach to partial label learning which conducts identification-based disambiguation;

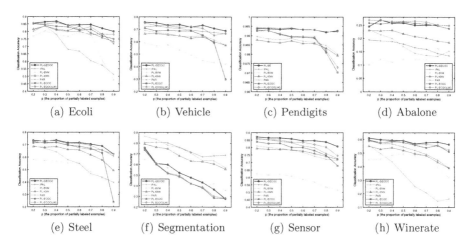

(a) Ecoli   (b) Vehicle   (c) Pendigits   (d) Abalone

(e) Steel   (f) Segmentation   (g) Sensor   (h) Winerate

Fig. 3. This figure illustrates the classification accuracy of each comparing method with varying p(the proportion of examples which are partially labeled) and fixed r and ϵ ($r = 2$ for *vehicle* and *sensor*, $r = 5$ for others).

- PL-SVM: A maximum margin approach to partial label learning which conducts identification-based disambiguation;
- PL-ECOC: A ECOC-based approach to partial label learning which conducts disambiguation-free method;
- PAR: A two-stage approach to partial label learning which conducts identification-based disambiguation;

Each comparative method is implemented with the default hyper parameters setup suggested in respective literature. For PL-GECOC, we use the implementations in LibSVM as the base regressor. The codeword length L for both PL-GECOC and PL-ECOC is set to be $\lceil 40 \cdot \log_2 q \rceil$ for fairness. Furthermore, the maximum number of iterations for PL-GECOC is set to be 30.

4.2 Experimental Results

Controlled UCI Datasets. Figure 3 illustrates the mean predictive accuracy of each comparing methods as p varies from 0.2 to 0.9 with step size 0.1 on controlled UCI datasets. Along with the ground truth label, r additional class labels in $\mathcal{Y}$ will be randomly chosen to instantiate the candidate label set of each partial label example. Figure 4 illustrates the mean predictive accuracy of each comparing method as ϵ varies from 0.1 to 0.7 with step-size 0.1 on each controlled UCI datasets($p = 1$, $r = 1$). For any ground truth label $y \in \mathcal{Y}$, one extra label $y' \neq y$ is selected as the coupling label which co-occurs with y with probability ϵ. As shown in Table 2, we also compare the performance with different candidate label sizes.

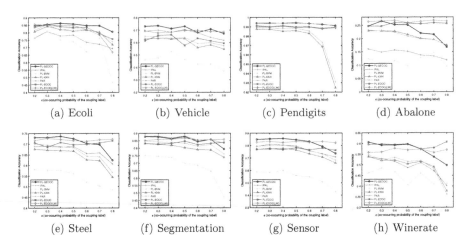

Fig. 4. This figure illustrates the classification accuracy of each comparing method with varying ϵ(the co-occurring probability of the coupling label) and fixed r and p ($r = 1$, $p = 1$).

As is shown in Fig. 3, the performance of PL-GECOC outperforms the comparing algorithms in most cases. An exception is *Segmentation*, where both PL-GECOC and PL-ECOC have poor performance, however PL-GECOC still outperforms PL-ECOC. In Fig. 4, the performance of PL-GECOC seems worse than that of identification-based methods, but PL-GECOC also outperforms PL-ECOC. In Table 2, PL-GECOC generally outperformed all comparing methods on t-test at 0.05 significance level. The accuracy of classification is set to be 0 ± 0 if an method fails to operate. As shown in this table, PL-ECOC fails to operate on difficult cases, meanwhile which is statically inferior to PL-GECOC on almost all easier cases. PL-GECOC achieves superior performance against PL-ECOC in 91% cases(105 out of 115). Additionally, comparing to averaging-based disambiguation approaches, PL-GECOC achieves superior performance against PL-KNN in 93% cases(107 out of 115). Comparing to identification-based disambiguation approaches, PL-GECOC achieves superior performance against IPAL in 82% cases(94 out of 115), PL-SVM in 95% cases(109 out of 115) and PAR in 90% cases(104 out of 115). Overall, the reported results show that PL-GECOC is competitive on not only easy cases but also under high ambiguity and large candidate label size settings.

Real-World Datasets. The classification accuracy of each comparing method on real-world datasets is reported in Table 3. PL-GECOC achieves superior performance against PL-ECOC on *Lost,MSRCv2* and *Malagasy* datasets, while achieves superior performance against both PAR and PL-KNN in 92% cases (24 out of 25) and PL-SVM in 72% cases (18 out of 25). Note that IPAL achieves

Table 2. Classification accuracy (mean±std) of each comparing method on the controlled UCI datasets with varying r (number of false positive candidate labels). In addition, •/○ indicates whether the performance of PL-GECOC is statistically superior/inferior to the comparing method on each dataset (pairwise t-test at 0.05 significance level).

dataset	r	PL-GECOC	PL-ECOC	IPAL	PL-SVM	PL-KNN	PAR
Ecoli	1	0.846 ± 0.025	0.661 ± 0.014•	0.760 ± 0.028•	0.744 ± 0.056•	0.808 ± 0.028	0.821 ± 0.025
	3	0.846 ± 0.040	0.669 ± 0.020•	0.732 ± 0.021•	0.793 ± 0.040	0.802 ± 0.028	0.810 ± 0.036
	5	0.745 ± 0.056	0.685 ± 0.011•	0.693 ± 0.030	0.411 ± 0.206•	0.611 ± 0.042•	0.611 ± 0.063•
Abalone	1	0.245 ± 0.009	0.265 ± 0.007○	0.157 ± 0.006•	0.120 ± 0.066•	0.225 ± 0.003•	0.232 ± 0.003•
	4	0.258 ± 0.011	0.055 ± 0.026•	0.122 ± 0.009•	0.090 ± 0.060•	0.190 ± 0.005•	0.190 ± 0.007•
	7	0.243 ± 0.006	0.000 ± 0.000•	0.105 ± 0.000•	0.073 ± 0.030•	0.151 ± 0.010•	0.150 ± 0.013•
	10	0.235 ± 0.015	0.000 ± 0.000•	0.091 ± 0.004•	0.158 ± 0.029•	0.117 ± 0.014•	0.116 ± 0.008•
Segmentation	1	0.881 ± 0.007	0.860 ± 0.008•	0.884 ± 0.009	0.753 ± 0.006•	0.828 ± 0.008•	0.848 ± 0.004•
	3	0.846 ± 0.025	0.000 ± 0.000•	0.838 ± 0.008	0.583 ± 0.057•	0.777 ± 0.013•	0.796 ± 0.016•
	5	0.234 ± 0.037	0.000 ± 0.000•	0.647 ± 0.016○	0.283 ± 0.159	0.448 ± 0.051○	0.421 ± 0.056○
Sensor	1	0.852 ± 0.007	0.822 ± 0.000•	0.815 ± 0.005•	0.592 ± 0.006•	0.776 ± 0.005•	0.797 ± 0.007•
	2	0.790 ± 0.008	0.000 ± 0.000•	0.675 ± 0.016•	0.440 ± 0.072•	0.683 ± 0.020•	0.707 ± 0.020•
Steel	1	0.734 ± 0.014	0.708 ± 0.006•	0.696 ± 0.010•	0.580 ± 0.020•	0.673 ± 0.014•	0.693 ± 0.016•
	3	0.691 ± 0.015	0.000 ± 0.000•	0.620 ± 0.033•	0.518 ± 0.033•	0.602 ± 0.021•	0.621 ± 0.019•
	5	0.394 ± 0.078	0.000 ± 0.000•	0.511 ± 0.020○	0.306 ± 0.126	0.332 ± 0.009	0.434 ± 0.036
Vehicle	1	0.724 ± 0.017	0.675 ± 0.019•	0.686 ± 0.017•	0.465 ± 0.039•	0.634 ± 0.029•	0.649 ± 0.029•
	2	0.654 ± 0.049	0.000 ± 0.000•	0.635 ± 0.017	0.298 ± 0.063•	0.535 ± 0.013•	0.551 ± 0.009•
Pendigits	1	0.994 ± 0.001	0.990 ± 0.003•	0.994 ± 0.001	0.810 ± 0.009•	0.988 ± 0.002•	0.990 ± 0.002•
	3	0.993 ± 0.001	0.000 ± 0.000•	0.993 ± 0.001	0.717 ± 0.024•	0.985 ± 0.001•	0.987 ± 0.001•
	5	0.992 ± 0.002	0.000 ± 0.000•	0.992 ± 0.002	0.465 ± 0.043•	0.966 ± 0.002•	0.969 ± 0.001•
Winerate	1	0.602 ± 0.013	0.568 ± 0.026•	0.538 ± 0.012•	0.474 ± 0.012•	0.536 ± 0.012•	0.556 ± 0.012•
	3	0.581 ± 0.014	0.000 ± 0.000•	0.383 ± 0.021•	0.464 ± 0.026•	0.443 ± 0.027•	0.439 ± 0.038•
	5	0.529 ± 0.028	0.000 ± 0.000•	0.269 ± 0.032•	0.291 ± 0.068•	0.319 ± 0.016•	0.279 ± 0.021•

Table 3. Classification accuracy (mean ± std) of each comparing method on the real-word datasets.

dataset	PL-GECOC	PL-ECOC	IPAL	PL-SVM	PL-KNN	PAR
Lost	0.598 ± 0.013	0.560 ± 0.022•	0.613 ± 0.000○	0.600 ± 0.063	0.262 ± 0.000•	0.310 ± 0.000•
MSRCv2	0.427 ± 0.010	0.394 ± 0.007•	0.501 ± 0.000○	0.316 ± 0.017•	0.408 ± 0.000•	0.395 ± 0.000•
Mirflickr	0.513 ± 0.223	0.469 ± 0.240	0.416 ± 0.183	0.410 ± 0.226	0.384 ± 0.145	0.400 ± 0.156
BirdSong	0.693 ± 0.012	0.688 ± 0.015	0.682 ± 0.000	0.488 ± 0.029•	0.624 ± 0.000•	0.648 ± 0.000•
Malagasy	0.616 ± 0.011	0.594 ± 0.008•	0.610 ± 0.000	0.554 ± 0.080	0.573 ± 0.000•	0.527 ± 0.000•

outstanding performance on real-world datasets. Nevertheless, PL-GECOC is competitive were it is inferior than IPAL in only 52% cases (13 out of 25).

Further Analysis. To validate the effectiveness of our proposed gradually induced coding strategy, we illustrate the classification accuracy on both controlled UCI datasets and real-world datasets at every iteration. For controlled UCI datasets, we set $p = 0.9$, $r = 2$ for *Sensor* and $r = 5$ for others. As shown in Fig. 5, the classification accuracy increases with the number of iterations on

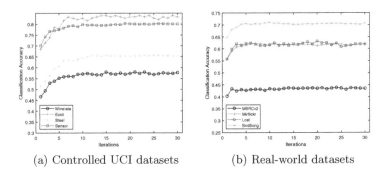

(a) Controlled UCI datasets   (b) Real-world datasets

Fig. 5. The classification accuracy of PL-GECOC during iterations.

both controlled UCI datasets and real-world datasets, which validates that the proposed strategy plays an important role in our proposed method.

5   Conclusion

In this paper, a novel partial label learning approach namely PL-GECOC is proposed. We propose to use the soft error-correction output coding strategy to transform the original partial label learning problem into a series of regression problems and further introduce the gradually induced coding strategy that iteratively trains the regressors and then use their own outputs to reduce the generated target noise. Extensive experiments on controlled UCI datasets and real world datasets show the effectiveness of our method especially on more difficult cases.

Acknowledgements. The authors wish to thank the anonymous reviewers for their helpful comments and suggestions. This work was supported by the National Science Foundation of China (62176055, 62225602).

References

1. Zhang, M.-L., Yu, F., Tang, C.-Z.: Disambiguation-free partial label learning. IEEE Trans. Knowl. Data Eng. **29**(10), 2155–2167 (2017)
2. Zhang, M.-L., Yu, F.: Solving the partial label learning problem: an instance-based approach. In: Proceedings of the 24th International Joint Conference on Artificial Intelligence (IJCAI 2015), Buenos Aires, Argentina, pp. 4048–4054 (2015)
3. Lei, F., Bo, A.: Partial label learning with self-guided retraining. In: Proceedings of the 33rd AAAI Conference on Artificial Intelligence (2019)
4. Hüllermeier, E., Beringer, J.: Learning from ambiguously labeled examples. In: Proceedings of the 6th International Conference on Advances in Intelligent Data Analysis, pp. 419–439 (2006)
5. Feng, L., An, B.: Leveraging latent label distributions for partial label learning. In: Twenty-Seventh International Joint Conference on Artificial Intelligence IJCAI-2018 (2018)

6. Nguyen, N., Caruana, R.: Classification with partial labels. In: Proceedings of the 14th ACM SIGKDD International Conference on Knowledge Discovery and Data Mining (KDD 2008), pp. 551–559. Association for Computing Machinery, New York (2008)

7. Zhang, M.-L., Fang, J.-P.: Partial multi-label learning via credible label elicitation. IEEE Trans. Pattern Anal. Mach. Intell. **43**(10), 3587–3599 (2021)

8. Cour, T., Sapp, B., Taskar, B.: Learning from partial labels. J. Mach. Learn. Res. **12**, 1501–1536 (2011)

9. Dietterich, T.G., Bakiri, G.: Solving multiclass learning problem via error-correcting output codes. J. Artif. Intell. Res. **2**(1), 263–286 (1995)

10. Zhou, Z.-H.: Ensemble Methods: Foundations and Algorithms. Chapman & Hall/CRC, Boca Raton (2012)

11. Lin, G., Liu, K., Wang, B., et al.: Partial label learning based on label distributions and error-correcting output codes. Soft Comput. **2020**(1) (2020)

12. Lin, G.-Y., Xiao, Z.-Y., Liu, J.-T., Wang, B.-Z., Liu, K.-H., Wu, Q.-Q.: Feature space and label space selection based on Error-correcting output codes for partial label learning, Inf. Sci. **589** (2022)

13. Pujol, O., Escalera, S., Radeva, P.: An incremental node embedding technique for error correcting output codes. Pattern Recogn. **41**()2), 713–725 (2008)

14. Escalera, S., Pujol, O., Radeva, P.: On the decoding process in ternary error-correcting output codes. IEEE Trans. Pattern Anal. Mach. Intell. **32**(1), 120–134 (2010)

15. Cour, T., Sapp, B., Jordan, C., Taskar, B.: Learning from ambiguously labeled images. In: Proceedings of the 20th IEEE Conference on Computer Vision and Pattern Recognition, pp. 919–926 (2009)

16. Zeng, Z., et al.: Learning by associating ambiguously labeled images. In: Proceedings of the 24th IEEE Conference on Computer Vision and Pattern Recognition, pp. 708–715 (2013)

17. Liu, L., Dietterich, T.G.: A conditional multinomial mixture model for superset label learning. In: Advances in Neural Information Processing Systems, pp. 548–556 (2012)

18. Briggs, F., Fern, X.Z., Raich, R.: Rank-loss support instance machines for miml instance annotation. In: Proceedings of the 18th ACM SIGKDD International Conference on Knowledge Discovery and Data Mining, pp. 534–542 (2012)

19. Garrette, D., Baldridge, J.: Learning a part-of-speech tagger from two hours of annotation. In: Proceedings of the 13th Conference of the North American Chapter of the Association for Computational Linguistics: Human Language Technologies, pp. 138–147 (2013)

20. Huiskes, M.J., Lew, M.S.: The mirflickr retrieval evaluation. In: Proceedings of the 1st ACM International Conference on Multimedia Information Retrieval, pp. 39–43. ACM (2008)

HMC-PSO: A Hamiltonian Monte Carlo and Particle Swarm Optimization-Based Optimizer

Omatharv Bharat Vaidya[1(✉)], Rithvik Terence DSouza[1(✉)],
Soma Dhavala[2(✉)], Snehanshu Saha[1(✉)], and Swagatam Das[3(✉)]

[1] BITS Pilani K.K. Birla Goa Campus, Sancoale, India
{f20180354,f20170200,snehanshus}@goa.bits-pilani.ac.in
[2] Wadhwani AI, Bangalore, India
soma@mlsquare.org
[3] Indian Statistical Institute, Kolkata, India
swagatam.das@isical.ac.in

Abstract. We introduce the Hamiltonian Monte Carlo Particle Swarm Optimizer (HMC-PSO), an optimization algorithm that reaps the benefits of both Exponentially Averaged Momentum PSO and HMC sampling. The coupling of the position and velocity of each particle with Hamiltonian dynamics in the simulation allows for extensive freedom for exploration and exploitation of the search space. It also provides an excellent technique to explore highly non-convex functions while ensuring efficient sampling. We extend the method to approximate error gradients in closed form for Deep Neural Network (DNN) settings. We discuss possible methods of coupling and compare its performance to that of state-of-the-art optimizers on the Golomb's Ruler problem and Classification tasks(HMC-PSO code and additional results are in the Github repository: https://github.com/rtdsouza/torchswarm).

Keywords: PSO · Monte Carlo · HMC-PSO · Gradient Descent

1 Introduction

Particle Swarm Optimization (PSO) techniques have been shown in the past to be analogous to Stochastic Gradient Descent (SGD) both qualitatively and quantitatively [1]. This has opened up the possibility of training deep neural networks (DNNs) with a diverse class of loss functions, since PSO techniques do not rely on gradients in the way that SGD-based optimizers do. Vanilla-PSO itself has evolved since its proposal in 1995. For instance, Exponentially Averaged Momentum PSO (EM-PSO) is a PSO variant with momentum that leads to faster convergence. Further, equivalence between EM-PSO and SGD can be used to approximate gradients by harvesting EM-PSO particle position information and fuse it with SGD. Such an approach was able to outperform the Adam

Supported by APPCAIR, BITS Pilani, India.
O. B. Vaidya and R. T. DSouza—Equal contribution.

optimizer on a variety of datasets and DNNs [1]. In the same spirit of coupling different families of optimization techniques, it has been recognized recently that learning rates can be transferred across training regimes. This grafting, facilitates decomposing and recomposing optimizers, and better hyperparameter tuning in DNN optimization tasks [2]. However, one of the potential drawbacks of EM-PSO, in general, is that it does not explore the search space well enough and it may settle in the local minima because of its first-order nature. In order to remedy this problem, we propose coupling PSO with a Markov Chain Monte Carlo (MCMC) sampling technique such as Hamiltonian Monte Carlo (HMC) [3]. The hope is that, this grafted optimizer-cum-sampler can better navigate the explore-exploit tradeoff that is prevalent in DNN training. Recall that HMC is used for sampling from probability distributions by simulating particles in a potential, subjected to Hamiltonian dynamics. Coupled with a conditional proposal acceptance step, this method produces a stationary distribution that approximates the original probability distribution fairly well. It explores the parameter space substantially and ensures that a global solution is found for the optimization problem. We discuss few possible ways of coupling HMC with PSO and report the findings on various multi-objective optimization as well as deep learning problems. We have observed that the modified optimizer makes the loss drop faster and leads to better testing accuracy on DNNs.

We term our proposed algorithm Hamiltonian Monte Carlo Particle Swarm Optimization (HMC-PSO). We demonstrate its effectiveness in solving a non-traditional yet important problem like the Golomb's ruler. Also, we compare the DNN version of HMC-PSO with other state-of-the-art (SOTA) optimizers like Adam, Adadelta, RMSprop, Adagrad, etc., on some important classification problems and show SOTA results. We begin with the background and motivation for the proposed HMC-PSO in Sect. 2. Required theoretical context, setup, and proposal for HMC-PSO are presented in Sect. 3. We showcase the experimental results and comparisons with current SOTA optimizers in Sect. 4. Finally, Sect. 5 states the conclusion.

2   Motivation and Background

The sources of inspiration to understand and improve optimization techniques in general, and Particle Swarm Optimization (PSO) in particular, come from several places. Said et al. [4] postulated that swarms behave similarly to classical and quantum particles. Their analogy helped in providing a dynamical system perspective, which was useful in unifying streams: Optimization and Markov Chain Monte Carlo sampling. In a seminal paper, Wellington and Teh [5], showed that a stochastic gradient descent (SGD) optimization technique could be turned into a sampling technique by just adding noise, governed by Langevin dynamics. Recently, Soma and Sato [6] provided further insights into this connection based on an underlying dynamical system governed by stochastic differential equations (SDE). These new results make a strong case for connections between optimization and sampling based on Stochastic Approximation and Finite Differences and made us wonder: Is there a larger, more general template of which the aforementioned approaches are special cases? Can we develop a unified dynamical system that can be made to operate either sampling mode (in the MCMC sense) or search model (in the SGD sense)? Precisely, can we (A) develop a unifying coupled-dynamical system that can couple existing paradigms such as an SGD and HMC, SGD and PSO, HMC and PSO, etc.? (B) Using the coupled-dynamical system, develop custom optimizers and samplers with better properties and apply them to non-smooth, non-convex problems from an optimization perspective and multi-modal distributions from a sampling perspective (C) develop an optimizer that specifically solves classification tasks on a DNN setting.

We provide an affirmative answer to these questions in a modest setting.

2.1 EM-PSO

Exponentially Averaged Momentum Particle Swarm Optimization (EM-PSO) [7] is a variant of PSO. It adopts PSO's major advantages, such as higher robustness to local minima. It provides more weight to the exploration part, which is an essential part of optimization problems. It has an additional tunable parameter i.e. exponentially averaged momentum, which adds flexibility to the task of exploration better than PSO or its vanilla momentum version.

The Adam optimizer [8] is a combination of SGD with momentum and RMSProp. It leverages the momentum by using the moving average of the gradient instead of the gradient itself like in SGD with momentum, and the squared gradients are used to scale the learning rate like in RMSProp. It calculates adaptive learning rates for different parameters from the estimates of the second and first moments of the gradients. The gradient of any function, differentiable or not, can be approximated by using the parameters of EM-PSO. We can leverage computed, approximated gradients from EM-PSO in the update rule for the Adam optimizer scheme. Experimental results show that this leads to lower execution time and comparable (and most times superior) performance to Adam [1].

2.2 Hamiltonian Monte Carlo (HMC)

Hamiltonian Monte Carlo (HMC) is an algorithm belonging to a class of algorithms known as Markov Chain Monte Carlo (MCMC) [3]. The original MCMC algorithm was devised in 1953 by Metropolis et al. to simulate the distribution of states for a system of idealized molecules [9]. In a 1987 paper by Duane et al., the stochastic MCMC algorithm was combined with the deterministic Hamiltonian algorithm to obtain Hybrid Monte Carlo, presently known as Hamiltonian Monte Carlo [10]. The goal of HMC is to sample from a given distribution in such a way that the distribution of the samples is equivalent to the original samples in the limit of samples drawn. It involves using Hamiltonian dynamics to produce more independent and distant proposals than the vanilla Metropolis algorithm with random walks [3]. A requirement of Hamiltonian dynamics, is that along with the position variable, there must be a momentum variable that stands for the momentum of the particle in the real world. However, since the momentum has no real analogue in a Monte Carlo simulation, it is simply considered a fictitious variable and is sampled from a random normal distribution.

2.3 Extending EM-PSO with HMC

One challenge with current MCMC samplers is that they don't scale well for large problems (in data and model size) - modern deep learning architectures being a case in point. HMC, based on Hamiltonian dynamics, is a very efficient MCMC sampler that exploits gradients. A Hamiltonian dynamical systems operates with a d-dimensional *position* vector q, and a d-dimensional *momentum* vector p [11]. This system is described by a the *Hamiltonian* $H(q,p)$. The partial derivatives of $H(q,p)$ determine how q and p change over time t, according to Hamilton's equations: $\frac{dq_i}{dt} = \frac{\delta H}{\delta p_i}$, $\frac{dp_i}{dt} = -\frac{\delta H}{\delta q_i}$ for $i = 1, 2, 3, ..., d$. We are seeking solutions to the above PDEs. For HMC, we typically write the Hamiltonian as: $H(q,p) = U(q) + K(p)$ where, $U(q)$ is called the *potential energy*, and can be defined to be negative the log probability density that we wish to sample from, plus any constant that is convenient. $K(p)$ is called the *kinetic energy*, and is usually defined as $K(p) = \frac{p^T M^{-1} p}{2}$. Here, M is a symmetric, positive-definite "mass matrix", which is typically diagonal, and is often a scalar multiple of the identity matrix. With these forms for H and K, Hamilton's equations of motion can be solved, which produce either the sampling paths or the search paths. Typically, the potential energy is the objective function being minimized in the optimization problem. The kinetic energy is somewhat an artificial construct and can be designed for convenience. It is exactly the kinetic energy we can play with. It may be plausible to couple the dynamics of HMC and PSO. Specifically, consider the PSO and say, we are tracking a particle, along the ith dimension: $p_i(t+1) = p_i(t) + c_1 r_1 (q^{best} - q(t))_i + c_2 r_2 (g^{best} - q(t))_i$ & $q_i(t+1) = q_i(t) + p_i(t+1)$. We can now construct the kinetic energy terms in the Hamiltonian, defined in terms of q. This is one way we can couple PSO with HMC. However, despite its appeal, in a DNN setting, the parameters can run into billions, and this variant of HMC-PSO may not be practical. Instead, we can consider a hybrid particle swarm, where some particles follow standard EM-PSO dynamics, and another set of particles follow HMC dynamics. The gradients of the potential function of the HMC can now be replaced with approximate gradients coming from the other particle set. We explore the latter approach in this paper.

3   Our Contribution: Hamiltonian Monte Carlo Particle Swarm Optimizer (HMC-PSO)

We have considered a few methods of coupling Hamiltonian Monte Carlo with PSO. The first question to ask was whether we would create an entirely new form of generic dynamics that used a "dial" to decide whether to optimize or sample, i.e., exploit or explore, or something in between? We have not used such a method since we wanted to stick with dynamics that had already been investigated, but we propose this as an avenue to be explored in the future. The methods we devised involved using particles that either used HMC dynamics or PSO dynamics.

3.1   HMC-PSO: Coupling HMC and PSO

An important challenge while integrating HMC with EM-PSO was to implement HMC dynamics without using explicitly computed gradients since we would otherwise lose the property of a gradient-free optimizer. As discussed earlier, such coupling may not scale to modern DNNs. The solution that we used was to use a modification of the approximation from EM-PSO for approximating the gradient in place of U_{grad} in the normal HMC. The approximation is given by:

$$U_{grad}(x) = -\frac{c_1 r_1 (p_{best} - x) + c_2 r_2 (g_{best} - x)}{\eta} \tag{1}$$

where $c_1 r_1$ and $c_2 r_2$ have the same meaning they do in EM-PSO, η is the total path length of a leapfrog step in HMC, g_{best} is the best point found by the swarm, and p_{best} is the best point found by the particle. While this would probably not be viable when requiring the samples to conform to a predefined distribution, we only need qualitative similarity between the sample distribution and the loss function. Thus, it appears reasonable to replace momentum and position updates in EM-PSO with HMC dynamics but leave the rest of the PSO machinery intact. The goal is to get a significantly better exploration of the space compared to EM-PSO, including a guarantee that the swarm would continue to explore as long as it runs. However, this happens at the cost of the local optimization capabilities of EM-PSO and speed of convergence (in comparison with EM-PSO). Slower convergence is expected since a single position-velocity-momentum update from EM-PSO is replaced by a significantly longer leapfrog update from HMC-PSO. As a remedy to over-correcting for EM-PSO's poor exploration capabilities, we reinstated EM-PSO while keeping HMC as a vital component. As a result, we devised a scheme to couple a single HMC particle with an EM-PSO swarm. Both these components are tightly coupled by sharing information about the best position found so far. Therefore, if the HMC particle, through its sampling dynamics, finds a better spot even after the swarm has settled at a local minimum, the swarm will migrate to this new minimum. In turn, the information about this g_{best} position will be used to guide the HMC particle via the EM-PSO approximation. The detailed HMC-PSO algorithm is given in Algorithm 1.

It might be noted that the EM-PSO approximation of the gradient leads to a loss of information for the HMC particle. Whereas earlier, the particle would be informed by the local landscape of the potential, it now receives information solely through g_{best}. As a result, upon convergence of the EM-PSO swarm, the HMC particle will behave as if it has been trapped in a potential centered around g_{best}. We can investigate the form of this pseudo-potential. The gradient update is given by Eq. 1. We can see that if the HMC particle comes sufficiently close to g_{best} since the gradient is directed towards it, we will have $p_{best} \approx g_{best}$. As a result, the approximation will have the form: $U_{grad}(x) = K(x - x_0)$, where, $x_0 = g_{best}$ and $K = \frac{c_1 r_1 + c_2 r_2}{\eta}$. Thus, the HMC particle may become trapped in a sort of randomized harmonic oscillator centered at g_{best}. However, this results in an exploration centered around the best known optimum. From elementary physics, we know that the resulting probability distribution for the HMC particle will be a gaussian centered at g_{best}. In fact, under such conditions, the draws from HMC can be used to perform a Laplce Approximation of the posteior density around a mode [12].

Algorithm 1: HMC-PSO

---

    input : Number of particles in the swarm, N
    input : Fitness function, fitness
    input : Parameters of PSO, c1 and c2
1 $g_{best}.value = \infty$
2 $g_{best}.position = $ undefined
3 swarm $\leftarrow$ [N EM particles]
4 hmc $\leftarrow$ HMC particle
5 **while** g_{best} has not converged **do**
6   **for** Particle p_i in swarm, including hmc **do**
7 $f = $ fitness(p_i)
8   **if** $f < p_i.p_{best}.value$ **then**
9 $p_i.p_{best}.value = f$
10 $p_i.p_{best}.position = p_i.position$
11   **if** $f < g_{best}.value$ **then**
12 $g_{best}.value = f$
13 $g_{best}.position = p_i.position$
14   **end**
15   **end**
16   **end**
17   **for** Particle p_i in swarm **do**
18 $p_i.move(g_{best})$; // move p_i using EM-PSO approximation with
           g_{best}
19   **end**
20   hmc.move(g_{best})
21 **end**

3.2 Complexity Analysis

From Algorithm 1, we see that the initialization steps 1 and 2 are $O(1)$ whereas the step 3 is $O(N)$, where N is the number of particles in the swarm. Similarly, step 4 i.e. initializing N EM particles takes $O(N)$ time [7]. Step 5 is $O(1)$. With regards to the while loop, let us assume that it takes t iterations to complete. We have two for loops on step 6 and step 17; they both take $O(N)$ time to complete, since all operations inside these loops are $O(1)$. Step 20 takes $O(L)$ time, where L is the number of steps per sample for Hamiltonian Monte Carlo. Hence, the overall complexity is $O(t(L + N))$.

3.3   HMC-PSO Solver with DNNs for Classification Tasks

There is an important advantage to utilizing HMC-PSO to train DNNs. PSO techniques do not rely on gradients. This means it is possible to use PSO on landscapes that do not rend well to gradient computation. Another advantage, something that has been found empirically, is that PSO allows the network to make better steps with the data given, so as to improve the loss faster than traditional methods of training. We shall elaborate on a method in which we can train DNNs using HMC-PSO.

We use the labels of the data to initialize a swarm of particles in such a way that they are close to a minimum of the loss function. The particles then follow HMC-PSO dynamics in order to find the best possible location in the vicinity of the initialization, g_{best}. The information about g_{best} is then used to make a first order estimation of the gradient of the loss function and make a step towards g_{best}. This is done because of an equivalence between EM-PSO and Stochastic Gradient Descent [1]. The estimate is given by:

$$grad = -\frac{(c_1 r_1 + c_2 r_2)}{\epsilon}(g_{best} - y) \tag{2}$$

where, $c_1 r_1$ and $c_2 r_2$ are the values obtained in the last iteration of Algorithm 1 and y is the value output by the last layer after forward propagation. This estimate is then used to update the weights of the rest of the network through backpropagation by the Adam optimizer. It's observed that this technique provides improved performance over Adam with numerically computed gradients [1].

It is important to note that the initialization of the swarm is an essential step since EM-PSO heavily relies on the initialization to produce a decent g_{best} since it almost always settles into the nearest local minimum. Each particle in the swarm has dimensions $batch\_size \times num\_classes$, where $batch\_size$ is the number of data items in a mini-batch, and $num\_classes$ is the number of neurons in the last layer. Each of the $batch\_size$ rows is initialized according to the index of the label. The values of the position matrix are set to 1 at the target index and -4 in all other columns of the row. Intuitively, we can understand how this incentivizes the network to learn the correct mapping from data to labels.

In this paper, we have utilized a Resnet-18 architecture for classification task on image-based datasets like: MNIST, Fashion MNIST, CIFAR-10, KMNIST, and CIFAR-100.

3.4   HMC-PSO Solver for Golomb's Ruler

A Golomb ruler is a set of distinct non-negative integers where the difference between every pair of integers is unique. This can be alternatively described by an imaginary ruler with integral markings such that no two pairs of marks are apart by the same distance. The order of the ruler is defined to be the number of markings, and the length as the largest marking. A Golomb Ruler is said to be *optimal* if no ruler of shorter length exists for the given order. An intuitive way of solving this problem would be by using powers of 2 as the markings. While this ensures a solution, it's length would be too large to be optimal. Furthermore, the computational complexity of the problem increases exponentially as the order of the ruler increases. The Traditional search-based algorithms like backtracking and branch-and-bound would not scale due to this computational complexity. Golomb's ruler finds application in several areas including error-correcting codes, selecting radio frequencies to reduce the effects of intermodulation interference with both terrestrial and extraterrestrial applications, design of phased arrays of radio antennas etc. Multi-ratio current transformers use Golomb's rulers to place transformer tap points.

The Golomb's Ruler problem is formulated by using a unconstrained optimization problem [13], which we solve by the HMC-PSO algorithm. Given the Golomb Ruler G, we measure the number of violations caused to the rule, i.e. a common differences obtained within markings. Let $V_G(d)$ be the violation score for a distance d. It is the number of times the distance d appears between any two markings of the ruler G. $V_G(d)$ is defined as: $V_G(d) = \max(0, |d_{ij} = d : 1 \leq j < i \leq m| - 1)$. The total violation of the ruler $V(G)$ can thus be given as:

$$V(G) = \sum_{d=1}^{n} V_G(d) \tag{3}$$

for a sufficiently large n. As closer we are to 0, closer we are to a Golomb Ruler. If $V(G) = 0$, it means we have successfully found a Golomb Ruler for the given order. For our experiment, we aim to achieve two important objectives:
1. For the given order n, find a Golomb Ruler (i.e. a ruler with $V(G) = 0$) &
2. Find the optimal or close to optimal Golomb ruler (i.e. a ruler with $V(G) = 0$ and smallest length).

It's not possible to use the total violation as the cost function for the optimization procedure directly as $V(G)$ can't differentiate between a valid ruler of smaller length and a valid ruler of larger length. We add a term proportional to the length of the proposed ruler. Let n be the order of G, the cost function for Golomb Ruler optimization i.e. $L(G)$ is defined as:

$$L(G) = V(G) + 10^{-k} \max(G) \tag{4}$$

where, k is the smallest positive integer such that $0 < 10^{-k} \max(G) < 1$. Intuitively, we pick a k such that the largest element obtained in the Golomb's Ruler in any iteration (while running HMC-PSO) when multiplied with 10^{-k} does not exceed 1. We do this to ensure priority of validity over length of G. Note that, $V(G)$ takes positive integer and 0 values. Thus $L(G)$ ensures that the loss via violation always remains greater than loss due to the ruler being large. This is essential since we need a valid Golomb's ruler rather than a small ruler with 1 violation.

Optimization Procedure: The space over which optimization is done is $\mathbb{R}^n$. Any position vector is given by $\boldsymbol{p} = (p_1, p_2, p_3, ..., p_n)$. The Golomb ruler corresponding to this position vector is given by: $\boldsymbol{p} = (p_1, p_2, p_3, ..., p_n) <=> G = \lfloor \|(p_1, p_2, p_3, ..., p_n)\| \rfloor$ where, $\|.\|$ is the absolute value function and $\lfloor . \rfloor$ is the floor function. A set of particles start at some initial position and follow the HMC-PSO rule to get to better positions based on $L(G)$. The results are mentioned in the next section.

4   Results

Before reporting the results of HMC-PSO, we note that: *1. EM-PSO couldn't solve the Golomb ruler problem 2. EM-PSO is just an approximation of first order derivative and was not exploited in the literature to solve Deep NN based classification problems 3. Exact methods such as Backtracking, Branch and Bound and inexact methods such as DQN yielded limited success in solving the Golomb ruler problem.*

Classification problems: We tested the optimizers HMC-PSO, Adam, Ada-Grad, AdaDelta, RMSProp, and NAdam in a DNN setting i.e. ResNet 18 architecture on the MNIST Digit Recognition, Fashion MNIST, CIFAR-10, KMNIST and CIFAR-100 datasets. We trained the model for 20 epochs for each of the optimizers and datasets (except CIFAR-100 for which we took 40 epochs). For the training dataset, *batch_size* is taken as 125, whereas for test, it is 100. For HMC-PSO experiments, hyperparameters taken are: *swarm_size* = 20, *dimensions* = *batch_size*, $c_1 = 2$, $c_2 = 2$, classes = 10 (for other datasets) or 100 (for CIFAR-100), and *true_y* = *targets*. These are determined from [1]. The results after 1 independent run are briefly mentioned in Table 1. The performance of HMC-PSO is better than or equivalent to other state-of-the-art optimizers.

Additionally, we performed a statistical inference test on HMC-PSO vs Adam using MNIST digit data and Cross-Entropy Loss. Let Null Hypothesis H_0 be given by H_0: HMC-PSO performs better than Adam. We ran HMC-PSO and Adam 10 different times to make a decision on H_0. The mean accuracy for 10 experiments using HMC-PSO was 98.96% with standard deviation of 0.29% and for Adam optimizer was 98.87% with standard deviation of 0.15%. 7 times out of 10, HMC-PSO performed better than Adam optimizer and hence H_0 is not rejected [1].

4.1   Golomb Ruler

P. Prudhvi et al. [13] utilized the use of Deep Q-Learning Networks (DQN) to solve the Golomb's Ruler problem. It used $-V(G)$ as the reward score. The agent repeatedly takes actions until all the markings are predicted and the violation score $V(G)$ is 0. This would mark the end of 1 episode and the agent is trained over a total of 100 episodes. The replay phase trains the agent to better predict the markings of a ruler. Results for [13] (for the 22nd episode) are mentioned in Table 2. It showcases the number of iterations required to converge to a solution

---

[1] Table of accuracies and loss obtained for each run is stored in the github repository: https://github.com/rtdsouza/torchswarm.

Table 1. Testing accuracy comparison of optimizers in a DNN framework on classification problems

Cross-Entropy Loss						
Dataset	HMC-PSO	Adam	RMSProp	NAdam	AdaGrad	AdaDelta
MNIST Digit	**99.25%**	98.63%	98.81%	98.85%	99.20%	97.24%
Fashion MNIST	93.62%	93.40%	**93.88%**	93.17%	92.89%	85.79%
CIFAR-10	**89.04%**	87.67%	87.90%	87.98%	80.39%	56.59%
KMNIST	**98.21%**	95.78%	97.04%	97.14%	97.69%	83.99%
CIFAR-100	**69.67%**	67.78%	64.46%	67.61%	58.98%	27.10%
Multi-Margin Loss						
Dataset	HMC-PSO	Adam	RMSProp	NAdam	AdaGrad	AdaDelta
MNIST Digit	98.19%	97.76%	98.24%	97.75%	98.54%	**98.88%**
Fashion MNIST	**93.43%**	88.53%	91.13%	91.17%	91.15%	93.31%
CIFAR-10	**82.33%**	78.19%	76.09%	80.26%	76.76%	78.24%
KMNIST	**98.06%**	96.86%	97.39%	96.65%	96.76%	76.25%
CIFAR-100	**69.32%**	64.39%	59.42%	64.97%	51.31%	21.37%

by various algorithms for the given ruler order: DQN shows computational savings as the order of the problem increases and a major improvement as compared to other traditional algorithms like Backtracking and Branch and Bound.

Table 2. Results for experiment by P. Prudhvi et al. comparing DQN to other methods for the Golomb's Ruler problem: The numbers represent the iterations taken by the algorithm to reach a solution for the given order.

Order	Backtracking	Branch and Bound	DQN
4	6	6	4
5	3615	13	5
6	170264	48	6
7	10095777	268	7

However, there is a major limitation for this technique. All methods analysed (including DQN) only arrive at a valid solution for the Golomb's ruler problem. It does not find an optimal Golomb's Ruler or rulers close to the optimal ruler.

Running HMC-PSO, we observe that uptill order $n = 50$, the HMC-PSO obtains a loss $L(G) < 1$ i.e. a valid Golomb Ruler within the first iteration itself. Hence, HMC-PSO achieves the same objectives that DQN and other algorithms aim in just 1 iteration and for a significantly higher order rulers. This shows that HMC-PSO is able to scale well on problems which require unimaginable number of computations (it's very difficult to estimate the number of iterations a backtracking algorithm would require for a Golomb ruler of order 50 if it takes 10095777 iterations for a ruler of order 7). Further, HMC-PSO then focuses on finding the optimal ruler, and is able to find rulers close to the optimal ruler. The number of iterations taken vs the loss obtained for Golomb ruler for different values of order n is given in Table 3. For these results, we notice that the length of Golomb Ruler never crosses 1000, and hence we take $k = 3$. We notice that HMC-PSO is able to find optimal Golomb ruler of order 7 and close to optimal but valid rulers for higher orders. The decimal digits of loss is the length of the ruler. For instance, for $n = 11$, a loss of 0.106 signifies that a valid Golomb ruler of length 106 is found after running 80 iterations. The optimal loss of 0.072 means that the optimal ruler for order $n = 11$ has a length of 72.

Table 3. HMC-PSO's results on the Golomb Ruler problem

Order	Iterations	Loss obtained	Optimal loss
5	3	0.011	0.011
7	65	0.025	0.025
9	80	0.060	0.044
11	80	0.106	0.072
15	80	0.395	0.151

5   Conclusion

In summary, we propose Hamiltonian Monte Carlo Particle Swarm Optimization (HMC-PSO) - an optimization algorithm that can either be directly used for any optimization procedure or can be utilized as a tool to approximate gradients of any differentiable or non-differentiable loss function. Because of coupling between HMC and PSO, the HMC-PSO algorithm reaps considerable benefits from both: extensively searching the parameter space and converging quickly to the minima. Apart from this, HMC-PSO acquires an essential benefit from EM-PSO: it is able to approximate gradients for non-continuous and non-differentiable loss functions which usual gradient-descent algorithms are unable to compute. This is one of the reasons why HMC-PSO stands out from other SOTA optimizers apart from better results. Apart from Back tracking, Branch and Bound, and DQN, it is also the 4th different optimizer technique for solving the Golomb's Ruler problem. In the future, it can also be used for solving other special problems like the N-Queens' problem. In this paper, we focused on the prior-work i.e. EM-PSO as the motivation for our research. We provided necessary motivation as to why we utilized HMC and how we successfully coupled HMC and PSO techniques. Further, we discussed two different uses of HMC-PSO: as a direct optimization procedure for solving the Golomb's Ruler problem and as a optimizer for DNN in classification problems. We obtained better results than other SOTA methods. However, an essential point of difference between EM-PSO and HMC-PSO is that HMC-PSO explores the search-space much more extensively and ensures that the global optimum is found. This was observed when we tested them on 9 different multi-modal Gaussian distributions. EM-PSO was only able to converge linearly to the local maxima, whereas HMC-PSO effectively explored and found all the maxima and converged to the global maxima. HMC-PSO produced sound test-results on 1-d benchmark optimization problems too [2].

References

1. Mohapatra, R., Saha, S., Coello, C.A.C., Bhattacharya, A., Dhavala, S.S., Saha, S.: AdaSwarm: Augmenting gradient-based optimizers in Deep Learning with Swarm Intelligence. IEEE Trans. Emerg. Top. Comput. Intell. **6**(2), 329–340 (2022). https://doi.org/10.1109/TETCI.2021.3083428
2. Agarwal, N., Anil, R., Hazan, E., Zhang, C.: Disentangling adaptive gradient methods from learning rates. ArXiv. https://doi.org/10.48550/arXiv.2002.11803
3. Brooks, S., Gelman, A., Jones, G., Meng, X.-L. (eds.) Handbook of Markov Chain Monte Carlo (1st ed.). Chapman and Hall/CRC (2011). https://doi.org/10.1201/b10905
4. Mikki, S.M., Kishk, A.A.: Particle swarm optimizaton: a physics-based approach. Morgan Claypool (2008). https://doi.org/10.2200/S00110ED1V01Y200804CEM020
5. Welling, M., Teh, Y.W.: Bayesian learning via stochastic gradient langevin dynamics. In: Proceedings of the 28th International Conference on Machine Learning, pp. 681-688 (2011). https://dl.acm.org/doi/abs/10.5555/3104482.3104568
6. Yokoi, S., Sato, I.: Bayesian interpretation of SGD as Ito process. ArXiv, vol. abs/1206.1901 (2012). https://doi.org/10.48550/arXiv.1911.09011
7. Jivani, U.N., Vaidya, O.B., Bhattacharya, A., Saha, S.: A swarm variant for the schrödinger solver. Int. Joint Conf. Neural Netw. (IJCNN) **2021**, 1–8 (2021). https://doi.org/10.1109/IJCNN52387.2021.9534221
8. Kingma, D.P., Ba, J.: Adam: a method for stochastic optimization. arXiv preprint arXiv:1412.6980 (2014). https://doi.org/10.48550/arXiv.1412.6980
9. Metropolis, N., Rosenbluth, A.W., Rosenbluth, M.N., Teller, A.H.: Equation of state calculations by fast computing machines. J. Chem. Phys. **21**, 1087–1092 (1953). https://doi.org/10.1063/1.1699114
10. Duane, S., Kennedy, A.D., Pendleton, B.J., Roweth, D.: Hybrid monte carlo. Phys. Lett. B **195**(2), 216–222 (1987). ISSN: 0370–2693. https://doi.org/10.1016/0370-2693(87)91197-X
11. Neal, R.M.: MCMC using Hamiltonian dynamics. ArXiv, vol. abs/1911.09011 (2019). https://doi.org/10.48550/arXiv.1206.1901
12. Guihenneuc-Jouyaux, C., Rousseau, J.: Laplace expansions in Markov chain monte Carlo algorithms. J. Comput. Graph. Stat. **14**, 75–94 (2005)
13. Prudhvi Raj, P., Saha, S., Srinivasa, G.: Solving the N-Queens and Golomb Ruler problems using DQN and an approximation of the convergence. In: Mantoro, T., Lee, M., Ayu, M.A., Wong, K.W., Hidayanto, A.N. (eds.) ICONIP 2021. CCIS, vol. 1517, pp. 545–553. Springer, Cham (2021). https://doi.org/10.1007/978-3-030-92310-5_63

---

[2] Additional results are stored here: https://github.com/rtdsouza/torchswarm.

Heterogeneous Graph Representation for Knowledge Tracing

Jisen Chen, Jian Shen, Ting Long, Liping Shen$^{(\boxtimes)}$, Weinan Zhang, and Yong Yu$^{(\boxtimes)}$

Shanghai Jiao Tong University, Shanghai, China
{chenjisen,r_ocky,longting,lpshen,wnzhang}@sjtu.edu.cn
yyu@apex.sjtu.edu.cn

Abstract. Knowledge tracing (KT) is a fundamental task of intelligent education, which traces students' knowledge states by their historical interactions. In KT, students, questions, concepts, and answers are four main types of entities, and they contain various relations, including student-question interactive relations, question-concept relations, and question-answer relations. Such rich knowledge in these heterogeneous relations could potentially improve the prediction of KT. However, it has not been sufficiently utilized in existing KT methods. In this paper, we propose a novel method, called Heterogeneous Graph Representation for Knowledge Tracing, to leverage these useful relations. Our method first models all the complex entities and relations in KT as a heterogeneous graph, and then uses a heterogeneous graph neural network to obtain entities' feature representations. After that, we feed the feature embeddings to a KT model in an end-to-end training manner. Due to the heterogeneous graph's high representational capacity, our method exploits the relations among students, questions, concepts, and answers in a concise and unified way. Experiments on four KT datasets show that our method achieves state-of-the-art performance.

Keywords: Knowledge Tracing · Heterogeneous Graph · Intelligent Education

1 Introduction

Currently, online education is developing rapidly and has gradually become a common way of learning. Many advanced algorithms are used to mine large-scale interactive data between students and the system to provide more intelligent educational services so that each student can have a better adaptive learning experience. *Knowledge tracing* (KT) [7] is a fundamental task of intelligent education, which uses students' historical learning interactive data to trace students' dynamic knowledge states and predict their performances in future interactions.

In KT, the collected historical data include relational information among entities of *students* (i.e., users), *questions* (i.e., problems/exercises), *concepts* (i.e., knowledge concepts/skills), and *answers* (i.e., responses). These entities

© The Author(s), under exclusive license to Springer Nature Switzerland AG 2023
M. Tanveer et al. (Eds.): ICONIP 2022, LNCS 13623, pp. 224–235, 2023.
https://doi.org/10.1007/978-3-031-30105-6_19

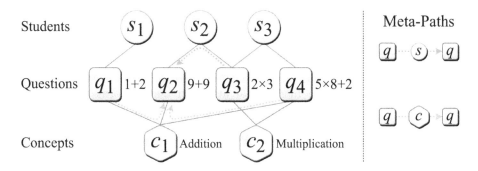

Fig. 1. An example of knowledge tracing data, student-question-concept relation graph, and meta-paths. Questions q_3 and q_4 are done by the same student s_3, and they have similar related concepts and difficulties. Questions q_1 and q_2 are related to the same concept c_1 but have different difficulties. Question q_2's meta-based neighbors include q_2, q_3 (via q-s-q), q_1, and q_4 (via q-c-q).

contain three main types of relations: student-question, question-concept, and question-answer. Existing KT methods utilize the knowledge in these entities and relations with varying degrees. However, they do not model all types of relations in a comprehensive and uniform way.

For question and concept entities, many KT models [19,33] only use concepts as the input to learn a student's concept mastery, and they ignore the specific information and the student's state of each question, causing the loss of latent information between them [1,30] (see q_1, q_2 in Fig. 1). For student entities, few KT models consider using student information as the input, and students are only indices of sequence. Using student information is meaningful, and the questions done by the same student can implicitly reflect that their learning stages and difficulties are similar. Some previous work partially utilizes relational information by introducing a question-concept graph [15,27,30] or student-specific parameters [20,31], but they do not fully use all types of relations.

We believe that the KT task can be better achieved by comprehensively and uniformly modeling all types of entities while retaining specific feature information. Since these entities with different types have unique effects and structural information, complete modeling of all entities and their relations can best extract the rich information contained therein.

Inspired by previous work [9,11], we represent all types of entities and relations in KT as vertices and edges in a heterogeneous graph. A heterogeneous graph [21,23] contains multiple types of vertices and edges, which can naturally express data with complex structures in the real world. As shown in Fig. 1 and 2, the heterogeneous graph demonstrates these entities and their relations. In addition, heterogeneous graph neural networks (HGNNs) [8,28] can obtain semantic relational information in a heterogeneous graph.

In this paper, we propose a deep KT model, namely **Heterogeneous Graph Representation for Knowledge Tracing** (HGRKT), to model various types of entities and relations in the KT task in a unified way. We first construct a *knowledge tracing heterogeneous graph* (KTHG), which describes the student-question-concept-answer relations. Then, we use a meta-path-based multi-layer HGNN to obtain heterogeneous relation-aware representations in KTHG. Specifically, we use Heterogeneous Graph Attention Network (HAN) to implement HGNN, which is based on hierarchical attention and performs well among HGNNs. After that, we feed the representations of questions into the KT model, and the whole model can be optimized through end-to-end backpropagation.

In summary, our main contributions are as follows:

1. We construct a KTHG to comprehensively and uniformly model the interactive and structural relations among students, questions, concepts, and answers, where each vertex has an embedding and represents an entity with a type. Our proposed method is simple and concise by exploiting the high representational capacity of the heterogeneous graph.
2. We apply HGNN to get representations for KT. HGNN can effectively extract the rich knowledge contained in relations into the generated question embeddings, and the embeddings are used as the input of the KT model.
3. Experiments on multiple benchmark datasets show our model's effectiveness. Our method improves AUC by 1% compared to the best baseline on average.

2 Related Work

2.1 Knowledge Tracing

Existing KT models can be roughly divided into non-deep models and deep models. In recent years, deep learning has been widely used in KT task research for its powerful ability to extract and represent features and discover complex structures [14]. DKT [19] first introduces deep learning to KT and uses a Recurrent Neural Network (RNN) [29] to model a student's knowledge state. DKVMN [33] uses Memory Augmented Neural Network to automatically discover basic concepts and trace the state of each concept. Since then, various deep KT methods have been proposed based on the two methods.

Many researchers introduce graph and question features to model knowledge structure in different ways, such as utilizing concept graphs [3,17,25], considering question features and combining them with concept features in prediction [16,27], and using question-concept relation graphs and graph neural networks (GNNs) [13] to learn embeddings for questions [15,30].

2.2 Heterogeneous Graph

A heterogeneous graph [21,23] is a graph with multiple types of vertices and edges, which can naturally describe complex relations and rich semantics in many data mining tasks [9,11]. Recently, researchers have proposed many Heterogeneous graph neural networks (HGNNs) models [8,28,32] to extract representations of vertices and obtain knowledge in heterogeneous graphs.

Our method mainly uses the Heterogeneous Graph Attention Network (HAN) [28]. HAN is an HGNN based on vertex-level and semantic-level attentions. Vertex-level attention uses a semantic-specific GAT [26] layer to learn the importance between vertices and their meta-path-based neighbors, while semantic-level attention measures the importance of meta-paths with an attention vector and fuses embeddings of all meta-paths to obtain the final embedding.

3 Preliminaries

3.1 Knowledge Tracing

In the KT task, a group of students $\mathcal{S}$ sequentially answer a series of questions from a set $\mathcal{Q}$. At the time step t, a student $s \in \mathcal{S}$ answers a question $q_t \in \mathcal{Q}$, and the correctness of the answer is $a_t \in \{0, 1\}$. The observed learning interaction sequence before the time step T is

$$\boldsymbol{X} = \{q_t, a_t\}_{t=1}^{T} = \{(q_1, a_1), \ldots, (q_t, a_t), \ldots, (q_T, a_T)\}. \tag{1}$$

In the time step $T + 1$, given the historical sequence $\boldsymbol{X}$ and a new question q_{T+1}, the goal of KT is to predict the probability of the student correctly answering the new question:

$$\hat{a}_{T+1} = P(a_{T+1} = 1 | \boldsymbol{X}, q_{T+1}). \tag{2}$$

Let ℓ be the loss function (e.g., binary cross-entropy). The predictive loss of the answer is $\ell(a_{t+1}, \hat{a}_{t+1})$, and the loss for a student is $\sum_t \ell(a_{t+1}, \hat{a}_{t+1})$.

3.2 Heterogeneous Graph

A *heterogeneous graph* [23] $\mathcal{G} = (\mathcal{V}, \mathcal{E})$ consists of a vertex set $\mathcal{V}$ and an edge set $\mathcal{E}$, and it is associated with a vertex type map and an edge type map so that each vertex and edge has its own type.

A *meta-path* [24] Φ is a path $A_1 \xrightarrow{R_1} A_2 \xrightarrow{R_2} \cdots \xrightarrow{R_l} A_{l+1}$ describing a composite relation of all intermediate edge types $R_1, \ldots, R_l$ between vertex types A_1 and A_{l+1}. Given a meta-path Φ, the *meta-path-based neighbors* $\mathcal{N}^\Phi$ is a map from vertex i to a set of vertices that connect with vertex i via meta-path Φ. Meta-path is a fundamental structure of a heterogeneous graph, which can express certain semantics and reveal diverse structure information.

As shown in Fig. 1, the meta-path question-concept-question represents questions with the same concept (e.g., q_1 and q_4), while the meta-path question-student-question represents questions done by the same student (e.g., q_2 and q_3). Question q_2's neighbors based on the meta-path question-student-question include q_2 (itself) and q_3, indicating that s_2 answers both. Its neighbors based on meta-path question-concept-question include q_2, q_1, and q_4, indicating that they share the same concept c_1.

3.3   Message Passing

To utilize the graph structural knowledge, we apply the message passing to get the new embeddings of vertices in graph $\mathcal{G}(\mathcal{V}, \mathcal{E})$. We can use a GNN layer to apply the message passing from source vertices $\{v_i | (v_i, v_j) \in \mathcal{E}\}$ to target vertices $\{v_j | (v_i, v_j) \in \mathcal{E}\}$ via all edges. We construct an embedding matrix $\mathbf{M} = (\mathbf{h}_1^T; \ldots; \mathbf{h}_N^T) \in \mathbb{R}^{N \times d}$, consisting of all vertices' embedding vectors, where $\mathbf{h}_i = \mathbf{M}_{i,*} \in \mathbb{R}^d$ is vertex v_i's embedding vector. The GNN transforms the input embedding matrix into a new embedding matrix $\mathbf{M}' = \text{GNN}(\mathbf{M}|\mathcal{G})$, where the new embedding of vertex i is $\mathbf{h}_i' = \mathbf{M}_{i,*}'$.

For a heterogeneous graph $\mathcal{G}'$, we should construct multiple embedding matrices for each type of vertices. For vertices with type A_k, its corresponding embedding matrix is M_k. The message passing is applied from source vertices to target vertices with various types. Let the source vertices have n_s different types $\{A_1^{\mathcal{S}}, \ldots, A_{n_s}^{\mathcal{S}}\}$ and target vertices have n_t different types $\{A_1^{\mathcal{T}}, \ldots, A_{n_t}^{\mathcal{T}}\}$, then the HGNN transforms the embedding matrices of sources into embedding matrices of targets, i.e., $\{M_1^{\mathcal{T}}, \ldots, M_{n_t}^{\mathcal{T}}\} = \text{HGNN}(M_1^{\mathcal{S}}, \ldots, M_{n_s}^{\mathcal{S}}|\mathcal{G}')$.

4   Method

4.1   KTHG

The data of knowledge tracing includes students, questions, concepts, answers, and their relations. We model them as vertices and edges with different types in a *knowledge tracing heterogeneous graph* (KTHG). Let $\mathcal{S}$, $\mathcal{Q}$, and $\mathcal{C}$ be the set of students, questions, and concepts separately. Let $\mathcal{A} = \{a | a = 0 \text{ or } a = 1\}$ be the set of answers, where the member 1 and 0 indicates whether the answer is correct or not. We define the KTHG $\mathcal{G}_{\text{KT}}(\mathcal{V}_{\text{KT}}, \mathcal{E}_{\text{KT}})$ as the combination of three simpler heterogeneous graphs $\mathcal{G}_{sq}, \mathcal{G}_{qc}, \mathcal{G}_{qa}$, i.e., $\mathcal{G}_{\text{KT}} = \mathcal{G}_{sq} \cup \mathcal{G}_{qc} \cup \mathcal{G}_{qa}$.

Student-Question Graph. The student-question graph $\mathcal{G}_{sq}(\mathcal{S} \cup \mathcal{Q}, E_{sq})$ is a bipartite graph, where E_{sq} contains the interactions between students and questions in the training data, and an undirected edge (s_i, q_j) means that student $s_i \in \mathcal{S}$ answers question $q_j \in \mathcal{Q}$, as is shown in the upper part of Fig. 2(b).

Question-Concept Graph. The question-concept graph $\mathcal{G}_{qc}(\mathcal{Q} \cup \mathcal{C}, E_{qc})$ is a bipartite graph, where E_{qc} contains the relations between questions and concepts. An undirected edge (q_i, c_j) means that question $q_i \in \mathcal{Q}$ is related to concept $c_j \in \mathcal{C}$, as is shown in the lower part of Fig. 2(b).

Question-Answer Graph. The question-answer graph $\mathcal{G}_{qa}(\mathcal{Q} \cup \mathcal{A} \cup \mathcal{Q}_A, E_{qq_a} \cup E_{aq_a})$ is a combination of two complete bipartite graphs $\mathcal{G}_{qq_a}(\mathcal{Q} \cup \mathcal{A}, E_{qq_a})$ and $\mathcal{G}_{aq_a}(\mathcal{A} \cup \mathcal{Q}_A, E_{aq_a})$, where $\mathcal{Q}_A$ is the set of questions combined with answers, and $\mathcal{Q}_A = \{q_{i0}, q_{i1}\}_{i=1}^{|\mathcal{Q}|}$ for $\mathcal{Q} = \{q_i\}_{i=1}^{|\mathcal{Q}|}$, and $|\mathcal{Q}_A| = 2|\mathcal{Q}|$. Concretely, to combine the question $q_i \in \mathcal{Q}$ with answer a, we use two new vertices $q_{i1}, q_{i0} \in \mathcal{Q}_A$ to

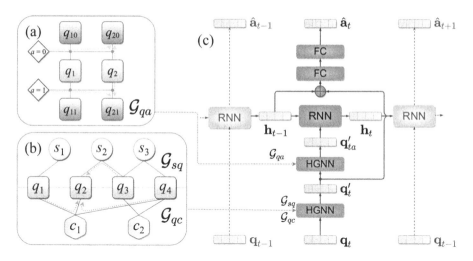

Fig. 2. The architecture of our method. (a) The question-answer graph $\mathcal{G}_{qa}$; (b) The student-question graph $\mathcal{G}_{sq}$ and the question-concept graph $\mathcal{G}_{qc}$; (c) The overview of our method. It shows the updating of a student's knowledge state in the time step t. Two parts of KTHG in (a) and (b) are used in two HGNN layers in (c) separately. Dotted lines show the message passing via meta-paths.

represent the question with the correct and incorrect answer respectively, called question-with-answer (qa for short) vertices.

As shown in Fig. 2(a), question q_1 and two answers, "$a = 0$" and "$a = 1$", are linked into q_{10} and q_{11} by two meta-paths, question-qa and answer-qa. We use these two meta-paths to fuse a question embedding and an answer embedding into a question-with-answer embedding.

4.2   Representation Learning on KTHG

We stack the HGNN layers to make up a multi-layer HGNN, and apply message passing on KTHG with a two-layer HGNN to generate representations of vertices with graph structure information about other vertices connected with them. Although we may get the new embeddings for any type of vertices, we only use generated embeddings about questions as the input feature of the KT model because students' performance is mainly related to them.

For all types of vertice sets $\mathcal{S}, \mathcal{Q}, \mathcal{C}, \mathcal{A}, \mathcal{Q}_A$, we construct embedding matrices $\mathbf{M}_S, \mathbf{M}_Q, \mathbf{M}_C, \mathbf{M}_A, \mathbf{M}_{QA}$. At first, We use an HGNN layer to transform the original question embeddings, student embeddings, and concept embeddings into aggregated question embeddings:

$$\mathbf{M}'_Q = \mathrm{HGNN}_{sqc}(\mathbf{M}_S, \mathbf{M}_Q, \mathbf{M}_C | \mathcal{G}_{sq} \cup \mathcal{G}_{qc}), \tag{3}$$

where $\mathbf{M}'_Q = (\mathbf{q}_1'^{\mathrm{T}}; \ldots; \mathbf{q}_N'^{\mathrm{T}})$, $\mathbf{q}'_i = (\mathbf{M}'_Q)_{i,*}$ and $N = |\mathcal{Q}|$. Here, we apply message passing on the student-question-concept graph and use meta-paths ended with the question vertex, and the aggregated question embeddings carry information about questions, students, and concepts.

When modeling the student's learning process, we need to combine the information about questions and answers. We use another HGNN layer to fuse an aggregated question embeddings $\mathbf{M}'_Q$ and an answer embeddings $\mathbf{M}_A$ into an aggregated question-with-answer embeddings $\mathbf{M}'_{QA}$:

$$\mathbf{M}'_{QA} = \mathrm{HGNN}_{qa}(\mathbf{M}'_Q, \mathbf{M}_A | \mathcal{G}_{qa}), \tag{4}$$

where $\mathbf{M}'_{QA} = \left(\mathbf{q}'^T_{10}; \mathbf{q}'^T_{11}; \ldots; \mathbf{q}'^T_{N0}; \mathbf{q}'^T_{N1}\right)$ and $\mathbf{q}_{ia} = \left(\mathbf{M}'_{QA}\right)_{2i+a-1,*}$. Here, we apply message passing on the question-answer graph and use the meta-paths ended with the question-with-answer vertex, so the aggregated question-with-answer embeddings carry the information of its corresponding aggregated question embeddings and answer embeddings.

4.3   KTHG Representations for KT Model

We trace a student's knowledge state by his/her learning interaction sequence $\{q_t, a_t\}^T_{t=1}$. Note that here q_t denotes the question ID in the time step t (instead of the question with ID t). The architecture of our KT model is shown in Fig. 2(c). After representation learning on KTHG, we get question embedding matrices $\mathbf{M}'_Q, \mathbf{M}_{QA}$ with graph structure information for the KT model. For question q_t, we can get its aggregated question embedding $\mathbf{q}'_t = (\mathbf{M}'_Q)_{q_t,*}$. When tracing the student's knowledge state, we can get the question-with-answer embedding $\mathbf{q}'_{ta} = (\mathbf{M}'_{QA})_{2q_t+a_t-1,*}$ for the current answer a_t.

Then, similar to DKT [19], we use an RNN to trace the student's knowledge state. For each time step t, we feed the embedding $\mathbf{q}'_{ta}$ into an RNN cell to update the RNN's hidden state vector $\mathbf{h}_t$:

$$\mathbf{h}_t = \mathrm{RNN}(\mathbf{q}'_{ta}, \mathbf{h}_{t-1}), t = 1, 2, \ldots, T. \tag{5}$$

In the prediction, we use the student's current knowledge state and question embedding to predict the answer. Specifically, we first concatenate RNN's current state $\mathbf{h}_{t-1}$ with the aggregated question embedding $\mathbf{q}'_t$. After that, we feed the concatenated vector into a fully connected layer to obtain the summary vector $\mathbf{h}'_t$, and then we feed the vector into a Sigmoid activation layer to calculate the probability of correctly answering, denoted as $\hat{a}_t$:

$$\mathbf{h}'_t = \tanh(\mathbf{W}_{fc}(\mathbf{h}_{t-1} \oplus \mathbf{q}'_t) + \mathbf{b}_{fc}), \tag{6}$$
$$\hat{a}_t = \sigma(\mathbf{W}_a \mathbf{h}'_t + \mathbf{b}_a), \tag{7}$$

where $\oplus$ denotes concatenation operator and σ is the Sigmoid function.

5   Experiment

5.1   Datasets

We evaluate our method on four open public datasets: ASSIST09, ASSIST12, EdNet, and Junyi, which are sampled from educational platforms [2,6,10].

Table 1. Dataset statistics.

Statistics	ASSIST09	ASSIST12	EdNet	Junyi
#Records	185,110	1,839,429	326,267	622,781
#Students	2,968	22,422	4,700	7,000
#Questions	15,003	45,543	11,060	1,978
#Concepts	121	99	189	39
#Concepts per Question	1–4	1	1–7	1
Avg. #Questions Per Concept	150.8	460.0	128.7	50.7
AVg. #Concepts Per Question	1.2	1.0	2.2	1.0
Avg. #Attempts Per Question	12.3	40.4	29.5	314.9
Avg. #Attempts Per Concept	1,914.2	18,580.1	4,026.9	15,968.7
Correct Rate (%)	63.8	69.6	59.7	67.3

We filter out records without concepts and students with less than 10 interactions and randomly sample students to make computing resources affordable. The statistics of the four datasets are shown in Fig. 1. We use the indices of questions, their related concepts, and students' answers from records. The maximum length of students' interaction sequence is set to 200. In the datasets, We split 4/5 of students for training and validation, and another 1/5 for testing, where no student appears in both training and testing sets.

5.2 Implementation Details

To evaluate the effectiveness of our method, we compare our method with three groups of 11 KT baseline methods. Single-state methods represent a student's knowledge state with a vector, which is often the hidden state of RNN; Multi-state methods represent the student's knowledge concept state with multiple vectors; State-free methods do not maintain a vector to represent the student's knowledge state. Note that our method is single-state.

To fairly compare the results of all methods, we uniformly set some shared parameters. The dimension of question embeddings and concept embeddings is 64. The batch size is 16. We initialize the parameters randomly, and the optimizer is Adam [12] with a learning rate of 0.001. We use a single-layer GRU [4] to implement RNN [4], and the dimension of its hidden state is 64. We run each method five times and show their average results of AUC.

We choose HAN [28] as the HGNN layer implementation, called **HAN-HGRKT**. In HAN, the dimension of the semantic-level attention vector is 128. The input dimension of the vertex embedding is set to 64. Besides, we only update part of useful vertices in the current batch to reduce memory usage. We pick five meaningful meta-paths for two HAN layers: (1) q-s-q, q-c-q, c-q for $\mathcal{G}_{sq}, \mathcal{G}_{qc}$; (2) q-$qa$, a-$qa$ for $\mathcal{G}_{qa}$, where "q", "s", "c", "a", and "qa" denote question, student, concept, answer, and question-with-answer respectively. The code of our implementation is available at https://github.com/chenjisen/HGR-KT.

Table 2. AUC results (%) on four datasets.

Group	Method	ASSIST09	ASSIST12	EdNet	Junyi
Single-state	DKT [19]	67.1	68.8	70.2	88.5
	EERNNA [22]	74.1	_74.1_	71.5	86.9
	DHKT [27]	74.2	_74.1_	71.6	_89.0_
	GIKT [30]	_75.5_	73.9	_72.5_	88.9
Multi-state	DKVMN [33]	65.6	68.6	68.9	88.5
	GKT [17]	75.1	68.5	69.1	85.8
	SKT [25]	71.3	66.4	66.4	86.1
State-free	SAKT [18]	68.9	69.5	70.5	79.5
	CKT [20]	70.3	70.0	64.9	87.7
	SKVMN [1]	64.7	68.2	68.9	88.2
	SAINT [5]	68.5	68.5	69.9	85.3
(Ours)	HAN-HGRKT *	**76.8**	**74.4**	**73.4**	**89.5**

Table 3. Attention weights of meta-paths in two HAN layers.

Graphs	Meta-path	ASSIST09	ASSIST12	EdNet	Junyi
$\mathcal{G}_{sq}, \mathcal{G}_{qc}$	q-s-q	0.39	0.64	0.53	0.52
	q-c-q	0.35	0.36	0.29	0.36
	c-q	0.26	0.00	0.18	0.12
$\mathcal{G}_{qa}$	q-qa	0.71	0.81	0.69	0.66
	a-qa	0.29	0.19	0.31	0.34

5.3 Results

We measure AUC to evaluate and compare the performance of methods. All decimals are rounded to the nearest tenth. Table 2 shows AUC results in experiments, and a higher AUC indicates that the method predicts student performances better. We observe that our method, HAN-HGRKT, outperforms all 11 baselines and achieves state-of-the-art performance in all four datasets, demonstrating that our method leverages all types of relations better with the HAN. Our method outperforms the best baseline (GIKT) by about 1% in AUC on average. In addition, there is no significant difference in performance between state-free, single-state, and multi-state methods.

However, our method has the least improvement on ASSIST12 and Junyi, which only outperforms the best baseline for no more than 1% in AUC. We assume that in these two datasets, a question is only related to a concept, so they are simpler than other datasets, and there is no place for the heterogeneous graph to show its powerful representation capacity.

Besides, we get the attention weight values of all meta-paths in HAN. Table 3 shows the results, where a mid-line separates the results from two HAN layers. We find that the HAN can automatically adjust the weight of each meta-paths on different datasets.

Table 4. AUC results (%) of HAN-HGRKT with different graphs and meta-paths in two HAN layers.

Graphs	Meta-paths	ASSIST09	ASSIST12	EdNet	Junyi
$\mathcal{G}_{sq}$	q-s-q	75.1	74.4	73.3	**89.5**
$\mathcal{G}_{qc}$	c-q	76.0	68.4	69.3	85.6
	q-c-q	76.3	**74.5**	71.7	89.1
	q-c-q, c-q	76.6	**74.5**	71.8	89.0
$\mathcal{G}_{sq}, \mathcal{G}_{qc}$	q-s-q, q-c-q	76.4	74.4	**73.4**	**89.5**
	q-s-q, q-c-q, c-q*	**76.8**	74.4	**73.4**	**89.5**
$\mathcal{G}_{qa}$	q-qa, a-qa *	**76.8**	74.4	**73.4**	**89.5**
	q-qa, aq-qa	71.4	71.7	70.9	86.8
(AE)	(AE)	76.7	74.4	73.2	89.4
(ZV)	(ZV)	76.5	74.3	73.2	89.1

5.4 Ablation Study

We design the ablation study to further analyze the effect of each sub-graph and meta-path in our method.

Student-Question Graph and Question-Concept Graph. We change the usage of $\mathcal{G}_{sq}$, $\mathcal{G}_{qc}$, and their related meta-paths in the first HAN layer, while not changing the second HAN layer. The upper part of Table 4 shows the results, where our original method performs the best, and removing any graphs or meta-paths decreases the performance. We notice that the meta-path c-q has the least effect, but it has the most significant improvement and the largest attention weight (see Table 3) on ASSIST09, indicating that the c-q relation contains irreplaceable semantic information on ASSIST09 compared to the other datasets, and our method can automatically catch it.

Question-Answer Graph. We compare our second HAN layer with other methods of combining questions with answers, while not changing the first HAN layer. The variants of the methods of combining answers with questions are listed as follows: (1) Using a new vertex type "aq", which means that we use two answer vertices for each question respectively, and the number of "aq" vertices is $2|\mathcal{Q}|$; (2) AE: concatenating new question embedding with answer embedding instead of using $\mathcal{G}_{qa}$, i.e., $\mathbf{q}'_{ta} = \mathbf{q}'_t \oplus \mathbf{a}_t$; (3) ZV: concatenating new question embedding with zero vectors in two orders instead of using $\mathcal{G}_{qa}$, i.e., $\mathbf{q}'_{ta} = \mathbf{q}'_t \oplus \mathbf{0}$ when $a_t = 1$ and $\mathbf{0} \oplus \mathbf{q}'_t$ when $a_t = 0$. The lower part of Table 4 shows the results, where our original method performs the best. We notice that if we construct two answer embedding for each question vertex individually, the performance decreases a lot, and the reason may be that it increases the complexity of the model and causes difficulty for training.

6   Conclusion

In this paper, we propose an end-to-end method, Heterogeneous Graph Representation for Knowledge Tracing (HGRKT), to learn question representations for the KT task. We introduce a KTHG to model heterogeneous relations among students, questions, concepts, and answers. Then, we use a multi-layer HGNN to obtain the question representations with rich relational knowledge. After that, we use a KT model to get the students' performances. We evaluate our method on four KT datasets and compare it with 11 knowledge tracing baseline methods. As our model gets better question representations, it performs state-of-the-art performance and outperforms the best baseline by 1% on average. The extensive ablation study shows the effectiveness of the whole KTHG with the meta-paths. In the future, we will apply different implementations of heterogeneous graph models to explore a better extraction method of heterogeneous relations.

Acknowledgments. This work is supported by Shanghai Municipal Science and Technology Major Project (2021SHZDZX0102) and National Natural Science Foundation of China (62177033).

References

1. Abdelrahman, G., Wang, Q.: Knowledge tracing with sequential key-value memory networks. In: Proceedings of SIGIR (2019)
2. Chang, H.S., Hsu, H.J., Chen, K.T.: Modeling exercise relationships in e-learning: a unified approach. In: Proceedings of EDM (2015)
3. Chen, P., Lu, Y., Zheng, V.W., Pian, Y.: Prerequisite-driven deep knowledge tracing. In: Proceedings of ICDM (2018)
4. Cho, K., et al.: Learning phrase representations using RNN encoder-decoder for statistical machine translation. In: Proceedings of EMNLP (2014)
5. Choi, Y., et al.: Towards an appropriate query, key, and value computation for knowledge tracing. In: Proceedings of L@S (2020)
6. Choi, Y., et al.: Ednet: a large-scale hierarchical dataset in education. In: Proceedings of AIED (2020)
7. Corbett, A.T., Anderson, J.R.: Knowledge tracing: modeling the acquisition of procedural knowledge. In: Proceedings of UMUAI (1994)
8. Dong, Y., Chawla, N.V., Swami, A.: metapath2vec: Scalable representation learning for heterogeneous networks. In: Proceedings of KDD (2017)
9. Fan, S., et al.: Metapath-guided heterogeneous graph neural network for intent recommendation. In: Proceedings of KDD (2019)
10. Feng, M., Heffernan, N., Koedinger, K.: Addressing the assessment challenge with an online system that tutors as it assesses. In: Proceedings of UMUAI (2009)
11. Gao, W., et al.: RCD: Relation map driven cognitive diagnosis for intelligent education systems. In: Proceedings of SIGIR (2021)
12. Kingma, D.P., Ba, J.: Adam: a method for stochastic optimization. In: Proceedings of ICLR (2015)
13. Kipf, T.N., Welling, M.: Semi-supervised classification with graph convolutional networks. In: Proceedings of ICLR (2017)
14. LeCun, Y., Bengio, Y., Hinton, G.: Deep learning. Nature (2015)

15. Liu, Y., Yang, Y., Chen, X., Shen, J., Zhang, H., Yu, Y.: Improving knowledge tracing via pre-training question embeddings. In: Proceedings of IJCAI (2020)
16. Minn, S., Desmarais, M.C., Zhu, F., Xiao, J., Wang, J.: Dynamic student classification on memory networks for knowledge tracing. In: Proceedings of PAKDD (2019)
17. Nakagawa, H., Iwasawa, Y., Matsuo, Y.: Graph-based knowledge tracing: modeling student proficiency using graph neural network. In: WI (2019)
18. Pandey, S., Karypis, G.: A self-attentive model for knowledge tracing. In: Proceedings of EDM (2019)
19. Piech, C., et al.: Deep knowledge tracing. In: Proceedings of NeurIPS (2015)
20. Shen, S., et al.: Convolutional knowledge tracing: modeling individualization in student learning process. In: Proceedings of SIGIR (2020)
21. Shi, C., Li, Y., Zhang, J., Sun, Y., Yu, P.S.: A survey of heterogeneous information network analysis. In: Proceedings of TKDE (2017)
22. Su, Y., et al.: Exercise-enhanced sequential modeling for student performance prediction. In: Proceedings of AAAI (2018)
23. Sun, Y., Han, J.: Mining heterogeneous information networks: a structural analysis approach. ACM SIGKDD Explor. Newsl. (2013)
24. Sun, Y., Han, J., Yan, X., Yu, P.S., Wu, T.: PathSim: Meta path-based top-k similarity search in heterogeneous information networks. Proc. VLDB Endow. (2011)
25. Tong, S., et al.: Structure-based knowledge tracing: an influence propagation view. In: Proceedings of ICDM (2020)
26. Veličković, P., Cucurull, G., Casanova, A., Romero, A., Liò, P., Bengio, Y.: Graph attention networks. In: Proceedings of ICLR (2018)
27. Wang, T., Ma, F., Gao, J.: Deep hierarchical knowledge tracing. In: Proceedings of EDM (2019)
28. Wang, X., et al.: Heterogeneous graph attention network. In: Proceedings of WWW (2019)
29. Williams, R.J., Zipser, D.: A learning algorithm for continually running fully recurrent neural networks. Neural Comput. (1989)
30. Yang, Y., et al.: GIKT: a graph-based interaction model for knowledge tracing. In: Proceedings of ECML-PKDD (2020)
31. Yudelson, M.V., Koedinger, K.R., Gordon, G.J.: Individualized Bayesian knowledge tracing models. In: Proceedings of AIED (2013)
32. Zhang, C., Song, D., Huang, C., Swami, A., Chawla, N.V.: Heterogeneous graph neural network. In: Proceedings of KDD (2019)
33. Zhang, J., Shi, X., King, I., Yeung, D.Y.: Dynamic key-value memory networks for knowledge tracing. In: Proceedings of WWW (2017)

Intuitionistic Fuzzy Universum Support Vector Machine

Anuradha Kumari$^{(\boxtimes)}$, M. A. Ganaie, and M. Tanveer

Department of Mathematics, Indian Institute of Technology Indore, Simrol 453552, Indore, India
{phd2101141007,phd1901141006,mtanveer}@iiti.ac.in

Abstract. The classical support vector machine is an effective classification technique. It solves a convex optimization problem to give a global solution. But it suffers from noise and outliers. To deal with this, an intuitionistic fuzzy number (IFN) is assigned to the training samples which reduces the effect of noise. In this paper, we propose intuitionistic fuzzy universum support vector machine (IFUSVM), where IFN is assigned to the training data points in presence of universum data. Universum points lead to prior knowledge about data distribution and assignment of IFN to the data points reduces the effect of outliers and noise. Thus, leading to the enhanced generalization property of the model. Numerical experimental results and statistical analysis over 17 binary benchmark UCI datasets show the superiority of the proposed model over the baseline models in terms of rank and accuracy.

Keywords: SVM · Intuitionistic Fuzzy Number · Universum · classification · outliers

1 Introduction

In the last few decades, the application of machine learning for the classification problems is growing rapidly. Two important learning algorithms for the classification problems are convolutional neural network (CNN) [1] and support vector machine (SVM) [2]. CNN has certain disadvantages over SVM, such as former leads to local minima through back propagation whereas the later leads to global solution by solving a quadratic programming problem (QPP). Along with that, CNN needs large training dataset whereas SVM does not require large scale dataset for training.

SVM is applicable in real world classification problems such as detection of face [3], remote sensing [4], speaker verification and identification [5] and so forth. The generalization performance of SVM for high dimensional data is good as its VC (Vapnik-Chervonenkis) dimension is low [6]. The optimization problem of SVM is convex which leads to a global solution. Also, it takes structural risk minimization (SRM) into consideration. The major issue with SVM is its time complexity of $O(l^3)$, which is very high (l being the total training samples). In order to decrease the complexity of SVM, methods such as, SVM light [7], generalized eigenvalue

M. Tanveer et al. (Eds.): ICONIP 2022, LNCS 13623, pp. 236–247, 2023.
https://doi.org/10.1007/978-3-031-30105-6_20

proximal support vector machine (GEPSVM) [8] and sequential minimal optimization (SMO) [9], have been introduced.

However, the construction of hyperplanes in SVM depends on training data known as support vectors. In some practical cases, the training data is corrupted by noise and outliers. As SVM is very sensitive to noise [10], the generalization performance of SVM decreases. SVM based on adaptive margin [11] has bounds given on generalization error which justify the robustness of the model against outliers. Classification in a normalized feature space using SVM [12] generate unit hypersphere. Huang et al. [13] introduced SVM with pinball loss which is insensitive to noise. Moreover, fuzzy support vector machine [14] (FSVM) is introduced to deal with the noise, where the contribution of each training point is different in the construction of the hyperplane. As a result, it is an effective method of mitigating the impact of noise and outliers. Different fuzzy based SVM models have different performances depending on fuzzy membership functions. In [14], the degree of membership is determined by the distance between the point and the class centre. Another fuzzy based SVM introduced by Zhou et al. [15], where the determination of degree of membership to the training points far and near to the class centers are different. This model has the catch that the data points lying away from the class center may get a higher membership value. Recently fuzzy based twin SVM variant, intuitionistic fuzzy weighted least squares twin SVMs [16] is introduced to diagnose Schizophrenia disease.

By using the concept of intuitionistic fuzzy number (IFN) and kernel function, the support vector machine based on intuitionistic fuzzy number and kernel function (KIFSVM) [17] is introduced. Each of the training points is given its corresponding IFN, which includes membership and nonmembership degree. The score value measures the contribution of each data point for hyperplanes and it is calculated using the membership and nonmembership values. In real cases, mostly the training datasets are inseparable. So, KIFSVM [17] transforms the training data points into a high dimensional space, and the decision surface is built in that space. Owing to the high dimensionality in the model [17], kernel functions are taken into account.

Weston et al. [18] introduced universum SVM (USVM). Universum data points are the samples not belonging to any class. It gives prior information about the data distribution leading to improved generalization performance. These variants of SVM are widely used in gender classification [19], text classification [20], EEG signal classification [21] and so on. Diagnosis of Alzheimer's disease using USVM based recursive feature elimination (RFE) [22] provides global information about the data in RFE process. This shows identification of important brain regions for feature selection and classification of MRI images.

By taking the motivation of having prior information about data [18] and reduced effect of noise and outliers [17], we propose a model called intuitionistic fuzzy universum support vector machine (IFUSVM) which assigns IFN to the training samples along with the consideration of universum data. The strength of the proposed model are as:

– Assignment of IFN allots degree of membership and nonmembership to the data point. If a data point has higher nonmembership degree, then it will be considered as noise. Thus, the proposed model has reduced consequence of noise and outliers.
– The presence of universum data leads to prior knowledge about the data distribution in the classification problem.

The rest of the paper is organized as follows: We briefly discuss the background in Sect. 2. In Sect. 3, we have discussed the proposed IFUSVM in detail. Numerical experiments and statistical analysis are discussed in Sect. 4. Section 5 includes conclusion.

2   Background

Suppose $\{(x_1, y_1), (x_2, y_2), \ldots, (x_m, y_m)\}$ be the input training set, having $x_i \in \mathbb{R}^n$ and $y_i \in \{1, -1\}$, for all $i = 1, 2, \ldots, m$. The decision hyperplane for the classification of the data is given by $f(x) = w^T \psi(x) + b$, $w \in \mathbb{R}^n$ and b is a real number. Here, ψ is a map from original space to higher dimensional space.

2.1   Support Vector Machine

The optimization problem for SVM [2] is:

$$\min_{w,b,\eta} \ \frac{1}{2}\|w\|^2 + C_1 \sum_{k=1}^{m} \eta_k$$
$$\text{subject to} \ \ y_k(w^T \psi(x_k) + b) \geq 1 - \eta_k, \tag{1}$$
$$\eta_k \geq 0 \ \ \forall \ k = 1, 2, \ldots, m.$$

Here, C_1 is positive regularization parameter and η_k's are nonnegative slack variables $\forall \ k = 1, 2, \ldots, m$. By imposing the Karush-Kuhn-Tucker (K.K.T.) condition, we obtain the dual as:

$$\min_{\gamma_k} \ \frac{1}{2}\sum_{k=1}^{m}\sum_{j=1}^{m} \gamma_k y_k \gamma_j y_j \psi(x_k)^T \psi(x_j) - \sum_{k=1}^{m} \gamma_k$$
$$\text{subject to} \ \ \sum_{k=1}^{m} \gamma_k y_k = 0, \tag{2}$$
$$0 \leq \gamma_k \leq C_1, \ \ \forall \ k = 1, 2, \ldots, m,$$

where γ_k's are non negative Lagrange multipliers. Applying K.K.T. condition to QPP (2), we get

$$w = \sum_{k=1}^{m} \gamma_k y_k \psi(x_k). \tag{3}$$

Determine the value of b with the help of support vectors. After determining unknowns, the labels are assigned to the test data depending on the function:

$$f(x) = \text{sign } (w^T \psi(x) + b), \tag{4}$$

where x is an arbitrary point from the test data.

2.2   Universum SVM

Universum SVM [18] is more effective than classical SVM for the classification problems. Universum data gives prior knowledge about the distribution of training data. Let U be the set of universum data points given by $U = \{x_1^*, x_2^*, \ldots, x_u^*\}$, where $x_i^* \in \mathbb{R}^n$. The QPP of universum SVM is given by:

$$\min_{w,b,\eta,\xi} \frac{1}{2}\|w\|^2 + C_1 \sum_{k=1}^{m} \eta_k + C_u \sum_{j=1}^{2u} \xi_j$$

$$\text{subject to } y_k(w^T\psi(x_k) + b) \geq 1 - \eta_k,$$

$$y_j(w^T\psi(x_j^*) + b) \geq -\epsilon - \xi_j, \tag{5}$$

$$\eta_k \geq 0, \ \xi_j \geq 0 \ \forall \ j = 1, 2, \ldots, 2u$$

$$\text{and } k = 1, 2, 3, \ldots, m.$$

Here, C_1, C_u are positive regularization parameters. η_k's and ξ_j's are the slack variables for $k = 1, 2, \ldots, m$ and $j = 1, 2, \ldots, 2u$, respectively. The number of universum data points is given by u. Also, $x_{m+i} = x_i^*$ with $y_{m+i} = 1$ and $x_{m+u+i} = x_i^*$ with $y_{m+u+i} = -1$ for $i = 1, 2, \ldots, u$.

The Wolfe dual of the above optimization problem using K.K.T. condition is obtained as

$$\min_{\gamma_k} \frac{1}{2} \sum_{k=1}^{m+2u} \sum_{j=1}^{m+2u} \gamma_k y_k \gamma_j y_j \psi(x_k)^T \psi(x_j) - \sum_{k=1}^{m+2u} \lambda_k \gamma_k,$$

$$\text{subject to } 0 \leq \gamma_k \leq C_1 \text{ for } k = 1, 2, \ldots, m,$$

$$0 \leq \gamma_k \leq C_u \text{ for } k = m+1, m+2, \ldots, m+2u,$$

$$\lambda_k = 1 \text{ for } k = 1, 2, \ldots, m, \tag{6}$$

$$\lambda_k = -\epsilon \text{ for } k = m+1, m+2, \ldots, m+2u,$$

$$\sum_{k=1}^{m+2u} \gamma_k y_k = 0,$$

where γ_k's are non negative Lagrange multipliers. The labels are assigned to the test data by Eq. (4). Applying K.K.T. condition to QPP (6), we get,

$$w = \sum_{k=1}^{m+2u} \gamma_k y_k \psi(x_k). \tag{7}$$

For test data point, the label is determined by Eq. (4).

2.3   SVM Using IFN and Kernel Function

KIFSVM [17] uses the concept of IFN on training points mapped to the high dimensional space. It assigns membership and nonmembership degree to the training points. The membership and nonmembership assignment to the data points decreases the effect of outliers and noise. The contribution of the training points is calculated in the high dimensional space.

Degree of Membership: It is given in terms of the distance between data points and the class center. Mathematically, it is defined as:

$$\mu(x_k) = \begin{cases} 1 - \frac{\|\psi(x_k) - C_+\|}{R_+ + a}, & \text{if } y_k = 1, \\ 1 - \frac{\|\psi(x_k) - C_-\|}{R_- + a}, & \text{if } y_k = -1, \end{cases} \tag{8}$$

where $a > 0$ is a small number. Radius of positive class is denoted by R_+ and that of negative class by R_-, defined as:

$$R_\pm = \max_{y_k = \pm 1} \|\psi(x_k) - C_\pm\|, \tag{9}$$

here, $C_\pm$ denote the centers of classes. Mathematically,

$$C_\pm = \frac{1}{m_\pm} \sum_{y_k = \pm 1} \psi(x_k). \tag{10}$$

Here, m_+ and m_- signify total positive and negative samples, respectively.

Degree of Nonmembership: It is the value assigned to the training points which measures the extent of a point not belonging to a class. It is defined as the proportion of cardinality of dissimilar points in some neighbourhood to the cardinality of total points in the neighbourhood. Mathematically,

$$\delta(x_k) = \frac{|\{x_j| \|\psi(x_k) - \psi(x_j)\| \le d, y_k \ne y_j|}{|\{x_j| \|\psi(x_k) - \psi(x_j)\| \le d\}|}, \tag{11}$$

where $|.|$ denotes the cardinality and $d > 0$ is a small real number. And degree of nonmembership

$$\sigma(x_k) = (1 - \mu(x_k))\delta(x_k).$$

By using the above concept such that $0 \le \mu(x_k) + \sigma(x_k) \le 1$, degree of membership and nonmembership are assigned to each training points, which are used to calculate the score value as shown:

$$s_k = \begin{cases} \mu_k, & \text{if } \sigma_k = 0, \\ 0, & \text{if } \mu_k \le \sigma_k, \\ \frac{1 - \sigma_k}{2 - \mu_k - \sigma_k} & \text{otherwise.} \end{cases} \tag{12}$$

IFN: For each input data point, the set

$$A = \{(x_k, \mu(x_k), \sigma(x_k)) \mid k = 1, 2, \ldots, m\}$$

is called intuitionistic fuzzy set and the tuple $(\mu(x_k), \delta(x_k))$ is called the IFN of kth data point. $\kappa(x_k) = 1 - \mu(x_k) - \sigma(x_k)$ denotes the degree of hesitation of x_k.

Formulation: The optimization problem for SVM with IFN [17] assigned to the data points is given by:

$$\min_{w,\eta} \frac{1}{2}\|w\|^2 + C_1 \sum_{k=1}^{m} s_k \eta_k$$

$$\text{subject to} \quad y_k(w^T \psi(x_k) + b) \geq 1 - \eta_k, \tag{13}$$

$$\eta_k \geq 0, \quad \forall\ k = 1, 2, \ldots, m,$$

where s_k denotes the score value of kth training data calculated using Eq. (12). Using K.K.T. condition, the dual of the optimization problem is obtained as:

$$\min_{\gamma_k} \frac{1}{2} \sum_{k=1}^{m} \sum_{j=1}^{m} \gamma_k y_k y_j \gamma_j \psi(x_k)^T \psi(x_j) - \sum_{k=1}^{m} \gamma_k$$

$$\text{subject to} \quad \sum_{k=1}^{m} \gamma_k y_k = 0, \tag{14}$$

$$0 \leq \gamma_k \leq s_k C_1, \quad \forall\ k = 1, 2, \ldots, m,$$

where γ_k's are non negative Lagrange multipliers for $k = 1, 2, \ldots, m$. Solving the QPP (14), we get γ_k. Obtain w by the relation (3) and determine b with the help of support vectors. For an arbitrary test data sample, the decision function is given by Eq. (4).

3   Proposed Intuitionistic Fuzzy Universum SVM

In the proposed IFUSVM model, we introduce the universum data to KIFSVM [17]. The presence of universum data points lead to prior knowledge about data distribution. The assignment of membership and nonmembership values determine the contribution of each data point for the construction of the hyperplanes, thereby reducing the consequences of noise and outliers. The score values are considered for the training data points only and it differentiates support vectors from noise and outliers. The optimization problem for the proposed IFUSVM is:

$$\min_{w,b,\eta,\xi} \frac{1}{2}\|w\|^2 + C_1 \sum_{k=1}^{m} s_k \eta_k + C_u \sum_{j=1}^{2u} \xi_j$$

$$\text{subject to} \quad y_k(w^T \psi(x_k) + b) \geq 1 - \eta_k,$$

$$y_j(w^T \psi(x_j^*) + b) \geq -\epsilon - \xi_j,$$

$$\eta_k \geq 0, \quad \forall\ k = 1, 2, \ldots, m \tag{15}$$

$$\xi_j \geq 0, \quad \forall\ j = 1, 2, \ldots 2u,$$

where C_1, C_u are the positive regularization parameters. s_k is the score value for the kth data point. η_k and ξ_j are slack variables which are the measure of hinge loss and ϵ−insensitive loss, respectively. The Lagrangian obtained for QPP (15) is given by:

$$L(w, b, \eta_k, \xi_j, \gamma_k, \beta_j, \alpha_j, \mu_k)$$

$$= \frac{1}{2}\|w\|^2 + C_1 \sum_{k=1}^{m} s_k \eta_k + C_u \sum_{j=1}^{2u} \xi_j + \sum_{k=1}^{m} \gamma_k(1 - \eta_k - y_k(w^T \psi(x_k) + b))$$

$$+ \sum_{j=1}^{2u} \beta_j(-\epsilon - \xi_j - y_j(w^T \psi(x_j^*) + b)) - \sum_{k=1}^{m} \mu_k \eta_k - \sum_{j=1}^{2u} \alpha_j \eta_j. \qquad (16)$$

Here, γ_k, β_j, μ_k, α_j are non negative Lagrange multipliers having k in the range 1 to m and j in the range 1 to $2u$. Also we will follow, $x_{m+i} = x_i^*$ with $y_{m+i} = 1$ and $x_{m+u+i} = x_i^*$ with $y_{m+u+i} = -1$ for $i = 1, 2, \ldots, u$. The Wolfe dual obtained for QPP (15) using the necessary sufficient condition of optimality is the following:

$$\min_{\gamma_k} \frac{1}{2} \sum_{k=1}^{m+2u} \sum_{j=1}^{m+2u} \gamma_k y_k \gamma_j y_j \psi(x_k)^T \psi(x_j) - \sum_{k=1}^{m+2u} \lambda_k \gamma_k$$

subject to $0 \leq \gamma_k \leq C_1 s_k$ for $k = 1, 2, \ldots, m,$

$\qquad\qquad 0 \leq \gamma_k \leq C_u$ for $k = m+1, m+2, \ldots, m+2u,$

$\qquad\qquad \lambda_k = 1$ for $k = 1, 2, \ldots, m,$

$$\lambda_k = -\epsilon \text{ for } k = m+1, m+2, \ldots, m+2u \text{ and } \sum_{k=1}^{m+2u} \gamma_k y_k = 0.$$

Using K.K.T. condition, we obtain the relation

$$w = \sum_{k=1}^{m+2u} \gamma_k y_k \psi(x_k). \qquad (17)$$

With the help of support vectors, determine b. A test data, say x is assigned to the class depending on the following:

$$f(x) = \text{sign } (w^T \psi(x) + b). \qquad (18)$$

4   Experimental Results

In this section, we discuss the numerical experiments that are performed to compare the proposed IFUSVM with the baseline models such as classical SVM [2], USVM [18] and KIFSVM [17].

4.1   Setup of the Experiment

All the experimental execution are performed on a machine having MATLAB 2019a, intel(R) core processor and 8 GB RAM. We employed ten-fold cross validation (cv) to compare the performance of the proposed IFUSVM with baseline methods. In ten-fold cv, each dataset is randomly partitioned into ten subsets where one subset is for testing (testing data) and the remaining subsets are for training (training data). We have represented the non linear case through our experiment by using the Gaussian kernel $exp(-\|x_1 - x_2\|^2/2\mu^2)$, where μ denotes the kernel parameter. The range to choose different hyperparameters μ, C_1, C_u is $[10^{-5}, 10^{-4}, \ldots, 10^4, 10^5]$ and range for ϵ is $[0.1, 0.2, 0.3, 0.5, 0.6, 0.7]$. To obtain the optimal parameters, we have used the grid search method [23]. The metrics used for the evaluation of the models are rank and accuracy. Universum data is generated by randomly averaging the data points of both the classes [24]. The binary class datasets for performing the experiments are obtained from UCI repository [25].

4.2   Results and Discussion

For the evaluation of the proposed IFUSVM in comparison to the baseline models, the results are given in Table 1 which shows the performance of each model on the testing data of the 17 UCI datasets. From Table 1, it is evident that the proposed IFUSVM model has the highest average accuracy. We assigned rank to each model across every dataset depending on their performance. Taking average of the assigned ranks across every dataset, we get average ranks of each model which are 1.41, 1.76, 3.47 and 3.35 for proposed IFUSVM, USVM, KIFSVM and SVM, respectively. The best performing model is the one with smallest rank and vice versa. Thus, the proposed IFUSVM model has best performance. For the analysis of the models, we conducted statistical tests such as Friedman test and Nemyeni post hoc test [26]. Friedman test has the assumption under null hypothesis that all the models have equal average rank. The Friedman statistics follows χ_F^2 distribution and it is given by

$$\chi_F^2 = \frac{12K}{N(N+1)} \left(\sum_{i=1}^{N} R_i^2 - \frac{N(N+1)^2}{4} \right), \tag{19}$$

where $K = 17$ is the number of datasets and $N = 4$ is the number of models. R_i signifies i^{th} classifier's average rank. The F_F statistic having F distribution with degree of freedom $((N-1), (N-1)(K-1))$ is:

$$F_F = \frac{(K-1)\chi_F^2}{K(N-1) - \chi_F^2}. \tag{20}$$

After calculation, we get $\chi_F^2 = 34.1608$ and $F_F = 32.4584$. Degree of freedom of F_F statistic is $(3, 48)$. The critical value of $F_{(3,48)} = 2.798$ at 5% level of significance. As $F_F > 2.798$, we reject the null hypothesis. Thus, statistically the

Table 1. Model classification performance on binary datasets having Gaussian kernel

Dataset	SVM [2] (ACC,Time(s)) (C_1, μ)	USVM [18] (ACC,Time(s)) $(C_1, C_u, \mu, \epsilon)$	KIFSVM [17] (ACC,Time(s)) (C_1, μ)	Proposed IFUSVM (ACC,Time(s)) $(C_1, C_u, \mu, \epsilon)$
breast-cancer	$(71.98, 0.0164)$ $(10^4, 0.1)$	$(73.05, 0.0781)$ $(1, 10, 0.1, 0.2)$	$(71.97, 0.0231)$ $(10^4, 0.1)$	$(75.1, 0.0672)$ $(10^3, 10^2, 0.1, 0.1)$
breast-cancer-wisc-prog	$(76.29, 0.0084)$ $(10^{-5}, 10^{-5})$	$(76.79, 0.0139)$ $(0.1, 10^{-5}, 0.1, 0.3)$	$(76.29, 0.0107)$ $(10^{-5}, 10^{-5})$	$(76.79, 0.0145)$ $(10^3, 10^{-5}, 0.1, 0.3)$
congressional-voting	$(62.24, 0.0249)$ $(10^5, 0.1)$	$(62.49, 0.0607)$ $(0.1, 10^{-5}, 1, 0.1)$	$(61.56, 0.0332)$ $(10^5, 0.1)$	$(62.73, 0.1184)$ $(10, 1, 0.1, 0.6)$
echocardiogram	$(74.78, 0.0073)$ $(10^4, 0.1)$	$(79.34, 0.0228)$ $(0.1, 10^5, 0.1, 0.5)$	$(74.78, 0.0055)$ $(10^4, 0.1)$	$(79.34, 0.0232)$ $(10^3, 10^5, 0.1, 0.1)$
fertility	$(88, 0.0068)$ $(10^3, 1)$	$(89, 0.0039)$ $(10, 0.1, 1, 0.1)$	$(90, 0.0061)$ $(10^3, 1)$	$(91, 0.0057)$ $(10^3, 1, 1, 0.1)$
heart-hungarian	$(82.63, 0.0129)$ $(10^4, 0.1)$	$(84.33, 0.0798)$ $(0.1, 1, 0.1, 0.5)$	$(82.63, 0.0133)$ $(10^4, 0.1)$	$(84.33, 0.0841)$ $(10^3, 10^2, 0.1, 0.3)$
hepatitis	$(83.96, 0.0062)$ $(10^5, 0.1)$	$(85.25, 0.0077)$ $(1, 0.01, 0.1, 0.2)$	$(83.96, 0.0088)$ $(10^5, 0.1)$	$(85.25, 0.0114)$ $(10^3, 10^3, 0.1, 0.1)$
molec-biol-promoter	$(76.55, 0.0073)$ $(10^4, 0.1)$	$(82.27, 0.0067)$ $(0.1, 10^3, 0.1, 0.6)$	$(76.55, 0.0059)$ $(10^4, 0.1)$	$(82.27, 0.0055)$ $(10^{-5}, 10, 0.1, 0.5)$
parkinsons	$(91.39, 0.0059)$ $(10^5, 1)$	$(92.87, 0.0125)$ $(10, 10^{-5}, 1, 0.1)$	$(91.39, 0.0107)$ $(10^5, 1)$	$(92.87, 0.0123)$ $(10^5, 10^{-5}, 1, 0.1)$
pittsburg-bridges-T-OR-D	$(87.18, 0.0052)$ $(10^5, 0.1)$	$(88.09, 0.0102)$ $(10, 0.1, 0.1, 0.1)$	$(87.18, 0.0069)$ $(10^5, 0.1)$	$(88.09, 0.009)$ $(10^5, 1, 0.1, 0.5)$
spect	$(69.17, 0.013)$ $(10^4, 0.1)$	$(70.66, 0.0928)$ $(0.1, 0.01, 0.1, 0.3)$	$(68.4, 0.0151)$ $(10^4, 0.1)$	$(72.89, 0.0604)$ $(10^3, 1, 0.1, 0.1)$
trains	$(60, 0.0051)$ $(10^4, 0.1)$	$(10^2, 0.006)$ $(0.01, 10, 10^{-4}, 0.5)$	$(60, 0.0049)$ $(10^4, 0.1)$	$(10^2, 0.0049)$ $(10^{-5}, 10^{-3}, 10^{-4}, 0.2)$
vertebral-column-2clases	$(83.87, 0.0121)$ $(10^5, 1)$	$(84.84, 0.022)$ $(1, 1, 1, 0.1)$	$(83.23, 0.0156)$ $(10^5, 1)$	$(85.81, 0.0252)$ $(10, 10, 1, 0.6)$
heart-switzerland	$(60.77, 0.005)$ $(10^{-5}, 10^{-5})$	$(64.1, 0.0071)$ $(10^4, 10^4, 10^5, 0.6)$	$(60.77, 0.0065)$ $(10^{-5}, 10^{-5})$	$(64.1, 0.0184)$ $(10^{-4}, 10^{-3}, 0.1, 0.5)$
ilpd-indian-liver	$(72.04, 0.0491)$ $(1, 10^2)$	$(72.04, 0.1695)$ $(1, 10^{-5}, 10^2, 0.1)$	$(72.04, 0.0648)$ $(1, 10^2)$	$(72.04, 0.1279)$ $(1, 10^{-5}, 10^2, 0.1)$
ionosphere	$(87.16, 0.0164)$ $(10^5, 1)$	$(88.87, 0.068)$ $(1, 10^2, 1, 0.2)$	$(87.16, 0.0198)$ $(10^5, 1)$	$(88.87, 0.0456)$ $(10^4, 10, 1, 0.5)$
pima	$(75, 0.0835)$ $(10^4, 0.1)$	$(75.52, 0.3994)$ $(0.1, 10^5, 0.1, 0.5)$	$(75, 0.0962)$ $(10^4, 0.1)$	$(75.52, 0.9315)$ $(10^3, 10^5, 0.1, 0.5)$
Average Accuracy	76.65	80.56	76.64	81
Average Rank	3.35	1.76	3.47	1.41

models are different. For pairwise comparison of the models, we use the Nemyeni post hoc test. In this test, pairs of models are compared. Models are considered statistically different, if the difference in average ranks of the compared models is greater than the critical difference (CD), where CD is determined by:

$$CD = q_\alpha \sqrt{\frac{N(N+1)}{6K}}, \tag{21}$$

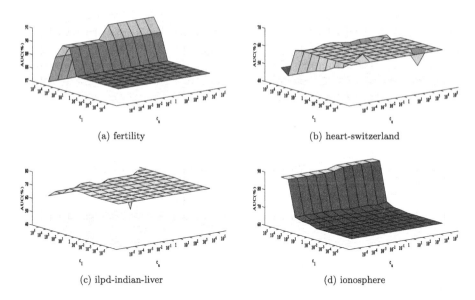

(a) fertility

(b) heart-switzerland

(c) ilpd-indian-liver

(d) ionosphere

Fig. 1. Sensitivity evaluation of the proposed IFUSVM classifier on UCI binary datasets using user-specified parameters and Gaussian kernel

Table 2. Win tie loss of the models in pair with proposed IFUSVM.

	SVM [2]	KIFSVM [17]	USVM [18]
KIFSVM	[1,12,4]		
USVM	[16,1,0]	[15,1,1]	
IFUSVM	[16,1,0]	[16,1,0]	[5,12,0]

Here, $[p, q, r]$ means p times wins, q times ties and r times loses of the row method over the column method

having $q_\alpha = 2.569$ for 4 classifiers. On calculating, we get $CD = 1.1376$. The average rank differences of the proposed IFUSVM model with respect to classical SVM, USVM and KIFSVM are 1.94, 0.35 and 2.06, respectively. The average rank differences of the proposed IFUSVM with respect to SVM and KIFSVM are greater than CD. Hence, the proposed method is better than SVM and KIFSVM. With USVM, though the rank difference is less than CD, we can see from the Table 1 that the overall rank and accuracy are better for the proposed model.

Further, we employed the win tie loss sign test [26]. Table 2 represents the win tie loss of baseline models in pair with the proposed IFUSVM. According to the this test, if each model of a pair wins on $K/2$ datasets, then under null hypothesis, the compared models are considered equivalent. If one of the models wins on $K/2 + 1.96\sqrt{K}/2$ datasets, then at a significant level of 5%, the compared models are statistically different. It is to be noted that if the number of ties between the models in a pair is even, then it is equally distributed among them.

If number of ties is odd, then ignoring one, it is equally distributed. For this case, we get, if number of wins of a model in a pair is greater than 12.54, then the null hypothesis fails. Thus, the proposed IFUSVM model performs comparatively better than SVM and KIFSVM. Figure 1 represents the effect of hyperparameters C_1, C_u on the generalization performance of the proposed IFUSVM classifier with Gaussian kernel.

5   Conclusion

In this paper, we proposed intuitionistic fuzzy universum support vector machine (IFUSVM) for the classification problems. In the proposed IFUSVM model, we have assigned membership and nonmembership values to the data points which signify their contribution to the construction of hyperplanes. Thus, diminishes the consequence of outliers and noise. Simultaneously, we have considered the concept of universum data which gives prior information about the distribution of the data. Results of numerical experiments and statistical analysis present that the proposed IFUSVM has superior performance having highest accuracy and lowest rank as compared to the baseline models. In future, one can explore different ways to reduce the complexity of the model, as SVM based models have very high complexity of $O(l^3)$, l being the number of samples. One can implement the sequential minimal optimization (SMO) so as to solve the optimization problem.

Acknowledgment. This work is supported and funded by science and Engineering Research Board (SERB) under Mathematical Research Impact-Centric Support (MATRICS) scheme grant no. MTR/2021/000787 and National Supercomputing mission under DST and Miety, Govt. of India with Grant number DST/NSM/RD HPC Appll/2021/03.29. Ms. Anuradha Kumari (File no - 09/1022 (12437)/2021-EMR-I) expresses her gratitude to Council of Scientific and Industrial Research (CSIR), New Delhi, India for the financial support provided as fellowship. We are grateful to Indian Institute of Technology Indore, India for providing the facilities and support.

References

1. O'Shea, K., Nash, R.: An introduction to convolutional neural networks. arXiv preprint arXiv:1511.08458 (2015)
2. Vapnik, V.: The Nature of Statistical Learning Theory. Springer, New York (1999). https://doi.org/10.1007/978-1-4757-3264-1
3. Osuna, E., Freund, R., Girosit, F.: Training support vector machines: an application to face detection. In: Proceedings of IEEE Computer Society Conference on Computer Vision and Pattern Recognition, pp. 130–136. IEEE (1997)
4. Pal, M., Mather, P.M.: Support vector machines for classification in remote sensing. Int. J. Remote Sens. **26**(5), 1007–1011 (2005)
5. Wan, V., Campbell, W.M.: Support vector machines for speaker verification and identification. In: Neural Networks for Signal Processing X. Proceedings of the 2000 IEEE Signal Processing Society Workshop (Cat. No. 00TH8501), vol. 2, pp. 775–784. IEEE (2000)

6. Vapnik, V.N.: An overview of statistical learning theory. IEEE Trans. Neural Netw. **10**(5), 988–999 (1999)
7. Joachims, T.: Making large-scale support vector machine learning practical, advances in kernel methods. Support Vector Learn. (1999)
8. Mangasarian, O.L., Wild, E.W.: Multisurface proximal support vector machine classification via generalized eigenvalues. IEEE Trans. Pattern Anal. Mach. Intell. **28**(1), 69–74 (2005)
9. Platt, J.: Sequential minimal optimization: a fast algorithm for training support vector machines (1998)
10. Song, Q., Hu, W., Xie, W.: Robust support vector machine with bullet hole image classification. IEEE Trans. Syst. Man Cybern. Part C (Appl. Rev.) **32**(4), 440–448 (2002)
11. Herbrich, R., Weston, J.: Adaptive margin support vector machines for classification (1999)
12. Graf, A.B., Smola, A.J., Borer, S.: Classification in a normalized feature space using support vector machines. IEEE Trans. Neural Netw. **14**(3), 597–605 (2003)
13. Huang, X., Shi, L., Suykens, J.A.: Support vector machine classifier with pinball loss. IEEE Trans. Pattern Anal. Mach. Intell. **36**(5), 984–997 (2013)
14. Lin, C.-F., Wang, S.-D.: Fuzzy support vector machines. IEEE Trans. Neural Netw. **13**(2), 464–471 (2002)
15. Zhou, M.-M., Li, L., Lu, Y.-L. Fuzzy support vector machine based on density with dual membership. in: 2009 International Conference on Machine Learning and Cybernetics, vol. 2, pp. 674–678. IEEE (2009)
16. Tanveer, M., Ganaie, M., Bhattacharjee, A., Lin, C.: Intuitionistic fuzzy weighted least squares twin SVMs. IEEE Trans. Cybern (2022). https://doi.org/10.1109/TCYB.2022.3165879
17. Ha, M., Wang, C., Chen, J.: The support vector machine based on intuitionistic fuzzy number and kernel function. Soft. Comput. **17**(4), 635–641 (2013)
18. Weston, J., Collobert, R., Sinz, F., Bottou, L., Vapnik, V.: Inference with the universum. In: Proceedings of the 23rd International Conference on Machine Learning, pp. 1009–1016 (2006)
19. Bai, X., Cherkassky, V.: Gender classification of human faces using inference through contradictions. In: 2008 IEEE International Joint Conference on Neural Networks (IEEE World Congress on Computational Intelligence), pp. 746–750. IEEE (2008)
20. Liu, C.-L., Hsaio, W.-H., Lee, C.-H., Chang, T.-H., Kuo, T.-H.: Semi-supervised text classification with universum learning. IEEE Trans. Cybern. **46**(2), 462–473 (2015)
21. Richhariya, B., Tanveer, M.: EEG signal classification using universum support vector machine. Expert Syst. Appl. **106**, 169–182 (2018)
22. Richhariya, B., Tanveer, M., Rashid, A.H., Initiative, A.D.N.: Diagnosis of Alzheimer's disease using universum support vector machine based recursive feature elimination (USVM-RFE). Biomed. Sig. Process. Control **59**, 101903 (2020)
23. Hsu, C.-W., Lin, C.-J.: A comparison of methods for multiclass support vector machines. IEEE Trans. Neural Netw. **13**(2), 415–425 (2002)
24. Richhariya, B., Tanveer, M.: A reduced universum twin support vector machine for class imbalance learning. Pattern Recogn. **102**, 107150 (2020)
25. Murphy, P., Aha, D.: UCI repository of machine learning databases (1992). http://www.ics.uci.edu/~mlearn. *MLRepository.html*
26. Demšar, J.: Statistical comparisons of classifiers over multiple data sets. J. Mach. Learn. Res. **7**, 1–30 (2006)

Support Vector Machine Based Models with Sparse Auto-encoder Based Features for Classification Problem

A. K. Malik[1]([✉]), M. A. Ganaie[1], M. Tanveer[1], and P. N. Suganthan[2,3]

[1] Department of Mathematics, Indian Institute of Technology Indore,
Simrol 453552, Indore, India
{phd1801241003,phd1901141006,mtanveer}@iiti.ac.in

[2] School of Electrical and Electronic Engineering, Nanyang Technological University,
Singapore, Singapore
epnsugan@ntu.edu.sg

[3] KINDI Center for Computing Research, College of Engineering, Qatar University,
Doha, Qatar

Abstract. Auto-encoder is a special type of artificial neural network (ANN) that is used to learn informative features from data. In the literature, the generalization performance of several machine learning models have been improved either using auto-encoder based features or high dimensional features (original + auto-encoder based features). Random vector functional link (RVFL) network also uses two type of features, i.e., original features and randomized features, that makes it a special randomized neural network. These hybrid features improve the generalization performance of the RVFL network. In this paper, we introduce the idea of using additional features into robust energy-based least squares twin support vector machines (RELS-TSVM) and least squares twin support vector machines (LSTSVM). We used sparse auto-encoder with L_1 norm regularization to learn the auxiliary feature representation from original feature space. These new additional features are concatenated with the original features to get the extended feature space. The conventional RELS-TSVM and LSTSVM are trained over new extended feature space. Experiments demonstrate that auto-encoder based features improve the generalization capability of the conventional RELS-TSVM and LSTSVM models. To examine the performance of the proposed classifiers, i.e., extended-RELS-TSVM (ext-RELS-TSVM) and extended LSTSVM (ext-LSTSVM), experiments have been conducted over 15 UCI binary datasets and the results show that the proposed classifiers have better generalization performance than the baseline classifiers.

Keywords: Robust energy based least squares twin SVM (RELS-TSVM) · SP-RVFL · Least squares twin SVM (LSTSVM) · Extended feature space · RVFL

1 Introduction

There are various machine learning models such as deep neural networks [1], support vector machines (SVMs) [2], and randomized neural networks [3] that have

M. Tanveer et al. (Eds.): ICONIP 2022, LNCS 13623, pp. 248–259, 2023.
https://doi.org/10.1007/978-3-031-30105-6_21

shown their strength in various domains such as bio-informatics [4] and computer vision. Among them, SVM has been implemented successfully in various fields such as classification [5], clustering [6] and so on, due to its better generalization ability and strong mathematical foundation. SVM generates an optimal hyper plane which separate the data samples into different classes and take the final decision there. To overcome the issue of high computational complexity in SVM, there have been many attempts by researchers to develop various algorithms such as Reduced SVM [7], Lagrangian SVM [8], and so on. In 2007, Jayadeva et al. [9] proposed twin SVM (TSVM) wherein, instead of solving a large QPP, two small QPPs are solved. There are several improvements over TSVM such as Improved sparse pinball twin SVM [10], robust general twin SVM with pinball loss function [11], large-scale pinball twin support vector machines [12]. To get more information about TSVM, we refer the readers to [13]. Kumar and Gopal proposed least squares twin support vector machines (LSTSVM) [14] that is very fast and efficient algorithm. Instead of solving QPPs (like in SVM and TSVM), LSTSVM solves a system of linear equations. Further, to improve the classification capability of LSTSVM, many variants of it have been developed such as Fuzzy least squares projection twin SVM [15] and an improved LSTSVM [16] where an extra regularization term was introduced to improve the generalization performance of LSTSVM. Recently, Ali et al. [17] proposed regularized least square twin SVM for multiclass classification.

Randomized neural networks (RNNs) like single hidden layer feed forward network (SLFN) [18,19] have been widely studied within machine learning community. Random vector functional link (RVFL) network [20,21] and one of its variants extreme learning machine (ELM) [22] are randomization based algorithms to train SLFN wherein parameters from input nodes to the hidden nodes are randomly generated from a predefined interval (such as $[-s, +s]$), where s is a scaling parameter, with uniform distribution (or other distributions) [23] and only the final parameters of the model are calculated via least squares regression method. The RVFL model has direct links which prevents it from overfitting [3]. RVFL utilizes original features via direct links and gets optimal parameters. RVFL model uses two types of features- original features and randomized features. This phenomena of dual features enhances the generalization capability of RVFL model. The ELM model has no such direct links and hence, it uses only randomized features which are calculated from original features via non-linear activation function. Recently, Wan et al. [24] proposed twin ELM, wherein, the twin SVM was discussed in ELM framework. Rastogi and Amisha proposed least squares twin ELM [25] for pattern classification. These algorithms utilize the advantage of random feature mapping mechanism. In RVFL, the original features are integrated with the randomized features so that the model can capture more complex non-linear hidden relationship within data. Recently, Zhang et al. [26] proposed sparse pre-trained RVFL (SP-RVFL) neural network, wherein unlike RVFL, sparse auto-encoder with L_1-norm regularization is used to get the more informative features which increase the generalization performance of RVFL. One can visit [27] to learn more about RVFL model. In this paper, having inspired by the success of randomized neural networks with hybrid features and least square twin SVM based models [28], we introduce the aforementioned idea of calculating the more informative features from the original feature space into robust energy-based least squares twin SVM (RELS-TSVM) and least squares twin SVM (LSTSVM) to see the impact of sparse auto-encoder based features. We obtained the sparse feature representation of the original features via sparse-auto encoder with L_1 norm regularization and then concatenated these sparse (additional) features with original features to get the extended features (original + additional features). Extensive experiments over 15 UCI binary datasets illustrate that the proposed idea improve the generalization capability of the baseline models. Note that the proposed methods (ext-RELS-TSVM and ext-LSTSVM) are different from RVFL+ [29]. In RVFL+, additional privileged information is available during training process and least square method is used to determine the latter parameters, whereas in proposed models, two non-parallel hyperplanes are generated for the classification problems. The remainder of this paper is structured as follows: Sect. 2 gives a brief description to SP-RVFL and RELS-TSVM model. Section 3 provides the detailed information about the proposed method. Section 4 discusses the obtained results and we conclude the paper in Sect. 5.

2   Related Works

In this section, we discuss the mathematical formulations of the models, i.e., SP-RVFL, LSTSVM and RELS-TSVM. Let $X = [x_1, x_2, ..., x_N]^T \subset \mathbb{R}^{N \times q}$ be the training samples where N, q represent the number of sample and features, respectively. Let $Y = [y_1, y_2, ..., y_N]^T \subset \mathbb{R}^{N \times m}$ be the original output matrix with m classes.

2.1   Sparse Pre-trained Random Vector Functional Link Network [26]

The sparse pre-trained random vector functional link (SP-RVFL) network is an efficient variant of RVFL. Unlike RVFL, in SP-RVFL the weights and biases between first layer to the second layer are calculated using L_1 norm based sparse auto-encoder to get more information from data. The optimization problem for the sparse auto-encoder with K hidden neurons can be written as:

$$\Omega_\theta = \min_\theta ||\hat{H}\theta - X||_2^2 + ||\theta||_1, \tag{1}$$

where $\hat{H} \in \mathbb{R}^{N \times K}$ represents the hidden layer output matrix, and $\theta \in \mathbb{R}^{K \times q}$ is the output weight matrix. The optimization problem in (1) is solved via fast iterative shrinkage thresholding algorithm (FISTA) [30]. Since the generalization capability of the neural network with random weights (NNRW) such as RVFL depends on random parameters and hence sub-optimal random parameters may degrade the performance of these models. Therefore, to alleviate the problem of generating random weights and biases, in SP-RVFL, the output weight matrix

θ calculated from (1) is employed as input weight matrix, i.e., θ^T is used as the input weight matrix and the hidden biases are calculated as:

$$\sigma_i = \frac{1}{q} \sum_{j=1}^{q} \theta_{ij}, \quad i = 1, 2, ..., K, \tag{2}$$

where σ_i represents the bias in the ith enhancement node of the hidden layer. The mathematical formulation for SP-RVFL is as follow:

$$y_i = \sum_{j=1}^{K} \bar{\beta}_j \phi(\theta_j^t \cdot x_i + \sigma_j) + \sum_{j=K+1}^{K+q} \bar{\beta}_j x_{ij}, \quad i = 1, 2, ...N, \tag{3}$$
$$= [\phi(\theta^t \cdot x_i + \sigma) \quad x_i] \, \bar{\beta},$$
$$= \bar{h}(x_i)\bar{\beta},$$

where $\phi(.)$ is an activation function and $\bar{\beta} \in \mathbb{R}^{(q+K) \times m}$ is the output weight matrix. The objective function of SP-RVFL in matrix form can be written as: $D\bar{\beta} = Y$, where $D \in \mathbb{R}^{N \times (q+K)}$ and $Y \in \mathbb{R}^{N \times m}$ are the concatenation matrix and target matrix, respectively. For avoiding overfitting, regularized least square method is used and the learning objective can be written as:

$$\min_{\bar{\beta} \in \mathbb{R}^{(q+K) \times m}} \frac{1}{2} \left\| D\bar{\beta} - Y \right\|_2^2 + \lambda \frac{1}{2} \left\| \bar{\beta} \right\|_2^2, \tag{4}$$

where λ is the hyper-parameter that needs to be tuned. The closed form solution of (4) can be obtained as:

$$\bar{\beta} = \begin{cases} (D^t D + \lambda I)^{-1} D^t Y, & \text{if} \quad (K+q) \leq N \\ D^t (DD^t + \lambda I)^{-1} Y, & \text{if} \quad N < (K+q), \end{cases} \tag{5}$$

where I is an identity matrix of appropriate dimension.

2.2 Robust Energy-Based Least Squares Twin Support Vector Machines (RELS-TSVM) [31]

Considering the binary classification problem, let N_1 data points (from X) belong to class +1 and rest N_2 data points belong to class -1. Let $F_1 \in \mathbb{R}^{N_1 \times q}$ matrix represents the patterns of class +1 and $F_2 \in \mathbb{R}^{N_2 \times q}$ matrix represents the patterns of class -1 with $N = N_1 + N_2$. The optimization problem of linear robust energy-based least squares twin support vector machines is as follows:

$$\min_{\bar{w}_1, \bar{b}_1} \frac{1}{2} \left\| F_1 \bar{w}_1 + e_1 \bar{b}_1 \right\|^2 + \frac{\lambda_1}{2} \left\| \eta_1 \right\|^2 + \frac{\lambda_3}{2} \left\| \begin{bmatrix} \bar{w}_1 \\ \bar{b}_1 \end{bmatrix} \right\|^2$$
$$s.t. \ -\left(F_2 \bar{w}_1 + e_2 \bar{b}_1 \right) + \eta_1 = E_1 \tag{6}$$

and

$$\min_{\bar{w}_2, \bar{b}_2} \frac{1}{2} \left\| F_2 \bar{w}_2 + e_2 \bar{b}_2 \right\|^2 + \frac{\lambda_2}{2} \left\| \eta_2 \right\|^2 + \frac{\lambda_4}{2} \left\| \begin{bmatrix} \bar{w}_2 \\ \bar{b}_2 \end{bmatrix} \right\|^2$$

$$s.t. \quad \left(F_1 \bar{w}_2 + e_1 \bar{b}_2 \right) + \eta_2 = E_2. \tag{7}$$

Now substitute the equality constraints into the objective function then the QPP (6) becomes:

$$L_1 = \frac{1}{2} \left\| F_1 \bar{w}_1 + e_1 \bar{b}_1 \right\|^2 + \frac{\lambda_1}{2} \left\| F_2 \bar{w}_1 + e_2 \bar{b}_1 + E_1 \right\|^2$$

$$+ \frac{\lambda_3}{2} \left\| \begin{bmatrix} \bar{w}_1 \\ \bar{b}_1 \end{bmatrix} \right\|^2. \tag{8}$$

Taking the partial derivatives of L_1 with respect to $\bar{w}_1$ and $\bar{b}_1$ and putting it equal to zero, the solution of (6) is given as:

$$\begin{bmatrix} \bar{w}_1 \\ \bar{b}_1 \end{bmatrix} = -\left(\lambda_1 P_2^t P_2 + P_1^t P_1 + \lambda_3 I \right)^{-1} \lambda_1 \ P_2^t \ E_1, \tag{9}$$

and similarly we can do the same calculations for solving the QPP (7) and gets the solution as:

$$\begin{bmatrix} \bar{w}_2 \\ \bar{b}_2 \end{bmatrix} = \left(\lambda_2 P_1^t P_1 + P_2^t P_2 + \lambda_4 I \right)^{-1} \lambda_2 \ P_1^t \ E_2, \tag{10}$$

where $P_1 = [F_1 \ e_1]$ and $P_2 = [F_2 \ e_2]$. For a testing sample x, the final decision is taken as given in [31].

3   Proposed Work

In this section, we explain the proposed binary classification algorithms, that is, extended LSTSVM (ext-LSTSVM) and extended RELS-TSVM (ext-RELS-TSVM). The RELS-TSVM and LSTSVM model solve two systems of linear equations that makes them very fast and efficient models as compared to twin SVM and twin bounded SVM that solves quadratic programming problems. Moreover, the linear RELS-TSVM and LSTSVM models use only original features for calculating the optimal classifier. In the literature, auto-encoder's based features have been used successfully to improve the generalization performance of the machine learning models. Auto-encoders can extract highly informative features and interesting structures from original dataset. Having additional features (calculated via hidden layer) in the training process of RVFL or SP-RVFL classifiers, it improves the generalization capability of these randomized neural networks (RNN) [3]. Therefore, we introduce the idea of hybrid features into LSTSVM and RELS-TSVM, and gets their improved variants. In this study, we concatenated sparse auto-encoder with L_1 norm regularization based features with original features to make the extended feature space (original + additional features) so that more informative features can be fed into RELS-TSVM and LSTSVM model to calculate the optimal parameters. Hence, we proposed extended RELS-TSVM (ext-RELS-TSVM) and extended LSTSVM (ext-LSTSVM). The final decision of the proposed methods are taken as in the conventional RELS-TSVM and LSTSVM methods. The process of constructing the proposed binary classification algorithm ext-RELS-TSVM can be seen in Algorithm 1 and the similar process is followed for constructing the proposed ext-LSTSVM with extended feature space. The flowchart of the proposed models has been given in Fig. 1.

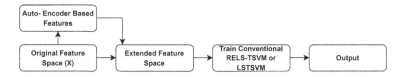

Fig. 1. The Flowchart of the proposed models.

Table 1. Average classification accuracy (%) of LSTSVM, RELS-TSVM, SP-RVFL and ext-LSTSVM and ext-RELS-TSVM and the best results are highlighted in bold.

Datasets	LSTSVM [14]	RELS-TSVM [31]	SP-RVFL [26]	ext-LSTSVM	ext-RELS-TSVM
breast-cancer	72.5352	72.8873	**75**	73.5915	69.7183
breast-cancer-wisc	97	**97.4286**	97	**97.4286**	**97.4286**
chess-krvkp	95.3692	95.4631	96.9962	96.9337	**97.0901**
congress-voting	58.945	58.945	**61.4679**	**61.4679**	58.945
credit-approval	87.064	86.7733	86.6279	**87.5**	86.6279
echocardiogram	84.8485	84.8485	**85.6061**	**85.6061**	84.8485
heart-hungarian	84.9315	85.6164	85.274	**86.3014**	82.8767
molec-biol-promoter	65.3846	64.4231	81.7308	56.7308	**89.4231**
oocytes_trisopterus_nucleus_2f	80.5921	80.5921	81.0307	82.5658	**84.7588**
ozone	**97.1609**	**97.1609**	**97.1609**	**97.1609**	97.082
parkinsons	88.7755	86.7347	90.8163	88.7755	**91.8367**
pima	76.6927	76.5625	76.8229	**77.6042**	76.4323
statlog-australian-credit	67.0058	66.8605	**68.0233**	67.0058	66.4244
statlog-german-credit	78.5	77.7	77.3	77.4	**79.4**
tic-tac-toe	97.5941	97.5941	**99.0586**	98.954	97.5941
Average accuracy	82.16	81.97	83.99	82.34	**84.03**

4 Experimental Setup and Results Analysis

In this section, we discuss the experimental setup details and illustrate the obtained results. All the experiments are carried out over the publicly available binary datasets given in UCI repository [32] and we setup the same environment and naming convention as in [33]. We used relu, selu and sigmoid activation functions and via grid search select the optimal function. The range of the number

Algorithm 1. Extended RELS-TSVM

1: **Input:** Let $X = [x_1, x_2, ..., x_N]^T \subset \mathbb{R}^{N \times q}$ be the matrix of original training samples. Let $\phi(.)$ and K be the given activation function and number of neurons in the hidden layer used in auto-encoder (1) and let the matrix $F_1 \in \mathbb{R}^{N_1 \times q}$ contains samples of class +1 (from X) and the rest samples of class -1 is contained in matrix $F_2 \in \mathbb{R}^{N_2 \times q}$.

2: Calculate the output weight matrix θ by solving 1 and then calculate the biases by (2). Apply the non-linear activation function $\phi(.)$ to calculate the additional features matrix from original feature matrix as:

$$\hat{X} = \phi(X\theta^T + \sigma) \tag{11}$$

$\hat{X}$ is the additional feature representation of the original feature space X.

3: Define the extended feature space as $X^* = [X, \hat{X}]$ and obtained $F_1^* = [F_1, \hat{F}_1]$ and $F_2^* = [F_2, \hat{F}_2]$, where $\hat{F}_1$ and $\hat{F}_2$ are the samples of class +1 and -1, respectively from $\hat{X}$.

4: After getting the extended feature space, define the objective function of ext-RELS-TSVM as:

$$\min_{\bar{w}_1, \bar{b}_1} \frac{1}{2} \left\| F_1^* \bar{w}_1 + e_1 \bar{b}_1 \right\|^2 + \frac{\lambda_1}{2} \left\| \eta_1 \right\|^2 + \frac{\lambda_3}{2} \left\| \begin{bmatrix} \bar{w}_1 \\ \bar{b}_1 \end{bmatrix} \right\|^2 \tag{12}$$
$$s.t. \quad -\left(F_2^* \bar{w}_1 + e_2 \bar{b}_1 \right) + \eta_1 = E_1$$

and

$$\min_{\bar{w}_2, \bar{b}_2} \frac{1}{2} \left\| F_2^* \bar{w}_2 + e_2 \bar{b}_2 \right\|^2 + \frac{\lambda_2}{2} \left\| \eta_2 \right\|^2 + \frac{\lambda_4}{2} \left\| \begin{bmatrix} \bar{w}_2 \\ \bar{b}_2 \end{bmatrix} \right\|^2 \tag{13}$$
$$s.t. \quad \left(F_1^* \bar{w}_2 + e_1 \bar{b}_2 \right) + \eta_2 = E_2.$$

5: Solve (12) and (13) as discussed in section 2 (2.2) and obtained the corresponding solutions. Let x' be the new sample, then the final decision of the proposed ext-RELS-TSVM is taken as in the conventional RELS-TSVM.

Table 2. Friedman rank of LSTSVM, RELS-TSVM, SP-RVFL and ext-LSTSVM and ext-RELS-TSVM and the best result are highlighted in bold.

Datasets	LSTSVM [14]	RELS-TSVM [31]	SP-RVFL [26]	ext-LSTSVM	ext-RELS-TSVM
breast-cancer	4	3	1	2	5
breast-cancer-wisc	4.5	2	4.5	2	2
chess-krvkp	5	4	2	3	1
congress-voting	4	4	1.5	1.5	4
credit-approval	2	3	4.5	1	4.5
echocardiogram	4	4	1.5	1.5	4
heart-hungarian	4	2	3	1	5
molec-biol-promoter	3	4	2	5	1
oocytes_trisopterus_nucleus_2f	4.5	4.5	3	2	1
ozone	2.5	2.5	2.5	2.5	5
parkinsons	3.5	5	2	3.5	1
pima	3	4	2	1	5
statlog-australian-credit	2.5	4	1	2.5	5
statlog-german-credit	2	3	5	4	1
tic-tac-toe	4	4	1	2	4
Average rank	3.5	3.53	2.43	**2.3**	3.23

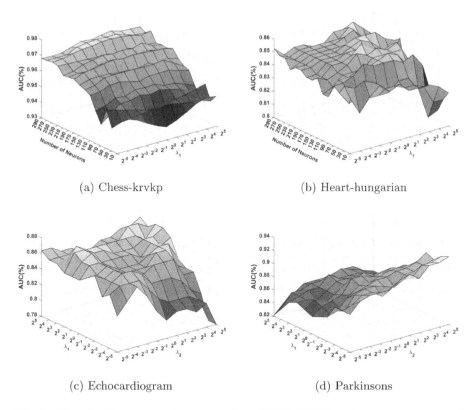

(a) Chess-krvkp

(b) Heart-hungarian

(c) Echocardiogram

(d) Parkinsons

Fig. 2. The classification performance of ext-RELS-TSVM with varying number of neurons, λ_1 and λ_2 for four data sets.

of neurons $(K) = 10 : 20 : 300$. The parameters $\lambda_{(1,2,3,4)}$ and the energy parameters $E_{(1,2)}$ are as follows: $\{2^j \mid j = -5, -4, ..., 5\}$ and $\{0.5, 0.6, 0.7, 0.8, 0.9, 1.0\}$, respectively. All experiments were conducted on a computer running Windows 10 with 16 GB of RAM and MATLAB R2021a.

Table 1 shows the average accuracy and average rank of the five models. One can see that the proposed model ext-RELS-TSVM has the highest accuracy (84.03%) that is almost 2% higher than RELS-TSVM (81.97%). Moreover, the other models have the following accuracy, i.e., ext-LSTSVM (82.34%), LSTSVM (82.16%) and SP-RVFL (83.99%). There is a small improvement of 0.16% in ext-LSTSVM as compare to standard LSTSVM. The proposed ext-LSTSVM model has better accuracy than LSTSVM on 10 out of 15 data sets. There are some datasets where one can see the impact of additional features approach such as the proposed ext-LSTSVM has winning performance with significant improvements over the following datasets, i.e., chess-krvkp, congress-voting, echocardiogram, heart-hungarian and pima with accuracy 96.93%, 61.47%, 85.61%, 86.30% and 77.60%, respectively. The proposed ext-RELS-TSVM has better accuracy (84.03%) than RELS-TSVM (81.97%). One can observe that on chess-krvkp dataset there is almost 2% improvement in accuracy in the proposed ext-RELS-TSVM model as compare to RELS-TSVM model. It shows that having additional features in the training process is an effective approach.

Moreover, we conducted Friedman rank test that ranks the models according to their performance on each dataset. Let N and A be the number of datasets and the number of models, respectively. Let A_i^j be the rank of the jth model out of A models on the ith dataset. Therefore, each model gets an average rank calculated as: $R_j = \frac{1}{N}\sum_i A_i^j$. Table 2 shows the average rank of each given models. The proposed ext-LSTSVM has the smallest average rank (2.3) among the compared models and the second average rank model is SP-RVFL (2.43) and third average rank model is ext-RELS-TSVM (3.23). The lower rank reflects that the model performance is good over maximum datasets. It shows that the proposed models have the comparable results with SP-RVFL. It can also be observed that the proposed ext-RELS-TSVM and ext-LSTSVM have lower rank than the conventional RELS-TSVM and LSTSVM, respectively. It reflects that there is advantage of using auto-encoder based features that improved the generalization performance of the models.

Figure 2 shows the classification performance of the proposed ext-RELS-TSVM model on chess-krvkp, heart-hungarian, echocardiogram and parkinsons data sets. In Fig. 2 (a) and (b), x-axis and y- axis represent λ_1 and number of neurons in the hidden layer parameters, respectively and z-axis shows the accuracy of the proposed ext-RELS-TSVM model. In Fig. 2 (a), one can see that the accuracy of the proposed ext-RELS-TSVM model is increasing as the number the neurons is increasing. The decreasing performance of the ext-RELS-TSVM can be seen in Fig. 2 (b) with higher λ_1 values and small number of neurons. It shows that having proper tuning of the neurons one can get the optimal performance of ext-RELS-TSVM model. In Fig. 2 (c) and (d), x-axis and y- axis represent λ_2 and λ_1 parameters, respectively and z-axis shows the classification performance of the proposed ext-RELS-TSVM model. Figure 2 (c) shows that the proposed ext-RELS-TSVM has lower performance with higher λ_2 values and small λ_1 values and as the value of the λ_2 parameter is decreasing the performance of ext-RELS-TSVM is increasing. The proposed ext-RELS-TSVM model has decreasing performance with higher λ_1 values on Parkinsons dataset in Fig. 2 (d). Therefore, for getting the best performance of the ext-RELS-TSVM model, one must carefully adjust the parameters according to the problems.

5   Conclusions and Future Work

In the literature, the auto-encoder based features have shown their strength in improving the generalization performance of the machine learning models [34,35]. In this paper, we propose the improved variants of RELS-TSVM and LSTSVM with extended feature space for binary classification problems. The proposed methods have two steps. First, the additional features are calculated via sparse auto-encoder with L_1-norm regularization from original feature space and second, the conventional RELS-TSVM and LSTSVM are trained over new extended feature space (original + additional features). The experiments over 15 UCI binary datasets demonstrate that the proposed idea improves the generalization performance of these baseline models. There are almost 2% improvement in the performance (average accuracy) in RELS-TSVM model (by using spare auto-encoder based features) and the proposed ext-LSTSVM has won over 10 out of 15 datasets as compared to LSTSVM model. It shows that sparse auto-encoder based features have significant information and improve the generalization performance of baseline models, i.e., LSTSVM and RELS-TSVM. There are some limitations also of this idea such as the dimension of features may resulted as dimension disaster problems. The proposed models need high computation cost to handle high-dimensional data as it works on feature space (matrix inversion problem). Both the proposed ext-RELS-TSVM and ext-LSTSVM model have higher computational cost as compared to the baseline models since they need to calculate the additional features. There is no statistically significant difference among the proposed models and the compared models. However, the proposed models achieve better generalization performance compared to baseline models. We design these experiments for binary classification problem. In future, we would like to extend this study for multi-class classification problem and regression problem as well. One can do the similar attempts with other support vector machine (SVM) based models and manifold learning based auto-encoders that generate efficient features.

Acknowledgment. This work is supported by National Supercomputing Mission under DST and Miety, Govt. of India under Grant No. DST/NSM/R&D HPC Appl/ 2021/03.29 and Department of Science and Technology under Interdisciplinary Cyber Physical Systems (ICPS) Scheme Grant no. DST/ICPS/CPS-Individual/2018/ 276 and Mathematical Research Impact-Centric Support (MATRICS) scheme grant no. MTR/2021/000787. Mr. Ashwani kumar Malik acknowledges the financial support (File no-09/1022 (0075)/2019-EMR-I) given as scholarship by Council of Scientific and Industrial Research (CSIR), New Delhi, India. We gratefully acknowledge the Indian Institute of Technology Indore for providing facilities and support.

References

1. Ganaie, M.A., Hu, M., Malik, A.K., Tanveer, M., Suganthan, P.N.: Ensemble deep learning: a review. Eng. Appl. Artif. Intell. **115**, 105151 (2022)
2. Cortes, C., Vapnik, V.: Support vector machine. Mach. Learn. **20**(3), 273–297 (1995)
3. Zhang, L., Suganthan, P.N.: A survey of randomized algorithms for training neural networks. Inf. Sci. **364**, 146–155 (2016)
4. Cao, Y., Geddes, T.A., Yang, J.Y.H., Yang, P.: Ensemble deep learning in bioinformatics. Nat. Mach. Intell. **2**(9), 500–508 (2020)
5. Shao, Y.-H., Deng, N.-Y., Yang, Z.-M.: Least squares recursive projection twin support vector machine for classification. Pattern Recogn. **45**(6), 2299–2307 (2012)
6. Tanveer, M., Gupta, T., Shah, M.: Pinball loss twin support vector clustering. ACM Trans. Multimedia Comput. Commun. Appl. (TOMM) **265** (2020)

7. Lee, Y.-J., Mangasarian, O.L.: RSVM: reduced support vector machines. In: Proceedings of the 2001 SIAM International Conference on Data Mining, pp. 1–17. SIAM (2001)
8. Mangasarian, O.L., Musicant, D.R.: Lagrangian support vector machines. J. Mach. Learn. Res. **1**(Mar), 161–177 (2001)
9. Jayadeva, R.K., Chandra, S.: Twin support vector machines for pattern classification. IEEE Trans. Pattern Anal. Mach. Intell. **29**(5), 905–910 (2007)
10. Tanveer, M., Rajani, T., Ganaie, M.A.: Improved sparse pinball twin SVM. In: 2019 IEEE International Conference on Systems, Man and Cybernetics (SMC), pp. 3287–3291. IEEE (2019)
11. Ganaie, M.A., Tanveer, M.: Robust general twin support vector machine with pinball loss function. In: Kumar, P., Singh, A.K. (eds.) Machine Learning for Intelligent Multimedia Analytics. SBD, vol. 82, pp. 103–125. Springer, Singapore (2021). https://doi.org/10.1007/978-981-15-9492-2_6
12. Tanveer, M., Tiwari, A., Choudhary, R., Ganaie, M.A.: Large-scale pinball twin support vector machines. Mach. Learn. 1–24 (2021). https://doi.org/10.1007/s10994-021-06061-z
13. Tanveer, M., Rajani, T., Rastogi, R., Shao, Y.-H., Ganaie, M.A.: Comprehensive review on twin support vector machines. Ann. Oper. Res. 1–46 (2022)
14. Kumar, M.A., Gopal, M.: Least squares twin support vector machines for pattern classification. Expert Syst. Appl. **36**(4), 7535–7543 (2009)
15. Ganaie, M.A., Tanveer, M., Initiative, A.D.N.: Fuzzy least squares projection twin support vector machines for class imbalance learning. Appl. Soft Comput. **113**, 107933 (2021)
16. Xu, Y., Xi, W., Lv, X., Guo, R.: An improved least squares twin support vector machine. J. Inf. Comput. Sci. **9**(4), 1063–1071 (2012)
17. Ali, J., Aldhaifallah, M., Nisar, K.S., Aljabr, A., Tanveer, M.: Regularized least squares twin SVM for multiclass classification. Big Data Res. **27**, 100295 (2022)
18. Hornik, K., Stinchcombe, M., White, H.: Multilayer feedforward networks are universal approximators. Neural Netw. **2**(5), 359–366 (1989)
19. Park, J., Sandberg, I.W.: Universal approximation using radial-basis-function networks. Neural Comput. **3**(2), 246–257 (1991)
20. Pao, Y.-H., Park, G.-H., Sobajic, D.J.: Learning and generalization characteristics of the random vector functional-link net. Neurocomputing **6**(2), 163–180 (1994)
21. Pao, Y.: Adaptive pattern recognition and neural networks (1989)
22. Huang, G.-B., Zhu, Q.-Y., Siew, C.-K.: Extreme learning machine: theory and applications. Neurocomputing **70**(1–3), 489–501 (2006)
23. Zhang, L., Suganthan, P.N.: A comprehensive evaluation of random vector functional link networks. Inf. Sci. **367**, 1094–1105 (2016)
24. Wan, Y., Song, S., Huang, G., Li, S.: Twin extreme learning machines for pattern classification. Neurocomputing **260**, 235–244 (2017)
25. Rastogi (nee Khemchandani), R., Bharti, A.: Least squares twin extreme learning machine for pattern classification. In: Deb, D., Balas, V.E., Dey, R. (eds.) Innovations in Infrastructure. AISC, vol. 757, pp. 561–571. Springer, Singapore (2019). https://doi.org/10.1007/978-981-13-1966-2_50
26. Zhang, Y., Wu, J., Cai, Z., Du, B., Philip, S.Y.: An unsupervised parameter learning model for RVFL neural network. Neural Netw. **112**, 85–97 (2019)
27. Malik, A.K., Gao, R., Ganaie, M.A., Tanveer, M., Suganthan, P.N.: Random vector functional link network: recent developments, applications, and future directions. arXiv:2203.11316 (2022)

28. Tanveer, M., Gautam, C., Suganthan, P.N.: Comprehensive evaluation of twin SVM based classifiers on UCI datasets. Appl. Soft Comput. **83**, 105617 (2019)

29. Zhang, P.-B., Yang, Z.-X.: A new learning paradigm for random vector functional-link network: RVFL+. Neural Netw. **122**, 94–105 (2020)

30. Beck, A., Teboulle, M.: A fast iterative shrinkage-thresholding algorithm for linear inverse problems. SIAM J. Imag. Sci. **2**(1), 183–202 (2009)

31. Tanveer, M., Khan, M.A., Ho, S.-S.: Robust energy-based least squares twin support vector machines. Appl. Intell. **45**(1), 174–186 (2016). https://doi.org/10.1007/s10489-015-0751-1

32. Frank, A.: UCI machine learning repository (2010). http://archive.ics.uci.edu/ml

33. Fernández-Delgado, M., Cernadas, E., Barro, S., Amorim, D.: Do we need hundreds of classifiers to solve real world classification problems? J. Mach. Learn. Res. **15**(1), 3133–3181 (2014)

34. Katuwal, R., Suganthan, P.N.: Stacked autoencoder based deep random vector functional link neural network for classification. Appl. Soft Comput. **85**, 105854 (2019)

35. Yang, Z., Xu, B., Luo, W., Chen, F.: Autoencoder-based representation learning and its application in intelligent fault diagnosis: a review. Measurement, 110460 (2021)

Selectively Increasing the Diversity of GAN-Generated Samples

Jan Dubiński[1]([✉])(iD), Kamil Deja[1](iD), Sandro Wenzel[2](iD), Przemysław Rokita[1](iD), and Tomasz Trzcinski[1,3,4](iD)

[1] Warsaw University of Technology, Warsaw, Poland
jan.dubinski.dokt@pw.edu.pl
[2] CERN, Geneva, Switzerland
[3] Jagiellonian University, Kraków, Poland
[4] Tooploox, Wrocław, Poland

Abstract. Generative Adversarial Networks (GANs) are powerful models able to synthesize data samples closely resembling the distribution of real data, yet the diversity of those generated samples is limited due to the so-called mode collapse phenomenon observed in GANs. Especially prone to mode collapse are conditional GANs, which tend to ignore the input noise vector and focus on the conditional information. Recent methods proposed to mitigate this limitation increase the diversity of generated samples, yet they reduce the performance of the models when similarity of samples is required. To address this shortcoming, we propose a novel method to selectively increase the diversity of GAN-generated samples. By adding a simple, yet effective regularization to the training loss function we encourage the generator to discover new data modes for inputs related to diverse outputs while generating consistent samples for the remaining ones. More precisely, we maximise the ratio of distances between generated images and input latent vectors scaling the effect according to the diversity of samples for a given conditional input. We show the superiority of our method in a synthetic benchmark as well as a real-life scenario of simulating data from the Zero Degree Calorimeter of ALICE experiment in LHC, CERN.

Keywords: Generative Adversarial Networks · Generative Models · Data Simulation

1 Introduction

Generative Adversarial Networks (GANs) [1] constitute a gold standard for synthesizing complex data distributions and they are, therefore, widely used across various applications, including data augmentation [3], image completion [5] or representation learning [6]. They are also employed in high energy physics experiments at the Large Hadron Collider (LHC) at CERN, where they allow to speed up the process of simulating particle collisions [4,7,12]. In this context, the generative models are used to generate samples of possible detectors' responses resulting from a collision of particles described with a series of physical parameters.

M. Tanveer et al. (Eds.): ICONIP 2022, LNCS 13623, pp. 260–270, 2023.
https://doi.org/10.1007/978-3-031-30105-6_22

For more controllable simulations, we can condition the generative models with additional parameters of the collision, using conditional GANs (cGANs) [15].

Although by conditioning GANs we obtain more context-dependent generations that are closer to the values observed in real experiments, these models are considered more vulnerable to the so-called mode collapse phenomenon [19], observed as a tendency to generate a limited number of different outputs per each conditional prior. This, in turn, significantly reduces the effectiveness of employing GANs for particle collision simulations, as alignment of generated samples with the real data distribution is fundamental for drawing correct conclusions from the performed experiments.

To address the above-mentioned limitations of cGANs, recent methods [2, 14, 21] attempt to increase the diversity of generated samples by modifying the associated cost function. However, they do not consider conditioning the diversity on the input conditioning values, assuming a uniform distribution of diversity across all of them. This assumption is rarely observed in practical applications, for instance in particle collision simulations at CERN diversity of generated samples highly depends on the set of conditioning variables.

In this work, we identify this shortcoming of existing models and propose a simple, yet effective method to selectively increase the diversity of GAN-generated samples, based on conditioning values. In principle, we introduce a regularization method that enforces GANs to follow diversity observed in the original dataset for a given conditional value. More exactly, we maximise the ratio of distances between latent vectors of generated images and inputs, scaling the effect accordingly to the diversity of samples corresponding to a given conditional input. Our approach, dubbed SDI-GAN, is readily applicable for conditional image synthesis models and does not require any modification of the baseline GAN architecture.

We evaluate our method on a challenging task of simulating data from the Zero Degree Calorimeter of the ALICE experiment in LHC, CERN. To better demonstrate the performance of our method we also include a synthetic dataset for 2D point generation. We compare our approach with competing methods and achieve superior results across all benchmarks.

The main contribution of this paper is a novel method for increasing the diversity of GAN-generated results for a selected subset of conditional input data while keeping the consistency of the results for the remaining conditional inputs.

2   Related Work

Generative Simulations: The need for simulating complex processes exists across many scientific domains. In recent years, solutions based on generative machine learning models have been proposed as an alternative to existing methods in cosmology [18] and genetics [17]. However, one of the most profound applications for generative simulations is in the field of High Energy Physics, where machine learning models can be used as a resource-efficient alternative to classic Monte Carlo-based [11] approaches.

Recent attempts [4,9,12] leverage solutions based on Generative Adversarial Networks [1] or Variational Autoencoders [13]. Although those methods offer considerable speed-up of the simulation process, they also suffer from the limitations of existing generative models. Controlling the diversity of simulated results while maintaining the high fidelity of the simulation is one of the challenges of using generative models for such applications.

Mode Collapse and Sample Diversity in cGAN: The authors of MS-GAN [2] address the mode collapse problem in cGANs by introducing mode-seeking loss. This additional regularization term added to the generator training function aims to improve generation diversity by maximizing the dissimilarity between two images generated from two different latent codes. During training, the generator tries to minimize the regularization term added to the loss function equal to the inverse of measured diversity.

DS-GAN [21] tries to tackle the problem with a similar approach. The main difference between the two methods is the fact that DS-GAN explicitly maximizes the measured diversity which is subtracted from the training loss of the generator.

In DivCo [14] the authors use contrastive learning to achieve diverse conditional image synthesis. They introduce a latent-augmented contrastive loss which encourages images generated from distant latent codes to be dissimilar and those generated from close latent codes to be similar. The similarity of images is measured using their latent representations extracted from the discriminator network.

Our approach shares a similar method of calculating the diversity of images with [2] and [21]. However, contrary to those approaches we do not base our measure of diversity on pixels of generated images. Instead we operate on image representation, similarly to [14]. Moreover, in principle, all previously described approaches do not account for different levels of variance of samples corresponding to different conditional inputs and instead maximize the diversity of the results generated for all conditional inputs.

3   Methodology

In this work we propose a novel selective diversity regularization method that improves alignment between generations from cGAN and data conditioned on available conditional values. For a dataset of images $\mathcal{X}$ with conditioning values $\mathcal{C}$ we want to learn a generator G that is able to produce realistic samples from the domain of $\mathcal{X}$ conditioned on $c \in \mathcal{C}$. Moreover, we want the synthesised images to be diverse or similar to each other depending on the variance of samples in $\mathcal{X}$ that correspond to a given c.

Traditional conditional GANs are trained using adversarial loss. Given a condition vector $c \in \mathcal{C}$ and a k-dimensional latent code $z \sim \mathcal{N}_k(0,1)$ the generator G takes both c and z as input and produces an output image $\hat{x} = G(z,c)$. The image $\hat{x}$ should be indistinguishable from real data by a discriminator D. During training generator and discriminator play a min-max game, in which D learns to distinguish real data from samples synthesised by G while G tries to generate samples that are considered by D as real.

$$\mathcal{L}_{adv}(G, D) = \mathbb{E}_{x \sim \mathcal{X}, c \sim \mathcal{C}}[\log D(x, c)] + \mathbb{E}_{c \sim \mathcal{C}, z \sim \mathcal{N}(0,1)}[\log(1 - D(G(z, c), c)] \tag{1}$$

The adversarial loss function encourages the generator to produce realistic data, but as observed by [14] it does not directly promote the diversity of synthesised samples.

To alleviate this problem Mao et al. [2] propose a regularization term that penalizes the low diversity of generated samples. More precisely, the introduced method maximizes the ratio of the distance between two images generated from two different latent codes z_1, z_2 and the same conditioning value c with respect to the distance between those latent codes. The proposed regularization term is added to the basic loss function from Eq. 1

$$\mathcal{L}_{ms} = \left(\frac{d_\mathbf{I}\left(G\left(c, \mathbf{z}_1\right), G\left(c, \mathbf{z}_2\right)\right)}{d_\mathbf{z}\left(\mathbf{z}_1, \mathbf{z}_2\right)} \right)^{-1} \tag{2}$$

Although, this approach successfully forces the generator to produce dissimilar examples it does not account for different levels of sample diversity for different conditioning input c. To address this issue we propose a simple yet effective modification of the regularization term.

As a data preprocessing step, for each unique conditioning input $c \in \mathcal{C}$ we calculate the diversity of samples from the dataset $\mathcal{X}$ that correspond to c denoted as $\mathcal{X}_c$. We base our measure of diversity f_{div} on the variance of samples. As denoted in Eq. 3, for each set of images $\mathcal{X}_c$ we sum the standard deviation of pixel values with the same coordinates for images in $\mathcal{X}_c$:

$$f_{div}(c) = \sum_{i,j} \sqrt{\frac{\sum_t (x_{ij}^t - \mu_{ij})^2}{|X_c|}} \tag{3}$$

where i and j are the pixel coordinates, t is the index of sample $x \in X_c$ and μ_{ij} is a mean value for a pixel ij from all samples from X_c. We normalize the obtained values of all sample diversity $f_{div}(c)$ to $<0, 1>$.

To account for varying levels of sample diversity for each conditioning input c we multiply the regularization term introduced before by $f_{div}(c)$.

$$\mathcal{L}_{div} = f_{div}(c) * \left(\frac{d_\mathbf{I}\left(G\left(c, \mathbf{z}_1\right), G\left(c, \mathbf{z}_2\right)\right)}{d_\mathbf{z}\left(\mathbf{z}_1, \mathbf{z}_2\right)} \right)^{-1} \tag{4}$$

This change forces the generator to better match the diversity observed in the original dataset with respect to conditional values. Intuitively, the generator produces more diverse images under conditions that allow for a higher variety of synthesised samples. At the same time, the generator is not forced to produce dissimilar results if the conditioning value corresponds to a set of similar images in the dataset. The overall objective of training SDI-GAN is the following:

$$\mathcal{L} = \mathcal{L}_{adv}(G, D) + \lambda_{div} \mathcal{L}_{div}(G) \tag{5}$$

where λ_{div} is a hyperparameter controlling the strength of the regularization.

Additionally, to better adapt our method to the task of applying cGANs as a fast simulation tool, we propose to base the distance d_g on the dissimilarity of latent representations of images rather than the dissimilarity of pixels. We measure the distance between two generated images by calculating the L_1 metric between their latent representations. We obtain those representations by treating the initial layers of the discriminator as an encoder E and extracting the latent features of the images from its penultimate layer during training.

$$e_{zc} = E(G(z, c)) \tag{6}$$

$$d_g(G(z_1, c), G(z_1, c)) = |e_{z_1 c}, e_{z_2 c}| \tag{7}$$

This change shifts the focus of the generator from the visual dissimilarity of images to the difference in their underlying characteristics extracted by the encoder.

4   Experiments

We evaluate our method on simulating data from the Zero Degree Calorimeter from the ALICE experiment in LHC, CERN. Additionally, we use a synthetic dataset as a benchmark. We compare our approach to a conditional DC-GAN [16], MS-GAN [2] and DivCo [14].

4.1   Synthetic Dataset

To clearly demonstrate the effects of our method and its comparison to competing approaches we provide a synthetic dataset of 2D points generation. Each point is conditioned on a class label (from 1 to 9) and a binary variable *spread*. The position of the generated point depends mainly on its class label. However, points with *spread* = False are generated very close to each other, while points with *spread* = True are heavily dispersed. The dataset is presented in Fig. 1.

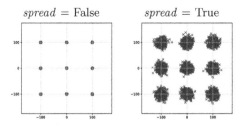

Fig. 1. Synthetic dataset. For each class the generated points form a cluster. For point with *spread* = False the variance of each cluster is equal to 1 and for points with *spread* = True the variance of each cluster is equal to 100.

4.2   Zero Degree Calorimeter Simulation

The task of simulating the response of the Zero Degree Calorimeter (ZDC) offers a challenging benchmark for generative models. The dataset consists of 295867 samples obtained from the GEANT4 [11] simulation tool. Each response is created by a single particle described with 9 attributes (mass, energy, charge, momenta, primary vertex).

During the simulation process, the particle is propagated through the detector for over 100 m while simulation tools must account for all of its interactions with the detector's matter. The end result of the simulation is the energy deposited in the calorimeter's fibres, which are arranged in a grid with 44×44 size. We treat the calorimeter's response as a 1-channel image with 44×44 pixels, where pixel values are the number of photons deposited in a given fibre. To create the dataset the simulation was run multiple times for the same input particles. For that reason, multiple possible outcomes correspond to the same particle properties. We refer to this dataset as HEP.

Although the process that governs the propagation of the particles is non-deterministic by nature, the majority of particles create consistent ZDC responses. However, a subset of particles produces highly diverse results and allows for multiple possible calorimeter responses. In Fig. 2 we present sampled simulations for two different conditional values. In the first row, we depict the calorimeter response for a high-energy proton, while for the second the input particle is a neutron, which has no electric charge, therefore responses observed in the calorimeter are much more consistent.

Fig. 2. Examples of ZDC calorimeter responses. We show 4 possible outputs of the simulation generated for 2 distinct particles.

Table 1. Results comparison on the synthetic and HEP datasets. Our solution outperforms competing approaches by generating data with variance close to the real data for both types of conditioning input in the synthetic case. We achieve the lowest Wasserstein distance between test and generated coordinates for the diverse points without any significant trade-off for the consistent points. The performance improvement introduced by SDI-GAN is further evaluated on the real-life HEP dataset, where SDI-GAN achieves the lowest Wasserstein distance between channels calculated from original and generated data.

	Synthetic dataset				HEP
	Variance		Wassesrstein ↓		Wassesrstein ↓
Spread	0	1	0	1	–
Real	1	100	–	–	–
DC-GAN	**0.2**	2.3	**0.7**	7.4	7.6
MS-GAN	115.5	280,7	9.0	6.7	21.7
DivCo	163.3	3.7	1.1	7.0	14.3
SDI-GAN (ours)	3.3	**127.4**	1.2	**2.1**	**4.5**

4.3 Results

We present the results of our experiments on the synthetic dataset using both qualitative and quantitative comparisons. As shown in Fig. 3 our method generates samples that are visually most similar to the real data. To confirm this observation we calculate the mean variance for all classes for points with *spread = 0* and *spread = 1*. The results in Table 1 show the superiority of our method in this scenario. DC-GAN is able to produce results with variance close to real data for conditional inputs with *spread* = 0, but fails to generate diverse results. MS-GAN properly reflect the diversity of possible results for conditional inputs with *spread* = 1, but does not generate consistent results. Although samples created by DivCo have different variance depending on the conditioning input, the data distribution generated under condition *spread* = 1 is distorted and does not match the real data, as presented in Fig. 3. Our approach is able to produce both diverse and similar results depending on the conditional information.

The most common method for evaluating GANs on real datasets utilizes Frechet Inception Distance (FID) [10]. However, for the HEP dataset, we propose a domain-specific evaluation scheme that better measures the quality of the simulation. Following the calorimeter's specification [8] we base our evaluation procedure on 5 channels calculated from the pixels of generated images. These channels reflect the physical properties of simulated collision and are used for analysing the output of the calorimeter. To measure the quality of the simulation we compare the distribution of channels for the original and generated data using Wasserstein distance [20].

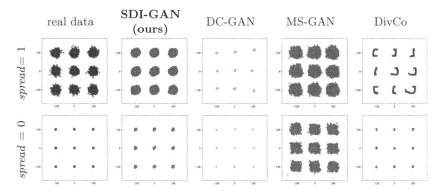

Fig. 3. Comparison of generated results for the synthetic dataset. Contrary to competing methods, our approach is able to properly synthesis both diverse and similar samples depending on the conditioning variable *spread*.

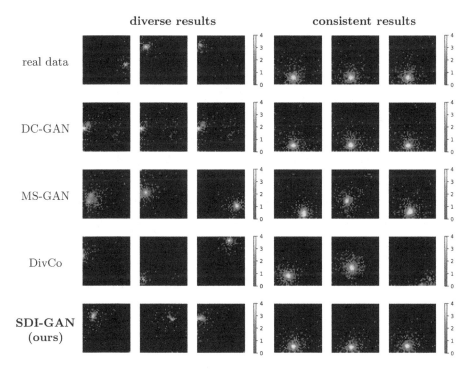

Fig. 4. Examples of calorimeters response simulations with different methods. DC-GAN works well for particles with consistent responses but fails to generate diverse outcomes when needed. Although MS-GAN and DivCo successfully increase the diversity of generated samples those models do not distinguish between particles that should produce diverse or consistent showers. Our method is able to generate diverse results while producing consistent responses for appropriate particles.

As presented in Table 1, our approach outperforms other solutions on the HEP datasets. In Fig. 4 we demonstrate that our method is able to generate diverse results for a specific subset of particles while keeping consistent responses for the remaining conditional inputs. The positive impact of this approach on the distribution of the generated samples is further confirmed by Fig. 5 where we compare channel distribution for SDI-GAN and competing approaches for 2 selected channels. Our method increases the fidelity of the simulation by smoothing the distribution of generated responses and covering the whole range of possible outputs.

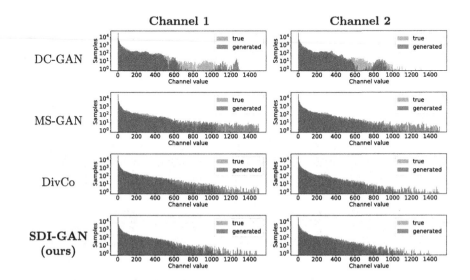

Fig. 5. Comparison of channel values distribution for selected channels. Our method decreases the differences between the distribution of original and generated data and smooths the distribution of the synthesised results. In the case of SDI-GAN the increased diversity of generated samples does not harm the fidelity of the simulation, contrary to competing approaches.

The additional regularization term for training of SDI-GAN does not influence the inference speed of the model. In our initial experiments, we observe a speed-up of simulations of two orders of magnitude when compared to the standard Monte-Carlo approach. With SDI-GAN this computation boost is observed without degradation in simulation quality. We leave the detailed analysis of this performance gain and the influence of fast simulations on physical experiments for future work.

5 Conclusions

In this work we introduce a simple, yet effective modification of the loss function for conditional generative adversarial networks. Our solution enforces increased sample diversity for a subset of conditional data without affecting samples that are characterised by conditional values associated with consistent responses.

We show that our solution outperforms other comparable approaches on the synthetic benchmark and the challenging practical dataset of calorimeter response simulations in the ALICE experiment at CERN.

Acknowledgments. This research was funded by National Science Centre, Poland grant no 2020/39/O/ST6/01478, grant no 2018/31/N/ST6/02374 and grant no 2020/39/B/ST6/01511.

References

1. Goodfellow, I., et al.: Generative adversarial nets. In: Advances in Neural Information Processing Systems, pp. 2672–2680 (2014)
2. Mao, Q., Lee, H.-Y., Tseng, H.-Y., Ma, S., Yang, M.-H.: Mode seeking generative adversarial networks for diverse image synthesis. In: IEEE Conference on Computer Vision and Pattern Recognition (2019)
3. Motamed, S., Khalvati, F.: Inception augmentation generative adversarial network. CoRR, abs/2006.03622 (2020)
4. Paganini, M., de Oliveira, L., Nachman, B.: CaloGAN: simulating 3D high energy particle showers in multi-layer electromagnetic calorimeters with generative adversarial networks. CoRR, abs/1705.02355 (2017)
5. Zhao, S., et al.: Large scale image completion via co-modulated generative adversarial networks. CoRR, abs/2103.10428 (2021)
6. Zieba, M., Semberecki, P., El-Gaaly, T., Trzciński, T.: BinGAN: learning compact binary descriptors with a regularized GAN. In: NeurIPS (2018)
7. Deja, K., Dubiński, J., Nowak, P., Wenzel, S., Spurek, P., Trzciński, T.: End-to-end Sinkhorn autoencoder with noise generator. IEEE Access **9**, 7211–7219 (2021)
8. Dellacasa, G., et al.: ALICE technical design report of the zero degree calorimeter (ZDC). In: CERN Document Server (1999)
9. Erdmann, M., Glombitza, J., Quast, T.: Precise simulation of electromagnetic calorimeter showers using a Wasserstein generative adversarial network. Comput. Softw. Big Sci. **3** (2019)
10. Heusel, M., Ramsauer, H., Unterthiner, T., Nessler, B., Hochreiter, S.: GANs trained by a two time-scale update rule converge to a local Nash equilibrium. In: Advances in Neural Information Processing Systems, pp. 6626–6637 (2017)
11. Incerti S., et al.: Geant4-DNA example applications for track structure simulations in liquid water: a report from the Geant4-DNA Project. Med. Phys. **45**, 722–739 (2018)
12. Kansal, R., et al.: Particle cloud generation with message passing generative adversarial networks. In: Annual Conference on Neural Information Processing Systems (NeurIPS) (2021)
13. Kingma, D.P., Welling, M.: Auto-encoding variational Bayes. CoRR, abs/1312.6114 (2013)
14. Liu, R., Ge, Y., Choi, C.L., Wang, X., Li, H.: DivCo: diverse conditional image synthesis via contrastive generative adversarial network. In: IEEE Conference on Computer Vision and Pattern Recognition (2021)
15. Mirza, M., Osindero, S.: Conditional generative adversarial nets. arXiv (2014). https://doi.org/10.48550/arXiv.1411.1784

16. Radford, A., Metz, L., Chintala, S., Unsupervised representation learning with deep convolutional generative adversarial networks (2015). https://doi.org/10.48550/arXiv.1511.06434

17. Rodríguez, A.C., et al.: Fast cosmic web simulations with generative adversarial networks. Comput. Astrophys. Cosmol. **5**(1), 1–11 (2018). https://doi.org/10.1186/s40668-018-0026-4

18. Rong, R., et al.: MB-GAN: microbiome simulation via generative adversarial network. GigaScience **10** (2021)

19. Salimans, T., Goodfellow, I., Zaremba, W., Cheung, V., Radford, A., Chen, X.: Improved techniques for training GANs. In: Advances in Neural Information Processing Systems (2016)

20. Tolstikhin, I., Bousquet, O., Gelly, S., Schoelkopf, B.: Wasserstein auto-encoders. arXiv (2017). https://arxiv.org/abs/1711.01558v4

21. Yang, D., Hong, S., Jang, Y., Zhao, T., Lee, H.: Diversity-sensitive conditional generative adversarial networks. In: Proceedings of the International Conference on Learning Representation (2019)

Cooperation and Competition: Flocking with Evolutionary Multi-Agent Reinforcement Learning

Yunxiao Guo[1], Xinjia Xie[2], Runhao Zhao[3], Chenglan Zhu[1], Jiangting Yin[4], and Han Long[1(✉)]

[1] College of Sciences, National University of Defense Technology, Changsha 410073, China
longhan@nudt.edu.cn
[2] College of Computer Science, National University of Defense Technology, Changsha 410073, China
[3] College of System Engineering, National University of Defense Technology, Changsha 410073, China
[4] Beijing International Studies University, Beijing 100000, China

Abstract. Flocking is a very challenging problem in a multi-agent system; traditional flocking methods also require complete knowledge of the environment and a precise model for control. In this paper, we propose Evolutionary Multi-Agent Reinforcement Learning (EMARL) in flocking tasks, a hybrid algorithm that combines **cooperation and competition** with little prior knowledge. As for **cooperation**, we design the agents' reward for flocking tasks according to the boids model. While for **competition**, agents with high fitness are designed as senior agents, and those with low fitness are designed as junior, letting junior agents inherit the parameters of senior agents stochastically. To intensify **competition**, we also design an evolutionary selection mechanism that shows effectiveness on credit assignment in flocking tasks. Experimental results in a range of challenging and self-contrast benchmarks demonstrate that EMARL significantly outperforms the full competition or cooperation methods.

Keywords: Flocking · Multi-Agent Reinforcement Learning · Evolutionary Reinforcement Learning · Swarming Intelligence

1 Introduction

Such as migratory birds and wasps, flocking in swarm intelligence refers to the behaviour of a large population that tend to gather in groups and move orderly [21]. It also be widely employed in numerous applications such as social evolution [4] multi-robotics control [5], and traffic model [16]. The research on the flocking

---

Y. Guo, X. Xie and R. Zhao—Equal Contribution.

model can help people understand the macroscopic dynamics of the population and its dependence on parameters [9] and also guide people in designing the control methods for the robotic population, such as UAVs [11].

The Traditional flocking method usually uses partial differential equations (PDEs) to describe the agent behaviour in the population [3,19,26], by solving the PDEs with numerical techniques, the flocking strategies can be found. However, the PDEs are generally difficult to solve, and when the agent number is large and the geometry is complex, the computation cost will increase. It also needs the model to know the full knowledge of the environment [20].

For the above reasons, the methods using deep reinforcement learning (DRL) to train the agents flocking self-organized have attracted lots of interest in recent years [6,11,20], especially the model-free multi-agent DRL (MADRL), which can handle the complicated tasks well without modelling the complex environment. Compared to the single-agent DRL algorithms, MADRL can also deal with more complicated problems [8], but also bring about new challenges like credit assignment, which is difficult to solve, particularly in full-cooperative tasks like flocking.

Besides, the reward design in the flocking task is essential. For the full-cooperative tasks, if the individual agent gets the same reward as the rest of the group no matter what it does, then the agent will find it hard to know their contribution to the group, leading to failed learning. In biological, the behaviour of agents to pursue higher individual rewards in the group can be seen as competition, and the lack of competition among the agents will constrain the performance of learning [2]. Nevertheless, we should not apply the competitive MARL methods such as the MARL based on Nash equilibrium. It will lead to the local optimal solution for the cooperative tasks because the agent would not sacrifice for the greater team reward.

In this paper, we proposed the Evolutionary Multi-Agent Reinforcement Learning (EMARL) algorithm, which introduces competition to solve the credit assignment in the full-cooperative flocking task. Our **main contributions** are as follows: (1) We redesign the reward of the flocking task based on the boids model and solve the credit assignment challenge in the flocking task. We propose EMARL that combines competition (Evolutionary RL) and cooperation (Multi-agent RL). (2) In the part of evolutionary RL, we propose evolutionary selection to sift agents with poor performance and prove the convergence of policy gradient with Evolutionary Selection which further demonstrates the effectiveness. (3) we illustrated our approach to the flocking task with two obstacles and demonstrated the results at different levels of competition and cooperation.

2   Background

2.1   Boids Model

To guide the agents to flock efficiently, literature [22] proposed the three basic rules in agents' observable space (inside the red dotted circle in Fig. 1:

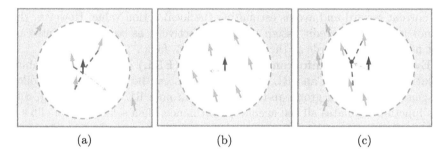

(a)   (b)   (c)

Fig. 1. The observable space of an agent (blue arrow) is inside the red dotted circle, the grey area is unobservable. (a) Separation. (b) Alignment. (c) Cohesion (Color figure online)

- **Separation:** To avoid the crash of agents during the flocking process, agents are required to keep a distance from others observed. For example, the agent (blue arrow) in Fig. 1 (a) will change it's toward to the orange dotted arrow.
- **Alignment:** It drives the agents to move in the same speed direction. Figure 1 (b) shows an agent who gets the acceleration to follow the other agents.
- **Cohesion:** To expand the scale gradually, agents who are nearby are attracted to join. The agent (blue arrow) in Fig. 1 (c) will get an acceleration along the orange dotted arrow to keep the flock close to the blue node.

Based on these rules, different flocking models were proposed adding natural population properties, such as [6,24]. We introduce the three rules into the action-value function estimation process of the multi-agent RL approach.

2.2 Multi-Agent and Evolutionary Reinforcement Learning

The flocking task can be considered a full-cooperation stochastic game so that it can be modelled as a partially observable Markov decision process (POMDP). POMDP also can be defined by a tuple $(\mathcal{S}, \mathcal{U}, \mathcal{P}, \mathcal{Z}, \mathcal{O}, R, \gamma)$. At time t, $s_t \in S$ represents the true state of the environment, the joint action of all agents is $\{u_t^1, u_t^2, \cdots, u_t^n\} = \mathbf{U_t} \in \mathbf{U} \subset \mathcal{U}^{\mathbf{n}}$, where $u_t^k \in \mathcal{U}$ represent kth agent action. The map $\mathcal{P} : \mathcal{S} \times \mathbf{U} \times \mathcal{S} \rightarrow [0, 1]$ represents the probability that agents take the joint action $\mathbf{U_t}$ at the state s_t then transit to s_{t+1}. The state that the kth agent gains from the environment $z_t^k \in \mathcal{Z}$ is partially observable. The map $\mathcal{O} : \mathcal{S} \rightarrow \mathcal{Z}$ is the observe function that maps the true state to the observed state. The observation-state history $\tau_t^k = \{z_0^k, u_0^k, z_1^k, u_1^k, \cdots, z_t^k\}$ recorded the state and action of kth agent from time 0 to t. After kth agent takes action at the observed state, it will receive a reward r_t^k defined by: $R : \mathcal{S} \times \mathcal{U} \rightarrow \mathbb{R}$.

To solve POMDP, we maximize the discount return: $R_t = \sum_{i=0}^{\infty} \gamma^i r_{t+i}$ by learning a joint policy decomposed of all n agent independent stochastic policies: $\boldsymbol{\pi}(\boldsymbol{u} \mid \boldsymbol{s}, \boldsymbol{\theta}) = \prod_{i=1}^{n} \pi_i \left(u^i \mid \tau^i, \theta_i \right)$.

A common MARL framework is Centralized Training with Decentralized Execution (CTDE) [8,17]. Each agent in the CTDE paradigm has a Critic for global information training and several actors for local information training, while COMA [8] tells an agent how much the selected action contributes to the current reward and avoids estimating the local action value. However, they cannot address the credit assign problem effectively, as paper [8] pointed out that using global rewards as the estimation of R_0 fails.

Evolutionary Reinforcement Learning (ERL) [13,14,18,29] combines evolutionary algorithm and RL. A common ERL can be summarized as follow: We generate an agent group and train them through RL. After training, a fitness function evaluates all the agents' performance, then select several elites by ranking their fitness value. Then, introduce the mutation mechanism like evolutionary algorithms for generating next-generation agents, and keep selecting the agent until the stop conditions. It shows highly competitive among the agents' learning process. During the process, the best agents will be retained.

3   Proposed Method

3.1   Cooperation: The Flocking Task

In this section, we introduce the agents' state and action space of the flocking tasks, and design the reward function.

State and Action Space
Considering a general flocking environment that consists of agents and obstacles. Each agent has the same velocity module $||v||$ and only observe the nearest k agents' state. The state of each agent includes its position and velocity (both are vectors). Set the action of agents is to rotate in the clockwise direction, then, given the rotation size ω and the action space size $|\mathcal{A}|$ of the individual agent, the action space can be generated as $\{-\omega, -\frac{(|\mathcal{A}|-1)\cdot\omega}{|\mathcal{A}|}, \cdots, 0, \cdots, \frac{(|\mathcal{A}|-1)\cdot\omega}{|\mathcal{A}|}, \omega\}$.

Reward Function
Three flocking rules of the Boids model: **separation, alignment, cohesion**, were modelled respectively by giving rewards to agents' behaviour in the environment. To avoid obstacles and focus on these rules, the reward of the whole Multi-agent flocking system at the time t is given as follows:

$$R(t) = R_{sep}(t) + R_{ali}(t) + R_{coh}(t) + R_{obs}(t) \tag{1}$$

where $R_{sep}, R_{ali}, R_{coh}$ represent the reward obtained from flocking rules, and R_{obs} is the reward for avoiding obstacles.

For the reward of **separation**, the farther the ith agent is from all observed agents, the higher the reward is. Therefore, we utilize the average module of position vectors' difference between ith agent and all n_i observed agents as the single agent's reward. Meanwhile, to avoid the distance between agents increasing infinitely by directly using the vectors' difference module as the reward, the penalty parameter d_s is given to punish the agent. The agent has received a negative reward for getting too close to other agents (less than d_s), but this reward does not increase with a distance greater than d_s:

$$R_{sep,i}(t) = \begin{cases} -\frac{1}{n_i}\sum_{j=1}^{n_i} e^{-\beta_{sep}\cdot||r_i-r_j||}, & ||r_i - r_j|| < d_s \\ 0, \text{otherwise} \end{cases} \tag{2}$$

For the second rule: **alignment**, we normalized all the agent's velocity vectors: $u_i = \frac{v_i}{||v_i||}$, and use vectors' differences like Eq. 2:

$$R_{ali,i}(t) = \frac{1}{n_i} \sum_{j=1}^{n_i} e^{\beta_{ali}||u_i - u_j||} \tag{3}$$

If we know the angle between each agent and the coordinate system: $\theta_1, \cdots, \theta_n$, where $\theta = \arctan(\frac{v_y}{v_x})$, the alignment reward can be rephrased as follow:

$$R_{ali,i}(t) = \frac{1}{n_i} \sum_{j=1}^{n_i} e^{-\beta_{ali} \cdot [2 - \cos(\theta_i - \theta_j)]} \tag{4}$$

Cohesion can be modelled as the distance between agent i and all n_i observed agents. As we didn't hope that all the agents get too close, a penalize parameter d_c like Eq. 5 is introduced:

$$R_{coh,i}(t) = \begin{cases} e^{-\beta_{coh} \cdot ||r_i - \frac{1}{n_i} \sum_{j=1}^{n_i} r_j||}, & ||r_i - \frac{1}{n_i} \sum_{j=1}^{n_i} r_j|| \geq d_c \\ 0, \text{otherwise} \end{cases} \tag{5}$$

For driving agents to avoid obstacles in the environment, construct the obstacle reward as a constant function determined by whether agents hit the obstacle:

$$R_{obs,i}(t) = \begin{cases} -\beta_{obs}, & \text{If agent hits obstacle} \\ 0, \text{otherwise} \end{cases} \tag{6}$$

Finally, we obtain the adding-up reward of an agent at time t. For the multi-agent system, the total reward is represented as the sum of all agents' rewards:

$$R(t) = \sum_{i=1}^{N} (R_{sep,i}(t) + R_{ali,i}(t) + R_{coh,i}(t) + R_{obs,i}(t)) \tag{7}$$

As a fully cooperative game, the flocking system hopes the overall return is as large as possible, and the target is to maximize the total reward Eq. 7 instead of each agent's reward. Thus we pursue the Pareto dominance (globally optimal solution) instead of Nash equilibrium (locally optimal solution) [10].

3.2 Competition: Evolutionary Multi-Agent Reinforcement Learning

We developed an algorithm named **Evolutionary Multi-Agent Reinforcement Learning (EMARL)**, which uses MARL to drive the agents to complete the flocking task full-cooperatively. Meanwhile, the trick of ERL is introduced simultaneously to encourage the agents to learn competitively and solve credit assignments in full-cooperatively MARL.

For a cooperative task, flocking require the agents to cooperate and maximize the team reward, and an agent must sacrifice its reward in exchange for the higher

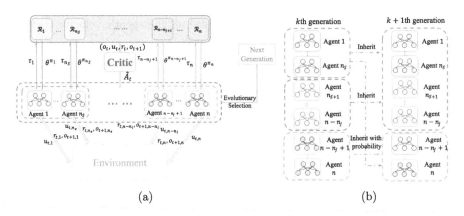

Fig. 2. (a) The framework of Evolutionary Multi-Agent Reinforcement Learning (EMARL); (b) The schematic diagram of Evolutionary Selection

team reward. To achieve cooperation, we use centralized critic to estimate the team reward to train decentralized actors. Randomly initialize the actors and critic's parameters $\{\theta^{\pi_1}, \cdots, \theta^{\pi_n}\}$ and $\theta^{\mathcal{Q}}$ respectively, all the actors and critic are neural networks. The actors interact with the environment simultaneously, storing their individual history $\tau_{i,t}$ into their individual buffer $\mathcal{R}_i$. After all the buffers are updated, the critic minimizes the loss $\mathcal{L}(\theta^{\mathcal{Q}}) = \frac{1}{T}\sum_{i=1}^{T}(y_i - \mathcal{Q}(s_i, u_i|\theta^{\mathcal{Q}}))^2$ by sampling the transition (o_t, u_t, r_t, o_{t+1}) from the total buffer $\mathcal{R} = \cup_{i=1}^{n}\mathcal{R}_i$. When updating the critic, because the train is centralized, we use the true global state s_t to replace the observation o_t. Then, update all the actors by sampling the transition from their individual buffer $\mathcal{R}_i$. Figure 2(a) illustrates all the setup. In addition, to reduce the variance, we introduce the state value function $V(s_t)$ as the baseline [15,25], with the policy gradient:

$$g_{\theta^{\pi_i}} = \mathbb{E}_\pi[\nabla_\theta log\pi_i(u|\tau, \theta^{\pi_i})(\mathcal{Q}(s, u|\theta^{\mathcal{Q}}) - V(s|\theta^{\mathcal{Q}}))] \tag{8}$$

There is an unovercome challenge of credit assignment on this method. Inspired by ERL that introduces evolutionary rules to enforce agents' competition with each other, and left high fitness one, eliminate low fitness one, we design an evolutionary selection stage that partially replaces low perform agents' parameters to solve this problem. As is shown in Fig. 2(b), after the kth updating, calculate every agent's ξ steps accumulative reward as its fitness value. Then, ranking the fitness value and selecting the first n_S agents as senior agents, the last n_J agents as junior agents, and using the percentage of the fitness value of the ith junior agent to the sum fitness of all senior agents as the probability of the ith agent parameter being replaced. We call it happens as *inherit with probability*. Other agents that are not included in junior agents will be kept and continue to be updated.

Unlike traditional ERL, **Evolutionary Selection** does not evolve all the agents. It only selects well-performed n_S agents to form senior agents and replace parameters of junior agents, which consist of worse-performed n_J agents because

not all agents that do not perform as well as others have poorer policies. In many cases, the policy did not show its advantage because the agent local on a bad position (e.g., the agent in the obstacle area like the blue arrows in Fig. 4), so the evolutionary selection only selects the worst n_J agents to let them inherit the parameters from senior agents. In the view of the evolutionary algorithm, a larger $n_S : n_J$ ratio indicates that the algorithm is more competitive (The experiments will be designed to explore it).

Next, we give the convergence analysis of policy gradient with evolutionary selection.

3.3   Convergence Analysis

Before we analyze the convergence of the EMARL, we give the following lemma:

Lemma 1. *The policy gradient of the actor-critic algorithm without evolutionary selection in ith iteration:*

$$g_{\theta_k} = \mathbb{E}_\pi[\nabla_{\theta_k} log\boldsymbol{\pi}(\boldsymbol{u}|\boldsymbol{\tau})A^\pi(s,\boldsymbol{u})] \tag{9}$$

will converge when:

$$\lim_{k\to\infty} \|\nabla_\theta \mathcal{J}\| = 0, w.p.1 \tag{10}$$

Lemma 2. *If $\pi_{J,i}$ is the policy of the junior agent, after enough updates, the $\pi_{J,i}$ will inherit the policy parameter from senior agents with probability 1.*

Using the lemmas, we give the convergence analysis of policy gradient with evolutionary selection:

Theorem 1. *The policy with evolutionary selection: $\pi_j^s(u_j|\tau_j), \forall j \in 1,2,...,n$, the gradient:*

$$g_{\theta_k}^s = \mathbb{E}_\pi[\sum_{j=1}^n \nabla_{\theta_k} log\pi_j^s(u_j|\tau_j)A^{\pi_j^s}(s,\boldsymbol{u})] \tag{11}$$

will converge as well:

$$\lim_{k\to\infty} \|g_{\theta_k}^s\| = 0, w.p.1 \tag{12}$$

Proof. After the kth update, collect the actors that have never been selected as junior agents, append their policy to P_{N_A}, and append the rest policies to P_{N_B}:

- $P_{N_A} = \{\pi_{A,1}^s, \pi_{A,2}^s, \cdots, \pi_{A,N_A}^s\}$, where $\pi_{A,i}^s$ is the policy that parameters never been inherit from before N iteration.
- $P_{N_B} = \{\pi_{B,1}^s, \pi_{B,2}^s, \cdots, \pi_{B,N_B}^s\}$, where $\pi_{B,i}^s$ is the policy that parameter was inherit in previous some iteration.

It is easy to know, $\forall \pi_j^s$, if $\pi_j^s \notin P_{N_A}$, then $\pi_j^s \in P_{N_B}$; so, we can assume P_{N_A} is not empty. $\forall \pi_{A,j}^s \in P_{N_A}$, they will never be selected as junior agents. Thus their parameters of them will not be inherited by others. Therefore, according to Lemma 1, we know: $\forall \varepsilon > 0, \exists N_{A,j} > 0$, when $k > N_{A,j}$, for $\pi_{A,j}$, the following equation hold:

$$\mathcal{P}\{||\mathbb{E}_\pi[\nabla_{\theta_k} log\pi_{A,j}^s(u_j|\tau_j, \theta_k^{\pi_{A,j}^s}) \cdot A^{\pi_{A,j}^s}(s,u)]|| < \varepsilon\} = 1 \tag{13}$$

So we only select $N = \max\{N_{A,1}, N_{A,2}, \cdots, N_{A,N_A}\}$, let $k \geq N$, the policies in P_{N_A} are convergent. For the policy $\pi_{B,i}$ in P_{N_B}, according to Lemma 2, $\exists N_{B,i} \geq 0$, when $n \geq N_{B,i}$, $\pi_{B,i}$ will inherit the parameter from senior agent.

So we only select $N_B = \max\{N, N_{B,i}\}$, after N_B update steps, there is some agent in senior agent convergent, once the $\pi_{B,i}$ inherit their parameters, the $\pi_{B,i}$ will convergent. Extend this result to all the policy in P_{N_B}, we have when $\max\{N, N_{B,1}, \cdots, N_{B,N_B}\}$ update steps, the policy in P_{N_B} convergent.

All the policies in P_{N_A}, P_{N_B} will converge $w.p.1$, therefore, the policy with evolutionary selection is converge $w.p.1$.

Based on Lemma 1, Theorem 1 shows that under the assumption [15], the policy gradient with **Evolutionary Selection** is also convergence with probability 1.

4   Experiments

4.1   Experimental Setup

Experiments of EMARL with different n_S and n_J were designed for validation. We implement the flocking environment by OpenAI gym, the basic of the codes we refer to [1]. We mainly compare our algorithm with different levels of competition and cooperation by adjusting the ratio of n_J and n_S.

Agents Setting: In the flocking visualization environment, we use a green arrow to represent the agent, and the arrow direction is the agent's toward. We test 15 and 30 agents in the environment, and these two group experiments include the different settings of n_S, n_J. The details of the environment setting can be seen in the Appendix.

Environment Setting: We construct the environment with two circle obstacles (shallow red area), one has a radius of 0.1 (lower-left corner) and the other 0.2 (center) (Fig. 4). Our environment allows the agent to swarm inside the obstacle to simplify the environment. We consider the obstacle as the forbidden zone and the agents who stay in the zone would suffer from the negative reward determined by Eq. 6. In our experiments, the alignment reward is not introduced because the relationship between performance and alignment reward is weak in the tentative experiments. Nevertheless, instead of denying the function of alignment reward, our target is to simplify the environment.

Baselines Setting: As for comparisons, we selected two training methods. **(a) Non-training:** In general, we need to choose non-training as the baseline to

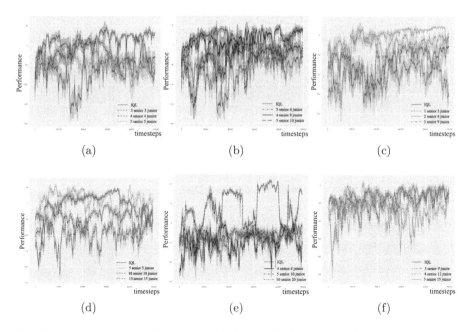

Fig. 3. Performance of Evolutionary Multi-Agent Reinforcement Learning in flocking task. Each row represents that the maximum number of agents is 15 and 30, respectively, while each column represents $n_S : n_J = 1, 2, 3$ respectively.

compare flocking and obstacle avoidance performance with other methods. **(b) IQL:** There are several representative methods according to MARL. [23] has demonstrated that IQL achieves state-of-the-art performance.

4.2 Evolutionary Reinforcement Learning Performance

The performance of the algorithms is shown in Fig. 3, where (a), (b), (c) represent the 15 agents with the number ratio of senior agents to junior agents $n_S : n_J$ equal to $1, 2, 3$ respectively. In terms of the group return obtained by agents in the training process, the environment of 15 agents and 30 agents performs excellently when $n_S : n_J = 1 : 2$ or $1 : 1$. In the view of the evolutionary algorithm, a larger $n_S : n_J$ ratio indicates that the algorithm is more competitive.

From the perspective of ERL, when $n_S : n_J = 1 : 3$, **Evolutionary Selection** is almost consistent with the conventional ERL algorithm. Nearly all agents are selected as senior agents or junior agents. Then, the parameter replacement will happen to them. Our experiment shows that although it can improve the algorithm compared with the traditional way, the extent of improvement is far less than that of the previous two ratios of n_S to n_J, which refer to appropriate levels of competition and cooperation.

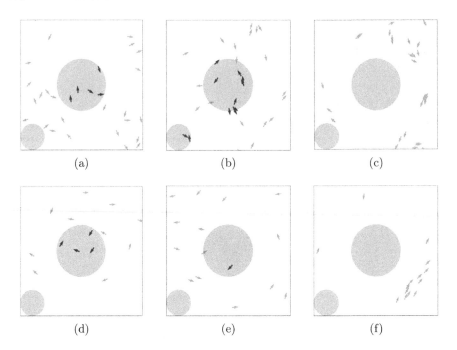

(a)   (b)   (c)

(d)   (e)   (f)

Fig. 4. The visualization of the agents flocking. (a): The flocking of 30 agents' envi-
ronmnet before training; (b): The flocking of 30 agents' environment trained with IQL.
(c): The flocking of 30 agents environment trained with EMARL.; (d): The flocking
of 15 agents' environment before training; (e): The flocking of 15 agents' environment
trained with IQL. (f): The flocking of 15 agents environment trained with EMARL.

4.3   Ablation Study

Flocking Visualization. In general, the proposed algorithm performs better
than the traditional one, and different n_S and n_J settings have different effects.
When $n_S : n_J = 1 : 2$, they perform best. Compared to the return curve of an
agent during training, we are more concerned with position distribution.

The visualized results are shown in Fig. 4, from which we see that before the
training, the distribution of all agents is irregular, and some agents even moved
into the obstacle. For the 30 agents' environment, after training with IQL, the
agents evolve into the flock, but almost one-third of the agents are in the area of
the forbidden zone; after training with EMARL, the agents develop into the obvious flock, with no agent inside the obstacle (forbidden zone), and the movement
of the agents tends to avoid obstacles. While for the 15 agents' environment,
after training with IQL, although there is only one agent in the forbidden zone,
agents do not evolve into any obvious flock; after training with EMARL, there is
no agent inside the forbidden zone and the agents evolve into the flock well, The
movement of agents tends to avoid the obstacle, which indicates the excellent
effectiveness of EMARL.

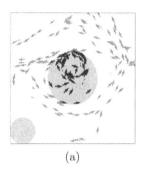
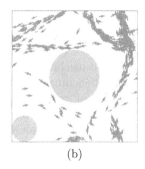

(a)   (b)

Fig. 5. The visualization of frame by frame superposition of agents flocking. (a): The flocking of 30 agents environment trained with IQL. (b): The flocking of 30 agents environment trained with EMARL.

Frame by Frame Superposition Visualization. The flocking visualization of frame by frame superposition shows the dynamic perspective's obstacle avoidance performance. From Fig. 5, we discover that after training with IQL, the agents moving into the forbidden zone are very intense and frequent, and the formation is not evolved into a good flock. While after training with EMARL, most agents avoided the obstacles and they went on orderly, which proves the advantages of our method more forcefully.

5   Related Work

Adopting RL to solve the flocking problem has received continuous attention. In literatures [11,12], the Q-learning based algorithms are introduced into the UAVs' flocking task, and the flocking reward is designed well; this work needn't know complete knowledge but still performs well when the UAV agent number is not too large. The paper [6] proposed a MARL framework for flocking tasks, and they designed a secondary reward that measures how an agent can perform complex cooperative learning driven by flocking. These works also pay more attention to cooperation than competition among the agents, and the algorithms can only deal with small-scale flocking tasks. The literature [5,27,28] introduced DRL and actor-critic architecture, which increase the number of agents the algorithm can handle. The literature [20] proposed a mean-field game-based method, which encourages the agents to reach the equilibrium solution by the Soft Actor-Critic algorithm. It is paying more attention to the competition, but the potential problem is that the equilibrium solution may not be the global optimal.

6   Conclusion

In this paper, we propose Evolutionary Multi-Agent Reinforcement Learning (EMARL) which first applies ERL in flocking tasks. We combine cooperation for further performance improvement and competition for the credit assignment problem. For cooperation, the agents' reward for flocking tasks is designed, and the target is to maximize the total reward. For the competition, we design a Roulette to make junior agents inherit the parameters of senior agents and an Evolutionary Selection mechanism. Experimental results in a range of challenging and self-contrast benchmarks indicate this algorithm with effectiveness. For multi-agent tasks similar to flocking [7], we would further study on the relationship between the number of agents, the ratio of senior and junior agents, and the degree of emergence making our method conducive to more complex environments.

References

1. "boid multi-agent rl environment & multi-agent rl agent" (2020). https://github.com/zombie-einstein/flock_env
2. Bansal, T., Pachocki, J., Sidor, S., Sutskever, I., Mordatch, I.: Emergent complexity via multi-agent competition. arXiv preprint arXiv:1710.03748 (2017)
3. Bardi, M., Cardaliaguet, P.: Convergence of some mean field games systems to aggregation and flocking models. Nonlinear Anal. **204**, 112199 (2021)
4. Bonabeau, E.: Agent-based modeling: methods and techniques for simulating human systems. Proc. Natl. Acad. Sci. 99(suppl_3), 7280–7287 (2002)
5. Chang, W., Lizhen, W., Chao, Y., Zhichao, W., Han, L., Chao, Y.: Coactive design of explainable agent-based task planning and deep reinforcement learning for human-UAVs teamwork. Chin. J. Aeronaut. **33**(11), 2930–2945 (2020)
6. Chen, C., Hou, Y., Ong, Y.S.: A conceptual modeling of flocking-regulated multi-agent reinforcement learning. In: 2016 International Joint Conference on Neural Networks (IJCNN), pp. 5256–5262. IEEE (2016)
7. Drugan, M.M.: Reinforcement learning versus evolutionary computation: a survey on hybrid algorithms. Swarm Evol. Comput. **44**, 228–246 (2019)
8. Foerster, J., Farquhar, G., Afouras, T., Nardelli, N., Whiteson, S.: Counterfactual multi-agent policy gradients. In: Proceedings of the AAAI Conference on Artificial Intelligence, vol. 32 (2018)
9. Grover, P., Bakshi, K., Theodorou, E.A.: A mean-field game model for homogeneous flocking. Chaos Interdiscip. J. Nonlinear Sci. **28**(6), 061103 (2018)
10. Hu, Y., Gao, Y., An, B.: Multiagent reinforcement learning with unshared value functions. IEEE Trans. Cybern. **45**(4), 647–662 (2014)
11. Hung, S.M., Givigi, S.N.: A q-learning approach to flocking with UAVs in a stochastic environment. IEEE Trans. Cybern. **47**(1), 186–197 (2016)
12. Hung, S.M., Givigi, S.N., Noureldin, A.: A dyna-q (lambda) approach to flocking with fixed-wing UAVs in a stochastic environment. In: 2015 IEEE International Conference on Systems, Man, and Cybernetics, pp. 1918–1923. IEEE (2015)
13. Khadka, S., et al.: Collaborative evolutionary reinforcement learning. In: International Conference on Machine Learning, pp. 3341–3350. PMLR (2019)
14. Khadka, S., Tumer, K.: Evolution-guided policy gradient in reinforcement learning. In: Advances in Neural Information Processing Systems, vol. 31 (2018)
15. Konda, V., Tsitsiklis, J.: Actor-critic algorithms. In: Advances in Neural Information Processing Systems, vol. 12 (1999)
16. Lighthill, M.J., Whitham, G.B.: On kinematic waves ii. a theory of traffic flow on long crowded roads. Proc. Roy. Soc. Lond. Ser. A. Math. Phys. Sci. **229**(1178), 317–345 (1955)

17. Lowe, R., Wu, Y.I., Tamar, A., Harb, J., Pieter Abbeel, O., Mordatch, I.: Multi-agent actor-critic for mixed cooperative-competitive environments. In: Advances in Neural Information Processing Systems, vol. 30 (2017)
18. Majumdar, S., Khadka, S., Miret, S., McAleer, S., Tumer, K.: Evolutionary reinforcement learning for sample-efficient multiagent coordination. In: International Conference on Machine Learning, pp. 6651–6660. PMLR (2020)
19. Mavridis, C.N., Tirumalai, A., Baras, J.S., Matei, I.: Semi-linear Poisson-mediated flocking in a Cucker-Smale model. IFAC-PapersOnLine **54**(9), 404–409 (2021)
20. Perrin, S., Laurière, M., Pérolat, J., Geist, M., Élie, R., Pietquin, O.: Mean field games flock! the reinforcement learning way. arXiv preprint arXiv:2105.07933 (2021)
21. Quera, V.Q.J., Salvador Beltrán, F., Dolado i Guivernau, R.: Flocking behaviour: agent-based simulation and hierarchical leadership. Jasss-J. Artif. Soc. Soc. Simul. **13**(2), 8 (2010)
22. Reynolds, C.: Boids background and update (2001). http://www.red3d.com/cwr/boids/
23. Tan, M.: Multi-agent reinforcement learning: independent vs. cooperative agents. In: Proceedings of the Tenth International Conference on Machine Learning, pp. 330–337 (1993)
24. Toner, J., Tu, Y.: Flocks, herds, and schools: a quantitative theory of flocking. Phys. Rev. E **58**(4), 4828 (1998)
25. Weaver, L., Tao, N.: The optimal reward baseline for gradient-based reinforcement learning. arXiv preprint arXiv:1301.2315 (2013)
26. Wu, J., Liu, Y.: Flocking behaviours of a delayed collective model with local rule and critical neighbourhood situation. Math. Comput. Simul. **179**, 238–252 (2021)
27. Yan, C., Xiang, X., Wang, C.: Fixed-wing UAVs flocking in continuous spaces: a deep reinforcement learning approach. Robot. Auton. Syst. **131**, 103594 (2020)
28. Yan, C., Xiang, X., Wang, C., Lan, Z.: Flocking and collision avoidance for a dynamic squad of fixed-wing UAVs using deep reinforcement learning. In: 2021 IEEE/RSJ International Conference on Intelligent Robots and Systems (IROS), pp. 4738–4744. IEEE (2021)
29. Zhu, S., Belardinelli, F., León, B.G.: Evolutionary reinforcement learning for sparse rewards. In: Proceedings of the Genetic and Evolutionary Computation Conference Companion, pp. 1508–1512 (2021)

Differentiable Causal Discovery Under Heteroscedastic Noise

Genta Kikuchi[1,2](✉)

[1] DENSO CORPORATION, Kariya, Aichi 448-8661, Japan
genta.kikuchi.j8g@jp.denso.com
[2] Shiga University, Banba Hikone, Shiga 522-8522, Japan

Abstract. We consider the problem of estimating directed acyclic graphs from observational data. Many studies on functional causal models assume the independence of noise terms. Thus, they suffer from the typical violation of model assumption: heteroscedasticity. Several recent studies have assumed heteroscedastic noise instead of additive noise in data generation, though most of the estimation algorithms are for bivariate data. This study aims to improve the capability of continuous optimization-based methods so that they can handle heteroscedastic noise under multivariate non-linear data with no latent confounders. Numerical experiments on synthetic data and fMRI simulation data show that our estimation algorithm improves the estimation of the causal structure under heteroscedastic noise. We also applied our estimation algorithm to real-world data collected from a ceramic substrate manufacturing process, and the results prove the possibility of using the estimated causal graph to accelerate quality improvement.

Keywords: Causal discovery · Structural equation models · Heteroscedasticity · Continuous optimization · Bayesian networks

1 Introduction

Understanding causal relationships between quantities of interest plays a fundamental role in most scientific fields, including neuroinformatics, bioinformatics, social science, and manufacturing [14,17,26,29]. Causality is also used in the area of machine learning research, such as fairness, model interpretation, and transfer learning [15,16,22]. While randomized controlled trials are highly recommended to obtain high-level evidence of a relationship between quantities, such experiments are in some cases unethical, technically difficult, or too expensive. To tackle this issue, many researchers have developed numerous methods to estimate causal relationships from observational data, termed causal discovery [7]. One important approach is the constraint-based approach, which estimates causal relationships based on conditional independencies between variables [27]. Constraint-based methods include well-known techniques such as the PC and FCI algorithms [27], and are widely applicable to various data types and distributions, though they cannot distinguish causal relationships that give the same sets of conditional independencies.

M. Tanveer et al. (Eds.): ICONIP 2022, LNCS 13623, pp. 284–295, 2023.
https://doi.org/10.1007/978-3-031-30105-6_24

Another class of approaches, termed functional causal models (FCM), make additional assumptions to the data generating process to exploit structural asymmetries [9,23]. Many studies assume independence of noise terms, hence estimation performance of FCM can easily be degraded by the violation of the model assumption: heteroscedasticity. Heteroscedasticity typically occurs in cases of time series data [1] and under the existence of latent confounders [13]. There is also heteroscedastic noise, in which the variance of the noise term is modulated by other variables; it is often observed in practice [6], but not sufficiently studied in the field of causal discovery.

Several recent studies have assumed heteroscedastic noise instead of additive noise. A method based on an asymmetric property of fourth-order moments [4] utilizes heteroscedasticity to determine causal direction. Causal autoregressive flow model (CAREFL) [12] estimates causal direction with the framework of normalizing flows, a methodology developed in deep learning. Heteroscedastic noise causal model (HEC) [30] divides data into segments exhibiting different noise variances and identifies causal direction using the Bayesian information criterion. Although it has been reported that fourth-order moments, CAREFL, and HEC perform well in bivariate data, applying those algorithms to data with more than two variables is not straightforward. Generalized root cause inference (GRCI) [28] transforms the regression residuals with the estimated mean absolute deviation (MAD), so that predictors and transformed residuals are independent when regressed in the correct causal direction. GRCI outputs causal order of variables, but it can be used to obtain an estimate of unique directed acyclic graphs. To the best of our knowledge, GRCI is the only algorithm that assumes heteroscedastic noise and can be applied to data with more than two variables. This motivated us to explore the possibility of other approaches.

In this paper, we propose an estimation algorithm that can also handle heteroscedastic noise under multivariate, non-linear data with no latent confounders. We extend the recently developed continuous optimization-based method, NOTEARS [31,32], to simultaneously estimate the conditional variance of noise terms as well as the conditional expectation of each variables, and we also define corresponding acyclicity constraints for our model. To mitigate the performance degradation of NOTEARS when standard Gaussian noise assumption does not hold [3,21], we do not assume any specific probability distribution of the noise. Instead, we leverage the estimation method of approximating log-probability. To verify the effectiveness of our approach, we conducted a numerical experiment to compare the proposed method with other functional causal models on synthetic data and fMRI simulation data. We also applied our algorithm to real-world data collected from ceramic substrate manufacturing.

2   Background

2.1   Structural Equation Models

Suppose we observe $p-$dimensional random vector $\boldsymbol{X} = (X_1, ..., X_p) \in \mathbb{R}^p$ with joint distribution $\mathcal{L}(\boldsymbol{X})$ generated from a system that can be represented by a directed acyclic graph (DAG) $\mathcal{G}$. Throughout this paper, we assume that all random variables have positive density and are absolutely continuous with respect to the Lebesgue measure. Denote the parent set of X_j as PA_j, a set of variables that have a direct connection to X_j. The following set of equations is called structural equation model (SEM) [18]:

$$X_j = f_j(\mathrm{PA}_j, N_j), \quad j = 1, ..., p \tag{1}$$

where N_j is a noise term. We assume that noise term $\boldsymbol{N} = (N_1, ..., N_p)$ is generated from product distribution $\mathcal{L}(\boldsymbol{N})$, which means there are no latent confounders. Although SEM (1) can represent a wide range of data-generating processes, it is too general to estimate the true graph $\mathcal{G}$ from the joint distribution $\mathcal{L}(\boldsymbol{X})$ [19]. By making additional restrictions to the causal structure, one can identify the graph $\mathcal{G}$ from the joint distribution.

2.2 Additive Noise Models

The additive noise model (ANM) [20] is a special case of SEM (1), where the noise term is restricted to be additive. ANM is defined as a collection of p equations:

$$X_j = f_j(\mathrm{PA}_j) + N_j. \quad j = 1, ..., p \tag{2}$$

Assuming additive noise, $\mathcal{G}$ is identifiable from $\mathcal{L}(\boldsymbol{X})$ if f_j are linear and N_j are non-Gaussian noises [23], or f_j are non-linear [9,20], or f_j are linear and N_j are Gaussian noises with equal error variances [19]. For non-linear cases, estimation of underlying graph structure is typically done by regression with a subsequent independence test (RESIT) algorithm [20]. RESIT recursively performs regression and independence tests to estimate the topological order of variables, starting from the sink variable.

Recently, the structure learning problem has been formulated as a continuous optimization problem using differentiable acyclicity constraint (NOTEARS) [31]. It was later extended to be capable of non-linear structure using multilayer perceptrons (MLP), named NOTEARS-MLP [32]. Let $\mathbf{X} = [\mathbf{x}_1|...|\mathbf{x}_d] \in \mathbb{R}^{n \times p}$ be data matrix consisting of n i.i.d. observations. Consider an MLP that consists of h hidden layers with m_l hidden units in each layer and an activation function σ, given by $\mathrm{MLP}(\mathbf{X}; A_j^{(1)}, ..., A_j^{(h+1)}) = A^{(h+1)}\sigma(\cdots A^{(2)}\sigma(A^{(1)}(\mathbf{X}))$, where $A^{(l)} \in \mathbb{R}^{m_l \times m_{l-1}}$ is a connectivity matrix with $m_0 = p, m_{h+1} = 1$. Let $\theta_j = \{A_j^{(1)}, ..., A_j^{(h+1)}\}$ be parameters of the j-th MLP which aims to predict $\mathrm{E}[X_j|\mathrm{PA}_j]$. NOTEARS-MLP is formulated by following optimization problem:

$$\min_\theta \frac{1}{n} \sum_{j=1}^p \ell(\mathbf{x}_j, \mathrm{MLP}(\mathbf{X}; \theta_j)) + \lambda\|A_j^{(1)}\|_1 \quad \text{s.t.} \quad h(W(\theta)) = 0, \tag{3}$$

where $\theta = (\theta_1, ..., \theta_p)$ and $h(W(\theta)) = \mathrm{tr}(e^{W(\theta) \circ W(\theta)}) - p$ is an acyclicity constraint that equals to zero only if no cycle exists in the weighted adjacency matrix $W(\theta)$. $W(\theta)$ is calculated based on the first layer of each MLP by

$[W(\theta)]_{kj} = \|[A_j^{(1)}]_{.k}\|_2$, where $[W(\theta)]_{kj} = 0$ if $\mathrm{MLP}(\mathbf{X}, \theta_k)$ is independent of x_j. Squared loss is used for loss function ℓ, and it has been seen that this is equivalent to assuming standard Gaussian noise, which the estimation is hindered when the assumption does not hold [3]. The violation of the assumption can easily occur by just scaling the variables [21].

2.3 Heteroscedastic Noise Models

The heteroscedastic noise model (HNM) [28] is also a special case of SEM (1), where the scale of the noise term is induced by other variables. HNM is defined as a collection of p equations in following form:

$$X_j = f_j(\mathrm{PA}_j) + s_j(\mathrm{PA}_j)N_j, \quad j = 1, ..., p \qquad (4)$$

where f_j and scaling function $s_j > 0$ can be non-linear. This type of heteroscedasticity $s_j(\mathrm{PA}_j)N_j$ is called multiplicative heteroscedasticity [8]. HNM is identifiable in linear and nonlinear cases, and the multivariate setting [28,30].

HEC [30] assumes that N_j is a standard Gaussian variable and the distributions of X_j have compact support. CAREFL [12] models the scaling functions by $e^{s(X)}$, where s represents autoregressive transformations. Both HEC and CAREFL are only applicable for $p = 2$. GRCI [28] does not assume specific distributions of N_j nor functional form of s_j. GRCI first estimates the skeleton of $\mathcal{G}$ using the PC-Stable algorithm [5], and then recursively finds the sink variable and obtains the causal order of the variables. Before evaluating the independence between the residual and the predictor, GRCI partials out the effect of PA_j by transforming the regression residuals using their estimated conditional MAD. We can get a unique DAG estimation using PC-Stable algorithm with conditioning sets restricted to preceding variables according to the causal order, followed by an orientation of directed edges according to the causal order.

Although GRCI considerably improves the estimation accuracy of recovering DAG under HNM, the results of our numerical experiment indicate that the performance degrades substantially under a linear multivariate setting. We suspect that estimating the conditional MAD of regression residuals with splines is too flexible, making the identification of causal direction based on independence test unstable in linear multivariate setting.

3 Estimation Algorithm Based on Continuous Optimization

In this section, we propose a continuous optimization-based estimation algorithm that also can handle heteroscedastic noise.

3.1 Log-Likelihood

Consider heteroscedastic noise model (4) and set $N_j^* := s_j(\mathrm{PA}_j)N_j$. We have:

$$X_j = f_j(\mathrm{PA}_j) + N_j^*. \quad j = 1, ..., p \qquad (5)$$

Given n i.i.d. observations, the log-likelihood of this model is in the form

$$\log L(\mathbf{X}) = -\frac{1}{n} \sum_{t=1}^{n} \sum_{j=1}^{p} \log \sigma_j^{(t)} + \frac{1}{n} \sum_{t=1}^{n} \sum_{j=1}^{p} \log \tilde{p}_j \left(\frac{x_j^{(t)} - f_j(\mathrm{PA}_j^{(t)})}{\sigma_j^{(t)}} \right), \quad (6)$$

where $\tilde{p}_j$ are the density functions of noises N_j^* standardized to unit variance, and $(\sigma_j^{(t)})^2$ are their conditional variance of t-th sample before standardization. $f_j(\mathrm{PA}_j^{(t)})$ and $\sigma_j^{(t)}$ describe the conditional expectation of X_j and the heteroscedasticity induced by s_j. Their estimations $\hat{x}_j^{(t)}$ and $\hat{\sigma}_j^{(t)}$ are given by MLP with h_A and h_C hidden layers:

$$\hat{x}_j^{(t)} = \mathrm{MLP}(\mathbf{x}^{(t)}; \theta_j^A), \quad (7)$$

$$\hat{\sigma}_j^{(t)} = \mathrm{MLP}(\mathbf{x}^{(t)}; \theta_j^C), \quad (8)$$

where $\theta_j^A = \{A_j^{(1)}, \cdots, A_j^{(h_A+1)}\}$ and $\theta_j^C = \{C_j^{(1)}, \cdots, C_j^{(h_C+1)}\}$ denote parameters of each MLP.

We do not assume any specific probability distributions for $\tilde{p}_j$. Instead, we use the estimation method of approximating log-probability $\log \tilde{p}_j$ and still obtain a satisfactory estimator by choosing from the two candidates according to whether the variable is super-Gaussian or sub-Gaussian [10,11]. During iteration, $\log \tilde{p}_j$ is determined from the following:

$$\log \tilde{p}_j(z) = \begin{cases} -2 \log \cosh(z) & \text{if } \gamma_j > 0, \\ -\left(z^2/2 - \log \cosh(z)\right) & \text{else} \end{cases}, \quad (9)$$

where scaler γ_j is calculated by

$$\gamma_j = \mathrm{E}\left[-\tanh(z)z + (1 - \tanh(z)^2)\right]. \quad (10)$$

γ_j is positive if z is super-Gaussian and negative if z is sub-Gaussian. We expect that using the approximated $\log \tilde{p}_j$ improves the estimation performance compared to the squared loss function, when the noise distribution is deviated from the standard Gaussian.

3.2 Acyclicity Constraint

Our estimation algorithm has an additional MLP which estimates conditional variances of a variable given its parents, compared to NOTEARS-MLP. Therefore, we can not directly use the well-known form $h(W(\theta)) = \mathrm{tr}(e^{W(\theta) \circ W(\theta)}) - p$. To obtain a DAG estimate, we have to consider an acyclicity constraint on θ_A as well as θ_C. Let the weighted adjacency matrix corresponding to θ_A and θ_C be $[W(\theta_A)]_{kj} = \|[A_j^{(1)}]_{\cdot k}\|_2$ and $[W(\theta_C)]_{kj} = \|[C_j^{(1)}]_{\cdot k}\|_2$, respectively. The acyclicity constraint for our model is given by:

$$h(W(\theta_A), W(\theta_C)) = \mathrm{tr}\left(e^{W(\theta_A) \circ W(\theta_A) + W(\theta_C) \circ W(\theta_C)}\right) - p = 0. \quad (11)$$

Acyclicity constraint (11) enforces $W(\theta_A) \circ W(\theta_A) + W(\theta_C) \circ W(\theta_C)$ to be acyclic. This means that, in addition to the acyclicity of $W(\theta_A)$ and $W(\theta_C)$ themselves, we do not allow cyclic relationships after combining them, for example:

$$X_1 = s_1(X_2)N_1, \tag{12}$$
$$X_2 = f_2(X_1) + N_2. \tag{13}$$

3.3   Objective Function

We propose solving the minimization problem of negative log-likelihood under acyclicity constraint. Using log-likelihood (6), acyclicity constraint (11), and by adding regularization term with respect to $A_1^{(1)}$ and $C_1^{(1)}$, we finally obtain our optimization problem:

$$\min_{\theta_A, \theta_C} \frac{1}{n} \sum_{t=1}^{n} \sum_{j=1}^{p} \log \mathrm{MLP}(\mathbf{x}^{(t)}; \theta_j^C) - \frac{1}{n} \sum_{t=1}^{n} \sum_{j=1}^{p} \log \tilde{p}_j \left(\frac{x_j^{(t)} - \mathrm{MLP}(\mathbf{x}^{(t)}; \theta_j^A)}{\mathrm{MLP}(\mathbf{x}^{(t)}; \theta_j^C)} \right)$$
$$+ \lambda_1 \sum_{j=1}^{p} \|A_j^{(1)}\|_1 + \lambda_2 \sum_{j=1}^{p} \|C_j^{(1)}\|_1 \tag{14}$$

$$\text{subject to} \quad \mathrm{tr}\left(e^{W(\theta_A) \circ W(\theta_A) + W(\theta_C) \circ W(\theta_C)} \right) - p = 0,$$

where scalers λ_1 and λ_2 denote regularization parameters. Following previous studies, L-BFGS [2] is used for optimization.

After the optimization, we calculate an estimation of weighted adjacency matrix by $\tilde{W} = 0.5(W(\theta_A) + W(\theta_C))$. Following previous studies, we round the numerical solution $\tilde{W}$ with a fixed small threshold $w > 0$. This removes redundant edges from the estimated DAG, which helps to reduce false positives [33] and remove cycles remaining in the graph.

4   Experiments

We report experiments to assess the effectiveness of our estimation algorithm. Baseline algorithms selected for comparison are NOTEARS [31] and DirectLiNGAM [24] for the linear cases, and NOTEARS-MLP [32], RESIT [20] and GRCI [28] for the non-linear cases. We also included the performance of the empty graph as a naive baseline. DirectLiNGAM and RESIT are included to investigate the performance of the independence-based estimation algorithms which do not assume heteroscedasticity. For each algorithm, we used the parameter set defined in the corresponding study and codes. We used structural Hamming distance (lower is better) as the evaluation metric.

We used for our model the same parameter settings as for NOTEARS and NOTEARS-MLP, where $\lambda_1 = 0.01$, $\lambda_2 = 0.01$, $w = 0.3$, and sigmoid function is used for σ. For $\mathrm{MLP}(\mathbf{x}^{(t)}; \theta_j^A)$, we set no hidden layer on the linear case denoted as Ours, and one hidden layer with 10 nodes on non-linear case denoted as Ours-MLP. For MLP($\mathbf{x}^{(t)}; \theta_j^C$), we set one hidden layer with 10 nodes for both linear and non-linear cases. Exponential activation function is applied to the output of MLP($\mathbf{x}^{(t)}; \theta_j^C$) to ensure positivity. To improve convergence, we initialized the weights of Ours and OursMLP with the fitted result of NOTEARS and NOTEARS-MLP, respectively.

4.1 Synthetic Data

In the experiments, ground truth DAGs were generated from Erdös-Rényi model with $2p$ edges (ER2). For the linear cases, each variable was generated by $X_j = \sum_{i \in \text{pa}(j)} W_{i,j} X_i + s_j(\text{PA}_j) N_j$, where pa($j$) denotes the index set of PA$_j$. For the non-linear cases, we used index models $X_j = \tanh(\sum_{i \in \text{pa}(j)} W_{i,j}^{(1)} X_i)$ $+ \cos(\sum_{i \in \text{pa}(j)} W_{i,j}^{(2)} X_i) + \sin(\sum_{i \in \text{pa}(j)} W_{i,j}^{(3)} X_i) + s_j(\text{PA}_j) N_j$. For the scaling function, we used $s_j(\text{PA}_j) = \exp(\sum_{i \in \text{pa}(j)} C_{i,j} X_i)$. Each non zero element of $p \times p$ weight matrix W and C were drawn randomly from $W_{i,j} \sim$ Uniform(-2.0, -0.5) $\cup$ Uniform(0.5, 2.0) and $C_{i,j} \sim$ Uniform(-0.8, -0.4) $\cup$ Uniform(0.4, 0.8), respectively. All variables were scaled to zero mean and unit variance.

We introduced an existence ratio of heteroscedasticity R_h, which controls how often heteroscedasticity occurs in the generated data. We first generated a binary matrix $B \in \{0,1\}^{p \times p}$ from ER2 graph, and then randomly selected each non-zero elements of B with probability R_h and obtained a binary matrix $\tilde{B}$. For example, if $R_h = 0$ we get $\tilde{B}$ with all zeros, and if $R_h = 1.0$, we get $\tilde{B} = B$. Thereafter, we generated each element of W and C according to B and $\tilde{B}$, respectively. Each experiment ran 10 times, and the average and the standard deviation of structural Hamming distance were collected.

We first conducted an experiment with different existence ratios of heteroscedasticity R_h. We changed $R_h = \{0, 0.2, 0.5, 0.8, 1.0\}$ with number of variables $p = 10$, sample size $n = 1000$ for the linear case and $n = 2000$ for the non-linear case. The results are given in Fig. 1. As R_h increases, Ours and Ours-MLP generally outperformed the others. Ours and Ours-MLP outperformed NOTEARS and NOTEARS-MLP under no heteroscedasticity ($R_h = 0$). This verifies the effect of using log-likelihood with variance estimation as an objective function, compared to squared loss function which is equivalent to the standard Gaussian noise assumption. GRCI matched Ours-MLP in the non-linear case, although its performance degraded in the linear case. DirectLiNGAM and RESIT did not perform well even in small R_h.

To evaluate the estimation performance under different sample sizes, we set $R_h = 0.5$, $p = 10$ and changed $n = \{50, 100, 200, 400, ..., 1000\}$ for the linear case and $n = \{50, 100, 200, 500, 1000, 2000, 3000\}$ for the non-linear case. As shown in Fig. 2, Ours and Ours-MLP generally outperformed others in different sample sizes. In this setting, Ours needed at least $n = 200$ in the linear case and $n = 1000$ for the non-linear case, thanks to the initialization using fitted NOTEARS and NOTEARS-MLP. From Fig. 2(a), the performance of GRCI on the linear case gradually decreases from $n > 400$, which implies that there is a case that GRCI converges to the wrong solution with large sample size.

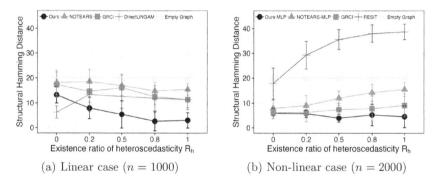

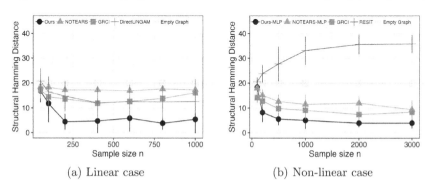

Fig. 1. Result of changing existence ratio of heteroscedasticity R_h (with $p = 10$)

(a) Linear case (b) Non-linear case

Fig. 2. Result of changing sample size n (with $p = 10$ and $R_h = 0.5$)

Finally, to investigate the change of estimation performance under different number of variables, we set $R_h = 0.5$ and changed $p = \{5, 10, 15, 20, 25\}$ with $n = 1000$ for the linear case and $n = 2000$ for the non-linear case. As shown in Fig. 3, Ours and Ours-MLP considerably outperformed the others. In Fig. 3(b), GRCI outperformed Ours-MLP only in the non-linear case with a small number of variables $p = 5$. Ours-MLP performed better on $p > 5$ compared to GRCI and the difference in performance got larger as p increased. This is likely to be the effect of the GRCI estimation procedure, which recursively finds sink nodes, where mistakes that occurred in the preceding iterations propagate through the whole procedure. Ours-MLP does not have that property, which is an advantage of continuous optimization-based methods.

4.2   fMRI Simulation Data

We used simulated fMRI data [25] to compare our estimation algorithm under a more realistic setting. We used datasets sim1, sim2, and sim3 which consist of 5, 10, and 15 variables generated from known ground-truth DAGs with 5, 11, and 18 edges, respectively. Each dataset consists of 50 time-series data subsets

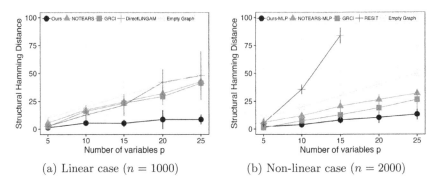

(a) Linear case ($n = 1000$)   (b) Non-linear case ($n = 2000$)

Fig. 3. Result of changing number of variables p (with $R_h = 0.5$)

with 200 data points, resulting in 10000 data points in total. The simulated data shows non-linearity and non-stationary in which the noise depends on the states of the neurons, similar to typical fMRI data. We compared Ours-MLP, NOTEARS-MLP, and GRCI with all the data points used to fit Ours-MLP and NOTEARS-MLP, whereas we randomly selected 5000 data points for GRCI, as we could not complete the calculation.

From the results given in Table 1, Ours-MLP succeeded in recovering the ground-truth DAG more accurately than the others. GRCI did not perform well in this dataset. This result implies that non-linearity in this dataset might not be strong, resulting in the weak performance of GRCI. We conclude that our estimation algorithm can improve estimation performance under realistic scenarios, regardless of the degree of non-linearity.

Table 1. Structural Hamming distance on each fMRI simulation data

Method	sim1 ($p = 5$)	sim2 ($p = 10$)	sim3 ($p = 15$)
Ours-MLP	**2**	**4**	**9**
NOTEARS-MLP	5	7	**9**
GRCI	3	13	17
Empty Graph	5	11	18

4.3   Ceramic Substrate Manufacturing Process Data

Finally, we applied Ours-MLP and GRCI to real-world data which we collected from a ceramic substrate manufacturing process. The goal was to identify the cause of cutting torque measured after the kneading process. Cutting torque is measured as an alternative characteristic of viscosity of ceramic, which is strongly related to the occurrence of crack failure. The ingredients of ceramic are fed to the mixing kneading machine which consists of two kneading screws: upper and medium, each cooled by separate chillers (denoted as U, M). We collected

temperatures (T), electricity (V, Hz), and pressure (P) from several positions of
the kneaders and chillers, resulting in 19 variables. We excluded obvious outliers
and fitted both models with 2000 data points. Knowing that cutting torque is the
sink variable, we used this prior knowledge for both method. Prior knowledge for
Ours-MLP is incorporated by adding constraints on the corresponding column
of $A_j^{(1)}$ and $C_j^{(1)}$ to be zero.

The estimated DAGs are shown in Fig. 4. The result of Ours-MLP matches
the domain knowledge in the sense that the cutter torque is affected by the temperature of the kneader (M_kneader_T), and the temperature of cooling water
from the chiller to the kneader (M_out_T) controls it. Variables of the upper
kneader and chillers (U_in_T, U_barrel_T) are placed on the upper stream, which
also meets the domain knowledge. As shown in Fig. 4(b), GRCI also correctly
placed the variables of the upper kneader, though the redundant arrow remained
hence hard to interpret. Moreover, there is no arrow from the temperature of
chillers (U_in_T, U_out_T) to the kneader, which does not meet the domain
knowledge. Consequently, we conclude that the result of Ours-MLP is preferable to that of GRCI. Red arrows in Fig. 4 indicate that there might be strong
heteroscedasticity in that causal relation. In the quality improvement activities,
we carefully consider how the variance of each quality measure changes after we
control the preceding process. Therefore, estimating the heteroscedasticity along
with DAG also helps to get an overview of the relationships we must consider.

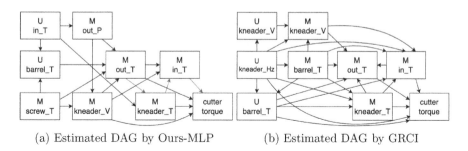

(a) Estimated DAG by Ours-MLP   (b) Estimated DAG by GRCI

Fig. 4. Estimated DAGs of ceramic substrate manufacturing process data. Red arrows
indicate strong heteroscedasticity, where $W(\theta_C) > 0.3$. Variables that were not estimated as an ancestor of cutter torque are excluded. (Color figure online)

5   Conclusion and Remarks

We proposed an estimation algorithm that can effectively deal with the heteroscedastic noise under multivariate, non-linear data with no latent confounders. We extended the recently developed continuous optimization-based
methods to estimate the conditional variance of noise terms and defined acyclicity constraints for our model. In contrast to the existing continuous optimization-
based methods, our estimation algorithm does not assume specific probability
distribution of the noise, which mitigates the performance degradation when the

standard Gaussian noise assumption does not hold. We verified the effectiveness of our estimation algorithm with numerical experiments on synthetic data, fMRI simulation data, and real-world data collected from a ceramic substrate manufacturing process. The estimation algorithm for estimating conditional variance which is robust to outliers remains in our future work. It will be also interesting to analyze the possibility of identifying causal direction under heteroscedastic noise and latent confounders, which both induce heteroscedasticity.

References

1. Bollerslev, T., Engle, R.F., Nelson, D.B.: Arch models. Handb. Econometrics **4**, 2959–3038 (1994)
2. Byrd, R.H., Lu, P., Nocedal, J., Zhu, C.: A limited memory algorithm for bound constrained optimization. SIAM J. Sci. Comput. **16**(5), 1190–1208 (1995)
3. Cai, R., Chen, W., Qiao, J., Hao, Z.: On the role of entropy-based loss for learning causal structures with continuous optimization. arXiv preprint arXiv:2106.02835 (2021)
4. Cai, R., Ye, J., Qiao, J., Fu, H., Hao, Z.: FOM: fourth-order moment based causal direction identification on the heteroscedastic data. Neural Netw. **124**, 193–201 (2020)
5. Colombo, D., Maathuis, M.H., et al.: Order-independent constraint-based causal structure learning. J. Mach. Learn. Res. **15**(1), 3741–3782 (2014)
6. Davidson, R., MacKinnon, J.G.: Econometric Theory and Methods, vol. 5. Oxford University Press, New York (2004)
7. Glymour, C., Zhang, K., Spirtes, P.: Review of causal discovery methods based on graphical models. Front. Genet. **10**, 524 (2019)
8. Harvey, A.C.: Estimating regression models with multiplicative heteroscedasticity. Econometrica: J. Econometric Soc. 461–465 (1976)
9. Hoyer, P., Janzing, D., Mooij, J.M., Peters, J., Schölkopf, B.: Nonlinear causal discovery with additive noise models. Adv. Neural Inf. Process. Syst. **21**, 689–696 (2008)
10. Hyvärinen, A., Karhunen, J., Oja, E.: Independent Component Analysis. Wiley, New York (2001)
11. Hyvärinen, A., Oja, E.: Independent component analysis by general nonlinear Hebbian-like learning rules. Sig. Process. **64**(3), 301–313 (1998)
12. Khemakhem, I., Monti, R., Leech, R., Hyvärinen, A.: Causal autoregressive flows. In: International Conference on Artificial Intelligence and Statistics, pp. 3520–3528. PMLR (2021)
13. Kupek, E.: Detection of unknown confounders by Bayesian confirmatory factor analysis. Adv. Stud. Med. Sci. **1**(3), 143–56 (2013)
14. Londei, A., D'Ausilio, A., Basso, D., Belardinelli, M.O.: A new method for detecting causality in fMRI data of cognitive processing. Cogn. Process. **7**(1), 42–52 (2006)
15. Ma, S., Tourani, R.: Predictive and causal implications of using Shapley value for model interpretation. In: Proceedings of the 2020 KDD Workshop on Causal Discovery, pp. 23–38. PMLR (2020)
16. Makhlouf, K., Zhioua, S., Palamidessi, C.: Survey on causal-based machine learning fairness notions. arXiv preprint arXiv:2010.09553 (2020)

17. Moneta, A., Entner, D., Hoyer, P.O., Coad, A.: Causal inference by independent component analysis: theory and applications. Oxford Bull. Econ. Stat. **75**(5), 705–730 (2013)
18. Pearl, J.: Causality. Cambridge University Press, Cambridge (2009)
19. Peters, J., Bühlmann, P.: Identifiability of Gaussian structural equation models with equal error variances. Biometrika **101**(1), 219–228 (2014)
20. Peters, J., Mooij, J.M., Janzing, D., Sch"olkopf, B.: Causal discovery with continuous additive noise models. J. Mach. Learn. Res. **15**, 2009–2053 (2014)
21. Reisach, A.G., Seiler, C., Weichwald, S.: Beware of the simulated DAG! Varsortability in additive noise models. arXiv preprint arXiv:2102.13647 (2021)
22. Rojas-Carulla, M., Schölkopf, B., Turner, R., Peters, J.: Invariant models for causal transfer learning. J. Mach. Learn. Res. **19**(1), 1309–1342 (2018)
23. Shimizu, S., Hoyer, P.O., Hyvärinen, A., Kerminen, A., Jordan, M.: A linear nongaussian acyclic model for causal discovery. J. Mach. Learn. Res. **7**(10) (2006)
24. Shimizu, S., et al.: DirectLiNGAM: a direct method for learning a linear nongaussian structural equation model. J. Mach. Learn. Res. **12**, 1225–1248 (2011)
25. Smith, S.M., et al.: Network modelling methods for FMRI. Neuroimage **54**(2), 875–891 (2011)
26. Spirtes, P., Glymour, C., Scheines, R., Kauffman, S., Aimale, V., Wimberly, F.: Constructing Bayesian network models of gene expression networks from microarray data (2000)
27. Spirtes, P., Glymour, C.N., Scheines, R., Heckerman, D.: Causation, Prediction, and Search. MIT Press, Cambridge (2000)
28. Strobl, E.V., Lasko, T.A.: Identifying patient-specific root causes with the heteroscedastic noise model. arXiv preprint arXiv:2205.13085 (2022)
29. Vuković, M., Thalmann, S.: Causal discovery in manufacturing: a structured literature review. J. Manuf. Mater. Process. **6**(1), 10 (2022)
30. Xu, S., Mian, O.A., Marx, A., Vreeken, J.: Inferring cause and effect in the presence of heteroscedastic noise. In: International Conference on Machine Learning, pp. 24615–24630. PMLR (2022)
31. Zheng, X., Aragam, B., Ravikumar, P., Xing, E.P.: DAGs with NO TEARS: continuous optimization for structure learning. In: Proceedings of the 32nd International Conference on Neural Information Processing Systems, pp. 9492–9503 (2018)
32. Zheng, X., Dan, C., Aragam, B., Ravikumar, P., Xing, E.: Learning sparse nonparametric DAGs. In: International Conference on Artificial Intelligence and Statistics, pp. 3414–3425. PMLR (2020)
33. Zhou, S.: Thresholding procedures for high dimensional variable selection and statistical estimation. IN: Advances in Neural Information Processing Systems, vol. 22 (2009)

IDPL: Intra-subdomain Adaptation Adversarial Learning Segmentation Method Based on Dynamic Pseudo Labels

Xuewei Li[1,2,3], Weilun Zhang[4], Jie Gao[1,2,3], Xuzhou Fu[1,2,3], and Jian Yu[1,2,3]($\boxtimes$)

[1] College of Intelligence and Computing, Tianjin University, Tianjin, China
{lixuewei,gaojie,fuxuzhou,yujian}@tju.edu.cn
[2] Tianjin Key Laboratory of Cognitive Computing and Application, Tianjin, China
[3] Tianjin Key Laboratory of Advanced Networking, Tianjin, China
[4] Tianjin International Engineering Institute, Tianjin University, Tianjin, China
zhangweilun@tju.edu.cn

Abstract. Unsupervised domain adaptation(UDA) has been applied to image semantic segmentation to solve the problem of domain offset. However, in some difficult categories with poor recognition accuracy, the segmentation effects are still not ideal. To this end, in this paper, Intra-subdomain adaptation adversarial learning segmentation method based on Dynamic Pseudo Labels(IDPL) is proposed. The whole process consists of 3 steps: Firstly, the instance-level pseudo label dynamic generation module is proposed, which fuses the class matching information in global classes and local instances, thus adaptively generating the optimal threshold for each class, obtaining high-quality pseudo labels. Secondly, the subdomain classifier module based on instance confidence is constructed, which can dynamically divide the target domain into easy and difficult subdomains according to the relative proportion of easy and difficult instances. Finally, the subdomain adversarial learning module based on self-attention is proposed. It uses multi-head self-attention to confront the easy and difficult subdomains at the class level with the help of generated high-quality pseudo labels, so as to focus on mining the features of difficult categories in the high-entropy region of target domain images, which promotes class-level conditional distribution alignment between the subdomains, improving the segmentation performance of difficult categories. For the difficult categories, the experimental results show that the performance of IDPL is significantly improved compared with other latest mainstream methods.

Keywords: unsupervised domain adaptation (UDA) · semantic segmentation · difficult category · dynamic pseudo labels · intra-subdomain adversarial learning

© The Author(s), under exclusive license to Springer Nature Switzerland AG 2023
M. Tanveer et al. (Eds.): ICONIP 2022, LNCS 13623, pp. 296–308, 2023.
https://doi.org/10.1007/978-3-031-30105-6_25

1   Introduction

The domain shift problem in UDA manifests itself as the inter-domain discrepancy problem between synthetic images and real images in semantic segmentation task, making difficult to generalize the model trained with synthetic data to real data. Previous UDA methods [1,2] have done a lot of works to reduce the domain shift, but there are still some problems with existing methods: First of all, most of the self-training methods based on pseudo labels select a fixed threshold to filter all pseudo labels with high confidence [3,4]. The model pay more attention to the conditional distribution alignment of the samples in easy categories between the two domains, resulting in the samples in difficult categories with high entropy may be discarded; Moreover, most methods adopt the strategy of global threshold [5,6], ignoring the separate consideration of different classes, which increases the risk of class imbalance in the generated pseudo labels.These problems lead to low accuracy of generated pseudo labels.

In addition, to alleviate the distribution gap between data within the target domain, the unsupervised intra-domain adaptation methods are proposed. For example, UIDA [7] considers the global feature map of samples during adversarial training between the easy and difficult subdomains. However, not all spatial regions maintain high entropy for difficult samples. Using global features for adversarial learning, the feature extraction of difficult samples in low-entropy regions may be affected, causing negative transfer. Therefore it is not reliable to use the global entropy of the predicted probability maps to divide easy/difficult subdomains. The results are shown in Fig. 1, for "difficult" categories (such as "pole" in purple boxes, "bike" in green boxes, and "sidewalk" in brown boxes), this method does not work well.

Inspired by UIDA [7], this paper proposes a more stable domain adaptation method to achieve intra-subdomain adversarial training, namely Intra-subdomain adaptation adversarial learning method based on Dynamic Pseudo Labels (IDPL). The method consists of 3 parts: Firstly, in order to improve the pseudo labels quality of intra-domain adversarial learning, the instance-level pseudo labels dynamic generation module is proposed. The threshold is dynamically adjusted for different semantic classes of each image, so that the model pays more attention to the high-entropy regions in the image; Then, the subdomain classifier module based on instance confidence is constructed to realize the hierarchical division from easy instances/difficult instances to easy subdomain/difficult subdomain; Finally, on the basis of the divided two subdomains, the subdomain adversarial learning module based on self-attention is constructed, and the multiple discriminator head structure is introduced to mine the class information contained in the high-entropy or entropy fluctuation regions, so as to more accurately narrow the intra-domain differences. As shown in Fig. 1, the pseudo labels of IDPL are significantly better than the comparison method, especially in these "difficult" categories.

The main contributions of this paper are summarized as follows:

(1) The instance-level pseudo labels dynamic generation module is proposed, which aims to dynamically adjust the global threshold of each class according

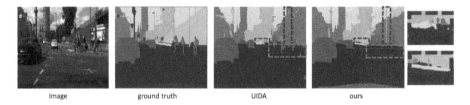

Fig. 1. Comparison of pseudo label results between IDPL and similar method. (Color figure online)

to the instance prediction of each class, fuse local and global information to guide the model to generate high-quality pseudo labels.

(2) The intra-subdomain adversarial learning module based on instance confidence is proposed. The confidence coefficient is constructed according to the relative proportion of easy and difficult instances, thus the target domain can be accurately divided into image-level easy and difficult subdomains; Guide the model to generate subdomain invariant features through intra-domain adversarial training at the semantic class level, so that the model pays more attention to the more difficult categories, and the influence of concentrated high-entropy regions on segmentation results can be alleviated more effectively.

2   Method

The overall framework is shown in Fig. 2. The method can be divided into 3 steps. Firstly, momentum is introduced for each instance to gradually update the threshold for each class, generating high-quality pseudo labels. Secondly, the confidence of each instance is calculated, and they are classified, then the target domain is split according to the relative proportion of the two types of instances. Thirdly, the self-attention heads are used to apply self-attention to each class of the two subdomains separately, and the self-attention maps are used as the weight of each class to conduct inter-class subdomain adversarial training.

2.1   Instance-level Pseudo Labels Dynamic Generation Module(PLDG)

First, the pre-trained segmentation model G is used for preliminary prediction of the target domain image, and the predicted probability maps $M_T = G(X_T)$ is obtained. Then, softmax is used to calculate the class probability prediction of each pixel position (h_t, w_t) in the image X_T, as shown in Eq. (1):

$$p_{i,h_t,w_t}^{x_t} = \frac{e^{m_t^{(h_t,w_t,i)}}}{\sum_{j=1}^{C} e^{m_t^{(h_t,w_t,j)}}}, m_t \in M_t \tag{1}$$

where i represents a specific semantic class, $i \in \{0, 1, \ldots, C\}$; The number with the largest value along the channel direction of a certain pixel position (h_t, w_t)

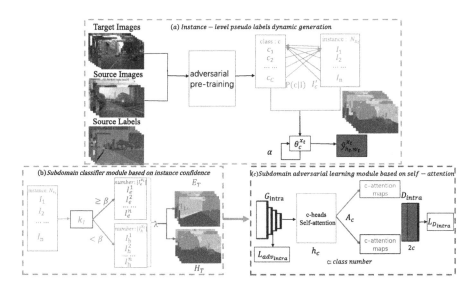

Fig. 2. The framework of the method proposed in this paper.

represents the class prediction probability of this pixel, and the corresponding channel subscript represents the predicted class. In the process of generating pseudo labels, the weight parameters of the generator network are fixed. The pseudo labels generation can be expressed by Eq. (2):

$$\min_{\hat{Y}_T} - \sum_{h_t, w_t} \sum_c \hat{y}_{h_t, w_t}^{x_t} \log \frac{p_{c, h_t, w_t}^{x_t}}{\theta_c^{x_t}} \tag{2}$$

where $\theta_c^{x_t}$ represents the threshold of the C-th class at the pixel position (h_t, w_t), and $\hat{y}_{h_t, w_t}^{x_t}$ indicates the pseudo labels of a vector of C classes.

Considering the problem of unbalanced sample number corresponding to different semantic classes, a threshold θ_c about class is introduced, which is used to provide a specific threshold for each of c different semantic classes in a batch, which is denoted as the global class thresholds. In order to generate high-quality pseudo labels map $\hat{y}_{h_t, w_t}^{x_t}$, this module aims to generate an adaptive threshold $\theta_c^{x_t}$ for each class c. The update process of the dynamic threshold of pseudo labels is described in detail below.

First, for the pseudo labels activation region $H_i = \left\{ (h, w) \mid \hat{y}_{h_t, w_t}^{x_t} = 1 \right\}, i \in \{0, 1, \ldots, C\}$ in a specific channel, the feature vector $F_i = \{f_{x,y} \mid (x, y) \in H_i\}$ of the corresponding region in the original feature map is selected. At this point, with the help of the clustering algorithm, the features of this region are separated into different instances, and the total number of instances sampled from a certain image sample x_t is denoted as N_{x_t}. Global average pooling is introduced to obtain the embedding vector of each different instance, denoted as I_n, where $n \in \{1, 2, \ldots, N_{x_t}\}$. Then, instances of all classes are more appropriately matched based on their similarity between the global and the local, in which all instances need to be dynamically classified. In addition, the method introduces a linear layer and Softmax activation function as a classifier for each instance feature I. The class prediction probability $p(c \mid I)$ of each instance is expressed by Eq. (3):

$$p(c \mid I) = \text{softmax}\left(W_c^T I\right) \tag{3}$$

where W_c is the trainable parameter of the class classifier. Then, a class sampling weight I_c' is set, and the sampling process is shown in Eq. (4):

$$I_c' = \frac{1}{h_t \times w_t} \sum_{h_t, w_t} p_{c, h_t, w_t}^{x_t} \times L_c, \forall x_t \in X_T \tag{4}$$

where L_c represents a set of several instance features, expressed as $L_c = \{I_n \mid n \in \{1, 2, \ldots, N_{x_t}\}\}$, $\times$ represents the sampling operation for matching and fusion.

Finally, when pseudo labels are generated, a global information is retained after iteration for each instance I; A local class threshold $\hat{\theta}_c^{x_t}$ is set for each image sample in the current batch, which is obtained by averaging the predicted classification of all instances in each class. The semantic class threshold at the image level can be obtained by fusing the local class threshold $\hat{\theta}_c^{x_t}$ and the global class thresholds θ_c initialized by the current batch in a certain proportion, thereby gradually generating high-quality pseudo labels. The process of threshold fusion is shown in Eq. (5):

$$\theta_c^{x_t} = \alpha \theta_c + (1 - \alpha)\hat{\theta}_c^{x_t} \tag{5}$$

α is the momentum factor used to hold the global class thresholds, with a range of (0,1). Within a certain range, as α gets larger, the dynamic update of $\theta_c^{x_t}$ becomes smoother and smoother. In this way, the global class thresholds and the class thresholds corresponding to the local instance are matched and fused to realize the adaptive matching between the global and local classes, so the dynamic iterative updating of the image-level class threshold is realized.

2.2 The Subdomain Classifier Module Based on Instance Confidence(SCIC)

In order to capture the high-entropy regions information and the difficult samples are accurately mined, this module can achieve more reliable image-level subdomain division of the target domain, thereby providing better guidance for reducing intra-domain differences.

Specifically, β is selected as the division threshold of the instances. Combined with the dynamic threshold $\theta_c^{x_t}$ of each class and the prediction probability $p(c \mid I)$ of each instance, the confidence of each instance can be calculated, as shown in Eq. (6):

$$k_I = p(c \mid I)\theta_c^{x_t} \tag{6}$$

Defining the set of easy instances as I_e and the set of difficult instances as I_h. When $k_I \geq \beta$, the instance is divided into I_e, otherwise into I_h.

λ is selected as the split ratio threshold of the target domain. Defining the easy subdomain as E_T and the difficult subdomain as H_T. Let $|I_e^{x_t}|$ and $|I_h^{x_t}|$ represent the number of instances of I_e and I_h in image x_t respectively. When $|I_e^{x_t}| / (|I_h^{x_t}| + |I_e^{x_t}|) \geq \lambda$, the image is divided into E_T; otherwise into H_T.

2.3 The Subdomain Adversarial Learning Module Based on Self-attention(SASA)

In order to reduce sample differences within the target domain, adversarial learning is introduced in this section to confuse the sample spatial distribution of the easy subdomain and the difficult subdomain. In addition, in order to further mine the correlation between pixel-level same semantic regions in target domain images, the self-attention mechanism is introduced for the adversarial learning of easy/difficult subdomains to train by class. Since the two subdomains share the same c semantic classes, the multi-head self-attention module introduces the self-attention mechanism for each class separately, achieve promoting the alignment of conditional distribution at the semantic class level.

The predicted probability maps obtained by the easy/difficult subdomains through the segmentation network are input into the self-attention module with c heads h_c, and the multi-head self-attention module generates c class self-attention maps $A_1, A_2, \cdots, A_C$ (each class corresponds to a self-attention map).

To focus on high-entropy regions, this paper improves the binary classification head of the traditional discriminator, and allocates a discriminator for each semantic class, namely c parallel binary discriminators, which forms a new class-level intra-domain discriminator D_{intra}.

The training purpose of D_{intra} is to distinguish whether the image belongs to the easy subdomain or the difficult subdomain, as shown in Eq. (7):

$$\min_{D_{intra}} L_{D_{\text{intra}}} (E_T, H_T) =$$
$$-\sum_c Q_{e,c}(1 - d) \log p (d = 0, c \mid f_e) - \sum_c Q_{h,c} d \log p (d = 1, c \mid f_h) \quad (7)$$

where $Q_{e,c}$ and $Q_{h,c}$ are the weights of class c of images in the "easy" subdomain and the "difficult" subdomain respectively, $Q \in R^{1 \times C}$. f_e and f_h represent the features extracted by F_{intra} from the images of two subdomains respectively, d represents the domain variable, where 0 represents the easy subdomain and 1 represents the difficult subdomain. $p(d \mid f)$ represents the probability of the discriminator D_{intra} output. Similarly, to obtain subdomain-invariant features, the discriminator D_{intra} is confused so that it cannot distinguish between easy and difficult subdomains. As shown in Eq. (8):

$$\min_{G_{intra}} L_{adv_{\text{intra}}} (H_T) = -\sum_c Q_{h,c} \log p (d = 0, c \mid f_T) \quad (8)$$

$L_{adv_{\text{intra}}} (H_T)$ aims to maximize the probability that the "difficult" subdomain feature is regarded as the "easy" subdomain feature without damaging the relationship between features and classes.

3 Experiment

3.1 Experimental Setup

Dataset: IDPL is evaluated on commonly used semantic segmentation tasks from synthetic domain to real domain: GTA5 [8] to Cityscapes [10] and SYN-THIA [9] to Cityscapes. The GTA5 dataset has 24,966 images from GTA5, and 19 classes with urban scenes. The SYNTHIA dataset includes 9400 images and 16 classes with urban scenes. The Cityscapes dataset is divided into training set, validation set and test set. Following the standard protocol in [1], using the training set of 2975 images as the target domain dataset and using the validation dataset to evaluate the model using IoU and mIoU.

Implementation Details: This paper uses PyTorch for training and inference on a single NVIDIA RTX TiTan/24GB. For fair comparison, DeepLabv2 [13] is used as the basic segmentation network, ResNet-101 [14] and VGG16 [15] as the backbone network. The training images are cropped randomly and resized to 1024×512. The initial learning rate is set to 2×10^{-4} and reduced according to the "poly" learning rate strategy with a power of 0.9. To train the discriminator, the Adam optimizer is used with $\beta_1 = 0.9$, $\beta_2 = 0.99$ and initial learning rate of 10^{-4}. Through experiments, the optimal hyperparameters obtained are set as $\alpha = 0.9$, $\beta = 0.6$, $\lambda = 0.7$.

3.2 Ablation Experiment

In order to understand the hyperparameter values and the impact of each module, we carry out ablation experiments using the ResNet-101 backbone in the GTA5 $\rightarrow$ Cityscapes task.

Module Ablation Experiment. Table 1 shows the results that verify the effectiveness of different modules in the method. Define PT as the model obtained by inter-domain adversarial pre-training using AdvEnt [2]. PT is the subsequent baseline model, mIoU=41.0%. PT+PLDG is 5.7% higher than PT. The method generates an adaptive semantic class threshold for each class by fusing the global and local information of each class instance by instance, achieving the clustering of the same class samples of inter-domain on the basis of edge alignment, which is feasible to improve the quality of pseudo labels. PT+PLDG+SASA, self-attention adversarial training is carried out by class in the two subdomains respectively, 7.5% higher than PT+PLDG. Subdomain adversarial training based on high-quality pseudo labels can focus on mining difficult categories in high-entropy regions, accurately guide the model to perform intra-domain adaptation in high-entropy and entropy fluctuation regions, and further improve the performance of the model.

Table 2 shows the results of ablation experiments by replacing each module in the framework with a module with similar function in other methods.

PLDG Module: This part verifies the pseudo labels generated by AdvEnt [2] proposed in UIDA [7], when these pseudo labels are used in the subsequent tasks of the proposed method, which is 4.4% lower than the final result of IDPL. This indicates that PLDG module sets a threshold for different classes respectively, and dynamically fine-tunes each semantic class threshold according to the confidence prediction of each instance with the cyclic training of the model, generating higher quality pseudo labels.

Table 1. Ablation experiments where each module is added individually.

PT	PLDG	SASA	mIoU	Δ
✓			41.0	–
✓	✓		46.7	+5.7
✓	✓	✓	**54.2**	+7.5

SCIC Module: This part randomly selects half of the images to be split into the "easy" subdomain and the rest into the "difficult" subdomain. The two subdomains are input into the subsequent module for experiments, which is 2.8% lower than the final result of ours. Therefore, this result indicates that selecting a fixed value as the threshold for domain separation cannot well perform targeted filtering for useful labels of each class. While SCIC module dynamically split the target domain into easy/difficult subdomains by using two hyperparameters for different datasets and tasks by considering the relative proportion of easy and difficult instances reasonably, implements hierarchical division of easy/difficult instances to easy/difficult subdomains, thus improve the model performance under different tasks. In particular, on the premise of not affecting the performance of "easy" categories, the goal of improving the segmentation accuracy of "difficult" categories is achieved by increasing the number of "difficult" categories as much as possible.

SASA Module: Compared with the improvement work of AdvEnt [2] in UIDA [7], that is, using the common generator and discriminator for intra-domain adversarial learning, which is 1.9% lower than the final result of ours. This part conducts intra-domain adversarial training for each class in the two subdomains with the help of the self-attention mechanism. The C self-attention maps generated by the multi-head self-attention mechanism assign different weights to different classes, which solves the problem that UIDA [7] ignores the class information contained in regions with high-entropy or entropy fluctuation. The final visualization results are shown in Fig. 3, compared with UIDA [7] and PLDG module, IDPL improves the quality of generated pseudo labels, especially for the "difficult" categories.(Such as "sign" and "light" in blue boxes in the first image; "sidewalk", "fence" and "mbike" in yellow boxes in the second image; "rider" and "bike" in brown boxes in the third image; "fence" in blue boxes in the fourth image.)

Model Parameter Analysis. In this section, sensitivity analysis will be conducted on the hyperparameters α, β, λ involved in Chap. 2.

α: In PLDG module, when the global iteration information momentum $\alpha = 0$, it means that there is only prior information for each class of the instance, and there is no global consideration by class. When iterating for each instance, with the increase of α, the weight of global information is gradually increasing, and the weight of local information is gradually decreasing. This causes the threshold for each class to update more and more slowly, as shown in Table 3. When $\alpha = 0.9$, on the basis of retaining the vast majority of the global information, the local information is slowly integrated, so as to better combine the global class information and the local instance information, and achieve the optimal threshold selection of each class at the image level. It has good adaptability to some difficult categories with few samples, and obtains more diversified information, which makes the model obtain the best performance.

Table 2. Ablation experiments of each module is replaced by a similar module.

Method	PT	PLDG	SCIC	SASA	mIoU	Δ
IDPL(ours)	✓	✓	✓	✓	**54.2**	–
General pseudo label	✓		✓	✓	49.8	−4.4
Random Select	✓	✓		✓	51.4	−2.8
UIDA [7]	✓	✓	✓		52.3	−1.9

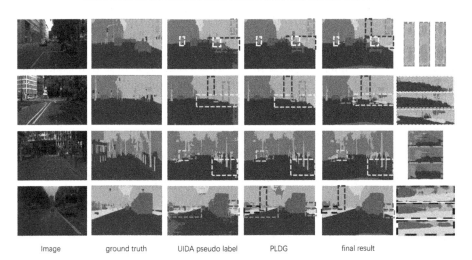

Image   ground truth   UIDA pseudo label   PLDG   final result

Fig. 3. Visualization results on the GTA5 → Cityscapes task. (Color figure online)

Table 3. The study of the hyperparameter α for preserving global information.

GTA5 ⟶ Cityscapes						
α	0	0.5	0.6	0.7	0.8	0.9	1.0
mIoU	41.7	47.7	49.3	50.8	53.2	**54.2**	53.8

β and λ are selected 10 values from $[0.1, 1.0]$ with a step size of 0.1, and the experimental results of 100 different combinations are presented in a three-dimensional stereogram. The intersection point is the mIoU of the corresponding combination. The darker the red is, the larger the value is.

β: When β is small, most of the instances are divided into "easy" instances set, and there are not enough "difficult" instances to train the model; As shown in Fig. 4, as β increases, the threshold for instances division into "easy" is increasing, the number of "difficult" instances is increasing, and the performance of the model is also increasing continuously; Until $\beta = 0.6$, the optimal relative allocation ratio of the two types of instances can be obtained, and when $\lambda = 0.7$, the best mIoU is reached; As β continues to increase, the proportion of "difficult" instances keeps increasing, resulting in the model overfitting to "difficult" instances. At the same time, the number of "easy" instances is decreasing, and there are not enough "easy" instances to train the model, resulting in the decline of mIoU .

λ: As λ increases, the number of images in the "easy" subdomain is gradually decreasing, more and more images are divided into the "difficult" subdomain. Since most of the classes in the "difficult" subdomain belong to "difficult" categories, the model performance is gradually improved; The optimal value is $\lambda = 0.7$, which realizes that the relatively difficult samples contained in the concentrated high-entropy region are divided into the "difficult" subdomain. At this point, the number of images in the "difficult" subdomain satisfies the training needs of the method for the "difficult" categories. As λ further increases, the number of images in the "difficult" subdomain is further increasing. Since most of the classes in the "difficult" subdomain are "difficult" categories, the model lacks a sufficient number of "easy" categories to make mIoU drop.

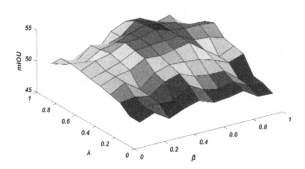

Fig. 4. The study of the hyperparameters β and λ.

3.3   Comparative Experiment

This section analyzes the quantitative results of IDPL and compares them with other mainstream methods. The results are shown in Tables 4 and 5. For fair comparison, all results are derived from single-scale inference.

IDPL outperforms other mainstream methods in more than half of the classes on both backbones for two tasks, and mIoU also reaches a high level, and mIoU also reaches a high level. Especially in the "difficult" categories such as "fence" and "pole", IDPL outperforms the others. On the one hand, these performance improvements show that the adaptive processing of classes is effective, and the dynamically generated pseudo labels can generate adaptive threshold for each class; On the other hand, when dividing the target domain, the method increases the number of images in the "difficult" subdomain as much as possible, and gives a higher weight to the "difficult" categories in the subdomain adversarial learning, adding more information of the "difficult" categories. Compared with the similar method UIDA [7] which adopts intra-domain adversarial learning, IDPL improved by 7.9% and 9.2% in mIoU in two tasks. In terms of specific classes, there is a significant improvement in almost all classes, especially in the difficult categories such as "fence" and "sign", the results of IDPL have been doubled. This proves that compared with some methods that only consider global feature alignment, IDPL considers the differences of different classes, realizes feature alignment at the class level, and solves the problem of class mismatch after domain alignment. Compared with CBST [3] and DAST [12], which also use the discriminator, IDPL assigns higher weights to difficult categories through self-attention mechanism at the class level and multiple processing for difficult categories, thus achieving a great performance improvement.

Table 4. IoU (%) comparison for each class on GTA5 → Cityscapes adaptation, evaluated on Cityscapes validation set.

Method	road	sidewalk	building	wall	fence	pole	light	sign	veg	terrain	sky	person	rider	car	truck	bus	train	mbike	bike	mIoU
VGG-16																				
Source Only	60.7	13.7	56.9	12.9	20.1	19.0	15.4	6.5	77.7	16.2	56.8	40.0	3.3	63.6	15.3	9.5	0.0	8.1	0.1	26.1
CBST [3]	90.4	**50.8**	72.0	18.3	9.5	27.2	28.6	14.1	82.4	25.1	70.8	42.6	14.5	76.9	5.9	12.5	1.2	14.0	**28.6**	36.1
AdvEnt [2]	86.9	28.7	78.7	28.5	25.2	17.1	20.3	10.9	80.0	26.4	70.2	47.1	8.4	81.5	26.0	17.2	18.9	11.7	1.6	36.1
FDA [16]	86.1	35.1	80.6	30.8	20.4	27.5	30.0	26.0	82.1	30.3	73.6	52.5	21.7	81.7	24.0	30.5	**29.9**	14.6	24.0	42.2
DAST [12]	90.5	49.2	**81.9**	34.0	27.0	26.5	26.6	21.5	83.0	**37.3**	76.3	52.0	23.1	**83.5**	29.9	42.0	12.1	19.8	25.8	44.3
IDPL(ours)	**90.7**	50.5	**81.9**	**35.8**	**31.0**	**30.9**	**32.1**	**29.8**	**83.5**	36.4	**81.1**	**55.4**	**28.3**	82.2	**31.2**	**42.6**	19.4	**22.1**	26.6	**46.9**
ResNet-101																				
Source-Only	75.8	16.8	77.2	12.5	21.0	25.5	30.1	20.1	81.3	24.6	70.3	53.8	26.4	49.9	17.2	25.9	6.5	25.3	36.0	36.6
AdvEnt [2]	89.4	33.1	81.0	26.6	26.8	27.2	33.5	24.7	83.9	36.7	78.8	58.7	30.5	84.8	38.5	44.5	1.7	31.6	32.4	45.5
CBST [3]	91.8	53.5	80.5	32.7	21.0	34.0	28.9	20.4	83.9	34.2	80.9	53.1	24.0	82.7	30.3	35.9	16.0	25.9	42.8	45.9
UIDA [7]	90.6	37.1	82.6	30.1	19.1	29.5	32.4	20.6	85.7	40.5	79.7	58.7	31.1	86.3	31.5	48.3	0.0	30.2	35.8	46.3
SUDA [20]	91.1	52.3	82.9	30.1	25.7	38.0	44.9	38.2	83.9	39.1	79.2	58.4	26.4	84.5	37.7	45.6	10.1	23.1	36.0	48.8
CaCo [19]	91.9	54.3	82.7	31.7	25.0	38.1	46.7	39.2	82.6	39.7	76.2	63.5	23.6	85.1	38.6	47.8	10.3	23.4	35.1	49.2
DAST [12]	92.2	49.0	84.3	36.5	28.9	33.9	38.8	28.4	84.9	41.6	83.2	60.0	28.7	87.2	**45.0**	45.3	7.4	33.8	32.8	49.6
RPLL [11]	90.4	31.2	85.1	36.9	25.6	37.5	**48.8**	**48.5**	85.3	34.8	81.1	64.4	**36.8**	86.3	34.9	52.2	1.7	29.0	44.6	50.3
PixMatch [17]	91.6	51.2	84.7	37.3	29.1	24.6	31.3	37.2	86.5	**44.3**	85.3	62.8	22.6	**87.6**	38.9	**52.3**	0.65	**37.2**	**50.0**	50.3
IDPL(ours)	**93.4**	**55.6**	**85.3**	**39.2**	**40.3**	**40.1**	41.7	41.2	**87.0**	42.3	**87.8**	**67.8**	33.1	85.1	42.2	52.2	**22.8**	33.1	40.6	**54.2**

Table 5. IoU (%) comparison for each class on SYNTHIA → Cityscapes adaptation, evaluated on Cityscapes validation set.

Method	road	sidewalk	building	wall*	fence*	pole*	light	sign	veg	sky	person	rider	car	bus	mbike	bike	mIoU	mIoU*
VGG-16																		
Source Only	4.7	11.6	62.3	10.7	0.0	22.8	4.3	15.3	68.0	70.8	49.7	6.4	60.5	11.8	2.6	4.3	25.4	28.7
CBST [3]	69.6	28.7	69.5	**12.1**	0.1	25.4	11.9	13.6	**82.0**	81.9	49.1	14.5	66.0	6.6	3.7	32.4	35.4	36.1
AdvEnt [2]	67.9	29.4	71.9	6.3	0.3	19.9	0.6	2.6	74.9	74.9	35.4	9.6	67.8	21.4	4.1	15.5	31.4	36.6
PyCDA [6]	80.6	26.6	74.5	2.0	0.1	18.1	**13.7**	14.2	80.8	71.0	48.0	19.0	72.3	22.5	12.1	18.1	35.9	42.6
DAST [12]	**86.1**	35.7	79.9	5.2	0.8	23.1	0.0	6.9	80.9	82.5	50.6	19.8	**79.7**	21.9	21.3	38.8	39.6	46.5
FDA [16]	84.2	35.1	78.0	6.1	0.44	27.0	8.5	**22.1**	77.2	79.6	**55.5**	19.9	74.8	24.9	14.3	**40.7**	40.5	-
IDPL(ours)	83.2	**37.4**	**80.1**	11.2	**0.83**	**27.9**	7.5	19.2	79.3	**82.8**	54.4	**21.6**	76.6	**31.4**	**22.8**	36.8	**42.1**	**48.7**
ResNet-101																		
Source-Only	36.30	14.64	68.78	9.17	0.20	24.39	5.59	9.05	68.96	79.38	52.45	11.34	49.77	9.53	11.03	20.66	29.45	33.65
AdvEnt [2]	87.0	44.1	79.7	9.6	0.6	24.3	4.8	7.2	80.1	83.6	56.4	23.7	72.7	32.6	12.8	33.7	40.8	47.6
UIDA [7]	84.3	37.7	79.5	5.3	0.4	24.9	9.2	8.4	80.0	84.1	57.2	23.0	78.0	38.1	20.3	36.5	41.7	48.9
CBST [3]	68.0	29.9	76.3	10.8	1.4	33.9	22.8	29.5	77.6	78.3	60.6	28.3	81.6	23.5	18.8	39.8	42.6	48.9
SUDA [20]	83.4	36.0	71.3	8.7	0.1	26.0	18.2	26.7	72.4	80.2	58.4	30.8	80.6	38.7	36.1	46.1	44.6	52.2
DAST [12]	87.1	44.5	82.3	10.7	0.8	29.9	13.9	13.1	81.6	**86.0**	60.3	25.1	83.1	40.1	24.4	40.5	45.2	52.5
CaCo [19]	87.4	48.9	79.6	8.8	0.2	30.1	17.4	28.3	79.9	81.2	56.3	24.2	78.6	39.2	28.1	48.3	46.0	53.6
PixMatch [17]	**92.5**	**54.6**	79.8	4.78	0.08	24.1	22.8	17.8	79.4	76.5	60.8	24.7	**85.7**	33.5	26.4	**54.4**	46.1	54.5
CVRN [18]	87.5	45.5	83.5	12.2	0.5	37.4	25.1	29.6	**85.9**	**86.0**	61.1	25.9	80.9	34.7	33.8	53.5	49.0	56.6
IDPL(ours)	85.1	42.2	**83.6**	**22.7**	**3.8**	**37.9**	**30.7**	**34.6**	80.3	85.5	**62.1**	**35.2**	85.4	**41.1**	**38.6**	51.5	**51.3**	**58.1**

4 Conclusion

In this paper, Intra-subdomain adaptation adversarial learning segmentation method based on Dynamic Pseudo Labels(IDPL) is proposed to improve the segmentation effect of difficult categories. Firstly, each instance is considered by iteration, and the threshold of each class is adjusted adaptively according to the class information in the global classes and local instances, thereby generating high-quality pseudo labels; Secondly, the target domain is dynamically split into two subdomains according to the relative proportion of instances; Finally, the class self-attention mechanism is applied between the two subdomains to enable intra-domain adaptive feature alignment at the class level with the help of high-quality pseudo labels. The overall performance of the proposed method outperforms other mainstream methods by significantly improving the performance of the difficult categories.

Acknowledgements. This work is supported by Tianjin Technical Export Project (Grant No. 21YDTPJC00090).

References

1. Tsai, Y.H., Hung, W.C., Schulter, S. et al.: Learning to adapt structured output space for semantic segmentation. In: CVPR, pp. 7472–7481 (2018)
2. Vu, T.H., Jain, H., Bucher, M. et al.: ADVENT: adversarial entropy minimization for domain adaptation in semantic segmentation. In: CVPR, pp. 2517–2526 (2019)
3. Zou, Y., Yu, Z., Vijaya Kumar, B.V.K., Wang, J.: Unsupervised domain adaptation for semantic segmentation via class-balanced self-training. In: Ferrari, V., Hebert, M., Sminchisescu, C., Weiss, Y. (eds.) ECCV 2018. LNCS, vol. 11207, pp. 297–313. Springer, Cham (2018). https://doi.org/10.1007/978-3-030-01219-9_18

4. Zou, Y., Yu, Z., Liu, X. et al.: Confidence Regularized Self-Training. In: ICCV, pp. 5982–5991 (2019)
5. Zheng, Z., Yang, Y.: Unsupervised scene adaptation with memory regularization in vivo. arXiv preprint arXiv:1912.11164 (2019)
6. Lian, Q., Lv̇, F., Duan, L.: Constructing self-motivated pyramid curriculums for cross-domain semantic segmentation: a non-adversarial approach. In: ICCV, pp. 6758–6767 (2019)
7. Pan, F., Shin, I., Rameau, F. et al.: Unsupervised intra-domain adaptation for semantic segmentation through self-supervision. In: CVPR, pp. 3764–3773 (2020)
8. Richter, S.R., Vineet, V., Roth, S., Koltun, V.: Playing for data: ground truth from computer games. In: Leibe, B., Matas, J., Sebe, N., Welling, M. (eds.) ECCV 2016. LNCS, vol. 9906, pp. 102–118. Springer, Cham (2016). https://doi.org/10. 1007/978-3-319-46475-6_7
9. Ros, G., Sellart, L., Materzynska, J. et al.: The synthia dataset: a large collection of synthetic images for semantic segmentation of urban scenes. In: CVPR, pp. 3234–3243 (2016)
10. Cordts, M., Omran, M., Ramos, S. et al.: The cityscapes dataset for semantic urban scene understanding.In: CVPR, pp. 3213–3223 (2016)
11. Zheng, Z., Yang, Y.: Rectifying pseudo label learning via uncertainty estimation for domain adaptive semantic segmentation. IJCV **129**(4), 1106–1120 (2021)
12. Yu, F., Zhang, M., Dong, H. et al.: DAST: unsupervised domain adaptation in semantic segmentation based on discriminator attention and self-training. In: AAAI, p. 10 (2021)
13. Chen, L.C., Papandreou, G., Kokkinos, I., et al.: DeepLab: semantic image segmentation with deep convolutional nets, atrous convolution, and fully connected CRFs. TPAMI **40**(4), 834–848 (2017)
14. He, K., Zhang, X., Ren, S. et al.: Deep residual learning for image recognition. In: CVPR, pp. 770–778 (2016)
15. Simonyan, K., Zisserman, A.: Very deep convolutional networks for large-scale image recognition. arXiv preprint arXiv:1409.1556 (2014)
16. Yang, Y., Soatto, S.: FDA: fourier domain adaptation for semantic segmentation. In: CVPR, pp. 4085–4095 (2020)
17. Melas-Kyriazi, L., Manrai, A.: PixMatch: unsupervised domain adaptation via pixelwise consistency training. In: CVPR, pp. 12435–12445 (2021)
18. Huang, J., Guan, D., Xiao, A. et al.: Cross-view regularization for domain adaptive panoptic segmentation. In: CVPR, pp. 10133–10144 (2021)
19. Huang, J., Guan, D., Xiao, A. et. al.: Category contrast for unsupervised domain adaptation in visual tasks. In: CVPR, pp. 1203–1214 (2022)
20. Zhang, J., Huang, J., Tian, Z. et al.: Spectral unsupervised domain adaptation for visual recognition. In: CVPR, pp. 9829–9840 (2022)

Adaptive Scaling for U-Net in Time Series Classification

Wen Xin Cheng[1]([✉])[ID] and Ponnuthurai Nagaratnam Suganthan[2][ID]

[1] School of Electrical and Electronic Engineering, Nanyang Technological University,
50 Nanyang Avenue, Singapore 639798, Singapore
wenxin001@ntu.edu.sg
[2] KINDI Center for Computing Research, College of Engineering, Qatar University,
Doha, Qatar
p.n.suganthan@qu.edu.qa

Abstract. Convolutional Neural Networks such as U-Net are recently getting popular among researchers in many applications, such as Biomedical Image Segmentation. U-Net is one of the popular deep Convolutional Neural Networks which first contracts the input image using pooling layers and then upscales the feature maps before classifying them. In this paper, we explore the performance of adaptive scaling for U-Net in time series classification. Also, to improve performance, we extract features from the trained U-Net model and use ensemble deep Random Vector Functional Link (edRVFL) to classify them. Experiments on 55 large UCR datasets reveal that adaptive scaling improves the performance of U-Net in time series classification. Also, using edRVFL on extracted features from the trained U-Net model enhances performance. Consequently, our U-Net-edRVFL classifier outperforms other time series classification methods.

Keywords: Ensemble Deep Random Vector Functional Link · Time Series Classification · U-Net

1 Introduction

Time series classification is one of the commonly-faced problems faced by researchers. Time series can be referred to as sequences that need to be presented in the exact order [9]. Any rearrangements of the values made within the time series will lose vital information required for classification. Time series problems have been explored in many different applications, including human activity recognition [35] and electronic health records [27].

Researchers have proposed many different types of Time Series Classification methods. Some of the methods search for the series that matches the input time series and assign the same label as the selected series in the database. These methods include Euclidean Distance (ED), Dynamic Time Warping (DTW) and Derivative DTW [12].

M. Tanveer et al. (Eds.): ICONIP 2022, LNCS 13623, pp. 309–320, 2023.
https://doi.org/10.1007/978-3-031-30105-6_26

Other algorithms extract useful features from time series before classifying them. Some methods such as Time Series Forest (TSF) [7] searches for a segment of the time series that contains the most useful information and compares them. Shapelet Transform (ST) [2] classify samples based on the presence of trained patterns found in time series. Other methods such as Bag of Symbolic Fourier Approximation (SFA) Symbols (BOSS) [29] classify samples based on the frequency of trained patterns. Some methods create alternate representations of a time series. In [18], time series are represented using Symbolic Aggregate Approximations to train neural networks. In [14], Minimal Description Length is proposed to optimize such methods to improve classification accuracy.

Ensembles are also getting popular among researchers. Elastic Ensemble (EE) [20] combines 11 different distance-based methods for time series problems. Shapelet Transform (ST) classifies samples using a set of classifiers on extracted features [2]. A more complex Collective Of Transformation Ensemble (COTE) [1] combines different feature extraction techniques and classifiers to create a large ensemble of classifiers. Researchers have proposed many different methods to improve performance, which include bagging, boosting, implicit/explicit ensembles and heterogeneous ensembles [10].

Deep Neural Networks are also getting popular among researchers due to their success in image-related tasks and time series problems. The performances of Fully Convolutional Neural Networks (FCN) and Residual Networks (ResNet) on time series classification problems have been explored in [36]. A more advanced Echo Memory Networks combines Echo State Networks and Convolutional Neural Networks to solve time series problems [22]. Some simpler methods require less computational effort. Some methods such as ROCKET [6] use randomized kernels for time series classification.

U-Net has been extensively explored in a variety of applications such as medical image segmentation [25], image synthesis [8] and concrete crack detection [21]. However, the performance of U-Net on time series classification has yet to be extensively explored. Therefore, we explore the use of U-Net on time series. We also introduce adaptive scaling to improve U-Net's performance. Finally, we explore the suitability of extracting features from U-Net by extracting features from trained U-Net models for ensemble deep Random Vector Functional Link (edRVFL).

The outline of the remaining sections is given in the next sentences. Section 2 describe U-Net, followed by Random Vector Functional Link (RVFL) and ensemble deep RVFL (edRVFL). Section 3 details proposed enhancements for U-Net in time series problems. Section 4 describes the dataset and experiment setup used in this work, followed by results and discussions. Finally, Sect. 5 concludes the paper.

2   Related Works

In this section, we briefly describe U-Net, followed by Random Vector Functional Link (RVFL) and ensemble deep RVFL (edRVFL).

2.1   U-Net

U-Net has shown excellent performance on Biomedical Images Segmentation Tasks [28]. In that work, the classifiers have to output a segmentation map of the input image. Each pixel of the segmentation map corresponds to a class.

For time series classification, we use one-dimensional convolutions are used to extract local patterns in time series. We construct U-Net as shown in Fig. 1. The network is made up of 9 double convolutional blocks and can be divided into two parts: the contracting part (left side) and the expensive part (right side). In each block, the input will undergo 2 convolution operations with kernels of length 3. Each convolutional operation will be followed by Batch Normalization [15] and Rectified Linear Units (ReLU) [24] operations. For this work, we apply zero-padding to maintain the length of the time series.

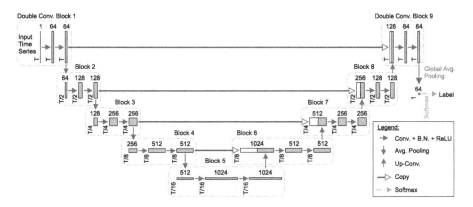

Fig. 1. Network structure of U-Net. B.N. refers to batch normalization. Each box represents a feature map. The number above the boxes indicates the number of maps. The expressions at the bottom left of each box indicates the relative length of the feature map.

In the contracting part (Blocks 1 to 4), each block is followed by an average pooling operation that halves the length of the time series. Here, we use average pooling instead of max pooling since we get better results during our experiments.

In the expanding part (Blocks 5 to 9), each block is followed by a transposed convolution layer that doubles the sequence length. The up-scaled features may be scaled down to match the length of corresponding output feature maps in the contracting part (See Fig. 1 for the connections). Finally, the features are concatenated with the corresponding output feature maps in the contracting part.

At the end of the network, we use global average pooling [19] to drastically reduce the feature size before feeding it into a fully connected layer (or Softmax Layer). This minimizes the number of parameters to train.

2.2   Random Vector Functional Link (RVFL)

Random Vector Functional Link (RVFL) is one of the popular algorithms that requires less computational effort than non-randomized neural networks. RVFL has shown exceptional performance for many tasks such as classification [16], regression [34] and forecasting [5]. Many types of RVFL have been proposed, ranging from shallow RVFL, deep RVFL, ensemble deep RVFL and ensembles involving bagging, boosting and stacking [23].

Random Vector Functional Link (RVFL) [26] is a randomized Feedforward Neural Network where input weights to the hidden layer are fixed during training. The basic version of RVFL comprises a randomized hidden layer and an output layer. In RVFL, direct links allow original features to propagate to the output layer for classification.

During training, the weights to the hidden layers are randomized and fixed. Then, randomized features are created by passing inputs to the hidden layer. The randomized features are concatenated with the original input features. Finally, the output layer takes the concatenated features for classification.

Output weights can be calculated using either Moore-Penrose pseudoinverse (given by $\beta = D^+Y$) if $\lambda_r = 0$ or ridge regression otherwise. Solution from Ridge regression is given in Eq. 1 and Eq. 2.

$$\text{Primal Space:}\quad \beta = (\mathbf{D}^T\mathbf{D} - \lambda_r\mathbf{I})^{-1}\mathbf{D}^T Y \qquad (1)$$

$$\text{Dual Space:}\quad \beta_s = \mathbf{D}^T(\mathbf{D}\mathbf{D}^T + \lambda_r\mathbf{I})^{-1}Y \qquad (2)$$

where $\mathbf{D} = [\mathbf{HX}]$ is the concatenation of hidden layer outputs $\mathbf{H}$ and original features $\mathbf{X}$ and Y is the target output. λ_r serves as a regularization hyperparameter to prevent overfitting.

Closed-form solutions can be computed easily as compared to back-propagation used in non-randomized neural networks [33].

2.3   Ensemble Deep Random Vector Functional Link (edRVFL)

Inspired by the success of RVFL and deep neural networks, researchers have extended the original RVFL into deep versions. In deep RVFL, multiple hidden layers are stacked back-to-back. All input weights to the hidden layers are randomized during training.

One popular variant is ensemble deep RVFL (edRVFL) [31]. edRVFL has also been used in short-term load forecasting [11]. In edRVFL, an ensemble is created using several output layers that are connected to different parts of the deep randomized neural networks. For an edRVFL of L hidden layers, there are L output layers. Each output layer in edRVFL is connected to the end of its corresponding hidden layer. In addition, there are direct links that connect to every output layer and deeper layers of edRVFL. Every output layer is trained independently of one another. Like shallow RVFL, Output weights can be estimated using closed-form solutions. Each output layer gives probability scores for every class. The final output can be obtained by either voting or averaging their probabilities. In this work, we average their probabilities.

More advanced variants includes Diversified edRVFL which includes multiple enhancements such as feature selection for direct links [4], Weighted edRVFL which combines adaptive boosting and edRVFL [30], Pruning-based edRVFL which removes redundant inputs in deeper layers [30], Jointly Optimized edRVFL and Semi-Supervised edRVFL for semi-supervised classification problems [32].

While edRVFL is powerful in classification tasks, edRVFL is not effective in extracting features in time series. Hence, we use U-Net as a powerful time series feature extractor for edRVFL to compensate for its shortcomings.

3   Proposed Enhancements

In this section, we outline adaptive scaling for U-Net and feature extraction from U-Net for training with edRVFL.

3.1   Adaptive Scaling for U-Net

To help U-Net to adapt to different lengths of time series in different datasets, we introduce adaptive scaling for U-Net. The scaling down operation is done using adaptive average pooling operation. Here, we use adaptive average pooling instead of adaptive max pooling since we get better results during our experiments. In adaptive average pooling, the algorithm will automatically compute the scaling factor and stride size such that the output length of the time series is given in Eq. 3 below:

$$T_{i+1} = T_i \times s \qquad (3)$$

where T_{i+1} is the target length of the output time series and T_i is the length of the input time series.

For our experiments, we use $s \in \{0.2, 0.25, 1/3, 0.4, 0.5, 0.6, 0.7, 0.8, 0.9, 0.95\}$. If s is set to 0.5, this operation becomes an average pooling operation with a neighbourhood size of 2 with strides of 2.

During U-Net's upscaling process, we use transposed convolutions with the following kernel size:

$$l_{uk} = \begin{cases} 1/s, & \text{if } s \in \{0.2, 0.25, 0.5\} \\ 3, & \text{if } s = 0.4 \\ 2, & \text{otherwise} \end{cases} \qquad (4)$$

where l_{uk} is the kernel length of the transposed convolutions and s is the selected scale. l_{uk} is always rounded up to the nearest integer.

The up-scaling process will expand the input time series by a factor of l_{uk}, which is always rounded up to the nearest integer. As such, the output feature maps can be longer than the corresponding output feature maps in the contracting part (See Fig. 1 for the connections). To prevent errors during copy and concatenate operations, we apply another adaptive average pooling layer to downscale the output feature maps to match the corresponding output feature maps in the contracting part.

3.2 Extracting Features from U-Net

In addition to U-Net, we also extract features from U-Net and use an advanced classifier to perform predictions. In this work, we use the simplest edRVFL [31].

For this method, we first train U-Net from scratch. Once U-Net is tuned and trained, we proceed to extract features from U-Net. Here, pass the input time series through the network and extract output features from every block (the features are taken before the pooling or up-scaling operation). This will give us 9 sets of feature maps. Then, for each set, we perform global average pooling to reduce the feature size. This is done so that we can minimize computation costs and overfitting. Finally, for each feature set, we train an edRVFL. This will give us an ensemble of 9 edRVFLs.

Each edRVFL in the ensemble will be tuned and trained independently. Once all edRVFLs are trained, we pass the extracted features through edRVFLs and the probability scores will be averaged. The class with the highest average probability score will be selected as the predicted class.

4   Experiments

In this section, we describe the datasets used in this study and the experiment setup. Next, the performance of adaptive scaling for U-Net will be discussed. Finally, we will compare our base U-Net, adaptive scaling for U-Net and edRVFL trained on extracted U-Net with other state-of-the-art time series methods.

4.1   Datasets

For this work, we use 55 UCR time series classification datasets found in the study of Echo Memory Networks [22]. We use the same training and testing splits used in [22] to ensure a fair comparison. Here, we select 55 datasets that have at least 100 training samples and 100 testing samples to ensure that there are sufficient samples to train U-Net. For more information on UCR time series classification datasets, refer to the data repository in [3].

For this work, we exclude datasets with small sample sizes (less than 100 training/testing samples) to avoid overfitting U-Net.

4.2   Experiment Setup

We use the same experimental setup as [22] for the evaluation of U-Net and edRVFL. In this work, we use the same training and testing splits. Classifiers are trained using only the data found in the allocated training set. The trained classifier models will then be used to predict samples from the allocated testing set. The performance of classifiers will be evaluated based on classification accuracy.

For tuning, we perform 4-fold cross-validation only on the training set. The validation folds are kept fixed during the entire tuning process. The best U-Net and edRVFL models are selected based on the best average classification accuracy across validation folds.

In U-Net, we only tune adaptive scaling with a wide range of scale $s \in \{0.2, 0.25, 1/3, 0.4, 0.5, 0.6, 0.7, 0.8, 0.9, 0.95\}$. The rest of the hyperparameters are fixed: fixed kernel length of 3, the number of feature maps is given in Fig. 1, batch size of 32 samples, Adam optimizer [17] with default parameters and a maximum of 500 epochs with early stopping based on training accuracy. To speed up tuning, we stop training when validation accuracy stops increasing. Once tuning is complete, we use the best-trained models are used to generate four sets of validation datasets for edRVFL (each validation dataset is generated for the corresponding training/testing fold).

Finally, we tune each edRVFL separately using the allocated feature sets. In edRVFL, we use the validation datasets generated during the U-Net tuning process. The number of neurons is tuned between $N \in \{K/2, K, 2K\}$, where K is the number of input features (also equivalent to the number of kernels in the corresponding U-Net convolutional layer). Regularization hyperparameter is tuned between $\lambda_r = 1/C$, where $C = 2^x$ and $x \in \{-8, -7, -6, ..., 17\}$. Number of layers are fixed at $L = 20$.

4.3   Performance of Adaptive Scaling

In this section, we report the performance of U-Net with adaptive scaling. In this experiment, only U-Net is trained without any hyperparameter tuning. The range of scales used in the downscaling operation is $s \in \{0.2, 0.25, 1/3, 0.5, 0.6, 0.7, 0.8, 0.9, 0.95\}$. We use $s = 0.5$ as a baseline since it is the closest implementation of our U-Net on time series arbitrary lengths.

We divide the datasets into 3 groups: Datasets with short time series ($T < 200$ points), datasets with medium time series ($200 \leq T < 500$ points), datasets with long time series ($T \geq 500$ points). For short and medium datasets, we use scales of $s \in \{0.5, 0.6, 0.7, 0.8, 0.9, 0.95\}$. For medium time series, we add $s = 0.4$ into the comparisons. For long time series, we add $s = 1/3$ and $s = 0.4$ into the comparisons. The results are shown in Table 1.

Based on the results, we observe that $s = 0.95$ is the best scale for short time series and $s = 0.5$ is the best scale for medium and long time series. Scale affects U-Net the most on medium time series. Although the extent of scaling down operations affects performance, we cannot infer the trend confidently.

Table 1. Adaptive U-Net Performance. Best values are marked in bold.

Scale	1/3	0.4	0.5	0.6	0.7	0.8	0.9	0.95
Short Time Series ($T < 200$ points)								
Avg. Acc. (%)			85.0	84.9	84.4	84.7	85.1	**85.4**
Avg. Rank			3.74	3.39	3.74	3.35	3.48	**3.30**
Medium Time Series ($200 \leq T < 500$ points)								
Avg. Acc. (%)		**83.2**	83.0	81.8	78.4	76.7	75.8	76.1
Avg. Rank		**2.10**	2.13	2.83	4.73	4.97	5.70	5.53
Long Time Series ($T \geq 500$ points)								
Avg. Acc. (%)	77.4	**77.5**	77.3	77.2	76.5	77.0	75.1	76.6
Avg. Rank	4.53	3.94	**3.85**	4.26	4.65	4.50	5.29	4.97

Setting the downscale hyperparameter to 0.5 generally works well in many datasets. Therefore, we proceed to tune s on every dataset. Table 2 shows the results.

Table 2. Test accuracy (in %) of U-Net with adaptive scaling.

Scale	0.5	0.6	0.7	0.8	0.9	0.95	Tuned
Avg. Acc.	82.09	81.67	80.33	80.15	79.48	80.15	**82.27**
Avg. Rank	3.40	3.64	4.46	4.21	4.76	4.53	**3.00**

Based on Table 2, our tuned configurations outperforms all other fixed configurations. Our tuned adaptive U-Net achieve the highest mean accuracy of 82.27% and the best average rank of 3.00. The baseline U-Net performs reasonably well, achieving a mean accuracy of 82.09% and an average rank of 3.40.

4.4   Benchmarking our Performance Against Other Classifiers

In order to verify the performance of our algorithm against other state-of-the-art methods, we compare our performance against the following time series classification methods: Bag of SFA Symbols (BOSS) [29], Shapelet Transform (ST) [2,13], Collection of Transformation Ensembles (COTE) [1], Fully Connected Networks (FCN) [36], Residual Networks [36], Echo Memory Networks (EMN) [22] and ResNet-Diverse edRVFL (R-DedRVFL) [4].

For our U-Net methods, we select the base U-Net. Also, we tune the scale for our U-Net with adaptive scaling (A. U-Net) to see how adaptive scaling helps in improving performance. Finally, we extracted the features from Adapt. U-Net and train an ensemble of edRVFL (AU-edRVFL) to see if we can further improve the overall performance. Table 3 shows the results.

Based on the results, we can tell that our U-Net with adaptive scaling improves performance. U-Net with a fixed scale of 0.5 scores 3rd best accuracy

Table 3. Test accuracy (in %) and average ranks among other time series classification methods. Bold values indicate best accuracy/rank. R-DedRVFL refers to ResNet-Diverse edRVFL. A. U-Net refers to U-Net with adaptive scaling. AU-edRVFL refers to U-Net-edRVFL.

Dataset	BOSS	ST	COTE	FCN	EMN	ResNet	R-DedRVFL	U-Net	A. U-Net	AU-edRVFL
Adiac	76.5	78.3	79.0	**85.7**	82.9	82.6	83.1	82.9	84.4	85.2
Chlorine	66.1	70.0	72.7	84.3	84.5	82.8	84.8	85.2	**85.5**	82.8
Computers	75.6	73.6	74.0	**84.8**	71.6	82.4	83.6	81.6	78.0	81.2
CricketX	73.6	77.2	80.8	81.5	78.2	82.1	80.0	81.8	**83.1**	79.5
CricketY	75.4	77.9	82.6	79.2	78.7	80.5	83.6	82.1	81.3	**84.6**
CricketZ	74.6	78.7	81.5	81.3	80.8	81.3	**84.4**	83.8	82.6	84.1
DistPhxAgeGp	74.8	77.0	74.8	83.5	**84.3**	79.8	81.8	76.2	81.2	84.0
DistPhxCorr	72.8	77.5	76.1	81.2	82.2	82.0	82.2	81.2	78.8	**83.2**
DistPhxTW	67.6	66.2	69.8	79.0	**79.5**	74.0	76.0	75.5	77.7	77.0
Earthquakes	74.8	74.1	74.8	80.1	81.1	78.6	81.1	**82.6**	82.0	82.3
ECG200	87.0	83.0	88.0	90.0	92.0	87.0	91.0	92.0	88.0	**93.0**
ECG5000	94.1	94.4	**94.6**	94.1	94.4	93.1	94.0	93.0	94.2	94.3
ElectricDevices	**79.9**	74.7	71.3	72.3	71.6	72.8	73.4	72.9	74.1	74.4
FaceAll	78.2	77.9	91.8	**92.9**	90.3	83.4	80.4	82.4	79.9	86.9
FacesUCR	95.7	90.6	94.2	94.8	94.7	**95.8**	95.4	95.4	94.2	95.1
FiftyWords	70.5	70.5	**79.8**	67.9	75.8	72.7	74.3	76.7	73.2	76.9
Fish	**98.9**	**98.9**	98.3	97.1	94.4	**98.9**	98.3	97.1	97.1	98.3
FordA	93.0	**97.1**	95.7	90.6	93.2	92.8	93.6	93.6	93.5	93.9
FordB	71.1	80.7	80.4	88.3	90.8	90.0	92.7	92.3	92.1	**92.8**
Ham	66.7	68.6	64.8	76.2	**78.1**	**78.1**	74.3	71.4	72.4	74.3
HandOutlines	90.3	**93.2**	91.9	77.6	89.1	86.1	85.5	87.5	92.0	90.0
Haptics	46.1	52.3	52.3	55.1	51.9	50.6	**55.2**	51.6	46.8	53.6
InlineSkate	**51.6**	37.3	49.5	41.1	46.0	36.5	35.6	45.6	42.2	49.8
InsWngSound	52.3	62.7	**65.3**	40.2	64.1	53.1	55.9	60.6	62.3	61.5
LrgKitApp	76.5	85.9	84.5	89.6	90.1	89.3	90.1	89.6	**90.7**	89.6
MedicalImages	71.8	67.0	75.8	79.2	77.5	77.2	**79.7**	74.6	72.5	75.8
MidPhxAgeGp	54.5	64.3	63.6	76.8	**80.0**	76.0	77.0	74.5	76.2	78.5
MidPhxCorr	78.0	79.4	80.4	79.5	81.5	79.3	**84.2**	79.7	57.2	83.3
MidPhxTW	54.5	51.9	57.1	61.2	**63.9**	60.7	61.9	57.4	60.9	59.9
NonInv_Thor1	83.8	95.0	93.1	**96.1**	93.3	94.8	95.0	94.2	93.6	95.9
NonInv_Thor2	90.1	95.1	94.6	95.5	93.9	95.1	95.1	94.0	94.7	**96.1**
OSULeaf	95.5	96.7	96.7	**98.8**	89.7	97.9	95.9	97.1	95.9	97.5
PhalCorr	77.2	76.3	77.0	82.6	83.2	82.5	83.8	83.0	83.3	**84.0**
Phoneme	26.5	32.1	**34.9**	34.5	23.9	32.4	**34.9**	29.7	31.8	34.2
Plane	100.0	100.0	100.0	100.0	100.0	100.0	100.0	100.0	100.0	100.0
ProxPhxAgeGp	83.4	84.4	85.4	84.9	85.4	84.9	85.9	82.9	85.9	**86.3**
ProxPhxCorr	84.9	88.3	86.9	90.0	89.0	91.8	**93.1**	90.4	91.8	92.8
ProxPhxTW	80.0	80.5	78.0	81.0	**83.0**	80.7	81.0	80.0	81.5	82.5
RefDev	49.9	**58.1**	54.7	53.3	56.0	52.8	56.3	53.1	56.3	53.9
ScreenType	46.4	52.0	54.7	66.7	55.5	**70.7**	64.8	62.7	61.3	60.0
ShapesAll	90.8	84.2	89.2	89.8	87.3	91.2	**92.3**	90.7	91.2	91.3
SmlKitApp	72.5	79.2	77.6	80.3	69.9	79.7	80.5	75.2	80.8	**81.1**
StarlightCurves	97.8	97.9	98.0	96.7	97.8	97.5	98.1	98.0	97.9	**98.2**
Strawberry	97.6	96.2	95.1	96.9	97.1	95.8	96.9	96.7	97.6	**97.7**
SwedishLeaf	92.2	92.8	95.5	96.6	94.1	95.8	95.0	96.8	**97.8**	97.4
Synth_Cntr	96.7	98.3	**100.0**	99.0	99.7	**100.0**	99.3	99.7	99.7	99.7
Trace	100.0	100.0	100.0	100.0	100.0	100.0	100.0	100.0	99.0	100.0
TwoPatterns	99.3	95.5	**100.0**	89.7	99.9	100.0	100.0	100.0	100.0	100.0
UWavGestAll	93.9	94.2	96.4	82.6	95.8	86.8	89.4	92.3	**97.7**	96.5
UWavGest_X	76.2	80.3	82.2	75.4	81.3	78.7	80.3	**83.7**	83.1	83.0
UWavGest_Y	68.5	73.0	75.9	72.5	73.6	66.8	71.1	76.5	76.8	**77.3**
UWavGest_Z	69.5	74.8	75.0	72.9	75.5	75.5	77.7	77.8	77.1	**78.9**
Wafer	99.5	100.0	100.0	99.7	99.8	99.7	99.7	99.8	99.8	99.8
WordSynonyms	63.8	57.1	**75.7**	58.0	66.3	63.2	61.8	70.1	68.2	72.3
Yoga	**91.8**	81.8	87.7	84.5	86.6	85.8	88.5	89.5	87.6	90.1
Avg. Acc	77.64	78.99	80.90	81.15	81.83	81.45	82.35	82.09	81.90	**83.50**
Avg. Rank	7.94	7.03	5.78	5.72	5.34	5.92	4.29	5.23	4.68	**3.08**

of 82.09% and 4th best average rank of 5.23. U-Net with adaptive scaling scores slightly lower average accuracy of 81.90% but achieve 3rd best rank of 4.68. Our U-Net-edRVFL improves the performance substantially, with the highest average accuracy of 83.50% and rank of 3.08. Our U-Net-edRVFL method outperforms other algorithms.

5   Conclusion and Future Works

In this paper, we propose a U-Net model for time series classification problems. For the contracting part, we down-scale features using average pooling layers. In the expanding part, we up-scale features using transposed convolutional layers and average pooling layers (if needed). For time series classification problems with varying series lengths, we propose adaptive scaling for U-Net which scale down the features by a suitable scale. In addition, we extract the features from U-Net and train edRVFLs from extracted features.

Based on the results, adaptive scaling affects U-Net performance substantially on time series classification, especially on series with medium lengths. The extent to which the features are down-scaled can be reduced for short time series. We got the best performance if we tune the extent of the down-scaling operations using 4-fold cross-validation on the training data. Our U-Net with adaptive scaling outperforms most of the classifiers. Training an ensemble of edRVFLs further improves classification. Our U-Net-edRVFL model outperforms all other algorithms.

Future works on U-Net can include using advanced methods such as Echo States Networks or Echo Memory Networks on extracted U-Net features. Also, a randomized U-Net can be explored to reduce computational effort. Finally, the performance of U-Net can be explored on other multivariate time series data, such as EEG data.

References

1. Bagnall, A., Lines, J., Hills, J., Bostrom, A.: Time-series classification with cote: the collective of transformation-based ensembles. IEEE Trans. Knowl. Data Eng. **27**(9), 2522–2535 (2015)
2. Bostrom, A., Bagnall, A.: Binary shapelet transform for multiclass time series classification. In: Hameurlain, A., Küng, J., Wagner, R., Madria, S., Hara, T. (eds.) Transactions on Large-Scale Data- and Knowledge-Centered Systems XXXII. LNCS, vol. 10420, pp. 24–46. Springer, Heidelberg (2017). https://doi.org/10.1007/978-3-662-55608-5_2
3. Chen, Y., et al.: The UCR time series classification archive, July 2015. www.cs.ucr.edu/~eamonn/time_series_data/
4. Cheng, W.X., Suganthan, P., Katuwal, R.: Time series classification using diversified ensemble deep random vector functional link and resnet features. Appl. Soft Comput. **112**, 107826 (2021). https://doi.org/10.1016/j.asoc.2021.107826
5. Dash, Y., Mishra, S.K., Sahany, S., Panigrahi, B.K.: Indian summer monsoon rainfall prediction: a comparison of iterative and non-iterative approaches. Appl. Soft Comput. **70**, 1122–1134 (2018). https://doi.org/10.1016/j.asoc.2017.08.055

6. Dempster, A., Petitjean, F., Webb, G.I.: ROCKET: exceptionally fast and accurate time series classification using random convolutional kernels. Data Min. Knowl. Disc. **34**(5), 1454–1495 (2020). https://doi.org/10.1007/s10618-020-00701-z

7. Deng, H., Runger, G., Tuv, E., Vladimir, M.: A time series forest for classification and feature extraction. Inf. Sci. **239**, 142–153 (2013). https://doi.org/10.1016/j.ins.2013.02.030

8. Esser, P., Sutter, E., Ommer, B.: A variational U-net for conditional appearance and shape generation. In: Proceedings of the IEEE Conference on Computer Vision and Pattern Recognition (CVPR), June 2018

9. Gamboa, J.C.B.: Deep learning for time-series analysis. arXiv preprint arXiv:1701.01887 (2017)

10. Ganaie, M., Hu, M., Malik, A., Tanveer, M., Suganthan, P.: Ensemble deep learning: a review. Eng. Appl. Artif. Intell. **115**, 105151 (2022). https://doi.org/10.1016/j.engappai.2022.105151

11. Gao, R., Du, L., Suganthan, P.N., Zhou, Q., Yuen, K.F.: Random vector functional link neural network based ensemble deep learning for short-term load forecasting. Expert Syst. Appl. **206**, 117784 (2022). https://doi.org/10.1016/j.eswa.2022.117784, https://www.sciencedirect.com/science/article/pii/S0957417422010545

12. Górecki, T., Łuczak, M.: Using derivatives in time series classification. Data Min. Knowl. Disc. **26**(2), 310–331 (2013)

13. Hills, J., Lines, J., Baranauskas, E., Mapp, J., Bagnall, A.: Classification of time series by shapelet transformation. Data Min. Knowl. Disc. **28**(4), 851–881 (2014)

14. Hu, B., Rakthanmanon, T., Hao, Y., Evans, S., Lonardi, S., Keogh, E.: Discovering the intrinsic cardinality and dimensionality of time series using mdl. In: 2011 IEEE 11th International Conference on Data Mining, pp. 1086–1091 (2011). https://doi.org/10.1109/ICDM.2011.54

15. Ioffe, S., Szegedy, C.: Batch normalization: accelerating deep network training by reducing internal covariate shift. arXiv preprint arXiv:1502.03167 (2015)

16. Katuwal, R., Suganthan, P., Zhang, L.: An ensemble of decision trees with random vector functional link networks for multi-class classification. Appl. Soft Comput. **70**, 1146–1153 (2018). https://doi.org/10.1016/j.asoc.2017.09.020

17. Kingma, D.P., Ba, J.: Adam: a method for stochastic optimization. arXiv preprint arXiv:1412.6980 (2014)

18. Lavangnananda, K., Sawasdimongkol, P.: Neural network classifier of time series: a case study of symbolic representation preprocessing for control chart patterns. In: 2012 8th International Conference on Natural Computation, pp. 344–349 (2012). https://doi.org/10.1109/ICNC.2012.6234651

19. Lin, M., Chen, Q., Yan, S.: Network in network. arXiv preprint arXiv:1312.4400 (2013)

20. Lines, J., Bagnall, A.: Time series classification with ensembles of elastic distance measures. Data Min. Knowl. Disc. **29**(3), 565–592 (2015)

21. Liu, Z., Cao, Y., Wang, Y., Wang, W.: Computer vision-based concrete crack detection using U-net fully convolutional networks. Autom. Constr. **104**, 129–139 (2019). https://doi.org/10.1016/j.autcon.2019.04.005, https://www.sciencedirect.com/science/article/pii/S0926580519301244

22. Ma, Q., Zhuang, W., Shen, L., Cottrell, G.W.: Time series classification with echo memory networks. Neural Netw. **117**, 225–239 (2019). https://doi.org/10.1016/j.neunet.2019.05.008

23. Malik, A.K., Gao, R., Ganaie, M.A., Tanveer, M., Suganthan, P.N.: Random vector functional link network: recent developments, applications, and future directions (2022). https://doi.org/10.48550/ARXIV.2203.11316

24. Nair, V., Hinton, G.E.: Rectified linear units improve restricted Boltzmann machines. In: Proceedings of the 27th International Conference on Machine Learning (ICML-10), pp. 807–814 (2010)

25. Oktay, O., et al.: Attention U-net: learning where to look for the pancreas. arXiv preprint arXiv:1804.03999 (2018)

26. Pao, Y.H., Takefuji, Y.: Functional-link net computing: theory, system architecture, and functionalities. Computer **25**(5), 76–79 (1992). https://doi.org/10.1109/2.144401

27. Rajkomar, A., et al.: Scalable and accurate deep learning with electronic health records. NPJ Digit. Med. **1**(1), 18 (2018)

28. Ronneberger, O., Fischer, P., Brox, T.: U-net: convolutional networks for biomedical image segmentation. In: Navab, N., Hornegger, J., Wells, W.M., Frangi, A.F. (eds.) MICCAI 2015. LNCS, vol. 9351, pp. 234–241. Springer, Cham (2015). https://doi.org/10.1007/978-3-319-24574-4_28

29. Schäfer, P.: The boss is concerned with time series classification in the presence of noise. Data Min. Knowl. Disc. **29**(6), 1505–1530 (2015)

30. Shi, Q., Hu, M., Suganthan, P.N., Katuwal, R.: Weighting and pruning based ensemble deep random vector functional link network for tabular data classification. Pattern Recogn. **132**, 108879 (2022). https://doi.org/10.1016/j.patcog.2022.108879, https://www.sciencedirect.com/science/article/pii/S0031320322003600

31. Shi, Q., Katuwal, R., Suganthan, P., Tanveer, M.: Random vector functional link neural network based ensemble deep learning. Pattern Recogn. **117**, 107978 (2021)

32. Shi, Q., Suganthan, P.N., Del Ser, J.: Jointly optimized ensemble deep random vector functional link network for semi-supervised classification. Eng. Appl. Artif. Intell. **115**, 105214 (2022). https://doi.org/10.1016/j.engappai.2022.105214, https://www.sciencedirect.com/science/article/pii/S0952197622002974

33. Suganthan, P.N., Katuwal, R.: On the origins of randomization-based feedforward neural networks. Appl. Soft Comput. **105**, 107239 (2021). https://doi.org/10.1016/j.asoc.2021.107239

34. Vuković, N., Petrović, M., Miljković, Z.: A comprehensive experimental evaluation of orthogonal polynomial expanded random vector functional link neural networks for regression. Appl. Soft Comput. **70**, 1083–1096 (2018). https://doi.org/10.1016/j.asoc.2017.10.010

35. Wang, J., Chen, Y., Hao, S., Peng, X., Hu, L.: Deep learning for sensor-based activity recognition: a survey. Pattern Recogn. Lett. **119**, 3–11 (2019). https://doi.org/10.1016/j.patrec.2018.02.010, Deep Learning for Pattern Recognition

36. Wang, Z., Yan, W., Oates, T.: Time series classification from scratch with deep neural networks: a strong baseline. In: 2017 International Joint Conference on Neural Networks (IJCNN), pp. 1578–1585, May 2017. https://doi.org/10.1109/IJCNN.2017.7966039

Permutation Elementary Cellular Automata: Analysis and Application of Simple Examples

Taiji Okano and Toshimichi Saito[✉]

Hosei University, Koganei, Tokyo 184-8584, Japan
tsaito@hosei.ac.jp

Abstract. This paper studies simple three-layer digital dynamical systems related to recurrent-type neural networks. The input to hidden layers construct an elementary cellular automaton and the hidden to output layers are one-to-one connection described by a permutation. Depending on the permutation, the systems generate various periodic orbits. Applications include walking robots, switching power converters, and reservoir computing. In order to analyze the dynamics, we introduce two feature quantities that evaluate complexity and stability of the periodic orbits. Calculating the feature quantities in simple example systems, we have clarified that the systems can generate various stable periodic orbits. Presenting an FPGA based hardware prototype, typical periodic orbits are confirmed experimentally.

Keywords: Elementary cellular automata · recurrent neural networks · permutation · periodic orbits · stability

1   Introduction

Elementary cellular automata (ECAs [1–3]) are simple digital dynamical systems where time, space, and state variables are all discrete. The dynamics is governed by rules of Boolean functions from three inputs to one output. The ECAs can generate various spatiotemporal patterns and the applications include signal processing [4], reservoir computing [5,6], and error correcting codes [7]. The ECAs are related deeply to discrete-time recurrent-type neural networks (RNNs, [8–10]) having various applications. Depending on the parameters, the ECAs and RNNs can generate a variety of periodic orbits of binary vectors (BPOs). Analysis of the BPOs is important not only as a basic study of nonlinear dynamics but also for engineering applications. However, the BPOs and their stability have not been studied sufficiently. In order to analyze the BPOs, we should reduce the number of parameters and should focus on simple examples.

This paper presents permutation elementary cellular automata (PECAs): simple three-layer digital dynamical systems with various BPOs. The input to hidden layers construct an ECA and the hidden to output layers are one-to-one connection described by a permutation. The dynamics is described by

© The Author(s), under exclusive license to Springer Nature Switzerland AG 2023
M. Tanveer et al. (Eds.): ICONIP 2022, LNCS 13623, pp. 321–330, 2023.
https://doi.org/10.1007/978-3-031-30105-6_27

autonomous difference equations of binary state variables. The PECAs bring benefits to FPGA based hardware implementation and precise analysis. In this paper, we select several ECA rules given by linearly nonseparable Boolean functions (non-LSBFs [2]) and consider the permutations as control parameters. In N-dimensional PECAs, the number of permutations is $N!$ that is much smaller than 2^{N^2}, the number of binary connection parameters in simple RNNs [10]. In order to analyze the PECA, we introduce two feature quantities that evaluate complexity and stability of BPOs. As a first step, we analyze simple/typical examples of PECAs. In the simple examples, we apply the brute force attack to calculate the feature quantities and clarify that the PECAs can generate various BPOs which are impossible in ECAs. Some of the BPOs are applicable to control of walking robots [11,12], control of switching power converters [10,13], and time-series approximation in reservoir computing [6]. Introducing an FPGA based hardware prototype, typical BPOs are confirmed experimentally.

The PECAs include permutation binary neural networks (PBNNs [14]) where the input to hidden layers are characterized by signum-type neurons that realizes linearly separable Boolean functions (LSBFs [2]). The PBNNs can be regarded as simplified systems of three-layer dynamics binary neural networks (DBNNs, [15]) with a large number of hidden neurons. As novelty of this paper, note that we have confirmed various BPOs from typical PECA examples with non-LSBF rules which are impossible in PBNNs with LSBF rules. Our basic results will be developed into more detailed analysis of PECAs and its applications.

2   Permutation Elementary Cellular Automata

First, we introduce ECAs on a ring of N cells. Let $x_i^t \in \{0,1\} \equiv B$ be the i-th binary state at discrete time t. The dynamics is described by

$$x_i^{t+1} = F(x_{i-1}^t, x_i^t, x_{i+1}^t), \quad i \in \{1, \cdots, N\}, \; N \geq 3 \tag{1}$$

where $x_0^t \equiv x_N^t$ and $x_{N+1}^t \equiv x_1^t$ for the ring topology. A Boolean function F transforms three binary inputs to one binary output, for example,

$$F(0,0,0) = 0 \; F(0,0,1) = 0 \; F(0,1,0) = 0 \; F(0,1,1) = 1$$
$$F(1,0,0) = 1 \; F(1,0,1) = 1 \; F(1,1,0) = 1 \; F(1,1,1) = 0$$

Decimal expression of the 8 outputs is referred to as the rule number (RN). In this example, the 8 outputs $(01111000)_2 = 120_{10}$ gives RN120. There exist $2^{2^3} = 256$ rules in the ECAs. Figure 1 (a) and (b) show ECA of RN120 and a BPO with period 6. This BPO is applicable to control hexapod walking robot [12].

Applying permutation connection, the PECA is constructed. The dynamics is described by the following autonomous difference equation.

$$x_i^{t+1} = y_{\sigma(i)}^t \qquad \sigma = \begin{pmatrix} 1 & 2 & \cdots & N \\ \sigma(1) & \sigma(2) & \cdots & \sigma(N) \end{pmatrix} \tag{2}$$
$$y_i^t = F(x_{i-1}^t, x_i^t, x_{i+1}^t),$$

where $y_i^t \in B$ is the i-th binary hidden state and σ is a permutation. Let $\boldsymbol{x}^t \equiv (x_1^t, \cdots, x_N^t) \in B^N$ and let $\boldsymbol{y}^t \equiv (y_1^t, \cdots, y_N^t) \in B^N$.

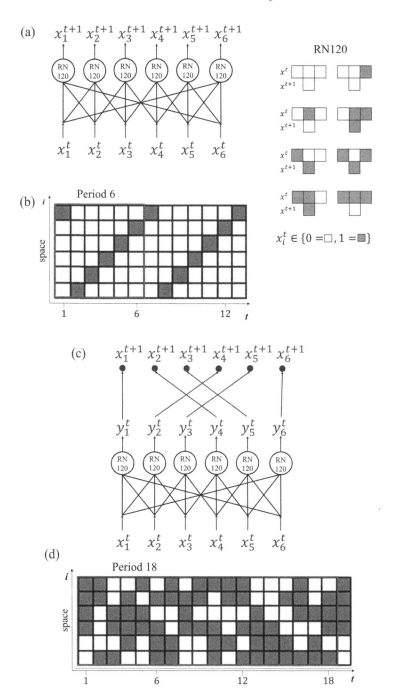

Fig. 1. ECA, PECA, and BPOs. (a) ECA network and rule table of RN120. (b) BPO with period 6 from ECA, RN120. (c) PECA with permutation connection. (d) BPO with period 18 from PECA, RN120.P145236.

As shown in Fig. 2 (c), the input to hidden layers construct an ECA and the hidden and output layers are connected by a permutation. The permutation connection transforms the binary hidden state vector y^t into the binary output vector x^{t+1}. In order to identify the permutation connection, we introduce permutation identifier $P\sigma(1)\cdots\sigma(N)$. A PBNN is identified by rule number the RN and the permutation identifier. In Fig. 2 (c), the PECA is identified by RN120 and P145236. Figure 2 (d) shows BPO with 18 from the PECA. It suggests that the permutation connection can make longer period BPOs.

3 Analysis of Binary Periodic Orbits

A BPO with period p is defined by.

$$
z^1, \cdots, z^p, \cdots \quad \begin{cases} z^{t_1} = z^{t_2} \text{ for } |t_2 - t_1| = np \\ z^{t_1} \neq z^{t_2} \text{ for } |t_2 - t_1| \neq np \end{cases} \tag{3}
$$

where $z^t = (z_1^t, \cdots, z_N^t)$, $z_i^t \in B$. A element of the BPO $z_p \in B^N$ is said to be a binary periodic point (BPP). A point $z_e \in B^N$ is said to be an eventually periodic point (EPP) if z_e is not a BPP but falls into a BPO. Figure 2 illustrates a BPO with period 6 and EPPs to the BPO.

In order to analyze the PECAs, we introduce two feature quantities. Since analysis of multiple BPOs is not easy, we consider one BPO with the maximum period (MBPO). The first feature quantity is defined by

$$
a = (\text{The period of an MBPO})/2^N, \ 1/2^N \le a \le 1. \tag{4}
$$

Roughly speaking, the quantity a evaluates complexity of the MBPO. The second feature quantity is defined by

$$
b = (\text{The number of EPPs into the MBPO})/2^N, \ 0 \le b \le 1 - a. \tag{5}
$$

Since stability of an MBPO becomes stronger as the number of EPPs increases, the quantity b evaluates stability of the MBPO. Note that the stability corresponds to bit error correction capability in digital dynamical systems. If multiple MBPOs exist, one MBPO with larger b is adopted.

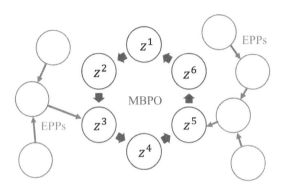

Fig. 2. MBPO and EPPs.

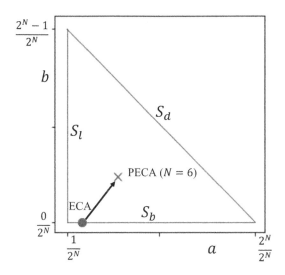

Fig. 3. Feature plane and three segments (S_b, S_l, S_d) Red circle: MBPO of ECA for $N = 6$, RN120. Red cross: MBPO of PECA for $N = 6$, RN120, P145236. (Color figure online)

Using the feature quantities a and b, we construct the feature plane as shown in Fig. 3. A PECA is represented by its MBPO and the MBPO gives one plot (point) in the feature plane. The plot exists in a triangle surrounded by the three segments each of which has the following meaning.

- $S_b (b = 0)$: There exists no EPP to the MBPO and no transient phenomenon to the MBPO.
- $S_d (a + b = 1)$: The PECA has unique BPO=MBPO. Except for the MBPO, all the initial points are EPPs to the MBPO.
- $S_l (a = 1/2^N)$: The MBPO is a fixed point (BPO with period 1).

In the figure, the red circle and red cross correspond to the MBPO of ECA in Fig. 1 (b) and the MBPO of PECA in Fig. 1 (d), respectively. The permutation P145236 makes the MBPO with longer period than that of the ECA.

Using the two feature quantities, we have analyzed typical 6-dimensional PECAs. In the 6-dimensional cases, the brute force attack is possible and some of MBPOs are applicable to engineering systems such as hexapod walking robots [11] and switching power converters with six switches [13]. First, as basic data, Fig. 4 (a) shows the feature plane for all the ECAs corresponding to P123456. The number of rules is 256 whereas the number of plots is 41: multiple rules give the same plots in the plane. The 41 plots are distributed in left region in the triangle.

The MBPO with the maximum period is given by RN45 ($a = 18/64, b = 36/64$). Next, after trial-and-errors, we have selected three rules (RN120, RN86, and RN154, red points in Fig. 4(a)). These rules are given by non-LSBFs which are impossible in the PBNN with LSBF rules [14]. Applying all the 6! permutations to ECAs of the three rules, we have calculated the feature quantities.

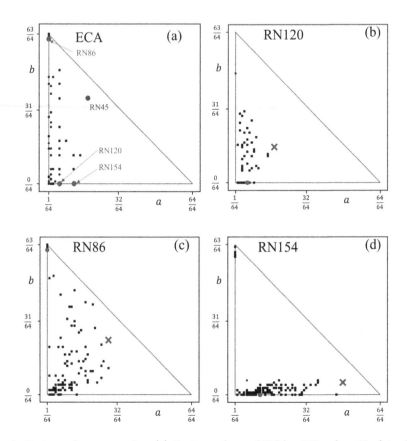

Fig. 4. Feature plane examples. (a) Feature plane of ECAs, 256 rules, 41 plots. Blue point: RN45 ($a = 18/64$, $b = 36/64$). Red pints (RN120, RN86, RN154). (b) RN120, 42 plots. Red point: ECA ($a = 6/64$, $b = 0/64$), Red cross: PECA(P145236, $a = 18/64$, $b = 15/64$). (c) RN86, 86 plots. Red point: ECA ($a = 1/64$, $b = 61/64$), Red cross: PECA(P315624, $a = 28/64$, $b = 23/64$). (d) RN154, 82 plots. Red point: ECA ($a = 12/64$, $b = 0/64$), Red cross: PECA(P146523, $a = 48/64$, $b = 5/64$). (Color figure online)

Figure 4 (b) shows 42 plots of 6! PECAs for the ECA with RN120. As stated in Fig. 3, applying the permutation P145236, the red circle of the ECA is moved into the red cross of the PECA: both a (period) and b (stability) become larger. The 42 plots distribute left bottom region in the triangle.

Figure 4 (c) shows 86 plots of PECAs for ECA with RN86. For example, applying the permutation P315624, a becomes larger whereas b becomes smaller (from red circle to red cross). The 86 plots distribute wider region in the triangle. The PECAs for RN86 can generate a variety of BPOs as compared with other examples.

Figure 4 (d) shows 82 plots of PECAs for ECA with RN154. For example, applying the permutation P146523, both a and b becomes larger (from red circle to red cross). Note that the red cross gives MBPO with the maximum period (48) in the figure. In [14], the maximum period in the 6-dimensional PBNN is 20 and that of 6-dimensional ECA is 18: the 6-dimensional PECA can generate MBPOs with longer period that are impossible in PBNNs and ECAs. The 82 plots distribute lower region in the triangle.

4   FPGA Based Hardware

In order to realize engineering applications, we introduce an FPGA based hardware prototype [14]. The FPGA is an integrated circuit designed to be configured by a designer after manufacturing. The advantages include high speed/precision operation and high degree of integration. Since the PECA uses binary signals, it is suitable for FPGA based hardware implementation. As a first step in engineering applications, we realize a BPO of ECA in Fig. 1 (b) and typical BPOs of PECA in Fig. 4. In the experiments, we have prepared the following tools:

- Vivado version: Vivado 2020.1 platform (Xilinx).
- FPGA: Xilinx Artix-7 XC7A35T-ICPG236C.
- Clock: 10[Hz]. The default frequency 100[MHz] is divided for clear measurement.
- Measuring instrument: ANALOG DISCOVERY2.
- Multi-instrument software: Waveforms 2015

Algorithms 1 and 2 show simplified SystemVerilog codes for the hardware design. In the Algorithm 1, as a rule number RN is given, the corresponding Boolean function is obtained and the ECA is realized. This design strategy is based on the Boolean functions from three input to one output. In the algorithm 2, as a permutation identifier is given, the permutation connection is realized. Varying the RN and permutation identifier, we obtain a desired PECA in $256 \times N!$ PECAs.

Using the algorithms, the ECA and PECA are implemented on the FPGA board experimentally. Figure 5(a) shows measured waveform of MBPO with period 6 from the ECA in Fig. 1 (b). This MBPO is applicable to control signal of hexapod walking robots [12]. Figure 5(b), (c), and (d) show measured waveforms of MBPOs (with period 18, 28, and 48) from the PECAs corresponding to red crosses in Fig. 4 (b), (c), and (d), respectively. Note that, in the PBNNs [14], the period of MBPO is at most 20. This FPGA prototype provides basic information to realize engineering applications of 6 or higher dimensional PECAs.

Algorithm 1. ECA

$N = 6$	// The number of cells

$\text{output } x^t[N]$
$\text{reg } x^{t+1}[N]$
$\text{wire } rule0, rule1, \cdots, rule7;$

$RN = 8\text{'b}120 \;;$	// Rule Number

$\textbf{for } j = 1; j <= N; j = j + 1 \textbf{ do}$
$\quad rule0 = RN[0]^*(\tilde{\,}x^t[j-1] \;\&\; \tilde{\,}x^t[j] \;\&\; \tilde{\,}x^t[j+1]);$
$\quad rule1 = RN[1]^*(\tilde{\,}x^t[j-1] \;\&\; \tilde{\,}x^t[j] \;\&\; x^t[j+1]);$
$\qquad \vdots$

$\quad rule7 = RN[7]^*(x^t[j-1] \;\&\; x^t[j] \;\&\; x^t[j+1]);$	// Boolean function

$\quad x^{t+1}[j] = (rule0)|(rule1)| \ldots |(rule7);$
$\textbf{end for}$

Algorithm 2. PECA

$N = 6$	// The number of cells

clk load rst

$\text{input } i[N]$	// Initial condition

$\text{output } x^t[N]$
$\text{reg } x^{t+1}[N]$

$y[N] = [1, 4, 5, 2, 3, 6];$	// Permutation identifier P145236

$\textbf{if } \text{load} == 1 \textbf{ then}$

$\quad x^t[N] \leftarrow i[N]$	//Initial value setting

$\textbf{else if } \text{rst} == 1 \textbf{ then}$

$\quad x^t[N] \leftarrow 0$	//Reset

$\textbf{else}$
$\quad \textbf{for } k = 1; k <= N; k = k + 1 \textbf{ do}$

$\qquad x^t[k] \leftarrow x^{t+1}[y[k]];$	// Permutation

$\quad \textbf{end for}$
$\textbf{end if}$
$\text{ECA}(x^t, x^{t+1});$

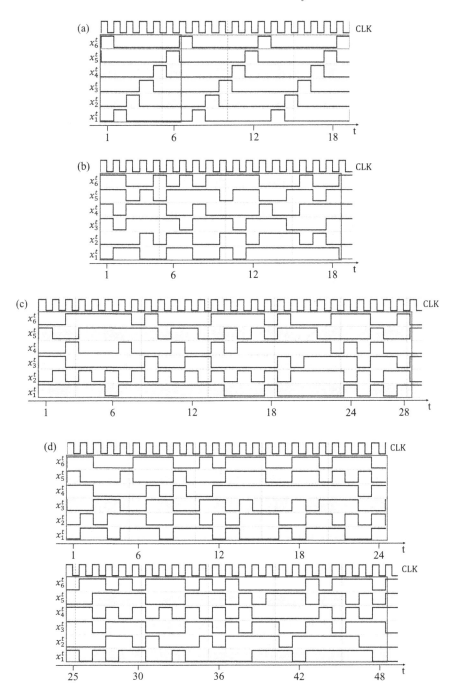

Fig. 5. Measured waveforms of MBPOs in the FPGA board. (a) MBPO with period 6 from ECA, RN120 in Fig. 4(a). (b) MBPO with period 18 from PECA, RN120, P145236 (red cross in Fig. 4(b)). (c) MBPO with period 28 from PECA, RN86, P315624 (red cross in Fig. 4(c)). (d) MBPO with period 48 from PECA, RN154, P146523 (red cross in Fig. 4(d)). (Color figure online)

5 Conclusions

Basic dynamics of the PECAs have been studied in this paper. In order to analyze the BPOs, we have introduced two feature quantities that evaluate complexity and stability of BPOs. Using the feature quantities, we have analyzed typical PECAs and have clarifies that the PECAs can generate various BPOs which are impossible in PBNNs and ECAs. In order to realize the engineering applications, we introduce an FPGA based hardware prototype. Using the hardware, typical MBPOs are confirmed experimentally. In our future works, we should consider various problems including detailed analysis of various BPOs, evolutionary learning algorithms of PECAs, FPGA based hardware implementation with learning function, and engineering applications such as control of switching circuits and time-series approximation in reservoir computing.

References

1. Wolfram, S.: Cellular automata and complexity: collected papers. CRC Press (2018)
2. Chua, L.O.: A nonlinear dynamics perspective of Wolfram's new kind of science. World Scientific (2006)
3. Schüle, M., Stoop, R.: A full computation-relevant topological dynamics classification of elementary cellular automata. Chaos **22**, 043143 (2012)
4. Wada, M., Kuroiwa, J., Nara, S.: Completely reproducible description of digital sound data with cellular automata. Phys. Lett. A **306**, 110–115 (2002)
5. Yilmaz, O.: Symbolic computation using cellular automata-based hyperdimensional computing. Neural Comput. **27**, 2661–2692 (2015)
6. Tanaka, G., et al.: Recent advances in physical reservoir computing: a review. Neural Netw. **115**, 100–123 (2019)
7. Chowdhury, D., Basu, S., Gupta, I., Chaudhuri, P.: Design of CAECC - cellular automata based error correcting code. IEEE Trans. Comput. **43**, 759–764 (1994)
8. Hopfield, J.J.: Neural networks and physical systems with emergent collective computation abilities. Proc. Nat. Acad. Sci. **79**, 2554–2558 (1982)
9. Michel, A.N., Farrell, J.A.: Associative memories via artificial neural networks. IEEE Control Syst. Mag. **10**, 6–17 (1990)
10. Sato, R., Saito, T.: Stabilization of desired periodic orbits in dynamic binary neural networks. Neurocomputing **248**, 19–27 (2017)
11. Minati, L., Frasca, M., Yoshimura, N., Koike, Y.: Versatile locomotion control of a hexapod robot using a hierarchical network of nonlinear oscillator circuits. IEEE Acess **6**, 8042–8065 (2018)
12. Suzuki, T., Saito, T.: Synthesis of three-layer dynamic binary neural networks for control of hexapod walking robots. In: Proceedings IEEE/CNNA (2021)
13. Holderbaum, W.: Application of neural network to hybrid systems with binary inputs. IEEE Trans. Neural Netw. **18**(4), 1254–1261 (2007)
14. Udagawa, H., Okano, T., Saito, T.: Permutation binary neural networks: analysis of periodic orbits and its applications. Discrete Contin. Dyn. Syst. Ser. B **28**(1), 748–764 (2023)
15. Anzai, S., Suzuki, T., Saito, T.: Dynamic binary neural networks with time-variant parameters and switching of desired periodic orbits. Neurocomputing **457**, 357–364 (2021)

SSPR: A Skyline-Based Semantic Place Retrieval Method

Jiamin Lu[1(✉)], Zhenyu Zhou[2], Jiahao Liu[2], and Jun Feng[1]

[1] Key Laboratory of Water Big Data Technology of Ministry of Water Resources,
Hohai University, Nanjing, China
{jiamin.luu,fengjun}@hhu.edu.cn
[2] College of Computer and Information, Hohai University, Nanjing, China
{zyzhou,jiahao.liu}@hhu.edu.cn

Abstract. With the introduction of spatial semantics in the knowledge base, semantic place retrieval on spatial RDF data has become a popular research topic in recent years. Most existing methods ignore the following two problems. First, exact matching leads to a large number of potential results being missed and ultimately returning limited results. Second, the Top-k linear ranking function transforms the multi-objective problem into the single-objective optimization, causing the results to be prone to extreme values. In this paper, we propose a new approach named SSPR, which replaces exact matching with fuzzy matching to make retrieval closer to the human experience of interpretation. In addition, inspired by skyline, we computation an efficient query algorithm to select places from both the semantic relevance and spatial distance, returning mutually non-dominated results. The experiments on different test sets demonstrate that our approach compared to the traditional kSP method balances spatial distance and semantic relevance while outperforming retrieval efficiency.

Keywords: Semantic place retrieval · Skyline query · Spatial RDF data · Fuzzy matching

1 Introduction

According to the relevant investigation and statistics, more than 28% of the information retrieved by users in the search engines is related to geographic information, such as attractions, restaurants, and city addresses [1]. The traditional retrieval method based on text keywords can no longer meet the semantic needs of users, so the information retrieval based on knowledge graph came into being. RDF is a widely used data model in knowledge graph to describe the relationships between entities and inter-entity. With the continuous growth of RDF data scale, some of its vertices introduce spatial coordinate information, i.e., RDF data contains geographic entities, which opens the road to semantic place retrieval.

© The Author(s), under exclusive license to Springer Nature Switzerland AG 2023
M. Tanveer et al. (Eds.): ICONIP 2022, LNCS 13623, pp. 331–342, 2023.
https://doi.org/10.1007/978-3-031-30105-6_28

At present, the mainstream methods of RDF data retrieval are roughly divided into two groups: (1) structured language (e.g., SPARQL) query; (2) keyword search. The former requires that query issuers fully understand the query language and the data domain, restricts common users to access RDF data, while the latter increases the applicability of common users by using only keywords. Given this, an RDF keyword search model [3,4] is proposed, which retrieves a set of (small) subtrees where the vertices of each subtree collectively cover all the query keywords. However, this model did not take into account the spatial information in the KG. In 2016, [2] proposed the Top-k Relevant Semantic Place (kSP) retrieval, which applies the RDF keyword search model to spatial RDF data retrieval with location-awareness. It takes as input a query location, a set of query keywords, and returns a set of Top-k subtrees according to a ranking function. This function combines the compactness of a subtree, measured by the *looseness score* [3], and the spatial distance between the query location and each candidate subtree. Each of the subtrees is rooted at a place entity and covers all query keywords. The lower the looseness, the more compact and relevant the subtrees are.

Many studies have proposed improvements to kSP, such as integrating diversity to facilitate retrieval of places with different features and orientations [5], and extending semantic place retrieval to semantic region retrieval to avoid situations where a single place cannot meet user demands [6]. However, none of the above improvements take into account the fuzzy matching of keywords. In the kSP retrieval, the looseness can be obtained only when all vertices of the subtree cover the query keywords. In this process, every keyword needs to be matched accurately. Nevertheless, there are often a large number of words with similar meanings but different expressions between the query keywords and the RDF data keyword sets in the actual retrieval scenario. This leads to the problem that the kSP retrieval will miss lots of potential results, and ultimately return limited results. For instance, on the YAGO2 [7] dataset, assume a top-2 query issued by a tourist with query keywords {*history, roman, ancient, catholic*}, the kSP retrieval returns the result *Montmajour Abbey, Roman Catholic Diocese*. However, when the synonyms of the above query keywords are entered but the query intent has not changed, such as {*history, roman, ancient, memorial*}, the kSP retrieval cannot return the results because the keyword *memorial* is not included in subtrees. Therefore, fuzzy matching is introduced in this paper to fundamentally increase the total number of eligible query results and return more suboptimal results to users.

On the other hand, the Top-k ranking function in the kSP retrieval transforms the multi-objective problem into single-objective optimization, which is insensitive to influencing factors, easily affected by weight parameters, and prone to extreme values. As shown in Fig. 1, consider the example kSP query issued by a user at location q with keywords {*childhood, scientific*}, The top-1 place is p_1 *Musee de I'Orangerie*. Although the place is spatially close to the query location, the looseness is simply too large to meet the needs of users. Driven by these problems, we propose a novel spatial RDF data retrieval method named

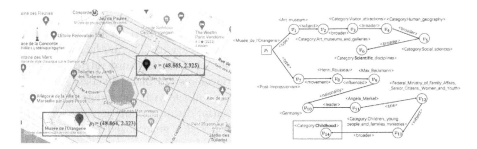

Fig. 1. kSP query example.

SSPR (Skyline-Based Semantic Place Retrieval) to support both accuracy and efficiency retrieval. We generate word embeddings of query keywords and keywords describing place vertices, and calculate the semantic relevance of the two to measure the extent to which place meets user semantic needs. We also propose a semantic place retrieval algorithm based on skyline, which finally returns the Top-k mutually uncontrolled solutions and improves the efficiency and accuracy of spatial RDF data retrieval.

The rest of this paper is organized as follows. Section 2 reviews the state-of-the-art RDF data keyword search method. We present the architecture of SSPR in Sect. 3. Section 4 describes the experimental setup and presents the evaluation results, compared to kSP. Finally, we conclude our contribution and the future work in Sect. 5.

2   Related Work

Keyword search on RDF data usually adopts the strategy of knowledge graph subgraph positioning to solve the problem. Le et al. [3] and Stefanidis et al. [4] proposed an RDF keyword search model that RDF graph is simplified by removing the outgoing edges from subjects which connect to attributes and literals, and by collecting all the keywords in the URIs, attributes, and literals to form a keyword document describing each vertex. A set of subgraphs are retrieved according to the query keywords, wherein the vertices of each subgraph collectively cover all the query keywords. The subgraphs are finally ranked based on statistical language models (LMs). However, this model does not take into account the spatial distance factor.

Shi et al. [2] proposed a new spatial RDF data retrieval method, namely Top-k Relevant Semantic Place Retrieval (kSP). The method takes a query location, a set of query keywords and the number of k results returned as input parameters, retrieves the R-tree based on the query location, obtains the place entities near the query location, and then traverses the spatial RDF graph from these vertices, finds the subtree containing all query keywords, and returns Top-k results according to the ranking function. In response to the problem that kSP retrieval results may be similar to each other, the Top-k Diversified Semantic

Places (kDSP) retrieval is proposed by [5], which integrates textual and spatial diversification into spatial RDF data retrieval, which facilitates retrieval with different characteristics relative to query locations. The method achieves a trade-off between relevance and diversity, enriches the keyword retrieval results of spatial RDF data, and makes the retrieval results more diverse. Although the kSP method consider the spatial distance along with the semantic requirements, it may return places with similar spatial distances but very low keyword relevance. Based on this, Wu et al. [6] proposed a generalized semantic place retrieval called the Semantic Region Retrieval (SR). It can retrieve multiple nearby related places to avoid situations where a single place cannot meet user demands, with each place being related to one or more query keywords. Also, Jin et al. [8] stated that a single subtree returned by kSP may not meet the user's needs. To this end, they propose a method to combine a set of subtrees together to collectively cover the query keywords, named CoSKQ-KB (Collective Spatial Keyword Query on a Knowledge Base).

3   The SSPR Architecture

3.1   Overview

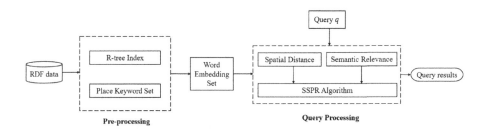

Fig. 2. Architecture of SSPR.

The main architecture of SSPR is illustrated in Fig. 2, mainly consisting of two components: the pre-processing module and query processing module.

First, the query q contains four parameters: query location q.λ, a set of query keywords q.ψ, the number k of requested places, and the semantic similarity threshold τ.

In the pre-processing module, the R-tree index is established for fast query of the desired place vertices within a certain range. For the purpose of keyword search, we simplify the RDF graph by eliminating outgoing edges from subject to types or literals and extracting all the keywords in the URI, types, and literals of such entities to form a general keyword set for each vertex. Then for each place vertex, we splice and de-duplicate the general keyword sets of this vertex and its children to form a place keyword set describing the entity. Finally, we generate word embedding sets based on the keyword sets to facilitate the calculation of semantic relevance.

In the query processing module, we consider the spatial distance and the semantic relevance factors. Firstly, we calculate the spatial distance between the place vertex and the query location. Then we calculate the semantic relevance of the place to the query keywords by using the word embedding sets. Lastly, we apply the SSPR algorithm to return the query results that are not dominated by each other in terms of semantic relevance and spatial distance.

3.2   Pre-processing

We define an RDF knowledge base as a directed graph $G = \langle V, E \rangle$, where V indicates a collection of vertices and E indicates a collection of edges. Figure 3(a) shows the graph representation of several triples extracted from the YAGO2. Both circles and squares are vertices in the RDF graph and represent entities. The general vertices denoted by n_i. Spatial coordinates are attached to some of the vertices and we call such vertices as place vertices, whose are denoted by np_i. The edges (labeled by predicates) model the relationships between entities, represented by square brackets. The pointed brackets denote the name of a vertex or edge, and double quotes denote types or literals a vertex has.

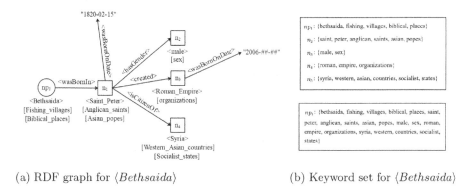

(a) RDF graph for $\langle Bethsaida \rangle$ (b) Keyword set for $\langle Bethsaida \rangle$

Fig. 3. RDF graph and keyword set example.

Keyword Set. Compared with triples, the representation of native graph supports general graph operations such as Random Walk, Reachability, and Community Discovery [9], which facilitates the measurement of geospatial and semantic relevance of spatial RDF data and can well improve the efficiency of data, and use adjacency lists to store data. The word embedding matrix is sparse if the types or literals of a vertex are not split, so according to previous keyword search efforts on RDF data, we built a general keyword set $n.\psi$ for each entity which is extracted from its URI, predicates, and literals. The upper part of Fig. 3(b) displays the general keyword set of all vertices in Fig. 3(a). For describing a place entity, we built a place keyword set $np.\lambda$ by stitching and de-duplicating the general keyword set of the node and its sub-nodes. The lower part of Fig. 3(b) displays the place keyword set of place entity $\langle Bethsaida \rangle$ in Fig. 3(a). In addition, to facilitate the subsequent computation of the place vertex looseness, we add the shortest distance from each keyword t_i to the place vertex np in its place keyword set.

Word Embedding Set. A word embedding set is established on the basis of the keyword set. Word embedding is a distributed representation of a word, generally consisting of a relatively low-dimensional dense vector. Simple mathematical operations can be performed between word embeddings to represent the similarity between words (we use the cosine similarity in this paper). In order to improve the computational efficiency of semantic relevance, we use the Word2vec [10] model to establish a static word embedding set and stores the word embedding representation of each keyword.

3.3   Query Processing

A query q consists of four parameters: the query location $q.\lambda$, the query keywords $q.\psi$, the number of requested places k, and the semantic similarity threshold τ. The query aims at finding the places that (i) are spatially close to the query location, (ii) the semantic similarity between the query keywords and the places greater than or equal to τ. The calculation method of semantic relevance is given in the following subsections.

Semantic Relevance. The semantic relevance is used to indicate relevance between the place and the query keywords. The higher the semantic relevance is, the more the places conform to the user's semantic requirements.

Definition 1. *(Semantic Similarity of Keyword Set, S) Given a query q and a RDF graph $G = \langle V, E \rangle$, query keywords set denoted by $q.\psi = \{t_1, t_2, \ldots, t_i, \ldots, t_n\}$, $1 \leq i \leq n$, and the place keyword set is $np.\psi = \{k_1, k_2, \ldots, k_j, \ldots, k_m\}, 1 \leq j \leq m$. $S(t_i, k_j)$ is used as the semantic similarity between keywords, and $S(np)$ is used as the semantic similarity between the query keywords $q.\psi$ and the place np, defined as follows:*

$$S(t_i, k_j) = \frac{E(t_i) \times E(k_j)}{\sqrt{(E(t_i))^2 \times (E(k_j))^2}} \tag{1}$$

$$S(np) = avg\left\{max[S(t_i, k_1), ..., S(t_i, k_j), ..., S(t_i, k_m)]\right\}, 1 \leq i \leq n \tag{2}$$

Among them, $E(t_i)$ and $E(k_j)$ represent the word embedding of the keywords t_i and k_j respectively. For each query keyword $t_i \in q.\psi$, calculate its semantic similarity with the keyword $k_j \in np.\psi$, and average the highest similarity to obtain the semantic similarity of the query keywords $q.\psi$ and the place np.

Considering that in the RDF graph, the distance of the keywords from the place vertex also affects the semantic relevance, we follow the definition of looseness and define the semantic relevance scoring function in Definition 2. Looseness aggregates the proximity of the query keywords at semantic places based on graph distance. The smaller the looseness, the more relevant the node where the query keyword is located is to the root (i.e. the place).

Definition 2. *(Semantic Relevance Score, f) Given a semantic place T_{np}, considering the two influencing factors of the semantic similarity of keyword set S and the looseness L, we give the semantic relevance scoring function as follows, which α is the weight parameter:*

$$f(np) = \frac{S(np) * \alpha}{L(T_{np} * (1 - \alpha)}\tag{3}$$

3.4   Query Algorithm

For a query, it needs to be sorted by combining semantic relevance and spatial distance. The traditional linear ranking model in kSP has many disadvantages, such as not taking into account the dominant relationship between the results. As one of the multi-objective decision methods, skyline query can efficiently filter out a set of data from a huge amount of data that cannot be dominated by other data, in which no single data will perform worse than all the data outside the skyline set in all aspects, and will be better than the other data in at least one aspect. In this paper, we propose SSPR (Skyline-Based Semantic Place Retrieval) algorithm to select the Top-k parallel and mutually uncontrolled results, which is suitable for the rapid processing of big data.

Following the idea of the Top-k Skyline query algorithm DFTS [11], we define the degree score function in Definition 3 to sort the dataset and filter out a large number of lower-ranked nodes. Then skyline query is made on the candidate set and Top-k results with high ranking will be selected as the final results.

Definition 3. *(Degree Score, ϕ) Given a d-dimensional data set $D(p_1, p_2, ..., p_n)$, $p_i[j]$ denotes p_i in the jth dimension, μ_j denotes the mean value of p_i in the jth dimension, $\lambda_{i,j}$ denotes p_i the degree score in the jth dimension, $\phi(p_i)$ denotes the sum of all the dimensions, as shown in the following:*

$$\phi(p_i) = \sum_{j=1}^{d} \lambda_{i,j}\tag{4}$$

where $\lambda_{i,j}$ and μ_j defined as follows:

$$\lambda_{i,j} = \frac{p_i[j] - \mu_j}{\mu_j}\tag{5}$$

$$\mu_j = \frac{\sum_{i=1}^{n} p_i[j]}{n}\tag{6}$$

The SSPR algorithm is shown in Algorithm 1. RSP_k is the queue that holds the Top-k results, and the elements in the queue are sorted in ascending order by degree score. $SkyList$ is used to store the candidate places that match the skyline query.

For a query q, according to the query location $q.\lambda$ and the spatial index $Rtree$, we first obtain at most $maxnum$ place vertices within the radius $maxdis$ of the origin as result candidates (line 2). These vertices are arranged ascending order of spatial distance from the $q.\lambda$. For each place vertex, calculate the semantic similarity S_{np} between the node and the query keyword list according to Eqs. 1 and 2 (line 4). If S_{np} higher than the semantic similarity threshold, then use the semantic relevance score f_{np} and spatial distance D_{np} of this node to construct a skyline query point $node(f_{np}, D_{np})$ and add it to skyline list (lines 5–8). Next, the nodes in the $skyList$ sorted in descending order of the degree score $\phi(node)$, and a large number of nodes are quickly filtered out, only a sufficiently small number of nodes are retained as the candidate set candList (lines 11–14). Then, a skyline query is made on the $candList$. Since the size of the $candList$ has been greatly reduced after filtering, we use the BNL algorithm for the skyline query (line 15). Finally, Top-k semantic results RSP_k are selected as the output according to the $\phi(node)$.

Algorithm 1. SSPR (Skyline-Based Semantic Place Retrieval)

Input: $q, Rtree, G, I$
Output: RSP_k
1: $skyList = \emptyset$
2: $list_{np} = GetByDis(R, q.\lambda, maxdis, maxnum)$
3: **while** $e = GetNext(list_{np})$ **do**
4:      Compute the semantic similarity S_e
5:      **if** $S_e \geq \tau$ **then**
6:          Compute the semantic relevance score f_e
7:          $D_e = D(q.\lambda, e)$
8:          $skyList.add(node(f_e, D_e))$
9:      **end if**
10: **end while**
11: **for** each node in $skyList$ **do**
12:      Compute $\phi(node)$ according to Equation 5
13: **end for**
14: $candList = skyList.top(n)$
15: Calling BNL do skyline query
16: $RSP_k = candList.top(k)$
17: **return** RSP_k

---

4   Evaluation

In this section, we compare our method with traditional kSP retrieval in the same environment. The experimental setup and a discussion of results are presented.

4.1   Experimental Setup

Experiments are carried out on a 3.4 GHz quad-core machine with Ubuntu 16.04 and 32G of memory, and the algorithm is implemented in Java.

We use two datasets to demonstrate our experimental results: YAGO2 and DBpedia [12]. The specific information of the datasets we use is shown in Table 1. We generate a query set A containing 100 queries. For each query of A, we randomly select a place p in the RDF graph and then randomly select the query location from a large range around this place. From p, we explore the RDF graph in BFS manner and randomly select at least $|q.\psi|/2$ and at most $|q.\psi| \times 2$ vertices that are reachable from p. If there are less than $|q.\psi|/2$ vertices reachable from p, p is discarded to avoid too few subgraphs around the query location. Among the selected $[|q.\psi|/2, |q.\psi| \times 2]$ vertices, at most $|q.\psi|$ vertices are randomly selected and $|q.\psi|$ different words are randomly extracted as query keywords from these vertices.

Table 1. Statistics of experimental datasets.

Dataset	Number of vertices	Number of sides	Keyword	Number of place vertices
YAGO2	809 ten thousand	50 million	3.7 million	477 ten thousand
DBpedia	809 ten thousand	72 million	2.9 million	88 ten thousand

Also, we created a query set B to simulate the possible user input keywords by randomly replacing one or two of the keywords with their most similar words in query set A.

4.2   Evaluation of Query Performance

Our methodology is evaluated by varying the (i) number k of requested places, (ii) the number of query keywords $|q.\psi|$. Table 2 lists the values of the parameters. The values in bold are the (fixed) default values. The semantic similarity threshold τ defaults to 0.7. In each experiment, we vary one parameter while fixing the remaining ones to their default values. For each setting, we run 100 queries and measure the average runtime, average spatial distance and average semantic relevance of top-k results.

Varying k. For query set A, the average runtime, average spatial distance and average semantic relevance of Top-k results for all methods were first evaluated, and the experimental results are shown in Fig. 4 and Fig. 5.

Table 2. Parameter settings.

Parameter	Values
k	1, 3, **5**, 8, 10, 15, 20
$|q.\psi|$	1, 2, **3**, 4, 5

The experimental results show that the runtime of kSP increases as k increases, while the runtime of SSPR is almost independent of k. This is because kSP request more places need to explore a larger search space, while SSPR need to traverse all the places within a given space, and the runtime is certain regardless of the returned results.

On the YAGO2, SSPR runs more than 2 times as fast as the kSP, and as k increases, SSPR becomes increasingly effective in optimizing the retrieval speed. This is because kSP requires the construction of TQSPs (Tightest Qualified Semantic Place), and the time to construct TQSPs dominates the running time, while the SSPR only needs to compute the similarity between word embeddings.

As shown in Fig. 4(b), unsurprisingly, the average spatial distance of the returned results of all methods increases with increasing k. The gap between SSPR and kSP remains stable as k increases, and the average spatial distance of SSPR is about 0.07–0.09 km farther than kSP. As shown in Fig. 4(c), the average semantic relevance of kSP remains stable and is almost unaffected by the value of k, while the average semantic relevance of SSPR increases with the increase of k value, which is 1.1–2.4 times higher than that of kSP, that is, the SSPR gets better in terms of semantic relevance compared with kSP as the k value increases. In a comprehensive analysis, SSPR balances spatial distance and semantic relevance to provide users with more diverse options, such as some results that are slightly more distant but more consistent with the user's semantic intent.

The experimental results on DBpedia are similar to YAGO2 (Fig. 5). Compared with YAGO2, DBpedia has a few less place vertices, kSP needs more time to explore the RDF graph for keyword exact matching, while SSPR has the feature of fuzzy matching and shows excellent performance on this large spatial RDF dataset.

Varying $|q.\psi|$. Varying $|q.\psi|$, the runtime of all methods is evaluated, and the results are shown in Fig. 6. The average spatial distance and average semantic relevance are almost not affected by $|q.\psi|$, so we do not show them. As $|q.\psi|$ increases, kSP needs to explore more vertices in the RDF graph to discover the TQSP covering all query keywords, and SSPR needs to generate more word

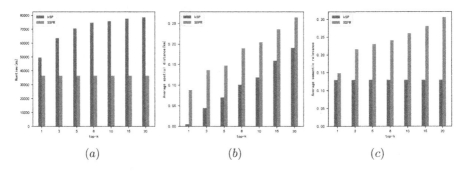

(a) (b) (c)

Fig. 4. Varying k on YAGO2. (a) Runtime; (b) Average spatial distance of Top-k; (c) Average semantic relevance of Top-k.

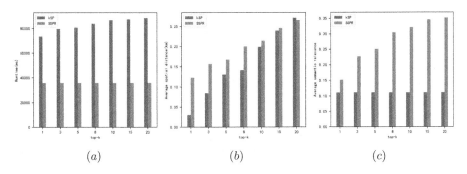

Fig. 5. Varying k on DBpedia. (a) Runtime; (b) Average spatial distance of Top-k; (c) Average semantic relevance of Top-k.

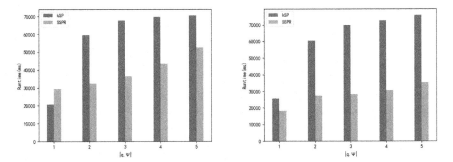

Fig. 6. Varying $|q.\psi|$ on YAGO2(left); Varying $|q.\psi|$ on DBpedia(right).

embeddings and calculate the semantic similarity of more words, so the runtime of both kSP and SSPR increases. The runtime of SSPR is smaller than kSP except for the case where $|q.\psi|=1$. With the increase of the number of keywords, SSPR is 1.3–1.8 times faster than kSP. Compared to YAGO2, the runtime gap between kSP and SSPR becomes larger on DBpedia because DBpedia has fewer place entities, so SSPR takes less time to visit fewer place entities for a given spatial extent.

For Query Set B. Finally, when we experimented with query set B, we found that kSP could return results for only 17 queries out of 100 queries, while SSPR could find results for 78 queries. This is because kSP requires exact keyword matching, and the actual keywords entered by the user may not exist in the RDF keyword set. This confirms that our approach can return results for more queries and is more user-friendly.

5   Conclusion

In this paper, we propose SSPR, a skyline-based semantic place retrieval method that takes as input a query location $q.\lambda$ and a set of query keywords $q.\psi$ and returns Top-k places on the RDF graph by their spatial distance from $q.\lambda$ and their semantic relevance to $q.\psi$. We utilize Word2vec to generate word embeddings, and calculate the semantic relevance between query keywords and places for fuzzy matching to return more possible results for users. In addition, we design a skyline-based efficient query algorithm to select places in terms of both semantic relevance and spatial distance. We compare the query performance of SSPR with traditional kSP. The experimental results show that our method outperforms in retrieval efficiency, and SSPR obtains the implied places that match the user's query intent, effectively improving the phenomenon of missing returned results.

For future work, we consider further improvements on the calculation of semantic relevance, by changing the weighting parameters.

Acknowledgements. The work is supported in part by the National Key R&D Program of China (Grant No. 2021YFB3900601).

References

1. Shan, X., Qiu, J., Wang, B., Dang, Y., Lu, T., Zheng, Y.: Place retrieval in knowledge graph. In: Scientific Programming (2020)
2. Shi, J., Wu, D., Mamoulis, N.: Top-k relevant semantic place retrieval on spatial RDF data. In: Proceedings of the 2016 International Conference on Management of Data, pp. 1977–1990 (2016)
3. Le, W., Li, F., Kementsietsidis, A., Duan, S.: Scalable keyword search on large RDF data. IEEE Trans. Knowl. Data Eng. **26**(11), 2774–2788 (2014)
4. Stefanidis, K., Fundulaki, I.: Keyword search on RDF graphs: it is more than just searching for keywords. In: Gandon, F., Guéret, C., Villata, S., Breslin, J., Faron-Zucker, C., Zimmermann, A. (eds.) ESWC 2015. LNCS, vol. 9341, pp. 144–148. Springer, Cham (2015). https://doi.org/10.1007/978-3-319-25639-9_28
5. Cai, Z., Kalamatianos, G., Fakas, G.J., Mamoulis, N., Papadias, D.: Diversified spatial keyword search on RDF data. VLDB J. **29**(5), 1171–1189 (2020)
6. Wu, D., Hou, C., Xiao, E., Jensen, C.S.: Semantic region retrieval from spatial RDF data. In: Nah, Y., Cui, B., Lee, S.-W., Yu, J.X., Moon, Y.-S., Whang, S.E. (eds.) DASFAA 2020. LNCS, vol. 12113, pp. 415–431. Springer, Cham (2020). https://doi.org/10.1007/978-3-030-59416-9_25
7. Hoffart, J., Suchanek, F.M., Berberich, K., Weikum, G.: Yago2: a spatially and temporally enhanced knowledge base from wikipedia. Artif. Intell. **194**, 28–61 (2013)
8. Jin, X., Shin, S., Jo, E., Lee, K.H.: Collective keyword query on a spatial knowledge base. IEEE Trans. Knowl. Data Eng. **31**(11), 2051–2062 (2018)
9. Zeng, K., Yang, J., Wang, H., Shao, B., Wang, Z.: A distributed graph engine for web scale RDF data. Proceed. VLDB Endow. **6**(4), 265–276 (2013)
10. Mikolov, T., Sutskever, I., Chen, K., Corrado, G.S., Dean, J.: Distributed representations of words and phrases and their compositionality. In: Advances In Neural Information Processing Systems 26 (2013)
11. Wei, L., Lin, Z., Lai, Y.: DTFS: A Top-k skyline query for large datasets. In: Computer Science, vol. 46, no. 5, pp. 150–156 (2019)
12. Lehmann, J., et al.: Dbpedia-a large-scale, multilingual knowledge base extracted from wikipedia. Semantic web **6**(2), 167–195 (2015)

Double Regularization-Based RVFL and edRVFL Networks for Sparse-Dataset Classification

Qiushi Shi[1][(✉)] [iD] and Ponnuthurai Nagaratnam Suganthan[1,2] [iD]

[1] School of Electrical and Electronic Engineering, Nanyang Technological University, 50 Nanyang Avenue, Singapore 639798, Singapore
qiushi001@e.ntu.edu.sg, EPNSugan@ntu.edu.sg
[2] KINDI Center for Computing Research, College of Engineering, Qatar University, Doha, Qatar

Abstract. In our previous work, the random vector functional link network (RVFL) and the ensemble deep RVFL network (edRVFL) have been proven to be competitive for tabular-dataset classification, and their sparse pre-trained versions (SP-RVFL and SP-edRVFL) perform well for sparse-dataset (datasets with a large number of features) classification. However, the sparse auto-encoder-based versions suffer from a time-consuming problem. Therefore, we need to find an alternative way to have similar performance and faster training time. In this paper, we propose the double regularization-based RVFL (2R-RVFL) and edRVFL networks (2R-edRVFL). Two different regularization parameters are assigned to the input and hidden features, respectively. The experiments on 12 sparse datasets show that the 2R-RVFL and 2R-edRVFL networks have similar performance as the SP-RVFL and SP-edRVFL networks, and the double regularized variants have a huge training time advantage. Thus, we believe the newly proposed 2R-RVFL and 2R-edRVFL networks are more suitable for sparse dataset classification.

Keywords: Random vector functional link network (RVFL) · Ensemble deep RVFL network (edRVFL) · Sparse auto-encoder · Double regularization · Training time advantage · Sparse-dataset classification

1 Introduction

In recent years, randomized neural networks are becoming more and more attractive because of their universal approximation ability and fast training speed [2,14,15]. Among them, Random Vector Functional Link (RVFL) network is the most popular one with various applications [1,3,5,6,9,17]. A typical RVFL network is a single hidden layer feed-forward network with a direct link converting the original features to the output layer. By concatenating the hidden features and the original features, the output weights can be solved by a closed-form solution [7,16].

M. Tanveer et al. (Eds.): ICONIP 2022, LNCS 13623, pp. 343–354, 2023.
https://doi.org/10.1007/978-3-031-30105-6_29

Inspired by the deep learning architectures, we propose the deep version of the shallow RVFL network named the ensemble deep RVFL (edRVFL) network in our previous paper [10]. And we propose some variants of the edRVFL network recently [11,12]. The edRVFL network achieves the best performance against other classifiers on benchmark datasets and it can be considered a competitive model for classification. We also develop the sparse pre-trained edRVFL network (SP-edRVFL) by extending the sparse pre-trained RVFL network (SP-RVFL) [18] to the ensemble deep version to deal with the sparse dataset (datasets with a large number of features) classification. The SP-RVFL and the SP-edRVFL use a sparse auto-encoder to pre-train the randomly generated hidden weights in the networks. And they can achieve higher accuracy than their original versions (RVFL and edRVFL) on the sparse dataset classification tasks.

However, these SP versions suffer from a time-consuming problem. Since we need to pre-train the hidden weights for every layer and this training procedure is an iterative learning method, the training time of the SP-edRVFL and the SP-RVFL is much more than their original versions'. This encourages us to find alternative ways to deal with the sparse datasets. Thus, in this paper, we propose the double regularization-based RVFL and edRVFL networks (2R-RVFL and 2R-edRVFL). In the original RVFL network, both the non-linear features (hidden nodes) and the linear features (direct-link) are concatenated and sent to the output layer together. In most cases, the output weights are calculated using ridge regression with the regularization term [4]. This means the two parts (linear and non-linear) of features share the same regularization degree λ. When the dataset has a small number of input features, this method can still work since the original features only take a small portion. However, when it comes to sparse datasets, it becomes necessary for us to give different regularization hyperparameters to these two parts, respectively. The same strategy can be applied to the edRVFL network because it contains a similar architecture as the RVFL network. We name these two new models double regularization-based RVFL and edRVFL networks (2R-RVFL and 2R-edRVFL).

The key contributions of this paper are summarized as follows:

- We introduce the double regularization to the RVFL and edRVFL networks and propose the 2R-RVFL and 2R-edRVFL networks to solve the time-consuming problem of the SP-RVFL and SP-edRVFL networks on sparse datasets.
- The 2R-RVFL and 2R-edRVFL networks have similar performance as the SP-RVFL and SP-edRVFL networks on 12 sparse datasets.
- The 2R-RVFL and 2R-edRVFL networks have a huge training time advantage compared to the SP-RVFL and SP-edRVFL networks.

2   Related Works

In this section, we first review the architecture of the RVFL network [7]. Then, we introduce its deep version - the edRVFL network [10].

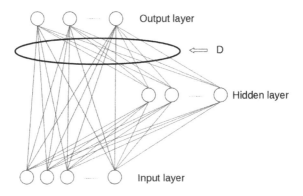

Fig. 1. The structure of RVFL network.

2.1 Random Vector Function Link

The RVFL network is a famous randomized neural network proposed in [7]. The standard framework of the RVFL network can be found in Fig. 1. A typical RVFL network has only one hidden layer. The uniqueness of this model lies in its direct link which conveys the input features to the output layer. Then these linear features are concatenated with the nonlinear hidden features. We denote this matrix $\mathbf{D}$. Therefore, the objective function of the RVFL network can be written as:

$$O_{RVFL} = \min_{\beta} ||\mathbf{D}\beta - \mathbf{Y}||^2 + \lambda ||\beta||^2 \tag{1}$$

Here β is the output weights we want to optimize, $\mathbf{Y}$ represents the labels of the samples, and λ is the regularization parameter so the term $\lambda||\beta||^2$ controls the complexity of the model.

Then, we can obtain the output weights β by ridge regression [4]. The solution is given below:

$$PrimalSpace : \beta = (\mathbf{D^T D} + \lambda \mathbf{I})^{-1}\mathbf{D^T Y} \tag{2}$$

$$DualSpace : \beta = \mathbf{D^T}(\mathbf{DD^T} + \lambda \mathbf{I})^{-1}\mathbf{Y} \tag{3}$$

2.2 Ensemble Deep Random Vector Functional Link

As shown in Fig. 2. The edRVFL network can be treated as the ensemble of several RVFL networks. The generation step of the first hidden of the edRVFL can be written as below:

$$\mathbf{H}^{(1)} = g(\mathbf{X}\mathbf{W}^{(1)}), \quad \mathbf{W}^{(1)} \in \mathbb{R}^{d \times n} \tag{4}$$

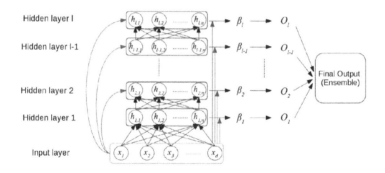

Fig. 2. The structure of edRVFL network.

where $\mathbf{X}$ is the input features, $\mathbf{W}^{(1)}$ represents the hidden weights matrix of the first hidden layer, and d, n are the dimension of the input features and the number of the hidden neurons, respectively. $g(\cdot)$ is the non-linear activation function used in each hidden neuron. When layer $l > 1$, both the features from the previous hidden layer and the input layer will be used to generate the next hidden layer, so the formula will change to:

$$\mathbf{H}^{(l)} = g([\mathbf{H}^{(l-1)}\mathbf{X}]\mathbf{W}^{(l)}), \quad \mathbf{W}^{(l)} \in \mathbb{R}^{(n+d)\times n} \tag{5}$$

Each hidden layer in the edRVFL network served as an independent classifier, or in other words, an RVFL classifier. For an edRVFL network that contains l hidden layers, we can have l different outputs. Thus, an ensemble method like majority voting [8] is employed to reach the final prediction.

3   Double Regularized RVFL and edRVFL Networks (2R-RVFL and 2R-edRVFL)

In this section, we propose the double regularization-based RVFL and edRVFL networks. For the direct link and the hidden features, two different regularization hyperparameters λ_X and λ_H are assigned to them, respectively. Instead of using a single λ to control the model complexity, this double regularization strategy makes the linear and non-linear parts have their preference of the regularization term. Therefore, the neural network can have better generalization ability for a specific dataset.

Firstly, we review the regularization scheme used in the standard RVFL network. As we can see in Fig. 3, the model complexity of the RVFL network is affected by two parts. One is the complexity of the output weights β_X for the original features which can be written as $\lambda||\beta_X||^2$, and the other one is the complexity of the output weights β_H for the hidden features which is represented by $\lambda||\beta_H||^2$. Since we concatenate the non-linear part and linear part together

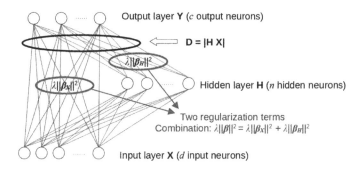

Fig. 3. The structure of RVFL network.

to calculate the whole output weights $\boldsymbol{\beta}$ (which is $\begin{bmatrix} \boldsymbol{\beta}_X \\ \boldsymbol{\beta}_H \end{bmatrix}$), the above two terms are usually combined and merged into $\lambda||\boldsymbol{\beta}||^2$. Thus, the optimization problem of the RVFL network is shown as:

$$O_{RVFL} = \min_{\boldsymbol{\beta}} ||\mathbf{D}\boldsymbol{\beta} - \mathbf{Y}||^2 + \lambda||\boldsymbol{\beta}||^2 \tag{6}$$

When the input feature number d is much smaller than the hidden neuron number n, using one regularization parameter λ is enough. Since the complexity of the hidden layer's output $\boldsymbol{\beta}_H$ is dominant in these cases, the regularization term $\lambda||\boldsymbol{\beta}_X||^2$ for the input features only has little effect on the whole model. Thus, tuning the hyperparameter λ is equal to tuning the importance of the model complexity of the non-linear part. Once we find a suitable $\lambda||\boldsymbol{\beta}_H||^2$ using cross-validation, the RVFL network is able to overcome the over-fitting problem on a specific dataset.

However, when we encounter the dataset with a large number of input features, which is named the sparse dataset by us, we cannot ignore the contribution of the regularization term $\lambda||\boldsymbol{\beta}_X||^2$. So it is easy for us to think about giving two different regularization parameters λ_X and λ_H to the linear and the non-linear part as in Fig. 4. Therefore, the optimization problem of the RVFL network is changed from Eq. (6) to the following expression:

$$O_{2R-RVFL} = \min_{\boldsymbol{\beta}} ||\mathbf{D}\boldsymbol{\beta} - \mathbf{Y}||^2 + \lambda_X||\boldsymbol{\beta}_X||^2 + \lambda_H||\boldsymbol{\beta}_H||^2, \quad \boldsymbol{\beta} = \begin{bmatrix} \boldsymbol{\beta}_X \\ \boldsymbol{\beta}_H \end{bmatrix} \tag{7}$$

For finding the optimal value of $\boldsymbol{\beta}$, we need to deform the formula of the two regularization terms $\lambda_X||\boldsymbol{\beta}_X||^2$ and $\lambda_H||\boldsymbol{\beta}_H||^2$. However, they cannot be easily combined as in Eq. (6) since we use different regularization parameters for the input layer and the hidden layer. So we try to replace $\boldsymbol{\beta}_X$ and $\boldsymbol{\beta}_H$ by $\boldsymbol{\beta}$ in a different way.

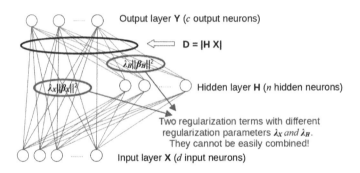

Fig. 4. The structure of 2R-RVFL network.

First of all, we can deal with the regularization parameters. Here λ_X and λ_H are two constant values. Can we put these two parameters in one matrix? The answer is Yes: We can create a diagonal matrix using these two regularization values to control the two parts' complexity in β separately. Denote the matrix sizes of β_X and β_H are $(d \times c)$ and $(n \times c)$, the diagonal regularization matrix λ_D is:

$$\begin{bmatrix} \lambda_1 & 0 & \cdots & 0 \\ 0 & \lambda_2 & \cdots & 0 \\ \vdots & \vdots & \ddots & \vdots \\ 0 & 0 & \cdots & \lambda_{d+n} \end{bmatrix} \tag{8}$$

where $\lambda_D \in \mathbb{R}^{(n+d)\times(n+d)}$. The first d diagonal values λ_1 to λ_d equal to $\sqrt{\lambda_X}$ and last n diagonal values λ_{d+1} to λ_{d+n} equal to $\sqrt{\lambda_H}$. The reason for this setting is to match β_X and β_H.

According to Eqs. (8) and $\beta = \begin{bmatrix} \beta_X \\ \beta_H \end{bmatrix}$, it can be easily proven that:

$$\begin{aligned} ||\lambda_D\beta||^2 &= || \begin{bmatrix} \sqrt{\lambda_X}\mathbf{I} & \\ & \sqrt{\lambda_H}\mathbf{I} \end{bmatrix} \begin{bmatrix} \beta_X \\ \beta_H \end{bmatrix} ||^2 \\ &= || \begin{bmatrix} \sqrt{\lambda_X}\mathbf{I}\beta_X \\ \sqrt{\lambda_H}\mathbf{I}\beta_H \end{bmatrix} ||^2 \\ &= ||\sqrt{\lambda_X}\beta_X||^2 + ||\sqrt{\lambda_H}\beta_H||^2 \\ &= \lambda_X||\beta_X||^2 + \lambda_H||\beta_H||^2 \end{aligned} \tag{9}$$

Thus, Eq. (7) can be rewritten as:

$$O_{2R-RVFL} = \min_{\beta} ||\mathbf{D}\beta - \mathbf{Y}||^2 + ||\lambda_D\beta||^2 \tag{10}$$

Finally, we can obtain the optimal β by the ridge regression:

$$\beta = (\mathbf{D^T D} + \lambda_D^T \lambda_D)^{-1}\mathbf{D^T Y} \tag{11}$$

Unfortunately, there is no dual solution for the 2R-RVFL network since the commutative law is no longer satisfied:

$$\lambda_D^T \lambda_D \mathbf{D}^T \neq \mathbf{D}^T \lambda_D^T \lambda_D \tag{12}$$

while it works when the λ is a constant value instead of a diagonal matrix in [15].

Now we complete the training part of the 2R-RVFL network through the above equations. Similarly, the same mechanism is used to generate the 2R-edRVFL network. We present the detailed training steps of the 2R-edRVFL network in Algorithm 1 (validation and hyperparameter tuning are omitted here).

Algorithm 1: 2R-edRVFL

Input: Giving m_1 training samples $\mathbf{X_1} \in \mathbb{R}^{m_1 \times d}$ with their labels $\mathbf{Y_1} \in \mathbb{R}^{m_1 \times c}$. And m_2 samples $\mathbf{X_2} \in \mathbb{R}^{m_2 \times d}$ with labels unknown.

Initialize the hidden weights and generate the hidden layers of 2R-edRVFL
The hidden nodes of the first hidden layer are generated as follows:

$$\mathbf{H}^{(1)} = g(\mathbf{X_1}\mathbf{W}^{(1)}) \tag{13}$$

while for every layer $l > 1$ it is defined as:

$$\mathbf{H}^{(l)} = g([\mathbf{H}^{(l-1)}\mathbf{X_1}]\mathbf{W}^{(l)}) \tag{14}$$

For the given combination of λ_X and λ_H, generate matrix λ_D according to Eq. (8)
for *Every Hidden Layer* **do**
 Calculate β according to

$$\beta = (\mathbf{D}^T\mathbf{D} + \lambda_D^T\lambda_D)^{-1}\mathbf{D}^T\mathbf{Y_1} \tag{15}$$

 where $\mathbf{D} = [\mathbf{H}\ \mathbf{X_1}]$
end
Save the model and feed it with the testing samples $\mathbf{X_2}$
for *Every Hidden Layer* **do**
 Get the prediction $\mathbf{Y_2'}$ for the testing data from the current layer

$$\mathbf{Y_2'} = \mathbf{D}\beta \tag{16}$$

end
Combine all the predicted $\mathbf{Y_2'}$ using ensemble methods to reach the final prediction $\mathbf{Y_2}$
Output: The predicted labels $\mathbf{Y_2}$

---

4   Comparison Between Sparse Pre-trained Networks and Double Regularized Networks

In the paper, we first review the previous works - the SP-RVFL network and the SP-edRVFL network. Then, we propose the 2R-RVFL and 2R-edRVFL networks using the double regularization scheme. Although these models are all developed for sparse datasets, the mechanism is quite different to deal with the problem. To avoid confusing the audience, we do a simple tabular comparison between the sparse auto-encoder-based networks (SP-RVFL and SP-edRVFL), double regularization-based networks (2R-RVFL and 2R-edRVFL), and their original versions (RVFL and edRVFL) in the following Table 1.

Table 1. Comparison of the networks

	Depth	Type of the hidden weights	Type of the regularization
SP-RVFL	Shallow	Pre-trained using the sparse auto-encoder	Only one λ for both linear and non-linear features
SP-edRVFL	Deep	Pre-trained using the sparse auto-encoder	Only one λ for both linear and non-linear features
2R-RVFL	Shallow	Randomly generated	λ_X and λ_H for linear and non-linear features, separately
2R-edRVFL	Deep	Randomly generated	λ_X and λ_H for linear and non-linear features, separately
RVFL	Shallow	Randomly generated	Only one λ for both linear and non-linear features
edRVFL	Deep	Randomly generated	Only one λ for both linear and non-linear features

5   Experiments

5.1   Datasets

In this paper, we conduct experiments on 12 sparse datasets from various domains. The feature dimensions of these datasets are all larger than 100. We present the details of these datasets in Table 2. It is also worth mentioning that we first separate the samples into two parts - training (75%) and testing (25%). Then, the training set is also divided into two groups for validation training (75%) and validation testing (25%) to tune the best hyperparameters.

5.2   Compared Methods

In this section, we compare the performance of 6 RVFL variants. Since the comparison of RVFL and edRVFL based methods with other popular machine learning models has already been down in our previous work [10], and the main purpose of this paper is to find more efficient methods to replace the SP-RVFL and SP-edRVFL networks, here we compare the performance of the sparse auto-encoder based networks, double regularization based networks, and their original versions in this section. The 6 algorithms are listed as follows:

Table 2. Sparse datasets

Dataset	#Patterns	#Features	#Classes
arrhythmia	452	263	13
BASEHOCK	1993	1000	2
bbc	2225	1000	2
bbcsport	737	1000	5
hill-valley	1212	101	2
low-res-spect	531	101	5
musk-1	476	167	2
musk-2	6598	167	2
RCV1	9625	1000	2
RELATHE	1427	1000	2
semeion	1593	257	10
TDT2	9394	1000	30

1. RVFL: Standard shallow RVFL network [7].
2. SP-RVFL: Sparse pre-trained RVFL network [18].
3. 2R-RVFL: Double regularized RVFL network.
4. edRVFL: Ensemble deep RVFL network [10].
5. SP-edRVFL: Sparse pre-trained edRVFL network [10].
6. 2R-edRVFL: Double regularized edRVFL network.

5.3   Hyperparameter Setting

In this work, we use Bayesian optimization [13] to find the best hyperparameter setting instead of the grid search. The maximum iteration number is set to 100 in this case. We present details of the hyperparameter considered for RVFL based methods in Table 3.

Table 3. Hyperparameters considered for RVFL based methods

Hyperparameter	Considered values
Regularization parameter λ (for RVFL, SP-RVFL, edRVFL and SP-edRVFL)	λ belongs to 2^x, $x \in [-12, 12]$
Regularization parameter λ_X and λ_H (for 2R-RVFL and 2R-edRVFL)	λ_X and λ_H both belong to 2^x, $x \in [-12, 12]$
Number of hidden neurons	[20,1000]
Maximum number of hidden layers (for edRVFL, 2R-edRVFL, and SP-edRVFL)	10
Activation function	Relu (0), Selu (1), and Sigmoid (2)

5.4   Experimental Results

The testing actuaries of 6 RVFL based methods are summarized in Table 4. The highest accuracy of each dataset is given in bold. As we can see from the table, the sparse pre-trained RVFL variants and the double-regularized variants are all performing better than the original RVFL versions. In addition, the double-regularized versions have similar performance as the sparse pre-trained versions. The 2R-RVFL network is slightly better than the SP-RVFL network (90.53% vs. 90.50%), and the 2R-edRVFL model is slightly worse than the SP-edRVFL model (90.80% vs. 90.99%).

Table 4. Comparison of 6 RVFL based methods on sparse datasets (%)

Dataset	SP-edRVFL [10]	SP-RVFL [18]	2R-edRVFL	2R-RVFL	edRVFL [10]	RVFL [7]
arrhythmia	71.68	71.90	71.90	**73.45**	69.91	71.46
BASEHOCK	97.99	97.49	**98.14**	97.69	97.04	96.59
bbc	**97.44**	97.17	96.81	97.17	96.81	96.63
bbcsport	98.10	**98.37**	97.15	**98.37**	97.15	97.28
hill-valley	70.96	**77.39**	73.93	73.93	67.00	69.64
low-res-spect	90.41	89.10	**91.73**	90.23	89.66	89.85
musk-1	90.13	84.03	**90.47**	84.45	88.24	80.67
musk-2	98.42	98.23	**98.58**	98.36	98.57	98.29
RCV1	**94.38**	93.33	93.33	93.18	93.18	93.21
RELATHE	89.28	88.16	87.35	**89.35**	86.33	86.62
semeion	**96.11**	94.60	93.84	93.84	92.96	92.90
TDT2	**96.98**	96.23	96.37	96.37	96.37	96.33
Average Accuracy	**90.99**	90.50	90.80	90.53	89.44	89.12
Mean Rank	**2.17**	3.50	2.67	2.88	4.71	5.08

5.5   Training Time Comparison

Although the SP-edRVFL achieves the best in Table 4, the time consuming is a serious problem for this method. Since the SP-edRVFL and SP-RVFL networks need to pre-train their hidden weights, they require more time to predict compared to the other methods. In this section, we perform experiments on the low-res-spect dataset to show the training time comparison of the 6 methods based on 100 Bayesian optimization iterations. The results can be found in Table 5.

Compare to the original edRVFL and RVFL networks, the sparse pre-trained and double-regularized versions are all slower during the training process. Furthermore, the sparse auto-encoder significantly increases the training complexity of the network. The training time for the SP-edRVFL network is almost 5 times of the 2R-edRVFL model and 13 times of the edRVFL model. Therefore, we think the double-regularized networks are more suitable for the sparse-dataset classification since the 2R-edRVFL network is only slightly worse than the SP-edRVFL network but has a huge training-time advantage.

Table 5. Training Time comparison on the low-res-spect dataset

Algorithm	Training Time (s)
SP-edRVFL	334.34
2R-edRVFL	70.01
SP-RVFL	30.95
edRVFL	25.53
2R-RVFL	16.76
RVFL	5.27

*Experiment environment:
Intel(R) Xeon(R) CPU E5-2620;
nVIDIA GeForce GTX-1080.

6   Conclusion

In this paper, we are focusing on the sparse-dataset (dataset with large feature number) classification. we first review the shallow sparse pre-trained RVFL (SP-RVFL) network and its ensemble deep version (SP-edRVFL). After that, we propose the double-regularized RVFL (2R-RVFL) and edRVFL (2R-edRVFL) networks. We assign two different regularization parameters to the original features and the hidden features since we believe the linear input features are also important to the sparse-dataset classification. Then, we conduct experiments on 12 sparse datasets where the 2R-RVFL and 2R-edRVFL networks have similar performance as the SP-RVFL and SP-edRVFL networks. At last, we compare the training time of these 6 RVFL based methods on the low-res-spect dataset. We can learn that the SP-edRVFL suffers from a time-consuming problem. And the training time for 2R-RVFL and 2R-edRVFL is much less than the sparse pre-trained versions. Thus, we believe the newly proposed 2R-RVFL and 2R-edRVFL are more suitable for sparse-dataset classification.

References

1. Cui, W., et al.: Received signal strength based indoor positioning using a random vector functional link network. IEEE Trans. Industr. Inf. **14**(5), 1846–1855 (2017)
2. Gallicchio, C., Scardapane, S.: Deep randomized neural networks. In: Recent Trends in Learning From Data, pp. 43–68 (2020)
3. Ganaie, M.A., Hu, M., et al.: Ensemble deep learning: a review. arXiv preprint arXiv:2104.02395 (2021)
4. Hoerl, A.E., Kennard, R.W.: Ridge regression: biased estimation for nonorthogonal problems. Technometrics **12**(1), 55–67 (1970)
5. Hu, M., Shi, Q., Suganthan, P.N., Tanveer, M.: Adaptive ensemble variants of random vector functional link networks. In: Yang, H., Pasupa, K., Leung, A.C.-S., Kwok, J.T., Chan, J.H., King, I. (eds.) ICONIP 2020. CCIS, vol. 1333, pp. 30–37. Springer, Cham (2020). https://doi.org/10.1007/978-3-030-63823-8_4
6. Malik, A.K., Gao, R., Ganaie, M., Tanveer, M., Suganthan, P.N.: Random vector functional link network: recent developments, applications, and future directions. arXiv preprint arXiv:2203.11316 (2022)
7. Pao, Y.H., Takefuji, Y.: Functional-link net computing: theory, system architecture, and functionalities. Computer **25**(5), 76–79 (1992)
8. Penrose, L.S.: The elementary statistics of majority voting. J. Roy. Stat. Soc. **109**(1), 53–57 (1946)
9. Ren, Y., Suganthan, P.N., Srikanth, N., Amaratunga, G.: Random vector functional link network for short-term electricity load demand forecasting. Inf. Sci. **367**, 1078–1093 (2016)
10. Shi, Q., Katuwal, R., Suganthan, P., Tanveer, M.: Random vector functional link neural network based ensemble deep learning. Pattern Recogn. **117**, 107978 (2021)
11. Shi, Q., Suganthan, P.N., Del Ser, J.: Jointly optimized ensemble deep random vector functional link network for semi-supervised classification. Eng. Appl. Artif. Intell. **115**, 105214 (2022)
12. Shi, Q., Suganthan, P.N., Katuwal, R.: Weighting and pruning based ensemble deep random vector functional link network for tabular data classification. arXiv preprint arXiv:2201.05809 (2022)
13. Snoek, J., Larochelle, H., Adams, R.P.: Practical Bayesian optimization of machine learning algorithms. arXiv preprint arXiv:1206.2944 (2012)
14. Suganthan, P.N., Katuwal, R.: On the origins of randomization-based feedforward neural networks. Appl. Soft Comput. **105**, 107239 (2021)
15. Suganthan, P.N.: On non-iterative learning algorithms with closed-form solution. Appl. Soft Comput. **70**, 1078–1082 (2018)
16. Zhang, L., Suganthan, P.N.: A comprehensive evaluation of random vector functional link networks. Inf. Sci. **367**, 1094–1105 (2016)
17. Zhang, L., Suganthan, P.N.: Visual tracking with convolutional random vector functional link network. IEEE Trans. Cybern. **47**(10), 3243–3253 (2016)
18. Zhang, Y., Wu, J., Cai, Z., Du, B., Philip, S.Y.: An unsupervised parameter learning model for RVFL neural network. Neural Netw. **112**, 85–97 (2019)

Adaptive Tabu Dropout for Regularization of Deep Neural Networks

Md. Tarek Hasan[✉], Arifa Akter, Mohammad Nazmush Shamael,
Md Al Emran Hossain, H. M. Mutasim Billah, Sumayra Islam,
and Swakkhar Shatabda

Department of Computer Science and Engineering, United International University,
Plot-2, United City, Madani Avenue, Dhaka, Badda 1212, Bangladesh
{tarek,swakkhar}@cse.uiu.ac.bd,
{aakter181254,mshamael181062,mhossain181144,hbillah181290,
sislam181123}@bscse.uiu.ac.bd

Abstract. Dropout is an effective strategy for the regularization of deep neural networks. Applying tabu to the units that have been dropped in the recent epoch and retaining them for training ensures diversification in dropout. In this paper, we improve the Tabu Dropout mechanism for training deep neural networks in two ways. Firstly, we propose to use *tabu tenure*, or the number of epochs a particular unit will not be dropped. Different tabu tenures provide diversification to boost the training of deep neural networks based on the search landscape. Secondly, we propose an adaptive tabu algorithm that automatically selects the tabu tenure based on the training performances through epochs. On several standard benchmark datasets, the experimental results show that the adaptive tabu dropout and tabu tenure dropout diversify and perform significantly better compared to the standard dropout and basic tabu dropout mechanisms.

Keywords: Online Learning & Bandits · Deep Neural Network Algorithms · Reinforcement Learning Algorithms · Heuristic Search · Local Search

1 Introduction

Deep neural networks are a powerful machine learning system that can obtain very effective results in many applications such as natural language processing, bioinformatics, computer vision and other similar fields. Deep neural networks containing a large number of parameters and multiple non-linear hidden layers can grasp many intricate relations in the data. However, while trying to understand these intricate relations, the model tends to perfectly fit the training data. In order to overcome this problem dropout can be used. Dropout [5,9] is an effective regularization technique that is designed to tackle the overfitting problem in deep neural network. During the training phase, we close some of the neurons in the network for each epoch. This allows us to construct a 'thinned' network for each epoch. The final model is a combination of this 'thinned' models. This method produces models with superior generalization for the test data.

Although dropout can provide better results to reduce overfitting, the methods used to determine which neurons gets dropped can affect the generalization for test

M. Tanveer et al. (Eds.): ICONIP 2022, LNCS 13623, pp. 355–366, 2023.
https://doi.org/10.1007/978-3-031-30105-6_30

data greatly. To resolve this problem, multiple approaches to dropout has been proposed over the years with varying level of effectiveness. The overlap and the difference of neurons between two epochs can affect the generalization of the data. Tabu dropout [12] tries to control this overlap and difference by not allowing a neuron to be dropped twice in a row. In tabu dropout, only the status of the neurons of the previous forward propagation needs to be stored. This allows the dropout method to increases the diversification of the neural network while being computationally efficient. In practice, Tabu Dropout outperforms many dropout techniques like AlphaDropout [8], Curriculum dropout [13] and standard dropout. However, further control of the overlap and difference of neurons between two epochs can be achieved by controlling how many epochs a single neuron is not allowed to be dropped after being dropped once. Thus, increasing the generalization for the test data.

In this paper, we propose two methods to further improve the tabu dropout method. The first method is a regularization technique that controls how many epochs a single neuron is prohibited from being dropped after being dropped once. We call this method Tabu Tenure Dropout. The Tabu Tenure method increases the diversity of a neural network model while reducing the error rate. It allows us to have additional control over the overlap and difference of neurons between two epochs. The second method is an Adaptive Tabu Dropout algorithm. Using this algorithm, we dynamically select which Tabu Tenure to use during the training stage to get the optimum results. This algorithm allows us to harness the full potential of the Tabu Tenure Dropout through adaptively choosing the best suited Tabu Tenure for the train data. We conducted our experiments on various standard dataset like MNIST, Fashion MNIST, CIFAR-10 and CIFAR-100 and got promising results. All of the Tabu Tenure dropout algorithms tested performed better than the standard Tabu dropout. Amongst the Tabu Tenure dropouts tested, tabu tenure 6 produced the best results in all experiments except the MNIST-CNN1 experiment. Using the Tabu Tenure dropout algorithm we reduced the error rate up to 62.06% when compared with the standard tabu dropout during our testing. The second method called Adaptive Tabu dropout is formulated as a multi-armed bandit problem where we calculated the reward based on the loss per epoch and arms are the tabu tenure (TT) values. We tested multiple policies to the Adaptive tabu tenure selection like random selection, greedy selection, epsilon (ϵ) greedy selection, probabilistic selection and softmax probabilistic selection. The Adaptive Tabu Dropout achieved similar or improved accuracy and error rate in general when compared to the Tabu Tenure dropout.

2   Related Work

Ma et al. [12] presented a new dropout diversification approach that seeks to generate more diverse neural network topology in fewer iterations. No neuron can drop twice in a row. The AS-Dropout [1] model drops most of the neurons based on the neurons' activation functions. Though the model controls the proportion of the active neurons, it cannot accurately control it at a particular value. So, the dropout technique learns both confidence and uncertainty. The contextual dropout [3] is an efficient alternative to data-independent dropouts, which learns the dropout probabilities. It is well suited with both Bernoulli dropout and Gaussian dropout. The model may be used on a wide range of models with just a minor increase in memory and computational cost.

In contrast to multi-layer perceptrons, Irsoy and Alpaydın [7] offer a form of dropout for hierarchical mixtures of experts that is true to the tree hierarchy described by the model, rather than a flat, unit-wise independent application of dropout. The model is overfitted only if the number of levels and leaves is too high. AutoDropout [14] finds the dropout patterns efficiently for both image recognitions as well as the language modeling. A controller learns to produce a dropout pattern at every channel and layer of a target network. The dropout pattern then utilized to train the target network, and the validation performance is then utilized as a learning signal for the controller. Li et al. [11] presented an adaptive dropout method in which neural network and Variational Auto-encoder (VAE) are used alternately in the training phase. To regularize its hidden neurons, the model adaptively sets activity to zero.

As the traditional binary dropout method is not that precise, Tang et al. [16] has proposed to merge distortions onto feature maps by utilizing the Rademacher complexity. Using the generalization error bound, randomly selected elements in the feature maps are changed with particular values during the training phase. Hu et al. [6] has presented a simple and efficient surrogate dropout as an alternative to the learning parameter in the Bernoulli distribution. It learns the parameters by using concrete distribution. There are two steps of the model technique. For measuring the significance of each neuron, the initial step trains a surrogate module that may be improved alongside the neural network. When the network converges, the surrogate module's output is used as a guiding signal for removing particular neurons, approximating the ideal pre-neuron drop rate.

3   Proposed Method

The tabu technique [4] is commonly used in local search algorithms, and it uses a memory structure (referred to as the tabu list) to prevent the local search from returning a previously visited candidate solution. In [12], the authors have presented a new dropout technique based on the tabu strategy named Tabu Dropout. Algorithm 1 shows how the forward propagation works with the Tabu Dropout technique described by [12]. It prevents a neuron to be dropped in the consecutive epoch. In the rest of the section we present two improvements of the standard tabu dropout: Tabu Tenure Dropout that generalizes the standard Tabu Dropout by extending it using a *tabu tenure* and an Adaptive Tabu Dropout that is able to select the tabu tenure automatically based on the learner's performance.

Algorithm 1 shows the standard Tabu Dropout Algorithm, where x represents a hidden layer of neural network architecture. Initially, Tabu is set to none and p is the dropout rate. At the time of training, mask is created using Bernoulli Distribution that represents which neurons should be dropped in this epoch. If nothing is set to Tabu, we do not have anything to compare with. But if the tabu is not none, then we need to check the neuron we are planning to drop whether it has already dropped in the previous epoch or not. If it has already dropped in the previous epoch then we would not consider dropping this neuron in this epoch also. After that, by doing Hadamard product of mask and x the value of x will be updated and normalize afterward. The mask value should be set to *tabu* to store the current state from that we can check whether a neuron has already dropped in the previous epoch or not.

3.1 Tabu Tenure Dropout

In this work, we propose a modified Tabu Dropout technique and named it Tabu Tenure Dropout. We generalize the idea of tabu not only for the consecutive epoch but for a certain number of epochs. We call this *tabu tenure* denoted by TT. In other words, a neuron cannot be dropped if it has been dropped in previous TT epochs. It thus results in increased diversity in the dropout of units. In the case of standard Tabu Dropout proposed in [12], the value of TT is 1.

Algorithm 1. Standard Tabu Dropout	**Algorithm 2.** Tabu Tenure Dropout
1: x : a hidden layer of neural network architecture	1: x : a hidden layer of neural network architecture
2: $Tabu \leftarrow$ **None**	2: $Tabu \leftarrow$ **None**
3: $p \leftarrow$ Dropout Rate	3: $p \leftarrow$ Dropout Rate
4:	4: $epoch \leftarrow 0$
5: **procedure** TABU_DROPOUT(x)	5: $TT \leftarrow$ Tabu Tenure Value
6: **if** not training **then**	6:
7: return x	7: **procedure** TABU_TENURE_DROPOUT(x)
8: **end if**	8: **if** not training **then**
9: ones $\leftarrow$ number of ones of size	9: return x
10: mask $\leftarrow$ Bernoulli(ones $\odot$ p)	10: **end if**
11: **if** $Tabu \neq$ **None then**	11: ones $\leftarrow$ number of ones of size
12: mask $\leftarrow$ mask $\|$ ($Tabu \wedge$ ones)	12: mask $\leftarrow$ Bernoulli(ones $\odot$ p)
13: **end if**	13: **if** $Tabu \neq$ **None then**
14: x $\leftarrow$ mask $\odot$ x	14: mask $\leftarrow$ mask $\|$ (($Tabu - epoch \leq TT$) & ($Tabu \neq 0$))
15: x $\leftarrow$ x / (1 - p)	15: **end if**
16: $Tabu \leftarrow$ mask	16: x $\leftarrow$ mask $\odot$ x
17: return x	17: x $\leftarrow$ x / (1 - p)
18: **end procedure**	18: Flip mask
	19: mask $\leftarrow$ mask $\odot$ $epoch$
	20: $Tabu \leftarrow$ maximum(mask, $Tabu$)
	21: $epoch \leftarrow epoch + 1$
	22: return x
	23: **end procedure**

Algorithm 2 shows how the forward propagation works with the Tabu Tenure Dropout. It tracks the epoch number and checks whether the neuron has already been dropped in the previous TT epochs, if it has already been dropped then it can not be dropped in the current epoch. This is how we can increase the diversification of the neural network architecture in a smaller number of epochs.

Algorithm 2 shows the Tabu Tenure Dropout where x represents a hidden layer of neural network architecture. Like Algorithm 1, here *Tabu* is set to none and p is the dropout rate. As the Tabu Tenure can consider more than one epoch while dropping a neuron that is why we need to track the epoch number and initially it's set to zero.

TT is the Tabu Tenure Value which means while dropping a neuron how many previous epochs we need to check whether that neuron has already dropped in previous TT epochs or not.

At the time of training, we create mask using Bernoulli distribution similar to Algorithm 1. If the *tabu* is not none, we need to check whether the neuron we are considering to drop has already dropped in the previous TT epochs or not. After that we update the value of x by Hadamard product with mask. Next, we flip the mask value and again Hadamard product with the current value of epoch. Now the mask represents the neurons those are dropped in the current epoch are set as epoch number and those have not dropped in the current epoch set as zero. Now the *tabu* will be updated by the element-wise maximum operation which represents the recent epoch where the neuron has dropped. After that, we increase the epoch value by 1. The only difference between Algorithm 1 and Algorithm 2 is Algorithm 1 considers only previous epoch while dropping a neuron. On the contrary, Algorithm 2 consider TT number of epochs while dropping a neuron where the value of TT is fixed before starting the training phase.

3.2 Adaptive Tabu Dropout

In Tabu Tenure Dropout, we consider TT previous epochs while dropping a neuron in hidden layers. Instead of using a fixed number of TT, we can change the TT value several times during training and make the algorithm to learn and set the TT value in an adaptive manner. Here we formulate Adaptive Tabu Dropout as a multi-armed bandit problem. The arms or actions of the bandit are the TT values taken from a set T containing several TT values for selection and the reward, R_i in epoch i will be decided based on the loss function. The value function $Q_t(a)$ at epoch t value of an arm a will be estimated an average or expected value of the rewards when an arm a is selected prior to epoch t, in our case an arm is the TT values selected. In this paper, we have considered a *sample-average* technique for the value function $Q_t(a)$. The equation for calculating $Q_t(a)$ is given in Eq. 1.

$$Q_t(a) = \frac{\sum_{i=1}^{t-1} R_i \cdot \phi(a)}{\sum_{i=1}^{t-1} \phi(a)} \tag{1}$$

Here, $\phi(a)$ is a function that returns 1 if TT value a is selected in epoch i or 0 if it is not selected and R_i is the reward returned by the selected TT value in epoch i. The default values of $Q_t(a)$ is set to 0 is case any action is yet to be selected.

Tabu Tenure Selection: There are several techniques for choosing arms in each epoch, for example: random, greedy, epsilon greedy, probabilistic, softmax probabilistic, and many others [15]. For adaptive tabu dropout, we propose to change the TT value after specific number of epochs in the training. Thus we introduce a new hyper-parameter for the algorithm named *adaption period*. After each *adaption period*, a new TT value is selected based on the TT selection strategy based on the value function $Q_t(a)$. Our adaptive tabu selection algorithm uses a policy, $\pi(a)$ to select the values of TT. The policy function $\pi(a)$ specifies a probability distribution over the possible TT values in the set $\mathcal{T}$. In our experiments, we used the TT values in the range $1, \ldots, TT_{max}$. In this paper we have experimented with five different policies. They are described in the following.

Random Policy: The first policy is a exploration based strategy that selects a TT value from the set $\mathcal{T}$ using a uniform random distribution. The policy is defined in the following equation.

$$\pi(a) = \frac{1}{|\mathcal{T}|}, \forall a \in \mathcal{T} \tag{2}$$

Greedy Policy: In greedy technique, we will choose the arm or TT value with the largest $Q_t(a)$ denoted TT^* defined in the following equation:

$$TT^* = arg \max_a Q_t(a) \text{ (with ties broken arbitrarily)} \tag{3}$$

The greedy policy which is an exploitation based strategy is given in the following equation:

$$\pi(a) = \begin{cases} 1, \text{ if } a = TT^* \\ 0, \text{ if } a \neq TT^* \end{cases} \tag{4}$$

Epsilon Greedy Policy: For balancing between exploration and exploitation, ϵ-greedy policy is used. Here, ϵ is the exploration parameter taking values from the range $[0, 1]$. The higher the value of epsilon the algorithm gives more emphasis to exploration and lower the value is it gives more emphasis to exploitation. A highest value of 1 makes the policy random and a lowest value 0 makes the algorithm greedy. In our experiments, we have set the value of epsilon, $\epsilon = 0.5$. The ϵ-greedy policy is defined in the following equation:

$$\pi(a) = \begin{cases} 1 - \epsilon + \frac{\epsilon}{|\mathcal{T}|}, \textbf{ if } a = TT^* \\ \frac{\epsilon}{|\mathcal{T}|}, \textbf{ if } a \neq TT^* \end{cases} \tag{5}$$

Probabilistic Policy: We also use a policy that learns a probability distribution based on the $Q_t(a)$ values. Here the $Q_t(a)$ values are normalized to convert into probability. The policy is defined in the following equation:

$$\pi(a) = \frac{Q_t(a)}{\sum_{a \in \mathcal{T}} Q_t(a)}, \forall a \in \mathcal{T} \tag{6}$$

(a)   (b)   (c)   (d)

Fig. 1. Samples from different datasets (a) MNIST, (b) Fashion-MNIST, (c) CIFAR-10 and (d) CIFAR-100.

Softmax Policy: We also use a softmax policy which is an extension of the probabilistic policy. It also converts the $Q_t(a)$ values to probability distribution, however it uses a softmax function for that, The following equation defines the softmax policy.

$$\pi(a) = \frac{e^{Q_t(a)}}{\sum_{a \in \mathcal{T}} e^{Q_t(a)}}, \forall a \in \mathcal{T} \tag{7}$$

Reward Modeling: In our experiments, we have used two approaches to model rewards in each epoch when a particular TT value is selected for an epoch. Since the reward is based on the loss function, the first model proposes to use the inverse of the loss function as reward. Formally

$$R_t(a) = \frac{1}{\mathcal{L}_t} \tag{8}$$

Here, $\mathcal{L}_t$ is the loss function when arm or tabu tenure a was selected at epoch t. This reward is then fed to Eq. 1 for value update. However, there is a problem with this approach, when the loss value is very low, then the reward will be a large value. In that case, the probability of choosing that selected TT value after the *adaption period* will dominate other TT values. We propose another reward model to improve over this issue. In the second model, we define the reward as the negative exponential of the loss function. Formally, it is defined as in the following:

$$R_t(a) = e^{-\mathcal{L}_t} \tag{9}$$

This function does not result into very large rewards for small losses and thus exploration is retained and allows to avoid the pitfall of the previous reward function. We have also experimented with differential loss functions, however the initial experiments did not show promising results and we did not report that in this paper.

4 Experimental Analysis

We used PyTorch, a deep learning framework for quick and flexible experimentation, to implement Tabu Dropout and Tabu Tenure Dropout. All the experiments were conducted on 3.60GHz Intel(R) Core(TM) i7-7700 CPU, 32 GB RAM and a NVIDIA TITAN XP with 12GB physical memory under Ubuntu 16.04.7 LIS. We run all of the algorithms five times each and on the average of the results for all of the experiments are reported in this paper.

4.1 Implementation Details

In this section, we provide the details of the neural network architecture and datasets used in this paper.

Neural Network Architectures: For the convenience of comparison we have used the same neural network architectures as Ma et al. [12] used in their work. We have used a multilayer perceptron (MLP) and two CNN architecture. LeNet-5, a newer version of LeNet, is the first CNN architecture [10] and the second one is named CNN1[1] as their work [12]. LeNet5 extracts features by intelligent design using convolution, parameter sharing, pooling, and other processes, eliminating a huge amount of processing expense, and then utilizes a fully connected neural network for classification and recognition. This network has lately used as the foundation for a significant variety of neural network topologies. The architecture of LeNet5, Conv2d (Convolutional layer 1, 3 channel input, 6 convolution kernels, kernel size 5×5) - ReLU - MaxPool2d (2×2 max pooling) - Conv2d (Convolutional layer 2, 6 input channels, 16 convolution kernels, kernel size 5×5) - ReLU - MaxPool2d (2×2 max pooling) - FC - ReLU - FC - ReLU - FC - Logsoftmax. The MLP model is a full-connected (FC) neural network with 1024 units in each hidden layer (FC-ReLU-FC-ReLU-FC-Logsoftmax). Between the two hidden layers, we introduce dropout variations.

Parameter Setting: For all of the trials, we utilized a dropout rate of 0.5. For each model, we trained 300 epochs using a learning rate of 0.01 and a batch size of 512. For all neural networks, Adam was used as the optimizer. The TT_{max} value in our experiment was 6. We have used 4 *adaption period* values, these are 10, 15, 20, 25.

4.2 Datasets

The experiments were carried out on four datasets, the specifics of which are listed below.

- MNIST[2]: It is a traditional handwritten digits dataset with 28×28 grayscale pictures of 10 digits (ranging from 0 to 9) which is frequently used in computer vision and machine learning. There are 60000 training images and 10000 images for testing. For this dataset, we have used MLP and CNN-1. Here the Fig. 1 (a) shows some samples of MNIST dataset.
- Fashion-MNIST[3]: It is identical to the MNIST dataset in terms of training, test, number of class labels and image dimensions. Here is also 60000 training set, 10000 test set, 10 class labels (T-shirt/top, Trouser/pants, Pullover shirt, Dress, Coat, Sandal, Shirt, Sneaker, Bag and Ankle boot) with 28×28 grayscale images. Here the Fig. 1 (b) shows some samples of Fashion-MNIST dataset.
- CIFAR-10[4]: It contains 60000 32×32 colored images where 50000 images for training set and 10000 for testing set. This dataset is classified into 10 different classes (airplanes, cars, birds, cats, deer, dogs, frogs, horses, ships, and trucks) and each class has 6000 images. Exactly 1000 randomly-selected images are in test set for every class but in training set there may contains more images from one class to another. Here the Fig. 1(c) shows some samples of CIFAR-10 dataset.

---

[1] https://github.com/pytorch/examples/blob/master/mnist/main.py.

[2] http://yann.lecun.com/exdb/mnist/.

[3] https://github.com/zalandoresearch/fashion-mnist.

[4] https://www.cs.toronto.edu/ kriz/cifar.html.

– CIFAR-100: It's almost identical to CIFAR-10 dataset, except it has 100 classes containing 600 images each. There are also 50000 images for training set and 10000 images for test set that means 500 for training and 100 for testing per class. This 100 classes of CIFAR-100 are grouped into 20 superclasses. Every images comes with two label "fine" for class and "coarse" for superclass. Here the Fig. 1(d) shows some samples of CIFAR-100 dataset.

4.3 Effectiveness of Tabu Tenure Dropout

The Standard Tabu dropout diversifies the neural network by dropping off the marked neurons in the successive epoch. The tabu tenure dropout further extends the tabu to a number of epochs controlled by the parameter TT value. Table 1 illustrates the comparison between the error rate or loss function of Standard Tabu dropout and the Tabu Tenure dropout on the testing datasets after 300 epochs. We have calculated the arithmetic mean of 5 runs to generate each value. Note that our experiments are limited to $TT = 6$. The bold values in each column of the table shows the best error rate achieved using different tabu tenures. The results are narrated dataset wise in the following.

Result on Different Datasets: Table 1 illustrates the result of mean error rate on different datasets. MNIST - CNN-1 obtains better performance until Tabu Tenure dropout 5. Though the Tabu Tenure dropout 6 shows a slightly higher error, the value is pretty close to the previous one. While using MLP and CNN-1, the increased value of Tabu Tenure outperforms the others. As per the fact that the dropping delay of neurons is expanding the variety of both MLP and CNN-1 gradually. Finally, the Tabu Tenure 6 dropout achieves the most favorable result among the others. We observe on the LeNet-5 model that in primary stages, the error was decreasing slowly. In the end, Tabu Tenure dropout 6 shows a significantly lower error than the Standard Tabu dropout. The LeNet-5 model with Tabu Tenure dropout 6 has an outstanding result in all the training epochs than the Standard Tabu dropout. For all the datasets, the Tabu Tenure Dropout outperforms the state-of-the-art Standard Tabu Dropout.

Table 1. The average error rate (after training 300 epochs) on the test datasets using Standard Tabu and Tabu Tenure dropout. The best results are marked in bold. (FMNIST is the short form of Fashion-MNIST, C10 is for CIFAR-10 and, C100 is for CIFAR-100).

	MNIST - MLP	MNIST - CNN1	FMNIST - MLP	FMNIST - CNN1	C10 - LeNet5	C100 - LeNet5
AlphaDropout	0.01162	0.00602	0.11989	0.03800	0.14851	0.74702
Non-dropout	0.01302	0.00729	0.12471	0.03856	0.15707	0.74258
Standard dropout	0.01076	0.00582	0.11242	0.03754	0.13904	0.73983
Standard Tabu Dropout	0.01083	0.00585	0.11886	0.03731	0.13744	0.73408
Tabu Tenure Dropout 1	0.00989	0.00625	0.11639	0.03765	0.14418	0.71176
Tabu Tenure Dropout 2	0.00811	0.00459	0.09762	0.03073	0.11643	0.48707
Tabu Tenure Dropout 3	0.00689	0.00316	0.08587	0.02312	0.10501	0.38576
Tabu Tenure Dropout 4	0.00768	0.00351	0.07747	0.02208	0.09340	0.34184
Tabu Tenure Dropout 5	0.00749	**0.00306**	0.07269	0.01936	0.08589	0.29113
Tabu Tenure Dropout 6	**0.00675**	0.00317	**0.06733**	**0.01885**	**0.08169**	**0.27844**

Table 2. The mean error rate (after training 300 epochs) on the test datasets using different arm choosing methods for Adaptive Tabu Dropout. (TD is the short form of Tabu Dropout, AP is for *Adaption Period*, SP is for Softmax Probabilistic, FMNIST is the short form of Fashion-MNIST, C10 is for CIFAR-10 and, C100 is for CIFAR-100. V1 refers to Eq. 8 and V2 refers to Eq. 9).

	MNIST - MLP	MNIST - CNN1	FMNIST - MLP	FMNIST - CNN1	C10 - LeNet5	C100 - LeNet5
TD Random (AP: 10)	0.00886	0.00372	0.08983	0.02759	0.11402	0.43774
TD Random (AP: 15)	0.00849	0.00385	0.08689	0.02489	0.10563	0.43635
TD Random (AP: 20)	0.00782	0.00332	0.08902	0.02533	0.11064	0.44873
TD Random (AP: 25)	0.00769	0.00394	0.08936	0.02597	0.10686	0.47377
TD Greedy (AP: 10)	0.00924	0.00369	0.08399	0.02271	0.09782	0.38999
TD Greedy (AP: 15)	0.00701	0.00409	0.07995	0.02257	0.10278	0.39555
TD Greedy (AP: 20)	0.00749	0.00374	0.08365	0.02509	0.10275	0.38272
TD Greedy (AP: 25)	0.00851	0.00439	0.09014	0.02318	0.10397	0.42020
TD Epsilon Greedy (AP: 10)	0.00824	0.00402	0.08504	0.02415	0.10667	0.41873
TD Epsilon Greedy (AP: 15)	0.00777	0.00414	0.08455	0.02163	0.10751	0.37835
TD Epsilon Greedy (AP: 20)	0.00759	0.00358	0.07988	0.02269	0.10184	0.45175
TD Epsilon Greedy (AP: 25)	0.00787	0.00392	0.08630	0.02232	0.09759	0.39445
TD Probabilistic (AP: 10)	0.00766	0.00336	0.08013	**0.01958**	0.09592	0.36828
TD Probabilistic (AP: 15)	0.00747	**0.00307**	0.09351	0.02248	0.10738	0.38269
TD Probabilistic (AP: 20)	0.00756	0.00339	0.08649	0.02370	0.09672	0.39072
TD Probabilistic (AP: 25)	0.00754	0.00360	0.08706	0.02005	0.09730	0.38556
TD SP (AP: 10) V1	0.00763	0.00398	0.08199	0.02630	0.10471	0.33584
TD SP (AP: 15) V1	0.00661	0.00378	0.07691	0.02558	0.09352	0.36606
TD SP (AP: 20) V1	0.00791	0.00361	0.10153	0.02356	0.10184	0.38102
TD SP (AP: 25) V1	0.00775	0.00349	0.07419	0.02211	0.10017	0.35419
TD SP (AP: 10) V2	0.00766	0.00332	**0.06729**	0.02493	**0.08722**	0.30933
TD SP (AP: 15) V2	0.00753	0.00321	0.06913	0.02559	0.09969	0.29918
TD SP (AP: 20) V2	0.00707	0.00392	0.06827	0.02497	0.09236	**0.29515**
TD SP (AP: 25) V2	**0.00651**	0.00386	0.06969	0.02144	0.09419	0.30444

4.4   Effectiveness of Adaptive Tabu Dropout

We have formulated the Adaptive Tabu dropout as a multi-armed bandit problem where the arms of the bandit are the TT values selected from a set $\mathcal{T}$. We have used a set of policies: random, greedy, epsilon greedy, probabilistic, and softmax probabilistic techniques for analysis. Two different reward models have been used. The Tabu Tenure values are changed after *adaption periods*. In the experiments we have used four different values of adaption periods (AP) of 10, 15, 20, and 25 epochs. Table 2 shows the error rate of various tabu tenure selection policies with different adaption periods. Each value reported in Table 2 is average of 5 runs. In case of the softmax policy, we have used two types of reward models defined by Eq. 8 and Eq. 9. Other policies are experimented with only with the reward model defined in Eq. 8. The bold values in each column represents the best values achieved by any policy.

From the values reported in the table, we note that softmax policy based tabu tenure selection achieves best values among 5 out of 6 combinations of dataset-architectures. In the case of MNIST dataset with CNN1 architecture, probabilistic policy performes better than softmax policy. Note that the greedy policy does not work well. In case of greedy we allowed an initial phase for exploration (150 epochs) and then followed the greedy policy. Results indicate that explorations should be encouraged in all stages of training.

If we compare the performances of adaptive selection or policy based tabu tenure with that of the fixed tabu tenure methods, we note that the results are very much similar. However, to test them further we have used AlexNet[5] architecture [9] which is one of the earliest successful deep neural network on the relatively larger datasets CIFAR-10 and CIFAR-100.

We have considered the average accuracy and error of the best-fixed tabu tenure version ($TT = 6$) and the best policy-based algorithm softmax policy with different adaptation periods. We have examined the average accuracy and error of the best-fixed tabu tenure version ($TT = 6$) and the best policy-based algorithm softmax policy with different adaptation periods. Softmax policy with adaption period 10 shows the minimum error on both CIFAR-10 and CIFAR-100 datasets. The highest accuracy is observed on softmax policy with an adaption period of 25 on the CIFAR-10 dataset. Note that all four combinations of dataset-architecture policy-based adaptive tabu tenure with softmax policy are achieving significantly superior values compared to the fixed tabu tenure version. This encourages us to conclude on the overall effectiveness of the adaptive tabu selection method.

4.5 Comparison with Other Methods

Note that we have reported the experimental results in Table 1 on the identical four datasets after training with 300 epochs using same neural network models as done in [12]. In their experiments, Ma et al. [12] showed the superiority of their tabu dropout over AlphaDropout [8], Curriculum Dropout [13], Standard Dropout [5,9], and non-dropout strategies. Table 1 implies a clear contrast with the other dropout approaches and shows that the Tabu Tenure Dropout outperforms all the dropout strategies mentioned in [12]. The Adaptive Tabu Tenure is very much similar in performance compared to the fixed tabu tenure method and outperforms fixed tabu tenure in terms of error and accuracy using AlexNet architecture on larger datasets. Therefore, we conclude that the Adaptive Tabu Tenure Dropout outperforms the rest of the dropout techniques with whom [12] has shown comparison in their paper.

5   Conclusion

In this paper, we have introduced adaptive tabu dropout based on the multi-arm bandit problem which is a simple mechanism that can diversify the neural network more and performs better compared to the standard Tabu dropout and Tabu Tenure dropout. While using Tabu Tenure dropout we need to select ta TT value wisely otherwise there may happen an overfitting problem. In our experiments, we have noticed that TT values beyond 6 overfits and also larger TT values run for longer epochs results in overfitting. This encourages to use multiple tabu tenures and adaptive selection policy. Also note that we have not explored much of the differential reward models and non-stationary modeling of the rewards for a better estimation of the $Q_t(a)$ values. The effectiveness of the methods proposed in this paper encourages us to assess Adaptive Tabu dropout in larger and different types of deep neural network models with longer training periods, such as RNN, Res-Net, etc. and bigger datasets, such as ImageNet [2].

---

[5] https://github.com/Lornatang/pytorch-alexnet-cifar100/blob/master/model.py.

References

1. Chen, Y., Yi, Z.: Adaptive sparse dropout: learning the certainty and uncertainty in deep neural networks. Neurocomputing **450**, 354–361 (2021)
2. Deng, J., Dong, W., Socher, R., Li, L.J., Li, K., Fei-Fei, L.: Imagenet: a large-scale hierarchical image database. In: 2009 IEEE Conference on Computer Vision and Pattern Recognition, pp. 248–255. IEEE (2009)
3. FAN, X., Zhang, S., Tanwisuth, K., Qian, X., Zhou, M.: Contextual dropout: an efficient sample-dependent dropout module. In: International Conference on Learning Representations. https://par.nsf.gov/biblio/10273825
4. Glover, F.: Tabu search-part i. ORSA J. Comput. **1**(3), 190–206 (1989)
5. Hinton, G.E., Srivastava, N., Krizhevsky, A., Sutskever, I., Salakhutdinov, R.R.: Improving neural networks by preventing co-adaptation of feature detectors. arXiv preprint arXiv:1207.0580 (2012)
6. Hu, J., Chen, Y., Zhang, L., Yi, Z.: Surrogate dropout: learning optimal drop rate through proxy. Knowl.-Based Syst. **206**, 106340 (2020)
7. İrsoy, O., Alpaydın, E.: Dropout regularization in hierarchical mixture of experts. Neurocomputing **419**, 148–156 (2021)
8. Klambauer, G., Unterthiner, T., Mayr, A., Hochreiter, S.: Self-normalizing neural networks. In: Proceedings of the 31st International Conference on Neural Information Processing Systems, pp. 972–981 (2017)
9. Krizhevsky, A., Sutskever, I., Hinton, G.E.: Imagenet classification with deep convolutional neural networks. Adv. Neural. Inf. Process. Syst. **25**, 1097–1105 (2012)
10. LeCun, Y., et al.: Backpropagation applied to handwritten zip code recognition. Neural Comput. **1**(4), 541–551 (1989)
11. Li, H., et al.: Adaptive dropout method based on biological principles. IEEE Trans. Neural Networks Learn. Syst. (2021)
12. Ma, Z., Sattar, A., Zhou, J., Chen, Q., Su, K.: Dropout with tabu strategy for regularizing deep neural networks. Comput. J. **63**(7), 1031–1038 (2020)
13. Morerio, P., Cavazza, J., Volpi, R., Vidal, R., Murino, V.: Curriculum dropout. In: Proceedings of the IEEE International Conference on Computer Vision, pp. 3544–3552 (2017)
14. Pham, H., Le, Q.V.: Autodropout: Learning dropout patterns to regularize deep networks. arXiv preprint arXiv:2101.01761 1(2), 3 (2021)
15. Sutton, R.S., Barto, A.G.: Reinforcement learning: An introduction. MIT press (2018)
16. Tang, Y., et al.: Beyond dropout: Feature map distortion to regularize deep neural networks. In: Proceedings of the AAAI Conference on Artificial Intelligence, vol. 34, pp. 5964–5971 (2020)

Class-Incremental Learning with Multiscale Distillation for Weakly Supervised Temporal Action Localization

Tianquan Chen⬡, Bairong Li⬡, Yusheng Tao⬡, Yuqing Wang⬡,
and Yuesheng Zhu$^{(\boxtimes)}$⬡

Shenzhen Graduate School, Peking University, Beijing, China
{chentianquan,ystao,wyq}@stu.pku.edu.cn, {lbairong,zhuys}@pku.edu.cn

Abstract. Despite recent works having made great progress in weakly supervised temporal action localization (WTAL), they still suffer from catastrophic forgetting. When only new-class videos can be utilized to update these models, their performance in old classes diminishes drastically. Even while some class-incremental learning methods are presented to assist models in continuously learning new-class knowledge, most of them focus on image classification but pay little attention to WTAL. To fill this gap, we propose a novel class-incremental learning method with multiscale distillation, which mines two separate scales of old-class information in incoming videos for updating the model. Precisely, we calculate class activation sequences (CAS) with frame-level spatio-temporal information to provide fine-grained old-class labels for the updated model. Moreover, since the high activation segments contain rich action information, we select them and construct video-level logits to constrain the updated model for maintaining the old-class knowledge further. The experimental results under various incremental learning settings on the THUMOS'14 and ActivityNet 1.3 datasets reveal that our method effectively alleviates the catastrophic forgetting problem in WTAL.

Keywords: Class-incremental learning · Weakly supervised temporal action localization · Knowledge distillation

1 Introduction

Temporal action localization (TAL) aims to identify human actions in untrimmed videos and has made great strides in video understanding and analysis. However, such fully supervised TAL methods [15,18,23] are limited by the high costs of obtaining frame-level labels and the subjectivity of manual labeling. Therefore, some weakly supervised temporal action localization (WTAL) methods [5,9–11,13,17,19,21] have been proposed to eliminate these limitations, requiring only

This work was supported in part by the National Innovation 2030 Major S&T Project of China under Grant 2020AAA0104203, and in part by the Nature Science Foundation of China under Grant 62006007.

Fig. 1. Catastrophic forgetting problem in WTAL. The green parts represent the action instances of the old class *Throw Discus*, and the yellow parts represent the new class *Clean and Jeck*. (**a**): The ground truth action instances. (**b**): The predicted action instances of the original model. (**c**): The predicted action instances of the updated model using fine-tuning, which has forgotten the old class *Throw Discus*. (Color figure online)

video-level class labels. However, many services in real-world scenarios receive continuous video streams. It leads to the following problems with these WTAL approaches: First, they have to train from scratch to adapt to new-class videos. Second, retraining requires old videos, raising privacy issues and storage stress. Therefore, we need to develop a method using only new-class videos for WTAL to permit effective continuous learning of new-class action instances.

Fine-tuning [3] is the simplest method but suffers the worst old-class knowledge forgetting, which adapts to the new data form by expanding the parameter shape of the network. Although the new network learns to localize new-class action instances with fine-tuning, it will gradually lose the capacity of old classes with the network's training. A phenomenon like this in which the network forgets the old knowledge when learning with only new data is called catastrophic forgetting. Figure 1 illustrates the catastrophic forgetting problem in WTAL. The original model can localize action instances of class *Throw Discus*. Using fine-tuning, the updated model can localize the action instances of the new class *Clean and Jerk*. However, it cannot detect the old class *Throw Discus*.

As an early exploration of incremental learning method for WTAL problems, we employ the most stringent definition of class-incremental learning, in which models must continuously learn new classes without utilizing any samples of old classes. Existing incremental learning methods typically reduce catastrophic forgetting using some of the three techniques. 1) parameter regularization [8], 2) knowledge distillation [4,12], or 3) rehearsal [14,22]. Regularized approaches measure parameter relevance for model adaption. Rehearsal-based methods store and replay representative examples while training new tasks. However, regularized techniques have limited effectiveness in class-incremental learning [16].

And rehearsal-based methods utilize a small number of old-class samples, making them unsuitable for the current scenario. Besides, despite researchers having proposed many incremental learning methods to solve catastrophic forgetting problems in other fields (image classification [8,12,14,22], object detection [4]), the catastrophic forgetting problem of WTAL has not been actively investigated.

Unlike the tasks mentioned above, WTAL must detect action instances under incomplete labeling. Existing class-incremental learning methods from other tasks are therefore inappropriate and difficult to apply to WTAL due to the following factors. (1) Both videos' spatial and temporal features must be memorized while learning new classes. (2) Since only video-level labels are available, single frames without labels tend to give high activation scores directly to the new classes rather than the old classes. (3) With the continuous learning of the network, a video like the example shown in Fig. 1 may contain action instances of both new and old classes, but the video-level class label will not regard this video as a positive sample of old classes. These reasons further exacerbate the catastrophic forgetting problem of WTAL.

In this paper, we fill this gap of continuous training for WTAL, proposing a novel class-**I**ncremental **L**earning method based on **L**ogits and class **A**ctivation sequences (ILLA) for WTAL. We save a frozen copy of the original model as the teacher model, using knowledge distillation [6] at different scales to guide the optimization direction of the old-class information on the updated model. Specifically, we calculate the class activation sequences (CAS), a frame-level class score matrix with spatio-temporal information, to provide strongly supervised old-class labels for the updated model. In addition, we also aggregate high activation segments with rich action information to form video-level logits. Finally, we simultaneously use these two different-scale old-class knowledge carriers to help the updated model maintain the old-class performance. The experimental evaluation on THUMOS'14 [7] and ActivityNet 1.3 [1] datasets shows the effectiveness of our approach.

In summary, our contributions are as follows:

- We study WTAL's catastrophic forgetting problem and propose ILLA with multiscale distillation to solve it. To the best of our knowledge, ILLA is the first class-incremental learning method for WTAL.
- Our method introduces the old-class information of video-level logits and frame-level CAS in incoming videos for the updated model by knowledge distillation at different scales to enhance performance. And the evaluation results on two datasets show that ILLA significantly mitigates the catastrophic forgetting problem in WTAL.

2   Approach

Figure 2 depicts the framework overview of our proposed approach ILLA. We first formulate the WTAL task before going into details.

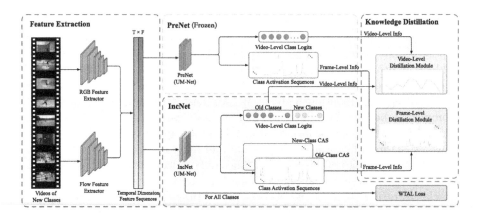

Fig. 2. The framework of our class-incremental learning method for weakly supervised temporal action localization.

2.1 WTAL Formulation

Given a dataset $\mathcal{D}$, It contains a video collection $\mathbf{V} = \{\mathbf{v}_k\}_{k=1}^{N^v}$. Each video $\mathbf{v}_k$ has N^ψ action instances which can be described by $\{\psi_i = (t_i^s, t_i^e, c_i)\}_{i=1}^{N^\psi}$, where t^s, t^e, c respectively represent the start time, end time and class of the action instance. For temporal action localization, It requires the framework to generate predicted action instances $\{\hat{\psi}_i = (\hat{t}_i^s, \hat{t}_i^e, \hat{p}_i, \hat{c}_i)\}_{i=1}^{N^{\hat{\psi}}}$ for each video $\mathbf{v}_k$, where $\hat{p}$ is expressed as the confidence score for the predicted action instance. Since the temporal labels are not visible in WTAL, there is only a video-level class label $\boldsymbol{y}_k \in \mathbb{R}^C$ for video $\mathbf{v}_k$, where C represents the number of action instance classes contained in the dataset. $\boldsymbol{y}_k[c] = 1$ if and only if video $\mathbf{v}_k$ contains at least one action instance of class c and class c is in the current training classes, otherwise $\boldsymbol{y}_k[c] = 0$. There may be multiple different types of action instances for each video, which means that $\sum_{i=1}^C \boldsymbol{y}_k[c] \geq 1$.

2.2 Feature Extraction

Follow the previous WTAL methods, We input the RGB stream and FLOW stream of each video $\mathbf{v}$ into the I3D network [2], which is used to extract the spatial feature set $\{\mathbf{x}_t^{\mathrm{RGB}}\}_{t=1}^{N^s}$ and temporal feature set $\{\mathbf{x}_t^{\mathrm{FLOW}}\}_{t=1}^{N^s}$. Each $\mathbf{x}_t \in \mathbb{R}^f$ is the feature vector of the segment composed of non-overlapping continuous frames on the video $\mathbf{v}$. Since the video length varies, we sample two feature sets and merge them in the feature dimension to create the fusion feature set $\mathbf{X} = \{\mathbf{x}_t^{\mathrm{BOTH}}\}_{t=1}^T \in \mathbb{R}^{T \times F}$, where T is a fixed number of sampled segments, and $F = 2f$ is the dimension of the segment feature.

2.3 UM-Net

Lee et al. developed a WTAL approach based on uncertainty modeling [10], which we call UM-Net. UM-Net is a typical, concise and effective WTAL model, so we employ it as the experimental network to build our class incremental learning framework. After inputting fusion feature $\mathbf{X}$, the feature embedding module of UM-Net employs the parameters $\boldsymbol{\Phi}$ to generate embedded feature $\mathbf{F} \in \mathbb{R}^{T \times F}$. For the convenience of description, we changed the classification module of UM-Net from one-dimensional convolution layer with kernel size 1 to fully connected layer (performance was not significantly affected, see Table 1), which was formulated as $\mathbf{A}^* = \mathbf{F}\mathbf{W}^T + \mathbf{b}$, where $\mathbf{A}^* \in \mathbb{R}^{T \times C}$ is the original CAS. In UM-Net, $\mathbf{A}^*$ will combine with the magnitudes of segments to generate the final class activation sequences $\mathbf{A} \in \mathbb{R}^{T \times C}$.

The loss function of UM-Net $\mathcal{L}_{\text{wtal}}$ consists of three parts. $\mathcal{L}_{\text{cls}}$ is a video-level classification loss. $\mathcal{L}_{\text{um}}$ tries to make pseudo-action segments and pseudo-background segments more differentiated. $\mathcal{L}_{\text{be}}$ encourages background segment scores to be evenly distributed in all classes. They are balanced by hyperparameters α, β:

$$\mathcal{L}_{\text{wtal}} = \mathcal{L}_{\text{cls}} + \alpha\mathcal{L}_{\text{um}} + \beta\mathcal{L}_{\text{be}} \tag{1}$$

During the interface phase, UM-Net uses different thresholds to generate proposed action instances on the CAS $\mathbf{A}$. For more implementation details about UM-Net, refer to the original paper [10].

2.4 Multiscale Knowledge Distillation

Our approach is founded on the concept of knowledge distillation. Iterative training of the incremental network using new-class video data changes parameters. However, Suppose the incremental network's video-level logits and class activation sequences are consistent with the previous network. In that circumstance, the incremental network can keep old-class action instances' recognition ability.

We first trained a network $\mathbf{N}^{\text{pre}}$ to localize the videos containing action instances belonging to the old-class set $\mathcal{C}^{\text{pre}}$. The goal now is to train a network $\mathbf{N}^{\text{inc}}$ using only the new-class videos, which can localize action instances of class set $\mathcal{C}^{\text{pre}} \cup \mathcal{C}^{\text{inc}}$. In order to adapt the new-class set $\mathcal{C}^{\text{inc}}$, we extend the parameters of the last fully connected layer of the previous network $\mathbf{N}^{\text{pre}}$ to obtain the incremental network $\mathbf{N}^{\text{inc}}$. For the weight $\mathbf{W}^{\text{pre}} \in \mathbb{R}^{|\mathcal{C}^{\text{pre}}| \times F}$ and bias $\mathbf{b}^{\text{pre}} \in \mathbb{R}^{|\mathcal{C}^{\text{pre}}|}$, we respectively extend them to $\mathbf{W}^{\text{inc}} \in \mathbb{R}^{|\mathcal{C}^{\text{pre}} \cup \mathcal{C}^{\text{inc}}| \times F}$ and $\mathbf{b}^{\text{inc}} \in \mathbb{R}^{|\mathcal{C}^{\text{pre}} \cup \mathcal{C}^{\text{inc}}|}$. And we saved a copy of the previous network with frozen parameters at the same time. When initializing the incremental network $\mathbf{N}^{\text{inc}}$, we loaded the parameters from $\boldsymbol{\Phi}^{\text{pre}}$, $\mathbf{W}^{\text{pre}}$ and $\mathbf{b}^{\text{pre}}$. For the expanded part of the new parameters $\mathbf{W}^{\text{inc;new}} \in \mathbb{R}^{|\mathcal{C}^{\text{inc}} - \mathcal{C}^{\text{pre}}| \times F}$ and $\mathbf{b}^{\text{inc;new}} \in \mathbb{R}^{|\mathcal{C}^{\text{inc}} - \mathcal{C}^{\text{pre}}|}$, We used the default initialization method to initialize them.

CAS is an essential carrier of localization information in WTAL because it comprises the class scores of each segment for a video. In CAS, different classes within the same segment reflect the segment's spatial information, whereas different segments within the same class indicate the class's temporal information. As shown in Fig. 2, we get the fusion feature $\mathbf{X}^{\text{new}}$ from the new-class video $\mathbf{v}^{\text{new}}$ through the feature extractor, and then input it to both $\mathbf{N}^{\text{pre}}$ and $\mathbf{N}^{\text{inc}}$ to obtain

$\mathbf{A}^{\text{pre}}$ and $\mathbf{A}^{\text{inc}}$. Then we select the old class activation sequences $\mathbf{A}^{\text{inc;old}}$ from $\mathbf{A}^{\text{inc}}$ and set $\mathbf{A}^{\text{pre}}$ as their optimization goal:

$$\mathcal{L}_{\text{dist\_cas}} = \frac{1}{N^b T |\mathcal{C}^{\text{old}}|} \sum_{k=1}^{N^b} \sum_{t=1}^{T} \sum_{c \in \mathcal{C}^{\text{old}}} (\mathbf{A}_k^{\text{inc}}[t, c] - \mathbf{A}_k^{\text{pre}}[t, c])^2 \tag{2}$$

where N^b represents the number of videos in this batch. t and c are used to traverse the old-class scores of each segment.

WTAL is generally modeled as a multi-instance learning task, so the highly activated segments in CAS usually contain more action frames. In order to make the incremental network tend to optimize the high activation segments with rich action information, we follow UM-Net to calculate the average raw output of high activation segments as video-level logit $\boldsymbol{o} \in \mathbb{R}^C$:

$$\boldsymbol{o}_k[c] = \frac{1}{k^{\text{act}}} \max_{\substack{\mathcal{A}_k[:,c] \subset \mathbf{A}_k^*[:,c] \\ |\mathcal{A}_k[:,c]| = k^{\text{act}}}} \sum_{\forall a \in \mathcal{A}_k[:,c]} a \tag{3}$$

where $\mathcal{A}_k[:, c]$ is the raw activation output vector of the highest k^{act} segments of $\mathbf{A}_k^*$ in class c. Thereafter, we select the old-class logits $\boldsymbol{o}^{\text{inc;old}}$ of $\mathbf{N}^{\text{inc}}$ and set the logits $\boldsymbol{o}^{\text{pre}}$ of $\mathbf{N}^{\text{pre}}$ as their optimization target:

$$\mathcal{L}_{\text{dist\_log}} = \frac{1}{N^b |\mathcal{C}^{\text{old}}|} \sum_{k=1}^{N^b} \sum_{c \in \mathcal{C}^{\text{old}}} (\boldsymbol{o}_k^{\text{inc}}[c] - \boldsymbol{o}_k^{\text{pre}}[c])^2 \tag{4}$$

Note that we do not utilize the modified softmax scores suggested by [12] since the original logits better reflected the proportion of high active segments in each class, whereas softmax altered the proportion value (see Table 1).

$\mathcal{L}_{\text{dist\_cas}}$ and $\mathcal{L}_{\text{dist\_log}}$ are balanced by the hyperparameter γ to obtain the distillation loss $\mathcal{L}_{\text{dist}}$:

$$\mathcal{L}_{\text{dist}} = \mathcal{L}_{\text{dist\_cas}} + \gamma \mathcal{L}_{\text{dist\_log}} \tag{5}$$

Finally, the incremental network $\mathbf{N}^{\text{inc}}$ can learn to localize new-class action instances using $\mathcal{L}_{\text{wtal}}$ and keep the old-class knowledge under the constraint of $\mathcal{L}_{\text{dist}}$. Its overall loss function $\mathcal{L}_{\text{total}}$ is as follows:

$$\mathcal{L}_{\text{total}} = \mathcal{L}_{\text{wtal}} + \sigma \mathcal{L}_{\text{dist}} \tag{6}$$

where σ is the hyperparameter for the weight of the distillation loss.

In the next stage of incremental training, the $\mathcal{C}_i^{\text{old}}$ will be updated to $\mathcal{C}_{i-1}^{\text{old}} \cup \mathcal{C}_{i-1}^{\text{new}}$ and $\mathcal{C}_i^{\text{new}}$ will be updated to the new-class set. The incremental network of the previous stage $\mathbf{N}_{i-1}^{\text{inc}}$ will become the previous network of that stage $\mathbf{N}_i^{\text{pre}}$. This step can be repeated for more classes.

3   Experiments

3.1   Experimental Settings

Datasets and Evaluation Metrics. The THUMOS'14 dataset contains 20 classes for WTAL. There are 200 validation videos for training and 213 test

Table 1. The average mAP (%) of the increment 10 classes setting on THUMOS'14, where *Fine-tuning* is the benchmark without incremental learning techniques.

Method	old classes(1–10)	new classes(11–20)	all classes(1–20)
N^{pre}	33.24	-	-
Fine-tuning	16.29(↓49.8%)	45.34(↓5.1%)	30.81(↓23.2%)
Train extended part only	5.33(↓83.6%)	19.51(↓59.1%)	12.42(↓69.0%)
Train FC layer only	7.87(↓75.7%)	20.60(↓56.9%)	14.23(↓64.5%)
Distill Score	17.39(↓46.4%)	44.54(↓6.7%)	30.96(↓22.8%)
Distill Logit	17.32(↓46.6%)	45.52(↓4.7%)	31.42(↓21.7%)
Distill CAS	24.95(↓23.1%)	46.24(↓3.2%)	35.60(↓11.2%)
Distill CAS + Score	27.34(↓15.7%)	44.16(↓7.5%)	35.75(↓10.9%)
ILLA[Conv]	**28.99**(↓10.7%)	**45.73**(↓4.3%)	**37.36**(↓6.9%)
ILLA[FC]	**29.03**(↓10.5%)	**45.30**(↓5.2%)	**37.16**(↓7.4%)
Joint training(upper bound)	32.45(−0%)	47.76(−0%)	40.11(−0%)

videos for evaluation. ActivityNet 1.3, which has 200 classes, contains 10,024 training videos and 4,926 validation videos. Following the standard protocol on temporal action localization, we report the average mean Average Precision (mAP) at different temporal Intersection-over-Union (tIoU) thresholds. The tIoUs are from 0.1 to 0.7 with step 0.1 for THUMOS'14 and from 0.5 to 0.95 with step 0.05 for ActivityNet 1.3. Note that only the action instances of the classes related to the evaluated model will participate in the calculation.

Implementation Details. The feature extractor I3D has been pre-trained by Kinetics [2] and is not fine-tuned. The TV-L1 algorithm [20] is employed to extract the optical flows. Each segment contains the feature of non-overlapping 16 frames of video. Except for modifying the one-dimensional convolution layer of the classification module to a fully connected layer, we keep the other parts and settings of UM-Net as same as [10]. In incremental training, we set the hyperparameter γ to 0.01 and σ to 50. As an exception, in the case of only increment one class, σ will be set to 1000 due to the low distillation loss caused by the lack of negative samples. We number all classes alphabetically. Since class-incremental learning for WTAL has not been actively explored before and those incremental methods specific to other tasks are difficult to migrate to WTAL tasks, we can only mainly compare ILLA with other universal transferable methods and our ablation settings.

3.2   Addition of Multiple Classes

Adding numerous classes at once is the most typical scenario for class-incremental learning. As shown in Table 1, we use the first 10 classes as old classes and the last 10 classes as new classes on THUMOS'14. *Fine-tuning* without incremental learning techniques suffers catastrophic forgetting, and its old-class performance is reduced to 16.29%. Another direct idea is to train only part parameters (*Train extended part only* and *Train FC layer only*). However, they can not maintain the

Table 2. The average mAP (%) of the increment 10 classes using different classes divisions setting on THUMOS'14.

Method	old classes(1–10)	new classes(11–20)	all classes(1–20)
ILLA	29.03(↓10.5%)	45.30(↓5.2%)	37.16(↓7.4%)
Joint training	32.45(−0%)	47.76(−0%)	40.11(−0%)
Method	**old classes(11–20)**	**new classes(1–10)**	**all classes(1–20)**
ILLA	45.63(↓4.5%)	30.87(↓4.9%)	38.25(↓4.6%)
Joint training	47.76(−0%)	32.45(−0%)	40.11(−0%)
Method	**old classes(6–7,13–20)**	**new classes(1–5,8–12)**	**all classes(1–20)**
ILLA	35.43(↓9.3%)	39.48(↓4.0%)	37.45(↓6.6%)
Joint training	39.07(−0%)	41.14(−0%)	40.11(−0%)

Table 3. The average mAP (%) of the increment 50 classes setting on ActivityNet 1.3.

Method	old classes(1–150)	new classes(151–200)	all (1–200)
N^{pre}	21.12	-	-
Fine-tuning	0(↓100%)	17.15(↓23.3%)	4.29(↓79.6%)
Distill Score	3.77(↓81.7%)	17.65(↓21.1%)	7.24(↓65.6%)
Distill Logit	1.13(↓94.5%)	17.09(↓23.6%)	5.12(↓75.7%)
Distill CAS	19.10(↓7.2%)	19.92(↓10.9%)	19.30(↓8.2%)
ILLA	**19.26**(↓6.4%)	**20.12**(↓10.0%)	**19.47**(↓7.4%)
Joint training(upper bound)	20.58(−0%)	22.36(−0%)	21.03(−0%)

old-class performance or even localize the new-class action instances due to most parameters being frozen. The transferred incremental learning method (e.g., *Distill Score*) from image classification has not achieved much effect. By distilling logits, the average mAP of old classes can be improved to 17.32%, and distilling CAS goes one step further, reaching 24.95%. Furthermore, ILLA[FC] can increase the average mAP to 29.03% by combining these two loss items. In addition, even with distillation loss constraints, the new-class performance of our method is still quite close to *Joint training*. Additionally, the experimental results of ILLA[FC] and ILLA[Conv] demonstrate that our method is not sensitive to the last layer form. We also compare the effects of distilling logits and distilling modified softmax scores (*Distill CAS + Score* VS ILLA[FC]) and prove that the original logits preserve relative activation information of critical high activation segments better than the modified softmax scores.

In order to explore the adaptability of ILLA to the recognition deviation of different classes, we have divided different old and new class groups many times according to the difficulty of recognition and carried out incremental training in the same steps. The results of Table 2 show that ILLA can flexibly adapt to this recognition deviation. And Table 3 analyzes the performance on ActivityNet 1.3 for increment 50 classes. As the number of classes increases, *Fine-tuning* and *Distill Score* forget almost all old-class knowledge. In contrast, ILLA can

Table 4. The average mAP (%) of the increment one class setting on THUMOS'14.

Method	old classes(1–19)	new class(20)	all classes(1–20)
N^{pre}	40.15	-	-
Fine-tuning	26.91($\downarrow$34.0%)	11.72($\downarrow$57.2%)	26.15($\downarrow$34.8%)
Distill Score	31.09($\downarrow$23.7%)	14.13($\downarrow$48.4%)	30.24($\downarrow$24.6%)
Distill Logit	28.48($\downarrow$30.1%)	15.77($\downarrow$42.4%)	27.84($\downarrow$30.6%)
Distill CAS	34.67($\downarrow$15.0%)	35.83($\uparrow$30.8%)	34.72($\downarrow$13.4%)
ILLA	**37.91**($\downarrow$7.0%)	**37.37**($\uparrow$36.4%)	**37.89**($\downarrow$5.5%)
Joint training(upper bound)	40.77(−0%)	27.40(−0%)	40.11(−0%)

Table 5. The average mAP (%) of the increment one class setting on ActivityNet 1.3.

Method	old classes(1–199)	new class(200)	all classes(1–200)
N^{pre}	20.81	-	-
Fine-tuning	0($\downarrow$100%)	2.17($\downarrow$97.09%)	0.01($\downarrow$99.95%)
Distill Score	0.03($\downarrow$99.86%)	2.17($\downarrow$97.09%)	0.04($\downarrow$99.81%)
Distill Logit	0($\downarrow$100%)	2.16($\downarrow$97.11%)	0.01($\downarrow$99.95%)
Distill CAS	20.76(-0%)	74.42($\downarrow$0.33%)	21.03(-0%)
ILLA	**20.81**($\uparrow$0.24%)	**75.90**($\uparrow$1.65%)	**21.09**($\uparrow$0.29%)
Joint training(upper bound)	20.76(−0%)	74.67(−0%)	21.03(−0%)

maintain old-class performance and localize new classes well since there are some similarities between the features of old and new classes.

3.3 Addition of One Class

In the extreme scenario, the new videos only include action instances of one class. Table 4 shows that *Distill Score* can retain old-class knowledge better than *Fine-tuning* indeed. However, they both perform badly in the new class due to a lack of negative samples. Compared with training from scratch, ILLA loses only 7.0% in the performance of old classes while increasing 36.4% in the new class. Given the multiscale distillation loss constraint, it demonstrates that the incremental network can overcome the disadvantage of an absence of negative samples by focusing on learning to localize new-class action instances to achieve better performance.

We also conducted the increment one class experiment on ActivityNet 1.3. Table 5 shows that the overall performance of ILLA is even better than *Joint training*. Because ActivityNet has a much lower average number of instances per video than THUMOS'14, making ILLA easier to retain old-class performance when only one new class needs to be identified. It indicates a new training paradigm for the WTAL frameworks. For example, the WTAL framework needs to improve its ability to recognize a specific class (such as the most difficult to localize) in some cases. According to the results of ILLA and *Joint training* in

Table 6. The average mAP (%) of the multi-stage incremental training setting on THUMOS'14. Each stage will add five classes.

Method	1–5	6–10	10–15	15–20	1–20
N^{pre}	35.53	-	-	-	-
Fine-tuning	2.49	0	0	10.73	3.31($\downarrow$91.7%)
Distill Score	0	1.64	36.90	19.70	14.56($\downarrow$63.7%)
Distill Logit	0	0	17.08	12.57	9.88($\downarrow$75.4%)
Distill CAS	17.88	17.35	41.42	25.05	25.42($\downarrow$36.6%)
ILLA	**19.99**	**20.07**	**48.20**	**26.29**	**29.64($\downarrow$26.1%)**
Joint training(upper bound)	31.85	33.05	63.74	31.79	40.11(-0%)

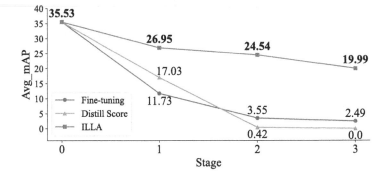

Fig. 3. The first five classes average mAP (%) at each stage of the multi-stage incremental training setting on THUMOS'14. After the third stage finishes, all classes have been learned.

Table 4 and Table 5, we can set this specific class as the new class. After the incremental training, this particular class's performance can be improved while the ability to localize the old classes won't be significantly affected.

3.4 Sequential Addition of Multiple Classes

In multi-stage incremental training, the incremental network in the current stage N_{i-1}^{inc} will become the previous network in the next stage N_i^{pre}. We set up this experiment by incrementing five classes at each stage in the THUMOS'14 dataset. Initially, the network can localize five classes of action instances. After three incremental training sessions, it can recognize 20 classes of action instances, which includes all classes and samples of the THUMOS'14 dataset, and finishes the increment sessions. We report the results in Table 6. Finally, ILLA still achieves a performance of 73.9% training from scratch for all classes. Figure 3 shows the ability of various methods to localize the action instances of the first five classes at each stage. Other methods drop to almost zero after two incremental training sessions, but ILLA can effectively preserve the capacity to localize original classes.

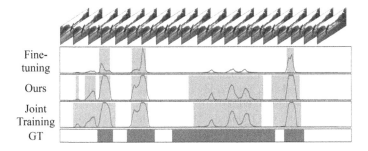

Fig. 4. Qualitative results on a video of THUMOS'14 that contains action instances of class *Golf Swing*. The lines depict the CAS of class *Golf Swing* in the temporal dimension, and the squares represent the proposed action instances.

3.5 Qualitative Results

Figure 4 illustrates the visualization results of localizing a video comprising the action instances of the old class *Golf Swing* after training with only the new-class videos. The results of ILLA are very similar to *Joint training*. By contrast, the CAS of *Fine-tuning* has been distorted, and it can no longer localize the third instance that has the most extended duration. It shows more intuitively the importance of transmitting localization information between the models by distilling CAS.

4 Conclusion

In this paper, we explore class-incremental learning for weakly supervised temporal action localization and propose ILLA with multiscale distillation to alleviate the catastrophic forgetting problem of WTAL. In order to memorize the old-class knowledge, we mine different scales of video-level logits and frame-level CAS, then utilize knowledge distillation to guide the optimization direction of the incremental network. The experimental results on two datasets show that the incremental network trained by ILLA can effectively keep the old-class knowledge when learning to localize new-class action instances. In the future, we will continue to develop the framework of ILLA to accommodate more types of WTAL models and investigate the potential of ILLA as a novel training paradigm for WTAL.

References

1. Caba Heilbron, F., Escorcia, V., Ghanem, B., Carlos Niebles, J.: Activitynet: a large-scale video benchmark for human activity understanding. In: CVPR, pp. 961–970 (2015)
2. Carreira, J., Zisserman, A.: Quo vadis, action recognition? a new model and the kinetics dataset. In: CVPR, pp. 6299–6308 (2017)

3. Girshick, R., Donahue, J., Darrell, T., Malik, J.: Rich feature hierarchies for accurate object detection and semantic segmentation. In: CVPR, pp. 580–587 (2014)

4. Hao, Y., Fu, Y., Jiang, Y.G., Tian, Q.: An end-to-end architecture for class-incremental object detection with knowledge distillation. In: ICME, pp. 1–6 (2019)

5. He, B., Yang, X., Kang, L., Cheng, Z., Zhou, X., Shrivastava, A.: Asm-loc: action-aware segment modeling for weakly-supervised temporal action localization. In: CVPR, pp. 13925–13935 (2022)

6. Hinton, G., Vinyals, O., Dean, J.: Distilling the knowledge in a neural network. arXiv preprint arXiv:1503.02531 (2015)

7. Idrees, H., Zamir, A.R., Jiang, Y.G., Gorban, A., Laptev, I., Sukthankar, R., Shah, M.: The thumos challenge on action recognition for videos "in the wild." Computer Vision and Image Understanding **155**, 1–23 (2017)

8. Kirkpatrick, J., et al.: Overcoming catastrophic forgetting in neural networks. Proc. Natl. Acad. Sci. **114**(13), 3521–3526 (2017)

9. Lee, P., Uh, Y., Byun, H.: Background suppression network for weakly-supervised temporal action localization. In: AAAI, vol. 34, pp. 11320–11327 (2020)

10. Lee, P., Wang, J., Lu, Y., Byun, H.: Weakly-supervised temporal action localization by uncertainty modeling. In: AAAI, vol. 35, pp. 1854–1862 (2021)

11. Li, B., Liu, R., Chen, T., Zhu, Y.: Weakly supervised temporal action detection with temporal dependency learning. IEEE Trans. Circuits Syst. Video Technol. **32**(7), 4474–4485 (2021)

12. Li, Z., Hoiem, D.: Learning without forgetting. IEEE Trans. Pattern Anal. Mach. Intell. **40**(12), 2935–2947 (2017)

13. Nguyen, P., Liu, T., Prasad, G., Han, B.: Weakly supervised action localization by sparse temporal pooling network. In: CVPR, pp. 6752–6761 (2018)

14. Rebuffi, S.A., Kolesnikov, A., Sperl, G., Lampert, C.H.: icarl: incremental classifier and representation learning. In: CVPR, pp. 2001–2010 (2017)

15. Shou, Z., Wang, D., Chang, S.F.: Temporal action localization in untrimmed videos via multi-stage cnns. In: CVPR, pp. 1049–1058 (2016)

16. Van de Ven, G.M., Tolias, A.S.: Three scenarios for continual learning. arXiv preprint arXiv:1904.07734 (2019)

17. Wang, L., Xiong, Y., Lin, D., Van Gool, L.: Untrimmednets for weakly supervised action recognition and detection. In: CVPR, pp. 4325–4334 (2017)

18. Xu, M., Zhao, C., Rojas, D.S., Thabet, A., Ghanem, B.: G-tad: Sub-graph localization for temporal action detection. In: CVPR, pp. 10156–10165 (2020)

19. Yu, J., Ge, Y., Li, Z., Chen, Z., Qin, X.: Context driven network with bayes for weakly supervised temporal action localization. In: ICME, pp. 1–6 (2021)

20. Zach, C., Pock, T., Bischof, H.: A duality based approach for realtime TV-L^1 optical flow. In: Hamprecht, F.A., Schnörr, C., Jähne, B. (eds.) DAGM 2007. LNCS, vol. 4713, pp. 214–223. Springer, Heidelberg (2007). https://doi.org/10.1007/978-3-540-74936-3_22

21. Zhang, C., Cao, M., Yang, D., Chen, J., Zou, Y.: Cola: weakly-supervised temporal action localization with snippet contrastive learning. In: CVPR, pp. 16010–16019 (2021)

22. Zhao, B., Xiao, X., Gan, G., Zhang, B., Xia, S.T.: Maintaining discrimination and fairness in class incremental learning. In: CVPR, pp. 13208–13217 (2020)

23. Zhu, Z., Tang, W., Wang, L., Zheng, N., Hua, G.: Enriching local and global contexts for temporal action localization. In: ICCV, pp. 13516–13525 (2021)

Nearest Neighbor Classifier with Margin Penalty for Active Learning

Yuan Cao[1,2,3], Zhiqiao Gao[2], Jie Hu[2], Mingchuan Yang[2],
and Jinpeng Chen[1,3(✉)]

[1] School of Computer Science (National Pilot Software Engineering School),
Beijing University of Posts and Telecommunications, Beijing, China
{caoyuanboy,jpchen}@bupt.edu.cn
[2] China Telecom Corporation Limited Research Institute, Beijing, China
{gaozhq6,hujie1,yangmch}@chinatelecom.cn
[3] Key Laboratory of Trustworthy Distributed Computing and Service (BUPT),
Ministry of Education, Beijing, China

Abstract. As deep learning becomes the mainstream in the field of natural language processing, the need for suitable active learning method are becoming unprecedented urgent. Active Learning (AL) methods based on nearest neighbor classifier are proposed and demonstrated superior results. However, existing nearest neighbor classifiers are not suitable for classifying mutual exclusive classes because inter-class discrepancy cannot be assured. As a result, informative samples in the margin area can not be discovered and AL performance are damaged. To this end, we propose a novel **N**earest neighbor **C**lassifier with **M**argin penalty for **A**ctive **L**earning (NCMAL). Firstly, mandatory margin penalties are added between classes, therefore both inter-class discrepancy and intra-class compactness are both assured. Secondly, a novel sample selection strategy is proposed to discover informative samples within the margin area. To demonstrate the effectiveness of the methods, we conduct extensive experiments on three real-world datasets with other state-of-the-art methods. The experimental results demonstrate that our method achieves better results with fewer annotated samples than all baseline methods.

Keywords: Active Learning · Text Classification · Bert

1 Introduction

Recently, Deep Learning (DL) has shown unparalleled ability in many areas especially in the field of natural language processing (NLP). DL-based [4,11,12] text classification methods has changed the landscape of text classification and achieved state-of-the-art performance. However, DL's superb learn capabilities heavily relies on large amount of labeled data. As a result, active learning (AL), which aims to maximize model performance while minimize labeling costs, is gradually receiving more attention [5,14,19,20,28], and may help ease the data shortage problems of DL.

© The Author(s), under exclusive license to Springer Nature Switzerland AG 2023
M. Tanveer et al. (Eds.): ICONIP 2022, LNCS 13623, pp. 379–392, 2023.
https://doi.org/10.1007/978-3-031-30105-6_32

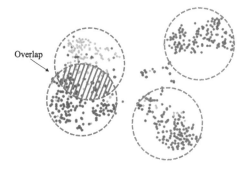

Fig. 1. Visualization of sample distribution results after 10 rounds of AL using NCENet on the **AGNEWS** dataset

So far, many AL methods analyse the output logits of the traditional softmax classifier for sample selection. The uncertainty-based method [5,6,21], a bunch of AL methods whose presence may date back to the era of machine learning, aims to calculate the uncertainty of the output logits to select the most uncertain samples for the model. Intuitively, those uncertainty-based methods are inherited in deep learning models. However, these ported methods didn't perform as well as they do on machine learning.

Fang et al. [26] pointed out that the problem encountered with deep models is actually a "false generalize" problem. DL models learn softmax classification boundaries form labeled samples, and incorrectly generalizes the classification boundaries to unlabeled samples. Fang then discards the traditional softmax classifier structure and utilizes a soft nearest neighbour classifier that classifies target samples by selecting prototype vectors on the feature space. NCENet [26] is proposed to avoid the "false generalize" problem by complete abandonment of the softmax classifier structure.

However, NCENet also has its shortcomings. The main structure of NCENet consists of n sigmoid functions instead of one softmax function. Although the sigmoid structure can be used in multi-classification scenario, it may encounter difficulty with classes that are mutually exclusive. For example, Fig. 1 shows a typical scene from a real AL training process on the **AGNEWS** dataset, visualised by t-SNE [16]. A clear overlap can be easily seen from the yellow class and blue class. The yellow class and the blue class are mixed together and no clear classification boundary can be established. In this way, two classes that were mutually exclusive become non-mutually exclusive. These are important indications that inter-class differences are not guaranteed by the sigmoid classifier. We define this phenomenon as the "non-exclusive problem". Solving the "non-exclusive problem" will enhance the performance of the model in classification and AL scenarios.

To this end, we propose **N**earest neighbor **C**lassifier with **M**argin penalty for **A**ctive **L**earning (NCMAL), which ensures class not overlapping by adding mandatory margins between classes so that the sigmoid classifier can be used in classifying mutual exclusive classes. In other words, inter-class discrepancy can be assured with the mandatory inter-class margin added. And at the same time, as we project the whole feature space onto a n-dimensional hyper-sphere, higher inter-class discrepancy brings higher intra-class compactness. As a result, the classification accuracy can be improved. Meanwhile, with margin area added, we believe that unlabeled samples within the margin area has a relatively high uncertainty, and have a high priority when labeling. We proposed a sample selection strategy that focuses high priority samples in the margin area.

Our contributions are summarized as follows:

- The proposed NCMAL effectively increases the inter-class discrepancy and make sigmoid-based classifier suitable for classifying mutual exclusive classes.
- We prove samples in the margin area tend to be more informative and propose a special sample selection strategy, which gives high priority to samples in the margin area.
- NCMAL outperforms state-of-the-art AL methods for text-classification on three real-world datasets.

2   Related Work

We focus on AL in pool-based scenarios. Pool-based AL methods can be roughly divided into three categories: uncertainty-based, representation-based and fusion methods which combines uncertainty-based method and representation-based method.

Uncertainty-Based Method. AL has been of interest to researchers since the days of machine learning, when one wanted to obtain better model performance with fewer labelled samples. In the most intuitive way of thinking, one determines whether a sample needs to be labelled by the uncertainty of the model on the sample. Different methods [17,18] have different measures of uncertainty, such as [2,24,25] based on least model confidence, [22] based on margin sampling, and [13,29] by measuring the entropy of the probability distribution to determine the uncertainty of this classification. In addition, [6] introduces Bayesian inference through the use of a Monte Carlo Dropout, which measures sample uncertainty more accurately by enabling the Dropout in the testing phase. However, the computational efficiency of this method is greatly limited by the need to perform multiple forward propagation.

Representation-Based Method. Representation-based methods aims to select the most important samples for labelling by analyzing the distribution of the unlabelled samples. As in the DAL (Discriminative Active Learning) [7] method, a binary classifier is trained to discriminate whether a sample comes from the labelled set or the unlabelled set, so that samples that best represent the entire data set can be selected. The EGL (Expected Gradient Length) [9] method measures the impact of a sample on the model by calculating the EGL of the labeled sample, and selects the labeled sample based on this criterion. Core-set [23] is also an emerging and very effective method that models the entire

AL process as a coreset problem. By solving the corresponding coreset problem in the learned representation space, samples that can best represent the entire dataset are selected to be labeled.

Fusion Method. In addition, there are many methods that combine the uncertainty-based method with the representation-based method, e.g. the BADGE [1] method can be considered as a combination of the BALD [8] method and the Coreset [23] method, and has been experimented on several models. For example, LL4AL [27] uses an additional network structure to predict the "loss" of a sample, which gives a more accurate measure of the diversity and uncertainty of the sample, and the top L labeled samples are obtained by sorting the loss values in descending order. The NCENet method uses a Nearest Neighbor Classifier to replace the traditional softmax classifier, thus solving the "false generalize" problem of the softmax classifier.

3   Methodology

In this section, we first introduce the NCMAL in detail. Then, the sample selection strategy, which aim to informative samples from the margin area, is described.

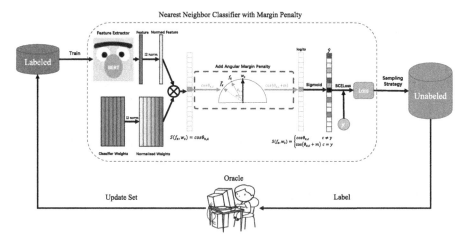

Fig. 2. One AL iteration of NCMAL

3.1   Overall Framework

Figure 2 shows the whole structure of NCMAL. This work is applied to pool-based active learning scenarios. Specifically, the algorithm is initialized with a small set of labeled samples $\mathcal{L}$ and a larger set of unlabeled samples $\mathcal{U}$. The samples $x_i \in \mathcal{L}$ all have corresponding labels y_i, while the unlabeled samples $x_i \in \mathcal{U}$ have no labels. Using $\mathcal{L}$ as training data, a text classifier $g(x|\Theta) : \mathbf{X} \rightarrow \mathbf{Y}$

is trained. The goal of the sample selection strategy is to select K samples from $\mathcal{U}$ by the classification result of the trained model $g(x|\Theta)$. The selected K samples are then annotated and added to $\mathcal{L}$, and used as the training data of the next round of training. The whole algorithm can be summarized as Algorithm 1.

Algorithm 1. *Nearest Neighbor Classifier with Margin Penalty for Active Learning(NCMAL)*

Input: Unlabeld set $\mathcal{U}$, initial budget K_{init}, budget K, margin factor m, deflation factor s, AL rounds r.

Output: Model parameters Θ, labeled set $\mathcal{L}$

1: Initialize Θ from Normal Distribution $\mathcal{N}(0, 0.01)$;
2: $\mathcal{L} \longleftarrow Random\_Select\_K\_Sample\_From(\mathcal{U}, K_{init})$
3: **for** $i = 1, 2, ..., r$ **do**
4:      **for** $x_i, y_i \in \mathcal{L}$ **do**
5:          Compute $o_{x_i,c}$ for every $c \in C$ according to Eq. 6;
6:          Compute loss L according to Eq. 7
7:          Update parameter Θ by gradient decent optimization with loss L
8:      **end for**
9:      **for** $x \in \mathcal{U}$ **do**
10:          Compute Margin Confidence score C_x^{Margin} according to Eq. 8
11:      **end for**
12:      $\mathcal{L}_i \longleftarrow Top\_K\_Sample\_Selection\_By\_Confidence\_Score(Conf(\mathcal{U}), K)$
13:      $\mathcal{L} \longleftarrow \mathcal{L} + \mathcal{L}_i$;
14:      $\mathcal{U} \longleftarrow \mathcal{U} - \mathcal{L}_i$
15: **end for**
16: **return** $\Theta, \mathcal{L}$;

3.2   Nearest Neighbor Classifier with Margin Penalty

In existing nearest neighbor classifier methods [10,26], take NCENet as an example, the classification result of an arbitrary sample mainly depends on the similarity between the feature vector $\boldsymbol{f}_x$ and the prototype vector $\boldsymbol{w}_c, c \in C$. The feature vector is extracted by a arbitrary feature extraction network (e.g. Bert, TextCNN). The output score function can be written as,

$$o_{x,c} = \sigma(\overline{S}(\boldsymbol{f}_x, \boldsymbol{w}_c)) \tag{1}$$

where $\overline{S}(\cdot, \cdot)$ is an arbitrary similarity function. As we can see from Eq. 1, the main structure of the NCENet consists of n sigmoid classifiers.

As we mentioned before, the transformation from softmax to n-sigmoid brings the "non-exclusive" problem. In order to solve this problem, intuitively, we refer to the approach in [3] and add an angular margin penalty between classes during training, which can increase the inter-class discrepancy and the intra-class compactness. In AL scenario, samples in the overlapping area contains much more information to better separate the overlapped classes. By adding mandatory margin, overlapping areas are now margin areas. Samples used to be in the overlapping area are now located in the created margin area, and can be better measured by the special sample selection strategy we describe later.

First, we utilize dot product to measure the similarity between vectors. We define the similarity between a feature vector $\boldsymbol{f}_x$ and prototype vector corresponding to class c as

$$S(\boldsymbol{f}_x, \boldsymbol{w}_c) = \boldsymbol{f}_x^{\mathrm{T}} \boldsymbol{w}_c = ||\boldsymbol{f}_x|| ||\boldsymbol{w}_c|| cos\theta_{x,c} \tag{2}$$

where $\theta_{x,c}$ is the angle between $\boldsymbol{f}_x$ and $\boldsymbol{w}_c$. We apply l_2 regularization to $\boldsymbol{w}_c$ so that $||\boldsymbol{w}_c|| = 1$. We also regularise $\boldsymbol{f}_x$ and rescale to s. With l_2 regularisation, we project $\boldsymbol{f}_x$ and $\boldsymbol{w}_c$ onto a feature space shaped as a hypersphere with radius s, making the multi-classification prediction dependent only on the angle between the sample vector and the prototype vector. The similarity can be then described as

$$S(\boldsymbol{f}_x, \boldsymbol{w}_c) = s * cos\theta_{x,c}. \tag{3}$$

Since the sample features as well as the prototype vectors are projected onto the same hypersphere with radius s, adding a angular margin becomes possible. We add a angular margin penalty m to $\theta_{x,y}$, where y is the ground-truth label of sample x. The new similarity function of $\boldsymbol{f}_x$ and $\boldsymbol{w}_y$ can be written as

$$S(\boldsymbol{f}_x, \boldsymbol{w}_c) = s * cos(\theta_{x,y} + m) \tag{4}$$

The whole similarity function can be written as

$$S(\boldsymbol{f}_x, \boldsymbol{w}_c) = \begin{cases} s * cos\theta_{x,c} & c \neq y \\ s * cos(\theta_{x,c} + m) & c = y \end{cases} \tag{5}$$

We then apply sigmoid classifier to the similarity score in order to calculate the probability of sample x belong to class c.

$$o_{x,c} = \sigma(S(\boldsymbol{f}_x, \boldsymbol{w}_c)) \tag{6}$$

A binary cross entropy loss function is applied. The loss function can be rewritten as

$$L = -\sum_c y \log o_{x,c} + (1 - y) \log(1 - o_{x,c})$$
$$= -\log \sigma(s * cos(\theta_{x,y} + m)) - \sum_{c \neq y} \log(1 - \sigma(s * cos\theta_{x,c})) \tag{7}$$

And in the testing phase, the sample will be predicted to be the class with maximum Eq. 3.

3.3 Sample Selection

In each round of active learning, we rely on the probability output $o_{x,c}$ to select the samples. As we mentioned in the previous section, the NCMAL creates a margin area between classes, and we believe that samples located in the margin area shall have higher priority when labeling. To best find the samples in the margin area, we proposed a confidence score function for NCMAL in order to give samples closer to the margin area a relatively high confidence score.

Margin Confidence

$$C_x^{Margin} = -|o_{x,c_0} - o_{x,c_1}|, \tag{8}$$

where c_0, c_1 are the classes with largest and second largest output probabilities respectively. It is worth noting that the Margin confidence score is closely related to the difference in hyperarc length from the sample point to the two nearest class centers after projected onto the feature hypersphere.

The higher the confidence score is, the higher priority the sample obtains when labeling. Samples with top-k confidence score will be queried and manually labeled for the next AL iteration. The effects of different sampling strategies will be discussed in the experimental subsection.

4 Experiments

NCMAL is tested on several datasets on a text classification task. In this section, we describe the implementation and results of the experiments in detail.

4.1 Experimental Settings

Datasets. Three different text classification datasets are used to prove the effectiveness of our method. The three datasets consists of two public datasets and one private dataset (**Telecom**). The **Telecom** dataset comprises of a total of 7,302 real-word messages from Chinese users. These messages were labeled and divided into 25 pre-defined mutually exclusive classes by a dedicated team. The dataset was divided into 5,841 training samples and 1,461 test samples. The Statistics for these three datasets are shown in Table 1.

Table 1. Statistics of datasets

Dataset	AGNEWS	IMDb	Telecom
#class	4	2	25
#train	120000	25000	5841
#test	7600	25000	1461
#init budget	50	100	500
#budget	10	20	20
#round	50	50	30

Feature Extraction Network. The commonly used pre-trained Bert [4] were chosen as the Feature Extraction Network. For all AL methods, hyperparameters were chosen consistently for fairness considerations. Necessary changes are made on the original Bert structure due to the requirements of both NCENet and NCMAL, while the number of parameters remains the same for fairness considerations.

Training Details. NCMAL is implemented on Pytorch and trained on 4* NVIDIA Tesla V100. Both init-budget and budget selection on different datasets is shown in Table 1. Batch size was set to 10 and the model trained on a learning rate of $2e^{-5}$ using the AdamW [15] optimizer. For each AL sampling method, 10 different random number seeds are used for testing and the final performance of the method was averaged over 10 experiments.

Baselines. We compare our approach to the following baselines.

– **Random.** It is the most commonly used baseline in active learning. The samples added to the labeled set in each round are randomly selected from the unlabeled set.
– **DBAL(Deep Bayesian Active Learning).** Monte Carlo Dropout was used to provide a more accurate measure of the uncertainty of the classifier. Both Confidence and Entropy were used in the final sample selection stage and the best performing of the two methods was selected as the performance of this method in the end.
– **Coreset.** Samples that best cover the entire feature space are selected. We chose two implementations of Coreset as described in [23], the greedy version of Coreset are implemented.
– **BADGE** [1]. It can be viewed as a combination of EGL and Coreset, and ensures diversity and uncertainty at the same time.
– **NCENet.** We implemented NCENet as described in [26].

The implementation[1] was based on the code[2] made available by [7].

4.2 Model Effect

Table 2. Accuracy performance on full training scenario. The best performing method in each row is boldfaced

Dataset	Classifier		
	Softmax	NCENet	NCMAL
AGNEWS	94.42%	94.67%	**94.87%**
IMDb	85.52%	85.57%	**85.89%**
Telecom	87.81%	89.11%	**89.45%**

---

[1] https://github.com/GhostAnderson/Nearest-Neighbor-Classifier-with-Margin-Penalty-for-Active-Learning.
[2] https://github.com/dsgissin/DiscriminativeActiveLearning.

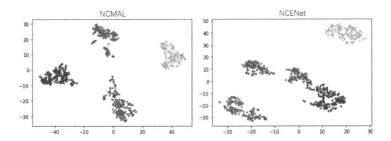

Fig. 3. Demonstration of sample distribution on **AGNEWS** dataset using NCMAL and NCENet

Performance on Full Training. We first tested the performance of our model under full training with all samples labeled. This result is also equivalent to the test result under AL scenario at 100% sample labeled. Table 2 illustrates the accuracy performance of our NCMAL and baselines fully trained on three datasets. Our NCMAL outperforms all baselines on all three datasets. This demonstrates the structural advantage of our classifier over both traditional softmax classifier and NCENet. Figure 3 visualises the sample distribution of NCMAL and NCENet after full training on **AGNEWS** dataset. The classes formed by NCMAL are more compact compared to NCENet, which can be easily seen from the red class. At the same time, we can see from the distribution of blue class and red class that, the inter-class discrepancy are better assured by the NCMAL.

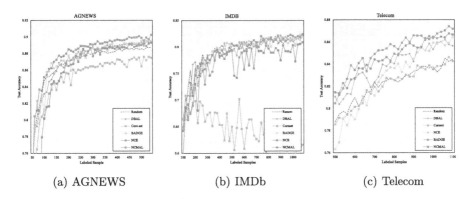

(a) AGNEWS   (b) IMDb   (c) Telecom

Fig. 4. Active learning performance and comparison with baseline methods

Performance on AL. Figure 4 illustrates test accuracy curves of our NCMAL and baselines under AL scenario on three datasets. From Fig. 4, we can draw three following critic conclusions.

1) In most cases except for Coreset, AL methods outperforms random selection, which shows the importance of AL methods. Meanwhile, our NCMAL outperforms all other baseline methods. Significant performance gaps can be observed on **Telecom** and **AGNEWS** dataset, in which NCMAL has a large performance gap with other methods from beginning to end. Under **IMDb** dataset, though advantages of all active learning methods over random selection are not very clear, our NCMAL still shows comparable performance over other baseline methods.

2) From Fig. 4 we can conclude that, the improvement are ascending as class number increases. Specifically, our method shows the greatest improvement over other methods on the **Telecom** dataset (25 classes) and marginal improvements on **IMDb** dataset (2 classes), which implies that, our methods is more suitable for classification with more classes. The reason may be that in classification with more classes, discrepancy between classes are even less guaranteed, and adding a margin term in such scenarios is more helpful in improving classification accuracy than in classification tasks where there are relatively few classes.

3) Compared with NCENet, our NCMAL shows constantly better performance especially in the front part of the learning curve. We believe that the addition of margin introduces a prior-knowledge to the model that classes are mutual exclusive, and thus the performance during early training period is improved.

4.3   Different Sample Selection Strategies

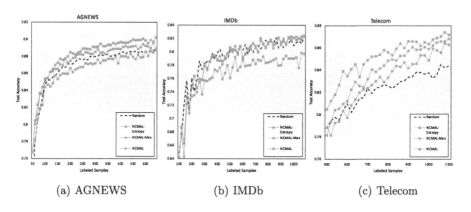

(a) AGNEWS   (b) IMDb   (c) Telecom

Fig. 5. Active learning performance and comparison with different variants of NCMAL

With the same NCMAL network structure, the effect of sample selection strategies other than Margin Confidence is also studied. Except for Margin Confidence (NCMAL for simplicity consideration), we bring out two variants of NCMAL with different sample selection strategies as comparison.

- **NCMAL-Entropy** To measure the confusion of the classifier, we propose a Entropy Confidence sample selection strategy.

$$C_x^{Entropy} = \sum_c \left(1 + o_{x,c} - \max_c o_{x,c}\right) \tag{9}$$

- **NCMAL-Max** As the traditional uncertainty-based methods do, we pick the sample with the lowest maximum probability to be labeled.

$$C_x^{Max} = -\max_c o_{x,c} \tag{10}$$

Those two sample selection strategies or their variants are often used in uncertainty-based methods.

Figure 5 shows the test accuracy curve of NCMAL with three different sample selection strategies. It is easy to see that the original NCMAL with Margin Confidence has a significant performance advantage over the other two methods (NCMAL-Entropy & NCMAL-Max) in most cases. Among them, NCMAL-Max performs the worst with weaker performance than random on both **AGNEWS** and **IMDb** data sets. Meanwhile, NCMAL-Entropy performs moderately, outperforming the random method on all three data sets and second only to original NCMAL.

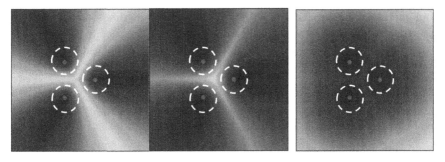

(a) Margin Confidence   (b) Entropy Confidence   (c) Max Confidence

Fig. 6. Demonstration of the confidence score distribution of the two sample selection strategies. The red points represent the class centroids. The white dashed line represents the class classification boundary. Yellower areas have a higher confidence score, and bluer areas represent lower confidence score. The higher the confidence score, the higher the priority when labeling (Color figure online)

We analyzed the reasons for this difference in performance. As shown in Fig. 6, the confidence score distribution of three sample selection strategies are demonstrated. The order of attention given to the margin area samples by different strategies is consistent with the order of their performance. The higher the confidence score given to samples from the margin area, the better the classification performance of the model obtains. The Margin Confidence in NCMAL gives the highest confidence score to samples from the margin area, thus more samples from the margin area will be selected and labeled. And as a result, NCMAL obtains the best performance. The Entropy Confidence gives moderate priority to samples in the margin area while Max Confidence gives nearly no priority to samples in the margin area, and as a result, NCMAL-Entropy gains moderate performance and NCMAL-Max obtains the worst performance. It is easy to see that the more samples from the margin area are selected, the better the model performs. This finding supports the hypothesis we presented in the previous section that, samples from the margin area tend to be more informative than samples from other areas and should have high priority when labeling. The combination of our NCMAL and Margin Confidence can best discover informative samples from the margin area, thus gains significant performance improvement.

5   Conclusion

In this paper, we propose a novel nearest neighbour classifier with margin for active learning (NCMAL). We add angular margin penalties so that the inter-class discrepancy can be assured thus the sigmoid classifier structure can be applied to mutual exclusive classification scenarios. This solves the problem of overlapping class boundaries that can occur in [26] and achieves both better classification results and active learning results. We demonstrate the effectiveness of our method by comparing it with several baseline methods on different real datasets. We also proposed a special sample strategy in order to discover informative samples which lies in the margin area. The experimental results proves the correctness of our hypothesis and demonstrate the superiority of our method for text classification tasks.

Acknowledgement. This work was supported by National Natural Science Foundation of China (Grant No. 61702043, No. 72274022).

References

1. Ash, J.T., Zhang, C., Krishnamurthy, A., Langford, J., Agarwal, A.: Deep batch active learning by diverse, uncertain gradient lower bounds. In: 8th International Conference on Learning Representations, ICLR 2020, Addis Ababa, Ethiopia, 26–30 April 2020. OpenReview.net (2020). https://openreview.net/forum?id= ryghZJBKPS
2. Culotta, A., McCallum, A.: Reducing labeling effort for structured prediction tasks. In: AAAI, vol. 5, pp. 746–751 (2005)
3. Deng, J., Guo, J., Xue, N., Zafeiriou, S.: Arcface: additive angular margin loss for deep face recognition. In: Proceedings of the IEEE/CVF Conference on Computer Vision and Pattern Recognition, pp. 4690–4699 (2019)

4. Devlin, J., Chang, M., Lee, K., Toutanova, K.: BERT: pre-training of deep bidirectional transformers for language understanding. In: Burstein, J., Doran, C., Solorio, T. (eds.) Proceedings of the 2019 Conference of the North American Chapter of the Association for Computational Linguistics: Human Language Technologies, NAACL-HLT 2019, Minneapolis, MN, USA, 2–7 June 2019, Volume 1 (Long and Short Papers), pp. 4171–4186. Association for Computational Linguistics (2019). https://doi.org/10.18653/v1/n19-1423

5. Dor, L.E., et al.: Active learning for BERT: an empirical study. In: Proceedings of the 2020 Conference on Empirical Methods in Natural Language Processing (EMNLP), pp. 7949–7962 (2020)

6. Gal, Y., Islam, R., Ghahramani, Z.: Deep Bayesian active learning with image data. In: International Conference on Machine Learning, pp. 1183–1192. PMLR (2017)

7. Gissin, D., Shalev-Shwartz, S.: Discriminative active learning. arXiv preprint arXiv:1907.06347 (2019)

8. Houlsby, N., Huszár, F., Ghahramani, Z., Lengyel, M.: Bayesian active learning for classification and preference learning. arXiv preprint arXiv:1112.5745 (2011)

9. Huang, J., Child, R., Rao, V., Liu, H., Satheesh, S., Coates, A.: Active learning for speech recognition: the power of gradients. arXiv preprint arXiv:1612.03226 (2016)

10. Kontorovich, A., Sabato, S., Urner, R.: Active nearest-neighbor learning in metric spaces. J. Mach. Learn. Res. **18**, 195:1–195:38 (2017). http://jmlr.org/papers/v18/16-499.html

11. Lai, S., Xu, L., Liu, K., Zhao, J.: Recurrent convolutional neural networks for text classification. In: Twenty-Ninth AAAI Conference on Artificial Intelligence (2015)

12. Lan, Z., Chen, M., Goodman, S., Gimpel, K., Sharma, P., Soricut, R.: ALBERT: a lite BERT for self-supervised learning of language representations. In: 8th International Conference on Learning Representations, ICLR 2020, Addis Ababa, Ethiopia, 26–30 April 2020. OpenReview.net (2020). https://openreview.net/forum?id=H1eA7AEtvS

13. Lewis, D.D., Gale, W.A.: A sequential algorithm for training text classifiers. In: Croft, B.W., van Rijsbergen, C.J. (eds.) SIGIR '94, pp. 3–12. Springer, London (1994). https://doi.org/10.1007/978-1-4471-2099-5_1

14. Li, C., et al.: Unsupervised active learning via subspace learning. In: Proceedings of the AAAI Conference on Artificial Intelligence, vol. 35, pp. 8332–8339 (2021)

15. Loshchilov, I., Hutter, F.: Decoupled weight decay regularization. In: 7th International Conference on Learning Representations, ICLR 2019, New Orleans, LA, USA, 6–9 May 2019. OpenReview.net (2019). https://openreview.net/forum?id=Bkg6RiCqY7

16. Van der Maaten, L., Hinton, G.: Visualizing data using t-SNE. J. Mach. Learn. Res. **9**(11) (2008)

17. Nafa, Y., et al.: Active deep learning on entity resolution by risk sampling. Knowl.-Based Syst. **236**, 107729 (2022)

18. Nguyen, C.V., Ho, L.S.T., Xu, H., Dinh, V., Nguyen, B.T.: Bayesian active learning with abstention feedbacks. Neurocomputing **471**, 242–250 (2022)

19. Nguyen, Q.P., Low, B.K.H., Jaillet, P.: An information-theoretic framework for unifying active learning problems. In: Proceedings of AAAI, pp. 9126–9134 (2021)

20. Prabhu, S., Mohamed, M., Misra, H.: Multi-class text classification using BERT-based active learning. In: Dragut, E.C., Li, Y., Popa, L., Vucetic, S. (eds.) 3rd Workshop on Data Science with Human in the Loop, DaSH@KDD, Virtual Conference, 15 August 2021 (2021). https://drive.google.com/file/d/1xVy4p29UPINmWl8Y7OospyQgHiYfH4wc/view

21. Ren, P., et al.: A survey of deep active learning. ACM Comput. Surv. **54**(9), 180:1–180:40 (2022). https://doi.org/10.1145/3472291
22. Scheffer, T., Decomain, C., Wrobel, S.: Active hidden Markov models for information extraction. In: Hoffmann, F., Hand, D.J., Adams, N., Fisher, D., Guimaraes, G. (eds.) IDA 2001. LNCS, vol. 2189, pp. 309–318. Springer, Heidelberg (2001). https://doi.org/10.1007/3-540-44816-0_31
23. Sener, O., Savarese, S.: Active learning for convolutional neural networks: a core-set approach. In: 6th International Conference on Learning Representations, ICLR 2018, Vancouver, BC, Canada, 30 April–3 May 2018, Conference Track Proceedings. OpenReview.net (2018). https://openreview.net/forum?id=H1aIuk-RW
24. Settles, B.: Active learning literature survey (2009)
25. Settles, B., Craven, M.: An analysis of active learning strategies for sequence labeling tasks. In: Proceedings of the 2008 Conference on Empirical Methods in Natural Language Processing, pp. 1070–1079 (2008)
26. Wana, F., Yuana, T., Fua, M., Jib, X., Yea, Q.H.Q.: Nearest neighbor classifier embedded network for active learning. In: Proceedings of the AAAI Conference on Artificial Intelligence, vol. 35, pp. 10041–10048 (2021)
27. Yoo, D., Kweon, I.S.: Learning loss for active learning. In: Proceedings of the IEEE/CVF Conference on Computer Vision and Pattern Recognition, pp. 93–102 (2019)
28. Zhou, B., Cai, X., Zhang, Y., Guo, W., Yuan, X.: Mtaal: multi-task adversarial active learning for medical named entity recognition and normalization. In: Proceedings of the AAAI Conference on Artificial Intelligence, vol. 35, pp. 14586–14593 (2021)
29. Zhu, J., Wang, H., Yao, T., Tsou, B.K.: Active learning with sampling by uncertainty and density for word sense disambiguation and text classification. In: Proceedings of the 22nd International Conference on Computational Linguistics (Coling 2008), pp. 1137–1144 (2008)

Factual Error Correction in Summarization with Retriever-Reader Pipeline

Weiwei Li, Junzhuo Liu, and Hui Gao$^{(\boxtimes)}$

University of Electronic Science and Technology of China, Chengdu, China
{202121080411,202121080401}@std.uestc.edu.cn, huigao@uestc.edu.cn

Abstract. Summarization models compress the source text without sacrificing the primary information. However, about 30% of summaries produced by state-of-the-art summarization models suffer from the factual inconsistencies between source text and summary, also known as hallucinations, making them less trustworthy. It has been challenging to reduce hallucinations, especially entity hallucinations. Most prior works use the entire source text for factual error correction, while input length is often limited. It prevents models from checking facts precisely and limits its applications on long document summarization. To address this issue, we propose a post-editing factual error correction method based on a Retriever-Reader pipeline. After the summarization model generates the summary, entities in the summary are examined and corrected iteratively by retrieving and reading only the most relevant parts of the source document. We validate the proposed approach on the CNN/DM and Gigaword datasets. Experiments show that the proposed approach outperforms all the baseline models in the consistency metric. Better results are also achieved on human evaluations.

Keywords: Summarization · Entity hallucination · Post-editing factual error correction

1 Introduction

Text summarization is a task of shortening the source text while preserving its meaning as much as possible. Summaries are expected to be grammatically correct, fluent, novel, and consistent with the source text. Large-scale text corpora enable state-of-the-art summarization models to produce fluent summaries. Though achieved excellent performance in metrics such as ROUGE [1], the faithfulness of generated summaries is still in doubt [2]. The reason is that those metrics have a low correlation with consistency. For example, ROUGE evaluates only the token overlap between human-written summaries and generated ones. A swap of the subject and object will not change its ROUGE score but result in a factual error. About 30% of summaries produced by state-of-the-art summarization models suffer from the inconsistency between the source text and

M. Tanveer et al. (Eds.): ICONIP 2022, LNCS 13623, pp. 393–405, 2023.
https://doi.org/10.1007/978-3-031-30105-6_33

Source text: Born at Alltnacaillich, Strathmore in 1714, Rob Donn could not read or write and dictated his poetry from memory in later life. A wall hanging depicting scenes from his life and work is to be created in a project led by Strathnaver Museum and Mackay Country Community Trust...	Gold	
	A Gaelic bard is to be remembered in needlework.	
	Extrinsic Hallucination	ROUGE-1: 44.44
	A celebration is to be held to **mark the 150th anniversary of the birth** of a Scottish poet .	

Source text: (...) the kind of horror represented by the blackwater case and others like it – from abu ghraib to the massacre at haditha to cia waterboarding – may be largely absent from public memory in the west these days, but it is being used by the islamic state in iraq and syria (isis) to support its sectarian narrative.(...)	**Intrinsic Hallucination**	ROUGE-1: 30.77
	(...)**Isis** has been used by the Islamic State in Iraq and Syria (...)	
	Corrected by Vanilla Reader	ROUGE-1: 30.77
	(...)**Isis** has been used by the Islamic State in Iraq and Syria (...)	
Gold	**Corrected by Ours**	ROUGE-1: 23.08
ISIS is using past Western transgressions in Iraq to justify its brutality. (...)	(...)**Horror** has been used by the Islamic State in Iraq and Syria (...)	

Fig. 1. Typical hallucinations in summaries, generated by BertSum on XSum Dataset and Bottom-up on CNN/DM Dataset. These examples show a weak correlation between ROUGE scores and summary consistency. It is also worth noting that corrections to factual errors can lead to a drop in ROUGE scores.

the summary, as known as hallucination [3,4]. Some generated summaries with typical hallucinations and their ROUGE scores are listed in Fig. 1. Hence, it is essential to reduce hallucinations in summaries in order to make text summarization systems more reliable.

A study of the hallucinations [4] in neural abstractive summarization systems shows that over 90% of extrinsic hallucinations are erroneous. Entity hallucinations are hallucinations of named entities or quantitative entities, which are frequent [5]. Previous works have pushed the frontier of alleviating the inconsistency of summaries. Some works [6–11] concentrate on improving the overall consistency of summarization by incorporating consistency enhancement mechanism in existing summarization models. Although they made some progress, these are not generic solutions for all summarization models as the consistency enhancement component is designed separately for each network. The explainability of these methods is also poor. And some works [5,8,12,13] focus more on post-editing factual error correction method. Though progress has been made in both boosting consistency metric scores and human evaluation scores on consistency in recent works, erroneous rates remain too high to convince us that generated summary is trustworthy. The limitation of input length also restricts those method's applications on long document summarization.

In this work, we propose a post-editing factual error correction method based on a Retriever-Reader approach, to reduce entity hallucinations in both short and long documents. We follow the work SpanFact [11], which adopts from a QA model and works in a generate-then-correct manner. Entity correction is performed by predicting the masked entity in summaries using the source text. Multiple studies [14,15] found that QA models may be biased and not really learn to understand texts, since they fail to give answers if presented in a different position of the context. This provides us with the intuition that, as only a tiny part of the source text contributes to the hallucination correction procedure, removing redundant sentences from the source text may improve the performance of the model. As illustrated in Fig. 1, The reader gives the wrong prediction *ISIS* with the full source text, but the correct prediction *Horror* with only the relevant context sentences. This result is in line with our intuition. It inspires us to extend it using a Retriever-Reader approach, a classical solution for open-domain QA problems. The redundant part is removed using a retriever, so that long document summarization can be checked, and the reader can check entities more precisely. It can be transferred to any summarization model. Our main contributions are as follows:

- We propose a more explainable post-editing factual error correction method based on a Retriever-Reader approach to reduce entity hallucinations. It can be applied to any summarization model, including long document summarization.
- We propose a context retrieval method to obtain shorter and more accurate context for entity examination.
- Factual error correction experiment is conducted on three datasets over multiple baselines. Experiments show an average increase of 3 points in the consistency metric FactCC compared to the best baseline while keeping a close ROUGE score compared to competitive methods.

2   Related Works

Adding consistency components to the summarization model and post-editing methods are two typical ways of improving summary consistency.

As a representative of modifying existing summarization models, Falke et al. [16] first introduced text entailment to improve consistency by ranking candidate summaries by correctness. Cao et al. [17] used contrastive learning for improving the faithfulness and factuality of the generated summaries of the state-of-the-art models, like BART [13] and PEGASUS [18]. Huang et al. [9]'s method benefits from structured information by using dual encoders: a sequential document encoder and a graph-structured encoder. Aralikatte et al. [6] introduced the focus attention mechanism and demonstrates its effect on promoting consistency and diversity of summaries. Nan et al. [7] proposed a QA-based evaluation metric and trained the model to maximize the proposed metric. Zhu et al. [19] integrated factual relations into the summary generation process via graph attention to produce summaries with higher factual consistency. Lee et al. [20] improved consistency in dialogue summaries through self-supervised learning.

Post-editing factual error correction method is to correct the summary after the summarization model generates it, so it is model-independent. Cao et al. [12] proposed a method based on the denoising ability of BART. The concatenation of source text and summary is the input, and the output at the corresponding position is the corrected token. However, it is hard for this model to handle the case when the number of tokens for the entity hallucination and the correct entity is different. Some works focus on entity hallucinations. Chen et al. [8] proposed a method that generates alternative candidate summaries by replacing named entities and quantities in summary with one compatible entity from the source document. Then the best candidate is selected as the final output summary. Ranking candidate summaries by correctness is similar to the idea of Falke et al., and their difference is that this work does not correct factual errors during summarization model inference, but after it is generated. Zhao et al. [5] focused on the quantitative entity inconsistency in summaries and verified each entity through sequence annotation. Dong et al. [11] proposed a suite of QA-based methods that correct factual errors in parallel and iteratively.

We follow the work SpanFact's idea to use a reader to validate and predict entities in summaries using the source text. However, due to the input token length limitation, the relevant context of the entities to be verified in the summaries may be truncated or cannot be input into the model. And redundant sentences may distract the Reader model. To address this issue, our proposed method uses a Retriever-Reader approach.

3   Proposed Method

The post-editing factual error correction task is, given a text-summary pair $< D, S >$, where D represents the original token sequence $(d_1...d_k)$ and S represents the summary sequence $(s_1...s_m)$, to generate a corrected summary $S' = (s_1...s_n)$, so that it has

$$F(D, S') \geq F(D, S), \tag{1}$$

where F represents the selected factual consistency metric [11].

3.1   Entity Hallucinations Correction

As illustrated in Fig. 2, there are four main steps in the proposed method. In the preprocessing step, summaries are split into sentences, and entities in each sentence are recognized. A query is built by masking one entity in the sentence. Context, which contains the possibly relevant information, is retrieved using the query from the source text in the Context Retrieval Step. Masked entities in a sentence are predicted in the Prediction step by feeding the concatenation of query and the context text to a Reader model. Finally, in the Updating Step, whether the entity will be updated is decided based on the prediction. The sentence will be updated if the entity prediction is present in the source text and is not a substring of the original entity or empty. The corrected sentence will be used to build the next query. And this procedure will be executed on all the entities in all sentences in summary. In the following sections, we introduce the details of Context Retrieval and Reader.

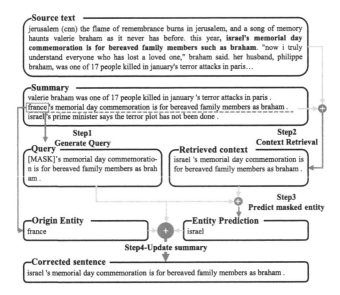

Fig. 2. Demonstration of entity hallucinations correction procedure. This is a running example selected from the outputs of Bottom-up Summarization on CNN/DM dataset. In our proposed method, Reader takes the retrieved context instead of the full source text as input.

3.2   Context Retrieval

Problem Formulation. To extract most relevant context sentences, we propose a context retrieval model. The problem is modeled as binary classification. Given an input (D, Q), where D is the source text, and Q is the query with one entity masked. The context retrieval model is expected to assign the label $L_i \in \{IRRELEVANT, RELEVANT\}$ for each D_i, where D_i is the i-th sentence of the source text, and $RELEVANT$ indicates one sentence may contain the masked entity.

Retrieve Model. The architecture is shown in Fig. 3. Given an input (D, Q), the input sequence I is formulated as,

$$I = ([CLS] \; Q \; [SEP] \; D_1 \; [SEP]... \; D_n \; [SEP]), \tag{2}$$

[SEP] is appended after each sentence. The BERT [21] Encoder calculates contextualized token-level encoding h_i,

$$h_i = BERT(I_i), \tag{3}$$

The encoded [CLS] is used as hidden representation h_i for the query and sentences in the source text. Sentence feature h'_i is obtained by concatenate current hidden representation with the previous one.

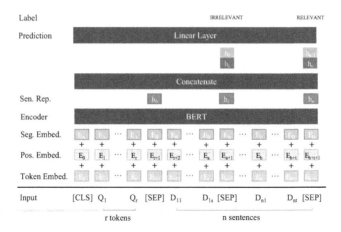

Fig. 3. Architecture of Context Retrieval Model

$$h'_i = [h_i, h_{i-1}], \tag{4}$$

Label sequence are generated by a linear layer with final hidden states

$$\hat{p}_i = Softmax(W * h'_i + b), \tag{5}$$

$$L_i = Argmax(\hat{p}_i), \tag{6}$$

where W and b are trainable parameters.

Loss Function. In most cases, only few sentences contains the masked entity, in others words, negative sample is dominant. So labels will be imbalance during training and inference. Simply applying cross-entropy loss may lead to failure, since model will be biased to mark every sample as negative.

Class-balanced cross-entropy loss [22]

$$L = \frac{1}{N}(\sum_{y_i=1}^{m} -\alpha log(\hat{p}) + \sum_{y_i=0}^{n} -(1-\alpha)log(1-\hat{p})) \tag{7}$$

is used as the loss function to solve this problem, where α is the class-balanced ratio, y_i is the ground-truth label, $\hat{p}$ is the network's output probability, and N is the batch size.

Additional Context. An additional context sentence that has the highest similarity score with the query is added to the retrieved context to balance the extractive and abstractive summarization. Levenshtein-distance and BLEU [23] are used to calculate the similarity score. Given $X, Y \in \Sigma^*$, the Levenshtein-distance $LD_{X,Y}$ is defined as

$$LD_{X,Y} = LD_{X,Y}(|X|, |Y|) \tag{8}$$

$$C_{X,Y}(i,j) = \begin{cases} 0 & X_i = Y_j \\ 1 & otherwise \end{cases} \tag{9}$$

$$LD_{X,Y}(i,j) = \begin{cases} \max(i,j) & i = 0 \ or \ j = 0 \\ \min \begin{cases} LD_{X,Y}(i, j-1) + 1 \\ LD_{X,Y}(i-1, j) + 1 \\ LD_{X,Y}(i-1, j-1) + C_{X,Y}(i,j) \end{cases} & otherwise \end{cases} \tag{10}$$

The $BLEU_n$ score is defined as

$$Count_{clip} = \min(Count, MaxRefCount), \tag{11}$$

$$BLEU_n = \frac{\sum_{C \in \{Candidates\}} \sum_{n-gram \in C} Count_{clip}(n-gram)}{\sum_{C' \in \{Candidates\}} \sum_{n-gram' \in C'} Count(n-gram')}, \tag{12}$$

where $Candidates$ are the candidate sentences.

Training Data Synthesis. The retriever model is trained using gold summaries in CNN/DM [24] and Gigaword [25] datasets. For each sentence in summary, we select entities that is contained in source text. And the summary sentence with one selected entity masked is the query, the context sentence that has the highest similarity score with the query and contains the entity is labeled as $RELEVANT$.

3.3 Reader Model

We use BERT [21] as the basis of our Reader model, a linear layer is added to predict the start position and end position of the answer span.

Given an input (C, S_i, E_{ij}), where C is the context, S_i is the i-th summary sentence, and E_{ij} is the j-th entity in S_i. The query Q_i is generated by replacing E_{ij} with [MASK]. The input sequence I, which is the concatenation of Q_i and C, is formulated as

$$I = ([CLS] \ Q \ [SEP] \ C). \tag{13}$$

Then the input sequence is embedded and fed to the BERT model. The answer span is predicted using the final hidden state:

$$(start\_pos, end\_pos) = H_0 * W + b, \tag{14}$$

where H_0 is the final hidden state of BERT, W and b are trainable parameters.

4 Experiments

In this section, we present the results of the proposed method on the factual error correction task. The effect of alleviating inconsistency of summarization is estimated using both automatic factual metrics and human evaluation.

4.1 Baselines and Datasets

We choose multiple baselines to compare the improvement of consistency on summarization models of different architectures. Seq2seq models such as Bottom-up [26], Pointer Generator Network [27], GenParse [28], models based on pre-trained language models such as BertSumAbs [29], BART [13], and unsupervised models such as SummaryLoop [30] are chosen. We compare our factual error correction method with Split Encoder(2020) [31], SpanFact(2020) [11], FC(2021) [19], and CLIFF(2021) [17]. CNN/DM [24] and Gigaword [25] are widely used summarization datasets. For each summarization model, we select the dataset on which the model achieves state-of-the-art performance as the test dataset. Details are listed in Table 1.

4.2 Metrics

We choose ROUGE [1] and FactCC [2] as metrics. ROUGE is chosen since it is a widely used metric of the summarization model. It calculates the overlap of generated summary and reference summary, so summaries with higher ROUGE scores are closer to human written summaries. Automatic consistency metrics, such as FactCC, show a high correlation with human evaluation results [2], and the consistency score of summaries can be computed without a gold summary. FactCC is a consistency metric based on BERT [21]. It takes the concatenation of the summary and the source text as the input. The output layer of the BERT model is connected to a binary classifier output layer, whose output indicates the consistency label of the summary.

4.3 Implementation Details

The proposed model is implemented using the Huggingface library [32] and PyTorch framework [33]. Parameters of model are initialized with bert-large-uncased checkpoint. We finetuned the proposed model using training data for 2 epochs. The reader model is finetuned on SQuAD [34] dataset. Batch size is set to 7. AdamW optimizer [35] with 1e–8 and an initial learning rate 3e-5 is used for training. Learning rate schedule follows a linear decay scheduler with warmup = 10,000. Class-balanced ratio is set to 2. The best model checkpoints are chosen based on performance on the validation set. Experiments are conducted using 2 RTX 3090 GPUs with 24 GB of memory. Threshold is set to 80 in the simple match context retrieval. We generate summaries with the published code and checkpoint of summarization models. If outputs on the test set of is published, we use the published one.

4.4   Experimental Result

Table 1. Results on CNN/DM Dataset

Method	FactCC	ROUGE		
		R-1	R-2	R-L
Baseline: Bottom-up				
FC	+0.09	−0.26	−0.12	−0.24
Split Encoders	−0.51	−1.46	−0.83	−1.14
SpanFact	+2.94	−0.11	−0.12	−0.11
Ours	**+5.69**	**+0.05**	**+0.09**	**+0.02**
Baseline: BertSumAbs				
Split Encoders	−0.33	−1.46	−1.08	−0.92
SpanFact	+1.93	−0.14	−0.16	−0.14
Ours	**+5.98**	**+0.48**	**+0.11**	**+0.46**
Baseline: BertSumExtAbs				
Split Encoders	−0.10	−1.32	−1.00	−0.49
SpanFact	+1.75	−0.12	−0.14	−0.13
Ours	**+5.44**	**+0.52**	**+0.13**	**+0.52**
Baseline: TransformerAbs				
Split Encoders	−0.97	−1.13	−1.12	−1.19
SpanFact	+2.31	−0.11	−0.13	−0.10
Ours	**+6.44**	**+0.69**	**+0.12**	**+0.71**
Baseline: BART-large				
CLIFF	+3.11	/	/	/
Ours	**+3.34**	−0.44	−0.37	−0.44
Baseline: Summaryloop				
Ours	**+4.89**	**+0.19**	**+0.10**	**+0.16**

Automatic Evaluation. Results on CNN/DM and Gigaword Dataset are listed in Tables 1 and 2. Each block lists the improvement in FactCC score and ROUGE score of the proposed method and competing methods compared to the same baseline. Performance comparisons with competing methods are performed only on published experimental data for these methods.

As shown in Table 1, on the CNN/DM dataset, the proposed method outperforms all competing methods in FactCC score and maintains a very close ROUGE score. It is worth noting that we only obtained a very slight performance improvement on FactCC when comparing with CLIFF. The reason is that the BART-large model obtained the highest score of 78.70 among all baselines.

The results on the Gigaword dataset are listed in Table 2. We get a similar performance boost on the FactCC score, but with a slight drop (−0.5 to −1.8)

Table 2. Results on Gigaword Dataset

Method	FactCC	ROUGE		
		R-1	R-2	R-L
Baseline: GenParseBase				
Split Encoders	+2.15	−0.91	−0.10	−0.90
SpanFact	+6.10	**−0.85**	**−0.61**	**−0.81**
Ours	**+10.21**	−0.89	−1.38	−1.02
Baseline: GenParseFull				
Split Encoders	+3.67	−1.60	−1.31	−1.36
SpanFact	+6.15	**−0.95**	**−0.84**	**−0.97**
Ours	**+9.98**	−1.05	−1.77	−1.21
Baseline: PointerGeneratorNetwork				
Split Encoders	+4.70	−1.08	−0.66	−1.14
SpanFact	+6.20	−1.33	**−1.07**	−1.30
Ours	**+8.24**	**−0.47**	−1.22	**−0.65**

in the ROUGE score. As illustrated in Fig. 1, in some cases, a correct entity correction may also lead to a drop in ROUGE score. The entity *ISIS*, which presents in the gold summary and the origin summary, should not have been there. This correction resulted in a drop of 7.7 in ROUGE-1 score. The reason that makes the ROUGE scores drop on the Gigaword dataset but not on the CNN/DM dataset is that the summary of the Gigaword dataset is usually only one sentence, which is much shorter compared to the CNN/DM dataset. This results in the replaced entity, which presents in the gold summary, being unlikely to be covered by other summary sentences. However, the ROUGE score only calculates the overlap of the generated summaries and human-written summaries. So the differences in the construction protocol of the datasets lead to variations in the performance of our proposed method on the two datasets.

Human Evaluation. We conduct a human evaluation with the proposed method FEC [12] and SpanFact [11] on the K2019 dataset [2]. As far as we know, it is the only annotated dataset for factual error correction. The K2019 dataset contains 508 context-claim pairs with the correctness of claim annotated but corrected claim absent. We check whether the corrected claim is consistent with the context. As shown in Table 3, the proposed method corrected 19 of 62 incorrect claims, and corrupted 5 of 446 correct claims. Other claims are kept unchanged or meaning unchanged. The proposed method shows better performance on the K2019 dataset than competitive methods.

Table 3. Results on K2019 Dataset

Method	Consistent		Inconsistent	
	before	after	before	after
SpanFact	446	3	62	18
FEC		5		11
Ours		5		**19**

5 Conclusion

In this work, we propose a factual error correction method based on a Retriever-Reader approach for summarization model. We focus on correcting factual errors in a more precise way. We build a BERT-based retriever model to label the relevance between the query and sentence in the source text and use that information to retrieve the most relevant context. Experiments are conducted on the CNN/DM and Gigaword datasets. Results show that the proposed method have improved consistency among multiple baseline summarization models of different architectures. In the future work, we plan to extend the method for correcting extrinsic hallucinations by incorporating external knowledge base.

References

1. Lin, C.Y.: Rouge: a package for automatic evaluation of summaries. In: Text Summarization Branches Out, pp. 74–81 (2004)
2. Kryściński, W., McCann, B., Xiong, C., Socher, R.: Evaluating the factual consistency of abstractive text summarization. In: EMNLP, pp. 9332–9346 (2020)
3. Cao, Z., Wei, F., Li, W., Li, S.: Faithful to the original: fact aware neural abstractive summarization. In: AAAI (2018)
4. Maynez, J., Narayan, S., Bohnet, B., McDonald, R.: On faithfulness and factuality in abstractive summarization. In: ACL, pp. 1906–1919 (2020)
5. Zhao, Z., Cohen, S.B., Webber, B.: Reducing quantity hallucinations in abstractive summarization. In: EMNLP(Findings), pp. 2237–2249 (2020)
6. Aralikatte, R., Narayan, S., Maynez, J., Rothe, S., McDonald, R.: Focus attention: promoting faithfulness and diversity in summarization. In: ACL-IJCNLP, pp. 6078–6095 (2021)
7. Nan, F., et al.: Improving factual consistency of abstractive summarization via question answering. In: ACL-IJCNLP, pp. 6881–6894 (2021)
8. Chen, S., Zhang, F., Sone, K., Roth, D.: Improving faithfulness in abstractive summarization with contrast candidate generation and selection. In: NAACL, pp. 5935–5941 (2021)
9. Huang, L., Wu, L., Wang, L.: Knowledge graph-augmented abstractive summarization with semantic-driven cloze reward. In: ACL, pp. 5094–5107 (2020)
10. Zhang, M., Zhou, G., Yu, W., Liu, W.: Far-ass: fact-aware reinforced abstractive sentence summarization. Inf. Process. Manage. **58**(3), 102478 (2021)

11. Dong, Y., Wang, S., Gan, Z., Cheng, Y., Cheung, J.C.K., Liu, J.: Multi-fact correction in abstractive text summarization. In: EMNLP, pp. 9320–9331 (2020)
12. Cao, M., Dong, Y., Wu, J., Cheung, J.C.K.: Factual error correction for abstractive summarization models. In: EMNLP, pp. 6251–6258 (2020)
13. Lewis, M., et al.: Bart: denoising sequence-to-sequence pre-training for natural language generation, translation, and comprehension. In: ACL, pp. 7871–7880 (2020)
14. Sen, P., Saffari, A.: What do models learn from question answering datasets? In: EMNLP, pp. 2429–2438 (2020)
15. Ko, M., Lee, J., Kim, H., Kim, G., Kang, J.: Look at the first sentence: position bias in question answering. In: EMNLP, pp. 1109–1121 (2020)
16. Falke, T., Ribeiro, L.F., Utama, P.A., Dagan, I., Gurevych, I.: Ranking generated summaries by correctness: an interesting but challenging application for natural language inference. In: ACL, pp. 2214–2220 (2019)
17. Cao, S., Wang, L.: Cliff: contrastive learning for improving faithfulness and factuality in abstractive summarization. In: EMNLP, pp. 6633–6649 (2021)
18. Zhang, J., Zhao, Y., Saleh, M., Liu, P.: Pegasus: pre-training with extracted gap-sentences for abstractive summarization. In: ICML, pp. 11328–11339. PMLR (2020)
19. Zhu, C., et al.: Enhancing factual consistency of abstractive summarization. In: NAACL, pp. 718–733 (2021)
20. Lee, D., et al.: Capturing speaker incorrectness: speaker-focused post-correction for abstractive dialogue summarization. In: EMNLP(newsum), pp. 65–73 (2021)
21. Devlin, J., Chang, M.W., Lee, K., Toutanova, K.: Bert: pre-training of deep bidirectional transformers for language understanding. In: NAACL, pp. 4171–4186 (2019)
22. Xie, S., Tu, Z.: Holistically-nested edge detection. In: Proceedings of the IEEE International Conference on Computer Vision, pp. 1395–1403 (2015)
23. Papineni, K., Roukos, S., Ward, T., Zhu, W.J.: Bleu: a method for automatic evaluation of machine translation. In: ACL, pp. 311–318 (2002)
24. Nallapati, R., Zhou, B., dos Santos, C., Gulçehre, Ç., Xiang, B.: Abstractive text summarization using sequence-to-sequence rnns and beyond. In: CoNLL, pp. 280–290 (2016)
25. Graff, D., Kong, J., Chen, K., Maeda, K.: English gigaword. Linguistic Data Consortium, Philadelphia 4(1), 34 (2003)
26. Gehrmann, S., Deng, Y., Rush, A.M.: Bottom-up abstractive summarization. In: EMNLP, pp. 4098–4109 (2018)
27. See, A., Liu, P.J., Manning, C.D.: Get to the point: summarization with pointer-generator networks. In: ACL, pp. 1073–1083 (2017)
28. Song, K., et al.: Joint parsing and generation for abstractive summarization. In: AAAI, vol. 34, pp. 8894–8901 (2020)
29. Liu, Y., Lapata, M.: Text summarization with pretrained encoders. In: EMNLP-IJCNLP, pp. 3730–3740 (2019)
30. Laban, P., Hsi, A., Canny, J., Hearst, M.A.: The summary loop: learning to write abstractive summaries without examples. In: ACL, pp. 5135–5150 (2020)
31. Shah, D., Schuster, T., Barzilay, R.: Automatic fact-guided sentence modification. In: AAAI. 34, pp. 8791–8798 (2020)
32. Wolf, T., et al.: Transformers: state-of-the-art natural language processing. In: EMNLP, pp. 38–45 (2020)
33. Paszke, A., et al.: Pytorch: an imperative style, high-performance deep learning library. NIPS 32 (2019)

34. Rajpurkar, P., Zhang, J., Lopyrev, K., Liang, P.: Squad: 100,000+ questions for machine comprehension of text. In: EMNLP, pp. 2383–2392 (2016)
35. Loshchilov, I., Hutter, F.: Fixing weight decay regularization in adam. arxiv 2017. arXiv preprint arXiv:1711.05101

Context-Adapted Multi-policy Ensemble Method for Generalization in Reinforcement Learning

Tingting Xu[1,2], Fengge Wu[1,2(✉)], and Junsuo Zhao[1,2]

[1] Institute of Software, Chinese Academy of Sciences, Beijing 100190, China
`fengge@iscas.ac.cn`
[2] University of Chinese Academy of Sciences, Beijing 100049, China

Abstract. Generalizability is a formidable challenge in applying reinforcement learning to the real world. The root cause of poor generalization performance in reinforcement learning is that generalization from a limited number of training conditions to unseen test conditions results in implicit partial observability, effectively transforming even fully observed Markov Decision Process (MDP) into Partially Observable Markov Decision Process (POMDP). To address such issues, we propose a novel structure, namely Context-adapted Multi-policy Ensemble Method (CAMPE), which enables the model to adapt to changes in the environment and efficiently solve implicit partial observability during generalization. The method captures local dynamic changes by learning contextual environment latent variables to equip the model with the ability of environment adaption. The latent variables and samples with contextual information are used as the input of the policy. Multiple policies are trained, combined in an integrated way to obtain a single policy to approximately solve the problem of partial observability. We demonstrate our method on various simulated robotics and control tasks. Experimental results show that our method achieves superior generalization ability.

Keywords: Deep reinforcement learning · Generalization · Partially Observable Markov Decision Process

1 Introduction

Real-world applications of reinforcement learning (RL) still face enormous challenges. Although reinforcement learning has been used in many fields, numerous studies have shown that agents trained with reinforcement learning are difficult to apply to new environments [1,2]. Even when the new environment simply replaces some visual features on the background, the agent achieves catastrophically low rewards [3,4]. This is mainly due to the partial observability of unseen environments during generalization. Therefore, the generalization problem caused by partial observability needs to be solved urgently. The generalization problem has been

CAS Project for Young Scientists in Basic Research, Grant No. YSBR-040.

extensively studied in supervised learning, and there are some proven methods such as regularization [5], data augmentation [6], etc. Some studies also introduce similar method to reinforcement learning, improving the generalization of reinforcement learning [7,8]. However, these methods focus more on the optimization aspect of the loss function, while ignoring a series of effects on generalization arising from the interaction of reinforcement learning with the environment. Due to the interaction between the agent and the environment, it is possible to affect the Markov decision process, thereby affecting the training model's generalization performance in the training environment once the environmental factors change slightly. Ghosh D [9] has shown that the root cause of poor generalization performance of reinforcement learning is that the current MDP is only one of the representations of the real MDP. Models trained with current MDP will have poor generalization performance due to partial observability.

We propose a new algorithm, CAMPE, that aims to solve this partial observability caused by generalization to unseen environments, significantly improving the generalization of reinforcement learning. CAMPE focuses on restoring the MDP of the environment as realistically as possible while adapting our model to changes in the environment. Our work is to encode the dynamic transition process and turn the dynamic transition process into a latent variable with environmental parameters as the input of the policy, so that the policy can dynamically pay attention to the changes of the environment, to realize the dynamic adaptation to the environment. For restoring the Markov decision process of the real environment, we sample multiple MDP samples from the empirical posterior distribution instead of directly learning the optimal policy in the posterior distribution, which would be normally inducing partial observability. In order to restore the MDP of the real environment, CAMPE uses a decomposition method to learn multiple policies from the cross-sampled MDP samples, and obtain a single policy through the weighted average of multiple policies. This policy approximates the policy in the true Markov decision process. Empirically, we demonstrate that CAMPE, which maximizes return in an approximation to the POMDP, achieves significant gains in test-time performance over standard RL methods on a simulated robot using the MuJoCo physics engine as well as on a classical control task from OpenAI Gym.

We thus formulate our main contributions with this work:

- We propose a new model structure, CAMPE, to improve the generalization of reinforcement learning, which solves the problem of poor generalization performance due to partial observability during testing.
- We achieve substantial adaptation to environmental changes, as the dynamic models can effectively capture environmental information during testing.
- With our flexible and straightforward approach, we predominantly see the effectiveness in methods where multiple policies can approximately represent all the information in the environment during testing.
- We show empirically that our framework is suitable for all reinforcement learning methods, not only model-based reinforcement learning methods, but also model-free reinforcement learning methods.

2   Related Work

The current research on reinforcement learning generalization mainly focuses on several aspects: enhancing the similarity between training data and test data, reducing the difference between training environment and test environment, and optimizing and improving methods for specific reinforcement learning problems [10]. Many previous works have proposed effective ways to enhance the similarity of training and test data, such as data augmentation [11], domain randomization [12], etc. Another main reason why reinforcement learning is prone to overfitting in the test environment is the difference between the training environment and the test environment. A strategy trained in the training environment is likely to memorize successful action sequences based on some deterministic factors in the training environment. This memorization will catastrophically bring about serious consequences when the training environment is different from the test environment [13]. A large number of scholars propose to use regularization to reduce the overfitting problem caused by the difference between the training environment and the testing environment [7,14,15]. The generalization solution of reinforcement learning described in this paper focuses on solving the POMDPs caused by the uncertainty of the MDP target, which the above methods cannot completely solve.

Partially Observable Markov Decision Process, or POMDP for short, is an extension of the Markov decision process when the state of the system is not necessarily completely observable [16]. In the real-world application, in many cases, using reinforcement learning to solve decision-making problems will eventually translate into finding the optimal solution in a partially observable Markov decision process. Therefore, a large amount of research has poured into the partially observable Markov decision process in various forms into the in-depth exploration [17–19]. POMDP also has more solutions. Among them, methods such as point-based methods [20], finite state controller policies [21], and compression techniques [22]are used to solve the approximate partially observable Markov decision process and has achieved good results. From the perspective of generalization, it is a new idea to solve the generalization problem caused by POMDP. That is what our work focus.

Our research work builds on research in Bayesian reinforcement learning. Bayesian uncertainty has been studied in many sub-fields of RL [23,24]. It considers that the agent can express some prior information about the specific problem according to the probability distribution [25]. However, this brings new problems, and it becomes a challenge to choose the correct representation to accurately express the prior information in a specific domain. This also provides a new idea for the generalization of reinforcement learning, that is, to more accurately infer the original MDP from the partially observable MDP in the real environment [9]. This is also the problem that our article focuses on solving. Therefore, our article aims to solve the POMDP in the generalization process in Bayesian reinforcement learning.

3 Setting

We formally describe the POMDP that appears in the generalization process in this paper as a tuple $<S_m, A, O, P, P_0, R>$. S_m denotes the state in the POMDP. A is the action space. O is an observation capable of observing the state under the current MDP. P can be expressed as $P_c((M', s')|(M, s), a)$, which refers to the probability of transitioning to s' after taking action a from the current state s under the current context c, formally expressed is $P_c((M', s')|(M, s), a) = p_c(M' = M)P(s'|s, a)$. P_0 is the initialization state. R refers to the reward function, which can be expressed as $R = R(s, a)$.

In this setting, the objective function is set to maximize the sum of expected rewards during an episode $J(\pi_\theta) = E_{\tau \sim p_\pi(\tau)}[\Sigma_{t=0}^H \gamma^t R(s_t, a_t)]$,where $\tau = (s_0, a_0, ...)$ is the trajectory of states s_t and actions a_t of the agent at time step t for an episode of length H. The initial state is sampled from distribution P_0, and agent j samples actions from its policy function $\pi_j(a_t|s_t)$. The next state is sampled from the transition dynamics function $s_{t+1} \sim P_c(s_{t+1}|s_t, a_t)$. The agent learns through many episodes. For each episode, a new MDP Mi is sampled from one MDP's family M. For different MDP Mi training different policies π_i, our goal is to restore the mdp of the real environment from the mdp family, that is, train a single policy π_0, integrate the characteristics of each policy π_i, and maximize the expected reward:

$$arg \max_\theta E_{p(M)}[J(\pi_\theta)].$$

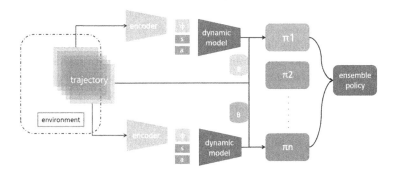

Fig. 1. The overall framework of the CAMPE. The trajectories are alternately sampled from the environment, which are not identical. After encoding, context-aware samples are added to the buffer to train different policies. The weighted average of multiple policies gets the final ensemble policy.

4 Method

In this chapter, we introduce the overall architecture and implementation of CAMPE. CAMPE is divided into three parts including: (1) estimation of posterior probability distribution, (2) local dynamic encoding, (3) policy ensemble and update. The architectures for the CAMPE are shown in Fig. 1. These three parts respectively explain how to restore the environment, how to make the model have the characteristics of adapting to the environment, and how to ensemble multiple policies. We discuss them separately in the following sections.

4.1 Estimating of the Posterior Probability Distribution

We collect n different MDPs, which we denote as $\{mdp_i\}_{i \in n}$. In order to enable the n mdp_i to restore the true MDP of the real environment as much as possible, we hope that the n MDP can approximately cover the entire environment. For the training sample C_{train} of the whole environment, we can get C_{train}^i by replacing it with sampling. Among the different sub-environments, there is a set of training environments with contexts sampled from $P_{C_{train}^i}(c)$. We define mdp_i as the empirical MDP in these sub-training environments. We argue that MDP_i is still a distribution over the context of the entire environment, rather than a single context, since we aim to sample posteriors from the entire context.

4.2 Local Dynamic Encoding

To make our policy more aware of environmental changes, we add a module that encodes the transition dynamics into a context-aware dynamic model. The traditional transition dynamics are fixed throughout the training period. When the environment changes during testing, the fixed transition dynamics cannot perceive and adapt to this change, which also causes many environmental factors to be ignored in the generalization. Several studies have shown that recent experience can reflect the current context [26]. So we create an encoder that takes the past trajectory $\tau_k = (s_0, a_0, ...)$ as input, and gets a vector $z_t = g(\tau_{t,K}^p; \phi)$ with environment parameters. To update the latent vector z_t, we train a dynamic model, given the current state s_t, the current action a_t and the current environment latent vector z_t go forward to predict the next state s_{t+1}. To optimize the dynamic model and latent vector z_t, our goal is to minimize the following loss:

$$L_i^{pred} = E_{(\tau_{t,M}^F, \tau_{t,K}^P) \sim B}[-\frac{1}{M} \sum_{i=t}^{t+M-1} log f(s_{i+1}|s_i, a_i, g(\tau_{t,K}^p; \phi))] \qquad (1)$$

where $\tau_{t,M}^F = (s_t, a_t), \ldots, (s_{t+M}, a_{t+M})$ is the future trajectory segment, $B = (\tau_{t,M}^F, \tau_{t,K}^P)$ is the training dataset, $\beta > 0$ is the penalty parameter, and $M > 0$ is the number of future samples. This means that we can obtain the maximum expected return using the current environment latent vector as well as the dynamic model. This is consistent with the goal of our entire reinforcement learning training.

4.3   Ensemble Policy

For the n mdps obtained by the posterior probability estimation, we get n different policies through interactive training. We hope that the n policies can be combined for the optimal policy π_0, which can reflect the maximum expected reward obtained by using this policy in the real Markov decision process, that is, the generalization is optimal. How to link n policies is a concern. We want to link multiple policies based on the ensemble idea. To avoid falling into a suboptimal solution due to greed, we propose a more balanced strategy, which we call the weighted average strategy. As the name suggests, weighted average will focus on the parts with large weights, but will also balance other parts. We want to slightly favor the well-behaved policy, i.e. the part with the largest expected reward at present, and we formulate π_0 as follows:

$$f(\{\pi_i\}_{i\in[n]}) = \pi_0(a|s) = \frac{max_{\alpha_i}\alpha_i\pi_i(a|s)}{\sum_{a'}\sum_i\alpha_i\pi_i(a'|s)} \tag{2}$$

$$\alpha_i = \frac{E_i}{E_1 + ... + E_n} \tag{3}$$

To update the parameters for π_i , we take gradient steps via the surrogate loss used for the policy gradient, augmented by a disagreement penalty between the policy and the combined policy $f(\{\pi_i\}_{i\in[n]})$ with a penalty parameter $\alpha > 0$, as in Eq. 4:

$$\pi_i \leftarrow \pi_i - \eta \nabla_i (\pi_i + \alpha E_{s\sim\pi_i}[D_{KL}(\pi_i(a \mid s) \parallel \pi_0(a \mid s))] + \beta E_{s\sim\pi_i}[D_{CE}\pi_i(a \mid s)]) \tag{4}$$

In this way, during the training process, each policy π_i is as close as possible to the linked policy π_0 to ensure that each policy can be as consistent as possible with the policy under the true MDP. Additional control over the amount of exploration is necessary when there are multiple policies. Imagine a scenario where one state space is small and is solved first, while the other tasks more complex and have much less reward. If there is no entropy term, and before the reward is encountered in the other sub-policies, the refined policy and all sub-policies can converge to a policy that solves very little state space. For single-policy RL, Mnih et al. [27] recently generalized the use of entropy regularization to combat premature convergence of greedy policies, which is especially severe in policy gradient learning. This also carries over to our multi-policy scenario, which is also the reason for the additional entropy regularization.

It is particularly difficult to update multiple policies for the policy update target proposed above. In order to optimize the above goals, there are many optimization techniques. We use the EM algorithm to maximize the true context policy π_0 and each context policy π_i. We can regard π_0 and π_i as the parameters of two sets of neural networks. Consider π_i as a parameter distributed with the latent variable π_0 to maximize π_i and vice versa. More intuitively, we can see that when π_0 is fixed, the gradient update of π_i becomes the following form:

$$\sum_i E_{\pi_i}[\sum_{t\geq 0}\gamma^t R_i(s_t,a_t)-\frac{\gamma^t}{\beta}log\pi_i(a_t|s_t)] \tag{5}$$

Simply given by the policy gradient with entropy regularization [27,28], which can be done in a framework like the advantage actor-critic [27].

When π_i is fixed, we can see that only one term is related to π_0:

$$\frac{\alpha}{\beta}\sum_i E_{\pi_i}[\sum_{t\geq 0}\gamma^t log\pi_0(a_t|s_t)] \tag{6}$$

The update of π_0 becomes the log-likelihood of the action-state mixture distribution. The maximum likelihood (ML) estimator can be derived from the state-action visit frequency, and the optimal ML solution is given by the state-dependent action distribution. In the optimization process, using stochastic gradient ascent will make π_i optimized to π_0, which also explains why using the above target update can make π_i gradually approach π_0.

The complete learning process is summarized in Algorithm 1. In our implementation, we alternately sample training samples to create $c_1, c_2, .., c_n$, where $c_i \in c_{train}$. Each iteration collects samples from one of the sub-training environments, updates the dynamic model, and updates the policy. The final policy is the weighted average set of multiple policies, according to Eq. 2.

5   Experiments

In the following experiments, we focus on two research questions. (i) Whether our method can effectively improve generalization and reduce the gap between train and test. (ii) Whether our method can be applied to multiple reinforcement learning basic models and improve their generalization ability. To answer these two questions, We demonstrate the effectiveness of our proposed method on a simulated robot using the MuJoCo physics engine as well as on a classical control task from OpenAI Gym.

5.1   Evaluation Setup

We demonstrate the effectiveness of our proposed method on simulated robots (i.e., HalfCheetah, Ant, and SlimHumanoid) using the MuJoCo physics engine and classic control tasks (i.e., CartPole and Pendulum) from OpenAI Gym which are widely used in reinforcement learning experiments [29,30]. The experimental environment is shown in the Fig. 2.

In these environments, We set up the training environment and the test environment separately in a way of partitioning the environment parameters. Therefore, at test time, we measure each model's performance in a unseen environment characterized by parameters outside the training range. We set different environmental parameters to distinguish training and testing including the length of the vehicle pole, the quality of the intelligent robot, damping and so on. For both training and testing, we sample environment parameters at the beginning of each episode.

Algorithm 1: CAMPE

Input: the number of past observations K and future observations M, learning
rate α, and batch size B, the number of ensemble members n
Output: $\pi_0(a \mid s)$ in Eq. 2
Initialize parameters of context encoder ϕ
Initialize n policies: 1, . . . n
Initialize dataset B $\leftarrow \varnothing$
for *each iteration* **do**
   // COLLECT TRAINING SAMPLES
   Sample $c_i \sim p_{seen}(c_i)$.
   for $t = 1$ *to Timesteps* **do**
      Collect samples $\{(s_t, a_t, s_{t+1}, r_t, \tau_{t,K}^P)\}$ from the environment with
      transition dynamics p_{c_i}
      Update B$\leftarrow$ B$\bigcup\{(s_t, a_t, s_{t+1}, r_t, \tau_{t,K}^P)\}$
   end
   // UPDATE DYNAMICS MODELS AND ENCODER
   for $t = 1$ *to N* **do**
      sample $\tau_{i,K}^P, \tau_{i,M}^F \sim B$
      Get context latent vector $z_i = g(\tau_{t,K}^P; \phi)$
      $L_i^{pred} \leftarrow L^{pred}(\tau_{i,M}^F; z_i)$ in Eq. 1
   end
   Update $\phi \leftarrow \phi - \alpha \bigtriangledown_\phi \frac{1}{N} \sum_{i=1}^N L_i^{pred}$
   //UPDATE POLICIES
   for *policy i=1,...,n* **do**
      Take gradient steps wrt πi on samples with augmented RL loss:
      $\pi_i \leftarrow \pi_i - \eta \bigtriangledown_i (\pi_i + \alpha E_{s \sim \pi_i, C_{train}^i} [D_{KL}(\pi_i(a \mid s) \parallel \pi_0(a \mid s))] + \beta E_{s \sim \pi_i, C_{train}^i} [D_{CE}\pi_i(a \mid s)])$
   end
end

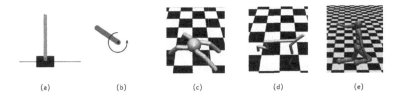

(a)   (b)   (c)   (d)   (e)

Fig. 2. Environments used in this work from left to right are (a) Cartpole, (b) Pendulum, (c) Ant, (d) Halfcheetah, and (e) SlimHumanoid. Among them, Cartpole and Pendulum are in the Gym. The latter three environments need the help of the Mujoco physics engine, and additional dependencies need to be installed.

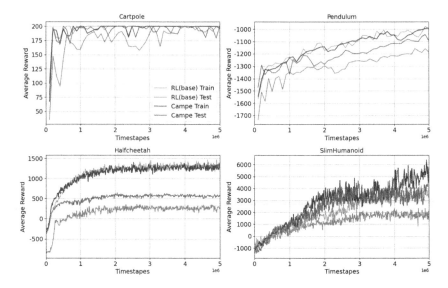

Fig. 3. Comparison of training and testing performance of traditional reinforcement learning and CAMPE methods in four environments. The RL(base) shown in the figure is the PPO algorithm, and the CAMPE marked in the figure combines the PPO and CAMPE structures.

5.2   Evaluation Results

To test the performance of our method, we compare with the baseline at test time. Our baseline is a basic PPO model that does not focus on environmental changes (i.e. the model does not change throughout the testing period) and has only a single policy. During the whole process, we train for 5×10^6 timesteps, using the average return during testing as the evaluation criterion. In all experiments, we choose the model with the highest average return during training, and report test performance as this model performs at test time. Figure 3 shows the experimental results. Compared with the baselines, CAMPE shows excellent performance in multiple tasks, greatly shortening the gap between train and test, and has higher average expected reward at test time. This implies that the generalization of the PPO algorithm is greatly improved by applying our structure.

Meanwhile, to demonstrate that our structure is applicable to multiple RL methods. We apply it to both model-free and model-based methods. The experimental results are shown in Fig. 4. We can see that applying the CAMPE structure in the model-based RL method also achieves a breakthrough generalization improvement.

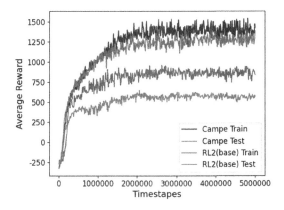

Fig. 4. Training and testing performance of CAMPE combined with model-base RL method on halfteechah environment. The RL2(base) shown in the figure is a model-based method, and the CAMPE marked in the figure is the combination of the model-based method and CAMPE structures.

5.3 Ablation

To verify the individual effects of each part of our architecture, we conducted three separate training: (i) Remove the context encoder part. (ii) Do not add cross-entropy in policy loss (iii) n = 1 in the part of the policy ensemble. We carried out experiments by removing three parts, and judged the influence of each part on the final result according to the performance during the test.

Table 1 shows the effect of removing the above parts on the results. According to the results, We can see that by removing the policy ensemble, the average return during testing will drop by about 30%. The main reason for the performance degradation is that a single strategy is prone to partial observability during testing. For removing the context encoder part, the average return at test time drops by about 15%, mainly because the model loses its adaptation to the environment at test time. As for removing the entropy regularization term, it also slightly reduces the average return during testing, showing that entropy regularization increases the exploration of the environment. Finally, CAMPE's highest average return also illustrates the positive effect of eliminating the partial observability caused by the generalization process and restoring the real environment on generalization.

Table 1. Ablation experiments on CAMPE.

	Pendulum	Halfcheetah	Ant	SlimHumanoid
ContextAdaption	$-1157.4\pm_{211.3}$	$493.3\pm_{158.6}$	$74.2\pm_{23.5}$	$3650.0\pm_{1350.6}$
CEM	$-1041.2\pm_{303.7}$	$536.8\pm_{200.5}$	$80.2\pm_{29.1}$	$3889.8\pm_{1211.9}$
MutiPolicy	$-1244.3\pm_{447.2}$	$411.7\pm_{178.1}$	$69.1\pm_{16.8}$	$2292.1\pm_{1028.7}$
RL(base)	$-1316.4\pm_{206.7}$	$261.4\pm_{121.5}$	$45.9\pm_{23.1}$	$1686.4\pm_{637.2}$
CAMPE	$-910.5\pm_{316.6}$	$674.2\pm_{378.3}$	$107.3\pm_{31.6}$	$4674.7\pm_{1378.3}$

6   Discussion

Overall, we succeeded in training an agent to generalize to unseen environments. We showed, on the one hand, that the model trained in our method is well capable of adapting to the environment. On the other hand, we showed that ensemble-based policy effectively addresses partial observability in the testing process. With CAMPE we drastically improve the generalizability at test time as it achieves a higher reward when evaluating the policy. Furthermore, we also demonstrate that our structure can be combined with multiple RL methods.

A limitation of our method is that it optimizes a rough approximation of the POMDP during generalization with a small number of posterior samples. It is a challenge to better approximate the true target. Better modeling of POMDP and optimization policies during generalization is an exciting avenue for future work, and we hope this direction will lead to further improvements in RL generalization.

References

1. Justesen, N., Torrado, R., Bontrager, P., Khalifa, A., Togelius, J., Risi, S.: Illuminating generalization in deep reinforcement learning through procedural level generation. arXiv: Learning (2018)
2. Zhang, C., Vinyals, O., Munos, R., Bengio, S.: A study on overfitting in deep reinforcement learning. arXiv, abs/1804.06893 (2018)
3. Gamrian, S., Goldberg, Y.: Transfer learning for related reinforcement learning tasks via image-to-image translation. In: International Conference on Machine Learning. PMLR (2019)
4. Farebrother, J., Machado, M.C., Bowling, M.: Generalization and regularization in DQN. arXiv preprint arXiv:1810.00123 (2018)
5. DeVries, T., Taylor, G.W.: Improved regularization of convolutional neural networks with cutout. arXiv preprint arXiv:1708.04552 (2017)
6. Cubuk, E.D., et al.: Autoaugment: learning augmentation policies from data. arXiv preprint arXiv:1805.09501 (2018)
7. Cobbe, K., et al.: Quantifying generalization in reinforcement learning. In: International Conference on Machine Learning. PMLR (2019)
8. Ponsen, M., Taylor, M.E., Tuyls, K.: Abstraction and generalization in reinforcement learning: a summary and framework. In: Taylor, M.E., Tuyls, K. (eds.) ALA 2009. LNCS (LNAI), vol. 5924, pp. 1–32. Springer, Heidelberg (2010). https://doi.org/10.1007/978-3-642-11814-2_1
9. Ghosh, D., Rahme, J., Kumar, A., et al.: Why generalization in RL is difficult: epistemic pomdps and implicit partial observability. In: Advances in Neural Information Processing Systems, vol. 34 (2021)
10. Kirk, R., et al.: A survey of generalisation in deep reinforcement learning. arXiv preprint arXiv:2111.09794 (2021)
11. Shanthamallu, U.S., et al.: A brief survey of machine learning methods and their sensor and IoT applications. In: 2017 8th International Conference on Information, Intelligence, Systems and Applications (IISA). IEEE (2017)
12. Tobin, J., et al.: Domain randomization for transferring deep neural networks from simulation to the real world. In: 2017 IEEE/RSJ International Conference on Intelligent Robots and Systems (IROS). IEEE (2017)

13. Bellemare, M.G., et al.: The arcade learning environment: an evaluation platform for general agents. J. Artif. Intell. Res. **47**, 253–279 (2013)
14. Zhang, A., Ballas, N., Pineau, J.: A dissection of overfitting and generalization in continuous reinforcement learning. arXiv, abs/1806.07937 (2018)
15. Liu, Z., Li, X., Kang, B., Darrell, T.: Regularization matters in policy optimization - an empirical study on continuous control. arXiv: Learning (2020)
16. Smallwood, R.D., Sondik, E.J.: The optimal control of partially observable Markov processes over a finite horizon. Oper. Res. **21**(5), 1071–1088 (1973)
17. Hausknecht, M., Stone, P.: Deep recurrent q-learning for partially observable MDPS. arXiv preprint arXiv:1507.06527 (2015)
18. Zhu, P., Li, X., Poupart, P., Miao, G.: On improving deep reinforcement learning for pomdps. arXiv preprint arXiv:1704.07978 (2017)
19. Qureshi, A.H., et al.: Robot gains social intelligence through multimodal deep reinforcement learning. In: 2016 IEEE-RAS 16th International Conference on Humanoid Robots (Humanoids). IEEE (2016)
20. Pineau, J., Gordon, G., Thrun, S.: Anytime point-based approximations for large POMDPs. J. Artif. Intell. Res. **27**, 335–380 (2006)
21. Amato, C., Bernstein, D.S., Zilberstein, S.: Optimizing fixed-size stochastic controllers for POMDPs and decentralized POMDPs. Auton. Agent. Multi-Agent Syst. **21**(3), 293–320 (2010)
22. Roy, N., Gordon, G., Thrun, S.: Finding approximate POMDP solutions through belief compression. J. Artif. Intell. Res. **23**, 1–40 (2005)
23. Ramachandran, D., Amir, E.: Bayesian inverse reinforcement learning. In: IJCAI (2007)
24. Lazaric, A., Ghavamzadeh, M.: Bayesian multi-task reinforcement learning. In: ICML (2010)
25. Ghavamzadeh, M., et al.: Bayesian reinforcement learning: a survey. Found. Trends® Mach. Learn. **8**(5-6), 359–483 (2015)
26. Zhou, W., Pinto, L., Gupta, A.: Environment probing interaction policies. In: ICLR (2019)
27. Mnih, V., et al.: Asynchronous methods for deep reinforcement learning. In: International Conference on Machine Learning (ICML) (2016)
28. Schulman, J., Abbeel, P., Chen, X.: Equivalence between policy gradients and soft Q-Learning. arXiv:1704.06440 (2017)
29. Todorov, E., Erez, T., Tassa, Y.: Mujoco: a physics engine for model-based control. In: IROS (2012)
30. Brockman, G., et al.: Openai gym. arXiv preprint arXiv:1606.01540 (2016)

Self-attention Based Multi-scale Graph Convolutional Networks

Zhilong Xiong[1] and Jia Cai[2]($^{(\boxtimes)}$) (iD)

[1] School of Statistics and Mathematics, Guangdong University of Finance
and Economics, Guangzhou 510320, Guangdong, China
[2] School of Digital Economics, Guangdong University of Finance and Economics,
Guangzhou 510320, Guangdong, China
`jiacai1999@gdufe.edu.cn`

Abstract. Graph convolutional networks (GCNs) have achieved remarkable learning ability for dealing with various graph structural data recently. In general, GCNs have low expressive power due to their shallow structure. In this paper, to improve the expressive power of GCNs, we propose two multi-scale GCN frameworks by incorporating self-attention mechanism and multi-scale information into the design of GCNs. The self-attention mechanism allows us to adaptively learn the local structure of the neighborhood, and achieves more accurate predictions. Extensive experiments on both node classification and graph classification demonstrate the effectiveness of our approaches over several state-of-the-art GCNs.

Keywords: Graph convolutional networks · Multi-scale · Self-attention

1 Introduction

Graph structural data related learning have drawn considerable attention recently. Graph neural networks (GNNs), particularly graph convolutional networks (GCNs), have been successfully utilized in recommendation systems [18], computer vision [4], molecular design [17], natural language processing [27] etc. In general, there are two convolution operations in the GCN models: spatial-based method and spectral-based approach. Spatial-based GCNs consider the aggregation method between the graph nodes. GAT [21] used the attention mechanism to aggregate neighboring nodes on the graph, and GraphSAGE [7] utilized random walks to sample nodes and then aggregated them. Spetral-based GCNs

The work described in this paper was supported partially by the National Natural Science Foundation of China (11871167, 12271111), Guangdong Basic and Applied Basic Research Foundation (2022A1515011726), Special Support Plan for High-Level Talents of Guangdong Province (2019TQ05X571), Foundation of Guangdong Educational Committee (2019KZDZX1023), Project of Guangdong Province Innovative Team (2020WCXTD011).

M. Tanveer et al. (Eds.): ICONIP 2022, LNCS 13623, pp. 418–430, 2023.
https://doi.org/10.1007/978-3-031-30105-6_35

focus on redefining the convolution operation by utilizing Fourier transform [3] or wavelet transform [24] to define the graph signal. However, the decomposition of Laplacian matrix is too exhausted. Therefore, ChebyNet [5] was introduced, which employed Chebyshev polynomial to approximate the convolution kernel. Based upon ChebyNet, to reduce computational cost, first-order Chebyshev polynomial [10] was used to approximate the convolution kernels. In addition, there are studies using ARMA filters [1] to approximate convolution kernels. He et al. [8] proposed an order-K Bernstein polynomial to approximate filters (BerNet).

Despite their great success, existing GCNs have low expressive power due to their shallow structure (with only two or three layers). Bronstein et al. [2] stated that there is no need for the label information to traverse the entire graph. Another intuitive reason is that graph convolution is a special form of Laplacian smoothing, which may make the representation of different nodes indistinguishable when the network is deepened. Several methods try to achieve an architecture scalable in depth. Sun et al. [19] developed a novel AdaGCN by incorporating AdaBoost into the design of GCNs. JKNet [25] used the information of each layer to improve the predictive ability of graph convolution. Rong et al. [15] suggested removing a few edges of the graph randomly, while [12,13] utilized multi-scale information to deepen GCNs and proposed two architectures, namely, Snowball and Truncated Krylov. However, the performance of their methods still drops severely when the layer of the network is increased.

The above-mentioned approaches ignore the importance of neighbors, may yield steep increasing of computation due to the accumulation of too many layers, although the structure of GCNs is deepened. In this paper, we propose two general GCN frameworks by employing the idea of self-attention and multi-scale information. On one hand, the introduction of multi-scale information will enhance the expressive power of GCNs by stacking many layers, which is scalable in depth. On the other hand, the self-attention mechanism captures the intuition that neighbors of the nodes might not be equally important. Hence, it can adaptively aggregate information from the neighbors, and discern the feature information of local structures, therefore improving the performance. Compared to GAT, our approach focuses more on adaptive exploitation of multi-scale information by employing self-attention graph pooling mechanism. We also notice that GSSA [29] and Structured Self-attention Architecture [30] utilized self-attention approach to enhance the expressive power of GCNs. However, GSSA aimed to automatically learn the hierarchical structures of feature importance through self-attention, while Structured Self-attention Architecture introduced an attention pooling approach to improve the performance of graph-level representation learning. In contrast, our scheme focuses more on the adaptive utilization of multi-scale information via self-attention. Moreover, the proposed frameworks are flexible and can be built on any GCN models. Extensive comparison against baselines in both node classification and graph classification is investigated. The experimental results indicate performance improvements over a variety of real-world graph structural data.

2   Preliminaries

Graph Convolutional Networks: Denote $\mathcal{G} = (\mathcal{V}, \mathcal{E})$ as an input undirected graph with $\mathcal{V}$ the node set and $\mathcal{E}$ the edge set, where $|\mathcal{V}| = N$, $|\mathcal{E}| = E$. Let $A \in \mathbb{R}^{N \times N}$ be a symmetric adjacency matrix, and D its corresponding diagonal degree matrix, i.e., $D_{ii} = \sum_j A_{ij}$. In conventional GCNs, the graph embedding of nodes with only one convolutional layer is depicted as

$$Z = \text{ReLU}(\hat{A}XW_0) \tag{1}$$

with $Z \in \mathbb{R}^{N \times K}$ the final embedding matrix (output logits) of nodes before softmax, K is the number of classes. Here $\hat{A} = \tilde{D}^{-\frac{1}{2}}\tilde{A}\tilde{D}^{-\frac{1}{2}}$ with $\tilde{A} = A + I$ (I stands for identity matrix), $\tilde{D}$ is the degree matrix of $\tilde{A}$, $X \in \mathbb{R}^{N \times d}$ is the feature matrix with d stands for the input dimension. Furthermore, $W_0 \in \mathbb{R}^{d \times H}$ denotes the input-to-hidden weight matrix for a hidden layer with H feature maps.

To design flexible deep GCNs for distinct multi-scale tasks, Luan et al. [13] generalized vanilla GCNs in block Krylov subspace forms, which is defined by introducing a real analytic scalar function g, and rewrite $g(A)X$ as

$$g(A)X = \sum_{n=0}^{\infty} \frac{g^{(n)}(0)}{n!} A^n X = [X, AX, \cdots, A^{m-1}X][(\Gamma_0^{(\mathbb{S})})^T, (\Gamma_1^{(\mathbb{S})})^T, \cdots, (\Gamma_{m-1}^{(\mathbb{S})})^T]^T,$$

where $\mathbb{S}$ is a vector subspace of $\mathbb{R}^{d \times d}$ containing I_d (the identity matrix), $\Gamma_i^{(\mathbb{S})} \in \mathbb{R}^{d \times d}, i = 1, \cdots, m - 1$ are parameter matrix blocks. m depends on A and X. To avoid choosing m and get richer representations, two architectures, namely Snowball and Truncated Krylov networks are developed.

Snowball, a densely-connected GCN, is depicted as the following

$$H_0 = X, \quad H_{\ell+1} = f(A[H_0, H_1, \cdots, H_\ell W_\ell]), \quad \ell = 0, 1, \cdots, n - 1$$
$$\mathbf{C} = g([H_0, H_\ell, \cdots, H_n]W_n)$$
$$\text{output} = \text{softmax}(L^p \mathbf{C} W_C), \quad p \in \{0, 1\},$$

where $W_\ell \in \mathbb{R}^{(\sum_{i=1}^{\ell} d_i) \times d_{\ell+1}}$, $W_n \in \mathbb{R}^{(\sum_{i=1}^{n} d_i) \times d_C}$, $W_C \in \mathbb{R}^{d_C \times d_0}$ are parameter matrices. $d_{\ell+1}$ is the number of output channels in the ℓ-th layer, f and g stand for pointwise activation functions. When $p = 0$, $L^p = I$ and when $p = 1$, $L^p = A$.

Truncated Krylov, unlike Snowball, stacks different scale information of $X, AX, \cdots, A^{m-1}X$ in each layer:

$$H_0 = X, \quad H_{\ell+1} = f([H_\ell, AH_\ell, \cdots, A^{m_\ell-1}H_\ell]W_\ell), \quad \ell = 0, 1, \cdots, n - 1$$
$$\mathbf{C} = g(H_n W_n)$$
$$\text{output} = \text{softmax}(L^p \mathbf{C} W_C), \quad p \in \{0, 1\},$$

where $W_\ell \in \mathbb{R}^{(m_\ell d_\ell) \times d_{\ell+1}}$, $W_n \in \mathbb{R}^{d_n \times d_C}$ are parameter matrices.

The iteration formulas of Snowball and Truncated Krylov demonstrate that multi-scale information are roughly concatenated, which means all the node features are stacked in the same way. Therefore the importance of features is ignored. We develop two architectures based on multi-scale information and self-attention mechanism in the sequel. Before addressing the detailed approaches, let us review self-attention first.

Self-attention: Attention mechanisms are a widely used deep learning technique, which aims at selecting important features. Especially, self-attention [20] or intra-attention is an attention mechanism that allows the input features to interact with each other and find out which features they should focus on. Self-attention mechanism can be described mathematically as

$$f = \tanh(\text{GNN}(X, A)) \tag{2}$$

where $\text{GNN}(X, A)$ stands for the graph convolution formula introduced by [10], i.e.,

$$H_{\ell+1} = \text{ReLU}(\hat{A} H_\ell W) \tag{3}$$

with $H_0 = X$, $W \in \mathbb{R}^{d_{h_\ell} \times d_{h_\ell}}$, where d_{h_ℓ} is the feature dimension of H_ℓ. In the sequel, we will observe that the introduction of self-attention mechanism can reflect the topology as well as the node features, and extract the important features of multi-scale information, thereby improving the performance of the models.

3   The Proposed Approaches

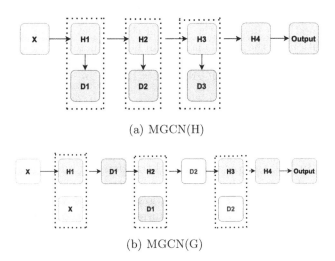

(a) MGCN(H)

(b) MGCN(G)

Fig. 1. The hierarchical GCN architecture and the global GCN architecture. H_i, $i = 1, \cdots$ represent the results of computing the graph convolutional layer, and D_i, $i = 1, \cdots$, denote the results of computing the self-attention mechanism.

We propose two architectures: MGCN(H) (H stands for hierarchical structure) and MGCN(G) (G stands for global structure) to amend the deficiency of Snowball and Truncated Krylov methods. These methods combine multi-scale feature information with self-attention mechanism, which both have the potential to be scalable in depth.

MGCN(H): In order to allow the model to use the information of different scales without increasing the computational cost, motivated by the PeleeNet [23] designed for object detection, we develop a multi-scale GCN (Fig. 1(a)), which adds an attention module in the middle of each layer. This module can expand the receptive field of the information achieved by the previous layer, combine the output of the previous layer and the obtained information from the attention module, and transfer them to the subsequent layer. Hence, this approach can not only enhance the discrimination ability of important features, but also improve the prediction accuracy of the designed model. Let us address the detailed structure of it. The first layer is the same as conventional GCN.

$$H_1 = \sigma(AXW_0),$$

where $W_0 \in \mathbb{R}^{d \times d_0}$ is the input-to-hidden weight matrix, σ is the activation function (e.g. tanh, ReLU). The attention module is computed as

$$D_\ell = \tanh(\text{GNN}(H_\ell, A)), \quad \ell = 1, 2, \cdots, n - 1,$$
$$C_\ell = \text{Concate}(H_\ell, D_\ell), \quad \ell = 1, 2, \cdots, n - 1, \tag{4}$$

where $C_\ell, \ell = 1, \cdots, n-1$ are used as input to the next layer, *Concate* stands for concatenation. Hence, C_ℓ conducts concatenation operation. Similarly, we can even perform self-attention calculations on the feature matrix X, and the result will be combined with itself as the input of the first layer. The computation of graph convolution for the middle layers can be conducted as

$$H_{\ell+1} = \sigma(AC_\ell W_\ell), \quad \ell = 1, 2, \cdots, n - 1,$$

$$\text{output} = \text{softmax}(AH_n W_C), \tag{5}$$

where $W_l \in \mathbb{R}^{(d_{\ell-1}+d_\ell) \times d_\ell}$, $W_C \in \mathbb{R}^{d_{n-1} \times d_C}$ are parameter matrices. d_ℓ is the number of output channels in the ℓ-th layer.

MGCN(G): In [13], the Snowball model transfers the information of previous layer to the next layer, so that the feature information of different scales could be well used, and alleviate the vanishing gradient problem, improve prediction accuracy. It ignores the importance of features however. Hence, we propose a self-attention based multi-scale global architecture (Fig. 1(b)) that not only inherits the merits of Snowball, but also overcome its shortcomings. The details are given as follows. The first layer is computed as

$$H_1 = \sigma(AXW_0),$$

$$C_1 = \text{Concate}(X, H_1),$$
$$D_1 = \tanh(\text{GNN}(C_1, A)), \tag{6}$$

The core idea for the computation of the subsequent layers is taking a self-attention module calculation after splicing the previous results, and passing it to the next layer.

$$C_{\ell+1} = \text{Concate}(H_{\ell+1}, D_l), \quad \ell = 1, 2, \cdots, n-1,$$
$$D_{\ell+1} = \tanh(\text{GNN}(C_{\ell+1}, A)), \quad \ell = 1, 2, \cdots, n-1, \tag{7}$$

$$H_{\ell+1} = \sigma(AD_\ell W_\ell), \quad \ell = 1, 2, \cdots, n-1,$$
$$\text{output} = \text{softmax}(AH_n W_C),$$

where $W_\ell \in \mathbb{R}^{(\sum_{i=0}^{\ell} d_i) \times d_l}$, $W_C \in \mathbb{R}^{d_{n-1} \times d_C}$ are learnable parameter matrices.

Our methods incorporate self-attention mechanism into GCNs by leveraging multi-scale information, which indeed enhance the expressive power of GCNs. Compared to Snowball and Truncated Krylov methods, the difference lies in that extra attention operations, i.e., Eqs. (4), (6) and (7), are considered in the proposed two architectures.

4 Experiments

To validate the proposed global architecture and hierarchical architecture for graph representation learning, we evaluate our two multi-scale GCN methods on both node classification and graph classification tasks. All the experiments are performed on a server running Ubuntu 16.04 (32 GB RAM).

4.1 Datasets

We evaluate the performance of the proposed MGCN (H) and MGCN (G) on 6 datasets with a large number of graphs ($>1k$) for the graph classification task, and 3 commonly used citation networks for the semi-supervised node classification task. Details of the data statistics about the citation networks (resp. graph datasets) are addressed in Table 1 (resp. Table 2)

Citation Networks: 3 benchmark citation networks, namely, Cora, Citeseer, and Pubmed [16] contain documents as node and citation links as directed edges, which stand for the citation relationships connected to documents. The characteristics of nodes are representative words in documents, and the label rate here denotes the percentage of node tags used for training. We use undirected versions of the graphs for all the experiments although the networks are directed. In this paper, we use public splits [26] approach for training.

Graph Datasets: Among TU datasets [14], we select six datasets including D&D, PROTEINS, MUTAGENICITY, NCI1, NCI109, and FRANKENSTEIN, where D&D, PROTEINS and MUTAGENICITY contain graphs of protein structures, while NCI1 and NCI109 are commonly used as benchmark datasets for graph classification task. FRANKENSTEIN is a set of molecular graphs with node features containing continuous values. For these 6 datasets, we randomly select 80%, 10%, and 10% of data for training, validation, and test.

Table 1. Summary of the citation networks.

Dataset	Cora	Citeseer	Pubmed
Nodes	2708	3327	19717
Edges	5429	4732	44338
Features	1433	3703	500
Classes	7	6	3
Label Rate	5.2%	3.6%	0.3%

Table 2. Summary of the datasets used for graph classification.

Dataset	D&D	PROTEINS	NCI1	NCI109	FRANKENSTEIN	MUTAGENICITY
Number of Graphs	1178	1113	4110	4127	4337	4337
Avg. # of Nodes Per Graph	284.32	39.06	29.87	29.68	16.90	30.32
Avg. # of Edges Per Graph	715.66	72.82	32.30	32.13	17.88	30.77
Number of Classes	2	2	2	2	2	2

4.2   Implementing Details

We compare against 3 classical GCNs: graph convolutional network (GCN) [10], graph attention network (GAT) [21], graph sample and aggregate (GraphSAGE) [7]. Moreover, our model is based on multi-scale tactic and self-attention mechanism. Hence, classical Chebyshev networks (ChebyNet) [5], Self-Attention Graph Pooling (SAGPoolg and SAGPoolh) [11], Graph U-Nets(U-nets) [6], DropEdge [15], JK-Net [25], Graph Statistical Self-Attention(GSSA) [29] and two famous multi-scale deep convolutional networks: Snowball and Truncated block Krylov network [13] are also considered as competitors.

Node Classification: We use RMSprop optimizer. There are several hyper-parameters, including learning rate and weight decay. The learning rate takes values from the set $\{1e-2, 4.7e-4, 4.7e-5\}$, while the weight decay is chosen from the set $\{3e-2, 5e-3, 5e-4, 5e-5\}$. We choose width of hidden layers from the set $\{200, 800, 1600, 1900, 3750\}$, number of hidden layers and dropout from the set $\{1, 2, \cdots, 8\}$ and $\{0, 0.2\}$, respectively. To achieve good training result, we utilize adaptive number of episodes (no more than 3000): the early stopping step is set to be 100, the same as those described in [13].

Graph Classification: Adam optimizer [9] is employed. We set learning rate $\gamma = 5e-4$, pooling ratio equals to 1, weight decay as $1e-4$, and dropout rate equals to 0.5. The model considered in this paper has 128 hidden units. We terminate the training if the validation loss does not improve for 50 epochs (the maximum of epochs is set as $100K$). In addition, for the two proposed architectures, we use MLP module in the last layer, and utilize the mean and the maximum readout, while the mean readout is employed for all the baselines.

4.3   Results

As demonstrated in Fig. 1, the MGCN (H) and the MGCN (G) stand for hierarchical GCN architecture and the global GCN architecture, respectively. We report the accuracy and training time. The comparison results for node classification and graph classification are summarized in Tables 3 and 4, respectively.

Experimental results in Tables 3 and 4 demonstrate that the proposed models obtain almost the best results over the baselines . Specifically, for 3 public citation networks and 6 graph datasets, MGCN (H) or MGCN (G) achieves satisfactory improvement in terms of accuracy on 3 public citation datasets, and outperforms the competitors on 4 out of 6 datasets. Intuitively, MGCN (H) and MGCN (G) are able to enhance the performance of GCNs, and provide better qualitative results for distinct types of graph structural data. To further illustrate the performance of the proposed model, we conduct paired t-tests for the accuracy. The p-value results are presented at the bottom of Tables 3 and 4, which indicate that our approaches are effective.

Table 3. Average accuracy over 10 runs for node classification.

Method	Cora	Citeseer	Pubmed
ChebyNet	78.0	70.1	69.8
GCN	80.5	68.7	77.8
GAT	83.0	72.5	79.0
GraphSAGE	74.5	67.2	76.8
JK-Net	81.1	69.8	78.1
DropEdge	82.8	72.3	79.6
GSSA	83.8	72.3	79.4
Snowball	83.6	72.6	79.5
Truncated Krylov	83.5	73.9	79.9
MGCN (H)	83.4	72.7	79.5
MGCN (G)	**84.0**	**74.0**	**80.0**
p-value	3.04e−5	2.36e−7	1.30e−5

Table 4. Average accuracy over 10 runs for graph classification.

Method	D&D	PROTEINS	NCI1	NCI109	FRANKENSTEIN	MUTAGENICITY
GCN	73.26	75.17	76.29	75.19	62.70	79.81
GraphSAGE	75.78	74.01	74.73	74.17	63.91	78.75
GAT	77.30	74.72	74.90	75.81	59.90	78.89
SAGPool (g)	76.19	70.04	74.18	74.06	62.57	74.51
SAGPool (h)	76.45	71.86	67.45	67.86	61.73	74.71
U-nets	77.02	80.71	76.25	76.61	61.46	80.30
Snowball	73.95	71.43	71.53	78.02	62.30	83.68
Truncated Krylov	79.54	78.85	**78.16**	78.16	78.16	79.54
MGCN (H)	82.53	**82.07**	78.10	79.31	74.48	**81.15**
MGCN (G)	**83.91**	81.84	77.24	**80.23**	**80.00**	**81.15**
p-value	2.58e-13	1.42e-11	8.42e-6	5.21e-12	8.01e-8	1.44e-11

To demonstrate whether the performance of the models degrades when the network is deepened. We add experiments on two datasets: PROTEINS, D&D, in which the number of layers is increased to 64. In Fig. 2, Tables 5 and 6, we can observe that the performance of the proposed MGCN (H) and MGCN (G) doesnot degrade too much. The accuracy achieved is still the highest even when the number of layers is stacked to 30 or 64. However, the accuracy achieved by Snowball and Truncated Krylov methods drops rapidly when the depth of the network is increased.

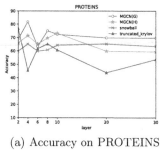

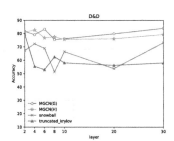

(a) Accuracy on PROTEINS dataset with different layers

(b) Accuracy on D&D dataset with different layers

Fig. 2. The accuracy on PROTEINS (a) and D&D (b) datasets with different layers.

Table 5. Accuracy on PROTEINS dataset using distinct number of layers.

Method	2-layer	4-layer	6-layer	8-layer	10-layer	20-layer	30-layer	64-layer
Snowball	59.82	65.18	59.82	60.71	64.29	65.18	63.39	61.61
Truncated Krylov	**76.78**	45.54	61.61	65.18	60.71	43.75	53.57	48.21
MGCN(H)	66.07	71.43	**65.18**	69.64	**73.21**	59.82	59.36	62.50
MGCN(G)	68.75	**81.84**	63.39	**75.00**	72.32	**69.75**	**69.64**	**70.54**

Table 6. Accuracy on D&D dataset using distinct number of layers.

Method	2-layer	4-layer	6-layer	8-layer	10-layer	20-layer	30-layer	64-layer
Snowball	67.23	72.27	68.91	51.26	66.39	53.78	73.11	56.30
Truncated Krylov	79.54	55.46	52.94	62.50	58.04	56.25	58.04	55.46
MGCN(H)	81.15	**82.53**	76.78	**77.47**	75.63	76.09	79.31	61.34
MGCN(G)	**81.61**	79.08	**83.22**	75.40	**76.09**	**79.77**	**83.91**	**63.87**

Table 7 displays the comparison of training time between MGCN (H), MGCN (G) and two recently proposed multi-scale deep convolutional networks: Snowball and Truncated Krylov. We can see that MGCN (G) runs faster on Cora and Citeseer datasets. We think the reason is that the self-attention mechanism allow the model to focus on using multi-scale information, instead of accumulating the information from the previous layer to the next layer mechanically, which can reduce computational cost.

Table 7. Average training time (s) comparison with Snowball and Truncated Krylov methods on 3 citation networks.

Method	Cora	Citeseer	Pubmed
Snowball	484.10	1228.57	265.41
Truncated Krylov	213.81	1194.94	254.85
MGCN (H)	**158.88**	304.53	219.92
MGCN (G)	184.90	**269.74**	**184.60**

4.4 Ablation Study

We conduct an ablation study on Cora, Citeseer and Pubmed to validate the effectiveness of the key components that contribute to the improvement of the proposed models. We utilize the following notations: w/o Multi-scale info means MGCN without multi-scale information, w/o Self-att stands for MGCN without the Self-attention mechanism while w/o both denotes MGCN without the multi-scale information and self-attention. We run each experiment ten times and report the averaged evaluation results on test data, which are presented in Table 8. We find that lack of self-attention or multi-scale information will reduce the performance of the models.

Table 8. Ablation study.

Method	Cora	Citeseer	Pubmed
MGCN(G)	84.0	74.0	80.0
w/o Multi-scale info	82.7	69.3	78.1
w/o Self-att	83.5	73.9	79.9
w/o both	80.5	68.7	77.8
MGCN(H)	83.4	72.7	79.5
w/o Multi-scale info	82.7	69.3	78.1
w/o Self-att	82.2	72.4	79.4
w/o both	80.5	68.7	77.8

5   Conclusion

In this paper, we propose two novel multi-scale graph convolutional networks based on self-attention mechanism. Our methods can not only improve the expressive power of GCNs, but also alleviate the problems of large scale calculation. We validate the performance of the proposed frameworks on both node classification and graph classification tasks. Our framework is universal, and one can replace the self-attention module with other graph pooling methods, for instance, gPool [6], DiffPool [28], Set2Set [22] etc.

References

1. Bianchi, F.M., Grattarola, D., Livi, L., Alippi, C.: Graph neural networks with convolutional arma filters. IEEE Trans. Pattern Anal. Mach. Intell. 1 (2021)
2. Bronstein, M.M., Bruna, J., LeCun, Y., Szlam, A., Vandergheynst, P.: Geometric deep learning: going beyond Euclidean data. IEEE Sig. Process. Mag. $\mathbf{34}(4)$, 18–42 (2017)
3. Bruna, J., Zaremba, W., Szlam, A., Lecun, Y.: Spectral networks and locally connected networks on graphs. In: 2nd International Conference on Learning Representations, pp. 1–14. ICLR, Canada (2014)
4. Casas, S., Gulino, C., Liao, R., Urtasun, R.: SpaGNN: spatially-aware graph neural networks for relational behavior forecasting from sensor data. In: 2020 IEEE International Conference on Robotics and Automation, pp. 9491–9497. IEEE, Paris (2020)
5. Defferrard, M., Bresson, X., Vandergheynst, P.: Convolutional neural networks on graphs with fast localized spectral filtering. In: Proceedings of the 30th International Conference on Neural Information Processing Systems, pp. 3844–3852. Curran Associates Inc., Red Hook (2016)
6. Gao, H., Ji, S.: Graph u-nets. In: Proceedings of the 36th International Conference on Machine Learning, pp. 2083–2092. ACM, California (2019)
7. Hamilton, W.L., Ying, R., Leskovec, J.: Inductive representation learning on large graphs. In: Proceedings of the 31st International Conference on Neural Information Processing Systems, pp. 1025–1035. Curran Associates Inc., Long Beach (2017)
8. He, M., Wei, Z., Huang, Z., Xu, H.: Bernnet: learning arbitrary graph spectral filters via Bernstein approximation. In: Advances in Neural Information Processing Systems, pre-proceedings, vol. 34, pp. 1–13. Curran Associates Inc., Virtual Conference (2021)
9. Kingma, D., Ba, J.: Adam: a method for stochastic optimization. In: 3rd International Conference for Learning Representations, pp. 1–12. ICLR, San Diego (2015)
10. Kipf, T., Welling, M.: Semi-supervised classification with graph convolutional networks. In: 5th International Conference on Learning Representations, pp. 1–14. ICLR, Toulon, France (2017)
11. Lee, J., Lee, I., Kang, J.: Self-attention graph pooling. In: Proceedings of the 36th International Conference on Machine Learning, pp. 3734–3743. ACM, California (2019)
12. Liao, R., Zhao, Z., Urtasun, R., Zemel, R.: Lanczosnet: multi-scale deep graph convolutional networks. In: 8th International Conference on Learning Representations, pp. 1–18. ICLR, New Orleans (2019)

13. Luan, S., Zhao, M., Chang, X.W., Precup, D.: Break the ceiling: stronger multi-scale deep graph convolutional networks. In: Proceedings of the 33rd International Conference on Neural Information Processing Systems, pp. 10945–10955. Curran Associates Inc., Vancouver (2019)

14. Morris, C., Kriege, N.M., Bause, F., Kersting, K., Mutzel, P., Neumann, M.: Tudataset: a collection of benchmark datasets for learning with graphs. In: ICML 2020 Workshop on Graph Representation Learning and Beyond (GRL+ 2020), pp. 1–11. ACM, Virtual Conference (2020)

15. Rong, Y., Huang, W., Xu, T., Huang, J.: Dropedge: towards deep graph convolutional networks on node classification. In: Eighth International Conference on Learning Representations, pp. 1–18. ICLR, Virtual Conference (2020)

16. Sen, P., Namata, G., Bilgic, M., Getoor, L., Gallagher, B., Eliassi-Rad, T.: Collective classification in network data. AI Mag. **29**, 93–106 (2008)

17. Stokes, J.M., Yang, K., Swanson, K., Jin, W., Collins, J.J.: A deep learning approach to antibiotic discovery. Cell **180**(4), 688-702.e13 (2020)

18. Sun, J., et al.: A framework for recommending accurate and diverse items using Bayesian graph convolutional neural networks. In: Proceedings of the 26th ACM SIGKDD International Conference on Knowledge Discovery and Data Mining, pp. 2030–2039. Association for Computing Machinery, Virtual Event (2020)

19. Sun, K., Lin, Z., Zhu, Z.: AdaGCN: adaboosting graph convolutional networks into deep models. In: The Ninth International Conference on Learning Representations, pp. 1–15. ICLR, Virtual Conference (2021)

20. Vaswani, A., et al.: Attention is all you need. In: Thirty-first Conference on Neural Information Processing Systems, pp. 5998–6008. Curran Associates Inc., Long Beach (2017)

21. Veličković, P., Cucurull, G., Casanova, A., Romero, A., Liò, P., Bengio, Y.: Graph attention networks. In: Sixth International Conference on Learning Representations, pp. 1–12. ICLR, Vancouver (2018)

22. Vinyals, O., Bengio, S., Kudlur, M.: Order matters: sequence to sequence for sets. In: The 4th International Conference on Learning Representations, pp. 1–11. ICLR, San Juan (2016)

23. Wang, R.J., Li, X., Ling, C.X.: Pelee: a real-time object detection system on mobile devices. In: Proceedings of the 32nd International Conference on Neural Information Processing Systems, pp. 1967–1976. Curran Associates Inc., Montréal (2018)

24. Xu, B., Shen, H., Cao, Q., Qiu, Y., Cheng, X.: Graph wavelet neural network. In: 8th International Conference on Learning Representations, pp. 1–13. ICLR, New Orleans (2019)

25. Xu, K., Li, C., Tian, Y., Sonobe, T., Kawarabayashi, K.I., Jegelka, S.: Representation learning on graphs with jumping knowledge networks. In: International Conference on Machine Learning, pp. 5453–5462. ACM, Stockholm (2018)

26. Yang, Z., Cohen, W.W., Salakhutdinov, R.: Revisiting semi-supervised learning with graph embeddings. In: Proceedings of the 33rd International Conference on International Conference on Machine Learning, pp. 40–48. JMLR.org, New York (2016)

27. Yao, L., Mao, C., Luo, Y.: Graph convolutional networks for text classification. In: Proceedings of the Thirty-Third AAAI Conference on Artificial Intelligence, pp. 7370–7377. AAAI Press, Honolulu (2019)

28. Ying, R., You, J., Morris, C., Ren, X., Hamilton, W.L., Leskovec, J.: Hierarchical graph representation learning with differentiable pooling. In: Proceedings of the 32nd International Conference on Neural Information Processing Systems, pp. 4805–4815. Curran Associates Inc., Montréal (2018)

29. Zheng, J., Wang, Y., Xu, W., Gan, Z., Li, P., Lv, J.: GSSA: pay attention to graph feature importance for GCN via statistical self-attention. Neurocomputing **417**, 458–470 (2020)
30. Fan, X., Gong, M., Xie, Y., Jiang, F., Li, H.: Structured self-attention architecture for graph-level representation learning. Pattern Recogn. **100**, 107084 (2020)

Synesthesia Transformer with Contrastive Multimodal Learning

Zhengxiao Sun[1], Feiyu Chen[1,2(✉)], and Jie Shao[1,2]

[1] University of Electronic Science and Technology of China, Chengdu 611731, China
sunzx@std.uestc.edu.cn, {chenfeiyu,shaojie}@uestc.edu.cn
[2] Sichuan Artificial Intelligence Research Institute, Yibin 644000, China

Abstract. Multi-sensory data, which exhibits complex relationships among modalities and temporal interactions, contains richer and more complex emotional representations for sentiment analysis. Yet, the effective integration of modalities remains a major challenge in the Multimodal Sentiment Analysis (MSA) task. We present a generalized model named Synesthesia Transformer with Contrastive learning (STC), which applies a synesthesia attention module enabling other modalities to guide the training of the input modality. It obtains a more natural and effective fusion and achieves competitive results on two widely used benchmarks CMU-MOSEI and CMU-MOSI.

Keywords: Transformer · Contrastive learning · Multimodal sentiment analysis

1 Introduction

Inspired by the increasing availability of multi-sensory data, the natural fusion and understanding of multimodal information on emotional artificial intelligent agents have been widely studied [1]. The task of sentiment analysis, initially conducted in a unimodal environment, has aroused strong interest in the research community, and multi-sensory data has prompted its development and gained a performance enhancement. The Multimodal Sentiment Analysis (MSA) task [13,23] aims at recognizing the sentiment of spoken utterances using multimodal information such as video, text and audio. Multimodal data can provide more cues for richer perception than unimodal data. In addition, with the development of the Internet, the practical application of multimodal scenarios is becoming increasingly widespread. Therefore, more researchers have focused on the multimodal aspects of sentiment analysis [14].

For better use of multi-sensory data, many attempts have been made on advanced fusion strategies of different modalities. Upgraded from simply concatenating modalities, dominant approaches focus on aggregation-based fusion

This work is supported by the National Natural Science Foundation of China (No. 61832001) and Sichuan Science and Technology Program (No. 2021JDRC0073).

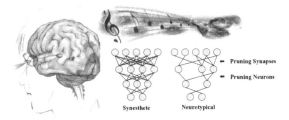

Fig. 1. When people with synesthesia receive information from a single modality, more neural units are receiving and processing the information, and a large number of synapses always bring about additional labelling of the other senses (figure is adapted from https://ircn.jp/en/pressrelease/20200422-hensch).

by respectively processing each modality first and then combining them [18]. However, these methods cannot effectively model the association of features between different modalities, and thus they cannot fully represent the intimate correlation of different modalities or be projected on the same high-dimensional space while containing enough features. These fusion methods often rely on artificial models made by manual splicing, without trying to exemplify the laws of modality fusion in reality. Therefore, for the purpose of fully modelling relationships of modalities, we take inspiration from the synesthesia phenomenon [8] for the human treatment of multimodal relationships. People with synesthesia will 'hear' sounds with their eyes or 'smell' music with their noses. Research evidence shows that synesthetic percepts are consistent and generic, and it is highly memorable with more synapses than normal neurotypical [8]. As Fig. 1 shows, neuroscientists have found out that people with synesthesia have more neurological connections than normal people while receiving one modality, which means that when one modality is processing, the neural units from other sensory organs would also participate in this activity to co-process the contents. These additional markings can keep a firm connection with the perception they accept across the sensory, and we believe that the MSA task can benefit from this phenomenon.

Most existing MSA methods focus on supervised learning since a large number of labelled data can be a good way to improve experimental results. The labelled data helps the model to show a more accurate recognition of sentiment. However, in practice, the complex interweaving of sentiment that does not occur in the dataset can make it difficult for the model to achieve the same recognition accuracy. According to this, many previous studies have taken unlabelled data into account and made unsupervised models [6,7]. However, unsupervised methods are essentially clustering algorithms, which improve predictions by making samples of the same category closer together. This will make the model vulnerable to those bad points which can easily shift the model in the wrong direction and thus affect the performance. Therefore, we suggest that labelled data can be seen as an anchor, which ensures that the output data would not cluster in the wrong space due to bad points, whilst unlabelled data can improve the robustness of models for real-world use. This is known as semi-supervised learning, which includes both labelled and unlabelled data for training and contains the advantages of supervised learning and unsupervised learning at the same time.

We therefore recommend not learning with full supervision or without supervision, but rather semi-supervised learning through contrast learning. A model with semi-supervised learning would perform a reasonable analysis based on the reality of the situation and identify more accurate results by processing both labelled and unlabelled data.

In this work, inspired by synesthesia, we seek to explore a mechanism that imitates the human processing of multimodal information with contrastive learning, which is called *synesthesia Transformer*. We combine each basic modality with other modalities and allow them to merge into a synesthesia unit by synesthesia attention, through which we imitate the process of synesthesia. By selecting a modality as the main body of the synesthesia unit, the other modalities will join neural networks together in an attention framework. As opposed to traditional methods, whose basic units are normally uni-modalities, synesthesia unit allows the model to mimic the appending of sentiments just like synesthesia, so that the labelling of sentiments is stable and effective among all modalities. We construct an anchor, a positive sample and a negative sample queue for contrastive learning, which is proven to achieve good performance in the MSA tasks.

The novel contributions of our work can be summarized as follows:

– We propose a Synesthesia Transformer with Contrastive learning (STC) - a multimodal learning framework that emphasizes multi-sensory fusion by semi-supervised learning. STC allows different modalities to join the feedforward neural network of each other to strengthen the multi-sensory connection, which leads to a more effective modality fusion and more powerful performance.
– Experiments on two datasets CMU-MOSI [22] and CMU-MOSEI [11] demonstrate that STC achieves competitive results.

2   Related Work

Over the years, the research community has come up with many creative solutions in the field of sentiment analysis through Transformers and some effective approaches to the contrastive learning of positive and negative samples. In this section, we will describe some related models on MSA and contrastive learning.

2.1   Multimodal Sentiment Analysis

In recent years, Transformer [16] has become increasingly used for the MSA task [13,15]. Inspired by the success of self-attention layers and Transformer architectures, several works have used self-attention layers to replace the popular Recurrent Neural Network (RNN) or Long Short-Term Memory (LSTM) such as EF-LSTM, LF-LSTM and RAVEN [10,17] for a better intra-modal connection through multi-head attention [4].

In MSA, better fusion of modalities can help model to perform better, and multimodal fusion has been widely studied such as Tensor Fusion Network (TFN) [20], Low-rank Multimodal Fusion (LMF) [9] and Memory Fusion Network (MFN) [21]. However, those multimodal fusion methods also cause some problems such as oversized framework or exploding gradient. In this work, by fully connecting different modalities together, we make use of the framework of self-attention, which is computed as a backbone to connect other innovative units. Our proposed STC model achieves a competitive accuracy than the corresponding traditional architectures on two popular datasets.

2.2 Momentum Contrast Architecture

Many efforts have been dedicated to the aspect of contrastive learning. Among these, the momentum contrast model [2] makes an encoder that can calculate a long negative queue and prevents it from fluctuating through training in gradient descent by using a momentum framework. In most contrastive learning models, negative samples are often selected by manual methods [5] and selected in limited numbers, which cannot prevent the models from being affected by choosing random similar positive samples as negative samples. However, the momentum contrast model maintains a dictionary as the queue of data samples: the encoded outputs of the current mini-batch are enqueued, and the oldest one is dequeued [3]. The queue decouples the dictionary size from the mini-batch size, allowing it to be larger. And the long queue negative sample model also succeeds in minimizing the effect of randomly similar positive samples. However, in this method negative samples from different mini-batches are not consistent and tend to make the gradient descent fluctuate drastically, making it difficult to converge properly. Therefore, the momentum contrast model uses a momentum optimization approach [19], where the negative queue encoder does not participate in gradient descent, but instead the parameters of the encoders are updated with momentum from the positive samples, successfully allowing the model to train smoothly.

3   Our Method

Three main modalities are typically involved in the synesthesia sequences, including visual (V), audio (A) and textual (T) modalities. In contrast to the previous approaches that use individual modalities as the basic unit, the proposed Synesthesia Transformer with Contrastive learning (STC) model chooses the synesthesia unit as the basic unit for processing. Specifically, each modality has a synesthesia unit $(V_{T,A}, A_{(V,T)}, T_{(A,V)})$. In this section, we will describe the details of the proposed STC.

3.1 Synesthesia Attention

Taking inspiration from the sense of synesthesia, for each modality, we use three basic modalities to form a synesthesia unit by a synesthesia attention framework,

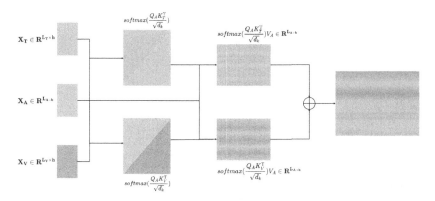

Fig. 2. An example of audio synesthesia attention. The output synesthesia unit in the diagram shows the composition of a basic unit and how other modalities guide the audio modality.

under the assumption that three modalities have synesthesia phenomenon on each other. For modality α, there are two other modalities β and γ adding aligned additional attentions on it. Firstly, modality β and γ respectively participate in the structure of attention as the key, so each modality message has access to the attentions of the other modalities α_β and α_γ. The attention results are then concatenated together as the output of synesthesia unit $\alpha_{(\beta,\gamma)}$ of modality α. Figure 2 shows an example of audio synesthesia Unit, where the basic inputs $X_\alpha \in R^{L_\alpha \times h}$, L_α is the length of modality α, and h is the size of hidden layers. Unlike the cross-modal attention module in the MULT model [15] that allows main modality to be trained as Query and other basic modalities to be trained as both Key and Value, our model is the exact opposite. We allow main modality α to be trained as both Query and Value and other modalities β and γ to be trained as Key to increase the weight of the main modality α in synesthesia attention.

Most of previous models create the positive samples through various data augmentation methods. In this work, we propose to adjust the degree of dropout parameters to construct positive samples. Based on the study of standard training on Transformer, it is known that in fully-connected layers there are dropout masks placed on it. By simply changing the rate of dropout mask η, we successfully build a positive sample pair $\alpha_{(\beta,\gamma)}$ and $\alpha^+_{(\beta,\gamma)}$ to a different dropout rate η^+. Formally,

$$\alpha_{(\beta,\gamma)} = concat\{softmax(\frac{Q_\alpha K_\beta^T}{\sqrt{d_k}})V_\alpha, softmax(\frac{Q_\alpha K_\gamma^T}{\sqrt{d_k}})V_\alpha, \eta\}, \tag{1}$$

$$\alpha^+_{(\beta,\gamma)} = concat\{softmax(\frac{Q_\alpha K_\beta^T}{\sqrt{d_k}})V_\alpha, softmax(\frac{Q_\alpha K_\gamma^T}{\sqrt{d_k}})V_\alpha, \eta^+\}. \tag{2}$$

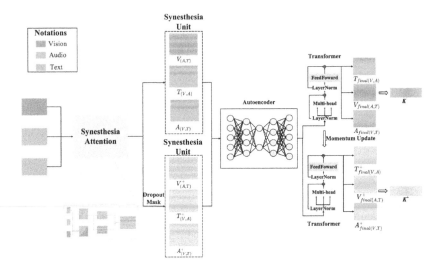

Fig. 3. Overall architecture for STC on modalities (V,T,A). Synesthesia units processed by the synesthesia attention module are selected as the basic unit and subsequently put in the autoencoder for processing. The positive samples are created by using different dropout masks on synesthesia attention.

As for the selection of negative samples, since there do not exist natural negative samples in the dataset, it is necessary to define or select negative samples. However, the selection of negative samples by the subjective method may have limitations. For instance, it is possible to choose a sample as the negative one which is very similar to the input sample. The previous demonstration of the loss function of contrastive learning [2] shows that models always tend to focus more on the least similar samples. Therefore, we select an extremely long queue $\{N_1, N_2, N_3, ...\}$ as the negative samples, as the effect of random similar positive samples can thus be reduced.

3.2 Autoencoder

To allow more effective multimodal fusion, we use an autoencoder to ensure that the output is consistent in features with the input while the features have an excellent fusion. From previous research [8], we know that when processing modalities, people with synesthesia will allow more synaptic connections to more nerve cells from different modalities. This neural connectivity gives other sensory organs access to the neural signals which do not belong to them, allowing multiple modalities to share a single neural network to process the information, thus effectively fusing the modalities. Therefore, we imitate this multi-sensory involvement by constructing an autoencoder through fully connected neural networks, as shown in Fig. 3. The output of the synesthesia unit $\alpha_{(\beta,\gamma)}$ along with its positive sample $\alpha_{(\beta,\gamma)}^+$ go through the autoencoder to produce $\alpha_{enhance(\beta,\gamma)}$

and $\alpha^+_{enhance(\beta,\gamma)}$. In this fashion, the features from modality β and γ participate to fuse the main modality α. Next, $\alpha_{enhance(\beta,\gamma)}$ and $\alpha^+_{enhance(\beta,\gamma)}$ are fed into the encoders of Transformers e_{nor} and e_{simi} to predict the final results $\alpha_{final(\beta,\gamma)}$ and $\alpha^+_{final(\beta,\gamma)}$.

3.3 Momentum Architecture

The samples in the queue are all from different mini-batches, and previous methods show that a large set of data from different batches will cause an intractable update of the encoder thorough back-propagation because all samples in the queue will be calculated by the gradient [2]. In order to prevent the autoencoder from jittering violently and losing feature consistency, we set that the training of the encoder of positive samples does not participate in gradient calculations, but instead changes with the parameters of the original encoder. This momentum architecture allows for stable convergence of models of positive samples. Formally,

$$\omega_{simi} \leftarrow k\omega_{simi} + (1-k)\omega_{nor}. \tag{3}$$

Here we set the parameters of e_{simi} as ω_{simi} and those of e_{nor} as ω_{nor}. Only the parameters ω_{nor} are updated by back-propagation. The momentum update in Eq. 3 makes ω_k evolve more smoothly than ω_q.

A slowly evolving key encoder allows the negative sample queue to successfully serve as an enhancement of the model.

3.4 Synesthesia Loss

There have been a lot of loss functions for contrastive learning. Contrastive loss often focuses on pushing away negative samples and pulling in positive samples. Hence, how to construct a function putting these two thoughts into account is a major concern. Inspired by NCEloss, we formulate a loss function for semi-supervised study. Mathematically,

$$\mathcal{L}_{syn} = -\log \frac{\exp\left(a \cdot K/\tau\right) + \exp\left(K \cdot K_+/\tau\right)}{\sum_{i=0}^{P} \exp\left(a \cdot K_i/\tau\right)}. \tag{4}$$

Here a is an anchor which represents the positive sample, K represents the final outputs, concatenating $T_{final(V,A)}, V_{final(A,T)}, A_{final(V,T)}$ together. K^+ represents the concatenation of the positive samples $T^+_{final(V,A)}, V^+_{final(A,T)}$ and $A^+_{final(V,T)}$. τ is defined as a hyper-parameter and P is the length of the negative sample queue.

4 Experiments

We compare the performance of the proposed STC model against two existing MSA works on two popular multimodal datasets, CMU-MOSEI [11] and CMU-MOSI [22], with semi-supervised training performed. To demonstrate the

Table 1. Performance comparison of the proposed STC model on CMU-MOSI and CMU-MOSEI.

Methods	CMU-MOSI				CMU-MOSEI			
	Acc-7	F1	MAE	Correlation	Acc-7	F1	MAE	Correlation
EF-LSTM	31.6	75.6	1.053	0.613	46.7	78.8	0.665	0.621
LF-LSTM	31.6	76.4	1.037	0.620	49.1	80.0	0.625	0.655
TFN	32.2	73.8	1.017	0.604	49.8	79.7	0.610	0.671
LMF	30.6	77.0	1.026	0.602	50.0	81.0	0.608	0.677
MFN	32.1	76.9	1.010	0.635	49.1	80.6	0.618	0.670
RAVEN	33.8	78.3	**0.968**	0.667	**50.2**	79.4	**0.605**	**0.680**
MULT	33.6	79.6	1.009	**0.667**	48.2	80.5	0.638	0.659
STC (ours)	**35.7**	**81.0**	1.018	0.649	49.1	**81.2**	0.606	0.678

effectiveness of the model, instead of running the model on word-aligned (by word, which is easier) multimodal language sequences, we choose to work on unaligned (which is more challenging) multimodal language sequences.

CMU-MOSEI is a sentiment and emotion analysis dataset made up of 23,454 movie review video clips taken from YouTube. The unaligned CMU-MOSEI sequences are extracted at a sampling rate 20 Hz for audio 15 Hz for visual signals. In the dataset, each sample is labelled by human annotators with a sentiment score from –3 (strongly negative) to 3 (strongly positive). We evaluate the model performance using various metrics, in agreement with those employed in prior works: 7-class accuracy (i.e., Acc-7: sentiment score classification in [–3, 3]), F1-score, Mean Absolute Error (MAE) of the score, and the correlation of the model's prediction with human beings.

CMU-MOSI is a collection of online videos in which a speaker is expressing his or her opinion about a film. Each video is split into multiple fragments, each containing the opinion expressed in one or more sentences. In total, CMU-MOSI consists of 2,199 short monologue video clips (each lasting the duration of a sentence). Audio and visual features of CMU-MOSI are extracted at a sampling rate of 12.5 Hz 15 Hz, respectively, while textual data are segmented per word and expressed as discrete word embeddings. Each segment has a sentiment label in [–3, +3], which is a continuous value representing the speaker's sentiment regarding some aspects of the film.

4.1 Baselines

We compare our model with the following methods:

- **Early Fusion LSTM (EF-LSTM)** [10] concatenates the input features of different modalities at word level, and then puts the concatenated features into an LSTM layer followed by a classifier to make prediction;

- **Late Fusion LSTM (LF-LSTM)** [10] uses an LSTM network for each modality to extract unimodal features and infer decision, and then combines the unimodal outputs by voting mechanism;
- **Tensor Fusion Network (TFN)** [20] applies outer product from unimodal embeddings to jointly learn unimodal, bimodal and trimodal interactions;
- **Low-rank Modality Fusion (LMF)** [9] leverages low-rank weight tensors to reduce the complexity of tensor fusion;
- **Recurrent Attended Variation Embedding Network (RAVEN)** [17] uses the features of the audio and visual modalities to model interactions by shifting language representations based on those two modalities;
- **Memory Fusion Network (MFN)** [21] employs a delta-attention module and multi-view gated memory network to discover intermodal interactions;
- **Multimodal Transformer (MULT)** [15] learns multimodal representation by using cross-modal Transformer to translate source modality into target modality.

The results reported in our experiments are processed in the same experimental environment according to the parameters in which they achieved the best results.

4.2   Comparison with Previous Results

We present our comparison results using the unaligned version of CMU-MOSI and CMU-MOSEI, which are shown in Table 1. Our proposed model achieves competitive results in main metrics. Specifically, it can be seen that the model achieves the best results in terms of F1 scores on both datasets and accuracy on CMU-MOSEI, demonstrating the effectiveness of the proposed STC.

4.3   Ablation Study

We perform comprehensive ablation studies on both datasets CMU-MOSEI and CMU-MOSI in unaligned versions. All results are shown in Tables 2 and 3.

First, we test the performance of STC in unimodal data, denoted as STC-Text, STC-Video and STC-Audio. We can notice that the performance with text modality only is better than the other two modalities. On the CMU-MOSI dataset, the F1-score of STC-Text is 6.7% higher than STC-Video and 9.9% higher than STC-Audio. On the CMU-MOSEI dataset, the F1-score of STC-Text is 7.4% higher than that of STC-Video and 9.1% higher than that of STC-Audio respectively. In terms of the metric of correlation, STC-Text performs better than other modalities with 0.206 higher than STC-Video and 0.171 higher than STC-Audio. Such a trend can also be found in prior works [12,15]. Moreover, the gaps between text modality and others are smaller than the results in the previous study [15], which in some ways demonstrate that the synesthesia attention and the framework of autoencoder can help unimodal data gain more information than normal unimodal inputs.

Then, we evaluate the performance of the model without autoencoder. We observe that without an autoencoder the results have a large decline. The performance of F1-score drops from 81.2% to 75.8% on the CMU-MOSEI dataset and

Table 2. Ablation studies on CMU-MOSI.

Methods	Accuracy-1	F1-score	MAE	Correlation
STC-Text	67.2	61.3	1.152	0.591
STC-Video	57.8	54.6	1.259	0.343
STC-Audio	56.3	51.4	1.264	0.310
STC-WA (w/o autoencoder)	72.5	70.9	1.191	0.643
STC-WS (w/o synesthesia)	76.9	77.0	1.148	0.670
STC-SL (w/o semi-supervised learning)	75.2	75.8	1.192	0.643
STC (ours)	77.3	81.0	1.018	0.649

Table 3. Ablation studies on CMU-MOSEI.

Methods	Accuracy-1	F1-score	MAE	Correlation
STC-Text	74.3	73.1	0.652	0.634
STC-Video	65.2	65.7	0.759	0.428
STC-Audio	64.1	64.0	0.764	0.463
STC-WA (w/o autoencoder)	76.3	75.8	0.691	0.643
STC-WS (w/o synesthesia)	77.1	79.2	0.648	0.670
STC-SL (w/o semi-supervised learning)	76.0	77.8	0.691	0.643
STC (ours)	79.1	81.2	0.606	0.678

delivers 10.1% decrease on the CMU-MOSI dataset. At the same time, MAE of STC-WA is nearly 9% higher than STC on the CMU-MOSEI dataset and nearly 18% higher than STC on CMU-MOSI. We suggest that this is because without an autoencoder, the simple concatenation of three units cannot deliver an effective fusion compared with using an autoencoder, so as to lead to less enhancement of the model. Hence, it is effective to use an autoencoder to make sure that different modalities can better fuse with others.

Next, we investigate the enhancement effect of the synesthesia attention module. From Tables 2 and 3 we can see that when just using unimodal data as input, the results show a small amount of decline for 2% in Accuracy-1, 2% in F1-score on the CMU-MOSEI dataset and 0.4% in Accuracy-1, 4% in F1-score on CMU-MOSI. Thus, it demonstrates the effectiveness of the proposed synesthesia attention module.

Finally, we compare the difference between using supervised learning and semi-supervised learning. When only using the anchor and the original outputs for contrastive learning, outputs are not as good as adding positive samples, and the model becomes supervised learning. The process is shown in Eq. 5:

$$\mathcal{L}_{supervised_{syn}} = -\log \frac{\exp\left(a \cdot k / \tau\right)}{\sum_{i=0}^{K} \exp\left(a \cdot k_i / \tau\right)}. \tag{5}$$

By testing, we find that Accuracy-1 of a model without positive samples declines to 76% on CMU-MOSEI and declines to 75.2% on CMU-MOSI. This shows the effectiveness of the semi-supervised setting.

5 Conclusion

In the paper, we proposed Synesthesia Transformer with Contrastive learning (STC) for the task of Multimodal Sentiment Analysis (MSA). The kernel of STC is the setting of synesthesia units and the usage of autoencoders. It allows for better integration of modalities by imitating a human approach - synesthesia - to process multimodality and obtains competitive results on two datasets in terms of unaligned human multimodal language sequences. While most previous MSA methods focus on supervised learning, STC constructs semi-supervised learning to ensure that the model training does not drift off by an anchor point and to enhance the performance of testing in real-world scenarios.

The performance of STC suggests many possibilities for future applications. The fusion of modalities is still not natural enough, and the additional annotation of the main modality by other basic modalities may not completely allow for a perfect fusion of modalities. It should also be considered that when a single modality is extremely dissimilar to other modalities but still participates in the synesthesia attention, it will interfere with the attention direction of the models. Subsequent work can explore how to create a better way of fusing modalities by allowing the unit vectors with higher weights to be noticed between modalities while avoiding the influence of irrelevant vectors on the fusion results.

References

1. Fedotov, D.: Contextual time-continuous emotion recognition based on multimodal data, Ph. D. thesis, University of Ulm, Germany (2022)
2. He, K., Fan, H., Wu, Y., Xie, S., Girshick, R.B.: Momentum contrast for unsupervised visual representation learning. In: 2020 IEEE/CVF Conference on Computer Vision and Pattern Recognition, CVPR 2020, pp. 9726–9735 (2020)
3. Heo, S., Lee, W., Lee, J.: mcBERT: momentum contrastive learning with BERT for zero-shot slot filling. CoRR abs/2203.12940 (2022)
4. Huddar, M.G., Sannakki, S.S., Rajpurohit, V.S.: Attention-based multi-modal sentiment analysis and emotion detection in conversation using RNN. Int. J. Interact. Multim. Artif. Intell. **6**(6), 112–121 (2021)
5. Jang, H., Choi, H., Yi, Y., Shin, J.: Adiabatic persistent contrastive divergence learning. In: 2017 IEEE International Symposium on Information Theory, ISIT 2017, pp. 3005–3009 (2017)
6. Jenckel, M.: Sequence learning for ocr in unsupervised training cases, Ph. D. thesis, Kaiserslautern University of Technology, Germany (2022)
7. Ji, X.: Unsupervised learning and continual learning in neural networks, Ph. D. thesis, University of Oxford, UK (2021)
8. Kann, K., Monsalve-Mercado, M.M.: Coloring the black box: what synesthesia tells us about character embeddings. In: Proceedings of the 16th Conference of the European Chapter of the Association for Computational Linguistics: Main Volume, EACL 2021, pp. 2673–2685 (2021)
9. Liu, Z., Shen, Y., Lakshminarasimhan, V.B., Liang, P.P., Zadeh, A., Morency, L.: Efficient low-rank multimodal fusion with modality-specific factors. In: Proceedings of the 56th Annual Meeting of the Association for Computational Linguistics, ACL 2018, Volume 1: Long Papers, pp. 2247–2256 (2018)

10. Mai, S., Zeng, Y., Zheng, S., Hu, H.: Hybrid contrastive learning of tri-modal representation for multimodal sentiment analysis. IEEE Trans. Affect. Comput. (2022). https://doi.org/10.1109/TAFFC.2022.3172360

11. Melanchthon, D.M.: Unimodal feature-level improvement on multimodal CMU-MOSEI dataset: uncorrelated and convoled feature sets. Proces. del Leng. Natural **67**, 69–81 (2021)

12. Pham, H., Liang, P.P., Manzini, T., Morency, L., Póczos, B.: Found in translation: learning robust joint representations by cyclic translations between modalities. In: The Thirty-Third AAAI Conference on Artificial Intelligence, AAAI 2019, pp. 6892–6899 (2019)

13. Qi, Q., Lin, L., Zhang, R., Xue, C.: MEDT: using multimodal encoding-decoding network as in transformer for multimodal sentiment analysis. IEEE Access **10**, 28750–28759 (2022)

14. Stappen, L., et al.: The muse 2021 multimodal sentiment analysis challenge: Sentiment, emotion, physiological-emotion, and stress. In: MuSe 2021: Proceedings of the 2nd on Multimodal Sentiment Analysis Challenge, pp. 5–14 (2021)

15. Tsai, Y.H., Bai, S., Liang, P.P., Kolter, J.Z., Morency, L., Salakhutdinov, R.: Multimodal transformer for unaligned multimodal language sequences. In: Proceedings of the 57th Conference of the Association for Computational Linguistics, ACL 2019, Volume 1: Long Papers, pp. 6558–6569 (2019)

16. Vaswani, A., et al.: Attention is all you need. In: Advances in Neural Information Processing Systems 30: Annual Conference on Neural Information Processing Systems 2017, pp. 5998–6008 (2017)

17. Wang, Y., Shen, Y., Liu, Z., Liang, P.P., Zadeh, A., Morency, L.: Words can shift: dynamically adjusting word representations using nonverbal behaviors. In: The Thirty-Third AAAI Conference on Artificial Intelligence, AAAI 2019, pp. 7216–7223 (2019)

18. Yang, B., Shao, B., Wu, L., Lin, X.: Multimodal sentiment analysis with unidirectional modality translation. Neurocomputing **467**, 130–137 (2022)

19. Yang, N., Wei, F., Jiao, B., Jiang, D., Yang, L.: xMoCo: cross momentum contrastive learning for open-domain question answering. In: Proceedings of the 59th Annual Meeting of the Association for Computational Linguistics and the 11th International Joint Conference on Natural Language Processing, ACL/IJCNLP 2021, (Volume 1: Long Papers), pp. 6120–6129 (2021)

20. Zadeh, A., Chen, M., Poria, S., Cambria, E., Morency, L.: Tensor fusion network for multimodal sentiment analysis. In: Proceedings of the 2017 Conference on Empirical Methods in Natural Language Processing, EMNLP 2017, pp. 1103–1114 (2017)

21. Zadeh, A., Liang, P.P., Mazumder, N., Poria, S., Cambria, E., Morency, L.: Memory fusion network for multi-view sequential learning. In: Proceedings of the Thirty-Second AAAI Conference on Artificial Intelligence, (AAAI-18), pp. 5634–5641 (2018)

22. Zadeh, A., Zellers, R., Pincus, E., Morency, L.: MOSI: multimodal corpus of sentiment intensity and subjectivity analysis in online opinion videos. CoRR abs/1606.06259 (2016)

23. Zheng, J., Zhang, S., Wang, X., Zeng, Z.: Multimodal representations learning based on mutual information maximization and minimization and identity embedding for multimodal sentiment analysis. CoRR abs/2201.03969 (2022)

Context-Based Point Generation Network for Point Cloud Completion

Lihua Lu[1,2,3], Ruyang Li[1,2], Hui Wei[1,2(✉)], Yaqian Zhao[1,2], and Rengang Li[1,2]

[1] Inspur Electronic Information Industry Co., Ltd., Jinan, China
{lulihua,weihui}@inspur.com
[2] Inspur (Beijing) Electronic Information Industry Co., Ltd., Beijing, China
[3] Shandong Massive Information Technology Research Institute, Jinan, China

Abstract. Existing sparse-to-dense methods for point cloud completion generally focus on designing refinement and expansion modules to expand the point cloud from sparse to dense. This ignores to preserve a well-performed generation process for the points at the sparse level, which leads to the loss of shape priors to the dense point cloud. To resolve this challenge, we introduce Transformer to both feature extraction and point generation processes, and propose a Context-based Point Generation Network (CPGNet) with Point Context Extraction (PCE) and Context-based Point Transformation (CPT) to control the point generation process at the sparse level. Our CPGNet can infer the missing point clouds at the sparse level via PCE and CPT blocks, which provide the well-arranged center points for generating the dense point clouds. The PCE block can extract both local and global context features of the observed points. Multiple PCE blocks in the encoder hierarchically offer geometric constraints and priors for the point completion. The CPT block can fully exploit geometric contexts existing in the observed point clouds, and then transform them into context features of the missing points. Multiple CPT blocks in the decoder progressively refine the context features, and finally generate the center points for the missing shapes. Quantitative and visual comparisons on PCN and ShapeNet-55 datasets demonstrate our model outperforms the state-of-the-art methods.

Keywords: Point Cloud Completion · Context-based Point Transformation · Center Point Generation · Point Context Extraction

1 Introduction

Point clouds have been widely used in 3D computer vision [3,12,14,25,27] since they are expressive and can be easily collected by sensors like depth cameras and LIDAR. Unfortunately, raw point clouds are usually sparse and incomplete due to limitations such as viewpoint occlusion, self-occlusion and poor sensor resolution. Therefore, point cloud completion is in need to infer the complete shape from the partially observed point clouds.

© The Author(s), under exclusive license to Springer Nature Switzerland AG 2023
M. Tanveer et al. (Eds.): ICONIP 2022, LNCS 13623, pp. 443–454, 2023.
https://doi.org/10.1007/978-3-031-30105-6_37

Previous methods [19, 24, 28–30] usually reconstruct the complete shape in a sparse-to-dense way. Typical methods like PCN [29] and Pointr [28] firstly infer a sparse point cloud and then increase the density of it to a dense one. SnowflakeNet [24] proposes a multi-stage generation architecture to progressively extend the number of points using the point deconvolution operations. However, these methods only employ the highly-encoded shape representation to infer the sparse point cloud via simple multi-layer perceptron (MLP) and reshape operations. This underuses geometric priors of the observed point clouds when inferring the missing shapes, and fails to comprehensively control the generation process of the sparse point cloud. Moreover, these methods heavily rely on well-performed refinement and expansion operations to expand the point cloud from sparse to dense.

To address these drawbacks, we propose a Context-based Point Generation Network (CPGNet) to control the point generation process at the sparse level, predicting the decent center points for the dense point cloud (depicted in Fig. 1). Our sight of generating the center points is to introduce Transformer in both feature extraction and shape completion processes, and design the Point Context Extraction (PCE) and Context-based Point Transformation (CPT) blocks.

The PCE block can learn geometric patterns of the observed points, providing shape priors for the missing point generation. It employs the EdgeConv layer [20] to extract local context features for each point from its neighbor points, and adopts the self-attention mechanism [17] to refine these features. The global receptive field of the self-attention mechanism enables the PCE block to integrate similar features together and learn each point with the global context features. Multiple PCE blocks hierarchically capture local and global context features of the observed regions at various scales, and offer shape priors and constraints to the generation of sparse point clouds for the missing regions.

The CPT block adopts the cross-attention mechanism to learn and refine the context features of the center points in missing regions. It summarizes the context features of the observed points, and transforms them into the context features of missing points. Consequently, the CPT block enables our model to concentrate on inferring the missing points at the sparse level, while preserving the observed points. Multiple CPT blocks in the decoder finally generate the center points facilitating the generation of the dense point cloud. In addition, the CPT block is plug-and-play, which can incorporate with the point expansion module in any sparse-to-dense point generation architecture. In our experiments, we adopt the point expansion module in [24] to expand the number of points and reconstruct a complete point cloud at the dense level.

In summary, our main contributions are three-fold:

- We propose a Context-based Point Generation Network (CPGNet), which focuses on controlling the generation processes of the point cloud at the sparse level. It can predict the well-arranged center points, providing shape priors and constraints for generating the dense point cloud.
- We introduce Transformer to both feature extraction and point generation processes and devise the point context extraction (PCE) and Context-based

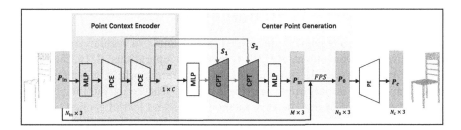

Fig. 1. The overall architecture of our Context-based Point Generation Network (CPGNet), which consists of the point context encoder, center point generation module and point expansion (PE) module. "FPS" denotes the farthest point sampling operation.

Point Transformation (CPT) blocks. The PCE block can learn global and local context features of the observed regions, while the CPT block can transform geometric contexts of the observed points to the context features of the missing points. Multiple PCE and CPT blocks can finally predict the center points representing the missing shapes.

– Qualitative and quantitative evaluations on PCN and ShapeNet-55 datasets demonstrate our CPGNet outperforms the state-of-the-art methods. Especially, the best completion results on the challenging ShapeNet-55 dataset prove our model can address challenging scenarios with more categories, different viewpoints and diverse incomplete patterns.

2   Related Work

2.1   Point Cloud Completion

Recently, point cloud completion has interested many researchers since point clouds are flexible and expressive. Inspired by PointNet [12] and PointNet++ [13], many works for point cloud completion [4,19,26,29,30] directly operate on points, and aim to predict complete point clouds from incomplete ones. Existing notable works like PCN [29] and FoldNet [26] usually follow an encoder-decoder architecture, and generate a complete point cloud only from a global shape representation derived from simply pooling local features of the whole discrete points. Works [19,24,28–30] provide better completion results by designing the sparse-to-dense decoders. Pointr [28] firstly infers a sparse point cloud and then increases the density of it to a dense one. Further, some works [8,19,24] propose a multi-stage generation architecture and achieve the state-of-the-art performance. These methods generally focus on the refinement and expansion modules to expand the number of points from sparse to dense. Though impressive performance can be achieved, these methods can't fully control the generation of the sparse point cloud. To address this challenge, we propose a model to control the point generation process at the sparse level, predicting the decent center points for the dense point cloud.

2.2   Feature Learning for 3D Point Clouds

Most existing methods convert unordered point clouds to regular voxels, which are intuitive and convenient for 3D convolution operations [1,10,15,23]. For instance, Atzmon et al. [1] present the PCNN framework in which the volumetric convolution is performed on voxels for the feature extraction. However, these methods suffer from expensive memory costs and loss of geometric details due to the quantization operation onto the voxels. Inversely, many works operate directly on discrete points to preserve geometric information. The Point-Net [12] and its variant PointNet++ [13] directly process the point clouds with feedforward neural networks. The subsequent works [20,22,25] concentrate on directly redefining convolution operations for irregular point clouds. EdgeConv [20] dynamically builds a local graph and learns features that describe the relationships between a point and its neighbors. However, these works focus on local features merely and lose sight of non-local relations among points. In our model, we use Transformer to refine the local features and learn context features of the points at both local and global scales.

2.3   Transformer in Computer Vision Tasks

Inspired by the success of Transformer in natural language processing (NLP) tasks, many researchers have introduced Transformer to computer vision tasks, such as image classification [2], object detection [18,33] and semantic segmentation [9,31,32]. When it comes to 3D computer vision, Transformer with self-attention and cross-attention mechanisms can naturally handle unordered and irregular point clouds and meet the demand of permutation invariance. And it has been widely proved to be highly effective in many 3D computer vision tasks like 3D point classification and segmentation [6] and 3D object detection [7,11].

Recent methods [24,28] apply transformer into the point cloud completion task. For example, SnowflakeNet [24] adopts the skip-transformer to design a snowflake point deconvolution module, which can expand the refine the sparse point cloud to a dense one with detailed geometries. However, these methods underuse geometric priors of the observed point clouds when inferring the missing shapes, and fails to comprehensively control the generation process of the sparse point cloud. To address these drawbacks, we introduce Transformer to both feature extraction and point generation processes, and propose a model to control the point generation process at the sparse level, predicting the well-arranged center points for the dense point cloud.

3   Method

3.1   Overview

Figure 1 illustrates the overall architecture of our Context-based Point Generation Network (CPGNet). There are three major parts: point context encoder, center point generation module and point expansion module. We firstly utilize the point context encoder containing multiple point context extraction (PCE) blocks to hierarchically learn local and global context features of observed regions. With these captured features, we employ our center point generation module consisting of multiple context-based point transformation (CPT) blocks to control the sparse point cloud generation, and generate the center points P_m for missing regions. We concatenate P_m with the input points P_{in}, and downsample them to form the sparse point cloud P_0 vis farthest point sampling. Finally, the point expansion module takes P_0 as seeds to increase the number of points and generate a dense point cloud P_c.

3.2 Point Context Encoder

The pioneering PointNet [12] and PointNet++ [13] extract features via pointwise operations on the raw points, which loses sight of contextual relations among points. Differently, EdgeConv [20] dynamically builds a local graph and learns features that describe the relationships between a point and its neighbors. Therefore, we employ the EdgeConv layer to learn local context features, which are lifted by the self-attention layer for the global context features. We design the point context extraction (PCE) block to learn context features for input points locally and globally (as shown in Fig. 1). Multiple PCE blocks constitute our point context encoder, which can capture local and global context features of the observed regions at multiple scales, and offer shape priors and constraints to the generation of the sparse point cloud.

The PCE block consists of an EdgeConv layer and a self-attention layer. In detail, given the incomplete point cloud $P_{in} \in \mathbb{R}^{N_{in} \times 3}$ with N_{in} points each having 3-dimensional coordinates, the multi-layer perceptron (MLP) and the farthest point sampling operations are adopted to obtain the subset $S_l = \{p_i\}_{i=1}^{N_l}$. An EdgeConv layer is attached to extract local context features for each point p_i from its k-nearest neighborhoods $k\text{-}NN = \{e_j\}_{j=1}^{k}$. The EdgeConv layer is implemented by the MLP and max-pooling operations, which can be conducted as follows:

$$p_i = \max_{j} \left\{ MLP([p_i; e_i^j - p_i]) \right\}. \tag{1}$$

Then a self-attention layer is added to compute global context features for each point, which also facilitates to compute the global feature of observed regions. Multiple PCE blocks build our point context encoder to extract local and global context features of points at different resolutions. The outputs of the last PCE block are maxpooled for the global feature of the observed regions. The point context encoder finally extracts the global feature $g \in 1 \times C$ and two context features S_1 and S_2 respectively with N_1 and N_2 point features.

As a contrast, we replace the EdgeConv layer with the PointNet feature extraction layer in our point context block. The comparison is shown in Table 3, and the quantitative results indicate the EdgeConv layer performs better in our PCE block.

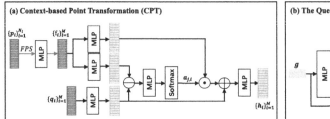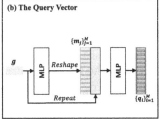

Fig. 2. (a) The detailed structures of the context-based point transformation (CPT) block. (b) The compute process of the query vector for the first CPT block. $\ominus$, $\oplus$ and $\odot$ respectively denote the element-wise subtraction, element-wise addition and Hadamard product.

3.3 Center Point Generation

We propose a context-based point transformation (CPT) block to control the generation process of the sparse point cloud, aiming to generate accurate center points for a dense point cloud. As depicted in Fig. 1, we stack multiple CPT blocks in the decoder to establish a center point generation module, which can progressively learn and refine features for missing regions, and finally generate a complete point cloud at the sparse scale.

The CPT block takes the context feature $S_l = \{p_i\}_{i=1}^{N_l}$ as the inputs, and outputs the features of points in the missing regions. It adopts a cross-attention [17] layer to adaptively capture the contributions from the observed regions to the missing regions, and refine the features for points in missing regions. The key step is to explicitly define the query, key and value vectors in the attention mechanism trimmed to the point generation task.

As depicted in Fig. 2(a), given the local features $S_l = \{p_i\}_{i=1}^{N_l}$, M points are sampled via a farthest point sampling operation, where each point feature is further transformed to f_i by the MLP. f_i is taken as the key vector. The value vector is the same as the key vector. Secondly, the output of the last CPT block is taken as the query vector q_j. The cross-attention layer aims to learn relations between the point q_j in missing regions and each point p_i in observed regions. Intuitively, it computes weight scores $a_{j,i}$ by measuring the similarity between the query vector and each key vector. The output vector is the sum of the value vectors weighted by the weight scores. Therefore, the point feature q_j for missing regions can be updated to h_j as follows:

$$a_{j,i} = \frac{exp(MLP((q_j) \ominus (f_i)))}{\sum_{i=1}^{N_l} exp(MLP((q_j) \ominus (f_i)))}, \tag{2}$$

$$h_j = MLP(q_j \oplus \sum_{i=1}^{N_l} a_{j,i} \odot f_i). \tag{3}$$

where $\ominus$, $\oplus$ and $\odot$ respectively denote the element-wise subtraction, element-wise addition and Hadamard product.

For the first CPT block, we compute the query vector from the global feature g (depicted in Fig. 2(b)). Specifically, we utilize the global feature g to generate the initial point features $\{m_j\}_{j=1}^{M}$ representing the missing regions via the simple MLP. Then each j-th point feature m_j is concatenated with the repeated global features, which are enhanced by the MLP to build the query vector $q_j = MLP(repeat(g); m_j)$.

Two CPT blocks are introduced to iteratively refine features for the missing regions, and ultimately generate the point cloud P_m with M points via the MLP for missing regions. Sequentially, P_m is merged with the input P_{in}, which are then down-sampled to P_0 with N_0 points. In this paper, we typically set $M = 256$ and $N_0 = 512$. We use the point expansion module [24] to extend P_0 to a dense point cloud for the missing regions, which is merged with the input point cloud P_{in} to form the final complete point cloud P_c. We adopt Chamfer Distance as our loss function. Given the ground-truth point cloud P_t, our final loss function is the sum of these two losses $\mathcal{L} = \mathcal{L}_s + \mathcal{L}_c$, where

$$\mathcal{L}_s = \frac{1}{|P_s|} \sum_{p \in P_s} \min_{g \in P_t} \|p - g\| + \frac{1}{|P_t|} \sum_{g \in P_t} \min_{p \in P_s} \|g - p\|, \tag{4}$$

$$\mathcal{L}_c = \frac{1}{|P_c|} \sum_{p \in P_c} \min_{g \in P_t} \|p - g\| + \frac{1}{|P_t|} \sum_{g \in P_t} \min_{p \in P_c} \|g - p\|. \tag{5}$$

4   Experiments

4.1   Datasets and Implementation Details

PCN Dataset. The PCN dataset [29] contains $30,974$ 3D CAD models derived from 8 categories. Each ground truth point cloud contains $16,384$ points, and the corresponding partial shape includes 2048 points. For a fair comparison, we use the same train/val/test splits as [25,28].

ShapeNet-55 Dataset. The ShapeNet-55 dataset [28] is collected from all the objects in ShapeNet [23], and contains 55 object categories. Point clouds are gathered from diverse viewpoints, with different incomplete patterns and various incompleteness levels. Every input point cloud contains $2,048$ points while each ground truth contains $8,192$ points. During training, a viewpoint is randomly chosen and n furthest points are removed from this viewpoint. From the remaining point cloud, $2,048$ points are downsampled as the input. n is randomly selected from 25% to 75%. We keep the standard protocol on this dataset as [28].

Implementation Details. Our model is trained on two RTX 3090Ti GPUs and optimized by an Adam optimizer. We adopt the mean Chamfer Distance (CD) as the evaluation metric. The CD-$\mathcal{L}1$ and CD-$\mathcal{L}2$ denote the distances in $\mathcal{L}1$-norm and $\mathcal{L}2$-norm. For the ShapeNet-55 dataset, during evaluation, 25%, 50% or

Fig. 3. Visual results on PCN dataset. Our model can generate the complete point cloud with less noises and more detailed geometrics.

Table 1. Completion results on PCN dataset in terms of $\mathcal{L}1$ Chamfer Distance $\times 10^3$. Lower is better.

Methods	Average	Plane	Cabinet	Car	Chair	Lamp	Sofa	Table	Watercraft
FoldNet [26]	14.31	9.49	15.80	12.61	15.55	16.41	15.97	13.65	14.99
TopNet [16]	12.15	7.61	13.31	10.90	13.82	14.44	14.78	11.22	11.12
AtlasNet [5]	10.85	6.37	11.94	10.10	12.06	12.37	12.99	10.33	10.61
PCN [29]	9.64	5.50	22.70	10.63	8.70	11.00	11.34	11.68	8.59
GRNet [25]	8.83	6.45	10.37	9.45	9.41	7.96	10.51	8.44	8.04
PMP-Net [21]	8.73	5.65	11.24	9.64	9.51	6.95	10.83	8.72	7.25
Pointr [28]	8.38	4.75	10.47	8.68	9.39	7.75	10.93	7.78	7.29
SnowflakeNet [24]	7.21	4.29	**9.16**	**8.08**	7.89	6.07	**9.23**	6.55	6.40
Ours	**7.11**	**4.09**	9.24	8.34	**7.62**	**5.72**	9.34	**6.39**	**6.14**

75% of the whole point cloud are removed representing three incomplete degrees: simple, moderate and hard. Besides, 8 viewpoints are fixed for evaluation.

4.2   Evaluation on PCN Dataset

Quantitative Comparison. The comparative results are shown in Table 1, demonstrating our CPGNet outperforms the state-of-the-art methods. Compared with the most recent investigations like Pointr [28] and SnowflakeNet [24], our model reduces the average CD to 7.11 since it can explicitly control the generation processes of the sparse point cloud, and provide the shape priors for a dense point cloud.

Visual Comparison. We also depict visualization results in Fig. 3. For the simple objects (listed in the first two lines), our CPGNet can generate the complete shape with geometric details, such as the plane engines. Then for the complicated objects (shown in the last three lines), no methods successfully generate the complete shape. But compared with the second-best SnowflakeNet [24], our model can generate a more clear geometry with less noisy structures because the proposed CPT block enables our model to generate more accurate center points for the dense point cloud.

Table 2. Completion results on ShapeNet-55 dataset in terms of $\mathcal{L}2$ Chamfer Distance $\times 10^3$. Lower is better.

Methods	Airplane	Car	Sofa	Birdhouse	Bag	Remote	Rocket	CD-S	CD-M	CD-H	CD-Avg
FoldNet [26]	1.43	1.98	2.48	4.71	2.79	1.44	1.48	2.67	2.66	4.05	3.12
PCN [29]	1.02	1.85	2.06	4.50	2.86	1.33	1.32	1.94	1.96	4.08	2.66
TopNet [16]	1.14	2.18	2.36	4.83	2.93	1.49	1.32	2.26	2.16	4.3	2.91
PFNet [8]	1.81	2.53	3.34	6.21	4.96	2.91	2.36	3.83	3.87	7.97	5.22
GRNet [25]	1.02	1.64	1.72	2.97	2.06	1.09	1.03	1.35	1.71	2.85	1.97
Pointr [28]	0.52	1.10	0.90	2.07	1.12	0.87	0.66	0.73	1.13	2.05	1.30
Ours	**0.43**	**0.94**	**0.88**	**1.66**	**0.83**	**0.54**	**0.46**	**0.56**	**0.80**	**1.45**	**0.93**

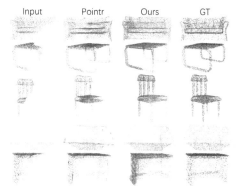

Fig. 4. Visual completion results on ShapeNet-55 dataset.

4.3 Evaluation on ShapeNet-55 Dataset

Quantitative Comparison. We evaluate our model under three incomplete degrees, where 25%, 50% or 75% of the whole point cloud are respectively removed. We report the average CD loss (CD-Avg) and CD losses at three incompleteness degrees: simple, moderate and hard (CD-S, CD-M and CD-H) in Table 2. Our model outperforms other related methods. In addition, the first three columns denote the completion results of categories with sufficient samples, and the following four columns report categories with insufficient samples. Our CPGNet performs better than the state-of-the-art methods, which proves its superiority in handling challenging scenarios with more categories, different viewpoints and diverse incomplete patterns.

Visual Comparison. We provides visualization results of our CPGNet and Pointr [28] under the moderate incomplete degree in Fig. 4. Both two methods can deal with the challenging inputs, but our model performs a little better. For example, the input of the chair (in the third line) loses many structures which seem hardly to be completed. Pointr [28] fails to predict missing regions, but our CPGNet generates a decent shape with geometric details like chair legs.

Fig. 5. Visual results of our center point generation module.

Table 3. Effects of the key parts in CPGNet on PCN dataset in terms of CD-$\mathcal{L}1$. Lower is better.

Methods	PointNet + Attention	EdgeConv + Attention	CPT blocks	CD-$\mathcal{L}1$
Model A	✓			7.29
Model B	✓		✓	7.17
Ours		✓	✓	**7.11**

4.4 Ablation Studies

Analysis of Our CPGNet. We deploy ablation experiments on PCN dataset to validate the effectiveness of the key components in CPGNet, and report the results in Table 3. All models employ the same point expansion module in [24]. Model A uses PointNet [12,13] feature extraction and self-attention layers to encode the features, and predicts the sparse point cloud from the global feature by a MLP. By contrast, model B utilizes two CPT blocks to generate the sparse point cloud. Our model replaces the PointNet [12,13] feature extraction layer in model B with the EdgeConv [20] layer. Model B reduces CD loss by 0.12 corresponding to model A, since the CPT blocks can predict more uniform and accurate seeds for the point expansion. Our model outputs a better result than model B, demonstrating the effectiveness of our point context encoder. We also plot visualization results of our center point generation module in Fig. 5. This module consisting of two CPT blocks can generate the complete point cloud at a sparse level (Ours-s), which serves as center points for a dense point cloud with detailed structures.

5   Conclusion

We propose the Context-based Point Generation Network which concentrates on controlling the point generation process at the sparse level. We introduce Transformer to both feature extraction and sparse point generation processes, and design the point context extraction and context-based point transformation blocks. The point context extraction block extracts local and global context features of the observed points, and the context-based point transformation block transforms them to the context features of the missing points. Multiple these two blocks are stacked to progressively refine the context features, and finally predict the well-arranged center points for the missing shapes. Massive experiments on PCN and ShapeNet-55 datasets demonstrate our model outperforms the state-of-the-art methods.

Acknowledgements. This work was supported by Shandong Provincial Natural Science Foundation under Grant ZR2021QF062.

References

1. Atzmon, M., Maron, H., Lipman, Y.: Point convolutional neural networks by extension operators. ACM Trans. Graph. **37**(4), 71 (2018)
2. Dosovitskiy, A., et al.: An image is worth 16x16 words: Ttansformers for image recognition at scale. arXiv preprint arXiv:2010.11929 (2020)
3. Fan, S., Dong, Q., Zhu, F., Lv, Y., Ye, P., Wang, F.Y.: SCF-Net: learning spatial contextual features for large-scale point cloud segmentation. In: CVPR, pp. 14504–14513 (2021)
4. Groueix, T., Fisher, M., Kim, V., Russell, B., Aubry, M.: AtlasNet: a papier-mâché approach to learning 3d surface generation. arxiv 2018. arXiv preprint arXiv:1802.05384 (1802)
5. Groueix, T., Fisher, M., Kim, V.G., Russell, B.C., Aubry, M.: A papier-mâché approach to learning 3d surface generation. In: CVPR, pp. 216–224 (2018)
6. Guo, M.H., Cai, J.X., Liu, Z.N., Mu, T.J., Martin, R.R., Hu, S.M.: PCT: point cloud transformer. Comput. Vis. Media **7**(2), 187–199 (2021)
7. He, C., Li, R., Li, S., Zhang, L.: Voxel set transformer: a set-to-set approach to 3D object detection from point clouds. In: CVPR, pp. 8417–8427 (2022)
8. Huang, Z., Yu, Y., Xu, J., Ni, F., Le, X.: PF-Net: point fractal network for 3D point cloud completion. In: CVPR, pp. 7662–7670 (2020)
9. Lai, X., et al.: Stratified transformer for 3D point cloud segmentation. In: CVPR, pp. 8500–8509 (2022)
10. Maturana, D., Scherer, S.: VoxNet: a 3D convolutional neural network for real-time object recognition. In: 2015 IEEE/RSJ International Conference on Intelligent Robots and Systems (IROS), pp. 922–928. IEEE (2015)
11. Pan, X., Xia, Z., Song, S., Li, L.E., Huang, G.: 3D object detection with pointformer. In: CVPR, pp. 7463–7472 (2021)
12. Qi, C.R., Su, H., Mo, K., Guibas, L.J.: PointNet: deep learning on point sets for 3D classification and segmentation. In: CVPR, pp. 652–660 (2017)
13. Qi, C.R., Yi, L., Su, H., Guibas, L.J.: Pointnet++: deep hierarchical feature learning on point sets in a metric space. arXiv preprint arXiv:1706.02413 (2017)

14. Tang, L., Zhan, Y., Chen, Z., Yu, B., Tao, D.: Contrastive boundary learning for point cloud segmentation. In: CVPR, pp. 8489–8499 (2022)
15. Tchapmi, L., Choy, C., Armeni, I., Gwak, J., Savarese, S.: SEGcloud: semantic segmentation of 3D point clouds. In: International conference on 3D vision (3DV), pp. 537–547. IEEE (2017)
16. Tchapmi, L.P., Kosaraju, V., Rezatofighi, H., Reid, I., Savarese, S.: TopNet: structural point cloud decoder. In: CVPR, pp. 383–392 (2019)
17. Vaswani, A., et al.: Attention is all you need. In: NeurIPS, pp. 5998–6008 (2017)
18. Wang, P., et al.: Omni-DETR: omni-supervised object detection with transformers. In: CVPR, pp. 9367–9376 (2022)
19. Wang, X., Ang Jr, M.H., Lee, G.H.: Cascaded refinement network for point cloud completion. In: CVPR, pp. 790–799 (2020)
20. Wang, Y., Sun, Y., Liu, Z., Sarma, S.E., Bronstein, M.M., Solomon, J.M.: Dynamic graph CNN for learning on point clouds. ACM Trans. Graph. **38**(5), 1–12 (2019)
21. Wen, X., et al.: PMP-Net: point cloud completion by learning multi-step point moving paths. In: CVPR, pp. 7443–7452 (2021)
22. Wu, W., Qi, Z., Fuxin, L.: Pointconv: deep convolutional networks on 3D point clouds. In: CVPR, pp. 9621–9630 (2019)
23. Wu, Z., et al.: 3D shapeNets: a deep representation for volumetric shapes. In: CVPR, pp. 1912–1920 (2015)
24. Xiang, P., et al.: SnowflakeNet: point cloud completion by snowflake point deconvolution with skip-transformer. In: ICCV, pp. 5499–5509 (2021)
25. Xie, H., Yao, H., Zhou, S., Mao, J., Zhang, S., Sun, W.: GRNet: gridding residual network for dense point cloud completion. In: Vedaldi, A., Bischof, H., Brox, T., Frahm, J.-M. (eds.) ECCV 2020. LNCS, vol. 12354, pp. 365–381. Springer, Cham (2020). https://doi.org/10.1007/978-3-030-58545-7_21
26. Yang, Y., Feng, C., Shen, Y., Tian, D.: FoldingNet: point cloud auto-encoder via deep grid deformation. In: CVPR, pp. 206–215 (2018)
27. Yew, Z.J., Lee, G.H.: REGTR: end-to-end point cloud correspondences with transformers. In: CVPR, pp. 6677–6686 (2022)
28. Yu, X., Rao, Y., Wang, Z., Liu, Z., Lu, J., Zhou, J.: PoinTr: diverse point cloud completion with geometry-aware transformers. In: ICCV, pp. 12498–12507 (2021)
29. Yuan, W., Khot, T., Held, D., Mertz, C., Hebert, M.: PCN: point completion network. In: International conference on 3D vision (3DV), pp. 728–737 (2018)
30. Zhang, W., Yan, Q., Xiao, C.: Detail preserved point cloud completion via separated feature aggregation. In: Vedaldi, A., Bischof, H., Brox, T., Frahm, J.-M. (eds.) ECCV 2020. LNCS, vol. 12370, pp. 512–528. Springer, Cham (2020). https://doi.org/10.1007/978-3-030-58595-2_31
31. Zheng, S., et al.: Rethinking semantic segmentation from a sequence-to-sequence perspective with transformers. In: CVPR, pp. 6881–6890 (2021)
32. Zhou, T., Li, L., Bredell, G., Li, J., Konukoglu, E.: Volumetric memory network for interactive medical image segmentation. Med. Image Anal. **83**, 102599 (2022)
33. Zhu, X., Su, W., Lu, L., Li, B., Wang, X., Dai, J.: Deformable DETR: deformable transformers for end-to-end object detection. arXiv preprint arXiv:2010.04159 (2020)

Temporal Neighborhood Change Centrality for Important Node Identification in Temporal Networks

Zongze Wu, Langzhou He, Li Tao$^{(\boxtimes)}$, Yi Wang, and Zili Zhang

College of Computer and Information Science, Southwest University,
Chongqing 400715, China
{tli,echowang,zhangzl}@swu.edu.cn

Abstract. In the field of complex networks, identifying important nodes is of great importance both in theoretical and practical applications. Compared with the important node identification of the static network, the important node identification of the temporal network is a more urgent problem to solve since most complex networks in reality change with time. The degree centrality method in static networks shows that the more nodes a node is connected to, that is, the more nodes it has in its neighborhood, the more influential and important this node is. Inspired by this method, our idea is that in a temporal network, as time changes, if a node's neighborhood keeps adding new nodes, the more nodes it affects, the more important it is. Therefore, we propose a new method for identifying important nodes in temporal networks, namely the temporal neighborhood change centrality(TNCC). The TNCC of a node is equal to its average neighborhood change rate over a period of time. The larger the TNCC of a node, the more important it is. We evaluate the proposed method against 7 baseline methods on 6 real temporal networks based on the infectious disease model SIR. Experimental results show that our method is more stable in identifying important nodes and has advantages in most cases.

Keywords: temporal networks · important nodes · centrality · neighborhood change

1 Introduction

Network science is increasingly important in numerous fields including physical, biological, financial, and social sciences. In fact, many complex systems can be properly represented as complex networks [3,4,6].

Identifying the most important nodes in large-scale networks, or evaluating the importance of a node relative to others, is one of the key issues in complex network research. Many real-world problems involve identifying multiple influential nodes in complex networks, such as finding some individuals that are critical to the dissemination of information on the Internet or, who will speed

© The Author(s), under exclusive license to Springer Nature Switzerland AG 2023
M. Tanveer et al. (Eds.): ICONIP 2022, LNCS 13623, pp. 455–467, 2023.
https://doi.org/10.1007/978-3-031-30105-6_38

up the process of spreading an infectious disease [21]. These nodes have a great influence on the structural connectivity and information dissemination process of complex networks, and are called important nodes.

If the nodes and edges in the network do not change over time, then such a network is a static network. Various important node identification methods have been developed for static networks [13]. There are methods based on local topology, such as degree centrality [2], which means that the more neighbors a node has, the more important the node is. There are methods based on shortest paths, such as betweenness centrality [8], which refers to the number of shortest paths that a node passes through between other nodes, and examines the ability of a node to control the dissemination of information from other nodes. For example, closeness centrality [19], which calculates the average shortest distance from a node to all other nodes, the smaller the distance, the closer the node is to other nodes in the network. There are methods based on eigenvectors, such as PageRank [9], which considers that the importance of a node depends on both the number of its neighbors and the importance of its neighbors.

But the real network is temporal, and its topology changes with time [16]. To identify important nodes in a temporal network, we need to consider both structural properties and temporal information. If we consider the time at which connections occur in a complex system, we can use temporal networks to represent the system.

The main idea of the existing methods for identifying important nodes in temporal networks is to build temporal network models based on time-series data, and then use network node sorting algorithms to identify important nodes based on the topology or dynamic characteristics of temporal networks [5]. Some of these methods have too high time complexity, and some may lose some time information.

We propose a new method, temporal neighborhood change centrality. The larger the average neighborhood change rate of a node over a period of time, the more important it is. Our proposed method focuses on the neighborhood changes of each node between different snapshots, avoiding high time complexity without losing too much temporal information, which is the greatest contribution of our work. We evaluate our method on 6 real temporal networks, based on the susceptible-infected-recovery (SIR) model, and select the top-ranked nodes of each method as the initial infection nodes to simulate the information propagation in the temporal network. Compared with some state-of-the-art methods, our method is more efficient in maximizing the size of the propagation and is more stable.

The rest of this paper is organized as follows. Section 2 introduces some basic concepts of temporal networks and work related to identifying important nodes in temporal networks. Section 3 proposes our method and Sect. 4 describes our experimental setup and dataset. Section 5 provides experimental results and discussion. Section 6 summarizes the paper.

2   Related Work

2.1   Preliminaries

Definition of Temporal Networks

$G^T=(V,E^T)$ represents the temporal network on the time interval $[0,T]$, the network consists of a set of nodes V and a set of temporal contact events E^T. Each temporal contact event is given by a triple (i, j, t), where i and j represent two nodes, and t represents the time when the two nodes interact.

Temporal Network Modeling

Temporal network modeling is the basis for identifying important nodes of temporal networks. The temporal network models we discussed in the paper have a common feature: the number of nodes in a temporal network is constant, and the edges appear and disappear with time. The main temporal network models can be roughly divided into three categories: temporal network models based on static graphs, temporal network models based on snapshots, and temporal network models based on explicit path flows [5].

We use a snapshot-based temporal network model which divides the entire time interval $[0, T]$ of the network into consecutive n snapshots $G1, G2, ...,Gn$ of length $w(n=T/w)$. The most common snapshot-based temporal network model is the time window graph model shown in Fig. 1. Figure 1(a) shows the aggregated graph of the entire temporal network. The set of numbers on the edge is the interaction time of two nodes. Figure 1(b) is the time window graph model corresponding to Fig. 1(a) with time window size setted to 1. The model divides the network into four snapshots, the interval of each snapshot is 1. Figure 1(c) is a time window graph model with time window size setted to 2 corresponding to Fig. 1(a). The model divides the network into two snapshots, and the time interval of each snapshot is 2.

Time-Respecting Path

The paths between nodes in a temporal network need to follow the time sequence of node interactions. For example, in the time aggregation graph of Fig. 2(a), there is a path $1 \rightarrow 2 \rightarrow 3$. But in Fig. 2(b), node 1 and 2 interact at time 2, and node 2 and 3 interact at time 1, so there is no path $1 \rightarrow 2 \rightarrow 3$. From above example we find that time information may be lost after the entire temporal network is aggregated into a static network. The path we find should be a time-respecting path.

2.2   Methods of Identifying Important Nodes in Temporal Networks

In temporal networks, methods for identifying important nodes are mainly divided into three categories. The first category is based on degree centrality, and Kim et al. [11] propose a sequential graph model that reduces sequential networks to static networks with path flows. The temporal degree centrality

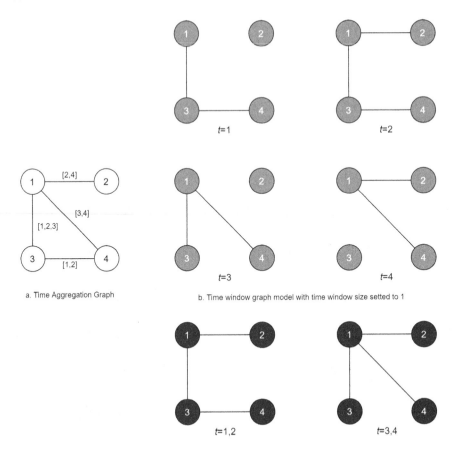

a. Time Aggregation Graph

b. Time window graph model with time window size setted to 1

c. Time window graph model with time window size setted to 2

Fig. 1. Time window graph model

of a node is equal to the average of its degree centrality at each time step in the time interval $[i,j]$. The larger the value, the more important the node is. However, they are refined to each time step, and the temporal network datasets we generally use have long spans, so the computational time of this method is unrealistic. Wang et al. [23] proposed the temporal degree centrality deviation value of nodes based on the time window graph model, which is equal to the standard deviation of node degree centrality within each snapshot. The larger the temporal degree centrality deviation value of the node, the more important the node is. But Wang et al. do not consider the connections between different snapshots, so this part of the time information is ignored. Both methods based on degree centrality focus on the fluctuation of the degree centrality of a node over time.

The second category is based on temporal paths. Kim et al. [11] also proposed temporal closeness centrality and temporal betweenness centrality based on the

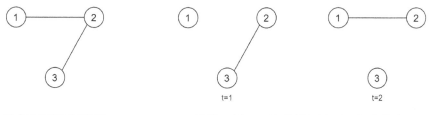

(a). Time Aggregation Graph   (b). Time window model with time window length setted to 1

Fig. 2. Temporal network with three nodes and T = 2 time steps.(a)time aggregated network. (b) Static snapshots.

temporal network model of path flow. In a time interval, the temporal closeness centrality of a node is equal to the sum of the inverses of the temporal shortest path distances to all other nodes in the network, and temporal betweenness centrality is equal to the number of temporal shortest paths through the node. Magnien et al. [15] considered an arbitrary time in the entire time interval as the starting time of a temporal path, and on this basis, the importance of nodes at any time can be evaluated. However, this method is relatively weak in identifying the global importance of nodes. Qu et al. [18] believed that the importance of a node depends on the importance of its neighbors, so they aggregated the initial information of the node and the information of neighbors with different distances to obtain the time-series information aggregation score of the node. Elmezain et al. [7] considered both the number of connections of nodes and the strength of connections between nodes, then proposed a new weighted network centrality measure based on temporal degree centrality and temporal closeness centrality. Bi et al. [1] proposed a time gravity model inspired by the physical gravity formula. The static and temporal properties of nodes in the network (i.e., static aggregate centrality and temporal centrality) are analogous to the mass in the gravitational formula, and the temporal distance from a node to other nodes is analogous to the distance in the gravitational formula. This type of method has the problem of high computational complexity because it needs to consider the actual temporal path.

The third category is based on eigenvectors, Taylor et al. [22] divide a temporal network with N nodes into T layers, and then couple the centrality matrices of all layers into a hypercentrality matrix of size $NT \times NT$. The principal eigenvectors of the hypercentricity matrix give the centrality of each node at each temporal layer. Yang et al. [10] believed that the coupling coefficient ω in the method [22] ignored the difference of the connection relationship between different temporal layers, so they proposed a neighbor topological overlap coefficient to represent the connection relationship between different layers. Lv et al. [14] believed that the existing feature vector-based centrality ignores the inter-layer interaction between different timestamps, so in order to better describe the characteristics of the multi-layer temporal network, they constructed a sixth-order tensor to Represents a multilayer temporal network. But the eigenvector-based method obtains the importance of nodes at a certain time layer, not the global importance.

3   Methods

We propose the temporal neighborhood change centrality (TNCC) method based on the snapshot temporal network model.

3.1   Temporal Neighborhood Change Centrality

The main idea of TNCC is to focus on the neighborhood change of a node between adjacent snapshots. The neighborhood change rate is defined as the rate of change of a node's neighbors on the current snapshot compared with its neighbors on the previous snapshot. The temporal neighborhood change centrality of a node is equal to the average neighborhood change rate of a node in all snapshots.

The temporal neighborhood change centrality (TNCC) of node v_i in the temporal network is defined as follows

$$TNCC^l(v_i) = \frac{1}{T}\left(\frac{|Nei_1^l(v_i)|}{N-1} + \sum_{t=2}^{T}\frac{|Nei_t^l(v_i) - Nei_{t-1}^l(v_i)|}{|Nei_t^l(v_i)|}\right) \tag{1}$$

where T represents the number of snapshots, l represents the neighborhood order, $Nei_t^l(v_i) = \{v_j|v_j \in e(v_i, v_j) \in E^t\}$ represents the l-level neighborhood nodes of the node v_i in the snapshot t, N represents the number of network nodes.

The temporal neighborhood change centrality method ranks nodes and selects the top k percent of nodes as spreaders in the network, as described in Algorithm 1.

Figure 3 illustrates the temporal neighborhood changed centrality approach. On the first snapshot, node 3 has a neighbor node 1, and may be connected to $(N-1)=4$ nodes at most, and its NCC is equal to $\frac{1}{4}$. On the second snapshot, the neighbor nodes of node 3 become node 2 and node 5. Compared with the neighbor nodes of the first snapshot, the neighbor nodes are all new nodes, so its NCC is $\frac{2}{2}$. On the third snapshot, the neighbor nodes of node 3 are node 4 and node 5, and the new node 4 is added, so the NCC is $\frac{1}{2}$.

4   Experiment

We use the epidemic spread simulation model SIR to compare the performance of the proposed method with 7 baseline methods on 6 real temporal networks: Temporal degree centrality(TDC) [11], Temporal closeness centrality(TCC) [11], Temporal betweenness centrality(TBC) [11], Temporal degree-degree(TDD) [7], Temporal gravity DC_s(TGDC) [1], Temporal gravity CC_s(TGCC) [1], Temporal gravity BC_s(TGBC) [1]. Our proposed method uses TNCC-1, TNCC-2, and TNCC-3 with neighborhood orders of 1, 2, and 3, respectively, since we consider the local changes of nodes, the order will not be set larger .

Algorithm 1. Temporal neighborhood change centrality.

Input: Temporal Network $G(V, E^T)$, the number of nodes N, the order of neighbors l, the number of time windows w

Output: top-k important nodes

1: Divide the network into $m = T/w$ snapshots
2: Nodes-List $\leftarrow$ [ ]
3: ImportantNodes-List $\leftarrow$ [ ]
4: Put nodes in the Temporal-List
5: **for** t $\leftarrow$ 1 to T **do**
6:     **if** $t = 1$ **then**
7:         $NCC^t(v_i) \leftarrow \frac{|neighbors_t^l(v_i)|}{N-1}$
8:     **else**
9:         $NCC^t(v_i) \leftarrow \frac{|neighbors_t^l(v_i) - neighbors_{t-1}^l(v_i)|}{|neighbors_t^l(v_i)|}$
10:    **end if**
11: **end for**
12: $TNCC_l(v_i) \leftarrow \sum\limits_{t=1}^{T} NCC^t(v_i)/T$
13: **for** nodes in Nodes-List **do**
14:     ImportantNodes-List $\leftarrow$ Sort nodes in ascending order based on $TNCC^l$
15: **end for**
16: Output top k percent of nodes in ImportantNodes-List

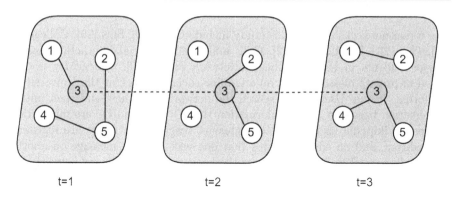

t=1   t=2   t=3

Fig. 3. Temporal neighborhood change centrality calculation for node 3 in a temporal network with 5 nodes and 3 snapshots.

4.1 SIR Model

In the SIR model, nodes have three states: susceptible, infected, and recovered. The first represents nodes that are not infected but potentially infected, the second represents nodes that are infected and are infectious, and the third represents nodes that have been removed from the infected. At the initial time step, we select the top k percent of nodes as the initial infected node , and the rest of the nodes are susceptible nodes. At each subsequent time step, the infected node randomly selects a neighbor node, and if the neighbor node is susceptible, it will be infected with probability β. The infected node also becomes a recovery node with probability γ. When the infected node recovers, it no longer has the ability to infect other susceptible nodes and will not be infected again. Finally, the iteration of the SIR model stops when there are no infected nodes in the network.

4.2   Performance Evaluation

We use the outbreak scale proportion in the network to evaluate the performance of our proposed method in identifying important nodes in temporal networks. The outbreak scale proportion at time step t is

$$F(t) = \frac{I(t) + R(t)}{N} \tag{2}$$

where $I(t)$ and $R(t)$ are the number of infected and recovered nodes at the end of time step t, respectively, and N is the number of network nodes.

4.3   Dataset

The 6 real networks used in this study include Calls [20], Sms [20], CollegeMsg [17], Cy [12] , Br [12], Oc [12]. Calls and Sms are social interactions among college students in the Copenhagen Network Study [20]. CollegeMsg is comprised of private messages sent on an online social network at the University of California, Irvine. Users could search the network for others and then initiate conversation based on profile information [17]. Cy, Br, and Oc are interactions between Wikipedia users in three languages, respectively, nodes are registered wiki editors, and an edge indicates that one user wrote a message on another user's talk page [12]. Table 1 presents the basic features of the six networks.

4.4   Parameters Settings

We analyze the outbreak scale proportion F(t) when changing k, where $k \in [0.05, 0.15]$, with a step of 0.05. β is fixed at 0.5. In addition, the infection rate β is a factor that may affect the outbreak size, so we also analyze the outbreak scale proportion F(t) when changing β, where $\beta \in [0.2, 0.6]$, with a step of 0.2. k is fixed at 0.10.

The recovery probability γ is fixed at 0.5 in all experiments. Experimental results are the average of 100 independent experiments.

Table 1. Basic features of 6 real networks. N is the number of nodes; M is the number of temporal edges; $< d >$ is the average degree; S is the time span in days.

Network	N	M	<d>	S
Calls	536	3600	6.7	28
Sms	568	24333	42.0	28
Cy	2333	10740	4.8	2320
Br	1181	13754	11.7	2320
Oc	3144	11059	3.5	2320
CollegeMsg	1899	59835	7.4	193

5   Results and Discussion

Table 2 gives the outbreak scale proportion $F(t)$ when k is 0.05. TNCC-3 has significant advantages and works best on the four networks of Calls, Sms, CollegeMsg and Cy. On Br, TNCC-2 outperforms other methods. On Oc, TGCC has obvious advantages over other methods.

Table 3 gives the outbreak scale proportion $F(t)$ when k is 0.10. TNCC-3 has the best effect on Sms, CollegeMsg and Cy, especially on Sms and CollegeMsg. TNCC-1 performs best on Calls and Br. TCC significantly outperforms other methods on Oc.

Table 4 gives the outbreak scale proportion $F(t)$ when k is 0.15. TNCC-3 is in the lead, with clear advantages on Calls, Sms and CollegeMsg. TNCC-2 outperforms other methods on Cy. TNCC-1 performs best on Br. TCC performs best on Oc.

From the above we can see that our proposed method has obvious advantages in performance on most networks when identifying different numbers of spreaders. Also, the more neighborhood orders our method considers, the better the performance may be. The reason for the poor performance on the Oc network may be that the OC network is sparse

Table 2. The outbreak scale proportion $F(t)$ when k is 0.05, β is 0.5.

	Calls	Sms	CollegeMsg	Cy	Br	Oc
TDC	0.3883	0.5148	0.8514	0.467	0.533	0.4076
TCC	0.3625	0.5002	0.8587	0.482	0.5518	0.4377
TBC	0.3477	0.4708	0.8526	0.4474	0.507	0.396
TDD	0.3746	0.5143	0.8511	0.4356	0.5287	0.402
TGDC	0.3449	0.4802	0.8511	0.4419	0.5235	0.4248
TGCC	0.3488	0.4756	0.8556	0.4639	0.5485	**0.4689**
TGBC	0.3376	0.4683	0.852	0.4372	0.5267	0.3808
TNCC-1	0.3977	0.4965	0.8636	0.5246	0.6286	0.4025
TNCC-2	0.3646	0.4932	0.8675	0.4945	**0.6433**	0.3998
TNCC-3	**0.4124**	**0.5156**	**0.8904**	**0.5584**	0.5917	0.416

Table 3. The outbreak scale proportion F(t) when k is 0.10, β is 0.5.

	Calls	Sms	CollegeMsg	Cy	Br	Oc
TDC	0.4353	0.5275	0.8377	0.4787	0.5137	0.4407
TCC	0.4202	0.5149	0.8455	0.4858	0.5443	**0.4672**
TBC	0.4018	0.4757	0.8352	0.4699	0.5502	0.4185
TDD	0.4396	0.5298	0.8335	0.4855	0.5114	0.4414
TGDC	0.3609	0.477	0.8378	0.4696	0.5246	0.4422
TGCC	0.369	0.4904	0.8424	0.4754	0.5231	0.4491
TGBC	0.4069	0.4561	0.8333	0.5045	0.5463	0.3971
TNCC-1	**0.4656**	0.5245	0.8442	0.4997	**0.6063**	0.4484
TNCC-2	0.398	0.4914	0.8461	0.5113	0.6016	0.442
TNCC-3	0.4588	**0.564**	**0.8839**	**0.5191**	0.5856	0.4214

Table 4. The outbreak scale proportion F(t) when k is 0.15, β is 0.5.

	Calls	Sms	CollegeMsg	Cy	Br	Oc
TDC	0.4785	0.5515	0.8224	0.4948	0.5259	0.4557
TCC	0.4679	0.5414	0.8345	0.5155	0.5589	**0.5072**
TBC	0.4613	0.4947	0.8237	0.4883	0.5407	0.4414
TDD	0.4635	0.5485	0.8267	0.4929	0.5351	0.4446
TGDC	0.3986	0.5182	0.8227	0.5076	0.5199	0.4702
TGCC	0.4101	0.5107	0.8364	0.5051	0.5588	0.4938
TGBC	0.441	0.4951	0.8259	0.4977	0.5243	0.4509
TNCC-1	0.4862	0.558	0.8347	0.5188	**0.6071**	0.4555
TNCC-2	0.4481	0.525	0.8356	**0.5224**	0.5987	0.4452
TNCC-3	**0.4912**	**0.5944**	**0.8869**	0.5069	0.5892	0.4863

Table 5 gives the outbreak scale proportion F(t) when β is 0.2. TNCC-1 performs better than other methods on Calls, Sms and Br. TNCC-3 has outstanding performance on CollegeMsg and outperforms other methods on Cy. TGCC has obvious advantages over other methods in Oc.

Table 6 gives the outbreak scale proportion F(t) when β is 0.4. TNCC-1 performs best on Calls and Br. TNCC-2 outperforms other methods on Cy. TNCC-3 has the best performance on Sms and CollegeMsg, especially on CollegeMsg. TCC performs best on Oc.

Table 7 gives the outbreak scale proportion F(t) when β is 0.6. TNCC-1 performs best on Calls. TNCC-2 outperforms other methods on Cy and Br. TNCC-3 has obvious advantages over other methods in Sms and CollegeMsg. TGCC significantly outperforms other methods on Oc.

From the above we can see that our proposed method can maintain its advantages on most networks regardless of whether β is small or large. The reason for the poor performance on the Oc network may be that the Oc network is sparse. How to deal with sparse networks is also the work that we need to improve later.

Table 5. The outbreak scale proportion F(t) when β is 0.2, k is 0.10.

	Calls	Sms	CollegeMsg	Cy	Br	Oc
TDC	0.224	0.3283	0.6806	0.292	0.3283	0.2426
TCC	0.2147	0.3307	0.6899	0.3165	0.3307	0.2767
TBC	0.2124	0.3376	0.6811	0.3026	0.3376	0.259
TDD	0.2262	0.3279	0.684	0.2987	0.3279	0.2629
TGDC	0.1925	0.3303	0.6815	0.2825	0.3303	0.2694
TGCC	0.2064	0.35	0.6901	0.3079	0.35	**0.3051**
TGBC	0.2182	0.3458	0.6802	0.3036	0.3458	0.2604
TNCC-1	**0.2301**	**0.3759**	0.6933	0.3106	**0.3759**	0.2665
TNCC-2	0.1904	0.3747	0.6961	0.3172	0.3747	0.264
TNCC-3	0.2184	0.3638	**0.7436**	**0.3242**	0.3638	0.2642

Table 6. The outbreak scale proportion F(t) when β is 0.4, k is 0.10.

	Calls	Sms	CollegeMsg	Cy	Br	Oc
TDC	0.366	0.4354	0.8036	0.4022	0.4718	0.399
TCC	0.356	0.428	0.8117	0.4563	0.4832	**0.4223**
TBC	0.3418	0.3851	0.8	0.4432	0.4926	0.3814
TDD	0.3725	0.4536	0.8011	0.421	0.4685	0.3376
TGDC	0.3031	0.3941	0.8048	0.4054	0.4722	0.3666
TGCC	0.3171	0.4022	0.8123	0.4514	0.5042	0.4032
TGBC	0.3497	0.3771	0.8046	0.3963	0.4694	0.3927
TNCC-1	**0.3982**	0.4392	0.8092	0.4465	**0.5442**	0.38
TNCC-2	0.3301	0.4039	0.8161	**0.4568**	0.5388	0.4177
TNCC-3	0.3824	**0.4738**	**0.8524**	0.4319	0.5154	0.3956

Table 7. The outbreak scale proportion F(t) when β is 0.6, k is 0.10.

	Calls	Sms	CollegeMsg	Cy	Br	Oc
TDC	0.4998	0.5962	0.8601	0.5163	0.5726	0.446
TCC	0.4899	0.5963	0.8731	0.55	0.5908	0.49
TBC	0.4686	0.5562	0.86	0.5198	0.5771	0.4544
TDD	0.5101	0.6067	0.8609	0.5051	0.5588	0.4339
TGDC	0.411	0.561	0.8622	0.525	0.5494	0.4686
TGCC	0.4284	0.5738	0.8697	0.5431	0.5911	**0.5019**
TGBC	0.466	0.5498	0.8595	0.5081	0.5805	0.4445
TNCC-1	**0.531**	0.5971	0.8717	0.5381	0.6507	0.4203
TNCC-2	0.4519	0.5736	0.8728	**0.5667**	**0.6508**	0.414
TNCC-3	0.5249	**0.6346**	**0.9098**	0.5593	0.6278	0.4676

6   Conclusion

Identifying important nodes in a temporal network has practical implications for message propagation or infectious disease control. We propose the temporal neighborhood change centrality method. In experiments on 6 real temporal networks, our proposed method is more effective than 7 baseline methods, and can better select important nodes that affect more nodes throughout the time interval.

However, the nodes in the temporal network may be in the multi-layer network and have various attributes. How to identify the important nodes in the multi-layer temporal network is a problem that needs to be solved in our future work.

References

1. Bi, J., Jin, J., Qu, C., Zhan, X., Wang, G., Yan, G.: Temporal gravity model for important node identification in temporal networks. Chaos Solitons Fract. **147**, 110934 (2021)
2. Bonacich, P.: Factoring and weighting approaches to status scores and clique identification. J. Math. Sociol. **2**(1), 113–120 (1972)
3. Carlos-Sandberg, L., Clack, C.D.: Incorporation of causality structures to complex network analysis of time-varying behaviour of multivariate time series. Sci. Rep. **11**(1), 1–16 (2021)
4. Charakopoulos, A., Karakasidis, T., Sarris, I.: Analysis of magnetohydrodynamic channel flow through complex network analysis. Chaos: Interdiscip. J. Nonlinear Sci. **31**(4), 043123 (2021)
5. Chen, S., Ren, Z., Liu, C., et al.: Identification methods of vital nodes on temporal networks. J. Univ. Electron. Sci. Technol. China **49**(2), 291–314 (2020)
6. Dorogovtsev, S.N., Mendes, J.F.: The Nature of Complex Networks. Oxford University Press (2022)
7. Elmezain, M., Othman, E.A., Ibrahim, H.M.: Temporal degree-degree and closeness-closeness: a new centrality metrics for social network analysis. Mathematics **9**(22), 2850 (2021)
8. Freeman, L.C.: A set of measures of centrality based on betweenness. Sociometry **40**, 35–41 (1977)
9. Gleich, D.F.: PageRank beyond the web. SAIM Rev. **57**(3), 321–363 (2015)
10. Jian-Nan, Y., Jian-Guo, L., Qiang, G.: Node importance identification for temporal network based on inter-layer similarity. Acta Phys. Sin. **67**(4), 048901 (2018)
11. Kim, H., Anderson, R.: Temporal node centrality in complex networks. Phys. Rev. E **85**(2), 026107 (2012)
12. Kunegis, J.: KONECT: the Koblenz network collection. In: Proceedings of the 22nd International Conference on World Wide Web, pp. 1343–1350 (2013)
13. Lü, L., Chen, D., Ren, X.L., Zhang, Q.M., Zhang, Y.C., Zhou, T.: Vital nodes identification in complex networks. Phys. Rep. **650**, 1–63 (2016)
14. Lv, L., et al.: Eigenvector-based centralities for multilayer temporal networks under the framework of tensor computation. Expert Syst. Appl. **184**, 115471 (2021)
15. Magnien, C., Tarissan, F.: Time evolution of the importance of nodes in dynamic networks. In: 2015 IEEE/ACM International Conference on Advances in Social Networks Analysis and Mining (ASONAM), pp. 1200–1207. IEEE (2015)

16. Michail, O.: An introduction to temporal graphs: an algorithmic perspective. Internet Math. **12**(4), 239–280 (2016)
17. Panzarasa, P., Opsahl, T., Carley, K.M.: Patterns and dynamics of users' behavior and interaction: network analysis of an online community. J. Am. Soc. Inform. Sci. Technol. **60**(5), 911–932 (2009)
18. Qu, C., Zhan, X., Wang, G., Wu, J.-L., Zhang, Z.-K.: Temporal information gathering process for node ranking in time-varying networks. Chaos: Interdiscip. J. Nonlinear Sci. **29**(3), 033116 (2019)
19. Sabidussi, G.: The centrality index of a graph. Psychometrika **31**(4), 581–603 (1966)
20. Sapiezynski, P., Stopczynski, A., Lassen, D.D., Lehmann, S.: Interaction data from the Copenhagen networks study. Sci. Data **6**(1), 1–10 (2019)
21. Su, Z., Gao, C., Liu, J., Jia, T., Wang, Z., Kurths, J.: Emergence of nonlinear crossover under epidemic dynamics in heterogeneous networks. Phys. Rev. E **102**(5), 052311 (2020)
22. Taylor, D., Myers, S.A., Clauset, A., Porter, M.A., Mucha, P.J.: Eigenvector-based centrality measures for temporal networks. Multiscale Model. Simul. **15**(1), 537–574 (2017)
23. Wang, Z., Pei, X., Wang, Y., Yao, Y.: Ranking the key nodes with temporal degree deviation centrality on complex networks. In: 2017 29th Chinese Control And Decision Conference (CCDC), pp. 1484–1489. IEEE (2017)

DOM2R-Graph: A Web Attribute Extraction Architecture with Relation-Aware Heterogeneous Graph Transformer

Jiali Feng[1,2], Cong Cao[1(✉)], Fangfang Yuan[1], Xiaoliang Zhang[1], Zhiping Li[1,2], Yanbing Liu[1,2], and Jianlong Tan[1,2]

[1] Institute of Information Engineering, Chinese Academy of Sciences, Beijing, China
{fengjiali,caocong,yuanfangfang,zhangxiaoliang,lizhiping,
liuyanbing,tanjianlong}@iie.ac.cn
[2] School of Cyber Security, University of Chinese Academy of Sciences,
Beijing, China

Abstract. Web attribute extraction refers to extracting structured entities with specific attributes (e.g. title, director, genre and mpaa rating for a movie) from HTML documents. Since each part of a web page corresponds to an unique node in the DOM tree, most of existing methods formulate web attribute extraction as a multi-class classification task of DOM tree nodes. However, they rarely focus on the multiple structural relations between DOM tree nodes, which will influence node semantic interactions. In this paper, we propose a novel web attribute extraction architecture called **DOM2R-Graph**, which integrates both node semantic information and heterogeneous structure information of DOM tree. Specifically, we first construct a heterogeneous graph by connecting DOM tree nodes and their contexts with edges indicating structural relations. Then, we propose a **R**elation-aware **H**eterogeneous **G**raph **T**ransformer (RHGT), to effectively capture the heterogeneous features of structural relations and learn representations of nodes on the graph at a fine-grained level. Extensive experimental results on the public SWDE dataset show that DOM2R-Graph outperforms the state-of-the-art methods.

Keywords: Web information extraction · Structured data extraction · Heterogeneous graph transformer

1 Introduction

Attribute extraction is to extract target attributes from text sources and form them into structured data. It is crucial for downstream tasks such as large-scale knowledge base/graph construction [6,21], personalized recommendation [18,19] and question answering systems [3,4]. Compared with plain texts [15], large-scale semi-structured vertical websites, such as IMDB with movie information, usually contain more comprehensive, real-time and high-quality information [16].

M. Tanveer et al. (Eds.): ICONIP 2022, LNCS 13623, pp. 468–479, 2023.
https://doi.org/10.1007/978-3-031-30105-6_39

Therefore, extracting attributes from semi-structured web pages (as shown in Fig. 1(a)) is an important problem worthy to be studied. However, the complex page contents and changeful page layouts make it a challenging task.

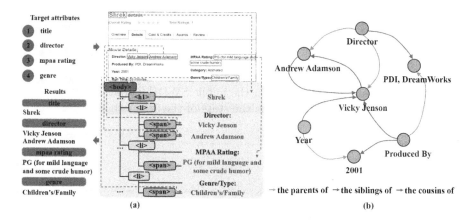

Fig. 1. (a) is an example of extracting target attributes from a semi-structured web page. (b) is a sub heterogeneous graph constructed from the DOM tree in (a). (b) shows that *parents* can provide description information, *siblings* can supplement attribute-related information since they are usually attribute values of the same type as current attribute value, *cousins* as the descriptions of other attribute values can provide co-occurrence information since important attributes are usually located in the same area.

Most of existing methods formulate web attribute extraction as a multi-class classification task of DOM tree nodes, predicting which attribute type a node belongs to using generalizable features. Some works [7,8,12] use visual features extracted from the rendered page, such as encoding of web page screenshots or the bounding box coordinates of nodes. However, the rendering process is computationally expensive. To alleviate this problem, recent works [14,22] try to exploit DOM tree features to avoid web page rendering. Lin et al. [14] propose to learn local representations of nodes and capture conditional dependencies between nodes, respectively. Zhou et al. [22] propose to concatenate contexts for nodes to enrich node representations. However, these methods fail to capture the complex heterogeneous structural relations between each node and its contexts. The contexts corresponding to different structural relations can make different types of contributions to the current node (as shown in Fig. 1(b)). Therefore, we argue that the semantics of each node and its contexts, as well as the heterogeneous structural relations together determine the node attribute type.

In this paper, we propose a novel web attribute extraction architecture, called DOM2R-Graph. Firstly, we generate a simplified DOM tree by retrieving the contexts of textual nodes based on structural relations. Then, to model both the texts and the heterogeneous structural relations, we construct a heterogeneous graph based on the simplified DOM tree. Furthermore, we propose a

Relation-aware Heterogeneous Graph Transformer, which captures the different importance of each structural relation to guide node semantic interactions.

In summary, the main contributions of our paper are as follows:

- We use a heterogeneous graph to model both node texts and heterogeneous structural relations, so as to establish more comprehensive contextual associations for each node.
- We propose a Relation-aware Heterogeneous Graph Transformer to learn fine-grained representations of nodes on the graph by considering the influence of relations on node semantic interactions.
- Extensive experimental results on the public SWDE [8] dataset demonstrate that DOM2R-Graph outperforms the state-of-the-art methods.

2   Methodology

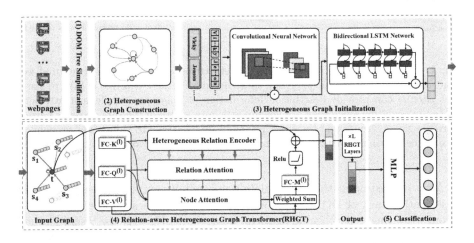

Fig. 2. The overall framework of DOM2R-Graph.

2.1   Problem Formulation

Like most of existing methods, we formulate web attribute extraction as a multi-class classification task of DOM tree nodes. We aim to learn an architecture (as shown in Fig. 2) that can classify each node into one of the pre-defined attribute collection (e.g. {title, director, genre, mpaa rating}) or none, where none means that this node doesn't contain any attribute values. And we follow the convention that one node can correspond to at most one attribute type [8].

Note that we only need to classify nodes with textual contents. Textual nodes in the DOM tree can be divided into fixed nodes N_f and variable nodes N_v [22]. Fixed nodes remain consistent across different web pages in the same website, while variable nodes often change in contents. Since attribute values usually vary in different web pages, we further narrow the range of nodes to be classified to variable nodes.

2.2   DOM Tree Simplification

In this module, our goal is to extract contexts for each variable node x, which consists of *parents*, *siblings* and *cousins*. A DOM tree is a hierarchical structure composed of a series of nodes. It originates from a node called *root* and extends down layer by layer. For each variable node x, all nodes on the path from *root* to x (excluding x) are its ancestors. Given a textual node set $X_n \subseteq N_f \cup N_v$ in which the distances from x and $x_n \in X_n$ to their lowest common ancestor doesn't exceed constant D. We define the *parents* of x as a fixed node set $X_p \subseteq X_n$ that contains at most one node x_p, where x_p is the only fixed node in the tree that originates from the lowest common ancestor of x_p and x. The *siblings* of x is denoted as a variable node set $X_s \subseteq X_n$, which consists of at most M variable nodes in X_n closest to x. The *cousins* of x is a fixed node set $X_c \subseteq X_n$, consisting of at most N fixed nodes in X_n closest to x.

2.3   Heterogeneous Graph Construction

To model both node texts and heterogeneous structural relations, we connect each variable node in the simplified DOM tree to its contexts and use the edges between them to indicate structural relations. Each edge on the graph is a directed edge starting from a contextual node. And we build a self-loop edge with "the siblings of" relation for each node.

We denote the heterogeneous graph as $\mathcal{G} = (\mathcal{V}, \mathcal{A}, \mathcal{E}, \mathcal{R})$, where the node $v \in \mathcal{V}$ and the edge $e \in \mathcal{E}$ can be mapped into their types by function $\tau(v) : \mathcal{V} \to \mathcal{A}$ and $\phi(e) : \mathcal{E} \to \mathcal{R}$, respectively. Here, we have only one type of node (i.e., textual node) and three types of edges, including "the parents of", "the siblings of" and "the cousins of". Hence, $|\mathcal{A}| = 1$ and $|\mathcal{R}| = 3$.

2.4   Heterogeneous Graph Initialization

The texts of node $v \in \mathcal{V}$ can be denoted as a sequence of words $\mathcal{S}_v = \{w_1, w_2, w_3, \cdots, w_{|\mathcal{S}_v|}\}$, and each word $w_i \in \mathcal{S}_v$ can be seen as a sequence of characters $\mathcal{S}_{w_i} = \{c_1, c_2, c_3, \cdots, c_{|\mathcal{S}_{w_i}|}\}$. We find that attribute values often have some morphological patterns at character level [14], which are helpful for node classification. For example, "yyyy-mm-dd" is a common pattern for *publication date* attribute in *book* vertical. To capture the useful morphological patterns, we encode the sequence of character embeddings of w_i using a convolutional neural network, and then get its character-level embedding $c^i \in \mathbb{R}^{d_c}$. To introduce richer external knowledge to assist the task, we retrieve the word-level embedding $g^i \in \mathbb{R}^{d_g}$ of w_i from the StanfordGloVE [17] word embeddings, and concatenate it with c^i to generate the final word embedding of w_i as follows:

$$h_{w_i}^v = \left[g^i; c^i\right] \tag{1}$$

where $h_{w_i}^v \in \mathbb{R}^{d_g + d_c}$. After that, we employ a bidirectional LSTM network to encode the word embedding sequence of node v and get its initial embedding by:

$$h_v^0 = \left[h_v^{forward}; h_v^{backward}\right] \tag{2}$$

where $h_v^0 \in \mathbb{R}^{d_t}$, $d_t = 2 * d_l$ and d_l is the dimension of the hidden layer of bidirectional LSTM network.

2.5   Relation-Aware Heterogeneous Graph Transformer

To learn fine-grained node representations, we exploit the Relation-aware Heterogeneous Graph Transformer to capture the influence of heterogeneous structural relations on node interactions. RHGT takes a heterogeneous graph $\mathcal{G}$ with node initial embeddings as the input, and outputs embeddings that contain contextual information. We denote the output of the $(l-1)$-th RHGT layer as $H^{(l-1)}$, which is also the input of the (l)-th layer. Initially, $H^{(0)} = \{h_1^0, h_2^0 \ldots h_{|\mathcal{V}|}^0\}$.

Given a target node $t \in \mathcal{V}$, its source node $s \in N_t$ and the edge $e = (s, t)$ between them, the operations of the (l)-th layer are as follows:

$$h_t^{(l)} \leftarrow \underset{\forall s \in N_t, \forall e \in E_{s,t}}{\textbf{Aggregate}} \left(\textbf{Attention}(s, e, t) \cdot \textbf{Message}(s)\right) \qquad (3)$$

where N_t and $E_{s,t}$ are the neighboring node set and neighboring egde set of t, respectively. In the above formulation, **Attention** estimates the importance of each source node, **Message** extracts the information of source nodes, and **Aggregate** aggregates messages from source nodes. Next we will describe their corresponding operations in detail.

Relation-Aware Attention. Each source node is connected to the target node by a specific relation, and different relations usually show different importance to the target node. Therefore, when measuring the importance of s, we consider the influence of relation $r = \phi(e)$. Specifically, we first project the target node t into a query vector $Q_t^{(l)}$ and the source node s into a key vector $K_s^{(l)}$ according to the following formulations:

$$Q_t^{(l)} = FC\text{-}Q^{(l)} \left(h_t^{(l-1)}\right) \qquad (4)$$

$$K_s^{(l)} = FC\text{-}K^{(l)} \left(h_s^{(l-1)}\right) \qquad (5)$$

where $FC\text{-}Q^{(l)} : \mathbb{R}^{d_t} \rightarrow \mathbb{R}^{d_k}$ and $FC\text{-}K^{(l)} : \mathbb{R}^{d_t} \rightarrow \mathbb{R}^{d_k}$ are linear projections, d_k is the dimension of query and key vectors. Then, we generate the representation of relation r and calculate its attention weight as follows:

$$H_{r,t}^{(l)} = \frac{1}{|N_t^r|} \sum_{s \in N_t^r} K_s^{(l)} \qquad (6)$$

$$W_{r,t}^{(l)} = \frac{\exp\left(\dfrac{Q_t^{(l)} \left[H_{r,t}^{(l)}\right]^T}{\sqrt{d_k}}\right)}{\displaystyle\sum_{r' \in \mathcal{R}} \exp\left(\dfrac{Q_t^{(l)} \cdot \left[H_{r',t}^{(l)}\right]^T}{\sqrt{d_k}}\right)} \qquad (7)$$

where the representation of relation r is calculated by averaging the key vectors of source nodes connected to t by relation r.

Next, we calculate the attention weight of s as follows:

$$W^{(l)}_{s,e,t} = \underset{\forall s \in N_t}{\text{softmax}} \left(\frac{Q^{(l)}_t \cdot W^{(l)}_{r,t} \cdot \left[K^{(l)}_s \right]^T}{\sqrt{d_k}} \right) \tag{8}$$

It can be seen that, in order to introduce the influence of relation r on node semantic interaction, we multiply the scaled dot product between $Q^{(l)}_t$ and $K^{(l)}_s$ by $W^{(l)}_{r,t}$ to calculate the attention score of s. And we normalize the attention scores of all source nodes in N_t to get the attention weight $W^{(l)}_{s,e,t}$ for each of them via the softmax function.

Message Passing. After obtaining the attention weights of all source nodes, we pass information from source nodes to target node. The message from s to t is calculated by:

$$V^{(l)}_s = FC\text{-}V^{(l)} \left(h^{(l-1)}_s \right) \tag{9}$$

where $FC\text{-}V^{(l)} : \mathbb{R}^{d_t} \rightarrow \mathbb{R}^{d_v}$ is a linear projection and d_v is the dimension of value vectors.

Target-Specific Aggregation. To obtain a rich representation for the target node, we aggregate the messages from its neighbors with attention weights, and get the updated embedding $\tilde{h}_t^{(l)}$ as:

$$\tilde{h}_t^{(l)} = \sigma \left(\sum_{s \in N_t} W^{(l)}_{s,e,t} \cdot V^{(l)}_s \right) \tag{10}$$

where $\sigma(\cdot)$ denotes the activation function. In order to make the training process more stable, we extend our relation-aware attention to multi-head attention. Specifically, we linearly project the queries, keys and values H times with different linear projections and perform the relation-aware attention in parallel. Therefore, the updated embedding can actually be calculated by:

$$\tilde{h}_t^{(l)} = \underset{head=1}{\overset{H}{\|}} \sigma \left(\sum_{s \in N_t} W^{(l)}_{s,e,t} \cdot V^{(l)}_s \right) \tag{11}$$

For each attention head, we set $d_k = d_v = \frac{d_t}{H}$, where d_t is the dimension of the initial embedding of each node.

Finally, we map the updated embedding back to the target feature distribution and conduct a residual connection [9] to get the output of t on (l)-th RHGT layer:

$$h_t^{(l)} = \sigma \left(FC\text{-}M^{(l)} \left(\tilde{h}_t^{(l)} \right) \right) + h_t^{(l-1)} \tag{12}$$

where $FC\text{-}M^{(l)} : \mathbb{R}^{d_t} \to \mathbb{R}^{d_t}$ is a linear projection. By stacking RHGT for L layers, we can get the final embedding $h_t^{(L)}$ for target node t.

2.6 Classification

To classify the attribute type of target node t, we connect its final node embedding to a multi-layer perceptron (MLP) as:

$$p = \text{MLP}(h_t^{(L)}) \tag{13}$$

where $p \in \mathbb{R}^{T+1}$, $T+1$ denotes the number of pre-defined attribute types (including none type). After that, we normalize p via the softmax function and select the attribute type with the highest score as the prediction $\hat{y}$:

$$s_i = \frac{\exp(p_i)}{\sum_{j=0}^{T} \exp(p_j)} \tag{14}$$

$$\hat{y} = \underset{i}{\arg\max}\, s_i \tag{15}$$

where $0 \leq i \leq T$. And we minimize the cross-entropy loss between the ground-truth y and the normalized probabilistic scores s over all variable nodes. The loss function is formalized as follows:

$$\mathcal{L} = -\sum_{v \in \mathcal{V}_{var}} \sum_{i=0}^{T} y_{v,i} \log s_{v,i} \tag{16}$$

where $\mathcal{V}_{var}$ denotes the set of variable nodes on the graph.

3 Experiments

3.1 Dataset

We evaluate DOM2R-Graph on the public SWDE [8] dataset. This dataset contains 8 verticals, each of which consists of 10 websites and has 3 to 5 attributes. The statistics of this dataset is shown in Table 1.

Fixing the order of all websites in each vertical, we take 10 cyclic permutations starting from the first website. Each time we select k ($1 \leq k \leq 5$) websites as the training set and the rest $10 - k$ website as the test set. Finally, 80 *"training set-test set"* pairs will be generated.

3.2 Evaluation Metrics

Following the evaluation metrics of [8,14,22], we calculate page-level F1 scores to evaluate the performance of DOM2R-Graph. Specifically, for each attribute, *precision* of a method is the number of pages for which the ground-truth attribute values are correctly extracted, called page hits, divided by the number of pages from which the method extracts values; while *recall* is the page hits divided by the number of pages containing ground-truth attribute values [8]. The page-level F1 score is the harmonic mean of *precision* and *recall*.

Table 1. The statistics of the public SWDE dataset.

Vertical	#Sites	#Pages	Fields
auto	10	17,923	model, price, engine, fuel economy
book	10	20,000	title, author, isbn, publisher, publish date
camera	10	5,258	model, price, manufacturer
job	10	20,000	title, company, location, date posted
movie	10	20,000	title, director, genre, mpaa rating
nbaplayer	10	4,405	name, team, height, weight
restaurant	10	20,000	name, address, phone, cuisine
university	10	16,705	name, phone, website, type

3.3   Implementation Details

In the stage of data pre-processing, we first use open-source LXML[1] library to process each page to get DOM tree structures. Then, we use a simple heuristic [14] to distinguish fixed nodes and variable nodes in the page. And we limit the text length of each node to no more than 5 words. When extracting the contextual nodes for each variable node, we limit the number of *parents* to no more than 1, the number of *siblings* to no more than 2, and the number of *cousins* to no more than 2. We set the size of word embedding d_g and the size of character embedding d_c as 100 when encoding the contents of DOM tree nodes. For the convolutional neural network, we use 50 filters and set kernel size as 3. And we set d_l, the hidden layer size of LSTM network as 64. For RHGT, we set the number of attention heads as 2 and stack two layers of RHGT.

3.4   Baseline Models

We compare the proposed DOM2R-Graph with five state-of-the-art methods, including SSM, Render-Full, FreeDOM-NL, FreeDOM-Full and SimpDOM.

- **SSM** [2]: The Stacked Skews Model (SSM) aligns the unseen web pages with seed web pages through handcrafted features and a tree alignment algorithm.
- **Render-Full** [8]: Render-Full utilizes the attribute-specific semantics and inter-attribute layout relationships to make the model can deal with other unseen sites in the same vertical after learning from seed sites.
- **FreeDOM-NL** [14]: FreeDOM-NL utilizes local text information and the markup information of HTML to learn node representations for classification.
- **FreeDOM-Full** [14]: FreeDOM-Full is a two-stage model. The FreeDOM-NL model in the first stage learns local representations of nodes. And the relational neural network in the second stage captures the distance and semantic correlation between node pairs.
- **SimpDOM** [22]: SimpDOM is a unified model which concatenates the textual features of DOM tree nodes and their contexts to obtain richer node representations.

---

[1] https://lxml.de/.

3.5   Overall Performance

Table 2. The F1 score comparison results (%) on the public SWDE dataset.

Model	#Seed Sites				
	$k = 1$	$k = 2$	$k = 3$	$k = 4$	$k = 5$
SSM	63.00	64.50	69.20	71.90	74.10
Render-Full	**84.30**	86.00	86.80	88.40	88.60
FreeDOM-NL	72.52	81.33	86.44	88.55	90.28
FreeDOM-Full	82.32	86.36	90.49	91.29	92.56
SimpDOM	83.06	88.96	91.63	92.84	93.75
DOM2R-Graph	83.26	**89.97**	**93.19**	**94.47**	**95.30**

Table 2 reports the comparison results between DOM2R-Graph and five baseline methods using different numbers of seed websites. We can see that when $k = 1$, DOM2R-Graph performs slightly worse than Render-Full, but when $k \geq 2$, DOM2R-Graph outperforms it significantly. This is because when the training data is extremely little, the visual features and rules carefully crafted by humans can capture more effective information. But as k increases, our neural architecture can quickly learn the more generalizable semantic and structural features.

Compared with the state-of-the-art method SimpDOM, DOM2R-Graph always maintains a significant advantage. We attribute the performance gain to two designs: (1) the heterogeneous graph constructed from the simplified DOM tree since it can preserve both node semantic information and DOM tree structure information, which are both crucial for inferring the attribute type of each node; (2) the Relation-aware Heterogeneous Graph Transformer as it considers the influence of structural relations on node semantic interactions to learn more fine-grained node representations.

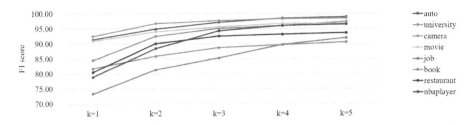

Fig. 3. The detailed performance (%) of DOM2R-Graph on the public SWDE dataset.

Furthermore, from Fig. 3, we can see that with the increase of training data, the overall performance is gradually improved, but the rate of performance improvement gradually levels off. The reason is that DOM2R-Graph has learned stable and effective knowledge from existing data, so continuing to increase the data can only bring a small improvement.

3.6   Ablation Study

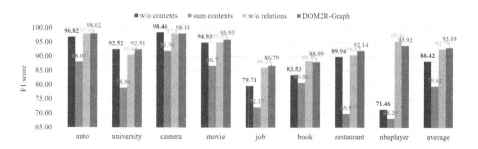

Fig. 4. The ablation study results (%) on the public SWDE dataset.

To verify the effectiveness of each design of DOM2R-Graph, we conduct ablation studies in this section. Specifically, we design tree variants: (1) *w/o contexts* predicts node attribute type using the texts of the node itself; (2) *sum contexts* obtains node representations by summing contexts; (3) *w/o relations* removes the relation-aware module, using a homogeneous graph transformer to learn node representations at semantic level. The ablation study results when using three seed websites as training set in each vertical (i.e., $k = 3$) are shown in Fig. 4. And similar results can be achieved with other k.

From Fig. 4, we can see that *sum contexts* performs much worse than *w/o contexts*. This is because it ignores the different contributions of contexts caused by the semantics of node texts, so weakly related contexts and noise will degrade the performance. Compared with *sum contexts*, *w/o relations* lifts up the average F1 score from 79.61% to 92.79%. The reason is that our node semantic interaction module can make useful contexts (such as the description words in front of attribute values) play a greater role and mitigate noise. The full architecture with RHGT further lifts up the average F1 score of *w/o relations* from 92.79% to 93.19%. This shows that the heterogeneous structural relations can strengthen the semantic interaction between nodes for better performance.

4   Related Work

4.1   Web Information Extraction

Early research of web information extraction mainly focus on wrapper induction [1,5,13]. They use a large number of manual rules and labels to generate extraction templates. However, these templates are only valid for specific websites.

To solve this problem, many recent studies [7,8,12,14,22] formulate web attribute extraction as a multi-class classification task of DOM tree nodes. Specifically, they use generalizable features such as visual features and DOM tree features to predict the attribute type of each DOM tree node. However, they ignore the heterogeneous structure information between nodes and their contexts, which is useful for inferring node semantics.

4.2 Heterogeneous Graph Neural Networks

Due to the high demand for processing heterogeneous data, many heterogeneous graph neural networks [10,11,20] have emerged in recent years to mine complex structure information and rich semantic information. Wang et al. [20] propose a hierarchical attention-based heterogeneous graph neural network, including node-level and semantic-level attentions. Hu et al. [10] propose a heterogeneous graph transformer architecture which maintains dedicated representations for different types of nodes and edges. Ji et al. [11] propose a heterogeneous graph propagation network to alleviate the semantic confusion. Inspired by the above works, we use heterogeneous graphs to model node texts and the structural relations between DOM tree nodes.

5   Conclusion

In this paper, we propose a novel web attribute extraction architecture, called DOM2R-Graph, which utilizes both node semantics and DOM tree structure to infer node attribute type. DOM2R-Graph first simplifies the DOM tree by retrieving useful contexts for nodes, and then constructs a heterogeneous graph based on the simplified DOM tree to preserve node texts and the structural relations between nodes. Furthermore, a Relation-aware Heterogeneous Graph Transformer is applied to learn fine-grained node representations by considering the influence of relations on node semantic interactions. Extensive experiments demonstrate the effectiveness of DOM2R-Graph.

References

1. Azir, M.A.B.M., Ahmad, K.B.: Wrapper approaches for web data extraction: a review. In: 2017 6th International Conference on Electrical Engineering and Informatics (ICEEI), pp. 1–6. IEEE (2017)
2. Carlson, A., Schafer, C.: Bootstrapping information extraction from semi-structured web pages. In: Daelemans, W., Goethals, B., Morik, K. (eds.) ECML PKDD 2008. LNCS (LNAI), vol. 5211, pp. 195–210. Springer, Heidelberg (2008). https://doi.org/10.1007/978-3-540-87479-9_31
3. Chen, L., et al.: WebSRC: a dataset for web-based structural reading comprehension. arXiv preprint arXiv:2101.09465 (2021)
4. Cui, W., Xiao, Y., Wang, H., Song, Y., Hwang, S.W., Wang, W.: KBQA: learning question answering over QA corpora and knowledge bases. arXiv preprint arXiv:1903.02419 (2019)

5. Dalvi, N., Kumar, R., Soliman, M.: Automatic wrappers for large scale web extraction. arXiv preprint arXiv:1103.2406 (2011)
6. Dong, X., et al.: Knowledge vault: a web-scale approach to probabilistic knowledge fusion. In: Proceedings of the 20th ACM SIGKDD International Conference on Knowledge Discovery and Data Mining, pp. 601–610 (2014)
7. Gogar, T., Hubacek, O., Sedivy, J.: Deep neural networks for web page information extraction. In: Iliadis, L., Maglogiannis, I. (eds.) AIAI 2016. IAICT, vol. 475, pp. 154–163. Springer, Cham (2016). https://doi.org/10.1007/978-3-319-44944-9_14
8. Hao, Q., Cai, R., Pang, Y., Zhang, L.: From one tree to a forest: a unified solution for structured web data extraction. In: Proceedings of the 34th International ACM SIGIR Conference on Research and Development in Information Retrieval, pp. 775–784 (2011)
9. He, K., Zhang, X., Ren, S., Sun, J.: Deep residual learning for image recognition. In: Proceedings of the IEEE Conference on Computer Vision and Pattern Recognition, pp. 770–778 (2016)
10. Hu, Z., Dong, Y., Wang, K., Sun, Y.: Heterogeneous graph transformer. In: Proceedings of the Web Conference 2020, pp. 2704–2710 (2020)
11. Ji, H., Wang, X., Shi, C., Wang, B., Yu, P.: Heterogeneous graph propagation network. IEEE Trans. Knowl. Data Eng. (2021)
12. Kumar, A., Morabia, K., Wang, J., Chang, K.C.C., Schwing, A.: Cova: context-aware visual attention for webpage information extraction. arXiv preprint arXiv:2110.12320 (2021)
13. Kushmerick, N.: Wrapper induction for information extraction. University of Washington (1997)
14. Lin, B.Y., Sheng, Y., Vo, N., Tata, S.: Freedom: a transferable neural architecture for structured information extraction on web documents. In: Proceedings of the 26th ACM SIGKDD International Conference on Knowledge Discovery and Data Mining, pp. 1092–1102 (2020)
15. Ling, X., Weld, D.S.: Fine-grained entity recognition. In: Twenty-Sixth AAAI Conference on Artificial Intelligence (2012)
16. Lockard, C., Dong, X.L., Einolghozati, A., Shiralkar, P.: Ceres: distantly supervised relation extraction from the semi-structured web. arXiv preprint arXiv:1804.04635 (2018)
17. Pennington, J., Socher, R., Manning, C.D.: GloVe: global vectors for word representation. In: Proceedings of the 2014 Conference on Empirical Methods in Natural Language Processing (EMNLP), pp. 1532–1543 (2014)
18. Sneha, Y., Mahadevan, G., Prakash, M.: A personalized product based recommendation system using web usage mining and semantic web. Int. J. Comput. Theory Eng. **4**(2), 202 (2012)
19. Wang, H., Zhang, F., Zhao, M., Li, W., Xie, X., Guo, M.: Multi-task feature learning for knowledge graph enhanced recommendation. In: The World Wide Web Conference, pp. 2000–2010 (2019)
20. Wang, X., et al.: Heterogeneous graph attention network. In: The World Wide Web Conference, pp. 2022–2032 (2019)
21. Wu, S., et al.: Fonduer: knowledge base construction from richly formatted data. In: Proceedings of the 2018 International Conference on Management of Data, pp. 1301–1316 (2018)
22. Zhou, Y., Sheng, Y., Vo, N., Edmonds, N., Tata, S.: Simplified DOM trees for transferable attribute extraction from the web. arXiv preprint arXiv:2101.02415 (2021)

Sparse Linear Capsules for Matrix Factorization-Based Collaborative Filtering

Xuan Li and Li Zhang[(⊠)]

School of Software, Tsinghua University, Beijing 100084, China
xuan-li15@mails.tsinghua.edu.cn, lizhang@tsinghua.edu.cn

Abstract. Collaborative filtering (CF) plays a key role in recommender systems, which consists of two basic disciplines: neighborhood methods and latent factor models. Neighborhood methods are most effective at capturing the very localized structure of a given rating matrix, while latent factor models are generally effective at capturing its global structure. However, two disciplines fail to capture these two structures simultaneously. Motivated by the sparse linear methods and the recently developed capsule networks, we propose a new matrix factorization model for collaborative filtering based on sparse linear capsule networks, which attempts to embed the neighborhood information into latent factors and finally get the very localized and global structure of a given rating matrix. Experiments on real-world datasets demonstrate that our model outperforms seven state-of-the-art matrix factorization-based CF methods in terms of rating prediction accuracy.

Keywords: Matrix Factorization · Collaborative Filtering · Capsule Networks

1 Introduction

Collaborative filtering (CF) plays a fundamental role in recommender system applications, which can be divided into two basic categories: neighborhood-based methods and latent factor models. Neighborhood-based methods are centered on exploring the neighborhood relationships between entities (users or items) and are most effective at capturing very localized structure of a sparse rating matrix. With low-rank assumption, latent factor models are centered on detecting the totality of week signals embedded in a sparse rating matrix and are generally effective at capturing its global structure, among which matrix factorization (MF) methods are the most successful ones. Both the two categories can not capture the local and global structure of a rating matrix simultaneously, while this is significantly important for more accurate predictions [9].

Recently, some clustering-based MF methods, *e.g.*, GLOMA [1] *etc.*, attempt to capture both local and global structure of a given rating matrix. They adopt

M. Tanveer et al. (Eds.): ICONIP 2022, LNCS 13623, pp. 480–491, 2023.
https://doi.org/10.1007/978-3-031-30105-6_40

clustering methods to divide the original rating matrix into submatrices, and then capture the local structure from these submatrices by MF method. However, the local structure captured by such way is in essence the global structure of these submatrices, which implies that the local structure of these submatrices cannot be captured effectively. In other words, these clustering-based MF-based CF methods cannot succeed in capturing the very localized relationships underlying the original rating matrix, while neighborhood-based methods can do best.

Since sparse linear methods (SLIM [18]) is an effective neighborhood learning method and the capsule networks can capture part-whole spatial relationships, we propose a new matrix factorization model called **S**parse **L**inear **Caps**ules **M**atrix **F**actorization (SLICapsMF), which attempts to ensure that the very localized structure of a rating matrix can be captured effectively. SLICapsMF tries to embed neighborhood information into latent factors by sparse linear capsules and finally get latent factors capturing both the local and global structure of the matrix. Experiments on real-world datasets demonstrate that our approach outperforms seven state-of-art MF-based CF methods in terms of rating prediction accuracy.

2 Related Work

2.1 Matrix Factorization

Matrix factorization (MF) and its extensions are among the most successful latent factor models in collaborative filtering. MF maps both users and items into a joint low-rank latent factor space to discover the underlying structure of the user-item rating matrix [10]. Indeed, it factorizes the user-item rating matrix R into the user latent factor matrix P and the item latent factor matrix Q. Accordingly, a vector p_i represents the latent factor vector of user i in joint latent factor space, and a vector q_j represents item j in the same space. To predict the unobserved user-item rating, matrix factorization computes $\hat{r}_{ij} = p_i q_j^T$.

The challenging problem is how to map users and items into the joint low-rank latent factor space. In collaborative filtering setting, the user-item rating matrix is highly incomplete. It is undefined to factorize the matrix by conventional SVD. Moreover, it is highly prone to overfitting for the relatively few observed entries in the matrix. It was proposed to fit directly observed entries in the matrix and to avoid overfitting through some regularizations [22]. To learn the latent factor vectors for users and items, the model fits the observed entries in the matrix by minimizing the regularized objective function:

$$\min_{P^*, Q^*} \sum_{(i,j) \in \mathcal{K}} [(r_{ij} - p_i q_j^T)^2 + \lambda(||p_i||^2 + ||q_j||^2)], \tag{1}$$

where $\mathcal{K}$ is the set of the (i, j) pairs that r_{ij} is observed in the training set. The regularizations penalize the magnitudes of the latent factor vectors. It plays a core role in the basic matrix factorization model. Probabilistic matrix factorization(PMF) [16] gives a probabilistic explanation of this regularization. The hyperparameter λ controls the extent of the regularizations. Usually, it is determined through cross-validation. The hyperparameters can also be determined automatically [20]. One can solve the minimum problem by the gradient descent methods [22].

Another challenging problem is that mapping users and items into the joint low-rank latent factor space with preserving the local structure of the user-item rating matrix. Weighted low-rank approximations [21] is a matrix factorization model, without regularization penalties, which attempts to capture the underlying structure by weighting the reconstruction error: $w_{ij}(r_{ij} - \boldsymbol{p}_i \boldsymbol{q}_j^T)^2$. [9] proposed a combined model that integrates both the neighborhood-based methods and the latent factor models to capitalize their advantages for more accurate predictions. [10] tried to incorporate the information of entities (users or items) into matrix factorization. With the assumption that the user-item rating matrix is locally low-rank, [2,11,14] divided the matrix completion problem into subproblems, solved these subproblems locally and combined the results to approximate the matrix. Instead of the inner product, [4] replaced the inner product with the function learned by neural network to construct the rating matrix. To capture more information of the matrix, deep learning techniques can be incorporated into matrix completion [24].

Different from these models aforementioned, SLICapsMF attempts to preserve both the local information and the global information about the underlying structure directly. SLICapsMF relies on matrix factorization to preserve global structure of the matrix and preserves the local structure by sparse linear capsules.

2.2 Sparse Linear Methods

SLIM is an effective neighborhood-learning method, which learns neighborhood relationships automatically from a sparse rating matrix $R = \{\boldsymbol{r}_1^I, \boldsymbol{r}_2^I, ..., \boldsymbol{r}_N^I\}$ [17]. In SLIM, neighborhood relationships for items can be learned by solving the following minimum optimization problem:

$$\min_{\boldsymbol{W}} \sum_j ||\boldsymbol{r}_j^I - \sum_k w_{kj} \boldsymbol{r}_k^I||^2 + \lambda_2 ||\boldsymbol{w}_j||_2^2 + \lambda_1 ||\boldsymbol{w}_j||_1^2 \tag{2}$$

$$s.t.\ \boldsymbol{w}_j \geq 0,\ w_{jj} = 0,\ for\ all\ j,$$

where $\boldsymbol{r}_j^I$ and $\boldsymbol{r}_k^I$ denote the rating vectors of item j and k respectively, and w_{kj} denotes the neighborhood relationship between item k and j. If w_{kj} is learned to be non-zero, it represents item k in item j's neighborhood and item k's contribution to the construction of item j's rating vector; otherwise, item k is not in item j's neighborhood. L_1 regularization is introduced to learn a sparse $\boldsymbol{W}$ and L_2 together prevents overfitting.

2.3 Capsule Networks

A capsule is a group of neurons that represent different properties of an entity [6]. Capsule Networks consist of different level layers, each of which contains

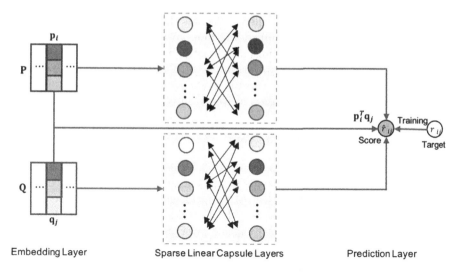

Fig. 1. An illustration of SLICapsMF model.

many capsules. Given the collection of representations of capsule i in layer l $\{\boldsymbol{u}_i \in \mathbb{R}^d | i = 1, 2, \cdots, N_l\}$. To get the presentations of capsule j in layer $(l+1)$ $\boldsymbol{v}_j$, the collection of representation votes of capsule i in layer l $\{\hat{\boldsymbol{u}}_{j|i} = \boldsymbol{W}_{ij}\boldsymbol{u}_i \in \mathbb{R}^D | i = 1, 2, \cdots, N_l\}$ are computed first, where $\boldsymbol{W}_{ij} \in \mathbb{R}^{D \times d}$ is a learnable transformation matrix that connects the capsule i in layer l to the capsule j in layer $(l+1)$. Subsequently, the representations of capsule j in layer $(l+1)$ is computed as a weighted sum of all votes: $\boldsymbol{v}_j = \sum_i c_{ij}\hat{\boldsymbol{u}}_{j|i}$, where the coupling coefficients c_{ij} is determined through a routing algorithm.

It was pointed out that the typical routing algorithms are in essence clustering algorithms to seek similar representations for high-level capsules [15], such as Dynamic Routing [19] and EM Routing [7]. To address high time consuming problem raised by the iterative process, some alternative strategies were proposed such as self-routing [5], ladder layer [8] and receptor skeleton [3].

3 Sparse Linear Capsules Matrix Factorization

As shown in Fig. 1, the proposed SLICapsMF model consists of three components: embedding layer, sparse linear capsule layers and prediction layer.

3.1 Embedding Layer

Given the user-item rating matrix $\boldsymbol{R} \in \mathbb{R}^{M \times N}$. Rows of the rating matrix represent user rating vectors $\{\boldsymbol{r}_1^U, \boldsymbol{r}_2^U, ..., \boldsymbol{r}_M^U\}$ and columns represent item rating vectors $\{\boldsymbol{r}_1^I, \boldsymbol{r}_2^I, ..., \boldsymbol{r}_N^I\}$. Matrix Factorization factorizes the rating matrix into the user latent factor matrix $\boldsymbol{P} \in \mathbb{R}^{d \times M}$ and item latent factor matrix $\boldsymbol{Q} \in \mathbb{R}^{d \times N}$. Specifically, a user i is described with an embedding vector $\boldsymbol{p}_i \in \mathbb{R}^d$ and similarly an item j is mapped to $\boldsymbol{q}_j \in \mathbb{R}^d$, where d denotes the embedding size. Therefore, $\boldsymbol{P}$ and $\boldsymbol{Q}$ can be seen as two embedding look-up tables. For the basic matrix factorization model, the embeddings are directly fed into the prediction layer to achieve the prediction score. However, in our SLICapsMF model, we feed these embeddings into the sparse linear capsule layers to embed neighbor-entity spatial relationships between entities (users or items) into latent factors.

3.2   Sparse Linear Capsule Layers

As shown in Fig. 1, we build capsule networks for users and items respectively in our model. Each capsule corresponds to one entity, the representation of which is d-dimensional. For example, the i-th capsule in each layer in user capsule network corresponds to the i-th user. Embedding layer propagates the initial embeddings to the low-level (first) capsule layer. Subsequently, we can get the representation of the i-th low-level capsule: $\boldsymbol{u}_i = \boldsymbol{p}_i$ in the user capsule network and $\boldsymbol{u}_i = \boldsymbol{q}_i$ in the item capsule network.

To synthesize the representation of the j-th high-level capsule, $\boldsymbol{v}_j$, we propose a new approach inspired by sparse linear methods. The premise behind latent factor models is that there are only a small number of factors influencing ratings given by users on items. With this premise, we assume that neighborhood relationships in rating vectors originate from their latent factors. Thus, the learned latent factors should encompass such neighborhood information. We also assume that learning latent factors and their neighborhood relationships simultaneously can embed neighborhood information into latent factors. SLIM is an effective neighborhood-learning method, which can learn neighborhood relationships from rating vectors automatically [18]. Inspired by these observations, we propose the sparse linear routing algorithm for entities (users or items) as follows:

We first adopt K nearest neighbors (KNN) method to build neighbor set $\mathcal{N}_j$ for entity j from entity rating vectors. Specifically, we use the *BallTree* algorithm to search K nearest neighbors as measured by cosine similarity:

$$cosine(j, i) = \frac{\boldsymbol{r}_j^T \boldsymbol{r}_i}{||\boldsymbol{r}_j|| \cdot ||\boldsymbol{r}_i||}, \tag{3}$$

where $\boldsymbol{r}_i$ and $\boldsymbol{r}_j$ denotes the rating vectors for entity i and j respectively (i.e., $\boldsymbol{r}_i = \boldsymbol{r}_i^U$, $\boldsymbol{r}_j = \boldsymbol{r}_j^U$ for users; $\boldsymbol{r}_i = \boldsymbol{r}_i^I$, $\boldsymbol{r}_j = \boldsymbol{r}_j^I$ for items). Then, the representation of j-th high-level capsule $\boldsymbol{v}_j$ can be computed by a weighted sum of these K neighbors from the outputs of the low-level capsule layer:

$$\boldsymbol{v}_j = \sum_{k \in \mathcal{N}_j} w_{kj} \boldsymbol{u}_k, \tag{4}$$

where $\boldsymbol{u}_k$ denotes the representation of the k-th capsule in low-level layer and w_{kj} denotes the neighborhood relationship to be learned between entity k and j.

The complete procedure of the sparse linear routing algorithm can be given in Algorithm 1. We may encounter missing data problem due to the sparsity of

---

Algorithm 1. Sparse Linear Routing Algorithm

1: Input: The representations of all the low-level capsules, $\{u_i|i \in \{1, 2, \cdots, A\}\}(A = M$ for users and $A = N$ for items); the rating vectors for all the corresponding entities, $\{r_i|i \in \{1, 2, \cdots, A\}\}$ $(r_i = r_i^U$ for users and $r_i = r_i^I$ for items); the neighborhood size, K.
2: Output: The j-th high-level capsule's representation v_j.
3: Compute the similarities between the entity j and all the other entities by Eq. (3).
4: Find the K nearest neighborhood set $\mathcal{N}_j$ for the entity j.
5: Combine the K nearest neighboring representations of the low-level capsules u_k $(k \in \mathcal{N}_j)$ with the learnable weight coefficients w_j to obtain the j-th high-level capsule's representation v_j by Eq. (4).

---

the rating matrix. The missing data problem can be solved by completing the training matrix with default values, e.g., the middle value of the rating range, etc. In this paper, we complete the training matrix with 0. The implementation of KNN follows the corresponding algorithm of Scikit-learn[1]. Note that KNN steps only need to be run once due to the fixed rating vectors and can be run in advance before training.

3.3 Prediction Layer

After propagating through the sparse linear capsule layers, we obtain the reconstruction embeddings containing the neighbor-entity spatial relationships in the rating matrix. Specifically, the i-th high-level user capsule's representation $\hat{p}_i$ is the reconstructed embedding for user i. Similarly, the j-th high-level item capsule's representation $\hat{q}_j$ is the reconstructed embedding for item j. Considering the excellent performance of the basic matrix factorization, we adopt the inner product of the embeddings of user i and item j to make the final prediction: $\hat{r}_{ij} = p_i^T q_j$. Training is performed by minimizing the point-wise loss between the score $\hat{r}_{ij}$ and its target value r_{ij} and also conducting regularization on the embeddings. To encourage neighborhood preserving embeddings, we add additional reconstruction loss on users embeddings and items embeddings respectively:

$$
\begin{aligned}
loss = &\sum_{(i,j)\in\mathcal{K}} [(r_{ij} - p_i^T q_j)^2 + \lambda(||p_i||^2 + ||q_j||^2)] \\
&+ \lambda_U \sum_i ||p_i - \hat{p}_i||^2 + \lambda_I \sum_j ||q_j - \hat{q}_j||^2,
\end{aligned}
\tag{5}
$$

where $\mathcal{K}$ denotes the set of entries in the training set, p_i and $\hat{p}_i$ denote the initial and reconstruction embeddings for user i, q_j and $\hat{q}_j$ denote the initial and reconstruction embeddings for item j. λ_U and λ_I control the extent of the sparse linear reconstruction for users and items respectively. We utilize the mini-batch gradient descent method to minimize the loss function.

---

[1] https://scikit-learn.org/.

4 Experiment

Experiments are conducted on the following four real-world datasets: Movie-Lens[2] 100K, 1M, 10M and Netflix, which contain 10^5, 10^6, 10^7 and 10^8 ratings respectively. The statistics of datasets are shown in Table 1. For the MovieLens 100K, we use the canonical five splits provided by GroupLens. For all rating prediction accuracy comparison, we randomly split the other three datasets into train-test sets by 90%/10%. All the experimental results are reported by the average of results on five different splits. We adopt the commonly used metric Root Mean Square Error(RMSE), which is computed as $RMSE = \sqrt{\sum_{(i,j)\in\mathcal{T}}(r_{ij} - \hat{r}_{ij})^2/|\mathcal{T}|}$, where $\mathcal{T}$ represents the indices of the observed entities in the test set.

Table 1. Statistics for datasets.

Dataset	Users	Items	Ratings	Density	Levels
MovieLens 100K	943	1,682	100,000	6.30%	1,2,...,5
MovieLens 1M	6,040	3,706	1,000,209	4.47%	1,2,...,5
MovieLens 10M	69,878	10,677	10,000,054	1.34%	0.5,1,...,5
Netflix	480,189	17,771	100,480,507	1.18%	1,2,...,5

4.1 Ablation Analysis

To verify the effectiveness of the individual components of our proposed SLI-CapsMF model, we conduct an ablation analysis on MovieLens 100K and 1M datasets.

Experimental Setting. In CF settings, there are two fundamentally different entities: users and items, which results in two capsule networks in SLICapsMF: one is applied on users and another applied on items. There are four variant models with different parameter settings: 1) SLICapsMF-U denotes the SLICapsMF model by setting $\lambda_U \neq 0$ and $\lambda_I = 0$, which only embeds the neighborhood information of users into the user latent factors; 2) SLICapsMF-I denotes the SLICapsMF model by setting $\lambda_U = 0$ and $\lambda_I \neq 0$, which only embeds the neighborhood information of items into the item latent factors; 3) SLICapsMF-UI denotes the normal SLICapsMF model by setting $\lambda_U \neq 0$ and $\lambda_I \neq 0$, which can embed both the user and item neighborhood information into the corresponding latent factors respectively; 4) SLICaspMF-None denotes the SLICapsMF model by setting $\lambda_U = 0$ and $\lambda_I = 0$, which does not embed any neighborhood information into the latent factors. It is in essence the basic matrix factorization (BasicMF).

---

[2] https://grouplens.org/datasets/movielens/.

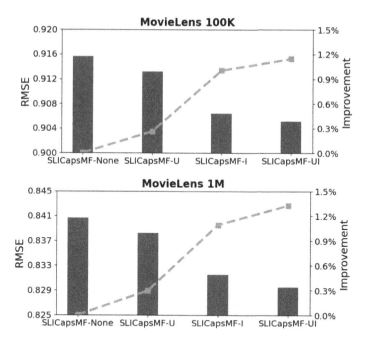

Fig. 2. RMSE comparison results on two benchmarks of MovieLens datasets.

The configuration of the experiments is as follows. The embedding size (rank) d is set to 64 for all models and the regularization parameter λ is set to 0.05 for MovieLens 100K and 0.02 for MovieLens 1M. The learning rate is set to 0.001. The batch size is set to 10K, 20K for MovieLens 100K and 1M respectively. For the three variant SLICapsMF models with capsule networks, the neighborhood size K is set to 6, and the reconstruction parameters (λ_U and λ_I) are chosen among $\{0.01, 0.02, 0.05, 0.1, 0.2, 0.5, 1, 2, 5\}$.

Experimental Results. As is shown in Fig. 2, experimental results validate the effectiveness of the sparse linear capsule layers: first, both SLICapsMF-I and SLICapsMF-U outperform SLICaspMF-None (BasicMF) in terms of rating prediction accuracy on both datasets; second, SLICapsMF-I is more effective than SLICapsMF-U, since SLICapsMF-I outperforms SLICaspMF-None by 1.0% and 1.1% respectively on MoveLens 100K and 1M, while SLICapsMF-U outperforms by 0.3% and 0.3% correspondingly; third, SLICaspMF-UI outperforms SLICaspMF-None by 1.1% and 1.3%. We can draw the following conclusions: applying the sparse linear capsules on items is more effective than that on users, likely since the average number of ratings of each item is more than those of user, and thus neighborhood information of items is more accurate than that of users; applying sparse linear capsules on both entities becomes more effective with the sparsity of the rating matrix increasing, since it can capture different localized structure of a rating matrix.

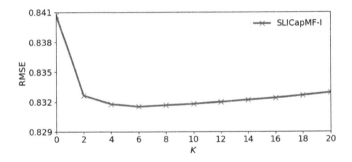

Fig. 3. Impact of neighborhood size on MovieLens 1M.

Time Complexity Analysis. The time complexity of the neighbor set construction is $O(N \log(K)M \log(M))$ for users and $O(M \log(K)N \log(N))$ for items, where M is the number of users and N is the number of items. It can be discovered that SLICapsMF-U consumes more time than SLICapsMF-I since M is usually larger than N in CF settings, e.g., MovieLens 1M, MovieLens 10M, Netflix, etc. Though SLICapsMF-UI performs best among the three variant capsule models, it consumes more time in all. With a trade off between the performance and efficiency, we only use SLICapsMF-I model in the following experiments.

4.2 Sensitive Analysis

In this section, we analyze the impact of two important parameters: neighborhood size K and the embedding size d.

Impact of the Neighborhood Size. The proposed SLICapsMF algorithm first searches K nearest neighbors for each entity, where K will control the complexity of the neighborhood information. When K is set to 0, SLICapsMF degrades to the BasicMF and no localized structure can be captured. We vary K in $\{0, 2, 4, 6, 8, 10, 12, 14, 16, 18, 20\}$ and evaluate the rating prediction performance on MovieLens 1M dataset. The other parameters are same as in Sect. 4.1. The results of SLICapsMF-I models with different neighborhood size K are shown in Fig. 3. As we can see from it, the RMSE of SLICapsMF-I drop off significantly when K increases from 0 to 2, and then its downward trend slows down and gets the minimum at around 6. When K exceeds 6, the performance get worse, because the unnecessary latent factor vectors may introduce confusion in delivering the necessary information as we mentioned in Sect. 3.

Impact of the Embedding Size. We evaluate the impact of the embedding size (rank) on MovieLens 10M with d varying from 50 to 600. Experimental results in Fig. 4 show that the embedding size has an important impact on the performance of SLICapsMF-I. We observe that the rate prediction accuracy is

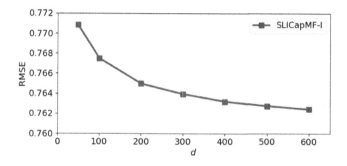

Fig. 4. Impact of embedding size on MovieLens 10M.

significantly improved with the increase of the embedding size. However, time consuming will increase at the same time. It is a tradeoff to determine the proper embedding size between the time complexity and result accuracy.

4.3 Rate Prediction Accuracy Comparison

In this section, we compare the rate prediction accuracy of SLICapsMF on two large datasets (MovieLens 10M and Netflix) with the state-of-the-art MF-based CF methods.

Baseline Methods. 1) DFC [14]: divides an user-item rating matrix into submatrices, and factorizes each submatrix in parallel, and combines submatrix estimates for matrix completion; 2) LLORMA [11]: reconstructs the rating matrix with a weighted sum of low-rank matrices, with assumption that the matrix is low-rank at local level instead of global level; 3) GSMF [23]: incorporates multiple behaviors into matrix facatorization by group sparsity regularization; 4) WEMAREC [2]: a weighted and ensemble matrix approximation method for recommendation; 5) SMA [13]: a stable matrix approximation that can achieve better generalization performance; 6) GLOMA [1]: a clustering-based matrix approximate methods which employs clustering techniques to capture global associations and local associations among users or items; 7) NORMA [12]: proposes an adaptive weighting strategy to decrease learning steps on noisy ratings.

Experimental Settings. The configuration of this experiment is as follows: the neighborhood size K is set to 6; λ_I is set to 0.05 for MovieLens 10M and 0.1 for Netflix; λ is set to 0.02; the learning rate is set to 0.001; the batch size is set to 20K and 100K for MovieLens 10M and Netflix respectively; the embedding size d is set to 500. The parameters of all the baselines are adopted from the original papers because they are all evaluated on the same datasets.

Experimental Results. The comparison results on MovieLens 10M and Netflix are shown in Table 2. The best results are in bold faces. We can observe

Table 2. RMSE comparison results between SLICapsMF and seven state-of-the-art MF-based methods on two large datasets. Note that our model statistically outperforms the other baselines with at least 95% confidence level.

Method	MovieLens 10M	Netflix
DFC [14]	0.8067 ± 0.0002	0.8453 ± 0.0003
LLORMA [11]	0.7855 ± 0.0002	0.8275 ± 0.0004
GSMF [23]	0.8012 ± 0.0011	0.8420 ± 0.0006
WEMAREC [2]	0.7775 ± 0.0007	0.8143 ± 0.0001
SMA [13]	0.7682 ± 0.0003	0.8036 ± 0.0004
GLOMA [1]	0.7672 ± 0.0001	0.8011 ± 0.0003
NORMA [12]	0.7641 ± 0.0008	0.7986 ± 0.0002
SLICapsMF-I	$\mathbf{0.7628 \pm 0.0008}$	$\mathbf{0.7973 \pm 0.0002}$

that SLICapsMF-I statistically outperforms seven state-of-the-art MF-based CF methods with at least 95% confidence level, since SLICapsMF-I can capture the global structure of a given rating matrix and its very localized structure based on sparse linear capsules simultaneously. For the GLOMA method, it divides the original rating matrix into entity submatrices first and then learn local structure of the original rating matrix from these submatrices by MF method. However the local structure captured by such way is in essence the global structure of these submatrices. Compared with these models, SLICapsMF can capture more effective localized structure of the rating matrix with the sparse linear item capsule layers.

5   Conclusion

In this paper, we proposed a novel matrix factorization model for collaborative filtering based on sparse linear capsules. Experimental studies on real-world datasets have validated the effectiveness of applying the sparse linear capsule networks on different entities and demonstrate that only embedding the neighborhood information of items into the item latent factors (SLICapsMF-I) outperforms seven state of-the-art MF-based CF methods. Sparse linear capsule networks can be constructed easily from a given rating matrix without additional information, and can also be employed by other MF-based CF methods since its idea is orthogonal to those methods.

Acknowledgement. The work described in this paper was supported by the National Key Research and Development Program of China (No. 2019YFB1707001) and the National Natural Science Foundation of China (Grant No. 62021002).

References

1. Chen, C., Li, D., Lv, Q., Yan, J., Shang, L., Chu, S.M.: Gloma: embedding global information in local matrix approximation models for collaborative filtering. In: Thirty-First AAAI Conference on Artificial Intelligence (2017)
2. Chen, C., Li, D., Zhao, Y., Lv, Q., Shang, L.: WEMAREC: accurate and scalable recommendation through weighted and ensemble matrix approximation. In: SIGIR, pp. 303–312 (2015)
3. Chen, J., Yu, H., Qian, C., Chen, D.Z., Wu, J.: A receptor skeleton for capsule neural networks. In: ICML (2021)
4. Dziugaite, G.K., Roy, D.M.: Neural network matrix factorization. arXiv preprint arXiv:1511.06443 (2015)
5. Hahn, T., Pyeon, M., Kim, G.: Self-routing capsule networks. In: NeurIPS (2019)
6. Hinton, G.E., Krizhevsky, A., Wang, S.: Transforming auto-encoders. In: ICANN (2011)
7. Hinton, G.E., Sabour, S., Frosst, N.: Matrix capsules with EM routing. In: ICLR (2018)
8. Jeong, T., Lee, Y., Kim, H.: Ladder capsule network. In: ICML (2019)
9. Koren, Y.: Factorization meets the neighborhood: a multifaceted collaborative filtering model. In: KDD (2008)
10. Koren, Y., Bell, R., Volinsky, C.: Matrix factorization techniques for recommender systems. Computer $42(8)$, 30–37 (2009)
11. Lee, J., Kim, S., Lebanon, G., Singer, Y.: Local low-rank matrix approximation. In: International Conference on Machine Learning, pp. 82–90 (2013)
12. Li, D.S., Chen, C., Gong, Z., Lu, T., Chu, S., Gu, N.: Collaborative filtering with noisy ratings. In: SDM (2019)
13. Li, D., Chen, C., Lv, Q., Yan, J., Shang, L., Chu, S.: Low-rank matrix approximation with stability. In: International Conference on Machine Learning, pp. 295–303 (2016)
14. Mackey, L.W., Talwalkar, A., Jordan, M.I.: Divide-and-conquer matrix factorization. arXiv abs/1107.0789 (2011)
15. Malmgren, C.: A comparative study of routing methods in capsule networks (2019)
16. Mnih, A., Salakhutdinov, R.R.: Probabilistic matrix factorization. In: Advances in Neural Information Processing Systems, pp. 1257–1264 (2008)
17. Ning, X., Desrosiers, C., Karypis, G.: A comprehensive survey of neighborhood-based recommendation methods. In: Recommender Systems Handbook (2015)
18. Ning, X., Karypis, G.: Slim: sparse linear methods for top-n recommender systems. 2011 IEEE 11th International Conference on Data Mining, pp. 497–506 (2011)
19. Sabour, S., Frosst, N., Hinton, G.E.: Dynamic routing between capsules. arXiv abs/1710.09829 (2017)
20. Salakhutdinov, R., Mnih, A.: Bayesian probabilistic matrix factorization using Markov chain Monte Carlo. In: ICML 2008 (2008)
21. Srebro, N., Jaakkola, T.: Weighted low-rank approximations. In: Proceedings of the 20th International Conference on Machine Learning (ICML-2003), pp. 720–727 (2003)
22. Webb, B.: Netflix update: Try this at home (2006). Blog post: https://sifter.org/simon/journal/20061211.html
23. Yuan, T., Cheng, J., Zhang, X., Qiu, S., Lu, H.: Recommendation by mining multiple user behaviors with group sparsity. In: AAAI (2014)
24. Zhang, S., Yao, L., Xu, X.: AutoSVD++: an efficient hybrid collaborative filtering model via contractive auto-encoders. In: SIGIR, pp. 957–960. ACM (2017)

PromptFusion: A Low-Cost Prompt-Based Task Composition for Multi-task Learning

Hetian Song[1,2], Hao He[2(✉)], Qingmeng Zhu[2], and Xiaoguang Xue[3]

[1] University of Chinese Academy of Sciences, Beijing, China
`songhetian20@mails.ucas.ac.cn`
[2] Institute of Software, Chinese Academy of Sciences, Beijing, China
`{hehao21,qingmeng}@iscas.ac.cn`
[3] Beijing Special Engineering and Design Institute, Beijing, China

Abstract. Prompt-tuning takes advantage of large-scale pretrained language models and achieves great performance while being more parameter-efficient. However, existing prompt-tuning methods require tuning different pretrained language models for each specific task, and fail to utilize information across different tasks, which limits their applicability in complex situations. To address above issues, we propose PromptFusion, a unique prompt-based multi-task transfer learning approach which learns knowledge from multiple tasks and incorporates for the target task at low cost. The proposed approach first learns task-specific parameters with prompts to extract information individually, then, a fusion module is designed to aggregate information. Our method is interpretable because it can explain which sources of tasks are the crucial factors to influence the model decision on the target task. We also examine a more effective way to encapsulate information by incorporating parallel adapter modules into transformer layers, and this makes a linkage between parameter-efficient transfer learning methods. We empirically evaluate our methods on the GLUE benchmark and a variety of hard NLU tasks. The results show that our approach outperforms full fine-tuning and other parameter-efficient multi-task methods.

Keywords: Prompt · Multi-task · Transfer learning · Parameter-efficient

1 Introduction

After the presence of GPT-3 [1], the "pre-train, fine-tune" paradigm in natural language processing (NLP) is gradually replaced by "pre-train, prompt, and predict" [14]. Prompt-tuning on pretrained language models (PLMs) [5,22,23] has become the most prevalent paradigm in NLP. However, training individual models on PLM per task usually leads to a huge cost of parameter training and storage due to multiple duplication of the same PLM, especially when it

M. Tanveer et al. (Eds.): ICONIP 2022, LNCS 13623, pp. 492–503, 2023.
https://doi.org/10.1007/978-3-031-30105-6_41

grows to trillions of parameters [6]. Compared with training one specialized model for each specific task, multi-task learning enables one model to handle multiple different NLP tasks, which not only allows for the re-using of repeated parameters and sharing of information among all tasks (especially related tasks), but also mitigates the catastrophic forgetting [7,19] while tuning the PLM.

Liu et al. [16] train the pretrained language model in an unsupervised pattern with all data. However, due to task interference, this training strategy usually has a negative transfer when the data for the target task is limited, and performs worse in high-resource data [16,18]. Traditional multi-task models also have the difficulty in implementing since they demand simultaneous access to all tasks during training, and can not freely add additional tasks without retraining. As a result, some works [11–13] try to improve by developing parameter-efficient methods, which adds a minimal number of parameters into the PLM while keeping the underlying PLM frozen during training. Mahabadi et al. [18] use Hypernetwork [8] to guide parameters generating according to different sources, and He et al. [10] use Hypernetwork to produce hyper-prompts in order to learn task-specific features. However, these methods lack sufficient interpretation to explain which sources contribute more to the target task, and they are difficult to reduce training sources which have negative impact on the target task. These methods hardly re-use the corresponding modules which contain specific information also.

In this paper, we attempt to address these limitations and propose Prompt-Fusion, a novel but natural two-stage approach for multi-task learning by extending prompt-tuning. Like previous works [12,13,15], we assume that prompts are likely to contain the particular knowledge of training data while the PLM is frozen. Therefore, we first introduce prompts to store specific information for each specific task, then a PromptFusion module is designed to assign attention weights to every prompts module. Additionally, we incorporate parallel adapter modules into transformer layers to improve performance. Our method is jointly trained across all tasks and performs well on the target task, while simply updating a small number of parameters and reducing re-training cost on the whole model. Besides, PromptFusion is interpretable to some extent because it can explain which sources of tasks are the crucial factors to influence the model decision on the target task.

Contributions. Overall, our main contributions are as follows:

(1) We propose a novel joint training approach PromptFusion, which is able to make NLP multi-task learning based on PLM with higher training and parameter efficiencies.
(2) To support proposed PromptFusion approach, we utilize different prompts to extract knowledge from multi-task resources and introduce a variant of adapter module to strengthen task-specific information.
(3) We evaluate our approach with fully fine-tuning and other state-of-the-art parameter-efficient methods in single-task or multi-task setup on the GLUE benchmarks and several hard NLU tasks using Roberta [17] across base and large model sizes. The experimental results show that our model achieves

matched or better performance compared with the competitive parameter-efficient models.

2 Related Work

2.1 Prompt-Tuning and Adapter-Tuning

Prompt-tuning is a new paradigm for adapting PLM to downstream tasks with the help of textual prompts [14]. To overcome the weakness of the hand-crafted design of prompt template, some works, such as Prefix-tuning [13], Prompt-tuning [12], and P-tuning, [15] optimize a series of embeddings which can be seen as prompts by inserting a small number of parameters into the last or all transformer self-attention layers. Similarity, adapter-tuning [11] is an effective way that inserts small extra modules named adapters into pretrained network layers. During training, the major body of the pretrained language model is frozen. Especially, the former shares the common module parameters across all transformer layers while adapter-tuning inserts adapter modules per transformer layer. Both of them can be considered as parameter-efficient methods.

2.2 Low-Cost Multi-task Learning

Multi-task learning is an inductive transfer method whose principle goal is to improve generalization performance by leveraging domain-specific information contained in the training data [4]. Multi-task learning is a challenging problem in parameter-efficient-tuning, because it is difficult to attain Nash Equilibrium and Pareto efficiency in terms of performance and efficiency across all tasks. Pfeiffer et al. [20] propose AdapterFusion that leverages the knowledge of several adapters from resources of tasks, and learns their aggregation for a single target task. HyperFormer [18] generates parameters of adapters by Hypernetwork [8]. More recently, He et al. [10] uses Hypernetwork to produce hyper-prompts in order to learn task-specific features, but they consider that tuning the whole model to perform better.

3 Methods

In this section, we first describe the problem formulation of multi-task for transfer learning [31]. We introduce our modification in Projection Network, which is used for reparameterizing prompts. The PromptFusion framework is shown in Fig. 1.

3.1 Problem Formulation

We assume that we are given a pretrained language model $f_\Theta(.)(e.g.$Roberta [17]) which is parameterized by Θ. The problem is considered as a general multi-task learning problem where the data for training from a set of tasks $\{\mathcal{D}_n\}_{n=1}^N$, where

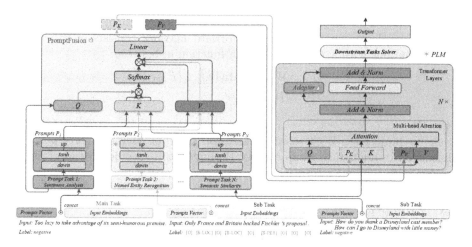

Fig. 1. PromptFusion framework: Every task-specific prompts $P_n, n \in \{1, ..., N\}$ learned in single-task setup are used to aggregate information in PromptFusion module, which generates P_k and P_v that are prepended to transformer key-value pairs K and V. For PromptFusionA, parallel adapters are inserted into transformer layers to strengthen the target-specific information.

N is the number of tasks. Specifically for each task, $\mathcal{D}_n = \{(x_n^i, y_n^i)\}_{i=1}^{T_n}$ means the training data for n-th task with T_n samples. Standard multi-task learning minimizes the following objective function:

$$\mathcal{L}(\Theta, \mathcal{D}) = \sum_{n=1}^{N} \sum_{i=1}^{T_n} Loss(f_\Theta(x_n^i), y_n^i) \tag{1}$$

where the $Loss(.,.)$ is typically the cross-entropy loss, and $f_\Theta(x_n^i)$ is the output of the PLM for training sample x_n^i. In this paper, our goal is to finetune the pretrained language model in multi-task setup at a low cost, and to leverage a set of N tasks to improve the target task n by sharing and composing parameters across tasks at the same time. Specifically, we wish to learn parameters Θ_n from Θ' which is expected to contain relevant information of all the N tasks:

$$\Theta_n = \arg\min_{\Theta'} Loss(\mathcal{D}_n; \Theta') \tag{2}$$

3.2 Prompt Projection Network

Prompt projection Network is used for reparameterizing prompts P and then transforming P into two sets of prefix embeddings P_k and P_v, which are concatenated with original key-value pairs K and V of transformer attention. The prompts P are series of trainable continuous embeddings and can be attended to as if they were virtual tokens. In our approach, we change the dimension in order to limit the number of parameters of PromptFusion module. In single task setup, we add a linear projection with parameters W_t to change the dimension for multi-head computation. The feed-forward network consists of two linear transformations with a $TANh$ activation function in between:

$$P' = W_{up}^T(\mathbf{Tanh}(W_{down}^T(P)))$$
$$P_k = P_v = W_t P' \tag{3}$$

where $W_{down} \in \mathcal{R}^{D_{hidden} \times D_{mid}}$ is a down projection and $W_{up} \in \mathcal{R}^{D_{mid} \times D_{hidden}}$ is an up projection. W_t is used to transform output into the corresponding dimension. In multi-task setup, we use PromptFusion module instead of linear projection to combine information.

3.3 PromptFusion Component

Inspired by [20,26], we propose PromptFusion to learn the composition of N prompts and flexibly share information by introducing a new set of weights Ψ. We anticipate that model can learn the useful information contained in multiple prompts through an attention mechanism. We use PromptFusion to polymerize prompts of each task and transform the output of PromptFusion O into prefix embeddings P_k and P_v in order to concatenated with the original K and V of the multi-head attention at every transformer layer. The parameters Ψ of PromptFusion are composed of Query, Key and Value matrices, denoted by Q_p, K_p, and V_p. The reparameterizing prompts $P'_{i \in 1,...,N}$ are used as input of both Key and $Value$, the n-th target task prompts P_n are taken as $Query$. We learn the fusion contextual representation of each prompts P_n by using dot-product attention:

$$S = softmax(P_n^T Q_p \otimes P_i'^T K_p), i \in \{1, ..., N\}$$
$$h_i = P_i'^T V_p, i \in \{1, ..., N\}$$
$$H = [h_1, h_2, ..., h_N] \tag{4}$$
$$O = W_h(S \otimes H)$$

where $\otimes$ indicates the dot product and $[.,.]$ means the concatenation of given embeddings. W_h is used to transform output into the corresponding dimension.

3.4 PromptFusionA

PromptFusion learns a parameterized mixer of the trained prompts to share multi-task information to improve the performance on the target task. To further mitigate the catastrophic forgetting and interference between other tasks, we use a variant of adapter to strengthen the task-specific information. We insert adapter module into every transformer layer, making it parallel to the FFN sub-layer. He et al. [9] apply this modification, but their method allocates more parameter budgets for bottleneck dimension and barely have an improvement. We reduce the dimension substantially to match with the length of the prompts, and add a scalar hyper-parameter α for scaling task-specific differences:

$$h' = \alpha W_{up}^T(\mathbf{ReLU}(W_{down}^T(h))) + FFN(h) \tag{5}$$

where h indicates the output of transformer layer. In this way, the variant of adapter can learn the specific information from the target task additionally.

3.5 Learning Algorithm

Algorithm 1. PromptFusion Learning Algorithm Framework

Step1: Single Prompts Extraction

1. For every n-th task in N tasks, train prompts P_n by minimizing the loss $LOSS(\mathcal{D}_n; \Theta, \Phi)$ individually, with frozen PLM parameters Θ. [**Eq.** 6]
2. Save trained Prompts P_n parameters Φ_n for re-using.

Step2: Multiple Prompts Aggregation

1. Select a set of prompts $[P_1, ..., P_n, ..., P_N]$ which are trained on different tasks, P_n is the prompts of target task.
2. Initialize PromptFusion module with parameters Ψ, optimize Ψ by minimizing the loss $LOSS(\mathcal{D}_n; \Theta, \Psi, \Phi_{1,...,N})$, with frozen PLM parameters Θ. [**Eq.** 7]

The overall algorithm is shown in Algorithm 1. PromptFusion algorithm has two stages. We train PromptFusion model for each of the N tasks individually at first, this also can be seen as common prompt-tuning. Specifically, we freeze the parameters of the pretrained language model Θ, and only the task prompts parameters Φ are trained. The target of each $n \in 1, ..., N$ is:

$$\Phi_n = \arg\min_{\Phi} Loss(\mathcal{D}_n; \Theta, \Phi) \tag{6}$$

where $Loss(.,.)$ is the cross-entropy loss. In this way, we save the collection of prompts weights from each task, and $\Phi \ll \Theta$. We believe that Φ has acquired task-specific knowledge.

Secondly, we then use PromptFusion to combine the collection of N prompts. We select the trained prompts that are useful for the target task. After that we freeze the parameters of the pretrained language model Θ and all prompts parameters Φ, only train the PromptFusion parameters Ψ.

$$\Psi_n = \arg\min_{\Psi} Loss(\mathcal{D}_n; \Theta, \Psi, \Phi_{1,...,N}) \tag{7}$$

where Ψ_n indicates the n-th task PromptFusion parameters. We use the n-th training source of $\mathcal{D}$ twice, for training task-specific prompts parameters Φ_n and PromptFusion parameters Ψ_n.

Additionally, to focus on the task-specific information, the variant of adapters are inserted in parallel with FFN sub-layers, and the objective turns to:

$$\Psi'_n, \Omega_n = \arg\min_{\Psi, \Omega} Loss(\mathcal{D}_n; \Theta, \Omega, \Psi, \Phi_{1,...,N}) \tag{8}$$

where Ω denotes the parameters of adapter, and Ω_n indicates the n-th task adapter parameters. By this learning algorithm, we can train multiple tasks separately, and the training weights of each task prompts can be easily re-used for other tasks.

4   Experiments

In this section, we evaluate the effectiveness of PromptFusion by conducting extensive experiments on the GLUE benchmarks and several hard natural language understanding (NLU) tasks.

4.1   Experiment Setup

Tasks and Datasets. We evaluate the performance of the models on the GLUE benchmarks [28], including sentiment analysis (SST-2), natural language inference (MNLI, QNLI, RTE), similarity and paraphrase (MRPC, STS-B, QQP), and linguistic acceptability (CoLA). Additionally, we test our model on several hard NLU tasks: sequence tagging and question answering, including named entity recognition (CoNLL03, CoNLL04 and OntoNotes5.0 [2,25,29]), semantic role labeling (CoNLL05 and CoNLL12 [3,21]), and extractive question answering (SQuAD1.1 and SQuAD2.0 [24]).

Experimental Details. We use the HuggingFace implementation [30] of the transformer model Roberta [17]. We set input length to 128 and training batch size to 64. The learning rate is a constant of 1e-4. We train single-task models in 60 epochs, multi-task models in 30 epochs, and save a checkpoint every epoch, then we report the best checkpoint for each task respectively. The default prompt length is 30 and the adapter bottleneck size is the same.

4.2   Results on the GLUE Benchmarks

As shown in Table 1, we compare our proposed PromptFusion and Prompt-FusionA with fine-tuning, two parameter-efficient single-task methods and two multi-task methods: Prompt-tuning, which is a widely used prompt-based method [15], MAM-adapter, which combines advantages of prompt and adapter [9], AdapterFusion [20], which is a adapter-based multi-task learning method, and HyperFormer++, which is a multi-task learning method with adapters and Hypernetwork.

PromptFusion model outperforms Prompt-tuning by 0.4/0.3% increase in average scores across base and large, this validates that extracting and aggregating information by prompt is effective. The reason why the performance is improving a little might be due to the interference between different tasks. Our PromptFusionA model, which inserts parallel adapter modules into every transformer layer, achieves the best performance on the GLUE benchmarks compared to other single-task or multi-task parameter-efficient models, especially better than Prompt-tuning by 1.0/1.3% improvement. This may benefit from the reinforcement of target resources by adapters. PromptFusionA model is better than full fine-tuning by 0.3/0.3%, while keeping the PLM frozen during training. Furthermore, due to the differences in datasets, not all tasks are able to benefit from multi-task learning. Tasks with smaller dataset typically benefits more.

Table 1. Results of all models on the GLUE validation set. For CoLA, we report Mattew's correlation. For MRPC, we report accuracy and F1-score. For STS-B, we report Pearson and Spearman correlation coefficients. We report accuracy for the rest. *: Our modification version of AdapterFusion by prompts. ◇: The results copy from [18] which based on T5 [23]. We report average performance and the best methods are bold.

Model	CoLA	SST-2	MNLI	RTE	QNLI	MRPC	QQP	STS-B	Avg
				Single-Task					
Fine-tuning$_{base}$	**62.6**	94.5	**87.6**	76.7	**92.8**	90.0/90.2	**91.3**	**90.3/90.0**	**85.7**
Prompt-tuning$_{base}$	60.7	94.4	86.8	76.2	92.4	90.0/87.1	89.6	89.8/89.7	85.0
MAM-adapter$_{base}$	61.8	**94.8**	87.4	**76.9**	92.7	**90.1/90.3**	91.1	90.1/89.9	85.6
Fine-tuning$_{large}$	68.0	96.4	90.2	86.6	94.5	**90.9**/91.9	**92.2**	**92.4/92.2**	88.9
Prompt-tuning$_{large}$	66.5	95.7	90.0	84.7	94.1	89.5/87.7	90.7	91.7/91.5	87.8
MAM-adapter$_{large}$	**68.7**	**96.5**	**90.3**	**86.9**	**94.8**	90.8/**92.3**	91.9	92.3/92.1	**89.0**
				Multi-Task					
AdapterFusion$^{*}_{base}$	63.4	**95.3**	86.9	**77.6**	92.5	**89.9**/88.0	90.6	**90.8/90.5**	85.9
HyperFormer++$^{◇}_{base}$	**64.0**	94.0	85.7	75.4	**93.0**	89.6/**92.6**	90.3	90.0/89.6	85.3
PromptFusion$_{base}$	63.5	94.3	86.6	76.8	92.3	89.2/86.7	90.4	89.8/89.8	85.4
PromptFusionA$_{base}$	63.9	**95.3**	**87.3**	77.3	92.8	89.7/87.2	**91.3**	90.3/90.0	**86.0**
AdapterFusion$^{*}_{large}$	67.9	96.4	89.8	87.2	94.3	**91.4**/90.0	92.5	**92.3**/91.9	89.0
HyperFormer++$^{◇}_{large}$	58.9	95.7	89.8	87.0	**94.5**	90.0/**92.7**	90.7	89.8/90.0	87.3
PromptFusion$_{large}$	68.2	95.1	89.4	86.2	92.8	89.7/88.0	91.4	91.9/91.7	88.1
PromptFusionA$_{large}$	**68.9**	**96.7**	**90.1**	**87.5**	**94.5**	91.2/**93.4**	**92.7**	92.3/92.0	**89.2**

4.3 Results on the Hard NLU Tasks

Table 2 presents the average performance of PromptFusion and PromptFusionA on hard NLU tasks such as name entity recognition (NER), extractive question answering (QA) and semantic role labeling (SRL). We use the same train and test split as [15]. The results show that PromptFusion performs better than prompt-tuning in NER and SRL datasets, but worse in QA datasets, and the overall effect is not as good as MAM-adapter and PromptFusionA. This also demonstrates that adapter contributes to strengthening information of the target task. PromptFusionA outperforms in general.

4.4 Ablation Study

To study the suitable inserted layers of PromptFusion module for prompts, and their effectiveness for low-resource, we conduct ablation experiments on RTE and SST-2. As shown in Fig. 2(a), we find that adding PromptFusion module to learn knowledge from frozen prompts in later layers is better than in previous layers, although the parameters of PromptFusion are shared across layers. The result is similar to original prompts [15]. We also show results on SST-2 for the various numbers of training samples in Fig. 2(b). Both PromptFusion and PromptFusionA models outperform fine-tuning while the training data has been limited.

Table 2. Results on hard NLU tasks.

Model	CoNLL03	CoNLL04	OntoNotes5.0	SQuAD1.1	SQuAD2.0	CoNLL05	CoNLL12
Prompt-tuning$_{base}$	91.7	85.8	88.6	90.7	79.8	82.1/88.1	82.5
MAM-Adapter$_{base}$	92.0	86.6	89.2	**91.7**	82.0	82.6/89.2	**85.0**
PromptFusion$_{base}$	91.8	86.6	89.2	89.8	79.3	82.3/88.5	82.7
PromptFusionA$_{base}$	**92.4**	**87.5**	**89.8**	91.6	**82.3**	**83.1/89.4**	84.3
Prompt-tuning$_{large}$	92.4	88.4	89.5	92.4	82.2	84.3/89.7	85.2
MAM-Adapter$_{large}$	92.9	88.7	90.4	93.1	83.3	85.4/90.5	85.7
PromptFusion$_{large}$	92.7	88.6	89.8	92.1	81.8	84.6/90.1	85.6
PromptFusionA$_{large}$	**93.1**	**89.2**	**91.0**	**93.2**	**83.7**	**86.2/90.7**	**86.1**

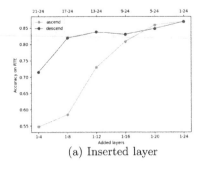

(a) Inserted layer

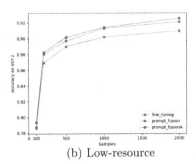

(b) Low-resource

Fig. 2. (a) We select k transformer layers to add PromptFusion module while using Roberta-large. [x-y] refers to the layer interval that we add. (b) We train models by using a small number of training samples (100,200,500,1000,2000) and models are based on Roberta-base.

5 Analysis

To further analyze our methods, we compare their efficiencies with other methods and conduct a PromptFusion Visualization.

5.1 Efficiency of PromptFusion

We analyze the efficiency of our multi-task methods as shown in Table 3. We test the trainable parameters and training time relative to fine-tuning and some parameter-efficient methods. For prompt setup, we calculate MLP parameters, despite the fact that they are generally ignored because they can be discarded at test time after training. We treat the parameters which are updated by gradient as the trainable ones.

The results from the efficiency test indicated that PromptFusion and Prompt-FusionA almost have the same training parameters and training time with the single-task methods such as prompt-tuning or MAM-adapter, and are better than fine-tuning. This is because they fix nearly all the model parameters, and almost only train the attention weights for multiple frozen prompts, so the parameter size almost depends on the size of Q_p, K_p and V_p. Also, while the

Table 3. Trainable parameters/Time relative to fine-tuning.

Single-task Model	Fine-tuning	prompt-tuning	MAM-adapter
Training parms	100.0%	**12.3%**	13.2%
Training times	100.0%	**72.1%**	76.8%
Multi-task Model	AdapterFusion	PromptFusion	PromptFusionA
Training parms	17.0%	**12.7%**	13.7%
Training times	100.3%	**68.1%**	73.1%

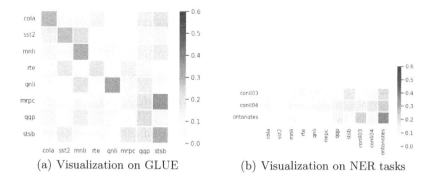

(a) Visualization on GLUE   (b) Visualization on NER tasks

Fig. 3. The heatmap of PromptFusion attention softmax scores. Rows indicate the target task and colunms indicate task prompts. We average softmax scores of Prompt-Fusion across 50 validation examples for each task to reduce the influence of samples.

adapter module simply requires a few parameters, it needs more cost in multi-task setup because the fusion module is present in each layer of the transformer.

5.2 Visualization of Attention

We analyze the attention weights of multiple prompts learned by PromptFusion. We assume that the softmax activation is higher if the target task benefits from other trained prompts, and we visualize the attention softmax scores with heatmap. We show the results on the GLUE benchmark and NER tasks. As shown in Fig. 3(a), we find that nearly every prompt contributes its own task's information to the target task, and PromptFusion tends to allocate more attention to its own task prompt. This reveals that PromptFusion is useful for extracting information from multiple prompts, and this can be seen as another view of SPoT [27], which uses one or more prompts trained in multiple resources to initialize the prompts of the target task. We also find that the target task benefits more from the similarity tasks, like the sentence similarity and para-phrase dataset STS-B and MRPC. To demonstrate this, we further conduct a visualization on NER datasets. As shown in Fig. 3(b), the prompts of NER tasks CONLL03, CoNLL04 and OntoNotes5.0 take a large proportion of soft-max scores, rather than other prompts which are trained on the GLUE datasets.

6   Conclusion

In this paper, we introduce PromptFusion, a novel prompt-based multi-task transfer learning approach which learns knowledge from multiple tasks and incorporates for the target task. PromptFusion re-uses prompts of each single-task to reduce re-training cost on the whole model. We also introduce a variant of adapter module to strengthen task-specific information. Our experiments show that our method matches or outperforms performance compared with the competitive parameter-efficient models. We also demonstrate that PromptFusion is able to select the most efficient trained prompts for the target task.

References

1. Brown, T., et al.: Language models are few-shot learners. Adv. Neural. Inf. Process. Syst. **33**, 1877–1901 (2020)
2. Carreras, X., Màrquez, L.: Introduction to the conll-2004 shared task: semantic role labeling. In: HLT-NAACL, pp. 89–97 (2004)
3. Carreras, X., Màrquez, L.: Introduction to the conll-2005 shared task: semantic role labeling. In: CoNLL-2005, pp. 152–164 (2005)
4. Caruana, R.: Mach. Learn. **28**(1), 41–75 (1997)
5. Devlin, J., Chang, M.W., Lee, K., Toutanova, K.: BERT: pre-training of deep bidirectional transformers for language understanding. In: NAACL-HLT, pp. 4171–4186 (2019)
6. Fedus, W., Zoph, B., Shazeer, N.: Switch transformers: scaling to trillion parameter models with simple and efficient sparsity. J. Mach. Learn. Res. **23**(120), 1–39 (2022)
7. French, R.M.: Catastrophic forgetting in connectionist networks. Trends Cogn. Sci. **3**(4), 128–135 (1999)
8. Ha, D., Dai, A.M., Le, Q.V.: Hypernetworks. In: ICLR (2017)
9. He, J., Zhou, C., Ma, X., Berg-Kirkpatrick, T., Neubig, G.: Towards a unified view of parameter-efficient transfer learning. In: ICLR (2022)
10. He, Y., et al.: Hyperprompt: prompt-based task-conditioning of transformers. In: International Conference on Machine Learning, pp. 8678–8690. PMLR (2022)
11. Houlsby, N., et al.: Parameter-efficient transfer learning for NLP. In: ICML, pp. 2790–2799 (2019)
12. Lester, B., Al-Rfou, R., Constant, N.: The power of scale for parameter-efficient prompt tuning. In: EMNLP, pp. 3045–3059 (2021)
13. Li, X.L., Liang, P.: Prefix-tuning: optimizing continuous prompts for generation. In: ACL/IJCNLP, pp. 4582–4597 (2021)
14. Liu, P., Yuan, W., Fu, J., Jiang, Z., Hayashi, H., Neubig, G.: Pre-train, prompt, and predict: a systematic survey of prompting methods in natural language processing. arXiv preprint arXiv:2107.13586 (2021)
15. Liu, X., et al.: P-tuning: prompt tuning can be comparable to fine-tuning across scales and tasks. In: ACL, pp. 61–68 (2022)
16. Liu, X., He, P., Chen, W., Gao, J.: Multi-task deep neural networks for natural language understanding. In: Korhonen, A., Traum, D.R., Màrquez, L. (eds.) ACL, pp. 4487–4496 (2019)
17. Liu, Y., et al.: Roberta: a robustly optimized BERT pretraining approach. arXiv preprint arXiv:1907.11692 (2019)

18. Mahabadi, R.K., Ruder, S., Dehghani, M., Henderson, J.: Parameter-efficient multi-task fine-tuning for transformers via shared hypernetworks. In: ACL/IJCNLP, pp. 565–576 (2021)
19. McCloskey, M., Cohen, N.J.: Catastrophic interference in connectionist networks: the sequential learning problem. In: Psychology of Learning and Motivation, vol. 24, pp. 109–165. Elsevier (1989)
20. Pfeiffer, J., Kamath, A., Rücklé, A., Cho, K., Gurevych, I.: AdapterFusion: non-destructive task composition for transfer learning. In: EACL, pp. 487–503 (2021)
21. Pradhan, S., Moschitti, A., Xue, N., Uryupina, O., Zhang, Y.: Conll-2012 shared task: modeling multilingual unrestricted coreference in ontonotes. In: EMNLP-CoNLL, pp. 1–40 (2012)
22. Radford, A., Wu, J., Child, R., Luan, D., Amodei, D., Sutskever, I., et al.: Language models are unsupervised multitask learners. OpenAI Blog 1(8), 9 (2019)
23. Raffel, C., et al.: Exploring the limits of transfer learning with a unified text-to-text transformer. J. Mach. Learn. Res. 21, 140:1–140:67 (2020)
24. Rajpurkar, P., Zhang, J., Lopyrev, K., Liang, P.: Squad: 100,000+ questions for machine comprehension of text. In: EMNLP, pp. 2383–2392 (2016)
25. Sang, E.F.T.K., Meulder, F.D.: Introduction to the conll-2003 shared task: language-independent named entity recognition. In: HLT-NAACL, pp. 142–147 (2003)
26. Vaswani, A., et al.: Attention is all you need. In: NIPS, pp. 5998–6008 (2017)
27. Vu, T., Lester, B., Constant, N., Al-Rfou', R., Cer, D.: SPoT: better frozen model adaptation through soft prompt transfer. In: ACL, pp. 5039–5059 (2022)
28. Wang, A., Singh, A., Michael, J., Hill, F., Levy, O., Bowman, S.R.: GLUE: a multi-task benchmark and analysis platform for natural language understanding. In: ICLR (2019)
29. Weischedel, R., et al.: Ontonotes release 5.0 ldc2013t19. In: LDC (2013)
30. Wolf, T., et al.: Transformers: state-of-the-art natural language processing. In: EMNLP, pp. 38–45 (2020)
31. Zhuang, F., et al.: A comprehensive survey on transfer learning. Proc. IEEE 109(1), 43–76 (2021)

A Fast and Efficient Algorithm for Filtering the Training Dataset

Norbert Jankowski[✉]

Department of Informatics, Nicolaus Copernicus University, Toruń, Poland
`norbert@umk.pl`

Abstract. The goal of this paper is to present a new algorithm that filters out inconsistent instances from the training dataset for further usage with machine learning algorithms or learning of neural networks. The idea of this algorithm is based on the previous state-of-the-art algorithm, which uses the concept of local sets. Sophisticated modification of the definition of local sets changes the merits of the algorithm. It is additionally supported by locality-sensitive hashing used for searching for nearest neighbors, composing a new efficient ($O(n \log n)$), and an accurate algorithm.

Results prepared on many benchmarks show that the algorithm is as accurate as previous but strongly reduces the time complexity.

Keywords: instance selection · prototype selection · classification · machine learning

1 Introduction

Learning from data characterizes all machine learning algorithms and the learning of neural networks. The dataset $\mathcal{D}$ can be defined as a set of pairs $\langle \mathbf{x}_i, y_i \rangle$ ($i = 1, \ldots, m$) of instances $\mathbf{x}_i \in R^n$ and class labels $y_i \in [1, \ldots, C]$.

The instance selection algorithms are a class of algorithms that reduce the number of instances in the original training dataset. The reduced set S should be enough to learn a classifier like the kNN [4] with similar or better accuracy.

Instance selection algorithms can be divided into prototype selection algorithms and filtering algorithms. The purpose of prototype selection algorithms is to reduce the size of the dataset as much as possible. For further information about the prototype selection algorithm, see [5,8]. The newest publications propose prototype selection in complexity $O(m \log m)$ see [1,9,12,14] In contrast, the filtering algorithms remove only inconsistent instances. Such a strategy is beneficial if a data cleaning is recommended before using the data by a different algorithm (like classifier learning). One well-known method is Edited nearest neighbors (ENN) proposed in [15]. However, currently, the best filtering method is the algorithm Local Set-based Smoother proposed in [10] and this algorithm will be a base for further considerations.

© The Author(s), under exclusive license to Springer Nature Switzerland AG 2023
M. Tanveer et al. (Eds.): ICONIP 2022, LNCS 13623, pp. 504–512, 2023.
https://doi.org/10.1007/978-3-031-30105-6_42

The main contribution of this article is the new instance filtering algorithm which has lower complexity than LSSm but keeps the same quality of classification accuracy. The FastLSSm can be parallelized.

Next section presents previous ideas and algorithms directly connected with the new algorithm presented in further section. The last section presents an empirical analysis of newly introduced algorithms. Section compares the new algorithm with others across several benchmark tests.

2   Local Sets and Instance Selection—Previous Work

First time the local sets were introduced in Iterative Case Filtering (ICF) algorithm in [3]. The *local set* in point $\mathbf{x}$ is defined as the set of points in the biggest hypersphere centered in $\mathbf{x}$ without any instance from the opposite class (without enemy):

$$LS(\mathbf{x}) = \{\mathbf{x}' \ : ||\mathbf{x}' - \mathbf{x}|| < ||\mathbf{x} - ne(\mathbf{x})||\} \tag{1}$$

where $ne(\mathbf{x}) = \arg\min_{\mathbf{x}'} ||\mathbf{x} - \mathbf{x}'|| \wedge y \neq y'$ is a nearest enemy (a nearest instance from opposite class).

The most successful application of local sets in selection algorithms was the Local set-based smoother (LSSm) as a filtering algorithm proposed in [10]. The LSSm uses the idea of local sets to define two properties that are very important in constructing the final algorithm. The first is the *usefulness* which is something like attractiveness of given instance $\mathbf{x}$ measured as the number of instances which has the $\mathbf{x}$ in their local sets:

$$u(\mathbf{x}) = |\{\mathbf{x}' \in \mathcal{D} \ : \ \mathbf{x} \in LS(\mathbf{x}')\}| \tag{2}$$

The second property used in LSSm is the *harmfulness* measured as the number of instances for which the $\mathbf{x}$ instance is the nearest enemy:

$$h(\mathbf{x}) = |\{\mathbf{x}' \in \mathcal{D} \ : \ ne(\mathbf{x}') = \mathbf{x}\}| \tag{3}$$

The difference between those two properties defines the strength of balance between usefulness and harmfulness, and the LSSm algorithm removes instances with greater harmfulness than usefulness. The final LSSm algorithm is presented in Algorithm 1.

Algorithm 1: LSSm($\mathcal{D}$)

Data: $\mathcal{D}$ — dataset
Result: S
1 $S = \{\}$
   compute local sets
2 **foreach** $\mathbf{x} \in \mathcal{D}$ **do**
3      $u(\mathbf{x}) = |\{\mathbf{x}' \in \mathcal{D} \ : \ \mathbf{x} \in LS(\mathbf{x}')\}|$
4      $h(\mathbf{x}) = |\{\mathbf{x}' \in \mathcal{D} \ : \ ne(\mathbf{x}') = \mathbf{x}\}|$
5   **if** $u(\mathbf{x}) \geq h(\mathbf{x})$ **then**
6          $S = S \cup \{\mathbf{x}\}$

The empirical analysis of the LSSm algorithm presented in [10] has proved its superiority—the LSSm had the highest accuracy among all analyzed algorithms. Of course, the amount of removed instances is small because this is a filtering algorithm, not the prototype selector algorithm. This means that, in the context of accuracy, it is a state of art algorithm among instance selection algorithms.

The complexity of LSSm is $O(m^2)$ because the computation of single usefulness or harmfulness is $O(m)$.

3   Firmly Local Sets and Fast Filtering Algorithm FastLSSm

The main goal of this paper is to propose an algorithm with the same quality (accuracy) but lower complexity. The main problem is that even with the support of locality-sensitive hashing (LSH) [6] the complexity will not be reduced because the cardinality of $LS(\mathbf{x})$ is $O(m)$. This means that LSH in such a case reduce potentially only a constant factor of complexity, not the complexity of LSSm.

The main observation is that the local sets are often massive, while more localized analysis could bring a similar effect. For example, let us assume that a given ($\mathbf{x}$) has 1000 instances. Is it necessary to base on so many neighbors of $\mathbf{x}$ to compute the accurate distance between usefulness and harmfulness? It is simple to observe that the local sets are massive for instances that lie on the outskirts, while much more localized analysis is enough to show that such cases are *clean*.

The idea of *firmly local set* is a localized version of a local set, and it is defined by surroundings of no more than k nearest neighbors of $\mathbf{x}$ without enemies:

$$FLS(\mathbf{x}) = \begin{cases} \{\mathbf{x}' \: : \: ||\mathbf{x}' - \mathbf{x}|| < ||\mathbf{x} - ne(\mathbf{x})||\}, & \text{if } \exists_{\mathbf{x}' \in N^k(\mathbf{x})} \: y \neq y' \\ N^k(\mathbf{x}), & \text{otherwise.} \end{cases} \quad (4)$$

This is very crucial in the context of complexity because if k is $O(1)$, then the $FLS(\mathbf{x})$ can be computed in average complexity $O(\log m)$ using locality sensitive hashing or better forests of locality sensitive hashing.

According to changes between local sets and firmly local sets the usefulness and harmfulness must be redefined by

$$u'(\mathbf{x}) = |\{\mathbf{x}' \in \mathcal{D} \: : \: \mathbf{x} \in FLS(\mathbf{x}')\}|, \quad (5)$$

and by

$$h'(\mathbf{x}) = |\{\mathbf{x}' \in \mathcal{D} \: : \: ne(\mathbf{x}') = \mathbf{x} \wedge \mathbf{x} \in N^k(\mathbf{x}')\}| \quad (6)$$

appropriately.

New definitions of the firmly local set, usefulness, and harmfulness enable us to propose a significantly faster version of LSSm algorithm, see Algorithm 2. First, it is necessary to construct a forest of LSH trees on $\mathcal{D}$. More information about the construction of the forest of LSH trees will be presented in a separate subsection below. The complexity of this step is $O(m \log m)$. The next step is to extract the nearest neighbors for each instance in $\mathcal{D}$ using a forest of LSH trees. The complexity of this step is $O(m)$ thanks to information deposed in LSH trees. After constructing the nearest neighbors, the enemies and the harmfulness are calculated. The complexity of this step is just $O(m)$ because k is assumed to be $O(1)$. The following loop is devoted to calculating usefulness with the same complexity as the last loop. The final loop collects all necessary instances in the final set in $O(m)$ complexity.

The above considerations mean that the final complexity of the whole algorithm is $O(m \log m)$.

The FastLSSm algorithm can be parallelized, even in each algorithm phase. The forest of trees can be built in parallel, and each following loop can be parallelized too.

Algorithm 2: FastLSSm($\mathcal{D}$)

Data: $\mathcal{D}$ — dataset
Result: S
1 $S = \{\}$
2 construct forest of balanced LSH trees
3 compute $N^k(\mathbf{x})$ for each $\mathbf{x} \in \mathcal{D}$ using forest of balanced LSH trees
4 **foreach** $\mathbf{x} \in \mathcal{D}$ **do**
5 | find nearest enemy $ne(\mathbf{x})$ in $N^k(\mathbf{x})$
6 | **if** *any enemy* **then**
7 | | $h'[ne(\mathbf{x})]++$

8 **foreach** $\mathbf{x} \in \mathcal{D}$ **do**
9 | **foreach** $\mathbf{x}' \in N^k(\mathbf{x})$ **do**
10 | | **if** *no enemy of* $\mathbf{x}$ *in* $N^k(\mathbf{x})$ $\lor$ $||\mathbf{x}' - \mathbf{x}|| < ||\mathbf{x} - ne(\mathbf{x})||$ **then**
11 | | | $u'[\mathbf{x}']++$

12 **foreach** $\mathbf{x} \in \mathcal{D}$ **do**
13 | **if** $u'(\mathbf{x}) \geq h'(\mathbf{x})$ **then**
14 | | $S = S \cup \{\mathbf{x}\}$

Of course, the h and h' are not the same; similarly, h and h' are not the same. Although, as it has been presented in Sect. 4 the u' and h' present very similar level of accuracy with significantly lower complexity.

3.1   Forest of Balanced Locality-Sensitive Hashing Trees

To compute nearest neighbors efficiently in the line 3 in Algorithm 2 an appropriate data structure are necessary. The best way is to use a forest of balanced locality-sensitive hashing trees. Hashing trees were proposed in [7], but in such cases, the space cuts created by random hyperplanes are pretty far from hyperspheres. This motivated the next step: constructing a forest of hashing trees presented in [2]. We extended the concept of the forest of hashing trees to the forest of balanced locality-sensitive hashing trees, which are more efficient in [13] . An example of sharp LSH trees and a smoother forest can be seen in Fig. 1.

Assuming that the k nearest neighbor has to be searched and that k is $O(1)$, then using the forest of balanced locality-sensitive hashing trees, the complexity reduces from $O(m)$ to $O(1)$. The cost of constructing a forest of balanced locality-sensitive hashing trees is $O(m \log m)$.

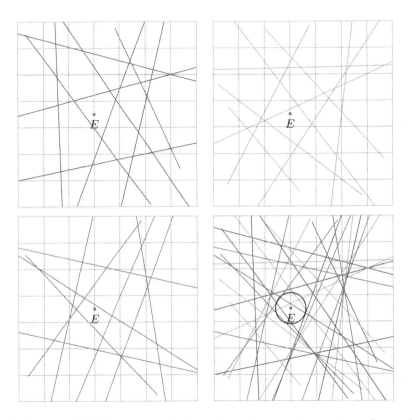

Fig. 1. Upper and left figures presents three trees of random hyperplanes. Lower right figure presents forest composed of those three trees.

4   Empirical Analysis of the LSSmLSH Algorithm

The [10] article has already shown the advantage of the LSSm algorithm (current state of the art) over a number of the most important instance selection algorithms. The results were presented not only in comparison with filtering algorithms but also with various prototype selection algorithms. Therefore, to show trustworthy the operation of the new FastLSSm algorithm, it was decided to simplify the test procedure to show its operation compared to the LSSm algorithm and the very well-known ENN and kNN algorithm.

In order to compare the above-mentioned algorithms, it was decided to select a number of different data sets from the UCI Machine Learning Repository [11]. In all tests, we used 10-fold stratified cross-validation and all learning machines were learning on the same sets of data partitions.

To visualize the performance of all algorithms we present average accuracy for each benchmark dataset and for each learning machine. Additionally, we present the average reduction of dataset size in separate tables. *Ranks* are calculated for each machine for a given dataset $\mathcal{D}$. The ranks are calculated as follows: First, for a given benchmark dataset $\mathcal{D}$ the averaged accuracies of all learning machines are sorted in descending order. The machine with the highest average accuracy is ranked 1. Then, the following machines in the accuracy order whose accuracies are not statistically different from the result of the first machine are ranked 1, until a machine with a statistically different result is encountered. That machine starts the next rank group (2, 3, and so on), and an analogous process is repeated on the remaining (yet unranked) machines. Notice that each cell of the main part of Table 1 is in a form: $acc + std(rank)$, where acc is average accuracy (for a given data set and given learning machine), std is its standard deviation and $rank$ is the rank described just above. If a given cell of the table is in bold it means that this result is the best for given data set or not worse than the best one (rank 1 = winners).

Table 2 presents information on the average reductions (%) for each algorithm and for each benchmark as in the previous table.

It can be seen that the accuracy of LSSm and LSSmLSH algorithms are very similar. All differences between both algorithms are shallow and are almost identical. The most important thing is that the LSSm and LSSmLSH algorithms, thanks to the better reduction of bad vectors, are more likely than ENN and kNN to have bigger ranks. This can be seen in the mean values of the ranks.

Similar conclusions can be seen in Table 2 about reduction of datasets. Again both algorithms (LSSm and LSSmLSH) have almost the same reduction level.

Because the accuracy and reduction properties of both algorithms are very similar, the algorithms can be seen as algorithms of the same quality except for the complexity—LSSmLSh has significantly lower complexity.

5   Summary

This article has presented a new instance selection algorithm (LSSmLSH), which belongs to the filtering algorithm group (removes inconsistent instances from training datasets).

The newly proposed algorithm LSSmLSH has the complexity $O(m \log m)$ while the previous LSSm algorithm has the complexity $O(m^2)$. The LSSm algorithm is based on *local sets* while the LSSmLSH uses a new concept of *firmly local sets*. Thanks to this idea, it was possible to reduce the complexity to $O(m \log m)$. Both algorithms, the LSSm and LSSmLSH have the same level of quality—the accuracy of both algorithms is almost the same, and the reduction is also the same, and it has been proven in the previous article that LSSm outperforms several best instance selection algorithms.

Table 1. Accuracy analysis

	LSSm	**LSSmLSH**	ENN	kNN
arrhythmia	**54.4 ± 20(1)**	**54.4 ± 20(1)**	51.1 ± 20(2)	49.1 ± 22(3)
autos	70.3 ± 9.8(2)	70.3 ± 9.8(2)	60.3 ± 11(3)	**80.7 ± 9.6(1)**
balance-scale	86.5 ± 2.9(3)	87.7 ± 2.5(2)	**88.1 ± 2.5(1)**	83.4 ± 3.2(4)
blood-transfusion	74.8 ± 3.6(2)	74.8 ± 3.8(2)	**76.7 ± 3.9(1)**	70.1 ± 4.2(3)
breast-cancer-diagnostic	**96.4 ± 2.4(1)**	**96.4 ± 2.3(1)**	95.9 ± 2.6(2)	95.2 ± 2.7(3)
breast-cancer-original	**96.5 ± 2(1)**	**96.6 ± 1.9(1)**	**96.6 ± 1.9(1)**	95.3 ± 2.1(2)
breast-cancer-prognostic	**77.8 ± 6.6(1)**	**77.8 ± 6.6(1)**	**77.9 ± 7.2(1)**	71.8 ± 8.2(2)
breast-tissue	**68 ± 12(1)**	**68 ± 12(1)**	65.9 ± 14(2)	**68.5 ± 13(1)**
car-evaluation	80.3 ± 2(3)	72.7 ± 1.3(4)	90.3 ± 1.5(2)	**93.1 ± 1.3(1)**
cardiotocography-1	75.8 ± 2.4(2)	75.8 ± 2.4(2)	74.4 ± 2.7(3)	**77 ± 2.7(1)**
cardiotocography-2	90.6 ± 1.9(2)	**90.8 ± 1.9(1)**	89.9 ± 1.8(3)	**90.7 ± 2(1)**
chess-rook-vs-pawn	90.1 ± 1.5(3)	90.2 ± 1.5(3)	91.3 ± 1.4(2)	**92.1 ± 1.4(1)**
cmc	**44.8 ± 3.8(1)**	**44.9 ± 4.1(1)**	45.3 ± 4(1)	43.9 ± 3.6(2)
congressional-voting	90.2 ± 5.7(2)	90.2 ± 5.7(2)	**90.9 ± 5.4(1)**	**90.8 ± 5.5(1)**
connectionist-bench-sonar	**87.1 ± 7.4(1)**	**87.1 ± 7.4(1)**	82.6 ± 7.7(2)	**87.2 ± 7.1(1)**
connectionist-bench-vowel	96.9 ± 2.4(2)	96.9 ± 2.4(2)	97.2 ± 2.8(2)	**99 ± 1.4(1)**
cylinder-bands	66.7 ± 8.4(2)	66.7 ± 8.4(2)	62 ± 8(3)	**72.2 ± 8.4(1)**
dermatology	**90.6 ± 3.9(1)**	**90.6 ± 3.8(1)**	**91 ± 4.1(1)**	**91.2 ± 3.9(1)**
ecoli	**86.3 ± 4.9(1)**	**86.3 ± 4.9(1)**	**86.3 ± 4.5(1)**	81 ± 5.1(2)
glass	**70.8 ± 8.2(1)**	**70.8 ± 8.2(1)**	69.4 ± 8.4(2)	**70.4 ± 9.1(1)**
habermans-survival	**70.4 ± 6.7(1)**	**70.5 ± 6.8(1)**	**71.4 ± 5.8(1)**	67.4 ± 8(2)
hepatitis	**84.9 ± 11(1)**	**84.9 ± 11(1)**	**85.6 ± 12(1)**	**85.3 ± 11(1)**
ionosphere	**87 ± 4.1(1)**	**87.1 ± 4.2(1)**	83.5 ± 4.7(2)	**86.9 ± 4.4(1)**
iris	**94.8 ± 5.3(1)**	**94.8 ± 5.3(1)**	**95 ± 5.3(1)**	**94.8 ± 5.6(1)**
libras-movement	82.1 ± 6.2(2)	82.1 ± 6.2(2)	78.9 ± 6.4(3)	**86.3 ± 5.5(1)**
liver-disorders	**64.1 ± 7.2(1)**	**64.1 ± 7.2(1)**	62 ± 7.5(2)	**63.3 ± 6.7(1)**
lymph	81.1 ± 10(2)	81 ± 10(2)	78.1 ± 10(3)	**83 ± 9.3(1)**
monks-problems-1	99.7 ± 1.2(2)	99.8 ± 0.98(2)	99.7 ± 1(2)	**100 ± 0(1)**
monks-problems-2	59.4 ± 5.6(2)	59.4 ± 5.6(2)	57.9 ± 5.2(3)	**68 ± 6.2(1)**
monks-problems-3	**98.9 ± 1.5(1)**	**98.9 ± 1.5(1)**	**98.9 ± 1.5(1)**	97.6 ± 1.9(2)
parkinsons	91.2 ± 6.1(2)	91.2 ± 6.1(2)	91.4 ± 5.8(2)	**94.5 ± 5.4(1)**
pima-indians-diabetes	**75.2 ± 4.8(1)**	**75.2 ± 4.8(1)**	74.4 ± 5.1(2)	70.9 ± 5.1(3)
sonar	**87.1 ± 7.4(1)**	**87.1 ± 7.4(1)**	82.6 ± 7.7(2)	**87.2 ± 7.1(1)**
spambase	**92.1 ± 1.3(1)**	92 ± 1.3(2)	91.1 ± 1.3(4)	91.5 ± 1.3(3)
spect-heart	**81.1 ± 7(1)**	**81.1 ± 7.1(1)**	78.1 ± 7.3(3)	79.6 ± 7.4(2)
spectf-heart	**74.3 ± 5.7(1)**	**74.3 ± 5.7(1)**	73 ± 6.4(2)	70.4 ± 8.2(3)
statlog-australian-credit	**78 ± 5.5(1)**	**78 ± 5.5(1)**	**77.7 ± 5.4(1)**	**78.2 ± 5.3(1)**
statlog-german-credit	**70.9 ± 3.6(1)**	**70.9 ± 3.6(1)**	**70.7 ± 3.4(1)**	67.8 ± 4.4(2)
statlog-heart	**78.5 ± 7.4(1)**	**78.5 ± 7.4(1)**	**78.6 ± 7.1(1)**	74.4 ± 8.2(2)
statlog-vehicle	69.8 ± 4(2)	69.8 ± 4(2)	**71.5 ± 3.9(1)**	70.2 ± 4.3(2)
teaching-assistant	51.8 ± 11(2)	51.6 ± 11(2)	46.1 ± 13(3)	**64.4 ± 11(1)**
thyroid-disease	**94.9 ± 0.41(1)**	94.8 ± 0.42(2)	94.7 ± 0.45(3)	94.4 ± 0.64(4)
vote	**91.3 ± 5.3(1)**	91.1 ± 5.5(1)	91.1 ± 5.4(1)	**91.7 ± 5.3(1)**
wine	**95.3 ± 4.9(1)**	**95.3 ± 4.9(1)**	95.2 ± 5(1)	**95.3 ± 4.9(1)**
zoo	50.8 ± 12(2)	50.9 ± 12(2)	47.5 ± 12(3)	**60.2 ± 12(1)**
Mean Rank	1.47 ± 0.094	1.49 ± 0.1	1.89 ± 0.13	1.62 ± 0.13

Table 2. Instance reduction analysis

	LSSm	**LSSmLSH**	ENN
arrhythmia	0.17 ± 0.02	0.17 ± 0.02	0.3 ± 0.04
autos	0.15 ± 0.02	0.15 ± 0.02	0.23 ± 0.02
balance-scale	0.088 ± 0.005	0.095 ± 0.004	0.088 ± 0.004
blood-transfusion	0.16 ± 0.006	0.14 ± 0.006	0.18 ± 0.007
breast-cancer-diagnostic	0.028 ± 0.003	0.025 ± 0.003	0.022 ± 0.003
breast-cancer-original	0.025 ± 0.002	0.022 ± 0.002	0.025 ± 0.002
breast-cancer-prognostic	0.14 ± 0.01	0.14 ± 0.01	0.17 ± 0.01
breast-tissue	0.17 ± 0.02	0.17 ± 0.02	0.16 ± 0.02
car-evaluation	0.15 ± 0.004	0.23 ± 0.004	0.047 ± 0.005
cardiotocography-1	0.11 ± 0.003	0.11 ± 0.003	0.14 ± 0.004
cardiotocography-2	0.064 ± 0.002	0.059 ± 0.002	0.063 ± 0.003
chess-rook-vs-pawn	0.061 ± 0.002	0.061 ± 0.002	0.027 ± 0.002
cmc	0.24 ± 0.005	0.21 ± 0.005	0.31 ± 0.007
congressional-voting	0.055 ± 0.006	0.052 ± 0.006	0.071 ± 0.007
connectionist-bench-sonar	0.047 ± 0.008	0.047 ± 0.008	0.11 ± 0.01
connectionist-bench-vowel	0.057 ± 0.005	0.057 ± 0.005	0.026 ± 0.006
cylinder-bands	0.15 ± 0.01	0.15 ± 0.01	0.24 ± 0.02
dermatology	0.06 ± 0.008	0.057 ± 0.007	0.041 ± 0.006
ecoli	0.095 ± 0.007	0.095 ± 0.007	0.1 ± 0.009
glass	0.15 ± 0.01	0.15 ± 0.01	0.2 ± 0.02
habermans-survival	$0.17 \pm\ \pm 0.009$	0.15 ± 0.008	0.21 ± 0.01
hepatitis	0.083 ± 0.01	0.083 ± 0.01	0.068 ± 0.02
ionosphere	0.036 ± 0.004	0.03 ± 0.004	0.13 ± 0.009
iris	0.066 ± 0.008	0.066 ± 0.008	0.035 ± 0.007
libras-movement	0.089 ± 0.007	0.089 ± 0.008	0.13 ± 0.01
liver-disorders	0.14 ± 0.01	0.14 ± 0.009	0.21 ± 0.02
lymph	0.091 ± 0.01	0.093 ± 0.01	0.11 ± 0.02
monks-problems-1	0.004 ± 0.003	0.0024 ± 0.002	0.00038 ± 0.001
monks-problems-2	0.18 ± 0.006	0.17 ± 0.007	0.3 ± 0.007
monks-problems-3	0.021 ± 0.004	0.014 ± 0.003	0.011 ± 0.002
parkinsons	0.066 ± 0.006	0.066 ± 0.006	0.053 ± 0.008
pima-indians-diabetes	0.13 ± 0.005	0.13 ± 0.005	0.18 ± 0.008
sonar	0.047 ± 0.008	0.047 ± 0.008	0.11 ± 0.01
spambase	0.052 ± 0.001	0.052 ± 0.001	0.067 ± 0.002
spect-heart	0.14 ± 0.007	0.13 ± 0.008	0.14 ± 0.02
spectf-heart	0.13 ± 0.009	0.12 ± 0.009	0.14 ± 0.01
statlog-australian-credit	0.12 ± 0.006	0.12 ± 0.006	0.15 ± 0.008
statlog-german-credit	0.13 ± 0.004	0.13 ± 0.004	0.19 ± 0.007
statlog-heart	0.1 ± 0.008	0.1 ± 0.008	0.11 ± 0.01
statlog-vehicle	0.14 ± 0.005	0.14 ± 0.005	0.15 ± 0.006
teaching-assistant	0.2 ± 0.01	0.18 ± 0.01	0.32 ± 0.03
thyroid-disease	0.043 ± 0.0007	0.042 ± 0.0007	0.045 ± 0.0008
vote	0.056 ± 0.008	0.06 ± 0.008	0.072 ± 0.009
wine	0.019 ± 0.005	0.019 ± 0.005	0.023 ± 0.006
zoo	0.23 ± 0.02	0.23 ± 0.02	0.35 ± 0.03
Mean Removed%	0.1 ± 0.007	0.1 ± 0.007	0.13 ± 0.01

The conclusion is that the LSSm and the LSSmLSH algorithms have the same accuracy and reduction, but the LSSmLSH has better complexity.

References

1. Arnaiz-González, A., Díez-Pastor, J.-F., Rodríguez, J.J., García-Osorio, C.: Instance selection of linear complexity for big data: Knowl.-Based Syst. **107**, 83–95 (2016)
2. Bawa, M., Condie, T., Ganesan, P.: LSH forest: self-tuning indexes for similarity search. In: Proceedings of the 14th International Conference on World Wide Web, pp. 651–660. Chiba, Japan (2005)
3. Brighton, H., Mellish, C.: Advances in instance selection for instance-based learning algorithms. Data Min. Knowl. Disc. **6**(2), 153–172 (2002)
4. Cover, T.M., Hart, P.E.: Nearest neighbor pattern classification. Instit. Electr. Electron. Eng. Trans. Inf. Theory **13**(1), 21–27 (1967)
5. Garcia, S., Derrac, J., Cano, J., Herrera, F.: Prototype selection for nearest neighbor classification: taxonomy and empirical study. IEEE Trans. Pattern Anal. Mach. Intell. **34**(3), 417–435 (2012)
6. Har-Peled, S., Indyk, P., Motwani, R.: Approximate nearest neighbor: towards removing the curse of dimensionality. Theory Comput. **8**, 321–350 (2012)
7. Indyk, P., Motwani, R.: Approximate nearest neighbor—towards removing the curse of dimensionality. In: The Thirtieth Annual ACM Symposium on Theory of Computing, pp. 604–613 (1998)
8. Grochowski, M., Jankowski, N.: Comparison of instance selection algorithms II. results and comments. In: Rutkowski, L., Siekmann, J.H., Tadeusiewicz, R., Zadeh, L.A. (eds.) ICAISC 2004. LNCS (LNAI), vol. 3070, pp. 580–585. Springer, Heidelberg (2004). https://doi.org/10.1007/978-3-540-24844-6_87
9. Jankowski, N., Orliński, M.: Fast encoding length-based prototype selection algorithms. Australian J. Intell. Inf. Process. Syst. **16**(3), 59–66 (2019). Special Issue: Neural Information Processing 26th International Conference on Neural Information Processing. http://ajiips.com.au/iconip2019/docs/ajiips/v16n3.pdf
10. Leyva, E., González, A., Pérez, R.: Three new instance selection methods based on local sets: a comparative study with several approaches from a bi-objective perspective. Pattern Recogn. **48**(4), 1523–1537 (2015). https://doi.org/10.1016/j.patcog.2014.10.001
11. Merz, C.J., Murphy, P.M.: UCI repository of machine learning databases (1998). http://www.ics.uci.edu/~mlearn/MLRepository.html
12. Olvera-López, J.A., Carrasco-Ochoa, J.A., Martínez-Trinidad, J.F.: A new fast prototype selection method based on clustering. Pattern Anal. Appl. **13**(2), 131–141 (2009)
13. Orliński, M., Jankowski, N.: Fast t-SNE algorithm with forest of balanced LSH trees and hybrid computation of repulsive forces. Knowl.-Based Syst. **206**, 1–16 (2020). https://doi.org/10.1016/j.knosys.2020.106318
14. Orliński, M., Jankowski, N.: $O(m \log m)$ instance selection algorithms–RR-DROPs. In: IEEE World Congress on Computational Intelligence, pp. 1–8. IEEE Press (2020). https://doi.org/10.1109/IJCNN48605.2020.9207158. http://www.is.umk.pl/~norbert/publications/20-FastDROP.pdf
15. Wilson, D.: Asymptotic properties of nearest neighbor rules using edited data. IEEE Trans. Syst. Man Cybern. **2**(3), 408–421 (1972)

Entropy-minimization Mean Teacher for Source-Free Domain Adaptive Object Detection

Xing Wei[1,2,3](✉), Ting Bai[1], Zhangling Duan[4], Ming Zhao[1], Chong Zhao[2,5], Yang Lu[1,3], and Di Hu[2]

[1] School of Computer and Information, Hefei University of Technology, Hefei, China
`weixing@hfut.edu.cn`
[2] Intelligent Manufacturing Institute of Hefei University of Technology, Hefei, China
[3] Engineering Research Center of Safety Critical Industrial Measurement and Control Technology, Ministry of Education, Hefei, China
[4] School of Internet, Anhui University, Hefei, China
[5] Engineering Quality Education Center of Undergraduate School, Hefei University of Technology, Hefei, China

Abstract. It is difficult to obtain source domain labeled samples in actual situations due to data privacy protection, limited storage space, high labor costs and other factors. Therefore, we propose a Entropy-minimization mean teacher for Source-free domain adaptive Object Detection(ESOD). We only need a pre-trained object detection model in the source domain and an unlabeled target domain dataset, the pseudo-label is used as the "ground-truth" to complete the pseudo-supervised learning process of the student model for training and generating a detector suitable for the target domain. Firstly, we load the pre-trained source domain detector into the teacher network and the student network, and the target domain samples preprocessed by random enhancement are sent into two networks respectively, the teacher network filters the generated pseudo-labels through the threshold tuning method of entropy minimization. Secondly, the Mean Teacher algorithm is adopted for training to maximize the acquisition of sample features in the target domain, and the consistent regularization strategy is used to minimize the differences between the teacher network and the student network. Experiments are carried out on multiple datasets such as Cityscapes, Foggy Cityscapes and SIM10K, and the results prove the effectiveness and robustness of the proposed method.

Keywords: Source-free Object Detection · Transfer Learning · Domain Adaptation

1 Introduction

Domain adaptive object detection is an essential direction in cross-domain object detection, formally proposed at the 2018 CVPR [1]. In this paper, the author

© The Author(s), under exclusive license to Springer Nature Switzerland AG 2023
M. Tanveer et al. (Eds.): ICONIP 2022, LNCS 13623, pp. 513–524, 2023.
https://doi.org/10.1007/978-3-031-30105-6_43

draws on the idea of transfer learning, and learns the common domain invariant features of the source and target domain by using labeled data in the source domain and unlabeled data in the target domain. However, due to differences in background, light intensity, weather conditions, etc., the image styles of the target domain and the source domain are very different, and the data distribution is quite different. Secondly, since the supervised samples of the object detection task need to label the object location and category at the same time, the labor cost is unaffordable when there are dozens or hundreds of types of objects.

The domain adaptive object detection method needs to access the source domain data in the cross-domain learning process. However, in some practical scenarios, or because of memory storage requirements, shared data, privacy issues, or other dataset processing issues, the source domain data cannot be accessed, and the feature space cannot be extracted, which hinders training. Therefore, the accuracy and robustness of the cross-domain object detection model [3,19] have encountered significant challenges.

The following are the main contributions of our work:

1. We provide a source-free domain adaptive object detection model and its training mechanism, that is, the pre-trained source domain model and target domain unlabeled data are trained to obtain a detector that achieves precise positioning and classification in the target domain.
2. We use the method of entropy minimization for threshold tuning, iteratively filter the generated pseudo-labels, reduce the entropy of the prediction results, complete the pseudo-supervised learning of the student model, and optimize the weight parameters of the mean teacher model.
3. We propose a weight regularization method to reduce the domain discrepancy between domains and optimize the network structure. The experimental analysis is used as the verification indicator to verify the feasibility and effectiveness of the proposed model.

2   Related Work

2.1   Domain Adaptation and Object Detection Domain Adaptation

As a branch of transfer learning, domain adaptive methods are widely used in deep learning, such as object detection [1,11], sentiment classification [12], natural language processing [5], semantic segmentation [15] and so on. Chen Y et al. [3] first proposed a cross-domain object detection method and designed two domain adaptive modules to eliminate the domain discrepancy between the image-level and instance-level. A consistent regularization method is proposed to learn domain invariant features. Zheng Y et al. [19] proposed a feature adaptive method from coarse-grained to fine-grained, gradually and accurately aligning the depth features to achieve two-stage cross-domain object detection. Inspired by the MMD method, our work uses regularization strategy to minimize the source and target domain distribution distances, achieve feature alignment and improve accuracy.

2.2   Source-Free Domain Adaptation

In the traditional domain adaptive method, a large number of works use source domain labeled datasets, the cost is massively for training datasets to be annotated, and private information is illegally accessed. So unsupervised DA(domain adaptation) applications are proposed. The related survey found that the application scenarios of unsupervised DA are primarily applied to classification networks. Liang J et al. [10] proposed a novel self-supervised pseudo-label method to enhance the representation learning of the target domain and introduced the idea of hypothesis transfer learning (HTL), shared classifier parameters in the target domain and the source domain. VK Kurmi et al. [8] provided a generative framework to solve the problem of the domain without source datasets, the classifier uses pseudo samples with labeled information to adapt to the target domain. Li R et al. [17] proposed a cooperative conditional generative adversarial network (3C-GAN) without the source domain dataset, while the generator and prediction model are collaboratively enhanced, and the classification performance improved.

As object detection is concerned, Lin Xiong et al. [16] designed an approximation method based on the Law of Large Numbers to obtain the domain perturbation, thereby constructing a super target domain, and using the learning alignment from the super target domain to the target domain to avoid the imbalance problem in cross-domain object detection to a certain extent. Dan Zhang et al. [18] proposed a new vision that utilizes the style information of batch normalization stored in a pre-trained source model, converted into source-like style features, to challenge the cross-domain object detection task. However, by analogy with the classification method, the biggest problem is that it cannot directly generate labeled source domain samples, the idea of source-free joining adversarial generation [4] is challenging to realize. Since each picture in object detection contains multiple labels, it is difficult to use the pre-trained model to generate labeled source domain pictures from noise through the GAN network. The research on domain adaptive object detection without source data is still in its infancy, and there is enormous room for performance improvement.

3   Proposed Method

The overall framework uses the pre-trained source model to initialize the training model, see Fig. 1. First, we provide a set of unlabeled target domain data samples, defined as $D_t = \{x_i\}_{i=1}^{N_t}$, where x_i represents the ith picture in the target domain, N_t represents the number of pictures in the target domain, two noisy pictures x_i^T and x_i^S are generated by Random Augmentation (RA) and sent to the teacher model and the student model respectively. The two models share the same region proposals generated by the teacher model's RPN to achieve feature extraction and generate feature maps f^T and f^S, where the backbone is based on the ResNet101 network.

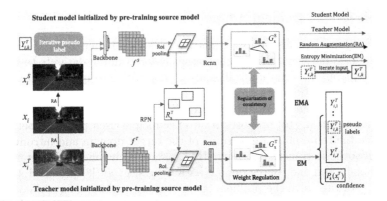

Fig. 1. Overview: The overall framework consists of two network models (the yellow part: Teacher model, the gray part: Student model), use the two noisy pictures x_i^T and x_i^S generated by the target domain image x_i as the input of the two networks respectively, where x_i^T is used as the input of the teacher model T_{model}, after threshold tuning of entropy minimization, the corresponding pseudo-label $Y_{i,k}^T$ of the kth iteration is generated. The x_i^S with the pseudo-label $Y_{i,k}^T$ is used as the input of the student model S_{model}, and the two network models generate feature maps f^T and f^S respectively after feature extraction. The weights of the teacher network are frozen, and the weight parameters are only updated by the exponential moving average(EMA) on the student side, and finally a detector suitable for the target domain is trained and generated.

3.1 Entropy Minimization Training

In object detection tasks, how identifying label errors and characterizing label noise is a necessary but easily overlooked task. The existence of label noise leads to the selection of too high or too low confidence threshold, which will affect the performance of training. Therefore, we use information entropy to evaluate the quality of pseudo-labels, the lower the entropy value, the lower the proportion of false-positive samples, and the higher the reliability value of the pseudo-label. During the iterative training process of the entire dataset, a reasonable lower entropy value is continuously selected, the confidence threshold is dynamically updated, and reliable pseudo-labels are generated to participate in the training.

Based on this task, we introduce the entropy minimization method in the teacher network, use the input sample $D_t = \{x_i\}_{i=1}^{N_t}$ after data augmentation to generate x_i^T as the input of the teacher model T_{model}, which T_{model} is initialized by the pre-trained model parameters, and then generates pseudo-labels and calculates the corresponding confidence, the formula is as follows:

$$\left\{ Y_{i,k}^T, P_k(x_i^T) \right\}_{i=1}^{N_t} = \left\{ T_{model}(x_i^T | h, \Re) \right\}_{i=1}^{N_t} \tag{1}$$

where $Y_{i,k}^T$ represents the pseudo-label generated by training the teacher model at the kth iteration, x_i^T represents the unlabeled target domain samples of the input teacher model, and represents the weight parameter of the teacher model at the *kth* iteration. $P_k(x_i^T)$ represents the confidence calculated by the *kth* iteration training, which is output by the softmax of the classification branch, the pseudo-label $Y_{i,k}^T$ is determined by the argmax of the foreground class probability, if the value is above the confidence threshold h, the corresponding box is assigned as the class label with the largest score. Otherwise, it is defined as the background class. h represents the confidence threshold, in which the first training uses a given confidence threshold to generate pseudo-labels, and then iterative training takes the local minimum entropy value to obtain the optimal threshold h.

$$H(D_t) = \sum_i^{N_t} \left(\frac{1}{n_c} \sum_c^{n_c} P_k^c(x_i^T) \log(P_k^c(x_i^T)) \right) \tag{2}$$

$$h_{optimal} = \arg\min(H(D_t)) \tag{3}$$

where $P_k^c(x_i^T)$ represents the confidence of a category at the *kth* iteration, n_c represents the total number of categories, and c represents the category.

Based on our model, the teacher outputs reliable pseudo-labels, and participates in the next iteration of training. Finally, the weight parameters of the teacher model are fine-tuned through EMA, new pseudo-labels are generated by filtering the confidence threshold obtained in the previous iteration, and the final labels and detectors suitable for the target domain are generated iteratively.

3.2   Weight Regulation Module

In this module, we introduce a weight regularization method to generate two relational graphs (G_x^S, G_x^T) respectively, propose two levels of regularization loss, guide the student model by backpropagation, and ensure the consistency of prediction between the teacher model and the student model to optimize our model.

Inter-graph Consistency. Firstly, the picture is input into the model and the relationship diagram $G_{x_i}^S = \{V_{x_i}^S, a_x^S\}$ is obtained through calculation, which $V_{x_i}^S$ represents the probability set of all regions, a_x^S represents the affinity matrix of the graph obtained by the student model. The similarity $(a_x^S)_{r_i r_j}$ between every two regions in the figure is obtained, where $r_i, r_j \in R_x^S$, which represents two different regions in the figure. Region features are extracted through the ROI-pooling layer and represented by $f_{r_i}^S$, and the corresponding feature is converted into a fixed-size dimension. The calculation formula of cosine similarity is introduced to obtain the similarity between every two regions, and generate an affinity matrix of the student model.

$$\left(a_x^S\right)_{r_i r_j} = f_{r_i}^S \cdot f_{r_j}^S / (\|f_{r_i}^S\|_2 \cdot \|f_{r_j}^S\|_2) \tag{4}$$

where $f_{r_i}^S$ and $f_{r_j}^S$ respectively represent the feature map with fixed dimensions obtained by the student model through the ROI-pooling layer processing, $(a_x^S)_{r_i r_j}$ represents the calculated similarity between every two regions.

The calculation method of the affinity matrix of the teacher model is the same as above, and two affinity matrices can be obtained through the above calculation. Therefore, the consistency difference between graphs can be defined as the mean square error between two affinity matrices, the calculation formula is as follows:

$$L_{Ite} = \frac{1}{|R_x^T|^2} \cdot \left\| a_x^S - a_x^T \right\|_2^2 \tag{5}$$

where a_x^S and a_x^T represent the affinity matrix of the student model and the teacher model respectively, R_x^T represents the region proposals generated by the teacher's RPN network.

Intra-graph Consistency. We hope to apply the principle of consistency to guide the student model through the pseudo-labels generated by the teacher model. First, we use the initial pre-training source model as benchmark to learn knowledge and generate pseudo-labels $l_r = \arg\max_{k \in C} \left(d_{r_k}^T \right)$ for each region proposal $r_x^T \in R^T$ on the teacher-side, which represents the total number of categories in the dataset, k represents a specific category, and $d_{r_k}^T$ represents the probability of the predicting. Then, the supervision matrix M_x^T of $(|R_x^T| \times |R_x^T|)$ can be generated in the following way to determine whether every region belongs to the same category:

$$(M_x^T)_{i,j} = \begin{cases} 1 \text{ if } l_{r_j} = l_{r_i} \\ 0 \text{ otherwise} \end{cases} \tag{6}$$

where l_{r_i} and l_{r_j} represent the pseudo-labels generated in a specific area, and $(M_x^T)_{i,j}$ represents the generated supervision matrix with i rows and j columns.

$$L_{Ita} = \frac{\sum\limits_{1 \leq i,j \leq |R_x^T|} (M_x^T)_{i,j} \cdot \left(1 - (a_x^S)_{r_i r_j} \right)}{MAX \left(1, \sum\limits_{1 \leq i.j \leq |R_x^T|} (M_x^T)_{i,j} \right)} \tag{7}$$

where $(a_x^S)_{r_i r_j}$ represents the calculated similarity between every two regions in the student model, $(M_x^T)_{i,j}$ represents the generated supervision matrix with i rows and j columns, $1 \leq i,j \leq |R_x^T|$ represents each region in the teacher model.

3.3   Overall Objective Function

Firstly, the weights of the teacher model are frozen, there is only one source for the weight update of the teacher model, which is the exponential moving average of the student network. Therefore, the teacher model's network weight can be considered a combination of the current student model and the earlier version. The network parameter update of the teacher model is obtained by Eq. 8:

$$\theta_t = \lambda \theta_{t-1} + (1 - \lambda) \theta_s \tag{8}$$

where θ_{t-1} and θ_t represent the upper training of the teacher model and the training parameters of this layer, respectively, θ_s represents the currently updated network parameters of the student model, λ represents the weight smoothing parameter, which is set to 0.99.

Secondly, the invariant features of the domain are learned from the perspective of the student model, and the network parameters are optimized to reduce the model deviation, including two regularization losses, to achieve inter-graph structure matching and intra-graph feature alignment respectively. The overall training loss is expressed as follows:

$$L = \alpha L_{Ite} + \beta L_{Ita} \tag{9}$$

where α, β is the tuning parameter.

4   Experiments

4.1   Dataset and Experimental Settings

In our experiment, the setup is consistent with [1]. We only use the pre-trained source model to participate in the training, and no longer use the source domain dataset. In all experiments, we use 4 GTX 1080 Ti with 11GB memory for training, batchsize is set to 1, the initial learning rate is set to 0.0001, the momentum is set to 0.9, the overall training loss α is set to 1.0, β is set to 0.99.

The Cityscapes [2], Foggy Cityscapes, SIM10K and other datasets are used for the training. In the experiment, several basic data augmentation strategies are used to process the input samples of the target domain, including techniques such as color jitter, gaussian blur and grayscale. This method adds noise by randomly adjusting the sample parameters. The input is the same image, but the input in different epochs is constantly changing, that is, the input-output mapping is not fixed, which is equivalent to the consistent regularization of the same unlabeled image between different epochs. The model does not easily overfit the noise in the pseudo-labels in this case.

4.2   Comparison Results

Normal Weather to Foggy Weather. In this experiment, the Cityscapes [2] dataset is used as the source domain, we use it to pre-train a source model, Foggy Cityscapes [14] is an unlabeled target domain dataset. The Foggy Cityscapes dataset is based on the Cityscapes image, it is a synthetic foggy dataset that simulates real scenes.

Table 1 obtained the object detection results in the adaptive process from Cityscapes to Foggy Cityscapes. The experimental results show that the baseline accuracy is only 22.6%, which is 9.3% higher than that of the baseline. Due to the similarity between the source domain and the target domain, the optimization effect of the model is remarkable, and its training effect exceeds some traditional domain adaptation methods, which proves the effectiveness of the method.

Table 1. *Cityscapes → Foggy Cityscapes adaptation* The average precision of all categories under different cross-domain object detection methods (mAP)

Methods	Person	Rider	Car	Truck	Bus	Train	Motor	Bicycle	mAP
baseline	24.2	23.0	34.2	15.0	26.4	14.3	15.7	28.1	22.6
DA-Faster [1]	25.0	31.0	40.5	22.1	35.3	20.2	20.0	27.1	27.6
BDC-Faster [13]	26.4	37.2	42.4	21.2	29.2	12.3	22.6	28.9	27.5
Selective-Faster [20]	33.5	38.0	48.5	16.5	39.0	23.3	28.0	33.6	33.8
MAF [6]	28.2	39.5	43.9	23.8	39.9	33.3	29.2	33.9	34.0
SFOD(SED) [9]	11.8	25.3	40.4	**34.3**	21.7	**34.5**	32.6	**44.0**	30.6
SFOD(Ideal) [9]	22.3	**44.0**	38.2	31.4	15.1	25.7	**34.6**	36.8	31.0
Ours	**28.3**	39.3	**49.3**	28.8	**31.5**	21.7	24.7	31.5	**31.9**
Oracle	33.0	45.1	49.3	32.0	41.9	38.2	32.5	49.1	40.1

Adaptation from Real to Artistic Images. In this experiment, the Pascal VOC dataset is used as the pre-training source model, which contains 20 common real-world categories. We use Pascal VOC's 5011 training set (2501 training set, 2510 validation set) in the experiment, The Watercolor2k dataset is used as the target domain to be tested. It contains six categories of watercolor-style artistic pictures, with 2000 pictures. In the experiment, we only use the 6 categories shared by the two datasets for training.

As shown in Table 2, we report the detection results of the dataset under six categories and compare them. Our model is 5.4% higher than the baseline, better than the traditional domain adaptive object detection method, and the source-free object detection accuracy of SOAP, proving the feasibility and applicability of our method in different application scenarios.

Table 2. The mean average precision (mAP) of the cross-domain detection of the *Pascal VOC → Watercolor adaptation.*

Methods	Bike	bird	car	cat	dog	person	mAP
baseline	74.6	48.4	45.1	28.9	22.0	53.1	45.4
DA-Faster [1]	75.2	40.6	48.0	31.5	20.6	60.0	46.0
BDC-Faster [13]	68.6	48.3	47.2	26.5	21.7	60.5	45.5
SOAP [16]	**77.7**	43.2	40.1	**48.2**	**38.8**	55.4	50.6
Ours	76.1	**48.7**	**45.2**	42.2	35.2	**57.4**	**50.8**

Adaptation from Virtual to Real Images. In this experiment, the SIM10K dataset is used as the pre-training source model, and Cityscapes is used as the unlabeled target domain dataset. The SIM10K dataset contains synthetic images in 10k computer game rendering. It contains 10,000 annotated images with the category "Car", so we only chose the category "Car" to participate in the experiment.

Table 3 compares the performance of cross-domain detection from the synthetic dataset to the real dataset. We only use the category "Car" to participate in the experiment. As can be seen from the table, our model's AP reaches 42.4%, which is 3.9% higher than DA-Faster, and 1.6% higher than the source-free object detection method SOAP. Therefore, experiments show that style transfer in virtual-reality can help the model better capture the discrepancies between graphs and improve the detection accuracy.

Table 3. The average precision (AP) of the "Car" category in the cross-domain detection of the $SIM10K \rightarrow Cityscapes$ adaptation

Methods	AP on Car
baseline	33.1
DA-Faster [1]	38.5
SW-Detection [13]	40.1
MAF [6]	41.1
AT-Faster [7]	42.1
SOAP [16]	40.8
SFOD(SED) [9]	42.3
Ours	**42.4**
Oracle	56.0

4.3 Ablation Experiment

We performed several ablation experiments to examine the contribution of each module to the performance of this object detector. Our model consists of 3 modules, while ensuring that all settings are exactly the same, we add each module in turn to study the effectiveness of each module. Table 4 shows the domain adaptive results of each module from Cityscapes to Foggy Cityscapes, where EM represents Entropy Minimization, Inter-G represents Inter-graph consistency and Intra-G represents Intra-graph consistency. We can observe that each module can improve the performance of the model. It can be seen from Table 4 that the baseline without any DA adjustment is only 22.6%, and the accuracy is increased to 4.4% after adding the EM module. Obviously, iteratively filtering pseudo-labels provides more accurate guidance for the mean teacher framework, prompting the generation of more valuable pseudo-labels in subsequent iterations and improving the model's overall performance.

4.4 Visualization of Results

Figure 2 shows two groups of cross-domain object detection results in 3 different scenarios. The three rows of images from top to bottom represent Cityscapes to Foggy Cityscapes, SIM10K to Cityscapes and Pascal VOC to Watercolor.

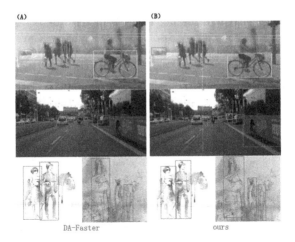

Fig. 2. Cross-domain object detection results of three different datasets, two sets of experiments(A and B) show the visualization results of the DA-Faster model and our model on the dataset under different scenarios, respectively.

Among the three sets of experimental scenarios we applied, it can be observed that the detection accuracy of each category is improved, and the probability of misprediction is reduced. Compared with the domain adaptive method, the experimental results are comparable, and can effectively demonstrate the effectiveness of our method on object detection under source-free domain conditions.

Table 4. Analysis of ablation experiments from Cityscapes to Foggy Cityscapes.

EM	Inter-G	Intra-G	mAP
			22.6
√			27.0
√	√		28.9
√		√	30.1
√	√	√	31.9

Figure 3 shows the visualizations using the t-SNE algorithm. The scene of the experiment is from Cityscapes to Foggy Cityscapes, with a total of 8 different colors to label the categories, the left is the source-only model, the right is the training result of our method, it can be seen the distribution gap between the source domain and the target domain. After model training, the problem of uneven sample distribution and unclear boundaries is optimized.

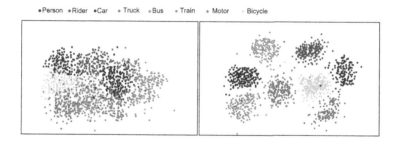

Fig. 3. Visualizations of feature map using t-SNE algorithm. The scene is from Cityscapes to Foggy Cityscapes. Left: the result obtained with the source-only model. Right: the result obtained with our proposed model.

5 Conclusion

In our work, in order to solve the problems of data privacy protection, limited storage space or high labor costs in the experimental process and training, we propose a model that uses entropy minimization for threshold tuning, iteratively filters the generated pseudo-labels, so that the model is more affected by the correct pseudo-labels during the training process, and uses weight regularization method to learn domain-invariant representations, and finally generate a detector suitable for target domain samples. Multiple datasets are used for training to obtain detection results, and experiments prove the effectiveness of our proposed method. In future, we will continue to study the domain adaptive problem based on object detection, and study the application of other domain adaptive methods in source-free domain.

Acknowledgements. This work was supported by Joint Fund of Natural Science Foundation of Anhui Province in 2020 (2008085UD08), Anhui Provincial Key R&D Program (202004a05020004), Open fund of Intelligent Interconnected Systems Laboratory of Anhui Province (PA2021AKSK0107), Intelligent Networking and New Energy Vehicle Special Project of Intelligent Manufacturing Institute of HFUT (IMIWL2019003, IMIDC2019002).

References

1. Chen, Y., Li, W., Sakaridis, C., Dai, D., Van Gool, L.: Domain adaptive faster R-CNN for object detection in the wild. In: Proceedings of the IEEE Conference on Computer Vision and Pattern Recognition, pp. 3339–3348 (2018)
2. Cordts, M., et al.: The cityscapes dataset for semantic urban scene understanding. In: Proceedings of the IEEE Conference on Computer Vision and Pattern Recognition, pp. 3213–3223 (2016)
3. Deng, J., Li, W., Chen, Y., Duan, L.: Unbiased mean teacher for cross-domain object detection. In: Proceedings of the IEEE/CVF Conference on Computer Vision and Pattern Recognition, pp. 4091–4101 (2021)

4. Ganin, Y., Lempitsky, V.: Unsupervised domain adaptation by backpropagation. In: International Conference on Machine Learning, pp. 1180–1189. PMLR (2015)
5. Gururangan, S., Marasović, A., Swayamdipta, S., Lo, K., Beltagy, I., Downey, D., Smith, N.A.: Don't stop pretraining: adapt language models to domains and tasks. arXiv preprint arXiv:2004.10964 (2020)
6. He, Z., Zhang, L.: Multi-adversarial faster-RCNN for unrestricted object detection. In: Proceedings of the IEEE/CVF International Conference on Computer Vision, pp. 6668–6677 (2019)
7. He, Z., Zhang, L.: Domain adaptive object detection via asymmetric tri-way faster-RCNN. In: Vedaldi, A., Bischof, H., Brox, T., Frahm, J.-M. (eds.) ECCV 2020. LNCS, vol. 12369, pp. 309–324. Springer, Cham (2020). https://doi.org/10.1007/978-3-030-58586-0_19
8. Kurmi, V.K., Subramanian, V.K., Namboodiri, V.P.: Domain impression: a source data free domain adaptation method. In: Proceedings of the IEEE/CVF Winter Conference on Applications of Computer Vision, pp. 615–625 (2021)
9. Li, X., et al.: A free lunch for unsupervised domain adaptive object detection without source data. arXiv preprint arXiv:2012.05400 (2020)
10. Liang, J., Hu, D., Feng, J.: Do we really need to access the source data? source hypothesis transfer for unsupervised domain adaptation. In: International Conference on Machine Learning, pp. 6028–6039. PMLR (2020)
11. Liu, Y.C., et al.: Unbiased teacher for semi-supervised object detection. arXiv preprint arXiv:2102.09480 (2021)
12. Sadr, H., Nazari Soleimandarabi, M.: ACNN-TL: attention-based convolutional neural network coupling with transfer learning and contextualized word representation for enhancing the performance of sentiment classification. J. Supercomput. **78**(7), 10149–10175 (2022)
13. Saito, K., Ushiku, Y., Harada, T., Saenko, K.: Strong-weak distribution alignment for adaptive object detection. In: Proceedings of the IEEE/CVF Conference on Computer Vision and Pattern Recognition, pp. 6956–6965 (2019)
14. Sakaridis, C., Dai, D., Van Gool, L.: Semantic foggy scene understanding with synthetic data. Int. J. Comput. Vision **126**(9), 973–992 (2018)
15. Tasar, O., Happy, S., Tarabalka, Y., Alliez, P.: ColorMapGAN: unsupervised domain adaptation for semantic segmentation using color mapping generative adversarial networks. IEEE Trans. Geosci. Remote Sens. **58**(10), 7178–7193 (2020)
16. Xiong, L., Ye, M., Zhang, D., Gan, Y., Li, X., Zhu, Y.: Source data-free domain adaptation of object detector through domain-specific perturbation. Int. J. Intell. Syst. **36**(8), 3746–3766 (2021)
17. Yang, S., Wang, Y., van de Weijer, J., Herranz, L., Jui, S.: Unsupervised domain adaptation without source data by casting a bait. arXiv preprint arXiv:2010.12427 (2020)
18. Zhang, D., Ye, M., Xiong, L., Li, S., Li, X.: Source-style transferred mean teacher for source-data free object detection. In: ACM Multimedia Asia, pp. 1–8 (2021)
19. Zheng, Y., Huang, D., Liu, S., Wang, Y.: Cross-domain object detection through coarse-to-fine feature adaptation. In: Proceedings of the IEEE/CVF Conference on Computer Vision and Pattern Recognition, pp. 13766–13775 (2020)
20. Zhu, X., Pang, J., Yang, C., Shi, J., Lin, D.: Adapting object detectors via selective cross-domain alignment. In: Proceedings of the IEEE/CVF Conference on Computer Vision and Pattern Recognition, pp. 687–696 (2019)

IA-CL: A Deep Bidirectional Competitive Learning Method for Traveling Salesman Problem

Haoran Ma, Shikui Tu$^{(\boxtimes)}$, and Lei Xu$^{(\boxtimes)}$

Department of Computer Science and Engineering, Shanghai Jiao Tong University,
Shanghai, China
{mahaoran,tushikui,lxu}@sjtu.edu.cn

Abstract. There is a surge of interests in recent years to develop graph neural network (GNN) based learning methods for the NP-hard traveling salesman problem (TSP). However, the existing methods not only have limited search space but also require a lot of training instances with ground-truth solutions that are time-consuming to compute. In this paper, we propose a deep bidirectional competitive learning method to address the above issues. The search space is expanded by training multiple weak but complementary models simultaneously, while the sample effiency is significantly improved by devising a gap-aware reweighting scheme over the TSP instances. Specifically, TSP is modeled in a one-by-one construction way by a GNN to assist the heuristic search. Weights are relatively increased for the instances with larger gap between the search algorithm's solution and the optimal one. The reweighted training set are pipelined to train the next TSP model with strength on the error part. With the error feedback from the search component, multiple complementary GNNs are obtained using this bidirectional alternations. Finally, we present a simple competing strategy by taking the minimum length of the predictions using the multiple TSP models. Experimental results indicate that our method achieves good generalization.

Keywords: Traveling Salesman Problem (TSP) · Graph Neural Network (GNN) · Deep Bidirectional Learning · Competitive Learning (CL)

1 Introduction

The traveling salesman problem (TSP) is a classic NP-hard combinatorial optimization problem [1]. For a given set of city coordinates, the goal of TSP is to find the shortest route that visits each city exactly once and back to the original city. Traditional heuristic algorithms rely on hand-crafted rules to efficiently search for an approximate solution.

Recently, there is a surge of interests to tackle TSP by developing deep learning methods. Generally, models give probabilities of feasible actions to direct

M. Tanveer et al. (Eds.): ICONIP 2022, LNCS 13623, pp. 525–536, 2023.
https://doi.org/10.1007/978-3-031-30105-6_44

the search of search algorithm. For example, Dai et al. [2] generate probabilities through a Graph Neural Network(GNN), called S2V, to guide the search of Greedy search algorithm, that is, choose the unvisited city with the highest probability every step. Many approaches have been devoted to devise a powerful model to provide more reliable probabilities [3–5], but they rarely noticed that the feedback of the search component are able to in turn guide the learning of the model. Most models in TSP are used to learning heuristics, rather than directly solving the objective function of TSP, so that the loss of the network is not enough to measure the error of the overall solution.

It is noted that the deviation of the solution by the search component from the optimal TSP tour can actually serve as a global feedback to guide the GNN representation learning. As suggested in [6–8], the neural network learning and the search algorithm are complementary and beneficial to each other. In this paper, we propose a deep bidirectional competitive learning method for TSP, called IA-CL, which learns the feedback from the search component in training a model set. It trains a model to join the model set in each IA alternation, and adjusts the training set so that the newly trained model can make up for the shortcomings of the previous models as much as possible. Gap-aware reweighting scheme is designed to properly define which instances are shortcomings and we assign higher weights to these instances so that the model trained by the next alternation will focus on them. This in turn leads to better results for the whole model set.

Our contributions mainly include the following aspects:

- We propose a deep bidirectional competitive learning framework for the TSP. It trains multiple TSP models which may be individually weak but have relative strength. This is realized by a new bidirectional alternation between GNN representation learning (over the TSP graph of selected or unvisited nodes) and the search algorithm (which constructs the TSP solution by selecting one node after another). In contrast, the existing methods pay little attention to the possible benefit from the search component to the GNN learning.
- We present a gap-aware reweighting scheme to enable the search algorithm to assist the deep GNN learning over the TSP graphs. The scheme adjusts the weights for each training instance according to gap between the predicted TSP solution and the ground-truth one. By focusing on learning the shortcomings of other models, the model set has achieved good performance, and in this process, sample effiency is improved by reweighting raw data.
- Empirical analysis demonstrates that our method can train a strong TSP solver. Our solver generalizes well to the TSP instances with more cities than those in the training set. It is better or at least comparable with the existing state-of-the-arts.

2   Related Work

2.1   Advances in Learning TSP

Learning-based TSP methods can be dated back to the Hopfield network in 1985 [9]. Recently, [10] proposed Pointer Net to learn the conditional probability of a

TSP tour. [11] improve the Pointer Net and train with reinforcement learning, but when the city number goes large, it becomes intractable to find the optimal tour. Later, learning methods were investigated on searching heuristics for two categories: constructing TSP tour by connecting one city after another, such as [2,3,12–14], or improving from an initial tour by continuously optimizing the current solution, such as [5]. The former commonly use Greedy search, Beam search and Sampling search as the search component, while the latter is often combined with 2-OPT [15]. There are also some works use advanced search algorithms such as MCTS and LKH to improve the performance [12–14,16].

The step-by-step construction paradigm attracts increasing attention, but they mainly focused on designing powerful model structures. [2] regard TSP instances as a graph, and GNN was applied to compute embedding of the current state of the intermediate solution and the structure of the graph. Transformers were also used as the network structure [3,5,17–19]. In [4], a residual graph convolutional network was devised to output an edge adjacency matrix denoting the probabilities of edges occurring on the TSP tour. There are also some works that exploit the symmetries in the representation [20].

Our method falls into the step-by-step construction paradigm. Different from the above methods, we not only apply deep GNN to provide construction heuristics to the search algorithm, but also consider the global feedback from the search component to enhance the GNN representation learning. We also train multiple weak models in a pipeline and compete them to achieve a strong TSP solver.

2.2 Bidirectional Learning for Combinatorial Problem

A bidirectional problem-solving scheme, called IA-DSM, was proposed for doubly stochastic matrix (DSM) featured combinatorial tasks including TSP [6]. The scheme was developed under the framework of deep bidirectional intelligence via yIng yAng (IA) system, which is featured by alternating between the A-mapping (for learning and perceiving) and the I-mapping (for solving and searching). The work in [21] implemented IA-DSM on graph matching problem, and indicate that this scheme is a promising direction. Our deep bidirectional competitive learning (CL) method for TSP also falls into the framework of IA-DSM, and thus we call it IA-CL. Our method is the first bidirectional problem-solving scheme for TSP.

3 Method

We consider symmetric 2D Euclidean TSP on a complete undirected graph $G(V, E)$, where V is the vertex set, and E is the edge set. A solution of TSP is a tour that visits each vertex exactly once and return to the original one, which can be defined as a permutation of vertices, and its length can be calculate by the sum of Euclidean distances of vertex coordinates. Our goal is to find the solution with the shortest length.

We propose a deep bidirectional competitive learning method, shortly called IA-CL, for TSP. The goal is to learn the feedback from search component in

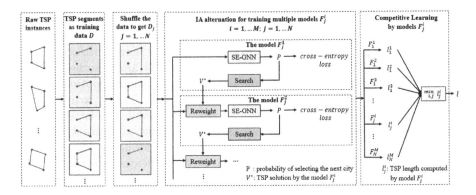

Fig. 1. Overview of IA-CL. It contrains five steps. The first two steps process raw TSP instances into training samples suitable for the network. Steps 3 and 4 describe how to train the model set through IA alternation, and the last step is Competitive Learning (CL), which is use to combine the models.

training the model set, so as to achieve high accuracy even with a small training set and simple model structure. Figure 1 shows the five steps of IA-CL. The first two steps process raw TSP instances into training samples suitable for the network. In next two steps, we train the model set through IA alternation. In each alternation, the training data is adjusted according to the performance of previous models. Training with the adjusted data allow new model focus on the error of previous models, thus make the model set achieve better results. The way to adjust data is called gap-aware instances reweighting. In the last step, the model set are generate solutions in a competitive manner for testing. In the following sections, we will explain each step in detail.

3.1 Data Preprocessing

Each model in the model set constructs solutions step by step. At each step, the model predicts next visited probability for all unvisited cities. In order to train in a supervised way, we collect raw *TSP instances* with the ground-truth tours, and for any TSP instance z, we define *TSP segments* by removing the vertices from the ground-truth tour one-by-one consecutively. For example, for a tour $z = (v_1, v_2, \ldots, v_K)$, the training data D would include all its segments, i.e., $\{(S_k, v_{k+1}), 1 \le k \le K\} \subseteq D$, where S_k denotes the partial tour $v_1, \ldots, v_k$, and v_{k+1} is used as the label.

3.2 Train the Model Set (IA)

After generating the training data D, we shuffle it for N times to get D_j, $j = 1, \ldots, N$. It is noted that D_j contain the same data but in different orders when they are fed into the model for training. Training processes on D_j would lead to different local optimums along separate trajectories, and obtain models with diversified abilities. Selecting N different random seeds to train the model can also achieve the same effect. Here we use RS as this part from the perspective of adjusting the data.

The IA Alternation. The goal of IA is to learn the feedback from the search component, which allows one model to focus on the shortcomings of other models, resulting in better performance for the whole model set. For each shuffled data set D_j, we perform M IA alternations. Each alternation consists of three parts: train one model with the current training set to join to the model set, calculate the performance of the current model set on the training set, and adjust the weights of training set for the next alternation.

In the first IA alternation, the raw training set D_j is used to train the first model F_j^1. We adopt the static edge graph neural network (SE-GNN) in [12] as the network backbone. Please refer to Eq. (3) (4) in [12] for model structure. SE-GNN predicts the next visited probability P for all unvisited cities, and we calculate cross entropy loss between the predicted P and the label v_{k+1} to train it in a supervised way. See Eq. (5) in [12] for the cross-entropy formula.

Next we need to know which TSP instances in the training set performs poorly with the current model set. Here we choose Greedy algorithm to assess it. For each sample in the training set D_j, network F_j^1 gives probability P of all unvisited cities, and Greedy algorithm selects the one with the highest probability, adds the city to partial route. Repeats this process until all cities are visited. Greedy algorithm is one of the simplest way to construct a TSP solution. Therefore, it can better reflect the ability of models themselves. Solutions generated in this way can reflect performance of the current model set.

Finally, in order to make the model trained in the next alternation better compensate for the shortcomings of the previous models, we calculate weights to adjust the training set using Gap-aware instances reweighting scheme. The adjusted training set contains the same samples, but samples performed poorly by the previous models are assigned higher weights.

In the next alternation, we use the adjusted training set to train the model F_j^2, and so on. With the help of the successive global feedback information, a collection of TSP models that have relative strengths would cover the TSP search space as much as possible, and finally work together in a competitive manner for the best output.

Gap-Aware Instances Reweighting. We expect the model to compensate for the shortcomings of other models. In TSP, the metric for evaluating shortcomings is gap as Eq. (1) shown. Therefore, we assign a weight to each TSP instances in the training set, which is calculated based on its gap, to focus on those TSP instances with relative large gap.

Figure 2 shows the process. The model (in green) assist the search algorithm (in blue) to predict the solutions, and the gap between the prediction and ground-truth in turn guide the model training by reweighting the data (in purple). This scheme of calculating weights based on gap, and reweighting the traing data called Gap-awawre instances reweighting.

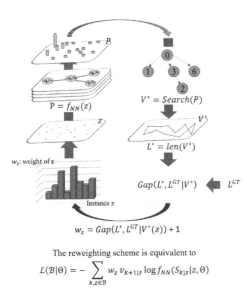

$$P = f_{NN}(z)$$

$$V^* = Search(P)$$

$$L^* = len(V^*)$$

$$Gap(L^*, L^{GT}|V^*) \Leftarrow L^{GT}$$

w_z: weight of z

$$w_z = Gap(L^*, L^{GT}|V^*(z)) + 1$$

The reweighting scheme is equivalent to

$$\mathcal{L}(\mathcal{B}|\Theta) = - \sum_{k,z \in \mathcal{B}} w_z v_{k+1|z} \log f_{NN}(S_{k|z}|z, \Theta)$$

Fig. 2. Gap-aware instances reweighting.

To measure the quality of the solution, we define the gap as the ratio between the TSP length of the predicted solution and the optimal solution, as follows:

$$Gap(L^*, L^{GT}|V^*) = (\frac{L^*}{L^{GT}} - 1) \times 100\%, \quad (1)$$

where L^* represents the length by the TSP solution V^* as indicated in Fig. 2, and L^{GT} denotes the length of the ground-truth (GT) solution. Since $L^* \geq L^{GT}$ always holds, the gap is nonnegative.

Based on Eq. (1), we define the weight w_z for the TSP instance z as below:

$$w_z = Gap(L^*, L^{GT}|V^*(z)) + 1. \quad (2)$$

Notice that if the gap is zero, the weighting on instance z is unchanged. More weight is placed on the instance with larger gaps, so that difficult instances for the current model set will get more attention when training the next model F_j^{i+1}. The models $\{F_j^i\}$ are trained with relative strengths for different TSP instances, and they work jointly to reduce the average gap between the predictions and the ground-truths.

The reweighting scheme on the TSP instances is equivalent to adding a weight to the loss function. Denote a batch of TSP segments as $\mathcal{B} = \{(S_{k|z}, v_{k+1|z})\}$, where $(S_{k|z}, v_{k+1|z})$ comes from the k-th segment of TSP instance z. Due to the Random Shuffle, the batch $\mathcal{B}$ may contain TSP segments from different instances, while the TSP segments of the same instance may not be all in the batch. Mathematically, the cross-entropy loss function for the batch $\mathcal{B}$ is

$$\mathcal{L}(\mathcal{B}|\Theta) = - \sum_{k,z \in \mathcal{B}} w_z v_{k+1|z} \log f_{NN}(S_{k|z}|z, \Theta). \quad (3)$$

We have abused the use of notation in Eq. (3) for simplicity: $k, z \in \mathcal{B}$ actually means $(S_{k|z}, v_{k+1|z}) \in \mathcal{B}$, $v_{k+1|z}$ is a one-hot vector indicating which vertex to select, f_{NN} is the neural network function by SE-GNN which outputs a probability vector, the multiplication between the two vectors is inner product, and w_z is computed by Eq. (2). The above loss function degenerates back to the one used by [12] for training SE-GNN, if fixing $w_z = 1$.

3.3 Combine the Model Set (CL)

The last step of Fig. 1 shows the process of Competitive Learning (CL). When testing the unknown TSP instances, every model F_j^i computes the tour length

ℓ^i_j respectively. The final prediction takes the minimum length, i.e., $\min_{i,j} \ell^i_j$, in a competitive learning manner [22]. The model F^i_j is weak as an individual, but the competing strategy which cover the weaknesses for each other leads to a strong TSP solver.

It is easy for IA-CL to employ other more delicate neural networks, e.g., the ones by [3,4], or other more powerful search algorithms, e.g., MCTS. Here, we demonstrate by a combination of SE-GNN with the simple greedy search that IA-CL is an effective paradigm for strong combinatorial solvers.

3.4   Convergence of IA-CL

We consider the convergence of training via IA alternations in Fig. 1. For a given set of raw TSP instances, the training data $\{D_j\}_{j=1}^N$ are fixed once the random shuffle is done. For each D_j, the first TSP model F^1_j would converge by minimizing the loss in Eq. (3) at $w_z = 1$. Then, the gap-aware reweighting scheme would also converge to generate the weighted loss for the second IA alternation. As the IA alternation proceeds, multiple TSP models $\{F^i_j, i = 1, \ldots, M; j = 1, \ldots, N\}$ are obtained to tackle different aspects of the difficulties in solving TSP. We provide an empirical evidence that the training process of IA-CL will converge in the experiments.

3.5   IA-CL vs Adaboost

IA-CL is similar to the classic AdaBoost [26], from the aspect of paying more attention to the difficult instances and combing complementary learners. AdaBoost trains a boosted classifier by subsequently tweaking the weak learners in favor of those instances misclassified by previous classifiers. IA-CL differs from AdaBoost in the following ways. First, the loss of the model only describe the error generated by a single step, however, the quality of the solution is composition of multi-step errors. The reweighting scheme in IA-CL is determined by the gap, rather than the loss of SE-GNN. Second, IA-CL directly and jointly trains the new models on the reweighted instances, while AdaBoost focuses on selecting and weighting a new learner from a given set of weak classifiers. Third, the weak models trained by IA-CL compete to win the right to output the solution, unlike AdaBoost in a weighted sum.

4   Experiments

In order to keep the number of TSP segments consistent, we train IA-CL on 100k, 40k and 20k TSP instances for TSP20, TSP50 and TSP100 respectively, that is, 2000k TSP segments are generated in each case. We use Gurobi [23] to compute the solutions as the ground-truth labels. For performance evaluation, we test the models on 10k TSP instances, and report the gap by Eq. (1) and inference time. The city coordinates of the TSP instances of training and testing are all randomly generated, and "k" means thousand. We split the training data into training and validation sets in a 9-to-1 ratio, and each batch contains 512 TSP segments.

We selected a variety of search algorithms to test IA-CL. Greedy search (G) selects one unvisited city with max probability provided by the network component at each step. Each step of the sampling search (S) selects an unvisited city according to the network probability. It will repeat this process many times (the times are called search width) and take the best solution among them. Beam search (BS) has similar performance to Sampling search when the search width is small (around 20), so we only report the performance of Sampling search.

Table 1. Comparisons with other methods. The results with [†] are taken from [4] and the results with [*] are taken from [12]. Symbols of the Type: H (Heuristic), G (Greedy), 2OPT (2-OPT local search), CL (competitive learning), S (Sampling), BS (Beam Search), MCTS (Monte Carlo Tree Search)

Methods	Type	TSP20			TSP50			TSP100		
		Len	Gap	Time	Len	Gap	Time	Len	Gap	Time
Gurobi [23]	Solver	3.83	0.00%	(7 s)	5.69	0.00%	(2 m)	7.76	0.00%	(17 m)
Concorde [24]	Solver	3.83	0.00%	(1 m)	5.69	0.00%	(2 m)	7.76	0.00%	(3 m)
LKH [25]	Solver	3.83	0.00%	(18 s)	5.69	0.00%	(5 m)	7.76	0.00%	(21 m)
Nearest Insertion	H	4.33	13.17%	(1 s)	6.78	19.11%	(2 s)	9.46	21.93%	(6 s)
Random Insertion	H	4.00	4.42%	(0 s)	6.13	7.70%	(1 s)	8.51	9.72%	(4 s)
Farthest Insertion	H	3.92	2.34%	(1 s)	6.00	5.54%	(2 s)	8.35	7.57%	(6 s)
Nearest Neighbor	H	4.49	17.25%	(0 s)	6.98	22.60%	(1 s)	9.69	24.83%	(5 s)
S2V [2]	G	3.89[†]	1.42%	–	5.99[†]	5.16%	–	8.31[†]	7.03%	–
GAT [17]	G	3.86[†]	0.66%	(2 m)	5.92[†]	3.98%	(5 m)	8.42[†]	8.41%	(8m)
GAT [17]	G, 2OPT	3.85[†]	0.42%	(4 m)	5.85[†]	2.77%	(26 m)	8.17[†]	5.21%	(3 h)
AM [3]	G	3.84	0.28%	(0 s)	5.79	1.68%	(2 s)	8.10	4.34%	(5 s)
GCN [4]	G	3.86	0.72%	(16 s)	5.90	3.65%	(1 m)	8.41	8.35%	(5 m)
SE-GNN [12]	G	3.87	1.00%	(5 s)	5.94	4.36%	(25 s)	8.46	9.04%	(2 m)
IA-CL(Ours)	G, CL (20)	**3.83**	**0.07%**	(1 m)	**5.75**	**1.06%**	(7 m)	**7.99**	**2.99%**	(32 m)
GAT [17]	S, CL	3.84[†]	0.11%	(5 m)	5.77[†]	1.28%	(17 m)	8.75[†]	12.70%	(56 m)
GAT [17]	S, 2-OPT	3.84[†]	0.09%	(6 m)	5.75[†]	1.00%	(32 m)	8.12[†]	4.64%	(5 h)
AM [3]	S, CL (1280)	3.83	0.06%	(10 m)	5.72	0.50%	(27 m)	7.94	2.31%	(1 h)
AM [3]	BS (1280)	**3.83**	**0.00%**	(10 m)	5.71	0.30%	(31 m)	7.95	2.48%	(1 h)
GCN [4]	BS (1280)	3.83	0.15%	(27 s)	5.71	0.30%	(3 m)	7.92	2.09%	(10 m)
GCN [4]	BS, CL (1280)	**3.83**	**0.00%**	(22 m)	**5.70**	**0.04%**	(33 m)	7.87	1.37%	(53 m)
SE-GNN [12]	MCTS	3.92[*]	0.01%	(3 h)	5.69[*]	0.20%	(10 h)	**7.81[*]**	**1.04%**	(77 h)
IA-CL(Ours)	S, CL (128 × 5)	**3.83**	**0.00%**	(15 m)	5.70	0.18%	(2 h)	7.88	1.53%	(7 h)
IA-CL(Ours)	S, CL (128 × 10)	**3.83**	**0.00%**	(29 h)	5.70	0.12%	(4 h)	7.86	1.33%	(15 h)

4.1 Comparisons with Other Approaches

We report the experimental results in Table 1 in four categories of methods, i.e., the exact solvers, the traditional heuristic algorithms, the combination of deep neural networks with greedy search, and the combination of deep neural networks with other more advanced search algorithms. It can be observed that traditional heuristic algorithms are fast but generally inferior to learning-based methods. Learning with advanced search approaches like MCTS, can further improve the performance but at the cost of large amount of computation. Here we choose 2 and 10 for M and N to train a total of 20 models to form a model set. When assisting the Sampling search to generate the results, we set sampling width to 128, and take top 5 and top 10 of 20 models for testing to compare with those methods with the width of 1280. According to the recommendation of [12], number of rollouts of MCTS is set to 800, 800 and 1200 for TSP20, TSP50, and TSP100, respectively.

Considering the learning methods with greedy search, advanced network structures like AM, GAT and GCN, are more capable of capturing the structural patterns of TSP graphs then S2V, SE-GNN, leading to better TSP solutions. However, it requires a huge number of TSP instances to train advanced deep networks to achieve superior performance. We compare the training sample size with GCN [4] that are also trained in a supervised way. It uses one million TSP instances for TSP20, TSP50 and TSP100, whereas the amount of data used to train SE-GNN in the original paper [12] for TSP100, is only 1/50 of GCN. It is noted that the proposed IA-CL is able to train multiple weak SE-GNNs that assist greedy search to greatly reduce the gap, using the same small training sample size as [12]. As the number of cities increases, IA-CL is still very robust. For TSP100, IA-CL achieves the gap at 2.99%, reducing the gap by 64.2% from 8.35% by GCN, which is trained on 50 times more instances.

4.2   Computational Time

Running time is also an important factor for TSP solvers, and it can be affected by equipment. We test on 2 TITAN Xp and 4 4-core Intel(R) Xeon(R) CPUs E5-2620 v4 @ 2.10GHz, and report the running time of all methods under the same resource in Table 1. The running time of IA-CL is acceptable. Since the code of S2V is written in C++, and IA-CL is written in Python, we do not compare the computational time of S2V. Actually, IA-CL can be further implemented more efficiently. Two main time-consuming parts of IA-CL is finding the route for each model separately and taking the route with the smallest length of all. Both the two parts can be implemented in parallel on multiple GPU resources, and the running time would reduce to be the same for a single SE-GNN. Because the test SE-GNN with MCTS is too time-consuming, we multiply the test time of a single TSP instance reported in the paper by 10000 as its overall test time.

4.3   Ablation Study and Hyperparameter Selection

To give guidance on the choice of hyperparameters, we test N and M from 1 to 5, respectively, on TSP10 with 100k samples. We repeat the experiment 5 times, and train a total of 50 models to draw Fig. 3. The purple line represents the set of models trained with IA alternation, and the green line represents the set of models trained randomly.

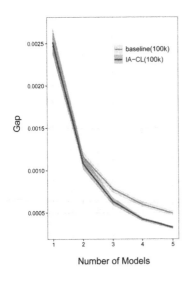

Fig. 3. IA-CL vs baseline

Here we remove the IA alternation in IA-CL, and only use the random shuffle as the green line. The horizontal axis represents the number of models in the model set, and is the value of M and N. The vertical axis is the gap of the model set on the test set. It can be seen that when the training set reaches 100k TSP instances, IA alternation can quickly make up for shortcomings in the training set and achieve better results. However, excessive IA-alternation will cause the model set overfitting on the training set, which requires to determined M based on the validation set. The tendency of overfitting can be slowed down by increasing N. The second consideration is the running time. It is better to set large N and M, but it is necessary to ensure that the running time of the model set is within an acceptable range.

In Fig. 3, we shaded intervals between the maximum and minimum values of 5 times. Because the variance of the model test set Gap obtained by SE-GNN training multiple times does not exceed 0.01, this also ensures the stability of IA-CL. Under 100k training samples, as the number of models increases, IA-CL shows a trend of convergence. This reflects that with a limited number of training samples, the effect that can be achieved is bound to be limited.

4.4 Generalization Performance

Table 2. Generalization to other scales.

SCALE	TSP20		TSP50		TSP100	
	LEN.	GAP.	LEN.	GAP.	LEN.	GAP.
TSP20	3.83	0.07%	5.81	2.11%	8.41	8.39%
TSP50	3.84	0.31%	5.75	1.06%	7.97	2.70%
TSP100	3.86	0.88%	5.80	1.90%	7.99	2.99%

Table 3. Performance on the TSPLib.

Num	AM	GCN	SE-GNN	IA-CL
50–99	6.97%	53.04%	9.70%	3.41%
100–399	18.35%	47.89%	19%	8.86%

*Ref:AM [3],GCN [4],SE-GNN [12].

The model set consisting of 20 models trained in Table 1 was used for examining the generalization performance of IA-CL. Rows in Table 2 represent scale of instances used to train the model set, and columns represent test scales. In general, IA-CL maintains well performance when tested under other scale TSP instances. Models trained at other scales are generally inferior to models trained at the corresponding scale. It is difficult for TSP20 to get good results on larger scale problems, and at the same time, the test results of models trained on TSP50 and TSP100 are not satisfactory on TSP20. This phenomenon improves as the number of cities increases, and the model set trained under TSP50 has outstanding generalization performance under three scales. Its gap on TSP20 is 0.31%, while that on TSP100 is 2.70%, which is similar to the performance of training on TSP100 itself, and even slightly better.

TSPLIB [27] is a library of sample instances for the TSP from various sources and of various types. We take TSP instances 50 to 399 from TSPLIB and normalize the coordinates to $[0, 1]$, and divide them into two data set: those with city numbers 50 to 99 and those with city numbers 100 to 399. For the former, we use the TSP50 model set of each method to generate solutions and calculate the average gap. For the latter case, we use the TSP100 model set. The results, shown in Table 3, again confirm that IA-CL is a strong learning-based TSP solver, and is promising in real applications.

5 Conclusion

In this paper, we have proposed a deep bidirectional competitive learning method, IA-CL, for approximately solving TSP. It gives an efficient way to train multiple weak solvers (IA), and a way to combine them to get strong solvers (CL), which can greatly improve the sample utilization, but increase the operation time. The application of IA-CL on larger instances should also be considered in the future.

Acknowledgement. This work was supported by the National Key R&D Program of China (2018AAA0100700), and Shanghai Municipal Science and Technology Major Project (2021SHZDZX0102).

References

1. Papadimitriou, C.H.: The Euclidean travelling salesman problem is NP-complete. In: Theoretical Computer Science, pp. 237–244 (1977)
2. Dai, H., Khalil, E., Zhang, Y., Dilkina, B., Song, L.: Learning combinatorial optimization algorithms over graphs. In: Advances in Neural Information Processing Systems, pp. 6348–6358 (2017)
3. Kool, W., van Hoof, H., Welling, M.: Attention, learn to solve routing problems! In: International Conference on Learning Representations (2019)
4. Joshi, C.K., Laurent, T., Bresson, X.: An efficient graph convolutional network technique for the travelling salesman problem (2019). arXiv preprint. arXiv:1906.01227
5. Wu, Y., Song, W., Cao, Z., Zhang, J., Lim, A.: Learning improvement heuristics for solving routing problems. In: IEEE Transactions on Neural Networks and Learning Systems, pp. 1–13 (2021)
6. Xu, L.: Deep IA-BI and five actions in circling. In: International Conference on Intelligent Science and Big Data Engineering, pp. 1–21 (2019)
7. Xu, L.: Deep bidirectional intelligence: AlphaZero, deep IA-search, deep IA-infer, and TPC causal learning. Appl. Inf. **5**(1), 1–38 (2018). https://doi.org/10.1186/s40535-018-0052-y
8. Xu, L.: An overview and perspectives on bidirectional intelligence: lmser duality, double IA harmony, and causal computation. In: IEEE/CAA Journal of Automatica Sinica, pp. 865–893 (2019)

9. Hopfield, J.J., Tank, D.W.: "Neural" computation of decisions in optimization problems. In: Biological Cybernetics, pp. 141–152 (1985)
10. Vinyals, O., Fortunato, M., Jaitly, N.: Pointer networks. In: Advances in Neural Information Processing Systems (2015)
11. Bello, I., Pham, H., Le, Q.V., Norouzi, M., Bengio, S.: Neural combinatorial optimization with reinforcement learning. In: International Conference on Learning Representations, pp. 1–8 (2017)
12. Xing, Z.H., Tu, S.K.: A graph neural network assisted Monte Carlo tree search approach to traveling salesman problem. In: IEEE Access, pp. 108418–108428 (2020)
13. Fu, Z.H., Qiu, K.B., Zha, H.: Generalize a small pre-trained model to arbitrarily large TSP instances. In: Proceedings of the AAAI Conference on Artificial Intelligence, pp. 7474–7482 (2021)
14. Xin, L., Song, W., Cao, Z., Zhang, J.: NeuroLKH: combining deep learning model with Lin-Kernighan-Helsgaun heuristic for solving the traveling salesman problem. In: Advances in Neural Information Processing Systems, pp. 7472–7483 (2021)
15. Mersmann, O., Bischl, B., Bossek, J., Trautmann, H., Wagner, M., Neumann, F.: Local search and the traveling salesman problem: a feature-based characterization of problem hardness. In: International Conference on Learning and Intelligent Optimization, pp. 115–129 (2012)
16. Kool, W., van Hoof, H., Gromicho, J., Welling, M.: Deep policy dynamic programming for vehicle routing problems. In: Integration of Constraint Programming, Artificial Intelligence, and Operations Research (2022)
17. Deudon, M., Cournut, P., Lacoste, A., Adulyasak, Y., Rousseau, L.M.: Learning heuristics for the tsp by policy gradient. In: Integration of Constraint Programming, Artificial Intelligence, and Operations Research, pp. 170–181 (2018)
18. Nazari, M., Oroojlooy, A., Snyder, L., Takác, M.: Reinforcement learning for solving the vehicle routing problem. In: Advances in Neural Information Processing Systems (2018)
19. Ma, Y., Li, J., Cao, Z., Song, W., Zhang, L., Chen, Z., Tang, J.: Learning to iteratively solve routing problems with dual-aspect collaborative transformer. In: Advances in Neural Information Processing Systems, pp. 11096–11107 (2021)
20. Kwon, Y.D., Choo, J., Kim, B., Yoon, I., Gwon, Y., Min, S.: POMO: policy optimization with multiple optima for reinforcement learning. In: Advances in Neural Information Processing Systems, pp. 21188–21198 (2020)
21. Zhao, K.X., Tu, S.K., Xu, L.: IA-GM: A deep bidirectional learning method for graph matching. In: Proceedings of the AAAI Conference on Artificial Intelligence, pp. 3474–3482 (2021)
22. Rumelhart, D.E., Zipser, D.: Feature discovery by competitive learning. In: Cognitive Science, pp. 75–112 (1985)
23. Gurobi Optimization, LLC (2022). Gurobi optimizer reference manual. Retrieved from http://www.gurobi.com
24. Applegate, D., Bixby, R., Chvatal, V., Cook, W. Concorde TSP solver (2006). Retrieved from http://www.math.uwaterloo.ca/tsp/concorde/
25. Helsgaun, K.: An effective implementation of the lin-kernighan traveling salesman heuristic. In: European journal of operational research, pp. 106–130 (2000)
26. Freund, Y. and Schapire, R.E.: A decision-theoretic generalization of on-line learning and an application to boosting. J. Comput. Syst. Sci., 119–139 (1997)
27. Reinelt, G.: TSPLIB-a traveling salesman problem library. ORSA J. Comput., 376–384 (1991)

Boosting Graph Convolutional Networks with Semi-supervised Training

Shuai Tang, Enmei Tu, and Jie Yang$^{(\boxtimes)}$

Institute of Image Processing and Pattern Recognition,
Shanghai Jiao Tong University, Shanghai, China
{tangshuai,tuen,jieyang}@sjtu.edu.cn

Abstract. Graph convolutional networks (GCN) suffer from the over-smoothing problem, which causes most of the current GCN models to be shallow. Shallow GCN can only use a very small part of nodes and edges in the graph, which leads to over-fitting. In this paper, we propose a semi-supervised training method to solve this problem, and greatly improve the performance of GCN. Firstly, we propose an integrated data augmentation framework to conduct effective data augmentations for graph-structured data. Then consistency loss, entropy minimization loss, and graph loss are introduced to help GCN make full use of unlabeled nodes and edges, which alleviates the excessive dependence of the model on labeled nodes. Extensive experiments on three widely-used citation datasets demonstrate our method can achieve state-of-the-art performance in solving the semi-supervised node classification problem. Especially, we get 85.52% accuracy on Cora with the public split.

Keywords: Graph Convolutional Networks · Semi-supervised learning · Node classification

1 Introduction

Nowadays, Graph convolutional networks(GCN) [11] and their variants [19] have been widely applied to many real-life applications, such as traffic prediction, recommender systems, and citation node classification. Compared with traditional algorithms for semi-supervised node classification, the success of GCN lies in the neighborhood aggregation scheme which simultaneously learns the node knowledge and edge knowledge in an end-to-end manner.

However, GCN suffers from over-smoothing [14] problem, which causes most of the current GCN models are shallow. Although the performance of a two-layer GCN has surpassed traditional algorithms by a large margin, it still suffers from over-fitting problems. For example, on Cora dataset, the accuracy on labeled nodes which can be seen as the training set will soon reach 100%, while the test accuracy is only 81.5%. According to the neighborhood aggregation scheme of GCN, the calculation of gradient backpropagation only involves the 2-hop neighbors of the labeled nodes, which is still only a very small part of the whole

© The Author(s), under exclusive license to Springer Nature Switzerland AG 2023
M. Tanveer et al. (Eds.): ICONIP 2022, LNCS 13623, pp. 537–548, 2023.
https://doi.org/10.1007/978-3-031-30105-6_45

dataset in semi-supervised setting. Likewise, in the training process of GCN, only the edges between the 2-hop neighbors of the labeled nodes are used.

In order to solve the over-fitting problem and further improve the performance of GCN on the semi-supervised node classification task, we propose helping GCN to make full use of unlabeled node knowledge and edge knowledge with semi-supervised training. Inspired by the success of semi-supervised training in the computer vision field, e.g. Fixmatch [17], MixMatch [2], we integrate DropOut [18], DropEdge [16], DropNode [7], and propagation [7] into a data augmentation framework, which conducts effective data augmentations for graph-structured data. Then we propose three semi-supervised losses to regularize the GCN backbone. Specifically, we use consistency loss to regular the GCN output consistent predictions over different augmented versions of the feature matrix. We use entropy minimization loss to push the model give high confidence prediction. Graph loss is adopted to guide the GCN output similar predictions for two linked nodes. During our training process, GCN will use all node feature information and edge information instead of only information from 2-hop neighbors.

In summary, our contributions are summarized as follows:

- We propose a integrated data augmentation framework, which could conduct effective data augmentations for graph-structured data.
- Three semi-supervised losses are introduced to help GCN make full use of unlabeled nodes and edges which alleviates the over-fitting problem of GCN and greatly improves the performance.
- Our method achieves state-of-the-art performance across three widely-used citation datasets for node classification task.

2   Method

In this section, we first formulate the semi-supervised node classification problem and briefly introduce the GCN model. We then give a detailed introduction to our proposed integrated data augmentation framework. Based on the data augmentation framework, we finally propose a regularized training method for improving the node classification performance under the semi-supervised setting.

2.1   Problem Formulation and Preliminaries

This work focuses on semi-supervised node classification on graph, which aims at predicting the labels of nodes, given only a portion of labeled nodes as guidance. Consider a graph $G = (X, A)$, where each row of X corresponds to a feature vector of a vertex in the graph and $X = [x_1, x_2, \ldots, x_n]^\top \in \mathbb{R}^{n \times d}$ is the feature matrix with n data points and d input channels. $A \in \{0, 1\}^{n \times n}$ denotes the adjacency matrix of G, with each element $A_{ij} = 1$ indicating there exists an edge between x_i and x_j, otherwise $A_{ij} = 0$. Let $Y = [Y^L; Y^U] \in \mathbb{R}^{n \times C}$ denote the labels of the nodes in the graph with C representing the number of classes. Each row $Y_i \in \mathbb{R}^C$ of label matrix Y denotes a C-dimensional one-hot vector of node i.

In semi-supervised node classification setting, what can be observed are the feature vectors of all nodes X and the full adjacencies A between them and the labels Y^L of a few of labeled nodes. Usually, the number of labeled nodes is much smaller than the number of unlabeled nodes. The final objective is to learn a predictive function $f : X, A, Y^L \rightarrow Y^U$ to infer the labels Y^U for unlabeled nodes.

2.2   A Brief Introduction to GCN

In this work, we take the GCN as the backbone network, due to its simple structure and powerful representation of graph-structured data. Formally, the node representation through a GCN layer can be obtained by the following formulation:

$$H^{(l+1)} = \sigma(\tilde{L}_{sym} H^{(l)} W^{(l)}), \tag{1}$$

where $W^{(l)}$ and $H^{(l)}$ are the weight matrix and the hidden node representation in the lth layer, $\sigma(.)$ represents the RELU activation function. $\tilde{L}_{sym}$ is the symmetric normalized adjacency matrix, whose eigenvalues is in $(-1, 1]$, which can effectively prevent the gradient disappearance or explosion phenomenon that occurs during multi-layer network optimization. $\tilde{L}_{sym}$ is defined as:

$$\tilde{L}_{sym} = \tilde{D}^{-\frac{1}{2}} \tilde{A} \tilde{D}^{-\frac{1}{2}}, \tag{2}$$

where $\tilde{A} = A + I$, $\tilde{D}$ is the degree matrix of $\tilde{A}$. From formula(1), we can see that GCN and MLP have the same weight matrix and GCN will degenerate into MLP without $\tilde{L}_{sym}$.

The formulation at the node level is as follows:

$$x_i = \sigma(\sum_{v_j \in \tilde{N}(v_i)} \tilde{L}_{sym}[i, j](W x_j)), \tag{3}$$

From the row vector perspective of matrix multiplication, $\tilde{L}_{sym} H^{(l)}$ equivalent to the aggregation operation on the feature vectors of neighbor nodes.

2.3   Integrated Data Augmentation Framework

In the field of computer vision, advanced data augmentation techniques have been proven to play a crucial role in improving the performance of semi-supervised learning algorithms. For example, fixmatch leverage CutOut [6] , CTAugment [1] , and RandAugment [4] for strong augmentation, which all produce heavily distorted versions of a given image and make it achieve state-of-the-art performance across a variety of standard semi-supervised learning benchmarks. The ICT [20] algorithm first applies Mixup [24] in semi-supervised learning, which greatly improves the generalization ability of the model.

Motivated by the success of data augmentation techniques in computer vision, we explore existing methods that may be used for graph-structured data augmentation and integrate them into a graph data augmentation framework. It is difficult to generate an augmented node in an existing graph because we don't know how to determine its adjacency to other nodes. Then, we cannot use graph neural networks, which need the adjacency matrix as input. This is the case with GraphMix [21], which use Mixup to generates new nodes in an existing graph and use MLP to train these nodes. We choose to augment the features of all nodes to generate a new feature matrix without changing the graph structure. Specifically, our framework integrates four techniques, i.e. DropNode, Propagation, DropEdge and DropOut. The overall process of our proposed integrated data augmentation framework is shown in Algorithm 1. Based on the homophily assumption [25] that adjacent nodes tend to have similar features and labels, we adopt propagation as the main augmentation method. What's more, in our framework, we inject three additional disturbance factors in the propagation process. DropNode perturbs the feature matrix from the node level to reduce the dependence on some nodes. DropOut perturbs the feature matrix from the feature level to reduce the dependence on some features. DropEdge perturbs the feature matrix from the graph level to reduce the dependence on some edges.

Algorithm 1. Framework of the integrated data augmentation system.

Input: The original feature matrix, X; The original adjacency matrix, A; The highest order of propagation, K; Drop rate for DropOut, δ_f; Drop rate for DropNode, δ_v; Drop rate for DropEdge, δ_e;
Output: A version of augmented feature matrix, $\overline{X}$;
  1: Generating a random integer between 0 and 1 recorded as r_1;
  2: Generating a random integer between 0 and 1 recorded as r_2;
  3: Generating a random integer between 0 and K recorded as $orders$;
  4: **if** $r_1 = 0$ **then**
  5:     $X \leftarrow dropnode(X, \delta_v)$;
  6: **end if**
  7: **if** $r_2 = 0$ **then**
  8:     $A \leftarrow dropdege(A, \delta_e)$;
  9: **end if**
10: Computing the symmetric normalized adjacency matrix $\tilde{L}_{sym}$ with A;
11: **for** $i = 1$ to $orders$ **do**
12:     $X \leftarrow dropout(X, \delta_f)$;
13:     $X \leftarrow \tilde{L}_{sym}X$;
14: **end for**
15: $X \leftarrow \overline{X}$;
16: **return** $\overline{X}$;

DropNode and Propagation. DropNode and Propagation were first raised in GRAND [7] as random propagation. Our use of dropnode is the same as GRAND, generating a perturbed feature matrix $\tilde{X}$ by randomly dropping out some nodes in X. Firstly, a binary mask vector $\epsilon = \{0, 1\}^n$ are sampled from the binomial distribution with $p(\epsilon_i = 0) = \delta$. Then, the perturbed feature matrix $\tilde{X}$

are obtained by multiplying each node's feature vector with its corresponding mask. Finally, we need to multiply $\tilde{X}$ by $\frac{1}{1-\delta}$ to guarantee the perturbed feature matrix is in expectation equal to X. We only use DropNode during training.

GRAND adopts the mixed-order propagation, formally:

$$\overline{X} = \sum_{k=0}^{K} \frac{1}{K+1} \tilde{L}_{sym}^k \tilde{X}, \tag{4}$$

Different from GRAND's use of propagation, we adopt random-order propagation for more diverse augmentations. Before each training epoch, we randomly sample a integer $k \in [0, K]$. The formulation of random-order propagation is as follows:

$$\overline{X} = \tilde{L}_{sym}^k \tilde{X}, \tag{5}$$

DropEdge. DropEdge [16] proposes a random edge removal technique. By randomly removing a certain ratio of edges, DropEdge increases the diversity of input data to prevent over-fitting and reduces message passing in graph convolution to alleviate over-smoothing. DropEdge converts 1 to 0 in the adjacency matrix A with a certain probability, generating a new adjacency matrix A_{drop}. We adopt DropEdge in propagation, i.e., using A_{drop} to compute the symmetric normalized adjacency matrix $\tilde{L}_{sym}$.

DropOut. DropOut [18] has been widely used for regularizing neural networks. In the training process , DropOut stops the activation of some neurons with a certain probability p, which reduces the model's dependence on some local features and improves the generalization ability of the model. We adopt DropOut in propagation, i.e., before each round of propagation, we pass the current feature matrix through the dropout module. Compared with DropNode, DropOut generates binary mask vectors for each element of each node, which belongs to the disturbance of feature level. One order of propagation with DropOut is as follows:

$$\overline{X}^{(l+1)} = \tilde{L}_{sym} dropout(\overline{X}^{(l)}), \tag{6}$$

where $dropout(.)$ represents the dropout module.

2.4   Semi-supervised Training

In the training process of the two-layer GCN model, the backpropagation of the gradient only uses the 2-hop neighbors of the labeled nodes, and the utilization of the graph only involves the edges between its second sister neighbors. GCN does not make full use of unlabeled nodes and the whole edge. Therefor, equipped with our proposed integrated data augmentation framework, we propose three additional loss functions which further improve the performance of GCN on semi supervised node classification tasks.

In our training method, we use the augmentation framework generate two augmented feature matrix, $\overline{X}_1$ and $\overline{X}_2$. Then, we get their logits Z_1 and Z_2 and their probability distributions P_1 and P_2,

$$Z_1 = f(\overline{X}_1, \tilde{L}_{sym}), Z_2 = f(\overline{X}_2, \tilde{L}_{sym}), \tag{7}$$

$$P_1 = softmax(Z_1), P_2 = softmax(Z_2), \tag{8}$$

where f is the GCN model, $softmax(.)$ is the softmax function.

We evaluate the cross-entropy loss over labeled nodes:

$$\ell_{sup} = \frac{1}{2|V_L|} \sum_{i \in V_L} (H(Y_i, P_{1i}) + H(Y_i, P_{2i})), \tag{9}$$

where $H(p, q)$ denotes the cross-entropy between two probability distributions p and q, P_{1i} is the ith-row vector of P_1, V_L is the labeled set.

For unlabeled nodes, we evaluate consistency loss and entropy minimization loss. According to the smoothness assumption, the model should output the same predictions for different augmented versions of the feature matrix. The loss function for label consistency is defined as:

$$\ell_{con} = \frac{1}{|V_U|} \sum_{i \in V_U} \|P_{1i} - P_{2i}\|_2^2, \tag{10}$$

where V_U is the unlabeled set. Based on the assumption of low-density segmentation, the decision boundary of the classifier should pass through the low-density region in the input space, which means that the model should output a probability distribution with high confidence. In semi-supervised learning, there are three ways to enforce this. Pseudo-Label [13] assigns "hard" labels to unlabeled nodes with the arg max of the model's output. MixMatch uses a "sharpening" function on the target distribution for unlabeled data to reduce entropy. Like VAT [15], we use a loss term called entropy minimization [8] to let the network output low-entropy predictions. The entropy minimization loss for unlabeled nodes is defined as:

$$\ell_{ent} = \frac{1}{2|V_U|} \sum_{i \in V_U} \sum_{j=1}^{C} (-P_{1ij} ln(P_{1ij}) - P_{2ij} ln(P_{2ij})), \tag{11}$$

where P_{1ij} is the (i,j) element in P_1. This loss minimizes the entropy of P_{1x} and P_{2x} for unlabeled data x .

In order to make further use of graph information, we propose graph loss, i.e. the graph $M = P_1 P_2^\top$ built from probability distribution of model output should be consistent with the original graph $\tilde{L}_{sym}$. The graph loss is defined as:

$$\ell_{graph} = \sum_{i=1}^{n} \sum_{j=1}^{n} -\tilde{L}_{sym_{ij}} ln(M_{ij}), \tag{12}$$

The graph loss pushes the model output similar predictions for connected nodes. What's more, using the symmetric normalized adjacency matrix $\tilde{L}_{sym}$ instead of A could prevent nodes with a large number of edges from leading this loss. Then, the overall loss is given by:

$$\ell_{overall} = \ell_{sup} + \lambda_1 \ell_{ent} + \lambda_2 \ell_{con} + \lambda_3 \ell_{graph}, \tag{13}$$

In short, firstly, we propose a powerful intergrated data augmentation framework for graph-structured data. Then with the consistency loss the graph loss and the entropy minimization loss, we could make full use of the knowledge from unlabeled nodes and the whole edge, which mitigates the over-fitting problem of GCN and greatly improve the performance on semi-supervised node classification task.

3   Experiments

In this section, we conduct experiments on three widely-used datasets for node classification task to validate the proposed augmentation framework and semi-supervised training method. We evaluate the performance by the metric of classification accuracy.

3.1   Datasets

The three benchmark datasets are Cora, Citeseer and PubMed, and the statistics of the datasets are shown in Table 1. These datasets are all citation network datasets, in which nodes correspond to documents, edges correspond to citation links, and each node has a sparse bag-of-words feature vector as well as a class label. We follow the same public data split method as the standard GNN settings on semi-supervised graph learning [11], which chose fixed 20 nodes per class as labeled set.

Table 1. Statistics of the datasets.

Datasets	Nodes	Edges	Classes	Features
Cora	2708	5429	7	1433
Citeseer	3327	4732	6	3703
PubMed	19717	44338	3	50

3.2   Experimental Setting

We use a two-layer GCN with 64 hidden units as the backbone network. The drop rate of dropout in the augmentation framework and GCN is 0.5. The drop rate of dropnode is 0.5 too. We use Adam [10] optimizer with learning rate 0.01, l_2-norm weight decay 5×10^{-4}, to train the model. We train the model for a maximum of 1500 epochs using early stopping with a window size of 200 according to the loss of the validation set. Due to the low accuracy in the early stage of training, we increase the weight of the unlabeled loss term and graph loss term from 0 to 1, over the training and the early stopping mechanism works after 600 epochs. Other hyper-parameters listed in Table 2 are searched on the validation set. We repeat the experiments 10 times and report the mean accuracy for different datasets.

Table 2. Statistics of the hyper-parameters.

Datasets	K	δ_e	λ_1	λ_2	λ_3
Cora	12	0.5	0.2	1.5	1
Citeseer	3	0.3	1	1	1
PubMed	2	0.3	0.01	1	0.1

3.3 Results

Table 3. Overall classification accuracy (%).

Categoriy	Method	Cora	Citeseer	Pubmed
Traditional Methods	MLP [22]	55.1	46.5	71.4
	LPA [25]	68.0	45.3	63.0
	Planetoid [23]	75.7	64.7	77.2
Advanced GNNs	GCN [11]	81.5	70.3	79.0
	GAT [19]	83.0±0.7	72.5±0.7	79.0±0.3
	DropEdge [16]	82.8	72.3	79.6
	APPNP [12]	83.8± 0.7	71.6 ±0.5	79.7± 0.3
	GCNII [3]	85.5± 0.5	73.4± 0.6	80.2± 0.4
GCN With Semi-Supervised Training	LC-GCN [22]	82.9± 0.4	72.3± 0.8	80.1± 0.4
	BVAT [5]	83.6± 0.5	74.0± 0.6	79.9± 0.4
	GraphMix [21]	82.7± 1.2	76.5± 1.6	80.7 ±1.1
	ours	85.5± 0.3	74.8± 0.3	80.3 ±0.3

To validate the effectiveness of our method on semi-supervised node classification, we choose three categories of methods for baselines. The first is the traditional methods including MLP [22], LPA [25], and Planetoid [23]. We take GNN and its improved variants as the second kind of method, such as two classical GNN: GCN [11], GAT [19], and three improved variants: DropEdge [16], APPNP [12], and GCNII [3]. The third type of method is a combination of the GCN and semi-supervised training, which is similar to our method. Specificity, we choose LC-GCN [22], BVAT [5], and GraphMix [21] as the third category. The overall results of classification accuracy are summarized in Table 3.

MLP only considers feature information X and LPA only considering edge information A. They both ignore part of the information in the graph, which leads to their poor performance. Planetoid also uses MLP as the model while it encodes the graph structure into embeddings through predicting context sampled according to edges and pseudo-labels. Since both feature and edge information are considered, the performance of Planetoid greatly exceeds that of MLP and LPA.

Compared with the methods in first part of Table 3, we can find that GCN is a strong model for graph-structured data which outperforms the traditional method extremely. Therefore, numerous research efforts have been devoted to developing new GNN architectures. GAT uses the attention mechanism to describe the importance of adjacent nodes to the central node which further increases the capacity of GCN. APPNP decouples feature transformation and propagation which help aggregate information from multi-hop neighbors. Inspired by ResNet [9], GCNII alleviates the over-smoothing problem for deep GCN by adopting Initial residual and Identity mapping and outperforms the state-of-the-art methods. Nevertheless, combining our semi-supervised training method with GCN can still achieve a great performance improvement that is even more effective than improving the network structure, which shows that our method is powerful. specifically, equipped with our training method, the accuracy of GCN was improved by 4.9% on Cora , 6.4% on Citeseer and 1.6% on Pubmed. In particular, our method based on a two-layer GCN beats a 64-layer GCNII on all the three datasets.

VBAT generates virtual adversarial perturbations for all nodes with an optimization process which needs large amount of computation. Equipped with mixup, GraphMix outperforms our method on Citeseer while it achieves poor performance on cora. In general, compared with the similar methods in the third part of the table, our method achieves state-of-the-art performance on all the three datasets. What's more, we can find that our method has the smallest variance, which means our method is robust.

3.4 Ablation Study

To explore the contributions of different components in our method, we conduct some ablation experiments on Citeseer dataset.

- **Without all of semi-supervised losses and augmentations:** We only use supervised loss.

- **Without all of semi-supervised losses:** We only use augmentations and supervised loss.

- **Without graph loss:** We set $\lambda_3 = 0$ to remove graph loss from our method.

Table 4. The ablation results.

augmentation	ℓ_{graph}	ℓ_{ent}	ℓ_{con}	Citeseer
✗	✗	✗	✗	70.3
✔	✗	✗	✗	70.79±0.44
✔	✗	✔	✔	74.28±0.35
✔	✔	✗	✔	74.62±0.37
✔	✔	✔	✗	74.55±0.40
✔	✔	✔	✔	74.82±0.30

- **Without entropy minimization loss:** We set $\lambda_2 = 0$ to remove entropy minimization loss from our method.

- **Without consistency loss:** We set $\lambda_1 = 0$ to remove consistency loss from our method.

As is shown in Table 4, Only using data augmentations for GCN, the improvement is not obvious. However, after adding any two additional auxiliary losses, the accuracy quickly increases from 70.79% to more than 74%, revealing that both our proposed data augmentation framework and semi-supervised training methods are indispensable. And we find that the absence of any semi-supervised loss will lead to a drop in accuracy.

4   Conclusions

In this work, we propose an integrated data augmentation framework, which could conduct rich perturbations for graph-structured data from the three levels of edges, nodes, and features. Equipped with the augmentation framework, we further propose three semi-supervised training losses to solve the over-fitting problem of GCN due to its insufficient utilization of unlabeled nodes and graph information. Our method achieves state-of-the-art performance across three widely-used datasets for the node classification task. The success of our work proves that combining graph neural networks and semi-supervised training methods is a feasible idea for graph semi-supervised tasks.

Acknowledgement. This research is partly supported by Ministry of Science and Technology, China (No. 2019YFB1311503) and Committee of Science and Technology, Shanghai, China (No.19510711200).

References

1. Berthelot, D., et al.: Remixmatch: semi-supervised learning with distribution matching and augmentation anchoring. In: ICLR (2020)
2. Berthelot, D., Carlini, N., Goodfellow, I.J., Papernot, N., Oliver, A., Raffel, C.: Mixmatch: a holistic approach to semi-supervised learning. arXiv abs/1905.02249 (2019)

3. Chen, M., Wei, Z., Huang, Z., Ding, B., Li, Y.: Simple and deep graph convolutional networks. arXiv abs/2007.02133 (2020)

4. Cubuk, E.D., Zoph, B., Shlens, J., Le, Q.V.: Randaugment: practical automated data augmentation with a reduced search space. In: 2020 IEEE/CVF Conference on Computer Vision and Pattern Recognition Workshops (CVPRW), pp. 3008–3017 (2020)

5. Deng, Z., Dong, Y., Zhu, J.: Batch virtual adversarial training for graph convolutional networks. ArXiv abs/1902.09192 (2019)

6. Devries, T., Taylor, G.W.: Improved regularization of convolutional neural networks with cutout. arXiv abs/1708.04552 (2017)

7. Feng, W., et al.: Graph random neural networks for semi-supervised learning on graphs. arXiv: Learning (2020)

8. Grandvalet, Y., Bengio, Y.: Semi-supervised learning by entropy minimization. In: CAP (2004)

9. He, K., Zhang, X., Ren, S., Sun, J.: Deep residual learning for image recognition. In: 2016 IEEE Conference on Computer Vision and Pattern Recognition (CVPR), pp. 770–778 (2016)

10. Kingma, D.P., Ba, J.: Adam: a method for stochastic optimization. CoRR abs/1412.6980 (2015)

11. Kipf, T., Welling, M.: Semi-supervised classification with graph convolutional networks. arXiv abs/1609.02907 (2017)

12. Klicpera, J., Bojchevski, A., Günnemann, S.: Predict then propagate: graph neural networks meet personalized pagerank. In: ICLR (2019)

13. Lee, D.H., et al.: Pseudo-label: the simple and efficient semi-supervised learning method for deep neural networks. In: Workshop on Challenges in Representation Learning, ICML (2013)

14. Li, Q., Han, Z., Wu, X.M.: Deeper insights into graph convolutional networks for semi-supervised learning. arXiv abs/1801.07606 (2018)

15. Miyato, T., Ichi Maeda, S., Koyama, M., Ishii, S.: Virtual adversarial training: a regularization method for supervised and semi-supervised learning. In: IEEE Transactions on Pattern Analysis and Machine Intelligence, vol. 41, pp. 1979–1993 (2019)

16. Rong, Y., Huang, W., Xu, T., Huang, J.: Dropedge: towards deep graph convolutional networks on node classification. In: ICLR (2020)

17. Sohn, K., et al.: FixMatch: simplifying semi-supervised learning with consistency and confidence. arXiv abs/2001.07685 (2020)

18. Srivastava, N., Hinton, G.E., Krizhevsky, A., Sutskever, I., Salakhutdinov, R.: Dropout: a simple way to prevent neural networks from overfitting. J. Mach. Learn. Res. 15, 1929–1958 (2014)

19. Velickovic, P., Cucurull, G., Casanova, A., Romero, A., Lio', P., Bengio, Y.: Graph attention networks. arXiv abs/1710.10903 (2018)

20. Verma, V., Lamb, A., Kannala, J., Bengio, Y., Lopez-Paz, D.: Interpolation consistency training for semi-supervised learning. Neural Netw. 145, 90–106 (2019)

21. Verma, V., Qu, M., Kawaguchi, K., Lamb, A., Bengio, Y., Kannala, J., Tang, J.: Graphmix: improved training of GNNs for semi-supervised learning. In: AAAI (2021)

22. Xu, B., Huang, J., Hou, L., Shen, H., Gao, J., Cheng, X.: Label-consistency based graph neural networks for semi-supervised node classification. In: Proceedings of the 43rd International ACM SIGIR Conference on Research and Development in Information Retrieval (2020)

23. Yang, Z., Cohen, W.W., Salakhutdinov, R.: Revisiting semi-supervised learning with graph embeddings. arXiv abs/1603.08861 (2016)

24. Zhang, H., Cissé, M., Dauphin, Y., Lopez-Paz, D.: Mixup: beyond empirical risk minimization. arXiv abs/1710.09412 (2018)

25. Zhu, X., Ghahramani, Z., Lafferty, J.D.: Semi-supervised learning using gaussian fields and harmonic functions. In: ICML (2003)

26. Zhu, X., Ghahramani, Z., Lafferty, J.D.: Semi-supervised learning using gaussian fields and harmonic functions. In: ICML (2003)

Auxiliary Network: Scalable and Agile Online Learning for Dynamic System with Inconsistently Available Inputs

Rohit Agarwal[✉][iD], Krishna Agarwal[iD], Alexander Horsch[iD], and Dilip K. Prasad[iD]

Bio-AI Group, UiT The Arctic University of Norway,
Hansine Hansens veg 18, 9019 Tromsø, Norway
agarwal.102497@gmail.com

Abstract. Streaming classification methods assume the number of input features is fixed and always received. But in many real-world scenarios, some features are reliable while others are unreliable or inconsistent. We propose a novel online deep learning-based model called Auxiliary Network (Aux-Net), which is scalable and agile and can handle any number of inputs at each time instance. The Aux-Net model is based on the hedging algorithm and online gradient descent. It employs a model of varying depth in an online setting using single pass learning. Aux-Net is a foundational work towards scalable neural network for a dynamic complex environment dealing ad hoc or inconsistent inputs. The efficacy of Aux-Net is shown on the Italy Power Demand dataset.

Keywords: Online Learning · Dynamic System · Inconsistent inputs

1 Introduction

Supporting varying number of input features can be a game changer in real life applications which deal with dynamic complex environments. Examples include a device in smart city environment, an autonomous vehicle, etc. To model such environment of inconsistent and scalable nature, we assume some reliable data channels and refer it as base input features, denoted by $\{x_1^B, \ldots, x_b^B, \ldots, x_B^B\}$. In addition, it may receive other information through auxiliary sensor arrays termed as auxiliary input features and denoted by $\{x_1^A, \ldots, x_a^A, \ldots, x_A^A\}$. Here, x denotes input features, B in superscript and subscript denotes base feature and the number of base features respectively. Similarly, A in superscript and subscript denotes auxiliary feature and the number of auxiliary features respectively. Due to the intermittent availability, only a subset of auxiliary features arrive along with the base features at any time instance t as shown in Fig. 1. This imparts a lot of challenges as mentioned in Table 1. This problem can be approached via minimalist (only using base features), maximalist (making ensemble of model, one for each possible combinations of auxiliary features) and imputation based approaches (imputing the values wherever possible) but all the solutions are inefficient to address the problem. More information is given in the Appendix.

© The Author(s), under exclusive license to Springer Nature Switzerland AG 2023
M. Tanveer et al. (Eds.): ICONIP 2022, LNCS 13623, pp. 549–561, 2023.
https://doi.org/10.1007/978-3-031-30105-6_46

An ideal solution would be an agile and scalable network architecture that adapts itself to the availability of auxiliary inputs without needing to maintain or train multiple networks or impute data. In this paper, we present a new paradigm of learning in the presence of inconsistently available auxiliary inputs, which we call auxiliary network (Aux-Net). The keystone of Aux-net is the separation of learning corresponding to auxiliary and base inputs into separate modules parallel to each other (see Fig. 2). The base features are processed as a chunk in base module while auxiliary module contains one independent layer per auxiliary input in parallel with other layers.

Table 1. Key differentiators of Aux-Net: ✓ Full support, ○ Partial support, × No support. Missing data: data for some time instances of an input feature is missing. Missing features: prior knowledge about their existence or distribution may be assumed (even if they arrive late). Obsolete features: features cease to exist after some time. Sudden unknown features: no prior knowledge of their existence is available and they arrive late unannounced (true ad hoc). Unknown no. of features: no information about the number of inputs.

Characteristics	Aux-Net	ODL	Impute based
Online	✓	✓	✓
Missing data	✓	○	✓
Missing features	✓	×	○
Obsolete features	✓	×	○
Sudden unknown features	✓	×	×
Unknown no. of features	✓	×	×

2 Related Works

Many methods based on Bayesian theory [14], k-nearest neighbour [1], support vector machines [15], decision tree [16], fuzzy logic [2,6] are proposed for streaming classification task. A brief study of all these techniques can be found in [17,18]. Furthermore, some incremental learning approaches are also proposed [7–11]. Other deep learning approaches for online learning include [19–22]. But these techniques assume the dimension of the input data is fixed. Hou et al. [12,13] proposed machine learning approaches for dynamic environments. However, they assumed the dimensionality of inputs is constant in batches and therefore batch wise learning can be used. Hou et al. [12] further assume that there are multiple sets of features, where one entire set is either available or unavailable in a given batch. [13] assumes an overlapping period between two batches when all the inputs from previous and next batches are available, which allows in supporting soft transition across batches. These methods are indeed more scalable than approaches that assume fixed input dimensionality. Nonetheless, they cannot handle as challenging situations as depicted in Fig. 1 and discussed in Table 1 where no assumption is made on the availability of auxiliary features in batches or sets. To the best of our knowledge, our work is a foundational work for problems of inconsistently available inputs. It has a general premise and the only assumption is that there are at least one base feature. We note that our framework is inspired from the concept of hedge algorithm [3] and online gradient descent (OGD) [4].

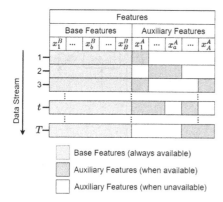

Base Features (always available)

Auxiliary Features (when available)

Auxiliary Features (when unavailable)

Fig. 1. Arrival of streaming data with all the base features and inconsistently available auxiliary features is demonstrated here.

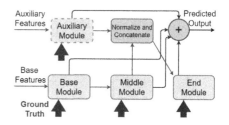

Fig. 2. Block diagram of Aux-Net. The green bold arrow represents the ground truth and the circle with a + sign calculates a final prediction from the weighted output of each classifier. All the modules are the combination of one or more hidden layers where the auxiliary module is scalable i.e. the number of layers keeps changing with time based on the needs of the application. (Color figure online)

3 Auxiliary Network (Aux-Net)

Problem Setting. Let's denote streaming classification data by $D = \{(x_1, y_1), ..., (x_t, y_t), ..., (x_T, y_T)\}$ where $x_t = \{x_t^B, x_t^{A_t}\}$ is input at time t. The base features are denoted by $x_t^B = \{x_{1,t}^B, ..., x_{b,t}^B, ..., x_{B,t}^B\}$, where B in superscript and subscript denotes base features and total number of base features respectively. $x_{b,t}^B$ denotes b^{th} base feature at t. The auxiliary features is represented by $x_t^{A_t} = \{x_{j,t}^A\}_{\forall j \in A_t}$ where $A_t \subseteq \{1, ..., a, ...A\}$ is subset of auxiliary features received at t. The notation A has denotation similar to B. The input $x_t \in \mathbb{R}^{d_t}$ where d_t is the dimension of x_t varying with t (Fig. 1). The output $y_t \in \mathbb{R}^C$ is the class label where C is the total number of classes. The Aux-Net learns a mapping $F : \mathbb{R}^{d_t} \to \mathbb{R}^C$. The prediction of the model is given by $\hat{y}_t = F(x_t)$. Model trains in an online setting where x_t arrives, it predicts $\hat{y}_t$, y_t is revealed and it is updated based on the loss.

Architecture. Consider a DNN with S number of base layers, one middle layer, A number of auxiliary layers and E number of end layers. The base layers, middle layer, auxiliary layers and end layers constitute the base module, middle module, auxiliary module and end module respectively. The base, middle and end modules are stacked sequentially and auxiliary module is placed in parallel to the base and middle module with a connection to the end module as shown in the Fig. 2. A softmax classifier is attached to each of the layer. The detailed architecture of the model is presented in Fig. 3. The output of the Aux-Net model is given as the weighted combination of all the classifiers by the equation:

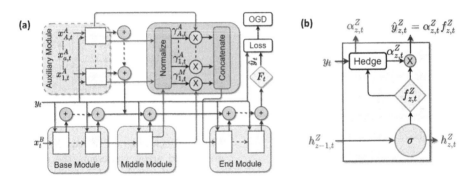

Fig. 3. (a) Detailed architecture of Aux-Net is presented here. The gray colored rectangular boxes represents a layer. (b) The functional diagram of a layer is shown here. (Color figure online)

$$F(x) = \sum_{Z \in U} \sum_{z=1}^{Z} \alpha_z^Z f_z^Z \tag{1}$$

where $U = \{S, M, E, A\}$ denotes all the modules, and S, M, E, A in superscript and subscript denotes module name and the total number of layers in module respectively. The notation f_z^Z and α_z^Z represents the output of the classifier associated with the layer z of module Z and weight of the classifier respectively.

Architecture of a layer is shown in Fig. 3(b). Each layer is attached to a classifier f parameterized by θ that gives an output $f_z^Z = \mathrm{softmax}(\mathrm{h}_z^Z \theta_z^Z)$, where h_z^Z is the hidden feature of layer. Each layer is parameterized by W and c and generates its hidden feature $h_z^Z = \sigma(W_z^Z h_{z-1}^Z + c_z^Z)$, where σ is the activation function, and θ, W and c are learnt using OGD. A hedge block is used to compute α based on the loss incurred by the classifier.

Now, we describe the inputs to different layers. The first base layer receives the complete x_t^B as the input i.e. $h_0^Z = x_t^B$. The subsequent base layers receive the hidden feature of its previous layer as its input. The input to middle layer is the hidden feature of the last base layer, i.e. $h_0^M = h_S^S$. The ath auxiliary layer receives the ath auxiliary feature, i.e., $h_a^A = x_a^A$. All the end layers, except the first end layer receive its previous layer features as the input. The first end layer is special since the input to it needs to support agility arising from only a subset of auxiliary features being available at time t. The input $h_{0,t}^E$ to the first end layer at time instance t is a vector derived by concatenating weighted hidden features h of the middle and the auxiliary layers corresponding to the currently available auxiliary inputs. It is given by $h_{0,t}^E = \left[\gamma_{1,t}^M h_{1,t}^M, \{\gamma_{j,t}^A h_{j,t}^A\}_{\forall j \in A_t}\right]$, where γ is importance of the layers connected to first end layer, denoting the fraction of the connected layer's output passed as an input to the first end layer.

Algorithm 1: Aux-Net algorithm

Inputs: Base Module: S; Middle Module: M; Auxiliary Module: A; End
    Module: E; Learning rate: η; Smoothing Parameter: λ; Discounting Parameter:
    β ;

Initialize: A DNN with $L = S + M + A + E$ layers and attach classifiers to each
    layer as shown in Figure 3(b); $\alpha_z^Z = 1/L \quad \forall Z \in \{S, M, A, E\}, \quad z \in \{1, ..., Z\}$; K_1
    using equation 7;

for $t = 1, ..., T$ **do**
    Receive input feature x_t;
    Create a list A_t of the auxiliary features received in x_t;
    Create the model M_t based on A_t using equation 8;
    Predict $\hat{y}_t$ on x_t using equation 1 based on M_t;
    Receive output label y_t;
    Calculate the loss of the model M_t based on y_t and $\hat{y}_t$ using equation 2;
    Update parameters of M_t based on the loss incurred and get M_t^* using 9;
    Update K_t based on M_t^* to get K_{t+1} using 10;
end

Parameters Learning. The learning of the model occurs in an online setting
through the use of a loss function defined as:

$$L(F(x), y) = \sum_{Z \in U} \sum_{z=1}^{Z} \alpha_z^Z L(f_z^Z(x), y) \tag{2}$$

where $L(f_z^Z(x), y)$ is the loss of classifier associated with layer z of module Z.
Based on the loss at each time step t, the values of $\gamma, \theta, W, c, \alpha$ are updated.

    Updating γ : The highlight of Aux-Net is the update of γ which allows for
soft handling of the asynchronous availability of auxiliary features. It depends
only on its classifiers weights and are calculated as follows:

$$\gamma_{p,t}^P = \frac{\alpha_{p,t}^P}{\alpha_{1,t}^M + \sum_{j \in A_t} \alpha_{j,t}^A} \quad \text{for} \quad \text{C1}:(P = M,\ p = 1) \text{ or } (P = A,\ p \in A_t) \tag{3}$$

    Updating θ : The parameter θ_z^Z is associated with only one classifier and
does not depend on the other classifiers. Therefore, its update will only be with
respect to the loss of its own classifier through OGD. After every time instance
t, θ_z^Z of classifier z of the module Z is updated as:

$$\theta_{z,t+1}^Z = \theta_{z,t}^Z - \eta \alpha_{z,t}^Z \Delta_{\theta_{z,t},z}^Z \quad \text{for} \quad \text{C2}:(Z \in U',\ z \in \{1, ..., Z\}) \text{ or } (Z = A,\ z \in A_t) \tag{4}$$

where, $\Delta_{\theta_{z,t},r}^R = \frac{\partial L(f_r^R(x_t), y_t)}{\partial \theta_{z,t}^Z}$, η is learning rate of parameters and $U' = \{S, M, E\}$.

    Learning W and c : The weights (W) and bias (c) of a layer are learned
by back propagation on the final loss similar to OGD. But, since each layer is
associated with a classifier unlike the traditional DNN where only last layer gives
a prediction, the gradient descent is different. Here, the parameters of a layer

depends on loss of all its successive layers that directly or indirectly influence it. The following equation shows update rule for W and same is applicable for c.

$$W_{a,t+1}^A = W_{a,t}^A - \eta \Big[\alpha_{a,t}^A \Delta_{W_{a,t}^A,a}^A + \sum_{e=1}^{E} \alpha_{e,t}^E \Delta_{W_{a,t}^A,e}^E \Big]$$

$$W_{z,t+1}^Z = W_{z,t}^Z - \eta \Big[\sum_{j=z}^{Z} \alpha_{j,t}^Z \Delta_{W_{z,t}^Z,j}^Z + \sum_{Q=set}^{} \sum_{q=1}^{Q} \alpha_{q,t}^Q \Delta_{W_{z,t}^Z,q}^Q \Big] \tag{5}$$

where $set = \{M, E\}, \{E\}, \phi$ if $Z \in \{S\}, \{M\}, \{E\}$ respectively, and $z \in \{1, ..., Z\}$.

Learning α : The value of α is learned through hedge algorithm. Initially, α is uniformly distributed i.e., $\alpha_z^Z = 1/L$, where L is the total number of layers, $L = S + M + A + E$. The loss incurred by classifier z of module Z at time t is $L(f_z^Z(x_t), y_t)$ and its weight is $\alpha_{z,t}^Z$. The weights of the classifier are updated as:

$$\alpha_{z,t+1}^Z = \alpha_{z,t}^Z \beta^{L(f_z^Z(x_t), y_t)} \qquad \text{for C2,} \tag{6}$$

where $\beta \in (0, 1)$ is the discount rate parameter. To avoid situation, $\alpha_z^Z \to 0$ (since we don't want to neglect any layer), a smoothing parameter $\lambda \in (0, 1)$ is introduced. It ensures minimum weight for each classifier by $\alpha_{z,t+1}^Z = \max(\alpha_{z,t+1}^Z, \lambda/L)$. The value of all α is then normalized such that $\sum_{Z=U'} \sum_{z=1}^{Z} \alpha_{z,t+1}^Z + \sum_{j \in A_t} \alpha_{j,t+1}^A = 1$.

Algorithm. Aux-Net is a test-then-train approach and since auxiliary features are changing, the trained model learned at t can't be used as it is for training or testing at $t + 1$. We define a knowledge base K (updated parameters represented by $'$) which is updated after every time t. K at any time instance t is given by

$$K_t = \{W_t', c_t', \theta_t', \alpha_t'\}, \quad \text{where} \quad G_t = \{G_t^S, G_t^M, G_t^A, G_t^E\} \text{ if } G \in \{W', c', \theta', \alpha'\} \tag{7}$$

Before training or testing, model needs to incorporate the incoming dynamic auxiliary features. We define a model M_t (eq. 8), that handles the asynchronous availability of auxiliary features (A_t) by introducing the variable γ. The model M_t predicts an output $\hat{y}_t$, given x_t and updates its parameter giving M_t^* based on the loss incurred. Before moving to the next instance, we update the final parameters of K_t based on M_t^*, giving knowledge base K_{t+1} (see Algorithm 1).

Creating Model (M_t): Based on the auxiliary features A_t received at time step t and knowledge base K_t, the model M_t is created before prediction and training. The auxiliary layers corresponding to A_t are kept active and all the other auxiliary layers are freeze. Freezing of layers means all the parameters associated with this layer will not be trained (or removing the layer from the model). Since, some of the auxiliary layers are removed, the value α of the model changes and a parameter γ is introduced. The model M_t is given by:

$$M_t = M(W_t, c_t, \theta_t, \alpha_t, \gamma_t) \tag{8}$$

where $G_t = \{G_t'^S, G_t'^M, \{G_{j,t}'^A\}_{\forall j \in A_t}, G_t'^E\}$ if $G = \{W, c, \theta\}$, $\alpha_t = \{\alpha_{z,t}^Z\}_{\forall \text{ C2}}$ where $\alpha_{z,t}^Z = \alpha_{z,t}'^Z / \Big[\sum_{Z=U'} \sum_{z=1}^{Z} \alpha_{z,t}'^Z + \sum_{j \in A_t} \alpha_{j,t}'^A \Big]$, $\gamma_t = \{\gamma_{p,t}^P\}_{\forall \text{ C1}}$, $\gamma_{p,t}^P = \alpha_{p,t}'^P / \Big[\alpha_{1,t}'^M + \sum_{j \in A_t} \alpha_{j,t}'^A \Big]$.

Obtaining Knowledge Base K_{t+1} **for Next Instance:** M_t is updated based on loss incurred at time t. The updated model, represented by M_t^* is given by:

$$M_t^* = M(W_t^*, c_t^*, \theta_t^*, \alpha_t^*) \tag{9}$$

where $W_t^*, c_t^*, \theta_t^*, \alpha_t^*$ are the parameters obtained by updating the parameters $W_t, c_t, \theta_t, \alpha_t$ of the model M_t by using Eq. 2, 4, 5 and 6. After training the model at time step t, we create the knowledge base K_{t+1} before moving to the next iteration. All the parameters updated at time step t and the parameters of the freezed layers $(A - A_t)$ are collected. Then, K_{t+1} is given by:

$$K_{t+1} = \{W'_{t+1}, c'_{t+1}, \theta'_{t+1}, \alpha'_{t+1}\} \tag{10}$$

where $G'_{t+1} = \{G_t^*, \{G'^A_{j,t}\}_{\forall j \in A - A_t}\}$ if $G \in \{W, c, \theta\}$, $\alpha'_{t+1} = \{\alpha'^Z_{z,t+1}\}_{Z \in U, z=\{1,...,Z\}}$ where $\alpha'^Z_{z,t+1} = \alpha''^Z_{z,t} / \left[\sum\limits_{Z \in U} \sum\limits_{z=1}^{Z} \alpha''^Z_{z,t} \right]$ and $\alpha''_{t+1} = \{\alpha_t^*, \{\alpha'^A_{j,t}\}_{\forall j \in A - A_t}\}$.

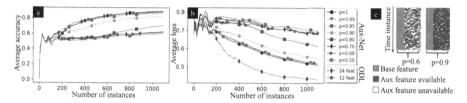

Fig. 4. Cumulative average accuracy (a) and loss (b) for different values of probability p of auxiliary inputs on Italy power demand dataset. ODL with 12 and 24 features is included for baseline. Snippet of data availability for $p = 0.6$ and $p = 0.9$ are shown in (c), analogous to Fig. 1.

4   Experimental Results

We show robust, agile and scalable performance on Italy power demand dataset [5]. It has 1096 data instances with 24 features. In all the studies, we retain the original order of features. To the best of our knowledge, no method incorporates the intermittently available input data in an online setting. Thus, we compare Aux-Net with ODL [21] (in minimalist approach). We train both the models in a purely online setting where after each instance the model predicts and trains.

Architecture Details The number of base layers (S) are 5, middle layer (M) is 1, and end layers (E) are 5 for Aux-Net. The number of auxiliary layers (A) are equal to number of auxiliary features. The number of layers for ODL is set as 11 $(S + E + M)$. For both Aux-Net and ODL, we used ReLU activation function, adam optimizer $(\eta = 0.01)$, cross-entropy loss, smoothing rate $(\lambda = 0.2)$, discount rate $(\beta = 0.99)$ and number of nodes in each layer was set as 50.

Varying Probability of the Availability of Auxiliary Inputs in Aux-Net. The first 12 features of Italy power demand dataset are considered as base features and remaining as auxiliary features. The availability of each auxiliary feature at a given time instance is modeled as a uniform distribution with probability p. The same value of p is used for all auxiliary features but the availability of each is computed independently. The results of Aux-Net with varying values of p, ODL with all 24 features, and 12 base features are presented in Table 2. We report the average of losses and accuracy observed across all time instances. The cumulative average loss and accuracy curves are shown in Fig. 4. We study the performance of Aux-Net and compare with ODL with the following aims:

Sensitivity of Aux-Net to p and its Performance: The average accuracy and loss for all the time instances in the dataset shows monotonic trend as a function of p, as noted in Table 2. This shows that Aux-Net is sensitive to the availability of the auxiliary inputs, as expected. Yet, the performance of Aux-Net degrades gracefully as p reduces. Moreover, Aux-Net still performs better compared to ODL with 12 features when $p < 1$ (as ODL can not work with inconsistent features). Further, the best case performance of Aux-Net when $p = 1$, is comparable to the scenario of ODL with 24 features.

This means that even though the knowledge base of Aux-Net supports for 2^{12} knowledge models, only the knowledge model with largest dimensionality is invoked and trained. In this case loss of Aux-Net is poorer than ODL, but the accuracy is better. In case of $p = 0.5$ which means no consistency in either availability or unavailability of the auxiliary inputs, the observed poorer performance of Aux-Net in comparison to ODL is only marginal, indicating robustness of Aux-Net to the extremely challenging scenario and its graceful degradation.

Agile Adaptation of Aux-Net: The demand on agility significantly enhances as p reduces. For example, for $p = 0.6$ in Fig. 4(c), not only a different knowledge model needs to be invoked at every instance but also the same knowledge model may not be invoked in next many instances. The situation is easier

Table 2. Average accuracy and loss of Aux-Net (for different values of probability(p) of availability of auxiliary features) and ODL (with different number of input features(feat)) in Italy Power Demand dataset. *ODL* is shown in italics.

Model	Accuracy	Loss
ODL(24 feat)	*0.8783*	*0.4297*
Aux-Net($p = 1.00$)	0.8884	0.5093
Aux-Net($p = 0.99$)	0.8811	0.5165
Aux-Net($p = 0.95$)	0.8637	0.5168
Aux-Net($p = 0.90$)	0.8243	0.5456
Aux-Net($p = 0.80$)	0.7054	0.6130
Aux-Net($p = 0.70$)	0.6240	0.6788
Aux-Net($p = 0.60$)	0.6167	0.6831
ODL(12 feat)	*0.6139*	*0.6868*
Aux-Net($p = 0.50$)	0.5956	0.6975

when $p = 0.9$ even though there are many instances when a different knowledge model is invoked. Nonetheless, Aux-Net remains stable in either case and adapts to the agility needs in an efficient manner, indicated in accuracy and loss plots in Fig. 4(a,b). Indeed, accuracy is better and loss decreases faster over time for $p=0.9$. Nonetheless, when $p=0.6$, the accuracy and loss curves closely follow ODL with 12 features, indicating that even though new knowledge models are being dynamically invoked every single instance, the performance of Aux-Net does not deteriorate in comparison to ODL and Aux-Net is indeed able to maintain agility over time, contributing to reduced loss and improved accuracy as time passes.

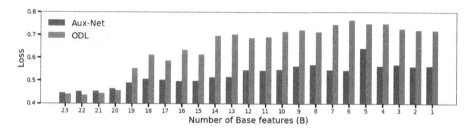

Fig. 5. Loss of Aux-Net as function of the number of base features B and ODL (trained using B number of features) in Italy power demand dataset. The probability of availability of the $(24 - B)$ auxiliary inputs is fixed at $p = 0.9$. Lower loss indicates better learning.

Decreased Loss and Improved Accuracy Over Time: For situation of 12 auxiliary inputs, support for 2^{12} knowledge models, and invocation of each knowledge model multiple times is needed to study the convergence of knowledge base over time. Yet, the decreasing loss (Fig. 4(b)) is a positive indicator of performance improvement over time and possible convergence.

Varying Number of Base Features. In this experiment, we fix p as 0.9, but vary the number of base features (B) from 1–23. The number of auxiliary features (A) are consequently (24-B). The first B features in the dataset are used as base features in Aux-Net and the only features in ODL. The average loss of Aux-Net and ODL are compared in Fig. 5 as a function of B. We observe the following:

Extreme Scalability: As expected, the performances of both Aux-Net and ODL deteriorate as B reduces. Nonetheless, the loss of Aux-Net is significantly smaller than ODL in the challenging scenarios when more than 4 inputs are inconsistently available. This clearly indicates that Aux-Net is able to leverage the auxiliary inputs for better learning. Especially, the extremely challenging scenarios ($B = 1$ for example) demonstrate that Aux-Net is indeed able to step up to the need of supporting several knowledge models of varying inputs and dimensionalities and provide better performance than the minimalist approach.

Poorer Performance than ODL when B $\in [20, 23]$**:** During initialization, Aux-Net assigns the same weights (α) to each classifier. However, the classifier corresponding to an auxiliary feature will be lossier as compared to the classifier of middle layer that uses base features. As time progresses, the value of α for each layer gets customized to suit its contribution towards accurate classification. Often, it means that α of auxiliary layer reduces in the first few time instances, indicating that Aux-Net has learnt that its inconsistent availability may cause increased loss if α corresponding to it is high.

Obsolete Features. In Fig. 6, we consider an example in which all the auxiliary features (12 out of 24 features of the original dataset) are available initially, but become obsolete after the 100th instances out of a total of 1096 instances (unavailable >90% of the time). Therefore, the remaining 12 (base) features represent the data trend over the long term. In this condition, since we have the information about the obsolete features from their previously received instances, we can perform imputations. In the first 100 instances when all the features are available, ODL (mean imputation) and Aux-Net generate lower loss than ODL (12 features). However, after the 12 auxiliary features become obsolete, Aux-Net quickly adapts to the new normal and converges to similar performance as ODL (12 features). On the other hand, ODL (mean imputation) performs robustly for sometime due to imputation, but very slowly converges thereafter towards the new normal over the long term.

Sudden Unknown Features. In Fig. 7, we consider an example where only 6 (base) features are available. Suddenly, 18 new features with no prior knowledge appear from 201st time instance. Aux-Net performs similar to ODL (6 features) till the 200th instance. Thereafter, Aux-Net quickly adapts to the availability of new features and its loss starts decreasing at a rapid rate.

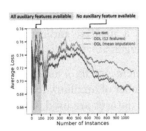
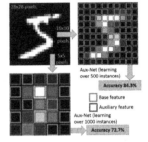

Fig. 6. All 12 auxiliary features become obsolete after 100th instance on Italy power demand dataset.

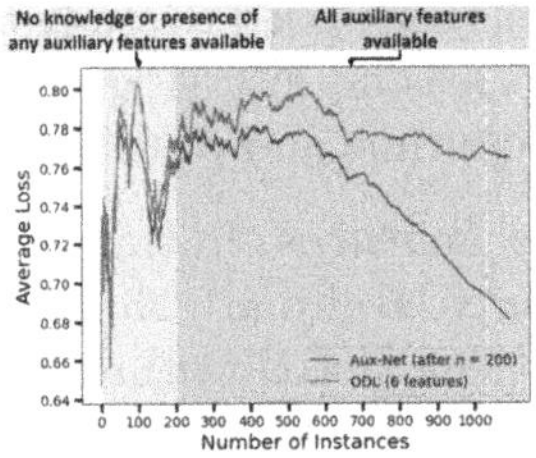

Fig. 7. 18 unknown auxiliary features starts appearing after 200th instance on Italy power demand dataset.

Fig. 8.   Downsampled MNIST data with feature unavailability (Color figure online)

MNIST Dataset. We applied Aux-Net on MNIST [23] dataset for classification of 2 similar looking digits, namely '5' and '6' (see Fig. 8) to demonstrate its performance on challenging situations. In our version, we consider availability of a down sampled versions of the original 28×28 pixels. We consider two cases, 10×10 pixels and 5×5 pixels. In either case, the intensity at each pixel is one feature and a series of images is provided as data stream. For the case of 10×10 pixels, we consider 75:25 ratio of base and auxiliary features (blue and magenta colored boxes in Fig. 8, respectively) and achieve classification accuracy of >84%. We further challenge Aux-Net with 12:13 ratio of base and auxiliary features for the case of 5×5 pixels, and achieve classification accuracy of >72%.

5   Conclusion

We have demonstrated scalability, agility, and stability of Aux-Net and its ability to deal with intermittently available inputs in an online setting. It supports scalability for the situations ranging from no auxiliary input being available to all auxiliary inputs being available. It incorporates knowledge models corresponding to all possible combinations of auxiliary inputs within a single knowledge base. The architectural support in Aux-Net for auxiliary inputs in the form of dedicated parallel layers is a critical feature for scalability. Agility in Aux-Net is characterized by its ability to dynamically invoke the relevant knowledge model without making the network unstable or unadaptive. A key factor that supports dynamic stability and agility is the importance parameter γ, which automatically adjusts the contributions of base inputs (through the middle layer) and the currently available auxiliary inputs so that neither the new auxiliary features introduce inordinate instability, nor are they suppressed. This, in our observation is not only the first such architecture, it is also a first demonstration of results on intermittently available input features. Having set a new paradigm, we hope that new datasets, frameworks, applications and more extensive studies are developed in future to exploit the possibility of learning in extremely dynamic and uncertain scenarios.

Acknowledgement. We acknowledge the following funding: ERC starting grant no. 804233 (Krishna Agarwal), Researcher Project for Scientific Renewal grant no. 325741 (Dilip K. Prasad), UiT's thematic funding project VirtualStain with Cristin Project ID 2061348 (Alexander Horsch, Dilip K. Prasad, Krishna Agarwal) and Horizon 2020 FET open grant OrganVision (id 964800). We thank Dr. Sk. Arif Ahmed (Assistant Prof., XIM University, Bhubaneswar) for his discussion and comments that greatly improved the manuscript.

Appendix

Minimalist: All uncertain inputs are dropped and a single knowledge model is trained using only base features. This model provides certain base accuracy, but does not utilize additional information from auxiliary inputs. The trade-off is the loss of opportunity for better performance.

Maximalist: An ensemble of 2^A networks can be formed to cater for all possible combinations of availability of auxiliary features. Therefore network with the smallest dimensionality caters to only the base features and network with the largest dimensionality caters to all the base and auxiliary features, where the number of inputs to a network is its dimensionality. However, learning the knowledge model in such ensemble of networks is cumbersome. Given A_t inputs features at time t, we have 2^{A_t} subsets, therefore the network corresponding to each subset needs to be trained. This results into long training duration. Another trade-off is that huge number of networks need to be maintained throughout.

Imputation: As shown in Table 1, it is possible to impute features whose prior information is known but it will introduce a lot of misinformation in cases where it is required to impute for many consecutive instances. Moreover, for the conditions of sudden unknown features, no prior information is available. So the approach can be to model noise (Gaussian or otherwise) but we don't know the total number of features hence making it impossible to impute.

References

1. Aggarwal, C.C., et al.: A framework for on-demand classification of evolving data streams. IEEE Trans. Knowl. Data Eng. **18**(5), 577–589 (2006)
2. Iyer, A.R., Prasad, D.K., Quek, C.H.: PIE-RSPOP: a brain-inspired pseudo-incremental ensemble rough set pseudo-outer product fuzzy neural network. Expert Syst. Appl. **95**, 172–189 (2018)
3. Freund, Y., Schapire, R.E.: A decision-theoretic generalization of on-line learning and an application to boosting. J. Comput. Syst. Sci. **55**(1), 119–139 (1997)
4. Zinkevich, M.: Online convex programming and generalized infinitesimal gradient ascent. In: Proceedings of the 20th International Conference on Machine Learning (ICML-03) (2003)
5. Dau, H.A., et al.: The UCR time series archive. IEEE/CAA J. Automatica Sinica **6**(6), 1293–1305 (2019)
6. Das, R.T., Ang, K.K., Quek, C.: ieRSPOP: a novel incremental rough set-based pseudo outer-product with ensemble learning. Appl. Soft Comput. **46**, 170–186 (2016)
7. Polikar, R., et al.: Learn++: an incremental learning algorithm for supervised neural networks. IEEE Trans. Syst., Man, Cybern. Part C (Appl. Rev.) **31**(4), 497–508 (2001)
8. Polikar, R., et al.: Learn++ .MF: a random subspace approach for the missing feature problem. Pattern Recogn. **43**(11), 3817–3832 (2010)
9. Muhlbaier, M.D., Topalis, A., Polikar, R.: Learn ++. NC: combining ensemble of classifiers with dynamically weighted consult-and-vote for efficient incremental learning of new classes. IEEE Trans. Neural Netw. **20**(1), 152–168 (2008)
10. Muhlbaier, M.D., Polikar, R.: Multiple classifiers based incremental learning algorithm for learning in nonstationary environments. In: 2007 International Conference on Machine Learning and Cybernetics, vol. 6. IEEE (2007)
11. Ditzler, G., Polikar, R., Chawla, N.: An incremental learning algorithm for non-stationary environments and class imbalance. In: 2010 20th International Conference on Pattern Recognition. IEEE (2010)
12. Hou, C., Zhou, Z.-H.: One-pass learning with incremental and decremental features. IEEE Trans. Pattern Anal. Mach. Intell. **40**(11), 2776–2792 (2017)
13. Hou, B.-J., Zhang, L., Zhou, Z.-H.: Learning with feature evolvable streams. In: Advances in Neural Information Processing Systems, vol. 30 (2017)
14. Seidl, T., et al.: Indexing density models for incremental learning and anytime classification on data streams. In: Proceedings of the 12th International Conference on Extending Database Technology: Advances in Database Technology (2009)
15. Tsang, I.W., Kocsor, A., Kwok, J.T.: Simpler core vector machines with enclosing balls. In: Proceedings of the 24th International Conference on Machine Learning (2007)
16. Domingos, P., Hulten, G.: Mining high-speed data streams. In: Proceedings of the Sixth ACM SIGKDD International Conference on Knowledge Discovery and Data Mining (2000)
17. Nguyen, H.-L., Woon, Y.-K., Ng, W.-K.: A survey on data stream clustering and classification. Knowl. Inf. Syst. **45**(3), 535–569 (2015)
18. Gama, J.: A survey on learning from data streams: current and future trends. Progress Artif. Intell. **1**(1), 45–55 (2012)
19. Das, M., et al.: FERNN: a fast and evolving recurrent neural network model for streaming data classification. In: 2019 International Joint Conference on Neural Networks (IJCNN). IEEE (2019)

20. Das, M., et al.: Muse-RNN: a multilayer self-evolving recurrent neural network for data stream classification. In: 2019 IEEE International Conference on Data Mining (ICDM). IEEE (2019)
21. Sahoo, D., et al.: Online deep learning: learning deep neural networks on the fly. arXiv preprint arXiv:1711.03705 (2017)
22. Ashfahani, A., Pratama, M.: Autonomous deep learning: continual learning approach for dynamic environments. In: Proceedings of the 2019 SIAM International Conference on Data Mining. Society for Industrial and Applied Mathematics (2019)
23. LeCun, Y.: The MNIST database of handwritten digits. http://yann.lecun.com/exdb/mnist/ (1998)

VAAC: V-value Attention Actor-Critic for Cooperative Multi-agent Reinforcement Learning

Haonan Liu, Liansheng Zhuang$^{(\boxtimes)}$, Yihong Huang, and Cheng Zhao

University of Science and Technology of China, Hefei 230027, China
phoenix_@mail.ustc.edu.cn, lszhuang@ustc.edu.cn

Abstract. This paper explores value-decomposition methods in cooperative multi-agent reinforcement learning (MARL) under the paradigm of centralized training with decentralized execution. These methods decompose a global shared value into individual ones to guide the learning of decentralized policies. While Q-value decomposition methods such as QMIX show state-of-the-art performance, V-value decomposition methods are proposed to obtain a reasonable trade-off between training efficiency and algorithm performance under the A2C training paradigm. However, existing V-value decomposition methods lack theoretical analysis of the relation between the global V-value and local V-values, and do not explicitly consider the influence of individuals on the total system, which degrades their performance. To address these problems, this paper proposes a novel approach called V-value Attention Actor-Critic (VAAC) for cooperative MARL. We theoretically derive a general decomposing formulation of the global V-value in terms of local V-values of individual agents, and implement it with a multi-head attention formation to model the impact of individuals on the whole system for interpretability of decomposition. Evaluations on the challenging StarCraft II micromanagement task show that VAAC achieves a better trade-off between training efficiency and algorithm performance, and provides interpretability for its decomposition process.

Keywords: Multi-agent reinforcement learning · Multi-agent policy gradients · Deep reinforcement learning

1 Introduction

Cooperative multi-agent reinforcement learning (MARL) has made significant progress in recent years [7,15,16], where a system of agents learns towards coordinated policies to optimize the accumulated global rewards. Many complex real-world tasks such as autonomous vehicle coordination [1] and sensor networks [20] can be modeled as cooperative MARL problems. One natural way to address cooperative MARL problems is the fully centralized approach that views the multi-agent system as a single-agent reinforcement learning task with

M. Tanveer et al. (Eds.): ICONIP 2022, LNCS 13623, pp. 562–573, 2023.
https://doi.org/10.1007/978-3-031-30105-6_47

a joint action space [13]. However, since the joint action space of the agents grows exponentially with the number of agents, the fully centralized approach has limited scalability. Besides, due to partial observability and communication constraint in practical environments, it is necessary to use decentralized policies that act only based on the local observation history of individual agents. The simplest approach to decentralized policies is independent learning which trains the agents independently, but suffers from non-stationarity since it views other agents as part of the environment [5].

To address these above issues, the paradigm of *centralized training with decentralized execution* (CTDE) [4] has attracted great attention of researchers, where decentralized policies are trained with access to additional global state information in a centralized fashion and executed only conditioned on local histories in a decentralized way. There is still a challenging problem of how to use a global shared value, such as joint action-value (Q-value) or global state value (V-value), for the training of decentralized policies. One approach is value-decomposition, which decomposes the global value into individual ones to guide the learning of decentralized policies. Many breakthroughs in Q-value decomposition methods have been made recently. *Value Decomposition Network* (VDN) [12] represents joint action-value Q_{tot} as a summation of individual Q-values that condition only on individual observations and actions. QMIX [8] extends to a broader class of monotonic functions using a mixing network of per-agent Q-values. QTRAN [10] proposes a provably more general value factorization method that avoids representation limitations. Qatten [18] theoretically derives a general formula of Q_{tot} and considers the impact of individuals on the global.

In real applications, how to improve the training efficiency of CTDE is a practical problem in cooperative MARL. A2C framework [6] is a popular training paradigm that promotes training efficiency by asynchronously executing multiple instances of the environment. However, Q-value decomposition methods such as QMIX do not perform well under the A2C paradigm, because these off-policy methods utilize replay buffers which are incompatible with multi-thread execution. As reported in [11], when using the A2C training paradigm, the performance of QMIX degrades on the StarCraft Multi-Agent Challenge (SMAC) [9]. On the other hand, on-policy actor-critic methods such as *counterfactual multi-agent* (COMA) [2] can exploit the A2C framework efficiently while having poor performance on SMAC [9].

To narrow the performance gap between on-policy actor-critic and Q-value decomposition methods, [11] extends value-decomposition to on-policy actor-critic methods and proposes a V-value decomposition framework called *value-decomposition actor-critic* (VDAC). VDAC represents the global state value V_{tot} as a monotonic function of local state values V^a, and introduces two V-value decomposition methods, i.e. VDAC-sum and VDAC-mix. The former method represents V_{tot} as a summation of V^a, while the later one generalizes the representation to a larger family of monotonic functions through a mixing network. However, both V-value decomposition methods impose certain assumptions which lack theoretical analysis of the relation between V_{tot} and V^a. Besides, they do not explicitly consider the influence of individuals on the total system, just viewing that each agent is equal or mixing local state values implicitly. These problems of existing V-value decomposition methods limit their performance.

To further achieve an acceptable trade-off between training efficiency and performance, this paper proposes a novel V-value decomposition approach called V-value Attention Actor-Critic (VAAC). We derive a decomposing formulation of V_{tot} in terms of V^a through theoretical analysis, and implement it with a multi-head attention formation to model the impact of agents on the whole system for interpretability of decomposition. Empirical results on SMAC show that VAAC outperforms other baselines under A2C. Next, we use ablation experiments to demonstrate the contribution of the multi-head attention formation. Moreover, we investigate the relationship between the weights for mixing V^a into V_{tot} and the properties of agents to interpret the decomposition process.

2 Background

2.1 Decentralized Partially Observable Markov Decision Process

We consider a fully cooperative multi-agent task that can be modeled as a *decentralised partially observable Markov decision process* (Dec-POMDP) consisting of a tuple $G = \langle S, U, P, r, Z, O, n, \gamma \rangle$, in which n agents identified by $a \in A \equiv \{1, ..., n\}$ choose sequential actions. The environment has a true state $s \in S$. Each agent simultaneously chooses an action $u^a \in U$ at each time step, forming a joint action $\mathbf{u} \in \mathbf{U} \equiv U^n$ which induces a transition probability function $P(s'|s, \mathbf{u}) : S \times \mathbf{U} \times S \rightarrow [0, 1]$ and a global reward function $r(s, \mathbf{u}) : S \times U \rightarrow \mathbb{R}$. We consider a partially observable setting, where each agent receives an individual partial observation $z \in Z$ from the observation function $O(S, A) : S \times A \rightarrow Z$. Each agent learns a stochastic policy $\pi^a(u^a|\tau^a) : T \times U \rightarrow [0, 1]$ conditioned on its local action-observation history $\tau^a \in T \equiv Z \times U$. We denote joint quantities over agents in bold and joint quantities over agents other than a given agent a with the superscript $-a$. All agents coordinate together to maximize the discounted return $R_t = \sum_{l=0}^{\infty} \gamma^l r_{t+l}$. The agents' joint policy induces a value function, i.e., the expected return for following the joint policy π from state s, $V^\pi(s_t) = \mathbb{E}[R_t|s_t = s]$, and an action-value function, i.e. the expected return for selecting joint action $\mathbf{u}$ in state s and following the joint policy π, $Q^\pi(s, \mathbf{u}) = \mathbb{E}[R_t|s_t = s, \mathbf{u}]$.

2.2 Single-Agent Policy Gradient Algorithms

Policy gradient methods optimise a single agent's policy parameterised by θ_π to maximize the objective $J(\theta) = \mathbb{E}_{s \sim p^\pi, u \sim \pi}[R(s, u)]$ by performing gradient ascent, where p^π is the state transition by following policy π. The gradient with respect to the policy parameters is $\nabla_\theta J(\theta) = \mathbb{E}_\pi[\nabla_\theta \log \pi_\theta(a|s) Q_\pi(s, u)]$. To reduce variations in gradient estimates, a baseline b is introduced. In *actor-critic* approaches, the actor, i.e., the policy, is trained by following a gradient that depends on a critic, which usually estimates a value function. This yields the advantage function $A(s_t, u_t) = Q(s_t, u_t) - b(s_t)$. $V(s_t)$ is commonly used as the baseline. Temporal difference (TD) error $r_t + \gamma V(s_{t+1}) - V(s_t)$, which is an unbiased estimate of $A(s_t, u_t)$, is a common choice for advantage functions.

2.3   IAC and COMA

Independent Actor-Critic (IAC) [2] is the simplest method to apply Policy Gradient Algorithms to multiple agents, which lets each agent learn its own actor and critic independently according to its own action-observation history. Each agent's critic estimates $V(o^a)$ to calculate TD error. IAC is straightforward and easy to implement but lacks information about other agents and global state during the training, which makes it difficult to learn coordinated strategies and estimate its contribution to the team's reward. To mitigate this issue, decentralized policies can be learned in the centralized training and decentralized execution (CTDE) paradigm. COMA [2] uses a centralized critic and applies the following counterfactual policy gradients:$\nabla_\theta J = \mathbb{E}_\pi [\sum_a \nabla_\theta \log \pi(u^a|\tau^a) A^a(s, \mathbf{u})]$, where $A^a(s, \mathbf{u}) = Q_\pi(s, \mathbf{u}) - \sum_{u^a} \pi_\theta(u^a|\tau^a) Q_\pi^a(s, (\mathbf{u}^{-a}, u^a))$ is the counterfactual advantage for agent a. COMA provides agents with tailored gradients to achieve credit assignment, but it becomes ineffective with complex cooperation behaviors.

2.4   Value Decomposition Actor-Critic

Value Decomposition Actor-Critic (VDAC) [11] is an actor-critic, on-policy framework that uses the paradigm of CTDE. VDAC has local critics for each agent to estimate the local state values V^a and a central critic to estimate the global state value V_{tot}. Inspired by *difference rewards* [17], VDAC decomposes the global state value $V_{tot}(s)$ into local states $V^a(o^a)$ through the following constraint:

$$\frac{\partial V_{tot}}{\partial V^a} \geq 0, \qquad \forall a \in \{1, ..., n\}. \tag{1}$$

With Eq. 1 enforced, given that the other agents stay at the same local states by taking $\mathbf{u}^{-a}$, any action u^a that leads agent a to a local state o^a with a higher value will also improve the global state value V_{tot}. In [11], they also prove the convergence of VDAC frameworks to a locally optimal policy. Two variants of value-decomposition that satisfy Eq. 1, VDAC-sum and VDAC-mix, are proposed in [11]. In VDAC-sum, $V_{tot}(s)$ is represented by a summation of local state values V^a, $V_{tot}(s) = \sum_a V^a(o^a)$. This linear representation satisfies the constraint. θ denotes the actors' parameters and θ_v denotes the distributed critics' parameters. The distributed critic is optimized by minibatch gradient descent to minimize the following loss $L_t(\theta_v) = (y_t - \sum_a V_{\theta_v}(o_t^a))^2$, where $y_t = \sum_{i=t}^{k-t-1} \gamma^i r_i + \gamma^{k-t} V_{tot}(s_k)$ is bootstrapped from the last state s_k, and k is upper-bounded by T. To generalize the representation to a larger family of monotonic functions, VDAC-mix uses a feed-forward neural network that takes input as local state values $V^a(o^a), \forall a \in \{1, ..., n\}$, and outputs the global state value V_{tot}. To enforce Eq. 1, the weights (not including bias) of the mixing network, which are produced by separate hypernetworks [3] that take s as an input, are restricted to be non-negative. The distributed critics are optimized by minibatch gradient descent to minimize the following loss $L_t(\theta_v) = (y_t - V_{tot}(s_t))^2 = (y_t - f_{mix}(V_{\theta_v}(o_t^1), ..., V_{\theta_v}(o_t^n)))^2$, where f_{mix} denotes the mixing network. The central critic is optimized by minimizing the same loss $L_t(\theta^c) = (y_t - V_{tot}(s_t))$, where θ^c denotes parameters in the hypernetworks. The policy network is trained using the following policy gradient $g = \mathbb{E}_\pi[\sum_a \nabla_\theta \log \pi^a(u^a|\tau^a)A(s, \mathbf{u})]$, where $A(s, \mathbf{u}) = r_t + \gamma V(s') - V(s)$ is a simple TD advantage.

3   Method

In this section, we propose a new V-value decomposition approach called V-value Attention Actor-Critic (VAAC). First, we perform the theoretical analysis of global and local V-values and derive a general decomposition formula. Then, we describe the architecture of VAAC that uses a multi-head attention formation to implement the decomposition formula.

3.1   Theoretical Analysis

Considering a stochastic policy π, the relationship between $V^\pi(s)$ and $Q^\pi(s, u)$ is $V^\pi(s) = \sum_u \pi(a|s)Q^\pi(s, u)$. Thus, V_{tot} and V^a can be formulated as:

$$V_{tot}(s) = \sum_{\mathbf{u}} \pi(\mathbf{u}|s)Q_{tot}(s, \mathbf{u}), V^a(o^a) = \sum_{u^a} \pi^a(u^a|\tau^a)Q^a(o^a, u^a), \quad (2)$$

where π denotes the joint policy and π^a denotes the individual policy of agent a. In [18], they theoretically derive a general decomposition formula of Q_{tot} by local state-action values Q^a:

$$Q_{tot}(s, \mathbf{u}) \approx c(s) + \sum_{a,h} \lambda_{a,h}(s)Q^a(o^a, u^a), \quad (3)$$

where $\mathbf{u} = (u^1, ..., u^n)$ and $\lambda_{a,h}$ is a linear functional of all partial derivatives $\frac{\partial^h Q_{tot}}{\partial Q^{a_1}...\partial Q^{a_h}}$ of order h, decaying super exponentially fast in h. Equation 3 appears to be a linear relationship between Q_{tot} and Q^a, yet contains the non-linear information due to the coefficient $\lambda_{a,h}$ that is a function of all partial derivatives of order h, and corresponds to all cross-terms $Q^{a_1}...Q^{a_h}$ of order h. Therefore, we could decompose V_{tot} to V^a as shown in the following Theorem 1.

Theorem 1. *Assuming that the joint policy π can be formulated as a product of independent actors: $\pi(\mathbf{u}|s) = \Pi_a \pi^a(u^a|\tau^a)$, then*

$$V_{tot}(s) \approx c(s) + \sum_{a,h} \lambda_{a,h}(s)V^a(o^a) \quad (4)$$

where c and $\lambda_{a,h}$ depend on the global state s, $\lambda_{a,h}$ is a linear functional of all partial derivatives $\frac{\partial^h Q_{tot}}{\partial Q^{a_1}...\partial Q^{a_h}}$ of order h, decaying super exponentially fast in h.

Proof. Since $\pi(\mathbf{u}|s) = \Pi_a \pi^a(u^a|\tau^a)$, we have $\pi(\mathbf{u}|s) = \pi(\mathbf{u}^{-a}|\tau^{-a})\pi^a(u^a|\tau^a)$, where $\pi(\mathbf{u}^{-a}|\tau^{-a})$ denotes a product of independent actors other than a given agent a, according to Eqs. 2 and 3 we have:

$$
\begin{aligned}
V_{tot}(s) &\approx \sum_{\mathbf{u}} \pi(\mathbf{u}|s)\Big(c(s) + \sum_{a,h} \lambda_{a,h}(s)Q^a(o^a, u^a)\Big)\\
&= c(s)\sum_{\mathbf{u}} \pi(\mathbf{u}|s) + \sum_{\mathbf{u}} \pi(\mathbf{u}|s)\sum_{a,h} \lambda_{a,h}(s)Q^a(o^a, u^a)\\
&= c(s) + \sum_{a,h} \lambda_{a,h}(s)\Big(\sum_{\mathbf{u}^{-a}} \pi(\mathbf{u}^{-a}|\tau^{-a})\sum_{u^a} \pi^a(u^a|\tau^a)Q^a(o^a, u^a)\Big)\\
&= c(s) + \sum_{a,h} \lambda_{a,h}(s)\Big(\sum_{\mathbf{u}^{-a}} \pi^a(\mathbf{u}^{-a}|\tau^{-a})V^a(o^a)\Big)\\
&= c(s) + \sum_{a,h} \lambda_{a,h}(s)\Big(V^a(o^a)\sum_{\mathbf{u}^{-a}} \pi^a(\mathbf{u}^{-a}|\tau^{-a})\Big)\\
&= c(s) + \sum_{a,h} \lambda_{a,h}(s)V^a(o^a)
\end{aligned}
\tag{5}
$$

3.2   Implementation

Following the above decomposition formula in Eq. 4, we propose VAAC based on the attention mechanism [14]. Figure 1 illustrates the overall architecture of VAAC. For each agent a, there is one agent network, which receives its action-observation history τ^a (last hidden states h_{t-1}^a and current local observation o_t^i) and outputs both $\pi^a(o^a)$ and $V^a(o^a)$ by sharing non-output layers between distributed critics and actors. The key to the mixing process is how to approximate different weights $\lambda_{a,h}$ corresponding to agent a and order h in Eq. 4. Thus, following the practice in Qatten [18], we feed local state values V^a and additional global state information (including the global state s and the agent's individual features μ^a) into the mixing network using a multi-head attention formation to model the individual impacts.

In Eq. 4, let the outer sum over h:

$$
V_{tot}(s) \approx c(s) + \sum_{h=1}^{H}\sum_{a=1}^{N} \lambda_{a,h}(s)V^a(o^a).
\tag{6}
$$

First, for each h, the inner weighted sum operation can be implemented by the differentiable key-value memory model [19], we compute the weights $\lambda_{a,h}$ as:

$$
\lambda_{a,h} \propto softmax\Big(\frac{e_a^T e_s}{\sqrt{d}}\Big),
\tag{7}
$$

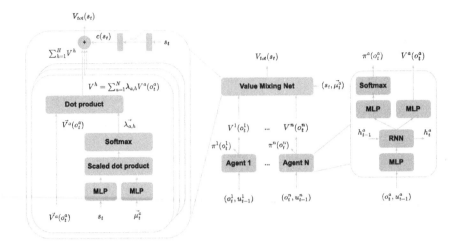

Fig. 1. The overall architecture of VAAC.

where e_s and e_a are obtained by a two-layer embedding transformation for s and μ^a, d is the embedding dim. Note that individual features μ^a are the part of the global state s related to agent a, hence $\lambda_{a,h}$ still only depends on s when we compute in Eq. 7. Then we compute the weighted sum of the local values V^a as $V^h = \sum \lambda_{a,h} V^a(o^a)$, where V^h denotes the output of a single attention. Next, for the outer sum over h, we adopt multiple attention heads to correspond to different orders of partial derivatives. Since $\lambda_{a,h}$ decays super exponentially fast in h, we stop at H for the feasibility of implementation, where H denotes the number of attention heads. Adding up the outputs of different heads and $c(s)$ which is produced by a two-layer network with the global state s as the input, we have

$$V_{tot}(s) \approx c(s) + \sum_{h=1}^{H} V^h. \tag{8}$$

Naturally, VAAC satisfies the constraint of VDAC in Eq. 1. The distributed critics and the value mixing network are optimized by minibatch gradient descent to minimize the following loss $L_t(\theta) = (y_t - V_{tot}(s_t))^2$. The policy network is trained using the following policy gradient $g = \mathbb{E}_\pi [\sum_a \nabla_\theta \log \pi^a(u^a|\tau^a) A(s, \mathbf{u})]$, where $A(s, \mathbf{u}) = r_t + \gamma V_{tot}(s_{t+1}) - V_{tot}(s_t)$ is a simple TD advantage.

4   Experiments

We evaluate VAAC against previous state-of-the-art multi-agent on-policy actor-critic methods such as VDAC-sum [12], VDAC-mix [12], COMA [2] and IAC [2], and multi-agent Q-learning method QMIX under A2C training paradigm (QMIX-A2C) [12] on the StarCraft Multi-Agent Challenge (SMAC) environment [9], in which each agent controls an individual allied army unit to beat the enemy. SMAC consists of various maps which have been classified as *easy*, *hard*, and *super hard*. We consider the following maps in our experiments: four *easy* maps (2s_vs_1sc, 2s3z, 3s5z, and 1c3s5z), three *hard* maps (2c_vs_64zg, bane_vs_bane and 3s_vs_5z) and one *super hard* map (MMM2). Note that all algorithms are trained under the A2C training paradigm where 8 episodes are rolled out independently during the training. Our method uses RMSprop with learning rate 0.0025, γ is set to 0.99 and λ is set to 0.8. For baseline algorithms, we use the same training setup as provided by their authors. The agent networks resemble a DRQN with a recurrent layer comprised of a GRU with a 64-dimensional hidden state, with a fully-connected layer before and after. The agent networks contain an additional layer to output local state values and the policy network outputs a stochastic policy. $c(s)$ is produced by a two-layer network with a 32-dimensional hidden state. We use ReLU for all activation functions. For the attention part, query (global state s) and key (agent's individual features μ^a) are obtained by two-layer embedding transformations, where hidden state dim is 64 and embedding dim is 32, and the number of heads H is 4.

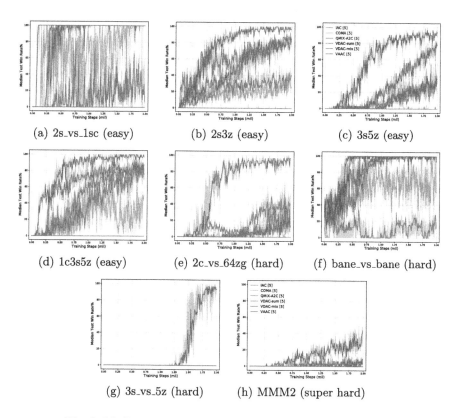

Fig. 2. Median win percentage on eight different SMAC maps

4.1   Main Results

We compare VAAC with other baselines on all maps mentioned above. The main evaluation metric is the median win percentage of evaluation episodes as a function of environment steps observed over the 2 million training steps [9]. The training is paused after every 10000 timesteps during which 32 test episodes are run with agents performing action selection greedily in a decentralized fashion. The median performance as well as the 25-75% percentiles are obtained by 5 independent training runs with different seeds. Figure 2 presents results.

In all scenarios, VAAC outperforms other baselines under A2C. On the *easy* and *hard* maps, VAAC can master these tasks and achieve competitive performance. Even on the *super hard* map, VAAC's win percentage can reach approximately 40%. VDAC-mix has good performance on *easy* maps but performs not well on *hard* and *super hard* maps due to the difficulty and complexity.

4.2   Ablations

We perform ablation experiments to investigate the influence and necessity of the multi-head attention formation. In our implementation, we use multiple attention heads to approximate different orders of partial derivatives in Eq. 4. Because $\lambda_{a,h}$ decays super exponentially fast in h, we stop at $H = 4$ which denotes the number of attention heads. We compare against VAAC without the multi-head attention formation by setting $H = 1$. We refer to this method as VAAC-H1. We test on the 3s5z (*easy*), 3s_vs_5z (*hard*), and MMM2 (*super hard*) maps. Figure 3 shows that VAAC outperforms VAAC-H1. It reveals that the multi-head attention formation could capture sophisticated relations between V_{tot} and V^a to improve performance.

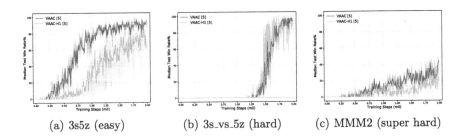

(a) 3s5z (easy)   (b) 3s_vs_5z (hard)   (c) MMM2 (super hard)

Fig. 3. Ablations of VAAC on three SMAC maps

4.3   Weights Analysis

We analyze the attention weights $\lambda_{a,h}$ to show how our method models the individual impact of agents, which provides interpretability for our method's decomposition process. In our implementation, the attention weights $\lambda_{a,h}$ are produced by the global state s, agent's individual features μ^a, and corresponding to different orders using multiple heads. We choose the 1c3s5z map as the representative for experiments since it contains three different types of units (including 1 Colossi, 3 Stalkers, and 5 Zealots).

First, we calculate the mean of $\lambda_{a,h}$ for different agents and heads. For clarity, we consider the difference between the mean of $\lambda_{a,h}$ and the average weight (Specifically, on the 1c3s5z map with nine agents, the average weight is $\frac{1}{9}$). Figure 4 presents the results. For each agent, the weights $\lambda_{a,h}$ of different heads are not the same, but in general, the higher value unit types of agents have higher weights. As shown in Fig. 4, Colossi's weights are much higher than others, the weights of Stalkers are roughly around the average weight and Zealots have the lowest weights. It demonstrates that our method could assign higher weights to more powerful and impactful agents.

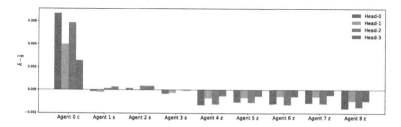

Fig. 4. The difference between the mean of $\lambda_{a,h}$ and the average weight (1/9 on this map) for different agents and heads on 1c3s5z map

Next, since the attention weights $\lambda_{a,h}$ are adaptive to the global state, we calculate the matrix of correlation coefficients for $\lambda_{a,h}$ during the training from the perspective of agents and attention heads respectively. Figure 5(a) shows the matrix of correlation coefficients from the perspective of Head-0. We find a strong positive correlation between the weights of agents of the same type, such as Stalkers (Agent 1, 2, and 3) or Zealots (Agent 4-8). Agents of the same type tend to perform the same function as a group. Interestingly, we notice that the weights of Stalkers and Zealots are strongly negatively correlated. For the 1c3s5z map, Stalkers and Zealots should maintain the formation to protect the Colossi that can deal a lot of damage to the enemy. Due to cooldown or other features, Stalkers and Zealots take turns taking a more important role in the team, reflected in the ebb and flow of their weights according to the team's needs. Figure 5(b) shows the matrix of correlation coefficients from the perspective of Agent 0. Generally, for a given agent, there is a certain degree of positive correlation between the weights $\lambda_{a,h}$ of different heads. According to Fig. 4 and 5(b), different attention heads may capture different features in sub-spaces to approximate the weights of different orders.

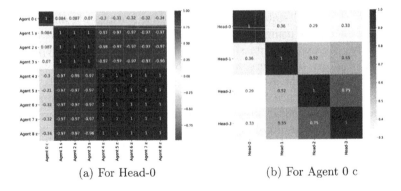

(a) For Head-0   (b) For Agent 0 c

Fig. 5. The matrix of correlation coefficients for $\lambda_{a,h}$ on 1c3s5z map

5 Conclusion

In this paper, we propose V-value Attention Actor-Critic (VAAC) for cooperative MARL. First, we derive a decomposition formulation of the global state value V_{tot} in terms of local state values V^a. To implement the decomposition formula, we adopt the multi-head attention formation in the mixing network, which can explicitly model the impact of individuals on the whole system for interpretability of decomposition. Experiments on the StarCraft II micromanagement task demonstrate that our method VAAC not only reaches better performance, but also provides interpretability for its decomposition process. In future work, we aim to further research on V-value decomposition methods.

Acknowledgement. This work was supported in part to Dr. Liansheng Zhuang by NSFC under contract No.U20B2070 and No.61976199.

References

1. Cao, Y., Yu, W., Ren, W., Chen, G.: An overview of recent progress in the study of distributed multi-agent coordination. IEEE Trans. Industr. Inf. **9**(1), 427–438 (2012)
2. Foerster, J., Farquhar, G., Afouras, T., Nardelli, N., Whiteson, S.: Counterfactual multi-agent policy gradients. In: Proceedings of the AAAI Conference on Artificial Intelligence (2018)
3. Ha, D., Dai, A., Le, Q.V.: Hypernetworks. arXiv preprint arXiv:1609.09106 (2016)
4. Kraemer, L., Banerjee, B.: Multi-agent reinforcement learning as a rehearsal for decentralized planning. Neurocomputing **190**, 82–94 (2016)
5. Lowe, R., Wu, Y.I., Tamar, A., Harb, J., Pieter Abbeel, O., Mordatch, I.: Multi-agent actor-critic for mixed cooperative-competitive environments. In: Advances in Neural Information Processing Systems 30 (2017)
6. Mnih, V., et al.: Asynchronous methods for deep reinforcement learning. In: International Conference on Machine Learning, pp. 1928–1937 (2016)
7. Qiu, W., et al.: RMIX: learning risk-sensitive policies for cooperative reinforcement learning agents. In: Advances in Neural Information Processing Systems 34 (2021)

8. Rashid, T., Samvelyan, M., Schroeder, C., Farquhar, G., Foerster, J., Whiteson, S.: QMIX: monotonic value function factorisation for deep multi-agent reinforcement learning. In: International Conference on Machine Learning, pp. 4295–4304 (2018)

9. Samvelyan, M., et al.: The starcraft multi-agent challenge. In: Proceedings of the 18th International Conference on Autonomous Agents and MultiAgent Systems, pp. 2186–2188 (2019)

10. Son, K., Kim, D., Kang, W.J., Hostallero, D.E., Yi, Y.: QTRAN: learning to factorize with transformation for cooperative multi-agent reinforcement learning. In: International Conference on Machine Learning, pp. 5887–5896 (2019)

11. Su, J., Adams, S., Beling, P.A.: Value-decomposition multi-agent actor-critics. In: Proceedings of the AAAI Conference on Artificial Intelligence, pp. 11352–11360 (2021)

12. Sunehag, P., et al.: Value-decomposition networks for cooperative multi-agent learning based on team reward. In: AAMAS (2018)

13. Tan, M.: Multi-agent reinforcement learning: independent vs. cooperative agents. In: Proceedings of the Tenth International Conference on Machine Learning, pp. 330–337 (1993)

14. Vaswani, A., et al.: Attention is all you need. In: Advances in Neural Information Processing Systems 30 (2017)

15. Wang, J., Ren, Z., Liu, T., Yu, Y., Zhang, C.: QPLEX: duplex dueling multi-agent Q-learning. In: International Conference on Learning Representations (2020)

16. Wang, T., Gupta, T., Peng, B., Mahajan, A., Whiteson, S., Zhang, C.: Rode: learning roles to decompose multi-agent tasks. In: Proceedings of the International Conference on Learning Representations (2021)

17. Wolpert, D.H., Tumer, K.: Optimal payoff functions for members of collectives. In: Modeling Complexity in Economic and Social Systems, pp. 355–369. World Scientific (2002)

18. Yang, Y., et al.: Qatten: a general framework for cooperative multiagent reinforcement learning. arXiv preprint arXiv:2002.03939 (2020)

19. Yun, C., Bhojanapalli, S., Rawat, A.S., Reddi, S., Kumar, S.: Are transformers universal approximators of sequence-to-sequence functions? In: International Conference on Learning Representations (2019)

20. Zhang, C., Lesser, V.: Coordinated multi-agent reinforcement learning in networked distributed POMDPs. In: Twenty-Fifth AAAI Conference on Artificial Intelligence (2011)

An Analytical Estimation of Spiking Neural Networks Energy Efficiency

Edgar Lemaire[1]($\boxtimes$), Loïc Cordone[1,2], Andrea Castagnetti[1],
Pierre-Emmanuel Novac[1], Jonathan Courtois[1], and Benoît Miramond[1]($\boxtimes$)

[1] Université Côte d'Azur, CNRS, LEAT, Nice, France
{edgar.lemaire,loic.cordone,andrea.castagnetti,pierre-emmanuel.novac,
jonathan.courtois,benoit.miramond}@univ-cotedazur.fr
[2] Renault Software Factory, Toulouse, France
loic.cordone@renault.com

Abstract. Spiking Neural Networks are a type of neural networks where neurons communicate using only spikes. They are often presented as a low-power alternative to classical neural networks, but few works have proven these claims to be true. In this work, we present a metric to estimate the energy consumption of SNNs independently of a specific hardware. We then apply this metric on SNNs processing three different data types (static, dynamic and event-based) representative of real-world applications. As a result, all of our SNNs are 6 to 8 times more efficient than their FNN counterparts.

Keywords: Spiking neural networks · Energy metrics · Computational metrics · Event-based processing · Low-power artificial intelligence

1 Introduction

Neuromorphic computing has been studied for many years as a game changer to address low-power embedded AI, assuming that the inspiration from the brain will natively come with a reduction in energy consumption. Neuromorphic computing mainly focuses on the encoding and the processing of the information with spikes. If this property takes an obvious place in the biological functioning, it is far from obvious that it is the only one to explain the efficiency of the brain. It is therefore necessary to ask the question whether considering this characteristic in isolation brings a gain compared to the classical neural networks used in deep learning. This is the question that this paper seeks to answer by restricting the study to standard machine learning tasks on three different types of data: static, dynamic and event-based data.

There already exist comparisons between Spiking Neural Networks (SNNs) and Formal Neural Networks (FNNs, *i.e.* non-spiking Artificial Neural Networks) in the literature. However, such comparisons are hardly generalizable since they

This research is funded by the ANR project DeepSee, Université Côte d'Azur, CNRS and Région Sud Provence-Alpes-Côte d'Azur.

focus on specific applications or hardware targets [11], [18]. Moreover, the considered applications are often toy examples not representative of real-world AI tasks. Another approach consists in producing metrics in order to evaluate the relative energy consumption between the two coding domains, based on their respective synaptic operations and activity. We thus propose a novel metric for energy consumption estimation taking synaptic operations, memory accesses and element addressing into account. Moreover, our metric is mostly independent from low-level implementation or hardware target to ensure its generality.

The proposed metric is described and applied to three datasets representative of the aforementioned data types: CIFAR-10 for the static case, Google Speech Commands V2 for the dynamic case, and Prophesee NCARS for the event-based case. Moreover, those datasets are closer to real-world applications than usual benchmarks of the neuromorphic community. The metric is used in conjunction with accuracy measurements to provide an in-depth evaluation of those three application cases and their relevance for spiking acceleration. We use the advanced Surrogate Gradient Learning technique and Direct Encoding spike conversion method, since they offer the best trade-off between prediction accuracy and synaptic activity [14].

Our code and trained SNN models are available upon request.

2   State of the Art

Data Encoding. In order to process data in SNNs, it must be encoded into spikes. Rate, Time and Direct encoding are three methods used to convert conventional data towards spiking domain. Rate coding [15] is the most notorious, since it provides state-of-the-art accuracy on most AI tasks. However, it generates a lot of spikes over a large number of timesteps, drastically impacting computational and energy efficiency. Time coding [12] intends to cope with this issue by encoding information in latency rather than rate, thus generating much fewer spikes. Yet, the temporal sparsity of latency-coded spikes causes long processing times, resulting in an energy overhead.

To cope with those limitations, we address a novel encoding scheme: Direct Encoding [21]. It should be noted that this term is a proposition of ours. In Direct Encoding, the first processing layer is made of hybrid neurons with analog inputs and spiking behaviour (IF, LIF...). The weights of this layer are learned during training, thus encoding can be tuned to reduce spiking activity in the network [17]. In the present work, we evaluate Direct Encoding on CIFAR-10 and GSC datasets. Additionally, we evaluate native spike encoding using event cameras [2]. In this method, each pixel of the sensor generates a spike whenever it detects a brightness variation, thus encoding movement into spikes. With such sensor, the input spiking activity is very low, since only the information of interest (*i.e.* moving objects) are returned by the camera. This property helps improving the computational and energy efficiency of SNNs. We evaluate native encoding using the Prophesee NCARS dataset.

Training of SNN in the Literature. Spiking neural networks cannot use the classical backpropagation training algorithm to learn their weights because its activations (spikes) are binary and thus non-differentiable. Encoding static data such as images using rate coding enables the conversion of an already trained FNN to a SNN. The most common way is to replace the ReLU neurons of the FNN by IF neurons [8]. However, the prediction accuracy obtained through conversion is systematically inferior to their FNN counterpart, while generating a lot of spikes over a large number of timesteps. Numerous works have studied how to train SNNs directly in spiking domain. The best results were obtained using backpropagation-based learning rules, such as the surrogate gradient [16]. To circumvent the non-differentiability of spikes in SNNs, the main idea of surrogate gradient learning is to use two distinct functions in the forward and backward passes: an Heaviside step function for the first, and a differentiable approximation of the Heaviside in the latter, such as a sigmoid function. Using surrogate gradient learning requires a fixed number of timesteps. As the number of computations performed by SNNs increases with the number of timesteps, being able to fix it beforehand and to tune it during the training is vital to increase the computational efficiency of SNNs.

Comparisons Based on Measurements. In the literature, a few comparisons of SNNs and FNNs have been produced based on hardware measurements. Some papers show competitive results for SNNs: in [3], the authors highlighted the influence of the spike encoding method on the accuracy and computational efficiency of the SNN. They compared the spiking and formal networks through a Resnet-18 like architecture on two classification datasets. They found that SNNs reached higher or equivalent accuracy and energy efficiency. In [11], the authors showed that an SNN could reach twice the power and resource efficiency of an FNN, with an MLP on MNIST dataset targeting ASIC. However, those encouraging results are still very specific and thus hardly generalizable. In a more holistic approach, the authors of [15] performed a design space exploration (including encoding, training method, level of parallelism...) and showed that the advantage of the SNN depended on the considered case, making it difficult to draw general rules. In [18], researchers showed that SNNs on dedicated hardware (Loihi) demonstrated better energy efficiency than equivalent FNNs on generic hardware (CPU and GPU) for small topologies, but observed the opposite using larger CNNs. Once more, the conclusions depended on the studied case and could not be generalized. Albeit encouraging, those results are not sufficient to draw general conclusions regarding the savings offered by event-based processing, since they depend on the selected application, network hyper-parameters and hardware targets. Therefore, another approach consists in comparing both coding domains through estimation metrics, taking a step back to produce more general conclusions.

Comparisons Based on Metrics. Most energy consumption metrics are based on the number of synaptic operations: accumulations (ACC) in the SNN and multiplication-accumulations (MAC) in the FNN. Those models have limitations: energy consumption is assimilated to the energy consumption of synaptic operations [13], thus other factors (such as neuron addressing in multiplexed architecture or memory accesses) are often neglected. Moreover, the models usually do not take into account some specific mechanisms, like membrane potential leakage, reset and biases integration. In [3], the authors proposed a metric based on synaptic operations only, and found great energy consumption savings for the SNN (up to 126× more efficient than the FNN baseline). In [15] the authors demonstrated that such simplistic metrics were not always coherent with actual energy consumption of circuits on FPGA. When taking memory into account, another team [7] found equivalent energy consumptions for SNNs and FNNs using various topologies on CIFAR10. Additionally, reference [6] measured a theoretical maximum spike rate of 1.72 to guarantee energy savings in the SNN based on a detailed metric, accounting for synaptic operations, memory accesses and activation broadcast. Those energy consumption models are enlightening, but still fail to settle whether event-based processing is sufficient to increase energy efficiency. That is mostly because those metrics are too hardware specific, or do not take all significant sources of energy consumption into account.

In the present work, we propose a metric intended to be independent from low-level implementation choices, based on three main operations: neuron addressing, synaptic operations and memory accesses.

3   Metrics

3.1   Operational Cost

In this section we define a metric to compute the number of ACC and MAC due to synaptic operations in SNNs and FNNs.

Convolutional Layers. For a convolution layer, the number of filters is defined by C_{out} and their size are noted $C_{\text{in}} \times H_{\text{kernel}} \times W_{\text{kernel}}$, where C, H and W stands for channel, height and width. The input and output of the layer are composed of a set of feature maps, with shapes $(C_{\text{in}} \times H_{\text{in}} \times W_{\text{in}})$ and $(C_{\text{out}} \times H_{\text{out}} \times W_{\text{out}})$ respectively. In the following we consider the padding mode "*same*" and a stride S. The number of timesteps is noted T. The equations describing the number of MAC and ACC operations in FNNs and SNNs, for convolution layers, are summarised in Eq. 1.

$$\text{MAC}_{\text{Conv-l}}^{\text{FNN}} = C_{\text{out}} \times H_{\text{out}} \times W_{\text{out}} \times C_{\text{in}} \times H_{\text{kernel}} \times W_{\text{kernel}}$$

$$\text{ACC}_{\text{Conv-l}}^{\text{FNN}} = C_{\text{out}} \times H_{\text{out}} \times W_{\text{out}}$$

$$\text{MAC}_{\text{Conv-l}}^{\text{SNN}} = T \times C_{\text{out}} \times H_{\text{out}} \times W_{\text{out}} \tag{1}$$

$$\text{ACC}_{\text{Conv-l}}^{\text{SNN}} = \theta_{1\text{-}1} \times (\lceil \frac{H_{\text{kernel}}}{S} \rceil) \times (\lceil \frac{W_{\text{kernel}}}{S} \rceil) \times C_{\text{out}}$$

$$+ T \times C_{\text{out}} \times H_{\text{out}} \times W_{\text{out}} + \theta_{1}$$

In FNNs, the integration of dense input matrixes requires a MAC operation for each element of the convolution kernels. That is described in the first row of Eq. 1. Additionally, the integration of synaptic biases as ACC operations, since it does not require multiplication with input activation. There is one bias per output neuron as shown in the second row of Eq. 1.

On the other hand, for an SNN the input activations are sparse binary matrices. The number of operations of the layer l, depends on the number of input and output spikes of that layer, noted θ_{l-1} and θ_l. Since spikes are binary and, they are integrated via ACC operations in contrast with FNNs. Each input spike causes one ACC operation per element of each filter, as shown in the first term of the fourth row ($\text{ACC}_{\text{Conv-}l}^{\text{SNN}}$) of Eq. 1. The second term accounts for the bias added to each membrane potential at each timestep. The third term accounts for the membrane potential reset whenever an output spike is generated. Additionally, SNNs may involve a membrane potential leakage (*i.e.* LIF neurons), which is modeled by an additional MAC operations for each output neuron, which is repeated at each timestep. This is depicted in the last two rows of Eq. 1.

Fully-Connected Layers. The same reasoning is applied to FC layers. For a given layer l the number of input and output neurons is noted N_{in} and N_{out} respectively. The equations for the number of MAC and ACC operations attributable to synaptic operations in FC layers are summarized in Eq. 2.

$$
\begin{aligned}
\text{MAC}_{\text{FC-}l}^{\text{FNN}} &= N_{\text{in}} \times N_{\text{out}} \\
\text{ACC}_{\text{FC-}l}^{\text{FNN}} &= N_{\text{out}} \\
\text{MAC}_{\text{FC-}l}^{\text{SNN}} &= N_{\text{out}} \times T \\
\text{ACC}_{\text{FC-}l}^{\text{SNN}} &= \theta_{l-1} \times N_{\text{out}} + T \times N_{\text{out}} + \theta_l
\end{aligned}
\tag{2}
$$

3.2 Memory Cost

In order to provide an evaluation of the energy used by both the FNN and the SNN, multiple assumptions have to be made. Without these assumptions, results could vastly vary between different unconstrained hardware implementations. Each layer of the FNN is assumed to have its own local (non-shared) memory. As a result, activations need to be kept in memory (*i.e.* I/O buffers) for all layers. The data flow between layers of the SNN is assumed to be sparse and asynchronous. Therefore, messages of incoming spikes must be buffered in a FIFO queue for each layer. Additionally, the SNN must keep the membrane potentials for all layers between timesteps. In both cases, we assume that all the memory is akin to local SRAM, including weights in order to support a reconfigurable architecture. Additionally, there is no local caching in a register bank. Only registers for the operands and a local accumulator are present and are excluded from this evaluation. All data is assumed to be represented with the same number of bits, including the messages describing a spike.

Memory Accesses. Data flowing from and to the memory is an important sink of energy. We attempt to describe each read or write operation of both the FNN and the SNN for each layer in order to evaluate the possible energy savings from using an SNN, which is mainly a result of its sparsity. Equations are provided for a single input for the FNN and a single timestep for the SNN.

Read Operations to Inputs. For a formal Conv layer, each output position matches with read operations from all input channels and all positions for which the kernel applies. For a formal FC layer, the number of read operations for the input data is equal to the number of inputs N_{in}.

$$RdIn_{\text{Conv}}^{\text{FNN}} = C_{\text{in}} \times C_{\text{out}} \times H_{\text{out}} \times W_{\text{out}} \times W_{\text{kernel}} \times H_{\text{kernel}} \tag{3}$$

$$RdIn_{\text{FC}}^{\text{FNN}} = N_{\text{in}} \tag{4}$$

For the SNN, the read operations in the queue directly depends on the number of incoming spikes $\theta_{\text{l-1}}$ and must be measured during inference:

$$RdIn^{\text{SNN}} = \theta_{\text{l-1}} \tag{5}$$

Read Operations to Parameters. In a formal Conv layer, each output position matches with read operations for all the weights in all filters associated to all input channels. The biases generate additional reads for all output positions and all filters. For an FC layer, every weight corresponding to all output neurons N_{out} and all inputs N_{in}. The biases cause additional read for all output neurons.

$$RdParam_{\text{Conv}}^{\text{FNN}} = (C_{\text{in}} \times W_{\text{kernel}} \times H_{\text{kernel}} + 1) \times C_{\text{out}} \times W_{\text{out}} \times H_{\text{out}} \tag{6}$$

$$RdParam_{\text{FC}}^{\text{FNN}} = (N_{\text{in}} + 1) \times N_{\text{out}} \tag{7}$$

In an spiking Conv layer, all received spikes $\theta_{\text{l-1}}$ will trigger a read for all output filters and all associated output positions (i.e. of the dimensions of the kernel). Biases for all output positions and filters must still be read. For an FC layer, the number of read operations for parameters is similar to an SNN except that weights are only read for all input spikes $\theta_{\text{l-1}}$. Biases must still be read for all output neurons N_{out}.

$$RdParam_{\text{Conv}}^{\text{SNN}} = \theta_{\text{l-1}} \times C_{\text{out}} \times W_{\text{kernel}} \times H_{\text{kernel}}$$
$$+ C_{\text{out}} \times W_{\text{out}} \times H_{\text{out}} \tag{8}$$

$$RdParam_{\text{FC}}^{\text{SNN}} = \theta_{\text{l-1}} \times N_{\text{out}} + N_{\text{out}} \tag{9}$$

Read Operations to Potentials. There is no membrane potential to update in an FNN so there is no associated read operation.

In a spiking Conv layer, the membrane potentials corresponding to all output positions affected by each input (i.e. of the dimensions of the kernel) in all filters must be read in order to update them. Biases need to be applied separately and therefore generate an additional read operation at each timestep for all output positions and all filters. For FC layers, the potentials of all output neurons are read for each input. Biases are applied separately and therefore generate an additional read operation at each timestep for all output neurons.

$$RdPot_{\text{Conv}} = \theta_{\text{l-1}} \times C_{\text{out}} \times W_{\text{kernel}} \times H_{\text{kernel}}$$
$$+ C_{\text{out}} \times H_{\text{out}} \times W_{\text{out}} \tag{10}$$

$$RdPot_{\text{FC}} = (\theta_{\text{l-1}} + 1) \times N_{\text{out}} \tag{11}$$

Write Operations to Outputs. In an formal Conv layer, each output position in all filters require a write operation. For an FC layer, each output neuron require a write operation. In both cases, the output is assumed to be fully computed in the local accumulator, including bias, before being written to RAM.

$$WrOut_{Conv}^{FNN} = C_{out} \times H_{out} \times W_{out} \qquad (12)$$

$$WrOut_{FC}^{FNN} = N_{out} \qquad (13)$$

For the SNN, the write operations to the queue directly depends on the number of generated spikes N_{output} and must be measured during inference:

$$WrOut = N_{output} \qquad (14)$$

Write Operations to Potentials. There is no membrane potential to update in an FNN so there is no associated write operation.

In an spiking Conv layer, the membrane potentials corresponding to all output positions affected by each input (i.e. of the dimensions of the kernel) in all filters must be written to in order to update them. Additionally, the biases must also be written separately to the potentials for all output positions and all filters at each new timestep. For an FC layer, the potentials of all output neurons must be written to for each input. Additionally, the biases must also be written to the potentials for all output neurons separately at each new timestep.

$$WrPot_{Conv} = \theta_{l-1} \times C_{out} \times W_{kernel} \times H_{kernel}$$
$$+ C_{out} \times H_{out} \times W_{out} \qquad (15)$$

$$WrPot_{FC} = \theta_{l-1} \times N_{out} + N_{out} \qquad (16)$$

3.3 Addressing in Sparse Vs. Dense Convolutions

In this subsection, we evaluate the cost of addressing in FNNs and SNNs. The first uses dense processing, in which all input synapses are stimulated at the same time. On the other hand, the second uses sparse processing, in which synapses are sparsely stimulated across time. Let us begin with convolution layers. In order to simplify the following matter, we consider convolutions with a *"same"* padding (input and output feature maps of same sizes) and a stride of S. In FNNs, a kernel scans all its possible positions (depending on padding, stride...) on the input sample and generates a dense output feature-map. In such dense convolutions, computation is performed sequentially and addresses can be computed by incrementing only an index (by 1 or S) assuming the memory is contiguous and ordered the same way it is processed. Thus, one index runs through the input, one index runs through the output and one index runs through the weights. In SNNs, sparse convolutions are performed asynchronously upon reception of input spikes, thus the kernel positions (*i.e.* output neuron addresses) must be calculated each time a spike is received. In a sparse representation, computation is performed non-sequentially with no prior knowledge of which output position is affected by an incoming spike. Computing the initial output position requires two multiplications. Thereafter, the computation of the remaining positions are computed by incrementing an index assuming the memory is contiguous and ordered as for FNNs. There is only one index running through the kernel weights. The cost of addressing in number ACC and MAC operations in spiking and formal convolution layer are summarized in Eq. 17.

$$
\begin{aligned}
\mathrm{ACC}^{\mathrm{FNN}}_{\mathrm{Addr\text{-}Conv\text{-}l}} &= C_{\mathrm{in}} \times H_{\mathrm{in}} \times W_{\mathrm{in}} + C_{\mathrm{out}} \times H_{\mathrm{out}} \times W_{\mathrm{out}} + C_{\mathrm{out}} \\
&\quad \times H_{\mathrm{kernel}} \times W_{\mathrm{kernel}} \\
\mathrm{MAC}^{\mathrm{SNN}}_{\mathrm{Addr\text{-}Conv\text{-}l}} &= \theta_{\mathrm{l\text{-}1}} \times 2 \\
\mathrm{ACC}^{\mathrm{SNN}}_{\mathrm{Addr\text{-}Conv\text{-}l}} &= \theta_{\mathrm{l\text{-}1}} \times C_{\mathrm{out}} \times H_{\mathrm{kernel}} \times W_{\mathrm{kernel}}
\end{aligned}
\tag{17}
$$

where C are the numbers of channels, W the widths and H the heights, of respectively the inputs when the index is in, outputs when it is out and kernels when it is kernel. $\theta_{\mathrm{l\text{-}1}}$ is the number of input spikes.

The same reasoning is applied to fully-connected layers. In FNNs, one index runs through the input, and another index runs through both the output. In SNNs, one single index runs through the output upon receiving an input spikes. This yields Eq. 18.

$$
\begin{aligned}
\mathrm{ACC}^{\mathrm{FNN}}_{\mathrm{Addr\text{-}FC\text{-}l}} &= N_{\mathrm{in}} + N_{\mathrm{out}} \\
\mathrm{ACC}^{\mathrm{SNN}}_{\mathrm{Addr\text{-}FC\text{-}l}} &= \theta_{\mathrm{l\text{-}1}} \times N_{\mathrm{out}}
\end{aligned}
\tag{18}
$$

where N_{in} and N_{out} are respectively the number of input and output neurons.

3.4   Energy Consumption Metric

In this section, we combine the equations obtained for computation, memory accesses and addressing in a global Energy evaluation metric. For this purpose, we multiply the energy cost of each operation by its number of occurrences, according to the metrics computed in Subsects. 3.1, 3.2 and 3.3. Our model can be summarized as shown in Eq. 19:

$$
E = E_{\mathrm{mem}} + E_{\mathrm{ops+addr}}
\tag{19}
$$

where E^{mem} is the energy consumption of memory accesses, E^{ops} that of synaptic operations, and E^{addr} that of addressing mechanisms.

Those elements are computed based on the metrics proposed in the above subsections, as shown in Eq. 20 for memory accesses, and Eq. 21 for addressing and synaptic operations. Equations are not repeated for those two last elements since they are identical.

$$E_{\text{mem}}^{\text{FNN}} = (RdIn^{\text{FNN}} + RdParam^{\text{FNN}}) \times E_{\text{RdRAM}}$$
$$+ WrOut^{\text{FNN}} \times E_{\text{WrRAM}}$$
$$E_{\text{mem}}^{\text{SNN}} = (RdIn^{\text{SNN}} + RdParam^{\text{SNN}} + RdPot) \times E_{\text{RdRAM}}$$
$$+ (WrOut^{\text{SNN}} + WrPot) \times E_{\text{WrRAM}} \tag{20}$$

With E_{RdRAM} and E_{WrRAM} the energy for a single read and a single write operation in RAM, respectively. In our computation, we assume that $E_{\text{RdRAM}} = E_{\text{WrRAM}}$ for simplicity purpose.

$$E_{\text{ops+addr}}^{\text{FNN}} = (E_{\text{ADD}} + E_{\text{MUL}}) \times \text{MAC}_{\text{ops+addr}}^{\text{FNN}} + E_{\text{ADD}} \times \text{ACC}_{\text{ops+addr}}^{\text{FNN}}$$
$$E_{\text{ops+addr}}^{\text{SNN}} = (E_{\text{ADD}} + E_{\text{MUL}}) \times \text{MAC}_{\text{ops+addr}}^{\text{SNN}} + E_{\text{ADD}} \times \text{ACC}_{\text{ops+addr}}^{\text{SNN}} \tag{21}$$

where E_{ADD} and E_{MUL} are the energy cost of single additions and multiplications respectively.

The energy consumption of single operations (addition, multiplication and memory accesses) are drawn from the literature [10] for 45 nm CMOS technology. For addition and multiplication with 32-bit integers, we use respectively 0.1 pJ and 3.1 pJ. For SRAM memory accesses, we compute a linear interpolation function based on 3 particular values: 8 kB (10 pJ), 32 kB (20 pJ) and 1 MB (100 pJ). This function enables to compute the energy cost of a memory access knowing the memory size (i.e. knowing the network hyper-parameters).

4   Methods

4.1   Spike Coding

The role of spike coding is to convert input pixels into spikes, which are in turn transmitted to an SNN for information processing. In the *Direct encoding* scheme, a spiking neuron (e.g. LIF or IF) located in the first layer (encoding) of the network is fed with a constant or dynamic input through T timesteps. The first layer thus converts the analog input into spike trains. In this paper, we propose three different uses of such encoding which are detailed below.

Static Frame-Based Data Encoding. The first method is Static Frame-based data encoding, in which the raw input data is directly broadcast to the first layer. The network input size is thus identical to that of the input sample, each of which is presented repeatedly during T timesteps. This method is adapted to every type of conventional data. The major drawback of this method is that the whole processing must be repeated over several timesteps (bias integration, membrane leakage...).

Dynamic Frame-Based Data Encoding. The second method is Dynamic Frame-based data encoding, adapted to temporal signals. The input data is split into several chunks along the temporal dimension. The network input size is the same as the size of a chunk, which are presented successively at the input, one per timestep. This approach has two main benefits: reducing the input size and no longer considering timesteps mechanisms as an additional cost.

Event-Based Data. Event-based data does not require a specific encoding since events can be interpreted as spikes. However, in order to process event-based data in modern deep learning models, we need to convert them into a dense representation. To process events with FNNs, we simply sum all events occurring during the sample duration d to reconstruct a single frame, containing integer values. On the other hand, when using SNNs, we accumulate events over T time windows (*i.e.* timesteps) lasting $\Delta t = \frac{d}{T}$ to reconstruct frames. This type of representation is called a *voxel grid* [4], where each voxel represents a pixel and a time interval. We added the constraint that the accumulation is a simple OR operation: if at least one event is present during the time window then its value will be 1 [5]. This way, all of our event frames stays binary in order to leverage the efficiency of spiking neural networks running on specialized hardware [1].

4.2 Organization of the Output Layer

Both our FNN and SNN for static frame-based data use a traditional final fully-connected layer. On the other hand, our models for dynamic frame-based and event-based data use a specific final classification layer as the feature maps are not sufficiently reduced to be flattened before the final fully-connected layer.

We followed the approach used in [5]. The output layer of our SNNs is simply composed of a batch normalization layer, a 1×1 convolution outputting *num_classes* channels and a final layer of LIF neurons. The final predictions are then obtained by summing all output spikes first in the spatial dimension and time dimension. We therefore obtain a tensor with a spatial size of 1×1 with *num_classes* channels, which is equivalent to the output of conventional fully-connected layers. The 1D convolution in the final layer enables to avoid the use of e.g. average pooling to reduce the spatial dimension as it would be incompatible with spikes computations. We used the same approach for the equivalent FNNs but without summation along the time axis since it does not exist.

5 Experiments and Results

5.1 Datasets and Models

Static Frame-Based Data. The CIFAR-10 dataset is made of 60000 32×32 RGB images representing 10 classes. For the SNN, CIFAR-10 samples are repeated as input over $T = 4$ timesteps, following the static frame-based data encoding described in Sect. 4.1. For this task, we use a VGG-16 architecture described in [19]. We dropped the max pooling layers by using a stride of 2 in their preceding convolution and we added batch normalization layers after each convolution.

Dynamic Frame-Based Data. The Google Speech Commands V2 dataset is a dataset of audio signals sampled at 16 kHz composed of 1-s recordings of 35 spoken keywords. We performed data augmentation by randomly changing the speed of the raw audio signals. The raw data are preprocessed to obtain images that can be fed to CNNs. We used 10 MFC Coefficients, FFT of size 1024, a window size of 640 with a hop of 320, and a padding of 320 on both sides. This results in a 48×10 image interpreted as 1D data truncated to 48 samples with 10 channels. For the SNN, we divided this temporal data in $T = 2$ timesteps, each of size 24. To tackle this classification task, we designed a 4-layers CNN with the following topology: 48c3 - 48c3 - 96c3 - 35c1. Each convolutional layer has a stride of 1 and was followed by a batch normalization layer.

Event-Based Data. The Prophesee NCARS dataset [20] is a classification dataset composed of 24k samples of length 100 ms captured with a Prophesee GEN1 event camera mounted behind the windshield of a moving car. The samples represent either a car or background. We resized all the samples to a size of 64×64 pixels using nearest neighbor interpolation to keep our inputs binary. For SNNs, we divided each sample in $T = 5$ timesteps, while all the events were summed into a single frame for the CNN. We proposed a variant of the classical Tiny VGG-11 architecture [19] that uses 4 times fewer channels in each convolution layers, reducing the number of parameters and calculations. Once again, we dropped the max pooling layers by using a stride of 2 in their preceding convolution and we added batch normalization layers after each convolution. Finally, we replaced the final 3 fully-connected layers by our output layer described in Sect. 4.2.

5.2   Results

We trained our FNNs using PyTorch, and our SNNs using SpikingJelly [9] with surrogate gradient learning. The models were trained over 50 epochs for GSC and NCARS, and 300 for CIFAR-10. All presented results represent the average over 5 runs. The performance of our networks were measured by their classification accuracy. We also measured the spike rate of our SNNs, corresponding to the average number of spikes per synapse in the network. Since computations are only performed when there is a spike, this has a direct impact on the SNN energy consumption within our metric. These results are summarized in Table 1.

For dynamic and event data, SNNs are able to reach equivalent or close accuracies to their FNN counterparts, a result never shown experimentally before on these datasets. On the CIFAR-10 dataset, our SNN reaches lower accuracy than the FNN. This is coherent with state of the art results, but should be improved in further works. Still, all of our SNNs reach these performance while having a very low spike rate, on average each neuron of a model spikes less than 0.14 times per inference for the three datasets.

Using the metrics proposed in Sect. 3.2, we were able to precisely estimate the energy consumption of our models. In our model, I/O buffers between SNN layers are FIFOs able to store 1000 32-bit elements. This assumption has been

Table 1. Accuracy and Activity comparisons between our proposed SNNs and CNNs on CIFAR-10, Google Speech Commands V2, and Prophesee NCARS.

Dataset	Network	#Params	Activation	Acc.	Spike Rate
CIFAR-10	VGG-16	15.2M	ReLU	0.951	–
			IF	0.884	0.10
GSC	4-layers CNN	29k	ReLU	0.936	–
			LIF	0.918	0.14
NCARS	Tiny VGG-11	356k	ReLU	0.934	–
			LIF	0.935	0.08

validated through hardware simulation using SPLEAT architecture [1]. On the other hand, the full feature maps are stored in SRAMs between FNN layers. The results are illustrated in Fig. 1 and detailed in Table 2.

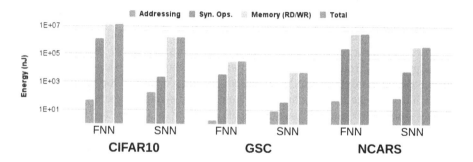

Fig. 1. Estimation of energy consumption for 45 nm CMOS technology.

The total energy consumption is dominated by the cost of memory accesses, which is yet unduly neglected in most metrics of the literature. While SNNs have additional memory accesses for updating the neuron potentials, it also requires fewer memory accesses for the weights and the I/O. Moreover, the size of I/O buffers are often much smaller in SNNs than in FNNs, since the first only requires FIFOs of 1000 elements whereas the second requires storing the full feature maps. The lower number of spikes combined with the absence of multiplication result in an energy consumption of synaptic operations two order of magnitude lower for SNNs than for FNNs. In the end, the total energy consumption of our SNNs is between 6.25 and 8.02 times lower than their FNN counterparts, a promising result for the implementation of SNNs on specialized hardware.

Table 2. Energy consumption estimations of our SNNs and their equivalent FNNs on the 3 datasets for 45nm CMOS technology using the metrics proposed in Sect. 3.2.

		CIFAR10		GSC		NCARS	
		SNN	FNN	SNN	FNN	SNN	FNN
Memory Accesses (nJ)	Potentials	1.12E+6	–	3.50E+3	–	2.69E+5	–
	Weights	4.60E+5	5.59E+6	1.73E+3	1.46E+4	8.10E+4	1.11E+6
	Bias	2.79E+2	6.97E+1	6.00E+0	3.00E+0	3.71E+1	7.43E+0
	In/Out	7.65E+2	6.09E+6	3.90E+1	1.50E+4	2.88E+2	1.54E+6
	Total	1.58E+6	1.17E+7	5.27E+3	2.96E+4	3.50E+5	2.64E+6
Synaptic Op. (nJ)		2.41E+3	1.29E+06	4.08E+1	3.53E+3	6.15E+3	2.67E+5
Addressing (nJ)		1.97E+2	4.94E+1	9.28E+0	1.93E+0	5.05E+1	7.65E+1
Total (nJ)		1.58E+6	1.24E+7	5.32E+3	3.32E+4	3.57E+5	2.86E+6
$E^{\mathrm{FNN}}/E^{\mathrm{SNN}}$		**8.19**		**6.22**		**8.17**	

6 Conclusion

Neuromorphic Engineering is based on the assumption that event-based processing is the key to mimic the unparalleled energy efficiency of the biological brain. However, this assumption remains to be proven. The goal of this work was to settle this question through a generic and accurate energy estimation metric, independent from low-level implementation choices and hardware targets. Our proposed analytical model is based on three types of operations occurring in hardware neural network implementations: synaptic operations, memory accesses and addressing mechanisms. This metric was applied to three datasets representative of three characteristic data types. In all three cases, spiking implementation could bring major energy savings compared to formal ones as our SNNs are respectively 8.19x, 6.22x and 8.17x more efficient on CIFAR-10, GSC and NCARS, while producing near state-of-the-art accuracy for the last two.

Future works will include a confrontation of those results with actual energy consumption measurements, using our own SNN hardware architecture (SPLEAT [1]) and other state-of-the-art deep learning accelerators. Moreover, further work is required on our CIFAR-10 model, in order to reach state-of-the-art accuracy and thus increase the fairness of the comparison. Additionally, we will study the impact of various quantization schemes on energy consumption.

References

1. Abderrahmane, N., Miramond, B., Kervennic, E., Girard, A.: Spleat: spiking low-power event-based architecture for in-orbit processing of satellite imagery. In: International Joint Conference on Neural Networks (2022)
2. Amir, A., et al.: A low power, fully event-based gesture recognition system. In: IEEE Conference on Computer Vision and Pattern Recognition, pp. 7243–7252 (2017)
3. Barchid, S., Mennesson, J., Eshraghian, J., Djéraba, C., Bennamoun, M.: Spiking neural networks for frame-based and event-based single object localization (2022). https://doi.org/10.48550/ARXIV.2206.06506

4. Bardow, P., Davison, A.J., Leutenegger, S.: Simultaneous optical flow and intensity estimation from an event camera. In: IEEE Conference on Computer Vision and Pattern Recognition, pp. 884–892 (2016). https://doi.org/10.1109/CVPR.2016.102
5. Cordone, L., Miramond, B., Thierion, P.: Object detection with spiking neural networks on automotive event data. In: International Joint Conference on Neural Networks (2022)
6. Davidson, S., Furber, S.B.: Comparison of artificial and spiking neural networks on digital hardware. Front. Neurosci. **15**, 651141 (2021)
7. Deng, L., et al.: Rethinking the performance comparison between SNNs and ANNs. Neural Netw. **121**, 294–307 (2020)
8. Ding, J., Yu, Z., Tian, Y., Huang, T.: Optimal ANN-SNN conversion for fast and accurate inference in deep spiking neural networks. In: International Joint Conference on Artificial Intelligence, pp. 2328–2336 (2021). https://doi.org/10.24963/ijcai.2021/321
9. Fang, W., et al.: Spikingjelly (2020). https://github.com/fangwei123456/spikingjelly. Accessed 29 July 2022
10. Jouppi, N.P., et al.: Ten lessons from three generations shaped Google's tpuv4i: industrial product. In: ACM/IEEE Annual International Symposium on Computer Architecture, pp. 1–14 (2021)
11. Khacef, L., Abderrahmane, N., Miramond, B.: Confronting machine-learning with neuroscience for neuromorphic architectures design. In: International Joint Conference on Neural Networks (2018). https://doi.org/10.1109/IJCNN.2018.8489241
12. Kheradpisheh, S.R., Masquelier, T.: Temporal backpropagation for spiking neural networks with one spike per neuron. Int. J. Neural Syst. **30**(06), 2050027 (2020)
13. Kundu, S., Datta, G., Pedram, M., Beerel, P.A.: Spike-thrift: Towards energy-efficient deep spiking neural networks by limiting spiking activity via attention-guided compression. In: Proceedings of the IEEE/CVF Winter Conference on Applications of Computer Vision, pp. 3953–3962 (2021)
14. Lemaire, E.: Modélisation et exploration d'architectures neuromorphiques pour les systèmes embarqués haute-performance. Ph.D. thesis, Univ. Côte d'Azur (2022)
15. Lemaire, E., Miramond, B., Bilavarn, S., Saoud, H., Abderrahmane, N.: Synaptic activity and hardware footprint of spiking neural networks in digital neuromorphic systems. ACM Trans. Embed. Comput. Syst. (2022)
16. Neftci, E., Mostafa, H., Zenke, F.: Surrogate gradient learning in spiking neural networks: bringing the power of gradient-based optimization to spiking neural networks. IEEE Sig. Process. Mag. **36**, 51–63 (2019). https://doi.org/10.1109/MSP.2019.2931595
17. Pellegrini, T., Zimmer, R., Masquelier, T.: Low-activity supervised convolutional spiking neural networks applied to speech commands recognition. In: 2021 IEEE Spoken Language Technology Workshop (SLT), pp. 97–103. IEEE (2021)
18. Rueckauer, B., et al.: NXTF: an API and compiler for deep spiking neural networks on intel Loihi (2021). https://doi.org/10.48550/ARXIV.2101.04261
19. Simonyan, K., Zisserman, A.: Very deep convolutional networks for large-scale image recognition. In: International Conference on Learning Representations (2015)
20. Sironi, A., Brambilla, M., Bourdis, N., Lagorce, X., Benosman, R.: Hats: Histograms of averaged time surfaces for robust event-based object classification. In: IEEE Conference on Computer Vision and Pattern Recognition, June 2018
21. Zimmer, R., Pellegrini, T., Singh, S.F., Masquelier, T.: Technical report: supervised training of convolutional spiking neural networks with PyTorch (2019)

Correlation Based Semantic Transfer with Application to Domain Adaptation

Florina Cristina Calnegru[1]([⊠]) [ID], John Shawe-Taylor[1] [ID], Iasonas Kokkinos[1] [ID], and Razvan Pascanu[2] [ID]

[1] Department of Computer Science, University College London, London, UK
`florina.calnegru.16@ucl.ac.uk`
[2] DeepMind, London, UK

Abstract. In this paper, we introduce a multifaceted contribution. First, we propose a definition, specific to convolutional neural networks (CNN's), for the notion of semantically similar features. Second, using this definition, we introduce a new loss, which semantically transfers features from one domain to another domain, where the features of both domains are learnt by two CNN's. Our transfer loss, named CBT, constrains the responses of the corresponding convolutional kernels of the two CNN's to correlate in similar contexts. When the features of the source domain are discriminative, with respect to a classifier, CBT helps to maintain in the target domain, the semantics of the feature space imposed by that classifier. Third, we show that CBT can be used for unsupervised domain adaptation (UDA) by proposing a novel approach for this problem.

Keywords: Semantically Similar Features · Transfer Loss · Transfer Learning · Unsupervised Domain Adaptation

1 Introduction

Deep Neural Networks have recently lead to great advances in a wide range of areas. However their performance is quite sensitive to changes in the input distribution. This means that a network that was trained on a source domain, but is employed with inputs from a target domain, can make a number of erroneous predictions, depending on how big the shift between the two domains is. In this case, if the label information is sparse, one needs to employ Domain Adaptation techniques. Domain Adaptation is concerned with leveraging labeled data in a source domain, to obtain learning mechanisms with good performance on a target domain, described by a different distribution. In this paper, we focus our attention on the unsupervised domain adaptation scenario, where we have no labels for the target domain. Different unsupervised domain adpatation (UDA) strategies were proposed through time. Some of the most succesfull UDA algorithms, try align the distributions of the source and target representations. A few examples are the methods that use adversarial training to generate target

M. Tanveer et al. (Eds.): ICONIP 2022, LNCS 13623, pp. 588–599, 2023.
https://doi.org/10.1007/978-3-031-30105-6_49

representations, that cannot be distinguished by a domain discriminator from the source representations, like [1], [2], [3], or the methods that minimize some measure of discrepancy between domains, like for instance the maximum mean discrepancy distance (MMD) [4]. Plenty of other UDA strategies are available, but despite the richness of their variety, some very important aspects still need to be tackled in order to better solve the problem of UDA. In the following, we present those aspects that represent a motivation for our work.

The first motivation for our work, is given by Theorem 2 from [5]. This theorem states that the target error of an hypothesis is upper bounded by the source error of that hypothesis, plus a measure of discrepancy between the two domains, plus the error of the optimal joint classifier for the two domains, plus a constant function. In the case of UDA the error of the optimal joint classifier cannot be computed, as there are no target labels. So, in order to minimize the target error of an hypothesis, most of the works on domain adaptation, following Theorem 2 of [5], have tried to minimize, in fact, a form of the discrepancy between the target and the source domain. But, as [6] has shown, when the target features are not discriminative, the error of the optimal joint classifier increases, and thus the target error increases. So, to achieve a minimization of the target error, it is not enough to minimize the discrepancy between the two domains. Rather, one also has to find target domain features that are discriminative. The second motivation for our work, can be found in Lemma 4.8 from [7], which states that when the distributions of the labels for the two domains are different, minimizing the discrepancy between the representations of the two domains, and using an invariant representation for the two domains, and the same classifier, actually leads to an increase in the target error. This means that, when the distributions of the labels are different, UDA methods that align the representations are theoretically shown by [7] to be sub-optimal.

The contributions of our work are the following. The first contribution, is that we propose a definition, specific to CNN's, for the notion of semantically similar features. The second contribution, is that, using this definition, we introduce a new loss, which semantically transfers features from one domain to another domain, and that we call Correlation Based Transfer (CBT). The third contribution, is that, based on the CBT loss, we propose a new approach to UDA, that we call Transferring Discriminative Hierarchies (TDH). TDH learns from the source domain which features are discriminative, and tries to promote in the target domain, features that emulate the discriminability of the source features. Unlike in the case of UDA methods that align the distributions of the representations, the features that we learn are not domain invariant. In our case, the target filters and the source filters are numerically different, but the features that they detect are semantically similar. By learning discriminative target features, we aim to diminish the first disadvantage of the current UDA algorithms, as presented in the first motivation of our work. By producing features that are not domain invariant, and that do not have the distributions of the representations aligned, we are trying to reduce the second downside of some of the existent UDA methods, as explained in the second motivation.

2   Semantically Similar Features

One way to obtain target domain representations that are discriminative is to semantically transfer discriminative source features to the target domain. In this way, since the target features are semantically similar with the source features, and the source features are discriminative, the target features will be discriminative. To be able to obtain via transfer, semantically similar features, we first need to be able to define them. As an example, at an intuitive level, of what we mean by 'semantically similar', if we use in the source domain features as head, torso, legs, etc. to recognize a body, we want to learn, in the target domain, features that also correspond to target specific heads, torso, legs, etc. We extend this intuition to CNN's. We define two features as semantically similar if they play the same role in the feature hierarchies corresponding to the two domains. For example, one can imagine that a CNN trained to do face recognition builds in intermediary layers, features that detect the eyes, the mouth, etc. The classifier decides whether an image is a face or not based on these features firing strongly enough in particular areas of the image. And this process is hierarchical. The eye detector relies on certain Gabor-like filters firing together, the same holds for the nose detector and so forth. Of course this is an idealized case. In practice, the semantics of the learnt features are not necessarily interpretable, and might not decompose around what we find meaningful. However their functional role is exactly the same. A feature on layer k corresponds to certain features on layer $k-1$ firing in a particular pattern. Taking this functional role into consideration, we define a feature on the target domain net as semantically similar to one in the source domain net, if it relies on the corresponding features on the layer below to fire together in roughly the same way. To achieve that, we introduce a new loss, the Correlation Based Transfer (CBT) loss, which constrains that the responses of the corresponding convolutional filters from the target and source nets correlate in similar contexts. This amounts to enforcing that the source and target filters play the same role in their features hierarchies.

3   Correlation Based Transfer Loss

Let $G_{i,j}^l$ be the i-th feature map, from the l-th layer of the target net, corresponding to the j-th target training point. Let $F_{i,j}^l$, be the i-th feature map, from the l-th layer on the source net, corresponding to the j-th source training point. Let γ_i^l be the kernel of the target net, used to compute the i-th feature map of layer l. ϕ_i^l, from the source network, is the corresponding kernel to γ_i^l from the target network. That is $G_{i,j}^l = a(G_j^{l-1} \odot \gamma_i^l)$, where $\odot$ represents the convolutional operator, a is our activation function, G_j^{l-1} are the target feature maps, from layer $l-1$, of the j-th target point, and for clarity we are ignoring the bias. Analogously $F_{i,j}^l = a(F_j^{l-1} \odot \phi_i^l)$. Let L be the number of layers, N_l the number of feature maps on l-th layer, R the number of data points in the source domain and M the number of data points in the target domain.

If we define the normalized cross correlation of random variables A and B as $NCC(A, B) = [(A - E(A)) * ((B - E(B))]/(\sqrt{Var(A)} * \sqrt{Var(B)})$, CBT is:

$$\mathcal{L}_{CBT} = 2 - \frac{\left(\sum_{l=1}^{L} \sum_{i=1}^{N_l} \sum_{j=1}^{R} NCC(a(G_j^{l-1} \odot \phi_i^l), F_{i,j}^l) \right)}{R(\sum_i^L N_i)} - \frac{\left(\sum_{l=1}^{L} \sum_{i=1}^{N_l} \sum_{j=1}^{M} NCC(a(F_j^{l-1} \odot \gamma_i^l), G_{i,j}^l) \right)}{M(\sum_i^L N_i)}$$

(1)

To explain what CBT does, we name $G_{i,j}^l$ the true feature maps of the target net and $F_{i,j}^l$ the true feature maps of the source net. We also name $a(G_j^{l-1} \odot \phi_i^l)$, the target hybrid feature maps. They are obtained on the l-th layer by correlating the i-th source filter from layer l, with the target feature maps from layer $l - 1$, corresponding to the j-th target point. Analogously we name $a(F_j^{l-1} \odot \gamma_i^l)$ the source hybrid feature maps. The CBT loss determines all the true feature maps and all the corresponding hybrid feature maps for the same domain, computes the NCC between every true feature map and every corresponding hybrid feature map, and then is averaging all these possible NCC-s. The intuition behind this is that a filter from the target net is semantically similar with its corresponding filter from the source net if, when it is used in the same context, it is capable to elicit a similar response. So, if for example, we have a filter in the source net capable to detect a hand in the source domain, the semantically similar feature from the target net must also be capable to detect a hand in the target domain. The hand in the target domain might look different from the hand in the source domain, but it is certainly more similar with the hand in the source domain than it is for example with a car. Thus when convolving both a 'source hand filter' and a 'target hand filter' with some feature maps that contain a hand at location q and a car at location p, we have in both convolution cases a strong response at location q and a weak response at location p. NCC actually measures the similarity of those responses. Note that one trivial solution to minimize CBT is to make the weights of the two networks identical. This however will not lead to good transfer on the target domain, and this is why we also use the reconstruction loss. As a consequence, the CBT loss will only play the role of a soft constraint, the association with the reconstruction loss allowing the target features to adapt to the target domain, while copying the semantics of the source net. The CBT loss allows for a large degree of freedom in changing the weights of a feature map while maintaining similar response profile. Compared with methods that just align marginal distributions, which might return the same score even when the relative correspondence between which features fire together is lost, we maintain this crucial information exploited by the classifier on top.

4   Transferring Discriminative Hierarchies

In this section, we present a new approach for UDA named Transferring Discriminative Hierarchies (TDH). TDH is concerned with transferring only those parts of the source representations that are relevant for the task at hand. We name those parts, the transferable (content) parts. We further assume that the style

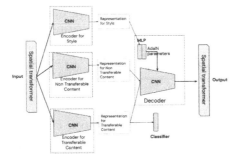

Fig. 1. Diagram depicting the architecture of both the source network and the target network.

is not relevant for the task at hand, and we choose to eliminate it. This means that to apply TDH, we have to disentangle the style from the content, and the transferable content from the non transferable content. We learn the representations, with the help of two CNN's, one for the target domain, and one for the source domain. The two CNN's have the same architecture, depicted in Fig. 1. We constrain, with CBT, the target transferable encoder to learn semantically similar features with the features learnt by the source transferable encoder. The architecture from Fig. 1 is inspired by the work of [8] for image to image translation. We borrow from them the idea of disentangling the style from content using Instance Normalization [9]. Like them, we add the style back to the content, in the decoder, using Adaptive Instance Normalization [10]. Unlike them we use two encoders for content, while they use only one. That is because [8] only consider images with one object, while our images can depict objects into an environment. Also their objects are all in the same positions in the images, while ours are not. To mediate that, we use two Spatial Transformers [11] in each domain. The aim of the first Spatial Transformer in the target domain is to eliminate the large affine differences between the transferable parts of the target domain and the transferable parts of the source domain. The role of the second Spatial Transformer in both domains is to recover for reconstruction, the inverse of the affine transformation, learnt by the first Spatial Transformer.

We denote by $\mathscr{E}_T$ and $\mathscr{E}_{NT}$ the encoders for the transferable respectively nontransferable content part, by $\mathscr{S}$ the style encoder and by $\mathscr{D}$ the decoder. For an input point x, we denote by $c_T = \mathscr{E}_T(x)$, $c_{NT} = \mathscr{E}_{NT}(x)$, and by $c = (c_T, c_{NT})$. We also write $c = \mathscr{E}(x)$. We denote by $\sigma = \mathscr{S}(x)$, the style representation, by $p(x)$ the marginal distribution over the inputs, by $p_T(c_T)$ and $p_{NT}(c_{NT})$ the distributions over the transferable, respectively nontransferable parts of the inputs. We add the superscript s or t, to signify that we refer exclusively to the source or the target domain. We denote by $\mathbb{E}$ the expected value, by $\mathcal{N}(0,1)$ the standard multivariate normal distribution and by $\|.\|$ the L2 norm. The **loss for disentangling the style and the content** is defined [8] as:

$$\mathcal{L}_{S/C} = \mathbb{E}_{x \sim p, \tilde{\sigma} \sim \mathcal{N}(0,1)}[\|\mathscr{E}(\mathscr{D}(c, \tilde{\sigma})) - c\| + \|\mathscr{S}(\mathscr{D}(c, \tilde{\sigma})) - \tilde{\sigma}\|] \tag{2}$$

$\mathcal{L}_{S/C}$ enforces that when we replace the old style with a random style, and when we decode the new representation and then encode the result, we can recover both the original content and the random style. **The loss for disentangling the transferable content from the nontransferable content** is defined as:

$$\mathcal{L}_{T/NT} = \mathbb{E}_{x \sim p, \tilde{c}_{NT} \sim p_{NT}} [\|\mathcal{E}_T(\mathcal{D}((c_T, \tilde{c}_{NT}), \sigma)) - c_T\| + \\ \|\mathcal{E}_{NT}(\mathcal{D}((c_T, \tilde{c}_{NT}), \sigma)) - \tilde{c}_{NT}\|] \qquad (3)$$

$\mathcal{L}_{T/NT}$ enforces that, when we replace the old non transferable part with a random non transferable part, and when we decode the new representation and then encode the result we can recover both the original transferable part, and the random nontransferable part. $\mathcal{L}_{T/NT}$ only ensures that the transferable and nontransferable content parts are independent. We can find an exponential number of ways to separate the features into independent parts, so this condition is not sufficient. To further identify the transferable content, we need to use the fact that the transferable part comprises the information useful for classification. So, in the source domain, the transferable part has to minimize the cross-entropy classification loss $\mathcal{L}_{classif} = \mathbb{E}_{x^s \sim p^s}[\mathcal{L}_{CE}(h(c_T^s), (x^s, y))]$, where h is the source classifier, and $\mathcal{L}_{CE}(P, (x, y)) = -\sum_{c=1}^{C} \delta_c \log(P_c)$, when P_c is the predicted probability that x belongs to class c, $\delta_c = 1$, if $c = y$, $\delta_c = 0$, if $c \neq y$, and C is the number of classes. In the target domain we have no labels, so we enforce that the target transferable part minimizes the conditional entropy, with respect to the source classifier h, defined as: $\mathcal{L}_{ent} = \mathbb{E}_{c_T^t \sim p_T^t}[-h(c_T^t) * ln((h(c_T^t)))]$. The conditional entropy is not enough to guarantee that all the information useful for classification is comprised into the transferable part. This is why, we also use the cyclic loss, which enforces that the transferable parts, of correspondent target and source examples, are interchangeable in controlled cases, from the point of view of their discriminative information. Although the cyclic loss is inspired from [12], we do not define it though as in [12], but we adapt it to our purpose. We define **the cyclic loss** as: $\mathcal{L}_{cyc} = \mathcal{L}_{rep}^{s-t} + \mathcal{L}_{rep}^{t-s} + \mathcal{L}_{sem}$

$$\mathcal{L}_{rep}^{t-s} = \mathbb{E}_{\tilde{c}_T^s \sim p_T^s, c_{NT}^t \sim p_{NT}^t} [\|\mathcal{E}_T^t(\mathcal{D}^t((\tilde{c}_T^s, c_{NT}^t), \sigma^t)) - \tilde{c}_T^s\| + \\ \|\mathcal{E}_{NT}^t(\mathcal{D}^t((\tilde{c}_T^s, c_{NT}^t), \sigma^t)) - c_{NT}^t\|] \qquad (4)$$

$\mathcal{L}_{rep}^{t-s}$ ensures that, if we have in the target transferable part the same information as in a source transferable part, we extract by disentanglement this information integrally. $\mathcal{L}_{rep}^{s-t}$ is defined by symmetry, changing the superscript t with s in Eq. 4. We define the semantic component of the cyclic loss as:

$$\mathcal{L}_{sem} = \mathbb{E}_{x^s \sim p^s, x^t \sim p^t}[\mathcal{L}_{CE}(h(\mathcal{E}_T^t(\tilde{x}_T^t)), (\tilde{x}_T^t, y)) + H^2(h(c_T^s), h(\mathcal{E}_T^t(\tilde{x}_T^t)))] \qquad (5)$$

where y is the label of x^s from the source domain, $\tilde{x}_T^t = \mathcal{D}^t((c_T^s, c_{NT}^t), \sigma^t)$ are synthetic target points, obtained from real target points, after we have replaced their transferable parts with transferable parts from random source points. $H^2(P, Q)$ is the Hellinger distance between two probabilities P, and Q, and is computed as $H^2((P, Q) = 1/\sqrt{2} * \|\sqrt{P} - \sqrt{Q}\|$, h is the classifier for both the source and the target domain. $\mathcal{L}_{sem}$ enforces that, for controlled cases, the target transferable parts contain information as useful for classification as their correspondent source transferable parts.

To train the source network we use $\mathcal{L}_{source} = \mathcal{L}_{classif} + \mathcal{L}_{rec} + \mathcal{L}_{S/C} + \mathcal{L}_{T/NT}$, where $\mathcal{L}_{rec}$ is the **reconstruction loss** that we define as $\mathcal{L}_{rec} = \mathbb{E}_{x \sim p}[\|x - \mathcal{D}(\mathring{\mathcal{E}}(x))\|]$, where $\mathring{\mathcal{E}}$ are all the encoders. To be noted that we actually transfer with TDH the discriminative hierarchies from the source domain, so we could have chosen to train the source transferable part only for classification and use no reconstruction, and thus no disentanglement in the source domain. However, in practice, we notice that training the source net with $\mathcal{L}_{source}$ leads to better transfer results. Since the CBT can only be used for the convolutional layers, we need two losses to train the target net. One loss is employed to train the convolutional layers of the target transferable encoders, and the other loss is used to train the bottleneck layers of the transferable encoder. The loss to train the convolutional layers is $\mathcal{L}_{target}^{conv} = \mathcal{L}_{ent} + \mathcal{L}_{CBT} + \mathcal{L}_{rec} + \mathcal{L}_{S/C} + \mathcal{L}_{T/NT} + \mathcal{L}_{cyc}$. The loss to train the bottleneck layers is $\mathcal{L}_{target}^{FC} = \mathcal{L}_{ent} + \mathcal{L}_{rec} + \mathcal{L}_{S/C} + \mathcal{L}_{T/NT} + \mathcal{L}_{cyc}$. In the costs for training the networks all the summed losses have weights that we do not include for the ease of presentation. The steps of our algorithm are:

1. Train the source net to minimize $\mathcal{L}_{source}$, and obtain classifier h.
2. Initialize the target net $\mathcal{E}_T$ and classifier with their source net counterparts.
3. Fix the parameters of the bottleneck layers of the target transferable encoder, and the target classifier, to their source value, and train for the rest of the parameters of the target net. The objective for training is $\mathcal{L}_{target}^{conv}$.
4. Keep the parameters of the target transferable convolutional layers fixed to their value from the previous step, use the same classifier, h, and fine tune the rest of the parameters of the target DNN. The objective for training the encoders is $\mathcal{L}_{target}^{FC}$. The objective for training the decoder is $\mathcal{L}_{rec}$.

5   Related Work

[13] have shown empirically that using *related* weights, instead of shared ones can improve the discriminability of the target features. We similarly do not share weights, but rather constrain them to stay close to each other. Unlike [13] we only constrain a subset of weights, the transferable part, they constrain all. They rely on MMD to align the distributions of the representations, while we do not. Also similar to us, [14] is using related encoders weights. Just like us, they employ a mechanism based on the feature maps to constrain that the weights of the encoders are close to each other. But while we are trying to maximize the correlation of the hybrid and true feature maps, they are trying to minimize the differences between corresponding attention maps. [14] combine their attention based alignment measure with a number of strategies that we do not use, like GAN's, expectation-maximization, and pseudolabels. Like [15] we are also using the correlation, but in a totally different way. Specifically, they are computing the correlation between the target representations and the correlation between the

source representations and then they are minimizing the log Euclidean distance between those correlations. Similar to [16] we are learning semantic representations, but their representations are domain invariant, while ours are not. Also, to be noted, that they obtain semantic representations by minimizing the distances between the centres of the corresponding classes representations in the source and the target domains. Like [1] we also use a separation of the domains into private (nontransferable) and shared (transferable) parts, but we differ substantially from them. They use weight sharing on the feature extractors of the transferable part, MMD, and adversarial training, while we do not. Similar to [17] we also have a reconstruction cost and a classification cost, but all our other cost functions are different. [12] define the cyclic and the semantic losses to verify that translating back and forth between the source domain and the target domain is consistent, both in a L_1 distance sense and semantically. We define those losses, differently, to ensure that the transferable target part is disentangled similarly with the source transferable part, when it contains semantically equivalent information.

6   Experiments

We first offer a proof of concept that CBT can be used for transferring features for UDA. We perform experiments on the UDA cases: MNIST [18]→SVHN [19], SVHN→MNIST, Syn-digits [20]→SVHN, MNIST→ USPS [21], CIFAR [22]→STL [23]. In our experiments on TDH the digits sets content encoders have 2, 5×5 convolutional layers (CL), with a number of filters of 100 and 150, 2, 3×3 CL of 150 and 50 filters, and 2 fully connected layers with 256 and 128 neurons. The content encoders for the object data set have 9, 3×3 CL, with a number of filters of 128, 128, 128, 128, 256, 256, 512, 512, 512, 512 and 2, 1×1 CL of 256 and 128 filters. The architecture of the decoder for a data set is the reverse of the architecture of a content encoder used for that data set. The digits data sets are converted to gray scale prior to applying TDH. We train TDH with RMSProp and use the learning rate (lr) $2*10^{-4}$ in step 3, and $0.5*10^{-4}$ in step 4 for digits sets. For the object set we use lr $0.5*10^{-3}$ in step 3, and 10^{-5} in step 4. We compare TDH against other methods and present the accuracies in Table 1, where NR stands for not reported by their authors. To show that CBT helps to preserve the semantics of the source classifier, we replace CBT in TDH, with MMD, and with the $\mathbb{L}_1$, and the $\mathbb{L}_2$ distances, between the corresponding transferable filters of the target and source DNN's. We compare CBT with MMD, because we want to show how CBT stands in relation with a metric that aligns the distributions, without putting an emphasis on preserving the semantics. We compare CBT with the $\mathbb{L}_1$ and the $\mathbb{L}_2$ distances, because CBT can be viewed as constraining the corresponding filters of the transferable encoders to elicit similar responses. So, we want to see if asking for similar responses is better than just asking that corresponding transferable filters have similar values. We present the accuracies in Table 2. This shows that CBT plays a part in promoting discriminative target features and thus, that CBT helps to preserves the semantics of

Table 1. Comparison with other methods

Datasets	MNIST →SVHN	SVHN →MNIST	Syn-digits →SVHN	MNIST →USPS	STL →CIFAR
DRCN[17]	40.05	80.2	NR	91.8	66.37
DSN[1]	NR	82.7	91.2	NR	NR
CyCleGAN+LWC[2]	NR	97.5	NR	97.1	NR
DANN[20]	60.6	68.3	90.3	NR	78.1
SHOT[24]	NR	98.9	NR	98.1	NR
VBDA[25]	NR	93.8	NR	96.0	NR
ATT[6]	52.8	86.2	92.9	NR	NR
SDDA-P[26]	43.6	76.3	NR	88.5	NR
DFA-ENT[3]	NR	98.5	NR	96.9	NR
IEDA[27]	78.8	98.9	NR	95.0	78.3
TDH after Step 2 i.e. Source Only	55.3	76.2	73.3	81.0	73.4
TDH after Step 3 (using CBT)	82.2	98.0	90.5	99.3	78.9
TDH after Step 4 (final results)	87.5	98.9	92.0	99.4	80.15

the source classifier on the target domain. It also shows that for more dissimilar domains (i.e. MNIST and SVHN) the improvement due to CBT is larger than the improvement for more similar ones (i.e. MNIST and USPS and Syn-digits and SVHN). We next present results of ablation studies, for MNIST→SVHN and MNIST→USPS, in Tables 3 and 4, where the column of Acc is for accuracies. In our ablation studies, we have excluded one by one, either the cost functions used to train the convolutional layers, or the Spatial Transformers. If on the column, corresponding to a loss, we have a 1, then this loss is employed to train the target net. If we have a 0, then this loss is not employed. The same applies to column STr, corresponding to the Spatial Transformers. Since we want to understand CBT, and it is only used when training the convolutional layers, we performed the ablations in Step 3 of TDH, and measured the accuracies after this step. When the domains strongly differ (Table 3), both CBT, that preserves the semantics, and the reconstruction loss, that forces the target features to be target specific, are necessary. When the nontransferable part is negligible, and the style differences are small (Table 4), all the losses with a role in disentanglement (i.e. $\mathcal{L}_{T/NT}$, $\mathcal{L}_{S/C}$ and $\mathcal{L}_{cyc}$) play a minor role for the system performance. This shows that once the transferable parts are isolated, using CBT for UDA is not dependent on employing these functions. All these results, point to the fact that the competitive acurracies of TDH, from Table 1, are in part determined by using CBT, and thus CBT is promising for UDA.

Table 2. Preserve semantics study

	MNIST SVHN	SVHN MNIST	Syndigits SVHN	MNIST USPS
MMD	60.0	59.3	84.4	98.3
L_1	62.05	41.2	84.7	94.8
L_2	67.0	69.0	89.3	95.7
CBT	82.0	98.0	90.5	99.3

Table 3. Ablation study MNIST− >SVHN

$\mathcal{L}_{CBT}$	$\mathcal{L}_{rec}$	$\mathcal{L}_{ent}$	$\mathcal{L}_{T/NT}$	$\mathcal{L}_{S/C}$	$\mathcal{L}_{cyc}$	STr	Acc
1	1	1	1	1	1	1	82.2
0	1	1	1	1	1	1	12.3
1	0	1	1	1	1	1	40.4
1	1	0	1	1	1	1	14.3
1	1	1	0	1	1	1	78.6
1	1	1	1	0	1	1	78.2
1	1	1	1	1	0	1	77.5
1	1	1	1	1	1	0	61.6

Table 4. Ablation study MNIST− >USPS

$\mathcal{L}_{CBT}$	$\mathcal{L}_{rec}$	$\mathcal{L}_{ent}$	$\mathcal{L}_{T/NT}$	$\mathcal{L}_{S/C}$	$\mathcal{L}_{cyc}$	STr	Acc
1	1	1	1	1	1	1	99.3
0	1	1	1	1	1	1	80.7
1	0	1	1	1	1	1	99.1
1	1	0	1	1	1	1	93.3
1	1	1	0	1	1	1	99.2
1	1	1	1	0	1	1	99.1
1	1	1	1	1	0	1	99.2
1	1	1	1	1	1	0	91.3

7   Conclusions

We have introduced the concept of 'semantically similar features', a new loss, CBT, that semantically transfers the hierarchies of features, and a new CBT based approach for UDA, named TDH. TDH represents a proof of concept that CBT can be used for UDA. Through our experiments, we give evidence that CBT is promising for UDA, and that the utilization of CBT for UDA is not dependent on the disentanglement cost functions that TDH employs. Although, in TDH, we availed ourselves of the reconstruction loss to make the target features target specific, other means to do that are also possible (E.g. by using self-supervised learning). This suggests that other UDA methods, leveraging CBT, can be devised. We hope to open thus, the path for a new class of UDA approaches, that employ CBT, or other future semantic transfer losses, like CBT, to transfer discriminative features in the target domain, and circumvent in this way the above-mentioned disadvantages of current UDA methods.

Acknowledgement. The second author is in part sponsored by the U.K. Engineering and Physical Sciences Research Council under grant number EP/R013616/1.

References

1. Bousmalis, K., Trigeorgis, G., Silberman, N., Krishnan, D., Erhan, D.: Domain separation networks. In: Lee, D., Sugiyama, M., Luxburg, U., Guyon, I., Garnett, R., editors, Advances in Neural Information Processing Systems, vol. 29. Curran Associates Inc., (2016)
2. Ye, S., et al.: Light-weight calibrator: a separable component for unsupervised domain adaptation. In: Proceedings of the IEEE/CVF Conference on Computer Vision and Pattern Recognition, pp. 13736–13745 (2020)
3. Wang, J., Chen, J., Lin, J., Sigal, L., de Silva, C.W.: Discriminative feature alignment: Improving transferability of unsupervised domain adaptation by gaussian-guided latent alignment. Pattern Recogn. **116**, 107943 (2021)
4. Long, M., Cao, Y., Wang, J., Jordan, M.: Learning transferable features with deep adaptation networks. In: International Conference on Machine Learning, pp. 97–105. PMLR (2015)
5. Ben-David, S., Blitzer, J., Crammer, K., Kulesza, A., Pereira, F., Vaughan, J.W.: A theory of learning from different domains. Mach. Learn. **79**(1), 151–175 (2010)
6. Saito, K., Ushiku, Y., Harada, T.: Asymmetric tri-training for unsupervised domain adaptation. In: International Conference on Machine Learning, pp. 2988–2997. PMLR (2017)
7. Zhao, H., Tachet Des Combes, R., Zhang, K., Gordon, G.: On learning invariant representations for domain adaptation. In: International Conference on Machine Learning, pp. 7523–7532. PMLR (2019)
8. Huang, X., Liu, M.-Y., Belongie, S., Kautz, J.: Multimodal unsupervised image-to-image translation. In: Ferrari, V., Hebert, M., Sminchisescu, C., Weiss, Y. (eds.) ECCV 2018. LNCS, vol. 11207, pp. 179–196. Springer, Cham (2018). https://doi.org/10.1007/978-3-030-01219-9_11
9. Ulyanov, D., Vedaldi, A., Lempitsky, V.: Improved texture networks: maximizing quality and diversity in feed-forward stylization and texture synthesis. In: Proceedings of the IEEE Conference on Computer Vision and Pattern Recognition, pp. 6924–6932 (2017)
10. Huang, X., Belongie, S.: Arbitrary style transfer in real-time with adaptive instance normalization. In: Proceedings of the IEEE International Conference on Computer Vision, pp. 1501–1510 (2017)
11. Jaderberg, M., Simonyan, K., Zisserman, A., Kavukcuoglu, K.: Spatial transformer networks. In: Cortes, C., Lawrence, N., Lee, D., Sugiyama, M., Garnett, R., editors, Advances in Neural Information Processing Systems, vol. 28. Curran Associates Inc., (2015)
12. Hoffman, J., et al.: Cycada: cycle-consistent adversarial domain adaptation. In International Conference on Machine Learning, pp. 1989–1998. PMLR (2018)
13. Rozantsev, A., Salzmann, M., Fua, P.: Beyond sharing weights for deep domain adaptation. IEEE Trans. Pattern Anal. Mach. Intell. **41**(4), 801–814 (2018)
14. Kang,G., Zheng, L., Yan, Y., Yang, Y.: Deep adversarial attention alignment for unsupervised domain adaptation: the benefit of target expectation maximization. In: Proceedings of the European Conference on Computer Vision (ECCV), pp. 401–416 (2018)
15. Morerio, P., Cavazza, J., Murino, V.: Minimal-entropy correlation alignment for unsupervised deep domain adaptation. arXiv preprint arXiv:1711.10288 (2017)
16. Xie, S., Zheng, Z., Chen, L., Chen, C.: Learning semantic representations for unsupervised domain adaptation. In: International Conference on Machine Learning, pp. 5423–5432. PMLR (2018)

17. Ghifary, M., Kleijn, W.B., Zhang, M., Balduzzi, D., Li, W.: Deep reconstruction-classification networks for unsupervised domain adaptation. In: Leibe, B., Matas, J., Sebe, N., Welling, M. (eds.) ECCV 2016. LNCS, vol. 9908, pp. 597–613. Springer, Cham (2016). https://doi.org/10.1007/978-3-319-46493-0_36
18. LeCun, Y., Bottou, L., Bengio, Y., Haffner, P.: Gradient-based learning applied to document recognition. Proc. IEEE **86**(11), 2278–2324 (1998)
19. Netzer, Y., Wang, T., Coates, A., Bissacco, A., Wu, B., Andrew Y.N.: Reading digits in natural images with unsupervised feature learning. In: NIPS Workshop on Deep Learning and Unsupervised Feature Learning (2011)
20. Ganin, Y., Lempitsky, V.: Unsupervised domain adaptation by backpropagation. In: Bach, F., Blei, D., editors, Proceedings of the 32nd International Conference on Machine Learning, volume 37 of Proceedings of Machine Learning Research, pp. 1180–1189, Lille, France, 07–09 Jul 2015. PMLR
21. Hull, J.J.: A database for handwritten text recognition research. IEEE Trans. Pattern Anal. Mach. Intell. **16**(5), 550–554 (1994)
22. Krizhevsky, A.: Learning multiple layers of features from tiny images. Technical Report (2009)
23. Coates, A., Andrew N., Lee, H.: An analysis of single-layer networks in unsupervised feature learning. In: Proceedings of the Fourteenth International Conference on Artificial Intelligence and Statistics, pp. 215–223. JMLR Workshop and Conference Proceedings (2011)
24. Liang, J., Hu, D., Feng, J.: Do we really need to access the source data? source hypothesis transfer for unsupervised domain adaptation. In: International Conference on Machine Learning, pp. 6028–6039. PMLR (2020)
25. Song, Y., et al.: Improving unsupervised domain adaptation with variational information bottleneck. In: ECAI (2020)
26. Kurmi, V.K., Subramanian, V.K., Namboodiri, V.P.: Domainkimpression: a source data free domain adaptation method. In: Proceedings of the IEEE/CVF Winter Conference on Applications of Computer Vision, pp. 615–625 (2021)
27. Choi, J., et al.: Visual domain adaptation by consensus-based transfer to intermediate domain. In: AAAI (2020)

Minimum Variance Embedded Intuitionistic Fuzzy Weighted Random Vector Functional Link Network

Nehal Ahmad[1,2(✉)], Mudasir Ahmad Ganaie[1], Ashwani Kumar Malik[1], Kuan-Ting Lai[3], and M. Tanveer[1]

[1] Department of Mathematics, Indian Institute of Technology Indore, Simrol, Indore 453552, India
{nehalahmad,phd1801241003,phd1901141006,mtanveer}@iiti.ac.in
[2] International Program of Electrical Engineering and Computer Science, National Taipei University of Technology, Taipei, Taiwan
[3] Department of Electronic Engineering, National Taipei University of Technology, Taipei, Taiwan

Abstract. Randomized neural networks such as random vector functional link network have been successfully employed in regression and classification problems. In real world, most of the data contaminated by outliers and noisy samples and hence, intelligent models are needed to classify such data. To make the model robust to noise and incorporate the geometric structure of the data, we propose minimum variance intuitionistic fuzzy RVFL (MVIFRVFL) model. In the proposed MVIFRVFL model, each sample is assigned an intuitionistic fuzzy number that is defined based on membership and non membership values. Membership value of a sample is considered according to its distance from the class center and the nonmembership value is assigned to each sample based on the ratio of the number of heterogeneous points to the total number of points in its neighborhood. Moreover, we incorporate the geometric aspect of the data, we minimise the variance among the data points of each class to improve the generalization performance. The performance of the classification models are analysed on 33 datasets taken from UCI and KEEL repository. From the experimental results, it is evident that the proposed MVIFRVFL model outperformed the state-of-the-art algorithms with best accuracy, highest number of wins among all the datasets and grabbed lowest rank amongst the baseline models.

Keywords: Intuitionistic fuzzy · RVFL · minimum variance · extreme learning machine

1 Introduction

Over the past decades, machine learning algorithms such as support vector machine (SVM) [1,2] decision tree (DT) [3] and artificial neural networks (ANNs) [4,5] have been successfully employed in several domain such as computer vision,

M. Tanveer et al. (Eds.): ICONIP 2022, LNCS 13623, pp. 600–611, 2023.
https://doi.org/10.1007/978-3-031-30105-6_50

health care, agriculture and so on. Among them, RNNs (randomised neural networks) [6] have demonstrated their effectiveness for classification and regression issues. Unlike back propagation algorithm based neural networks, RNNs [6] have closed form solution and hence avoid the shortcomings of back propagation based networks [7]. RNNs have a variety of traits, including a straightforward architecture, universal approximation capability [8] and higher data modelling capabilities [9]. Algorithms based on randomization, like the random vector functional link (RVFL) network, have the universal approximation capability [10,11]. The RVFL has proven its enormous capability in a number of notable areas such as forecasting [12], ensemble learning [13,14], biomedical engineering [15] and various other fields of machine learning [16]. RVFL is a unique variety of single hidden layer feed forward neural network (SLFN) in which both the hidden layer and output layer are directly coupled to the input layer. In RVFL model, the weights from input nodes to hidden nodes are randomly initialized with some probability distribution [17] from a predefined interval and kept constant throughout the learning process. The hidden layer consist of a non linear activation function that performs the transformation of input features into randomized feature space. The RVFL network's output weights are calculated using the closed-form method [7,18]. In the past decades, several improvements have taken place in the architecture of RVFL based models [19] [20] to improve its generalization performance. Zhang and Yang [21] proposed RVFL+ with LUPI paradigm and its kernalized version known as KRVFL+. Kernel based RVFL+ model is more robust than RVFL+ and RVFL network. Two variants of RVFL model proposed by [20] known as Total-Var-RVFL and Class-Var-RVFL model wherein, total variance and intra class variance are minimized, respectively and have better generalization performance. L_1 norm based sparse auto-encoder is employed in SP-RVFL [22] to calculate the hidden layer weights. Topological characteristics of the data are considered in sparse Laplacian regularized RVFL (SLapRVFL) [23] that consists of L_{21} norms in terms of regularization and hence, compared to traditional RVFL, it is more reliable and performs better in terms of generality. A novel robust RVFL model proposed by [24] that uses weighted error term in the optimization problem to handle noisy samples in the data. For more details about RVFL model, one can see the RVFL review paper [25].

In order to reduce the negative influences of the outliers or noisy samples, fuzzy theory has been successfully employed with several machine learning models such as SVM [26]. As compared to classical SVMs, the fuzzy based variants of SVMs have more capability in reducing noise effect and outliers [27]. A fuzzy SVM proposed by [28], wherein each sample of data is assigned a degree of fuzzy membership to decide its contribution in the learning process. However, fuzzy membership based models [28] have some drawbacks, i.e., it might not be able to distinguish the outliers from support vectors which lie on the edge [29]. Another approach has been proposed to enhance the aforementioned fuzzy membership scheme i.e., intuitionistic fuzzy membership [30,31]. Several other models such as intuitionistic fuzzy kernel ridge regression (IFKRR) [32] and intuitionistic fuzzy RVFL (IFRVFL) [33] have employed intuitionistic fuzzy theory to develop robust models compared to the base line models. The RVFL model assigns uniform weights to each pattern for determining the output parameters of the classifier and thus, it is not a robust model. Although, IFRVFL model employed intuitionistic fuzzy theory to handle the outlier and noise samples, however, it doesn't consider the aspect of variance of data . Moreover, MVRVFL model minimizes the within-class variance of the data and uses L_2 norm that is sensitive to noise and hence, doesn't has better generalization performance in noisy environment. Motivated by minimum variance embedded RVFL [20] and IFRVFL [33], in this article, minimum variance embedded intuitionistic fuzzy weighted RVFL (MVIFRVFL) model is proposed. The proposed MVIFRVFL model uses membership and nonmembership function to handle outlier and noisy sample in the data and hence, is robust compared to [20].

The remaining part of this research article is as follows: Section II discusses related work such as standard RVFL, intuitionistic fuzzy membership scheme. the proposed work is presented in Section III and the experimental analysis is performed in Section IV. Conclusion with some future directions is given in Section V.

2   Related Work

Let $X = \{(x_1, t_1), \ldots, (x_N, t_N) \mid (x_i, t_i) \in R^n \times R^c, i = 1, 2, \ldots, N\}$ be the set of input data with N samples, n is the total number of features and c is the total number of classes.

2.1   Random Vector Functional Link (RVFL) Network [34]

The architecture of RVFL has three layers (input layer, hidden layer and output layer). The model's ability to generalise is enhanced by the direct linkages of RVFL [35]. The output (weights) parameters in the RVFL network are calculated while taking into account both the input data and the nonlinear data learned by the hidden nodes. The mathematical formulation of RVFL can be written as: $g : R^n \rightarrow R^c$ where

$$g(x_i) = \sum_{j=1}^{n} \beta_j x_{ij} + \sum_{j=n+1}^{h_l} \beta_j \Theta \left(v_j, x_i, b_j\right), i = 1, 2, \ldots, N, \tag{1}$$

where β_j represents weight of jth node, v_j is randomly generated weight, b_j is the bias and Θ is an activation function. The RVFL approach optimization problem with h_l hidden nodes can be given by:

$$\min_{\beta \in R^{(n+h_l) \times c}} \frac{1}{2} \|\beta\|_2^2 + \frac{1}{2} \lambda \|D\beta - T\|_2^2, \tag{2}$$

here β represents weight matrix of the output and D represents the concatenated matrix contains D_1 and D_2, moreover, λ is the tunable parameter. T represents the target matrix. The matrix (D) and β are defined as: $D = [D_1 \ D_2]$. Where,

$$D_1 = \begin{bmatrix} x_{11} & \cdots & x_{1n} \\ \vdots & \ddots & \vdots \\ x_{N1} & \cdots & x_{Nn} \end{bmatrix}_{N \times n} \tag{3}$$

and

$$D_2 = \begin{bmatrix} \Theta\left(v_1 \cdot x_1 + b_1\right) & \cdots & \Theta\left(v_{h_l} \cdot x_1 + b_{h_l}\right) \\ \vdots & \ddots & \vdots \\ \Theta\left(v_1 \cdot x_N + b_1\right) & \cdots & \Theta\left(v_{h_l} \cdot x_N + b_{h_l}\right) \end{bmatrix}_{N \times h_l} \tag{4}$$

$\beta = [\beta_1, \beta_2, \cdots, \beta_{n+h_l}]^T_{(n+h_l) \times c}$ and $T = [t_1, t_2, \cdots, t_N]^T_{N \times c}$.

Here, the output vector $\beta_j = [\beta_{j1}, \beta_{j2}, \cdots, \beta_{jc}]$ connects the hidden and output nodes, $v_j = [v_{j1}, v_{j2}, \cdots, v_{jn}]$ is the weight vector connecting the input nodes to the hidden node.

Moreover, b_i and $\Theta(\cdot)$ are the bias and activation function of ith hidden node, respectively. Finally, the solution of optimization problem (2) can be calculated as:

$$\beta = \begin{cases} (D^t D + \frac{1}{\lambda} I)^{-1} D^t T, & (n + h_l) \leq N, \\ D^t (D D^t + \frac{1}{\lambda} I)^{-1} T, & N < (n + h_l), \end{cases} \tag{5}$$

where the identity matrix of the relevant dimension is represented by I.

2.2 Intuitionistic Fuzzy Membership Scheme [29]

Here the descriptions are given about membership function, score function and nonmembership function.

Membership Function The relevance of each sample in the data is determined by the membership function in accordance with the sample's distance from the appropriate class centre in the feature space. The importance (weights) of samples that are close to the class centre are higher than those of samples that are far from the class centre. The membership value (degree) of each sample is calculated as:

$$f(x_i) = \begin{cases} 1 - \frac{\|\theta(x_i) - c^+\|}{r^+ + \epsilon}, & c_i = +1, \\ 1 - \frac{\|\theta(x_i) - c^-\|}{r^- + \epsilon}, & c_i = -1. \end{cases} \tag{6}$$

Here, $c^+ (c^-)$ represent the center points of positive (negative) class and $r^+ (r^-)$ represents the radius of positive (negative) class. Moreover, $\epsilon > 0$ denotes the tunable parameter and θ is a nonlinear map which project samples from input space to higher dimensional space. The definition of center points of every class is given as: $c^+ = \frac{1}{n_+} \sum_{c_i = +1} \theta(x_i)$ and $c^- = \frac{1}{n_-} \sum_{c_i = -1} \theta(x_i)$ where n_+ and n_- represent the number of data samples in the positive and negative class, respectively. The radius of positive and negative classes are calculated as: $r^{\pm} = \max_{c_i = \pm 1} \|\theta(x_i) - c^{\pm}\|$.

The Nonmembership Function Each sample is also assigned a nonmembership value (degree) that considers the ratio of the number of heterogeneous points to all its neighbouring data points in the calculation. Hence, the non-membership function can be defined as: $f^*(x_i) = (1 - f(x_i))\Phi(x_i)$, where $0 \leq f(x_i) + f^*(x_i) \leq 1$ and $\Phi(x_i)$ is calculated as:

$$\Phi(x_i) = \frac{|\{x_j : \|\theta(x_i) - \theta(x_j)\| \leq \eta, c_j \neq c_i\}|}{|\{x_j : \|\theta(x_i) - \theta(x_j)\| \leq \eta\}|}, \tag{7}$$

where η represents the tunable parameter, $|\cdot|$ reflects the set's cardinality.

Score Function After the membership value calculation and nonmembership value calculation corresponding to each sample in the data, a score function is defined that finally assigns an specific value to each data sample to decide their contribution in the learning process. The transformed training data set is represented as:
$X^* = \{(x_1, t_1, f_1, f_1^*), (x_2, t_2, f_2, f_2^*), \cdots, (x_N, t_N, f_N, f_N^*)\}$.

The score function can be given as:

$$s_i = \begin{cases} f_i, & f_i^* = 0 \\ 0, & f_i \leq f_i^* \\ \frac{1 - f_i^*}{2 - f_i - f_i^*}, & \text{others.} \end{cases} \tag{8}$$

Here, membership, nonmembership and score functions are defined in higher dimensional space using θ map. Without knowing the θ map explicitly, there is a relation between θ map and kernel function.

Theorem 1 [29] : Assume that a kernel function is $K(x, y)$. The distance can then be calculated as: $\|\theta(x) - \theta(y)\| = \sqrt{K(x, x) + K(y, y) - 2K(x, y)}$.

Corollary 1 [29]: The following formula is used to determine how far the samples are from the relevant class centre: Let $d^\pm = \|\theta(x_i) - c^\pm\|$ then,

$$d^\pm = \sqrt{K(x_i, x_i) + \frac{1}{n_\pm^2} \sum_{c_m = \pm 1} \sum_{c_n = \pm 1} K(x_m, x_n) - \frac{2}{n_\pm} \sum_{c_j = \pm 1} K(x_i, x_j)}. \tag{9}$$

3 Proposed Minimum Variance Embedded Intuitionistic Fuzzy Weighted Random Vector Functional Link Network

The main goal of the proposed algorithm is to minimize the intra-class variance of the data and develop robust model by employing the intuitionistic fuzzy theory. The optimization problem of the proposed MVIFRVFL model can be formulated as:

$$\min \frac{1}{2}\lambda \left\| S^{\frac{1}{2}}\xi \right\|_2^2 + \frac{1}{2}\|\beta\|_2^2 + \frac{1}{2}\alpha \left\| S_w^{\frac{1}{2}}\beta \right\|_2^2$$

subject to

$$D\beta - T = \xi \tag{10}$$

here, $S = diag(s_1, s_2, \cdots, s_N)$ is the diagonal matrix, wherein s_i is the score value calculated via (8) and ξ is the error matrix. Moreover, S_w is the within-class scatter matrix of the training data and it is defined as:

$$S_w = \sum_{k=1}^{m} \sum_{i \in C_k} (x_i - m_k)(x_i - m_k)^t, \tag{11}$$

where m is the number of classes, m_k is the mean vector of the kth class and $x_i \in R^{n+h_l}$. The optimization problem (10) is categorized as the convex quadratic problem, which means that there exist a unique or only one solution. The first term in (10) is the weighted error term that handle the noise and outliers in the data, second term is the regularization and the third term is minimizing intra class variance of the data. We get the Lagrange's function of (10) as:

$$L = \frac{1}{2}\lambda \|S^{\frac{1}{2}}(D\beta - T)\|_2^2 + \frac{1}{2}\|\beta\|_2^2 + \frac{1}{2}\alpha\|S_w^{\frac{1}{2}}\beta\|_2^2. \tag{12}$$

The Karush-Kuhn-Tucker (K.K.T.) condition is as follows: $\frac{\partial L}{\partial \beta} = \lambda D^t S(D\beta - T) + \beta + \alpha S_w \beta = 0$. After calculations, the obtained output parameter is given as:

$$\beta = (D^t S D + \frac{1}{\lambda}I + \frac{\alpha}{\lambda}S_w)^{-1}D^t S T. \tag{13}$$

Corresponding to test sample x, the final decision is taken as follows: $g(x) = [x, h(x)]\beta$.

4   Experimental Results

The experimental findings of the baseline models and the proposed MVIFRVFL model are given in this section. The performance evaluation is done on 33 datasets [36].

4.1   Experimental Setup

In this research, the simulations are performed on Windows-10 operating system, Intel(R) Xeon(R) CPU E5−2697 v4 2.30 GHz, 128−GB RAM and MATLAB R2017b. We employed a Gaussian kernel, $K(x, y) = \exp(-(\|x - y\|^2)/\mu^2)$. Here, μ denotes the kernel parameter. The experiments are conducted with UCI and KEEL datasets [36]. The dataset is randomly divided into 70 : 30 ratio for training and testing sets, respectively. Grid search is applied for the optimization of hyperparameters of different models with a five fold cross validation. There are five disjoint sets that have been partitioned randomly in which four sets are used for training and one set is utilized for testing. The adaptable parameters corresponding to each model are taken from the following range: $\lambda = \alpha \in \{10^{-5}, 10^{-4}, \cdots, 10^4, 10^5\}$. The hidden neurons (h_l) varies as $3 : 20 : 203$ and relu activation function is employed. For the performance evaluation, we have employed several metrics such as model accuracy, average rank, pairwise sign test, win-tie loss and Nemenyi test.

Table 1. Experimental results of the baseline models and the proposed MVIFRVFL model.

Dataset	IFTWSVM [37] (Acc., Time(s))	KRR [38] (Acc., Time(s))	IFKRR [32] (Acc., Time(s))	ELM [39] (Acc., Time(s))	RVFL [40] (Acc., Time(s))	IFRVFL [33] (Acc., Time(s))	MVRVFL [20] (Acc., Time(s))	MVIFRVFL (Acc., Time(s))
Aus	0.8482, 0.0846	0.842, 0.0138	0.8023, 0.0445	0.8484, 0.0032	0.8492, 0.001	0.8526, 0.0306	0.8518, 0.0014	0.8536, 0.0489
Brwisconsin	0.9876, 0.0978	0.9938, 0.0189	0.5, 0.0558	0.995, 0.0011	0.9907, 0.0007	0.9938, 0.0278	0.9907, 0.002	0.9938, 0.017
Bupa-or-liver-disorders	0.6965, 0.0297	0.6789, 0.0031	0.5945, 0.0321	0.7189, 0.001	0.6989, 0.0012	0.6496, 0.064	0.7051, 0.0008	0.6867, 0.0068
Checkerboard_Data	0.8482, 0.0809	0.842, 0.0112	0.8023, 0.1005	0.8484, 0.0033	0.8492, 0.0007	0.8526, 0.0314	0.8518, 0.0015	0.8536, 0.0274
Cleve	0.8216, 0.0198	0.8162, 0.0022	0.7422, 0.0114	0.8371, 0.0024	0.851, 0.0016	0.7725, 0.0097	0.8192, 0.0008	0.8314, 0.0053
Cmc	0.6881, 0.3283	0.6969, 0.0659	0.5, 0.1751	0.6791, 0.0068	0.6877, 0.0075	0.6708, 0.0863	0.6892, 0.0024	0.7083, 0.0725
Crossplane130	1, 0.0201	1, 0.0029	0.5133, 0.0928	1, 0.0002	1, 0.0004	1, 0.0021	0.996, 0.0003	1, 0.0017
Crossplane150	1, 0.0216	0.9643, 0.0015	0.5873, 0.0192	0.9893, 0.0003	0.9643, 0.0004	0.9821, 0.0023	0.9893, 0.0006	0.9821, 0.0043
Ecoli-0-1_vs_2-3-5	0.8277, 0.0284	0.7143, 0.0017	0.8498, 0.0061	0.7113, 0.0004	0.7143, 0.0003	0.7143, 0.0803	0.7798, 0.0007	0.7784, 0.0041
Ecoli2	0.8571, 0.0328	0.7218, 0.0025	0.8087, 0.0077	0.6926, 0.0011	0.8072, 0.0007	0.9341, 0.0086	0.7213, 0.0009	0.9066, 0.009
Ecoli-0-1_vs_5	0.8868, 0.0286	0.8259, 0.0024	0.8259, 0.009	0.8303, 0.0003	0.8259, 0.0013	0.8333, 0.0042	0.8137, 0.0004	0.8333, 0.0035
Ecoli-0-1-4-6_vs_5	0.9321, 0.0329	0.9753, 0.0024	0.9938, 0.0083	0.9963, 0.0016	1, 0.0006	0.9753, 0.0075	0.9102, 0.0019	0.9938, 0.0066
Ecoli4	0.8691, 0.0435	0.8948, 0.0065	0.8897, 0.0082	0.8959, 0.001	0.8845, 0.0011	0.8948, 0.0583	0.8928, 0.0011	0.8948, 0.0066
Ecoli-0-3-4-6_vs_5	0.8083, 0.0278	0.825, 0.0046	0.825, 0.0046	0.795, 0.0004	0.8333, 0.0012	0.8333, 0.0038	0.8283, 0.0007	0.8333, 0.005
Ecoli-0-1-4-7_vs_5-6	0.8642, 0.0282	0.875, 0.061	0.8696, 0.0113	0.875, 0.001	0.875, 0.0024	0.875, 0.0107	0.875, 0.0011	0.8696, 0.0509
Ecoli-0-6-7_vs_5	0.7336, 0.0218	0.7418, 0.0096	0.7172, 0.0089	0.8169, 0.0004	0.75, 0.0011	0.7923, 0.0059	0.8251, 0.0009	0.8169, 0.0036
Ecoli-0-4-6_vs_5	0.9286, 0.0233	0.8571, 0.0012	0.8571, 0.0051	0.8605, 0.0011	0.8571, 0.0004	0.9013, 0.0063	0.8392, 0.0011	0.8571, 0.0043
Ecoli0137vs26	0.9872, 0.0321	0.9706, 0.0024	0.9321, 0.0114	0.9529, 0.0006	0.9706, 0.0015	0.9449, 0.0145	0.9532, 0.0012	0.9936, 0.0065
Ecoli3	0.8571, 0.0283	0.7218, 0.004	0.7218, 0.004	0.6926, 0.0008	0.8072, 0.0103	0.9341, 0.0092	0.7213, 0.0012	0.9066, 0.007
Glass5	0.4922, 0.0275	0.7266, 0.0014	0.7031, 0.0086	0.7453, 0.0015	0.7188, 0.0008	0.7188, 0.007	0.775, 0.0029	0.9844, 0.0073
Haber	0.5588, 0.0312	0.6138, 0.0022	0.5891, 0.0129	0.5822, 0.0016	0.5253, 0.0007	0.6097, 0.0056	0.5455, 0.0022	0.6191, 0.0059
Haberman	0.5588, 0.0311	0.6138, 0.0079	0.5891, 0.0177	0.5822, 0.0015	0.5253, 0.0005	0.6097, 0.0388	0.5455, 0.0043	0.6191, 0.0043
Heart-stat	0.8398, 0.0238	0.8717, 0.0035	0.7866, 0.0679	0.8653, 0.0019	0.8574, 0.0013	0.8647, 0.0065	0.8753, 0.0004	0.8754, 0.0051
Led7digit-0-2-4-5-6-7-8-9_vs_1	0.9016, 0.0535	0.9097, 0.0043	0.5, 0.0255	0.9073, 0.0009	0.9419, 0.003	0.9258, 0.0178	0.9008, 0.0013	0.9379, 0.014
New-thyroid1	0.9444, 0.0256	0.9912, 0.0013	0.9825, 0.0169	0.9947, 0.0004	1, 0.0003	0.9825, 0.0043	1, 0.0009	0.9912, 0.003
Vehicle1	0.7859, 0.1416	0.7766, 0.0181	0.7366, 0.0782	0.7626, 0.0029	0.7683, 0.0011	0.7954, 0.0379	0.7297, 0.0061	0.7766, 0.0325
Vehicle2	0.9819, 0.1389	0.9974, 0.0246	0.9738, 0.1828	0.9889, 0.0046	0.9893, 0.0066	0.9758, 0.0558	0.9802, 0.005	0.9155, 0.0422
Ripley	0.9203, 0.2319	0.9204, 0.0369	0.5, 0.1321	0.915, 0.0038	0.915, 0.0026	0.9133, 0.0749	0.9077, 0.0064	0.9155, 0.0524
Shuttle-6_vs_2-3	0.75, 0.0262	0.9924, 0.0028	0.9394, 0.0072	0.9985, 0.0003	1, 0.0002	0.9924, 0.0221	1, 0.0004	1, 0.0033
Votes	0.9558, 0.0373	0.9558, 0.0046	0.5, 0.011	0.9665, 0.0013	0.9715, 0.0017	0.943, 0.0114	0.9709, 0.0013	0.9494, 0.0081
Shuttle-c0-vs-c4	1, 0.6015	0.9865, 0.2812	1, 0.5837	0.9892, 0.0007	0.9865, 0.0013	0.9865, 0.2233	0.9865, 0.0019	0.9865, 0.1867
Transfusion	0.6146, 0.1027	0.6281, 0.018	0.5, 0.0683	0.606, 0.002	0.6257, 0.0003	0.6555, 0.0347	0.6056, 0.0009	0.6754, 0.0211
Yeast-0-2-5-7-9_vs_3-6-8	0.9345, 0.2167	0.9297, 0.0286	0.5, 0.0929	0.9036, 0.003	0.881, 0.0033	0.9188, 0.0769	0.8921, 0.0021	0.8948, 0.0435
Average Accuracy	0.8388	0.8446	0.7339	0.8437	0.8462	0.8575	0.8414	**0.873**
Average Rank	4.6818	4.3333	6.5152	4.2273	4.303	4.1364	4.8636	**2.9394**
Overall Win-Tie-Loss	[4,0,5]	[2,0,0]	[1,0,17]	[3,0,4]	[4,0,2]	[3,0,0]	[1,0,4]	**[9,0,0]**

Table 2. Significance difference among the models based on the Nemenyi test

	IFTWSVM [37]	KRR [38]	IFKRR [32]	ELM [39]	RVFL [40]	IFRVFL [33]	MVRVFL [20]	MVIFRVFL
IFTWSVM			✓					
KRR			✓					
IFKRR	✓	✓		✓	✓	✓		✓
ELM			✓					
RVFL			✓					
IFRVFL			✓					
MVRVFL								✓
MVIFRVFL			✓			✓		

Here, ✓ represents the significant difference exist between the row and the column method. Empty entry represents that no significance difference exists between the row and the column method.

Table 3. Win-tie-loss count (pairwise)

	IFTWSVM [37]	KRR [38]	IFKRR [32]	ELM [39]	RVFL [40]	IFRVFL [33]	MVRVFL [20]	MVIFRVFL
KRR	[17,2,14]							
IFKRR	[10,1,22]	[5,3,25]						
ELM	[18,1,14]	[16,2,15]	[26,0,7]					
RVFL	[18,1,14]	[13,8,12]	[24,2,7]	[18,2,13]				
IFRVFL	[19,1,13]	[12,8,13]	[29,1,3]	[14,2,17]	[16,6,11]			
MVRVFL	[14,0,19]	[13,2,18]	[23,0,10]	[14,2,17]	[13,5,15]	[13,2,18]		
MVIFRVFL	[22,1,10]	[21,7,5]	[28,3,2]	[19,2,12]	[19,6,8]	[20,7,6]	[24,2,7]	

Here, win tie loss [x,y,z] represents that the row method has x-times wins, y-times ties and z-times loses corresponding to the column method.

Table 4. Win-tie-loss (pairwise): Sign test

	IFTWSVM [37]	KRR [38]	IFKRR [32]	ELM [39]	RVFL [40]	IFRVFL [33]	MVRVFL [20]
MVIFRVFL		✓	✓				✓

Here, ✓ represents the significant difference exist between the row and the column method. Empty entries represent that there is no
significance difference exists between the row and the column method.

4.2 Experiments Analysis

The experimental findings in Table 1 show how the proposed MVIFRVFL model and baseline models performed. One can observe from the results that the average accuracy of the proposed MVIFRVFL model is higher than the baseline models. From Table 1, it is very clear that the proposed model outperformed with the average accuracy of 87.3%, and IFRVFL model stands at the second position with the average accuracy of 85.75% and rest baseline algorithms, i.e. RVFL, IFKRR, KRR, IFTWSVM, ELM, and MVRVFL model has the average accuracy 84.62%, 73.39%, 84.46%, 83.88%, 84.37% and 84.14%, respectively. Comparing the models based on average accuracy can be a biased measure because a model can have very low accuracy on few datasets that can be compensated by high accuracy on single dataset. As a result, we also performed statistical tests to verify the models' effectiveness on a statistical level. Here, we considered the Friedman test [41]. Let us consider k models that need to be evaluated on N_0 datasets. Therefore, we assign rank $r_{i,j}$ to the jth method on the ith database. The avg. rank of each model is obtained as: $R_j = \frac{1}{N_0} \sum_{i=1}^{N_0} r_{i,j}$. All classifiers are equally effective under the null hypothesis, and as a result, their average ranks are equal. The average rank of the proposed MVIFRVFL model is 2.9394 which is the minimum amongst the ranks of the base line models. Minimum model's rank implies better performance of the algorithm. Moreover, the average ranks of the baseline algorithms, i.e. MVRVFL, IFTWSVM, IFKRR, ELM, KRR, RVFL, IFRVFL and are 4.8636, 4.6818, 6.5152, 4.2273, 4.3333, 4.303, 4.1364 respectively.

The overall performance of models are equal under the null hypothesis and hence, their average rank are equal. In Friedman test, χ_F^2 is statistically defined as: $\chi_F^2 = \frac{12N_0}{k(k+1)} \left[\sum_j R_j^2 - \frac{k(k+1)^2}{4} \right]$, it is distribution with the degree of freedom

(dofs) $(k-1)$. Furthermore, a better statistical analysis, i.e. F_F can be defined as: $F_F = \frac{(N_0-1)\chi_F^2}{N_0(k-1)-\chi_F^2}$, here F_F follows the F-distribution with $(N_0 - 1)$ and $(N_0 - 1) \times (k-1)$. After performing simple calculations, χ_F^2 is 38.2107 and $F_F(7, 224)$ is 6.3424 for $k = 8$ models with $N_0 = 33$ datasets. The critical difference $F_F(7, 224)$ is 2.055 from the F distribution table at $\alpha = 0.05$ significant level. Here, the value $F_F = 6.3424 > 2.055$. Therefore, we reject the null hypothesis and it indicates that there are statistical difference among the models. In addition, for pairwise comparison between the models, Nemenyi post hoc [41] test is performed. The models are considered to be considerably different if their average rank differs by a critical difference (CD) or more. The CD is calculated as: $CD = q_\alpha \left[\frac{k(k+1)}{6N_0} \right]^{1/2}$.

After calculation, we gets $CD = 1.8300$. The Table 2 demonstrates that there are considerable differences between the proposed MVIFRVFL network and the baseline networks based on Nemenyi test. Moreover, we also examine the win-tie-loss (pairwise) sign test for the statistical analysis of the models. In this test, the evaluation is done on the basis of the datasets on which the classifiers are winner, tie or loses. If model has the number of wins $\frac{N_0}{2}$ out of N_0 number of datasets then, two models perform equal under the null hypothesis. The performance of two algorithms are said to be statistically unalike, only if the number of total wins of a algorithm is at least $\frac{N_0}{2} + 1.96\frac{\sqrt{N_0}}{2}$ which is equal to 22.12. Table 3 shows the total count of pairwise win-tie-loss of baselines models as well as the proposed MVIFRVFL model. One can observe that the proposed MVIFRVFL model has wins between 19 to 28 out of 33 datasets with respect to baseline models. It shows the superiority of the proposed MVIFRVFL model over the baseline models. Table 4 shows that the proposed MVIFRVFL network is different than KRR and MVRVFL network based on sign test.

5   Conclusion

In this article, we proposed minimum variance intuitionistic fuzzy RVFL (MVIFRVFL) model that utilize the strength of both IFRVFL and MVRVFL model. The optimization problem of the proposed MVIFRVFL model minimizes three terms, i.e. the weighted error, within-class variance of the data and the regularization term, simultaneously, to generates the final output parameters of the model. The proposed MVIFRVFL network can pay closer attention to each sample in order to enhance the model's classification performance. Therefore, the proposed MVIFRVFL model gets the higher generalization performance compared to the baseline models. The effectiveness of the proposed MVIFRVFL model is shown through experimental findings and statistical analyses. Experimental analysis reveals that the proposed MVIFRVFL model beat other models in term of average accuracy and average rank. In future, we will try to extend the proposed work in deep and ensemble deep frameworks.

Acknowledgment. This work is supported by National Supercomputing Mission under the department of science and technology, Miety and Indian Govt. under Grant No. DST/NSM/R&D HPC Appl/2021/03.29 and SERB under Mathematical Research Impact-Centric Support scheme grant no. MTR/2021/000787. Mr. A. K. Malik, a recipient of CSIR fellowship (New Delhi, India) File no-09/1022 (0075)/2019-EMR-I, acknowledges economic support. The grateful acknowledgement is given to the Indian Institute of Technology Indore for providing facilities and support.

References

1. Cortes, C., Vapnik, V.: Support-vector networks. Mach. Learn. **20**(3), 273–297 (1995)
2. Tanveer, M., Rajani, T., Rastogi, R., Shao, Y.: Comprehensive review on twin support vector machines. Ann. Oper. Res. (2022). https://doi.org/10.1007/s10479-022-04575-w
3. Breiman, L.: Random forests. Mach. Learn. **45**(1), 5–32 (2001)
4. Abiodun, O.I., Jantan, A., Omolara, A.E., Dada, K.V., Mohamed, N.A., Arshad, H.: State-of-the-art in artificial neural network applications: A survey. Heliyon **4**(11), e00938 (2018)
5. Ganaie, M.A., Hu, M., Malik, A.K., Tanveer, M., Suganthan, P.N.: Ensemble deep learning: A review. Eng. Appl. Artif. Intell. **115**, 105151 (2022)
6. Park, J., Sandberg, I.W.: Universal approximation using radial-basis-function networks. Neural Comput. **3**(2), 246–257 (1991)
7. Suganthan, P.N.: On non-iterative learning algorithms with closed-form solution. Appl. Soft Comput. **70**, 1078–1082 (2018)
8. Scarselli, F., Tsoi, A.C.: Universal approximation using feedforward neural networks: a survey of some existing methods, and some new results. Neural Netw. **11**(1), 15–37 (1998)
9. Chakravorti, T., Satyanarayana, P.: Non linear system identification using kernel based exponentially extended random vector functional link network. Appl. Soft Comput. **89**, 106117 (2020)
10. Wang, N., Er, M.J., Han, M.: Generalized single-hidden layer feedforward networks for regression problems. IEEE Trans. Neural Netw. Learn. Syst. **26**(6), 1161–1176 (2014)
11. Igelnik, B., Pao, Y.-H.: Stochastic choice of basis functions in adaptive function approximation and the functional-link net. IEEE Trans. Neural Netw. **6**(6), 1320–1329 (1995)
12. Gao, R., Du, L., Yuen, K.F., Suganthan, P.N.: Walk-forward empirical wavelet random vector functional link for time series forecasting. Appl. Soft Comput. **108**, 107450 (2021)
13. Malik, A.K., Ganaie, M.A., Tanveer, M., Suganthan, P.N.: A novel ensemble method of RVFL for classification problem, In: 2021 International Joint Conference on Neural Networks (IJCNN). IEEE, pp. 1–8 (2021)
14. Katuwal, R., Suganthan, P.N., Zhang, L.: An ensemble of decision trees with random vector functional link networks for multi-class classification. Appl. Soft Comput. **70**, 1146–1153 (2018)
15. Wang, W., Peng, Y., Kong, W.: EEG-Based Emotion Recognition via Joint Domain Adaptation and Semi-supervised RVFL Network. In: Li, X. (ed.) IASC 2021. LNDECT, vol. 80, pp. 413–422. Springer, Cham (2022). https://doi.org/10.1007/978-3-030-81007-8_46

16. Shi, Q., Katuwal, R., Suganthan, P.N., Tanveer, M.: Random vector functional link neural network based ensemble deep learning. Pattern Recogn. **117**, 107978 (2021)

17. Cao, W., Gao, J., Ming, Z., Cai, S., Zheng, H.: Impact of Probability Distribution Selection on RVFL Performance. In: Qiu, M. (ed.) SmartCom 2017. LNCS, vol. 10699, pp. 114–124. Springer, Cham (2018). https://doi.org/10.1007/978-3-319-73830-7_12

18. Rao, C. R.: Generalized inverse of a matrix and its applications, Vol. 1 Theory of Statistics. University of California Press. 6, pp. 601–620 (1972)

19. Ganaie, M. A., Tanveer, M., Suganthan, P.: Co-trained random vector functional link network. In: 2021 International Joint Conference on Neural Networks (IJCNN). IEEE, pp. 1–8 (2021)

20. Ganaie, M.A., Tanveer, M., Suganthan, P.N.: Minimum Variance Embedded Random Vector Functional Link Network. In: Yang, H., Pasupa, K., Leung, A.C.-S., Kwok, J.T., Chan, J.H., King, I. (eds.) ICONIP 2020. CCIS, vol. 1333, pp. 412–419. Springer, Cham (2020). https://doi.org/10.1007/978-3-030-63823-8_48

21. Zhang, P.-B., Yang, Z.-X.: A new learning paradigm for random vector functional-link network: RVFL+. Neural Netw. **122**, 94–105 (2020)

22. Zhang, Y., Wu, J., Cai, Z., Du, B., Philip, S.Y.: An unsupervised parameter learning model for RVFL neural network. Neural Netw. **112**, 85–97 (2019)

23. Guo, X., Zhou, W., Lu, Q., Du, A., Cai, Y., Ding, Y.: Assessing dry weight of hemodialysis patients via sparse Laplacian regularized RVFL neural network with l2, 1-norm, BioMed. Res. Int. vol. 2021, (2021)

24. Dai, W., Chen, Q., Chu, F., Ma, X., Chai, T.: Robust regularized random vector functional link network and its industrial application, IEEE Access **5**, pp. 162–172 (2017)

25. Malik, A.K., Gao, R., Ganaie, M.A., Tanveer, M., Suganthan, P.N.: Random vector functional link network: recent developments, applications, and future directions. (2022) arXiv preprint arXiv:2203.11316

26. Rezvani, S., Wang, X., Pourpanah, F.: Intuitionistic fuzzy twin support vector machines. IEEE Trans. Fuzzy Syst. **27**(11), 2140–2151 (2019)

27. Ganaie, M.A., Tanveer, M., Initiative, A.D.N.: Fuzzy least squares projection twin support vector machines for class imbalance learning, Appl. Soft Comput. **113**, p. 107933 (2021)

28. Lin, C.-F., Wang, S.-D.: Fuzzy support vector machines. IEEE Trans. Neural Netw. **13**(2), 464–471 (2002)

29. Ha, M., Wang, C., Chen, J.: The support vector machine based on intuitionistic fuzzy number and kernel function. Soft. Comput. **17**(4), 635–641 (2013)

30. Laxmi, S., Gupta, S.K.: Intuitionistic fuzzy proximal support vector machines for pattern classification. Neural Process. Lett. **51**(3), 2701–2735 (2020)

31. Rezvani, S., Wang, X.: Class imbalance learning using fuzzy ART and intuitionistic fuzzy twin support vector machines. Inf. Sci. **578**, 659–682 (2021)

32. Hazarika, B.B., Gupta, D., Borah, P.: An intuitionistic fuzzy kernel ridge regression classifier for binary classification. Appl. Soft Comput. **112**, 107816 (2021)

33. Malik, A.K., Ganaie, M., Tanveer, M., Suganthan, P.N., Initiative, A.D.N.I.: Alzheimer's disease diagnosis via intuitionistic fuzzy random vector functional link network. IEEE Trans. Comput. Soc. Syst. (2022). https://doi.org/10.1109/TCSS.2022.3146974

34. Pao, Y.-H., Takefuji, Y.: Functional-link net computing: theory, system architecture, and functionalities. Computer **25**(5), 76–79 (1992)

35. Zhang, L., Suganthan, P.N.: A comprehensive evaluation of random vector functional link networks. Inf. Sci. **367**, 1094–1105 (2016)
36. Dua, D., Graff, C.: UCI machine learning repository (2017). http://archive.ics.uci.edu/ml
37. Liang, Z., Zhang, L.: Intuitionistic fuzzy twin support vector machines with the insensitive pinball loss. Appl. Soft Comput. **115**, 108231 (2022)
38. Saunders, C., Gammerman, A., Vovk, V.: Ridge regression learning algorithm in dual variables. In: ICML-1998 Proceedings of the 15th International Conference on Machine Learning 04, 515-521 (1999)
39. Huang, G.-B., Zhou, H., Ding, X., Zhang, R.: Extreme learning machine for regression and multiclass classification Man, and Cybernetics. IEEE Trans. Syst. Part B (Cybernetics) **42**(2), 513–529 (2011)
40. Pao, Y.-H., Park, G.-H., Sobajic, D.J.: Learning and generalization characteristics of the random vector functional-link net. Neurocomputing **6**(2), 163–180 (1994)
41. Demšar, J.: Statistical comparisons of classifiers over multiple data sets. J. Mach. Learn. Res. **7**, 1–30 (2006)

Neural Network Compression by Joint Sparsity Promotion and Redundancy Reduction

Tariq M. Khan[1]([⊠]), Syed S. Naqvi[2], Antonio Robles-Kelly[3], and Erik Meijering[1]

[1] School of Computer Science and Engineering, UNSW, Sydney, Australia
tariq045@gmail.com
[2] Department of Electrical and Computer Engineering, CUI, Islamabad, Pakistan
[3] Faculty of Science, Engineering and Built Environment, Deakin University,
Waurn Ponds, VIC 3216, Australia

Abstract. Compression of convolutional neural network models has recently been dominated by pruning approaches. A class of previous works focuses solely on pruning the unimportant filters to achieve network compression. Another important direction is the design of sparsity-inducing constraints which has also been explored in isolation. This paper presents a novel training scheme based on composite constraints that prune redundant filters and minimize their effect on overall network learning via sparsity promotion. Also, as opposed to prior works that employ pseudo-norm-based sparsity-inducing constraints, we propose a sparse scheme based on gradient counting in our framework. Our tests on several pixel-wise segmentation benchmarks show that the number of neurons and the memory footprint of networks in the test phase are significantly reduced without affecting performance. MobileNetV3 and UNet, two well-known architectures, are used to test the proposed scheme. Our network compression method not only results in reduced parameters but also achieves improved performance compared to MobileNetv3, which is an already optimized architecture.

Keywords: Convolutional Neural Networks · Neural Network Compression · Joint Sparsity Promotion · Redundancy Reduction

1 Introduction

One of the key enablers of the unprecedented success of deep learning in recent years is the availability of very large neural network models [1]. Modern deep neural networks (DNNs), in particular convolutional neural networks (CNNs), typically consist of many cascaded layers totaling millions to hundreds of millions of parameters (weights) [2–4]. The larger-scale neural networks tend to enable the extraction of more complex high-level features and therefore lead to a significant improvement of the overall accuracy [5–7]. On the other side, the layered deep structure and large model sizes also demand increasing computational and memory resources [8,9]. The problem with many modern neural networks is that they suffer from parameter explosion [10,11]. Due to the large number of parameters, training can take weeks on a central processing unit (CPU) or days on a graphics processing unit (GPU) [12]. For larger and higher-dimensional image sizes, even a GPU is not able to train a network with a large number of parameters.

M. Tanveer et al. (Eds.): ICONIP 2022, LNCS 13623, pp. 612–623, 2023.
https://doi.org/10.1007/978-3-031-30105-6_51

The aim of the work presented in this paper is to design sparse, low-complexity neural networks by reducing the number of parameters while keeping performance degradation negligible. Memory and computational requirements in particular complicate the deployment of deep neural networks on low-power embedded platforms as they have limited computing and power budget. The energy efficiency challenge of large models motivates model compression. Several algorithm-level techniques have been proposed to compress models and accelerate DNNs, such as quantization to 16-bit [13], group L1 or L2 regularization [14], node pruning [15–17], filter pruning for CNNs ([17–20], weight pruning using magnitude-based methods [21], weight quantization [3,22], connection pruning [21], and low-rank approximation [23].

Pruning—which removes entire filters, or neurons, that make little or no contribution to the output of a trained network—is a way to make a network smaller and faster. There are two forms in which structural pruning is commonly applied: i) using a predefined per-layer pruning ratio, or ii) simultaneously over all layers. The second form allows pruning to automatically find a better architecture [24]. An exact solution for pruning will be to minimize the l_0 norm of all neurons and remove those that are zeroed out. However, l_0 minimization is impractical as it is non-convex, NP-hard, and requires combinatorial search. Therefore, prior work has tried to relax the optimization using Bayesian methods [25,29] or regularization terms.

Motivated by the success of sparse coding, several methods relax l_0 minimization with l_1 or l_2 regularization, followed by soft thresholding of parameters with a predefined threshold. These methods belong to the family of iterative shrinkage and thresholding algorithms (ISTA) [3]. Han et al. [7] applied a similar approach for removing individual weights of a neural network to obtain sparse non-regular convolutional kernels. Li et al. [22] extended this approach to remove filters with small l_1 norms.

Due to the popularity of batch-normalization [17] layers in recent networks [10,15], several approaches have been proposed for filter pruning based on batch-norm parameters [23,33]. These works regularize the scaling term (γ) of batch-norm layers and apply soft thresholding when the value falls below a predefined threshold. Furthermore, floating-point operations (FLOPS) based penalties can also be included to directly reduce computational costs [5]. A more general scheme that uses an ISTA-like method on scaling factors [16] can be applied to any layer.

While these approaches can offer a reasonable parameter reduction (e.g. by 9× to 13× in [21]) with minor accuracy degradation, they suffer from three drawbacks: 1) the sparsity regularization and pruning typically result in an irregular network structure, thereby undermining the compression ratio and limiting performance and throughput [24]; 2) the training complexity is increased due to the additional pruning process [21] or low-rank approximation step [23]; and 3) the compression ratios depending on network are heuristic and cannot be precisely controlled.

The main contributions of this work are threefold. First, we propose a novel training scheme based on composite constraints that removes redundant filters during training. Second, our experiments on several benchmarks demonstrate a significant reduction in the number of neurons and the memory footprint of networks in the test phase without affecting their accuracy. And, finally, we demonstrate online filter redundancy removal in encoder-decoder networks for the task of pixel-wise segmentation.

2 Methodology

2.1 Loss Function

Our goal is to formulate the loss function with added constraints given as

$$f = \alpha \left(y, y' \right) + g(\boldsymbol{W}), \tag{1}$$

where α is the standard loss, y and y' are the true and predicted labels respectively, and $g(\boldsymbol{W})$ represents the set of added constraints on the network parameters $\boldsymbol{W}$. Correspondingly, the l-th iteration of the back-propagation can be defined as:

$$\boldsymbol{W}_l^{t+1} = \boldsymbol{W}_l^t + \gamma \nabla_{\boldsymbol{W}_l} \alpha + \gamma \nabla \left(\sum_{i \in l} g\left(\boldsymbol{W}_i \right) \right), \tag{2}$$

where $\boldsymbol{W}_l$ represents the parameters learned in the l-th iteration, t and $t + 1$ represent the current and next states, and γ is the learning rate. According to (2), a new term $\nabla \left(\sum_{i \in l} g\left(\boldsymbol{W}_i \right) \right)$ must be added to each constrained layer during back-propagation to compute the overall gradient. In our case, the two terms that define $g(\boldsymbol{W})$ are differentiable and therefore can be easily incorporated in the back-propagation as shown in later sections.

2.2 Composite Constraints

We aim to remove the effect of redundant filters during training and effectively minimize network parameters without compromising network performance. To this end, we propose the following two composite constraints for $g(\boldsymbol{W})$ in (1).

Inter-Filter Orthogonality Constraint: The idea of inter-filter orthogonality has been explored in the past [25] as a measure to guide the filter pruning process, which is generally performed in an offline manner. In contrast, we employ it as a regularization term in our loss function to remove the effect of redundant filters during online training. The proposed inter-filter orthogonality constraint R is given as:

$$R\left(\boldsymbol{W} \right) = \left\| \boldsymbol{W} \boldsymbol{W}^T - I \right\|_2^2. \tag{3}$$

Here, instead of using the absolute difference [25], we use the squared l_2 norm, which we found to perform well in our experiments.

Sparse Regularization Constraint: Prior works have explored group-sparse regularization including l_2 or l_1 norms [17,26]. Instead, we propose to use the l_0 norm to further eliminate the contribution of redundant filters towards network learning. The proposed sparsity constraint is defined as:

$$H\left(\boldsymbol{S}_l \right) = \left\| \boldsymbol{S}_l \right\|_0. \tag{4}$$

Here, $\boldsymbol{S}_l$ represent the layer-wise sparse gradients corresponding to the weights.

2.3   Overall Update Equations

Having defined the composite constraints constituting $g(W)$, we now define the overall update equations during online training. The updated loss function incorporating the composite constraints is defined as:

$$f = \alpha(y, y') + \lambda \sum_{i \in l} R(W_i) + \beta \sum_{i \in l} H(S_i), \tag{5}$$

where λ and β are the parameters controlling the two regularization terms. Correspondingly, the weight update in the l-th iteration incorporating the gradient terms for the constraints is given as:

$$W_l^{t+1} = W_l^t + \gamma \nabla_{W_l^t} \alpha + \lambda \gamma \nabla \left(\sum_{i \in l} R(W_i) \right) + \beta \gamma \nabla \left(\sum_{i \in l} H(S_i) \right). \tag{6}$$

As the contribution of the sparsity term approaches zero,

$$\beta \gamma \nabla \left(\sum_{i \in l} H(S_i) \right) = 0, \tag{7}$$

the update equation (6) becomes

$$W_l^{t+1} = W_l^t + \gamma \nabla_{W_l^t} \alpha + \tau \nabla_{W_l^t} R(W_l), \tag{8}$$

where the gradient for the inter-filter orthogonality constraint is calculated as:

$$\nabla_{W_l} R(W_l) = W \left\| WW^T - I \right\|_2. \tag{9}$$

The update for the sparsity regularization term is obtained as:

$$S_l^{t+1} = S_l^t + \gamma \nabla_{S_l^t} \alpha + \zeta \nabla_{S_l^t} H(S_l^t). \tag{10}$$

3   Experiments

3.1   Benchmark Datasets

We have evaluated our proposed training strategy on three publicly available benchmark datasets across diverse applications including medical image segmentation, saliency detection, and scene understanding. Specifically, we employed the retinal vessel segmentation dataset DRIVE [27], the salient object dataset DUT-OMRON [28], and the video recognition dataset CamVid [29] for these applications, respectively. The DRIVE [27] dataset consists of 40 retinal scans from a population of 400 diabetic patients from the Netherlands, out of which only 7 exhibited signs of mild diabetic retinopathy. The color images (RGB) were acquired using a Canon 3CCD camera at a resolution of 768×584 pixels at 8 bits per pixel per color channel and are stored in JPEG format. Half of the images constitute the training set, while the other half were employed as test images. Manual segmentations of the vasculature verified by experts are available as gold standard for both training and test images. The standard training and test sets were employed in our experiments. The CamVid [29] dataset includes 700 color images (RGB) with pixel-wise reference masks for 32 semantic classes. As the name suggests, the images were derived 30 Hz footage of driving scenarios acquired from CCTV cameras from the perspective of a driving automobile. In this work, 307 training images, 60 validation images, and 101 test images with labels corresponding to three semantic classes were employed. The included classes feature the "sky" class and "building" class while all other semantic objects are grouped into a single "other" class. The images have a resolution of 360×480 pixels and are stored in PNG format. Finally, the DUT-OMRON [28] dataset consists of a collection of 5168 natural color images (RGB) with various resolutions available in JPEG format. The images include one or more salient objects with a challenging background. Binary reference masks are also available. For our experiments, we randomly sampled 420 images from the DUT dataset, with 70–30 ratio for training and test images.

3.2   Network Models

To demonstrate the prunability of the proposed scheme, we selected U-Net [30] as a large-scale model and MobileNetV3-Small [31] as a lightweight model for our experiments. For U-Net, we employed the implementation from [32][1], whereas the sub-classing-based implementation of [33][2] was adopted for MobileNetV3-Small. We used weight-based pruning in our experiments, where 10% and 50% weights are magnitude pruned. To evaluate the efficacy of the proposed scheme on a wide range of weight percentages, we used iterative pruning. Specifically, we used three iterations of 10% and 50% pruning, and after each iteration the model was trained for an additional 5 epochs to update the weights. The experimental results include performance results for different iterations specified by the keyword "iter". For training the networks, stochastic gradient descent with a learning rate of 0.2 and a momentum of 0.9 was used, where the learning rate was reduced on plateaus. The models were first trained for 100 epochs before pruning with early stopping. The intersection-over-union (IOU) score was employed for monitoring the learning rate and early stopping.

3.3   Evaluation Measures

The available gold-standard reference segmentation maps are binary images indicating for each pixel whether it corresponds to a feature (object) or non-feature (background). Hence there are four types of pixels for each output image: feature pixels correctly predicted as features (true positives: TP), non-feature pixels correctly predicted as non-features (true negatives: TN), non-feature pixels incorrectly predicted as features (false positives: FP), and feature pixels incorrectly predicted as non-features (false negatives: FN). From these, we calculated several common performance metrics [34]:

---

[1] https://github.com/JanMarcelKezmann/TensorFlow-Advanced-Segmentation-Models.

[2] https://github.com/xiaochus/MobileNetV3.

$$\text{Sensitivity} = \frac{TP}{TP + FN}, \tag{11}$$

$$\text{Specificity} = \frac{TN}{TN + FP}, \tag{12}$$

$$\text{Accuracy} = \frac{TP + TN}{TP + FN + TN + FP}, \tag{13}$$

$$\text{Balance Accuracy} = \frac{\text{Sensitivity} + \text{Specificity}}{2}, \tag{14}$$

$$F1 = \frac{2TP}{2TP + FP + FN}, \tag{15}$$

$$\text{Jaccard} = \frac{TP}{TP + FP + FN}, \tag{16}$$

$$\text{Error} = 1 - \text{Jaccard}. \tag{17}$$

The Jaccard measure (16) is the ratio between the intersection and the union of the segmented image and the gold-standard mask and is also known as the intersection-over-union (IOU) measure [35]. For all measures, higher values imply better performance, except for the error measure (17) where lower values imply better performance.

3.4   Experiment I

In the first experiment, we used the DRIVE dataset to evaluate the efficacy of the proposed loss function and the corresponding prunability (potential parameter saving) on the lightweight network MobileNetV3-Small [31]. For a thorough evaluation, variants of the MobileNetV3-Small architecture with unique training settings were compared. These included training from scratch, training from scratch with additional filters added to the backbone, and pretrained backbone. The lightweight network was trained with both the binary cross-entropy loss (default) and the proposed loss (termed as custom loss) to obtain six variants. From the results (Table 1) it is evident that the proposed custom loss achieved performance improvements over the cross-entropy loss in most cases. This confirms the sparsification property of the proposed loss and its ability to suppress unwanted information and attend to important regions of interest. The proposed loss

Table 1. Performance comparison of the standard loss and the proposed custom loss for the MobileNetV3-Small network on the DRIVE dataset. Bold indicates best performance per metric.

Method	Sens.	Spec.	Acc.	BAcc.	F1	Jacc.	Error
MobileNetV3-Scratch	**0.6054**	0.9486	0.9186	**0.7770**	0.5652	0.3945	0.6055
MobileNetV3-CustomLoss-Scratch	0.5860	**0.9554**	**0.9231**	0.7707	**0.5709**	**0.4001**	**0.5999**
MobileNetV3-Scratch-AddFilters	0.6264	0.9521	0.9236	0.7892	0.5885	0.4175	0.5825
MobileNetV3-CustomLoss-Scratch-AddFilters	**0.6381**	**0.9585**	**0.9305**	**0.7983**	**0.6158**	**0.4453**	**0.5547**
MobileNetV3-Pretrained	**0.6480**	0.9593	0.9322	0.8037	0.6250	0.4550	0.5450
MobileNetV3-CustomLoss-Pretrained	0.6477	**0.9614**	**0.9341**	**0.8046**	**0.6315**	**0.4619**	**0.5381**

Table 2. Parameter saving ability of the proposed loss. Test times were computed without removing the saved parameters.

Method	Parameters Backbone	Parameters Saved	Train Time	Test Time
MobileNetV3-Pretrained	0.445M		777	0.107
MobileNetV3-Pretrained-Optimized	0.433M	11,290	766	0.105
MobileNetV3-Scratch-AddFilters	0.531M		760	0.109
MobileNetV3-Scratch-AddFilters-Optimized	0.517M	13,035	753	0.104

incited parameter savings and train/test time savings (Table 2). Visual comparison of sample results with the different losses (Fig. 1) illustrates overall superior performance of the proposed loss, in that it resulted in less false pixel predictions.

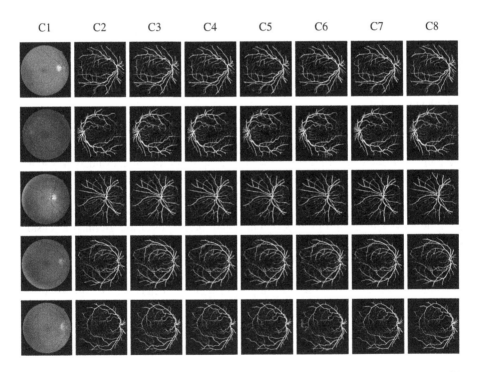

Fig. 1. Example visual results for the DRIVE dataset. C1: Input images. C2: Corresponding gold-standard reference segmentations. C3-C5: Output of the model with standard loss and, respectively, training from scratch, pretrained backbone based training, and using additional filters. C6-C8: Output of the model with custom loss and, respectively, training from scratch, pretrained backbone based training, and using additional filters. In C3-C8, white pixels indicate true positives, red pixels indicate false positives, and yellow pixels indicate false negatives. (Color figure online)

3.5   Experiment II

In the second experiment, we evaluated the performance of the proposed pruning scheme using U-Net [30] and MobileNetV3-Small [31] on the CamVid and DUT-OMRON datasets in terms of mean IOU (mIOU) and the number of model parameters.

The results on the CamVid dataset (Table 3) show a decrease in mIOU for both 10% and 50% weight pruning. The drop in performance is relative to the degree of pruning, which could be attributed to training-based damage and network forgetting. For 10% pruning, there is about 14% decrease in mIOU after three iterations, with a decrease of about 18% (4.6 M) parameters. For 50% pruning, while there is a considerable decrease of about 25% in mIOU, this is achieved at a significant decrease of about 88% (22.7 M) parameters. Example visual results (Fig. 2) show that the segmentation outputs of the 10% pruned U-Net (iter 3) have considerable overlap with the gold standard and closely follow the original U-Net for the "building" class with slight misclassification for the "sky" class. These segmentations are achieved with around 18% fewer parameters compared with U-Net. The U-Net (iter 3) 50% model still remembers the semantic level segregation between the "building" and "sky" class but slightly forgets the representation for the background class. It is noteworthy that the 50% pruned model predicts the building occluded by the trees which is also neglected in the reference maps.

Results of the proposed pruning scheme for the DUT-OMRON dataset (Table 4) exhibit a slight decrease of about 4% in mIOU performance of U-Net for 10% pruning with a 3 M parameter saving. For the 50% pruned U-Net model we observe about 16% loss in mIOU at almost 59% parameter saving. It is noteworthy that the performance of the 10% pruned MobileNetV3-Small network is improved with a decrease in the number of parameters. Similar to U-Net, there is a considerable decrease in the mIOU for the 50% pruned MobileNetV3-Small model, but with only a slight decrease of 4.4% parameters in this lightweight network. Representative visual example results of MobileNetV3-Small on the DUT-OMRON dataset (Fig. 3) show that the pruned model can obtain better coverage of objects of interest. This can be attributed to the prunability of the model and the weight selection of the proposed scheme, which enable the network to better attend to regions of interest.

Table 3. Pruning results on the CamVid dataset.

Dataset	Weights (%)	Methods	mIOU	#Params (M)
CamVid	10%	U-Net	0.8798	25.8
		U-Net (iter 1)	0.8714	24.3
		U-Net (iter 2)	0.7822	22.8
		U-Net (iter 3)	0.7575	21.2
	50%	U-Net	0.8798	25.8
		U-Net (iter 1)	0.6838	18.2
		U-Net (iter 2)	0.6087	10.6
		U-Net (iter 3)	0.6553	3.1

Fig. 2. Example visual results of the proposed scheme on the CamVid dataset. Columns from left to right: input images, gold standard reference images, U-Net results, U-Net (iter 3) 10% results, and U-Net (iter 3) 50% results.

Table 4. Pruning results on the DUT-OMRON dataset.

Dataset	Weights (%)	Methods	mIOU	#Params (M)
DUT-OMRON		U-Net	0.8145	25.8
	10%	U-Net (iter 1)	0.8036	24.3
		U-Net (iter 2)	0.7804	22.8
	50%	U-Net (iter 1)	0.6773	18.2
		U-Net (iter 2)	0.6843	10.6
		MobileNet-V3-Small	0.4951	0.45
	10%	MobileNet-V3-Small (iter 1)	0.5113	0.44
		MobileNet-V3-Small (iter 2)	0.5154	0.44
	50%	MobileNet-V3-Small (iter 1)	0.3005	0.43
		MobileNet-V3-Small (iter 2)	0.3231	0.43

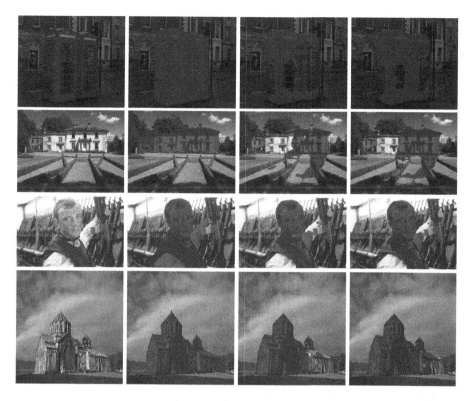

Fig. 3. Example visual results of the proposed scheme on the DUT-OMRON dataset. Columns from left to right: input images, gold standard reference images, MobileNetV3-Small results, and MobileNetV3-Small (iter 1) 10% results.

4   Conclusion

In this paper, a novel training scheme based on composite constraints is presented. The scheme removes redundant filters and reduces the impact that these filters have on the overall learning of the network by promoting sparsity. In addition, in contrast to previous works that make use of pseudo-norm-based sparsity-inducing constraints, the framework that we have developed includes a sparse scheme that is based on gradient counting. The proposed strategy is evaluated on three publicly available dataset across diverse application such as DRIVE (medical image segmentation), DUT-OMRON (saliency detection) and CamViD (scene understanding). The efficacy of the proposed filter pruning is tested on UNet and several variants of MobileNetV3.

References

1. Khan, T.M., Robles-Kelly, A.: Machine learning: Quantum vs classical IEEE Access.**8**, pp. 275–294 (2020)

2. Krizhevsky, A., Sutskever, I., Hinton, G. E.: ImageNet classification with deep convolutional neural networks, In Advances in Neural Information Processing Systems, **25**, pp. 1–9 (2012)

3. Lin, D., Talathi, S., Annapureddy, S.: Fixed point quantization of deep convolutional networks, In: International Conference on Machine Learning, pp. 2849–2858 (2016)

4. Khan, T.M., Naqvi, S.S, Meijering, E.: Leveraging image complexity in macro-level neural network design for medical image segmentation, (2021) arXiv preprint arXiv:2112.11065

5. Le, Q.V.: Building high-level features using large scale unsupervised learning In: IEEE International Conference on Acoustics, Speech and Signal Processing, pp. 8595–8598 (2013)

6. Ciregan, D., Meier, U., Schmidhuber, J.: Multi-column deep neural networks for image classification. In: IEEE Conference on Computer Vision and Pattern Recognition, pp. 3642–3649 (2012)

7. Schmidhuber, J.: Deep learning in neural networks: an overview. Neural Netw. **61**, 85–117 (2015)

8. Khan, T. M., Robles-Kelly, A., Naqvi, S. S.: T-net: a resource-constrained tiny convolutional neural network for medical image segmentation. In: Proceedings of the IEEE/CVF Winter Conference on Applications of Computer Vision, pp. 644–653 (2022)

9. Khan, T.M., et al.: Width-wise vessel bifurcation for improved retinal vessel segmentation. Biomed. Signal Process. Control **71**, 103169 (2022)

10. Khan, T.M., Robles-Kelly, A., Naqvi, S.S., Arsalan, M.: Residual Multiscale Full Convolutional Network (RM-FCN) for High Resolution Semantic Segmentation of Retinal Vasculature. In: Torsello, A., Rossi, L., Pelillo, M., Biggio, B., Robles-Kelly, A. (eds.) S+SSPR 2021. LNCS, vol. 12644, pp. 324–333. Springer, Cham (2021). https://doi.org/10.1007/978-3-030-73973-7_31

11. Khan, T.M., Robles-Kelly, A., Naqvi, S.S.: Rc-net: a convolutional neural network for retinal vessel segmentation. In: Digital Image Computing: Techniques and Applications (DICTA). IEEE **2021**, 01–07 (2021)

12. Khan, T.M., Robles-Kelly, A.: A Derivative-Free Method for Quantum Perceptron Training in Multi-layered Neural Networks. In: Yang, H., Pasupa, K., Leung, A.C.-S., Kwok, J.T., Chan, J.H., King, I. (eds.) ICONIP 2020. CCIS, vol. 1333, pp. 241–250. Springer, Cham (2020). https://doi.org/10.1007/978-3-030-63823-8_29

13. Gupta, S., Agrawal, A., Gopalakrishnan, K., Narayanan, P.: Deep learning with limited numerical precision. In: International Conference on Machine Learning, pp. 1737–1746 (2015)

14. Alemu, H.Z., Zhao, J., Li, F., Wu, W.: Group $l_{1/2}$ regularization for pruning hidden layer nodes of feedforward neural networks. IEEE Access **7**, 9540–9557 (2019)

15. Castellano, G., Fanelli, A., Pelillo, M.: An iterative pruning algorithm for feedforward neural networks. IEEE Trans. Neural Netw. **8**(3), 519–531 (1997)

16. Zhang, Z., Qiao, J.: A node pruning algorithm for feedforward neural network based on neural complexity. In: International Conference on Intelligent Control and Information Processing, pp. 406–410 (2010)

17. Wen, W., Wu, C., Wang, Y., Chen, Y., Li, H.: Learning structured sparsity in deep neural networks. In: International Conference on Neural Information Processing Systems, pp. 2082–2090 (2016)

18. Li, H., Kadav, A., Durdanovic, I., Samet, H., Graf, H.P.: Pruning filters for efficient convnets. (2017) arXiv:1608.08710

19. He, Y., Zhang, X., Sun, J.: Channel pruning for accelerating very deep neural networks. In: IEEE International Conference on Computer Vision, pp. 1398–1406 (2017)

20. Liu, Z., Li, J., Shen, Z., Huang, G., Yan, S., Zhang, C.: Learning efficient convolutional networks through network slimming, In: IEEE International Conference on Computer Vision, pp. 2755–2763 (2017)

21. Han, S., Pool, J., Tran, J., Dally, W.: Learning both weights and connections for efficient neural network. In: Advances in Neural Information Processing Systems, vol. 28, pp. 1–9 (2015)
22. Wu, J., Leng, C., Wang, Y., Hu, Q., Cheng, J.: Quantized convolutional neural networks for mobile devices. In: IEEE Conference on Computer Vision and Pattern Recognition, pp. 4820–4828 (2016)
23. Tai, C., Xiao, T., Zhang, Y., Wang, X., Ee, W.: Convolutional neural networks with low-rank regularization. (2016) arXiv:1511.06067
24. Yu, J., Lukefahr, A., Palframan, D., Dasika, G., Das, R., Mahlke, S.: Scalpel: customizing DNN pruning to the underlying hardware parallelism In: ACM/IEEE Annual International Symposium on Computer Architecture, pp. 548–560 (2017)
25. Prakash, A., Storer, J., Florencio, D., Zhang, C.: RePr: improved training of convolutional filters. In: IEEE/CVF Conference on Computer Vision and Pattern Recognition, pp. 10 658–10 667 (2019)
26. Zhou, H., Alvarez, J. M., Porikli, F.: Less is more: towards compact CNNs. In: European Conference on Computer Vision, pp. 662–677 (2016)
27. Staal, J., Abramoff, M., Niemeijer, M., Viergever, M., van Ginneken, B.: Ridge-based vessel segmentation in color images of the retina. IEEE Trans. Med. Imaging **23**(4), 501–509 (2004)
28. Yang, C., Zhang, L., Lu, H., Ruan, X., Yang, M.-H.: Saliency detection via graph-based manifold ranking. In: IEEE Conference on Computer Vision and Pattern Recognition, pp. 3166–3173 (2013)
29. Brostow, G.J., Shotton, J., Fauqueur, J., Cipolla, R.: Segmentation and recognition using structure from motion point clouds. In: European Conference on Computer Vision, 2008, pp. 44–57 (2008)
30. Ronneberger, O., Fischer, P., Brox, T.: U-Net: Convolutional networks for biomedical image segmentation. In: Medical Image Computing and Computer-Assisted Intervention, pp. 234–241 (2015)
31. Howard, A., et al.: Searching for MobileNetV3 In: IEEE/CVF International Conference on Computer Vision, pp. 1314–1324 (2019)
32. Kezmann, J.-M.: Tensorflow advanced segmentation models (2020). https://github.com/JanMarcelKezmann/TensorFlow-Advanced-Segmentation-Models
33. Xiaochus, L.: A Keras implementation of MobileNetV3 and lite R-ASPP semantic segmentation (2020). https://github.com/xiaochus/MobileNetV3,
34. Maier-Hein, L., et al.: Metrics reloaded: Pitfalls and recommendations for image analysis validation (2022) arXiv:2206.01653
35. Everingham, M., Eslami, S.M.A., Gool, L.V., Williams, C.K.I., Winn, J.M., Zisserman, A.: The PASCAL visual object classes challenge: A retrospective. Int. J. Comput. Vision **111**, 98–136 (2014)

Author Index

M. Tanveer et al. (Eds.): ICONIP 2022, LNCS 13623, pp. 625–627, 2023.
https://doi.org/10.1007/978-3-031-30105-6

Printed in the United States
by Baker & Taylor Publisher Services